Crop Production and Horticulture

Crop Production and Horticulture

Edited by **Shirley Doy**

Callisto Reference
New York

Published by Callisto Reference,
106 Park Avenue, Suite 200,
New York, NY 10016, USA
www.callistoreference.com

Crop Production and Horticulture
Edited by Shirley Doy

International Standard Book Number: 978-1-63239-683-9 (Hardback)

Printed in China.

Contents

Preface

Crop production and horticulture are interrelated disciplines focused on harvesting plants or crops for food, fuel, medicine, or other commercial uses. Recent advances in the field of agriculture have resulted in advancement and expansion of these fields. This book discusses the current methods and techniques for cultivation of plants. The researches included in this book have been contributed by an international group of authors which are bound to provide incredible insights to the readers. It is a beneficial text for agriculturists, horticulturists, professionals and students engaged in these fields.

The researches compiled throughout the book are authentic and of high quality, combining several disciplines and from very diverse regions from around the world. Drawing on the contributions of many researchers from diverse countries, the book's objective is to provide the readers with the latest achievements in the area of research. This book will surely be a source of knowledge to all interested and researching the field.

In the end, I would like to express my deep sense of gratitude to all the authors for meeting the set deadlines in completing and submitting their research chapters. I would also like to thank the publisher for the support offered to us throughout the course of the book. Finally, I extend my sincere thanks to my family for being a constant source of inspiration and encouragement.

Editor

1

The IDA Peptide Controls Abscission in *Arabidopsis* and *Citrus*

Leandro H. Estornell[1†], Mari Wildhagen[2†], Miguel A. Pérez-Amador[3], Manuel Talón[1], Francisco R. Tadeo[1] and Melinka A. Butenko[2*]*

[1] Centre de Genómica, Institut Valencià d'Investigacions Agràries, Montcada, Spain, [2] Section for Genetics and Evolutionary Biology, Department of Biosciences, University of Oslo, Oslo, Norway, [3] Instituto de Biología Molecular y Celular de Plantas, Universidad Politécnica de Valencia–Consejo Superior de Investigaciones Científicas, València, Spain

Edited by:
Timothy John Tranbarger,
Institut de Recherche pour le Développement, France

Reviewed by:
Lijun Chai,
Huazhong Agricultural University, China
Sara E. Patterson,
University of Wisconsin, USA

***Correspondence:**
Franscisco R. Tadeo
tadeo_fra@gva.es;
Melinka A. Butenko
m.a.butenko@ibv.uio.no

† These authors are joint first authors.

Organ abscission is an important process in plant development and reproduction. During abscission, changes in cellular adhesion of specialized abscission zone cells ensure the detachment of infected organs or those no longer serving a function to the plant. In addition, abscission also plays an important role in the release of ripe fruits. Different plant species display distinct patterns and timing of organ shedding, most likely adapted during evolution to their diverse life styles. However, it appears that key regulators of cell separation may have conserved function in different plant species. Here, we investigate the functional conservation of the citrus ortholog of the *Arabidopsis* peptide ligand INFLORESCENCE DEFICIENT IN ABSCISSION (AtIDA), controlling floral organ abscission. We discuss the possible implications of modifying the citrus IDA ortholog for citrus fruit production.

Keywords: IDA, HAESA, receptor-like kinases, *Citrus*, floral abscission, fruit production, crop yield

INTRODUCTION

Small peptides are used as important signaling ligands to control plant growth and development and more than 1000 genes encoding putative peptides have been discovered in *Arabidopsis thaliana* (*Arabidopsis*; Lease and Walker, 2006; Murphy et al., 2012; Grienenberger and Fletcher, 2015). One such peptide ligand, INFLORESCENCE DEFICIENT IN ABSCISSION (AtIDA), was found to be important for the regulation of floral organ abscission in *Arabidopsis* as the *ida* mutant failed to abscise its floral organs (Butenko et al., 2003). In wild type (wt) plants the abscission process takes place at the boundary between the organ to be shed and the main plant body, in cell files constituting the abscission zone (AZ). After the formation of an AZ, which occurs early and simultaneously with the development of lateral organs from the apical meristem, the AZ cells can be distinguished from their neighbors by being small, densely cytoplasmic and lacking vacuoles (Addicott, 1982; Sexton and Roberts, 1982; Osborne, 1989; Roberts et al., 2000; Liljegren, 2012). Once the abscission process is initiated these cells go through three sequential developmental stages; (i) cell fate determination and acquisition of competence to respond to abscission signals; (ii) cell wall loosening and expansion by cell wall remodeling (CWR) enzymes followed by organ separation; and (iii) differentiation of a protective lignified layer (Patterson, 2001; Aalen et al., 2013; Gubert et al., 2014).

In the *ida* mutant, the cell separation event fails to take place (Butenko et al., 2003). In accordance with this, an IDA overexpression line (35S:AtIDA) exhibited ectopic abscission and displayed an enlarged floral AZ in *Arabidopsis* (Stenvik et al., 2006). Additionally, organ loss was observed at the bases of the pedicel, branches of the inflorescence, and cauline leaves; places where abscission normally does not occur in *Arabidopsis*. AtIDA signaling and the 35S:AtIDA phenotypes are dependent on the two leucine-rich repeat (LRR) receptor-like kinases (RLKs) HAESA (HAE) and HAESA-LIKE 2 (HSL2; Jinn et al., 2000; Cho et al., 2008; Stenvik et al., 2008). Upon receptor activation a MAP kinase signaling event is turned on ultimately leading to the induction of genes encoding CWR enzymes (Cai and Lashbrook, 2008; Cho et al., 2008; Kumpf et al., 2013). Recently, it was shown that a hydroxyprolinated AtIDA peptide of 12 amino acids was sufficient to mediate AtIDA signaling. Furthermore, this same peptide was shown to bind the HSL2 receptor with high affinity (Butenko et al., 2014).

The gene encoding the AtIDA peptide belongs to a family of five *AtIDA-LIKE* (*AtIDL*) genes in *Arabidopsis* that are expressed in a variety of tissues in the *Arabidopsis* plant body, including the base of the pedicel, in the floral tissue AZ, in the funicle AZ and in the main root tip (Stenvik et al., 2008). Interestingly, some of the regions of expression represent cell files where abscission takes place in other plants such as in *Citrus* (citrus) species (Lewis et al., 2006; Estornell et al., 2013), where, unlike for *Arabidopsis*, leaf abscission does take place. Flower and fruit abscission in citrus occurs at the pedicel AZ (called AZ-A), located close to the boundary between the pedicel and the twig, and at the ovary/fruit AZ (called AZ-C), located in the calyx between the pericarp and the nectary or floral disk, respectively (Tadeo et al., 2008). Cultivated citrus trees usually display heavy flowering. However, a high number of flowers and young fruits are shed during the fruit set period (physiological drop) thus maintaining only those fruits that can be nurtured until maturity. The attachment force by which a fruit is held to the calyx shows a large reduction in value at the end of the maturation phase of fruit development in almost all citrus species. However, early and mid-season varieties of sweet orange are especially prone to a premature decline in the attachment force of the fruit leading to pre-harvest fruit abscission. This problem has a serious economic impact especially in those citrus producing areas dedicated to the fresh fruit market as the Mediterranean Basin. The premature decline in the attachment force of the fruit prevents on-tree storage of fruit which shortens the harvesting season and hinders the orderly fruit marketing. Therefore, understanding the mechanisms controlling fruit abscission in citrus, with the prospect of genetic engineering (transgenic or CRISPR gene editing technologies), is of importance for two reasons; one for controlling fruit loss during the physiological drop and pre-harvest abscission, and two to facilitate shedding of unmarketable fruits remaining on the tree once the harvesting season is over.

INFLORESCENCE DEFICIENT IN ABSCISSION and the IDL peptides are evolutionary conserved across the plant kingdom and they exhibit sequence similarity in their conserved C-terminal domain (PIP domain) containing the highly active AtIDA peptide (Butenko et al., 2003; Stenvik et al., 2008; Butenko et al., 2014). Furthermore, the HSL receptors are found across the plant kingdom and in dicots the HSL receptor sequences are conserved (Stø et al., 2015). It has previously been proposed that the presence of *IDL* and *HSL* transcripts at sites where cell separation occur in *Arabidopsis* may indicate that these genes play a more general role in cell separation (Butenko et al., 2009). Here, we explore on the possibility that this may not only be the case in *Arabidopsis* but also in citrus.

We show that the C- terminal domain of AtIDA is highly conserved in the IDA and IDL citrus orthologs. We also provide evidence that the citrus IDA (CitIDA) most similar to AtIDA, CitIDA3, has a function in abscission. We discuss the potential agriculture implications of modifying CitIDA3 and additional downstream signaling components of the IDA pathway identified in *Arabidopsis*.

MATERIALS AND METHODS

Plant Material and Generation of Transgenic Lines

The *CitIDA3* CDS was amplified from clementine cDNA, obtained from total RNA isolated from post-anthesis floral buds, by amplification of a 270 bp fragment with primers: 5′ CACCATGGCTTCTTCTTCTTCTTC 3′, 5′ TCAATTTTGAGT AGAATCCACAACAGA 3′. In parallel, screening of a clementine BAC collection (Terol et al., 2008) was carried out, and as a result BAC clones CCL005I18 and CCH3006E10 were identified to contain the genomic region of *CitIDA3*. A 3 kb *CitIDA3* promoter fragment was amplified from BAC clone CCL005I18 using 5′ CACCGAATTTGTAATTAACTTGTCTTCTT 3′ and 5′ ATAAA TTGTTTGTTTTGGGTTGGC 3′ and cloned into pKGWFS7.0 (Plant Systems Biology, Ghent, Belgium). *35S:CitIDA3* lines were made by cloning the CDS fragments into pK2GW7.0 (Plant Systems Biology, Ghent, Belgium). Sixteen independent lines were obtained and two independent lines were investigated in detail. The *Agrobacterium tumefaciens* C58 strain was used for *Arabidopsis* transformation by flower dipping (Clough and Bent, 1998).

Plant Material for Complementation of the *ida* Mutation

Two independent *35S:CitIDA3* lines were crossed into the *ida-2* mutant background (Cho et al., 2008). The F_2 progeny of both lines were phenotypically scored based on their ability to abscise and genotyped by the *ida-2* genotyping primers 5′ CGGTGTTGGTGGATCCAAGTC 3′ and 5′ CCCTCATTTCCG CCACACTTA 3′ and the T-DNA LBb1 primer 5′ ATTTTGC CGATTTCGGAAC 3′. The presence of the *35S:CitIDA3* trans-gene was verified with primers: 5′ CGCACAATCCCACTATCC TT 3′, 5′ TCAATTTTGAGTAGAATCCACAACAGA 3′.

Petal Breakstrength Measurements

The measurements were performed using a petal breakstrength meter, as previously described (Stenvik et al., 2008). The petal breakstrength was quantified as the force, in gram equivalents, required to remove a petal from a flower (Butenko et al., 2003).

Identification of IDL and HSL Citrus Orthologs

We employed two different strategies to identify IDL citrus orthologs. We first performed TBLASTN searches against the citrus (sweet orange and clementine) genome at the Join Genome Institute (JGI[1]) using the amino acid sequences corresponding to the variable region and the EPIP motif of the *Arabidopsis* IDL proteins. Second, we used only the amino acid sequences of the EPIP motifs of the *Arabidopsis* IDL proteins as query to perform TBLASTN searches in the NCBI GenBank dbEST database for the expressed sequence tag (EST) set of citrus and *Poncirus trifoliata*. Selected EST sequences (query coverage > 85%; max identity > 55%) were downloaded from the GenBank database and assembled using the CAP3 program (Huang and Madan, 1999). We used the sequences of the unigenes to query against the citrus (sweet orange and clementine) genome at the JGI.[1] Signal peptide predictions of all putative citrus IDL proteins were carried out using the SignalP 4.0 server[2].

To identify the citrus HSL orthologs TBLASTN (six-frame translation) searches against the citrus (sweet orange and clementine) genome were performed at the JGI[1] using the amino acid sequences of the *Arabidopsis* HAE (AT4G28490), HSL1 (AT1G28440), and HSL2 (AT5G65710) proteins.

Protein Alignments and Phylogenetic Analysis

Multiple sequence alignments of the IDL and HSL proteins were performed using ClustalW tools offered by GenomeNet[3] with default parameters and displayed with GENEDOC. Based on the aligned sequences of the citrus IDL and HSL proteins, Neighbor Joining trees were constructed using MEGA version 6.0 (Tamura et al., 2013) with a bootstrap of 1000 replicates.

RESULTS

Identification of the Citrus Members of the IDA-HAE/HSL2 Signaling Module

In the citrus (sweet orange and clementine) genome five *IDL* genes were identified. They were named *CitIDA1* to *CitIDA5* depending on the location of their sequences in the citrus chromosomes or scaffolds (Supplementary Table S1). The phylogenetic relationship between the six *Arabidopsis* and five citrus IDL proteins were investigated, showing CitIDA3 to be the closest related to AtIDA (**Figure 1A**). This gave rise to the possibility that CitIDA3 could have a function in regulating cell separation during organ abscission. Supporting this idea the complete nucleotide sequence of *CitIDA3* can be reconstructed from two ESTs derived from AZ libraries (Supplementary Table S3).

All of the citrus *IDL* genes showed uninterrupted open reading frames encoding translation products with a predicted N-terminal signal peptide sequence followed by a variable region and a highly conserved C-terminal PIP domain (Stenvik et al., 2008; Butenko et al., 2014; **Figure 1B**). Within the twelve amino acid residues that make up the PIP domain, there were invariant amino acid residues (positions 3, 5, 7, 8, and 12) and others in which the substitutions were conservative (positions 1, 2, and 6). CitIDA3 was the IDA-like protein with highest similarity to the PIP motif of AtIDA differing only in one amino acid (**Figure 1B**). Given these similarities we hypothesized that CitIDA3 has a conserved function similar to that of AtIDA.

Our genome-wide analysis identified three HSL proteins in the citrus (sweet orange and clementine) genome highly similar to each of the *Arabidopsis* proteins (**Figure 1A** and Supplementary Table S2). Therefore, the IDA-HAE/HSL2 signaling module could also work in citrus.

Function and Expression of *CitIDA3* Phenocopies *AtIDA*

To investigate whether *CitIDA3* has a conserved function to *AtIDA* in regulating cell separation during organ abscission, *Arabidopsis* plants were transformed with a construct driving *CitIDA3* expression with the strong constitutive cauliflower mosaic virus 35S promoter. Fully developed *35S:CitIDA3* plants showed reduced stature and shorter siliques in comparison to wild-type plants (**Figures 2A,B**). The *35S:CitIDA3* transgenic plants abscised their floral organs at an earlier stage than wild-type plants. Petal breakstrength (pBS), the force needed to remove petals from a flower, decreases along the length of the inflorescence. Changes in progression of abscission are determined by pBS (Bleecker and Patterson, 1997). In wt pBS reaches zero at floral position 8, while for *35S:CitIDA3* this occurs at position 4, comparable to that observed for *35S:AtIDA* (**Figure 2C**; Stenvik et al., 2006). Furthermore, the premature floral organ abscission observed for *35S:CitIDA* was associated with an increase in the size of the AZ similar to that observed for *35S:AtIDA* (**Figure 2D**; Stenvik et al., 2006; Shi et al., 2011).

In an additional experiment performed to investigate the spatial and temporal activity of the *CitIDA3* promoter, we monitored the expression of a $Pro_{CitIDA3}$*:GUS* construct in *Arabidopsis* plants. In the flowers, *CitIDA3* was expressed in the style and in floral organ AZs (Supplementary Figure S1A). GUS expression was absent in flowers from positions 1–4 but detected in the AZ of flowers from positions 5–8 (**Figure 2G**). *CitIDA3* was also expressed in cotyledons, developing leaves and roots of 1-week-old transgenic *Arabidopsis* seedlings (Supplementary Figure S1B). The expression of *CitIDA3* in *Arabidopsis* flowers is similar to that of *AtIDA* (Butenko et al., 2003).

35SCitIDA3 Complements the Abscission Deficiency of the *ida* Mutant

In order to investigate to what extent CitIDA3 could replace the function of AtIDA, *35S:CitIDA3* plants were crossed into the *ida-2* mutant background. The F_2 progeny of two independent lines was genotyped in order to identify *ida-2* plants. 6 out of 29 and 5 out of 27 plants were homozygous for the *ida-2*

[1]http://phytozome.jgi.doe.gov/
[2]http://www.cbs.dtu.dk/services/SignalP/
[3]http://www.genome.jp/tools/clustalw/

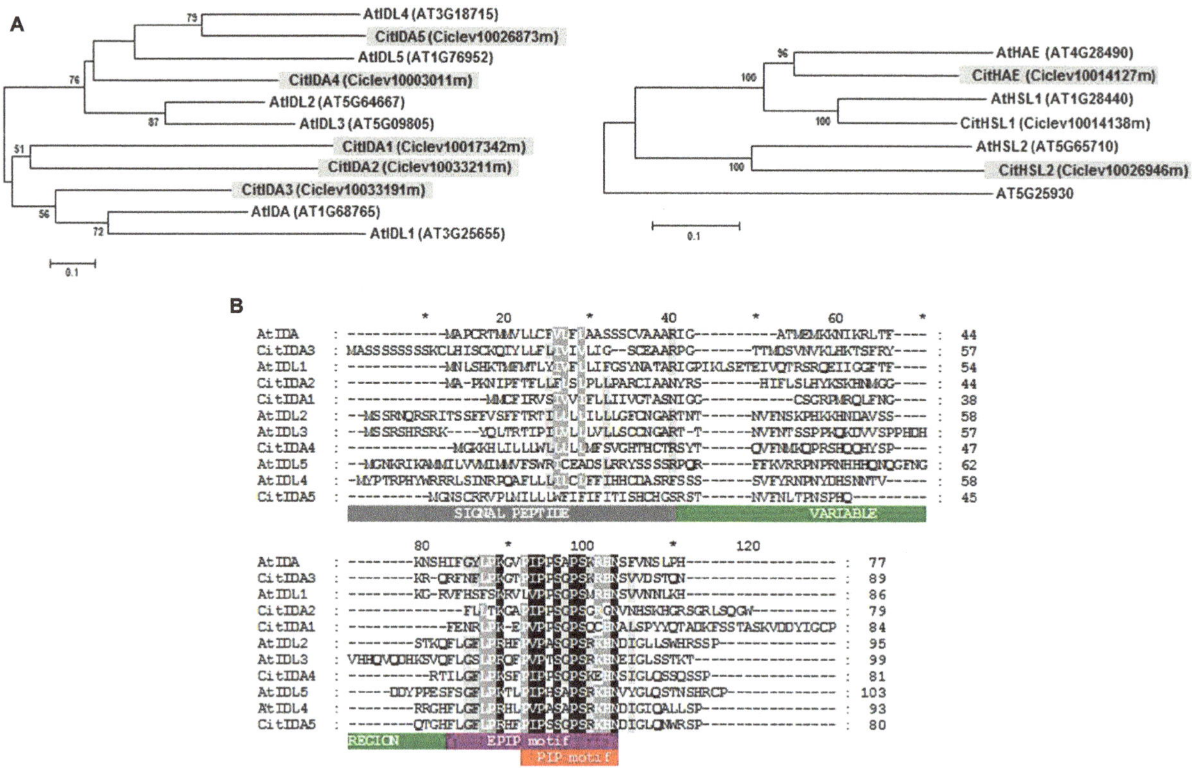

FIGURE 1 | Proteins encoded by *IDL* and *HSL* genes from *Arabidopsis* and citrus. (A) Phylogenetic relationships between IDL and HSL proteins. At5g25930 is a protein kinase not related to HSL proteins. **(B)** Alignment of the IDL full-length proteins showing the N-terminal secretion signal, the variable region, and the short proline-rich EPIP and PIP motives.

allele. Furthermore, the plants were scored for their abscission phenotype. Two and three of the *ida-2* plants from each cross, respectively, showed the *ida-2* phenotype, with attached petals after anthesis (**Figures 2E,F**), The remaining *ida-2* homozygotes showed wt abscission (**Figures 2E,F**). All of these plants contained the *35S:CitIDA3* transgene indicating that the presence of *35S:CitIDA3* is sufficient to induce abscission in *ida-2* mutant plants.

DISCUSSION

Pre-harvest abscission is a problem in citrus producing areas dedicated to the fresh fruit market, preventing on-tree storage of fruit and thereby shortening the harvesting and marketing season. On the other hand, the force required to remove mature fruits from the trees in citrus varieties dedicated to the processing market (juices and concentrates) is too large for effective mechanical harvesting. Therefore, new citrus genotypes with different abscission behavior or plants derived from genetic engineering (transgenic or CRISPR gene editing technologies) aiding in timing or execution of citrus fruit specific abscission would be favorable.

In *Arabidopsis*, the small peptide IDA forms a ligand-receptor module with the two RLKs HAE and HSL2 to initiate a signaling pathway in floral organ abscission (Stenvik et al., 2008). Even though IDA was first found to regulate floral organ abscission, it has later been shown to play a role in cell separation during lateral root emergence, providing this peptide with a more general function in the control of cell separation (Kumpf et al., 2013). The discovery that IDA, HSL and additional downstream components of the IDA signaling system are found across the plant kingdom (Butenko and Simon, 2015; Stø et al., 2015) makes it plausible that orthologs in citrus have a conserved function. Indeed, several auxin response factor (ARF) genes conserved between citrus and *Arabidopsis* have been shown to have a role in fruit abscission in citrus and floral abscission in *Arabidopsis* (Ellis et al., 2005; Tadeo et al., 2015). Thus, different cell separation processes appear to share common signaling elements that are conserved across plant species.

The results shown here indicate that *CitIDA3* plays a role in floral abscission when overexpressed in *Arabidopsis*. The *Arabidopsis 35S:CitIDA3* transgenic plants were characterized by reduced plant and silique growth and precocious floral organ abscission together with an increased size of the AZ (**Figures 2B–D**). Overexpression of *CitIDA3* resulted in a highly similar phenotype to that observed in *Arabidopsis 35S:AtIDA* plants (Stenvik et al., 2006). Furthermore expression of the *CitIDA3* transgene in an *ida-2* mutant background was

FIGURE 2 | CitIDA3 phenocopies the function of AtIDA. (A) Phenotype of fully developed *Arabidopsis* wt (Col-0) plants and *Arabidopsis* plants ectopically overexpressing the citrus IDA ortholog, *35S:CitIDA3*. Scalebar = 5 cm. **(B)** Comparison between main shoot (upper panel) and silique (lower panel) length in wt and *35S:CitIDA3* plants. Values represent mean ± SD, $n = 10$ (main shoot length) $n = 25$ (silique length). **(C)** Petal breakstrength measurements of wt and *35S:CitIDA3* petals. Values represent mean ± SD, $n = 15$. **(D)** Comparison of wt and *35S:CitIDA3* AZ. Scalebar = 2 mm. **(E)** Comparison of *ida-2* and *ida-2 35S:CitIDA3* floral abscission. Scalebar = 3 mm. **(F)** Comparison of the inflorescence of *ida-2* and *ida-2 35S:CitIDA3*. Arrow head indicates attached petal. Scalebar = 3 cm. **(G)** *Pro_{CitIDA3}:GUS* expression in *Arabidopsis* floral AZ. The numbers indicate flower position along the inflorescence, where position 1 refers to the flower at anthesis. Black arrow heads indicate the AZ. Scalebar = 1 mm.

sufficient to rescue the abscission defect of the mutant and the $Pro_{CitIDA3}$:*GUS* expression pattern in *Arabidopsis* plants is also consistent with a role of this gene in organ abscission (**Figures 2D–G**). Additionally, we identified two LRR-RLKs highly homologous in sequence to the *Arabidopsis* HAE and HSL2 in the citrus (sweet orange and clementine) genome (**Figure 1A** and Supplementary Table S2). All these results strongly suggest that the IDA-HAE/HSL2 abscission-signaling pathway characterized in *Arabidopsis* is conserved in citrus and therefore it could be feasible to manipulate *CitIDA3* in order to prevent or stimulate organ abscission in this fruit crop.

Downstream components of the IDA-HAE/HSL2 abscission-signaling pathway could also be potential candidates for manipulation in order to control abscission in citrus. There is genetic evidence for several KNOTTED LIKE HOMEOBOX (KNOX) transcription factors working downstream of the activated IDA-HAE/HSL2 pathway. These include *BREVIPE DICELLUS (BP)/KNOTTED-LIKE FROM ARABIDOPSIS THALIANA1* (*KNAT1*), *KNAT2*, and *KNAT6* (Shi et al., 2011). There are two genes in the citrus (sweet orange and clementine) genome (LOC102615394/Ciclev10001508m and LOC102628261/Ciclev10001779m, respectively) highly homologous to *KNAT1* and *KNAT2/KNAT6* suggesting that the known downstream components of the IDA-HAE/HSL2 abscission-signaling pathway also exist in citrus.

The MADS-domain transcription factor AGAMOUS-like 15 (AGL15) that previously was reported to play a role in regulating floral abscission in *Arabidopsis* has recently been shown to act downstream of the IDA-HAE/HSL2 and MAP kinase signaling module to regulate *HAE* expression (Fernandez et al., 2000; Patharkar and Walker, 2015). Phosphorylation of AGL15 relieves repression of *HAE* expression leading to production of *HAE* transcript in a positive feedback loop, thereby increasing the expression of the receptor (Patharkar and Walker, 2015). Manipulation of a citrus version of AGL15 (LOC102612882/Ciclev10032519m) could be a target to control abscission as could the CWR enzymes induced by IDA. Several members of different classes of CWR enzymes have been found in *Arabidopsis*, tomato and citrus (Estornell et al., 2013). By investigating the enzymes that are expressed in the citrus AZs, fruit abscission prevention or stimulation could be further modified.

Considering all the above, we can glimpse the possibilities available to manipulate the expression of *CitIDA3* and/or the citrus orthologs of the downstream components of the IDA-HAE/HSL2 abscission-signaling pathway as an agronomic tool that would affect positively on the economic benefit of growers and the citrus industry in general.

AUTHOR CONTRIBUTIONS

LE, MW, MP-A, MT, FT, and MB designed the research. LE, MW, MP-A, MT, FT, and MB performed the research. MW, FT, and MB analyzed the data. MW, FT, and MB wrote the article.

ACKNOWLEDGMENTS

Work at the Centre de Genómica was supported by Ministerio de Economia e Innovación grants PSE-060000-2009-8, IPT-010000-2010-43, and AGL2011-30240 and by Grants PSE-060000-2009-8, IPT-010000-2010-43 and AGL2011-30240 to MT and BIO2011-26302 to MP-A from the Ministerio de Economia e Innovación of Spain. Work at UiO was supported by Grants 13785/F20 and 230849/F20 to MB, 348256/F20 to MW from the Research Council of Norway. The help and expertise of Clara Fuster, Isabel Sanchís, and Ángel Boix are gratefully acknowledged.

REFERENCES

Aalen, R. B., Wildhagen, M., Sto, I. M., and Butenko, M. A. (2013). Ida: a peptide ligand regulating cell separation processes in *Arabidopsis*. *J. Exp. Bot.* 64, 5253–5261. doi: 10.1093/jxb/ert338

Addicott, F. T. (1982). *Abscission*. Berkeley, CA: University of California Press.

Bleecker, A. B., and Patterson, S. E. (1997). Last exit: senescence, abscission,and meristem arrest in *Arabidopsis*. *Plant Cell* 9, 1169–1179. doi: 10.1105/tpc.9.7.1169

Butenko, M. A., Patterson, S. E., Grini, P. E., Stenvik, G.-E., Amundsen, S. S., Mandal, A., et al. (2003). INFLORESCENCE DEFICIENT IN ABSCISSION controls floral organ abscission in *Arabidopsis* and identifies a novel family of putative ligands in plants. *Plant Cell* 15, 2296–2307. doi: 10.1105/tpc.0 14365

Butenko, M. A., and Simon, R. (2015). Beyond the meristems: similarities in the CLV3 and IDA peptide mediated signalling pathways. *J. Exp. Bot.* 66, 5195–5203. doi: 10.1093/jxb/erv310

Butenko, M. A., Vie, A. K., Brembu, T., Aalen, R. B., and Bones, A. M. (2009). Plant peptides in signalling: looking for new partners. *Trends Plant Sci.* 14, 255–263. doi: 10.1016/j.tplants.2009.02.002

Butenko, M. A., Wildhagen, M., Albert, M., Jehle, A., Kalbacher, H., Aalen, R. B., et al. (2014). Tools and strategies to match peptide-ligand receptor pairs. *Plant Cell* 26, 1838–1847. doi: 10.1105/tpc.113.120071

Cai, S., and Lashbrook, C. C. (2008). Stamen abscission zone transcriptome profiling reveals new candidates for abscission control: enhanced retention of floral organs in transgenic plants overexpressing *Arabidopsis* ZINC FINGER PROTEIN2. *Plant Physiol.* 146, 1305–1321. doi: 10.1104/pp.107.1 10908

Cho, S. K., Larue, C. T., Chevalier, D., Wang, H., Jinn, T. L., Zhang, S., et al. (2008). Regulation of floral organ abscission in *Arabidopsis thaliana*. *Proc. Natl. Acad. Sci. U.S.A.* 105, 15629–15634. doi: 10.1073/pnas.0805539105

Clough, S. J., and Bent, A. F. (1998). Floral dip: a simplified method for *Agrobacterium*-mediated transformation of *Arabidopsis thaliana*. *Plant J.* 16, 735–743. doi: 10.1046/j.1365-313x.1998.00343.x

Ellis, C. M., Nagpal, P., Young, J. C., Hagen, G., Guilfoyle, T. J., and Reed, J. W. (2005). AUXIN RESPONSE FACTOR1 and AUXIN RESPONSE FACTOR2 regulate senescence and floral organ abscission in *Arabidopsis thaliana*. *Development* 132, 4563–4574. doi: 10.1242/dev.02012

Estornell, L. H., Agusti, J., Merelo, P., Talon, M., and Tadeo, F. R. (2013). Elucidating mechanisms underlying organ abscission. *Plant Sci.* 19, 48–60. doi: 10.1016/j.plantsci.2012.10.008

Fernandez, D. E., Heck, G. R., Perry, S. E., Patterson, S. E., Bleecker, A. B., and Fang, S. C. (2000). The embryo MADS domain factor AGL15 acts postembryonically. Inhibition of perianth senescence and abscission via constitutive expression. *Plant Cell* 12, 183–198. doi: 10.1105/tpc.12.2.183

Grienenberger, E., and Fletcher, J. C. (2015). Polypeptide signaling molecules in plant development. *Curr. Opin. Plant Biol.* 23, 8–14. doi: 10.1016/j.pbi.2014.09.013

Gubert, C. M., Christy, M. E., Ward, D. L., Groner, W. D., and Liljegren, S. J. (2014). ASYMMETRIC LEAVES 1 regulates abscission zone placement in *Arabidopsis* flowers. *BMC Plant Biol.* 14:195. doi: 10.1186/s12870-014-0195-5

Huang, X., and Madan, A. (1999). Cap3: a DNA sequence assembly program. *Genome Res.* 9, 868–877. doi: 10.1101/gr.9.9.868

Jinn, T. L., Stone, J. M., and Walker, J. C. (2000). Haesa, an *Arabidopsis* leucine-rich repeat receptor kinase, controls floral organ abscission. *Genes Dev.* 14, 108–117.

Kumpf, R. P., Shi, C. L., Larrieu, A., Sto, I. M., Butenko, M. A., Peret, B., et al. (2013). Floral organ abscission peptide IDA and its Hae/Hsl2 receptors control cell separation during lateral root emergence. *Proc. Natl. Acad. Sci. U.S.A.* 110, 5235–5240. doi: 10.1073/pnas.1210835110

Lease, K. A., and Walker, J. C. (2006). The *Arabidopsis* unannotated secreted peptide database, a resource for plant peptidomics. *Plant Physiol.* 142, 831–838. doi: 10.1104/pp.106.086041

Lewis, M. W., Leslie, M. E., and Liljegren, S. J. (2006). Plant separation: 50 ways to leave your mother. *Curr. Opin. Plant Biol.* 9, 59–65. doi: 10.1016/j.pbi.2005.11.009

Liljegren, S. J. (2012). Organ abscission: exit strategies require signals and moving traffic. *Curr. Opin. Plant Biol.* 15, 670–676. doi: 10.1016/j.pbi.2012.09.012

Murphy, E., Smith, S., and De Smet, I. (2012). Small signaling peptides in *Arabidopsis* development: how cells communicate over a short distance. *Plant Cell* 24, 3198–3217. doi: 10.1105/tpc.112.099010

Osborne, D. J. (1989). Abscission. *Crit. Rev. Plant Sci.* 8, 103–129. doi: 10.1080/07352688909382272

Patharkar, O. R., and Walker, J. C. (2015). Floral organ abscission is regulated by a positive feedback loop. *Proc. Natl. Acad. Sci. U.S.A.* 112, 2906–2911. doi: 10.1073/pnas.1423595112

Patterson, S. E. (2001). Cutting loose. Abscission and dehisence in *Arabidopsis*. *Plant Physiol.* 126, 494–500. doi: 10.1104/pp.126.2.494

Roberts, J. A., Whitelaw, C. A., Gonzalez-Carranza, Z. H., and Mcmanus, M. T. (2000). Cell separation processes in plants- models, mechanisms and manipulation. *Ann. Bot. (Lond.)* 86, 223–235. doi: 10.1006/anbo.2000.1203

Sexton, R., and Roberts, J. A. (1982). Cell biology of abscission. *Ann. Rev. Plant Physiol.* 33, 133–162. doi: 10.1146/annurev.pp.33.060182.001025

Shi, C. L., Stenvik, G. E., Vie, A. K., Bones, A. M., Pautot, V., Proveniers, M., et al. (2011). *Arabidopsis* class I KNOTTED-like homeobox proteins act downstream in the Ida-Hae/Hsl2 floral abscission signaling pathway. *Plant Cell* 23, 2553–2567. doi: 10.1105/tpc.111.084608

Stenvik, G. E., Butenko, M. A., Urbanowicz, B. R., Rose, J. K., and Aalen, R. B. (2006). Overexpression of INFLORESCENCE DEFICIENT IN ABSCISSION activates cell separation in vestigial abscission zones in *Arabidopsis*. *Plant Cell* 18, 1467–1476. doi: 10.1105/tpc.106.042036

Stenvik, G. E., Tandstad, N. M., Guo, Y., Shi, C. L., Kristiansen, W., Holmgren, A., et al. (2008). The EPIP peptide of INFLORESCENCE DEFICIENT IN ABSCISSION is sufficient to induce abscission in *Arabidopsis* through the receptor-like kinases HAESA and HAESA-LIKE2. *Plant Cell* 20, 1805–1817. doi: 10.1105/tpc.108.059139

Stø, I. M., Orr, R. J. S., Fooyontphanich, K., Jin, X., Knutsen, J. M. B., Fischer, U., et al. (2015). Conservation of the abscission signaling peptide IDA during Angiosperm evolution: withstanding genome duplications and gain and loss of the receptors HAE/HSL2. *Front. Plant Sci.* 6:931. doi: 10.3389/fpls.2015.00931

Tadeo, F. R., Agustí, J., Merelo, P., Estornell, L. H., Cercós, M., Terol, J., et al. (2015). "To fall or not to fall, that's the question!" Molecular mechanisms underlying organ abscission in Citrus. *Acta Hortic.* 1065, 1189–1195.

Tadeo, F. R., Cercós, M., Colmenero-Flores, J. M., Iglesias, D. J., Naranjo, M. A., Ríos, G., et al. (2008). Molecular physiology of development and quality of citrus. *Adv. Bot. Res.* 47, 148–202.

Tamura, K., Stecher, G., Peterson, D., Filipski, A., and Kumar, S. (2013). Mega6: molecular evolutionary genetics analysis version 6.0. *Mol. Biol. Evol.* 30, 2725–2729. doi: 10.1093/molbev/mst197

Terol, J., Naranjo, M. A., Ollitrault, P., and Talon, M. (2008). Development of genomic resources for *Citrus clementina*: characterization of three deep-coverage BAC libraries and analysis of 46,000 BAC end sequences. *BMC Genomics* 9:423. doi: 10.1186/1471-2164-9-423

Conflict of Interest Statement: The authors declare that the research was conducted in the absence of any commercial or financial relationships that could be construed as a potential conflict of interest.

2

Identification of Putative Candidate Genes for Water Stress Tolerance in Canola (*Brassica napus*)

Jing Zhang[1,2], **Annaliese S. Mason**[3,4], **Jian Wu**[5], **Sheng Liu**[5], **Xuechen Zhang**[2†], **Tao Luo**[1], **Robert Redden**[6], **Jacqueline Batley**[2,4], **Liyong Hu**[1*] **and Guijun Yan**[2*]

[1] Ministry of Agriculture (MOA) Key Laboratory of Crop Ecophysiology and Farming System in the Middle Reaches of the Yangtze River, College of Plant Science and Technology, Huazhong Agricultural University, Wuhan, China, [2] Centre for Plant Genetics and Breeding, School of Plant Biology, Faculty of Science and The UWA Institute of Agriculture, The University of Western Australia, Perth, WA, Australia, [3] Plant Breeding Department, IFZ Research Centre for Biosystems, Land Use and Nutrition, Justus Liebig University, Giessen, Germany, [4] School of Agriculture and Food Sciences and Centre for Integrative Legume Research, The University of Queensland, Brisbane, QLD, Australia, [5] National Key Laboratory of Crop Genetic Improvement, College of Plant Science and Technology, Huazhong Agricultural University, Wuhan, China, [6] Australian Grains Genebank, Department of Economic Development Jobs Transport and Resources, Horsham, VIC, Australia

Edited by:
Vasileios Fotopoulos,
Cyprus University of Technology,
Cyprus

Reviewed by:
Panagiotis Madesis,
Institute of Applied Biosciences,
Greece
Costas Delis,
Technological Educational Institute of
Peloponnese, Greece

***Correspondence:**
Liyong Hu
liyonghu@mail.hzau.edu.cn;
Guijun Yan
guijun.yan@uwa.edu.au

†Present Address:
Xuechen Zhang,
School of Land and Food, University
of Tasmania, Launceston, TAS,
Australia

Drought stress can directly inhibit seedling establishment in canola (*Brassica napus*), resulting in lower plant densities and reduced yields. To dissect this complex trait, 140 *B. napus* accessions were phenotyped under normal (0.0 MPa, S0) and water-stressed conditions simulated by polyethylene glycol (PEG) 6000 (−0.5 MPa, S5) in a hydroponic system. Phenotypic variation and heritability indicated that the root to shoot length ratio was a reliable indicator for water stress tolerance. Thereafter, 66 accessions (16 water stress tolerant, 34 moderate and 16 sensitive lines) were genotyped using 25,495 *Brassica* single nucleotide polymorphisms (SNPs). Genome-wide association studies (GWAS) identified 16 loci significantly associated with water stress response. Two *B. napus* accessions were used for RNA sequencing, with differentially-expressed genes under normal and water-stressed conditions examined. By combining differentially-expressed genes detected by RNA sequencing with significantly associated loci from GWAS, 79 candidate genes were identified, of which eight were putatively associated with drought tolerance based on gene ontology of *Arabidopsis*. Functional validation of these genes may confirm key drought-related genes for selection and breeding in *B. napus*. Our results provide insight into the genetic basis of water stress tolerance in canola.

Keywords: canola, water stress tolerance, genome-wide association study, RNA sequencing, drought-related genes

INTRODUCTION

Canola (*Brassica napus*) is one of the largest oil crops in the world together with soybean and oil palm. However, it is very sensitive to drought stress, a major yield-limiting factor in this crop (Wan et al., 2009). Drought stress can directly inhibit seedling establishment, resulting in lower plant densities and reduced yields. Increased drought conditions are predicted worldwide due to the predicted long-term effects of global climate change (Cook et al., 2007). An important solution to this challenge is the development of canola cultivars which can tolerate water stress in order to maintain oil production. A key step toward genomics-assisted breeding for water stress tolerance in canola involves characterization of functional water-stress-tolerant genes or markers closely linked to these genes (Xue et al., 2013).

Genetic improvement of crops for drought tolerance requires investigation of possible mechanisms including water stress tolerance at the seedling stage and exploration of genetic variation for drought tolerance within a species (Dhanda et al., 2004; Lu et al., 2011). Previous studies have revealed considerable variation in water stress tolerance in wheat (Dhanda et al., 2004), sorghum (Bibi et al., 2012) and soybean (Bouslama and Schapaugh, 1984; Kpoghomou et al., 1990). In wheat seedlings, Dhanda et al. (2004) indicated benefits from selection for osmotic membrane stability of leaf segments and root to shoot length ratio. In sorghum seedlings, root length had the highest proportional contribution to drought tolerance (Bibi et al., 2012). In soybean, plant height stress index was a reliable parameter for predicting cultivar growth performance in early stages of development (Kpoghomou et al., 1990). Thus, seedling characteristics can often be used to study the effects of water stress on plants and to identify underlying functional genes.

Genome-wide association studies (GWAS) have proven powerful in revealing the complex genetic basis of important traits in sequenced genomes of *Arabidopsis thaliana*, rice and maize (Atwell et al., 2010; Huang et al., 2010, 2012; Zhao et al., 2011; Wang et al., 2012; Li et al., 2013; Chen et al., 2014; Wen et al., 2014; Matsuda et al., 2015; Ogura and Busch, 2015). Single nucleotide polymorphisms (SNPs) are currently recognized as the marker of choice in most species for GWAS (Raman et al., 2014). Few GWAS-based studies had been reported in canola until recently, due to limited marker numbers, low-throughput genotyping technologies and few available lines (Cai et al., 2014; Li et al., 2014a). However, the recent release of the Illumina Infinium Brassica 60K SNP array produced by the International Brassica SNP Consortium, and the sequenced and annotated genome of *B. napus* (http://www.genoscope.cns.fr/brassicanapus) (Chalhoub et al., 2014), now provide a low-cost and efficient method for high-density, sequence-based, genome-wide polymorphism screening in *B. napus* populations (Liu et al., 2013; Raman et al., 2014). Since these advances, GWAS has been used to dissect the genetic control of seed weight and quality, seed germination and seedling vigor in *B. napus* (Li et al., 2014a; Hatzig et al., 2015). Although the genetic mechanisms of drought tolerance are complex, GWAS has been successfully used to identify functional variation in both known and unknown genes associated with drought tolerance in maize (Hao et al., 2011; Xue et al., 2013). Previous drought studies in canola only used low marker density QTL mapping, which may be unable to capture global genetic diversity for drought tolerance (Li et al., 2014b; Fletcher et al., 2015).

In this study, 140 canola accessions were phenotyped under normal and water stress (simulated by PEG 6000) conditions; of these, 66 accessions ranging from water stress tolerant through to sensitive were selected to perform GWAS using the Illumina Infinium Brassica 60 K SNP array (http://illumina.com; 52157 SNPs). Two selected accessions were used to conduct RNA sequencing (RNA-seq). The objective of this study was to identify SNP markers and candidate genes significantly associated with water stress tolerance during early seedling growth in canola.

MATERIALS AND METHODS

Plant Materials

In this study, 140 accessions originating from 17 countries were screened. Most of the lines came from the Australia Temperate Field Crops Collection (ATFCC), while three came from Huazhong Agricultural University (HZAU) and six came from the Oil Crops Research Institute, Chinese Academy of Agriculture Science (OCRI-CAAS) (Supplementary Table 1).

Experimental Design and Phenotyping

Seeds were germinated in petri dishes on filter paper (Whatman™, Malaga, WA, Australia) soaked with 5 ml deionized water and maintained in a dark incubator at 25°C. When the radicles were 3–5 mm long, seeds from each accession were transferred to two hydroponic boxes with deionized water. The seedlings were grown under controlled conditions in a constant temperature room (25°C) with 12 h light (93.7 $\mu mol \cdot m^{-2} s^{-1}$) and 12 h dark at 50% relative humidity. After 5 days, seedlings were exposed to either 0.0 MPa (normal condition, S0) or −0.5 MPa (water stress, S5) for 7 days using ¼ strength Hoagland's solution or ¼ strength Hoagland's solution supplemented with polyethylene glycol (PEG) 6000 (Michel and Kaufmann, 1973; Michel, 1983), respectively. Deionized water was added to the solution to compensate for evaporation and the solution was stirred to ensure aeration. The experiment was arranged in a completely randomized design with five replications for each line in both treatments. Root length (RL), shoot length (SL), root to shoot length ratio (R/S) and fresh weight (FW) were recorded. Drought stress index (DSI) was calculated for each accession, $DSI = \frac{\text{FW under S5}}{\text{FW under S0}} \times 100\%$ (Kpoghomou et al., 1990).

SNP Genotyping and Filtering

Based on the DSI, 66 inbred lines classified as water stress tolerant (>50%), moderate response (30–50%) or sensitive (<30%) were selected and genotyped using the Illumina Infinium Brassica 60 K SNP array (Supplementary Table 2). DNA was extracted according to methodology detailed in Fulton et al. (1995). All DNA samples were hybridized to an Illumina Infinium Brassica 60 K SNP array released for the *B. napus* genome (http://illumina.com; 52157 SNPs) according to the manufacturer's instructions. SNP chips were scanned using an Illumina HiScanSQ and data visualized using Genome Studio V2011.1 (Illumina, Inc., San Diego, CA, USA) (Mason et al., 2015). The source sequences of the SNP array were used to perform a BLAST (Altschul et al., 1990) search against the public *B. napus* genome sequence (Chalhoub et al., 2014). Only the top blast hits were considered, with those matched to multiple loci not considered (Li et al., 2014a). SNPs with a call frequency <0.8 or minor allele frequency (MAF) <0.05 were excluded from further analysis. In addition, SNPs with >0.25 heterozygous calls were removed.

Genome-Wide Association Studies (GWAS)

The 66 genotyped lines were used for the GWAS association panel. The 4794 SNPs which were evenly distributed across the whole genome were selected to perform the relative kinship analysis using the software package SPAGeDi (Hardy and Vekemans, 2002). The relative kinship matrix (K) comparing all pairs of 66 lines was calculated. The software STRUCTURE was used to infer the population structure (Q) (Pritchard et al., 2000). Principal components analysis (PCA) was performed using GCTA tools (Yang et al., 2011). GWAS was performed using six models: naïve, Q, PCA, K, Q+K and PCA+ K in Tassel 4.0 (Bradbury et al., 2007; Li et al., 2014a). The model which fitted the data best was used to conduct GWAS with 25,495 SNPs (Supplementary Table 2) for data from each environment (S0 and S5). Significance was defined at a uniform threshold of $P < 3.92 \times 10^{-5}$ ($P = 1/N$, N = total markers used: Bonferroni-correction) (Wen et al., 2014).

RNA Extraction and Sequencing

Whole seedlings of two drought sensitive inbred lines according to the drought stress index (Supplementary Table 1)—L72 (MUTU) and L106 (Alku)—were sampled after being subjected to S0 and −1.0 MPa (S10) for 6 h (methods as per Section Experimental Design and Phenotyping). Total RNA was extracted using a Bioline Isolation II RNA Plant Kit (Bioline, Sydney, NSW, Australia) and the genomic DNA contamination was removed by on-column digestion with DNase I which was included in the kit following the manufacturer's protocols. The yield was assessed using a NanoDrop™ ND-1000 spectrophotometer. The RNA quality was also assessed using agarose gel electrophoresis and Agilent 2100 to check the integrity and the residual DNA.

RNA from four samples was used to construct cDNA libraries separately with the Illumina® TruSeq™ RNA Sample Preparation Kit v2. The libraries were sequenced on an Illumina Hiseq™ 2000 platform following the manufacturer's protocols, which produced 50 bp paired-end reads. The raw reads were cleaned using the NGS QC toolkit to remove reads containing primer/adaptor sequences, low-quality reads (sequences whose number of bases with PHRED-like score (Q-score) less than 20 accounted for more than 30%) and reads less than 50 bp in length. Sequences with the first 10 base pairs of reads showing unstable base composition according to the percentages of four different nucleotides (A, T, C, and G) and low-quality reads (Q-score < 20) from 3′ end were also trimmed (Patel and Jain, 2012). All high-quality reads from each sample were separately mapped to the *B. napus* genome sequence (Chalhoub et al., 2014) by Tophat v2.0.11 using default parameters (Trapnell et al., 2009). Only uniquely-mapped reads were considered for gene expression analysis. The transcript abundance of each gene was estimated by FPKM (Fragments per Kilobase of transcript per Million fragments mapped) with a cutoff value of 1.

Gene Ontology and Differential Gene Expression Analysis

All *B. napus* genes (101,040) were searched against the National Center for Biotechnology Information (NCBI) non-redundant (Nr) protein database using BlastP with *E*-value ≤ 1E-05. Gene ontology (GO) terms associated with each BLAST hit were annotated using Blast2go (Conesa et al., 2005), and *B. napus* genes were searched against the InterPro database (http://www.ebi.ac.uk/interpro/) using InterProScan5 (Jones et al., 2014). The GO terms for the *B. napus* genes were annotated by merging Blast2go and InterPro annotation results. Furthermore, homologs of *Arabidopsis* genes in the *B. napus* genome were identified based on BlastP program with *E*-value ≤ 1E-05, identity ≥50% and coverage ≥50%. Genes that had a different transcript level between S0 and S10 samples were identified at $|\log_2(S10/S0)| \geq 1$ and false discovery rate (FDR) ≤ 0.01 across the two plant accessions assessed, and were subsequently regarded as differentially-expressed genes (DEGs).

Real-Time Quantitaitve RT-PCR

The same RNA for RNA sequencing was used to obtain first-strand cDNA with the EasyScript First-strand cDNA Synthesis SuperMix kit (TransGen Biotech, Beijing, China) according to the manufacturer's instructions. Specific primers were designed for the target genes (Supplementary Table 5) using Primer 3 online software version 4.0.0 (Untergasser et al., 2012) and adjusted by oligo software version 7.56 (Rychlik, 2007). Real-time quantitative PCR (qRT-PCR) was performed in a 20 μL reaction vessel containing 10 μL *TransStart*® Tip Green qPCR Supermix (TransGen Biotech, Beijing, China), 10 pmoL of forward and reverse gene-specific primers, and 1 μL cDNA. PCR amplification was performed in an IQ5 machine (Bio-Rad, Hercules, CA, USA) according to the manufacturer's instructions with three independent technical replicates for each sample. Following a denaturation step at 95°C for 5 min, the amplification step comprised 40 cycles at 95°C for 15 s and 60°C for 30 s. A melting curve was constructed to determine the specificity of each PCR primer by maintaining the reaction at 95°C for 1 min, cooling the sample to 55°C for 1 min and finally heating to 95°C at a rate of 0.5°C per 6 s. The samples were initially normalized to a selected internal control gene (serine/threonine-protein phosphatase PP2A-1 catalytic subunit, *PP2A-1*), and the relative gene expression levels were determined using the $2^{-\Delta Ct}$ method (An et al., 2015).

Data Analysis

Analyses of variance and correlation coefficients were conducted using Statistix 8.1. Mean values of different characters were used to compute heritability (Burton and Devanez, 1953). QQ plots and Manhattan plots, Venn diagrams and heatmaps were drawn with gplots package, Vennerable and the pheatmap library, respectively, using R version 3.1.1 (http://www.r-project.org/) (R Foundation for Statistical Computing, 2012).

RESULTS

Phenotypic Variation and Heritability of Measured Traits Under Normal (S0) and Water Stress (S5) Conditions

Analysis of variance indicated significant differences among accessions, between environments and their interactions ($P <$

0.01). Significant differences were observed between the means for all traits under normal (S0) and water stress (S5) conditions. Under water stress, RL and R/S significantly increased by 14.3% and 91.0%, respectively, but SL and FW significantly decreased by 40.1% and 60.1%, respectively, compared to the control (**Table 1**). Medium to high heritability estimates were obtained for all tested traits, ranging from 68.2 (SL under S5) to 88.6 (R/S under S5). R/S showed high heritability (88.1 under S0 and 88.6 under S5) compared to RL (87.8 under S0 and 86.1 under S5) (**Table 1**). The level of phenotypic variation detected among all accessions was affected by water stress. Relative to the average variation of all measured traits under normal conditions, variation in RL and R/S increased while variation in SL and FW decreased under water stress conditions (**Figure 1**).

TABLE 1 | Mean values and broad-sense heritability of seedling traits observed under normal (S0) and water stress (S5) conditions.

Trait[a]	Treatment	Mean	*SD*	Reduction[b] (%)	Broad-sense heritability (h^2_B)	Accessions (A) (139 d.f.)[c]	Treatments (T) (1 d.f.)	A × T (1112 d.f.)
RL (cm)	S0	7.8	2.3	−14.3	87.8	**	**	**
	S5	8.9	2.8		86.1	**	**	**
SL (cm)	S0	4.7	1.1	40.1	77.7	**	**	**
	S5	2.8	0.7		68.2	**	**	**
R/S	S0	1.8	0.7	−91.0	88.1	**	**	**
	S5	3.4	1.8		88.6	**	**	**
FW (mg/seedling)	S0	147.5	37.4	60.1	86.5	**	**	**
	S5	58.9	20.8		83.4	**	**	**

[a] *RL, root length; SL, shoot length; R/S, root to shoot length ratio; FW, fresh weight.*

[b]
$$\text{Reduction}(\%) = \frac{(\text{Trait under S0} - \text{Trait under S5})}{\text{Trait under S0}} \times 100\ \%.$$

[c] *d.f., degrees of freedom; **Significant at the 1% level of significance among accessions (A), between treatments (T) and interaction effects of accessions by treatments (A × T).*

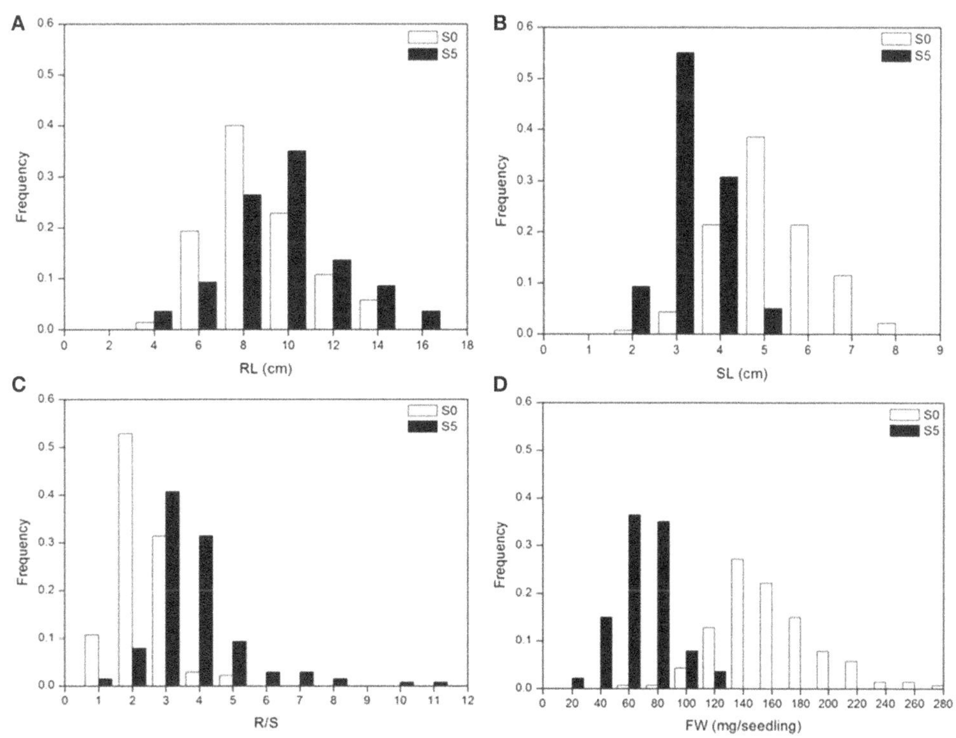

FIGURE 1 | Distribution of seedling traits phenotyped under normal (S0) and water stress (S5) conditions. (A), root length (RL); **(B)**, shoot length (SL); **(C)**, root to shoot length ratio (R/S); **(D)**, fresh weight (FW).

Correlation Coefficients Among Seedling Traits Measured Under Normal (S0) and Water Stress (S5) Conditions

All correlation coefficients for pairs of measured traits between control and water stress conditions were significant and positive ($P < 0.01$). R/S was significantly and positively correlated to RL, but negatively correlated to SL under both control and water stress conditions. FW was significantly and positively ($P < 0.01$) correlated to RL and R/S under both control and water stress conditions, but relatively lower correlation coefficients were observed in the control than under water stress. Also, significant and positive ($P < 0.01$) correlations were observed between FW and SL under water stress conditions. DSI was significantly and negatively correlated to SL and FW in the control, but positively correlated to RL, R/S and FW under water stress (**Table 2**).

Genome-Wide Association Studies

Of the six models tested, the PCA model fitted the data best for all measured traits except R/S under S5 and FW under S0 according to QQ plots (**Figure 2**). Therefore, the PCA model (generalized linear model (GLM) with 10 principal components for population structure) was used to perform GWAS and identify significant associations. A total of 36 SNP-trait associations with $P < 3.92 \times 10^{-5}$ were identified based on Manhattan plots (**Figure 3**); all significant SNPs within 200 kb were consolidated into one highest *P*-value SNP to account for linkage disequilibrium. Hence, a total of 17 associations involving 16 SNPs were identified, explaining 25.8–35.8% of the phenotypic variance (**Table 3**). However, more associations were identified under water stress than control conditions, and 10 associations were detected for R/S under S5 (**Table 3**).

Exploration of Degs in Response to Water Stress

A total of 2566 genes in L72 and 2155 genes in L106 were differentially regulated between water stress and control conditions. Of these, 1387 genes in L72 and 1125 genes in L106 were up-regulated and 1179 genes in L72 and 1030 genes in L106 were down-regulated (**Table 4**). A total of 726 up-regulated DEGs and 626 down-regulated DEGs were found in both L72 and L106 in the Venn diagram (**Figure 4**). Further analysis of these genes identified 21 drought-responsive genes in *B. napus* including five *BnLEA4* (Dalal et al., 2009) and five *BnECP63* (Singh et al., 2009) genes, which were significantly up-regulated in both accessions under water stress (Supplementary Table 3).

Candidate Drought-Response Genes

Based on both GWAS and RNA-seq analyses, 79 DEGs in both cultivars were identified within the vicinity of 1.0 Mb of the 16 SNP locations. Of these, 77 genes were homologs of 64 *A. thaliana* genes, with 58/64 annotated. There were 42 up-regulated genes and 37 down-regulated genes (Supplementary Table 4).

In this study, 11 genes were homologs of annotated drought-related genes in *A. thaliana*, matched with six RL, SL and R/S trait-associated genomic regions under water stress. Moreover, these 11 genes were up-regulated in both tested accessions (**Figure 5**). One gene, *BnaA03g27910D*, was located 568 kb from the marker Bn-A03-p15141821 (associated with both RL under S5 and R/S under S5) and encoded for an ortholog of *A. thaliana* gene "Late embryogenesis abundant protein (LEA) family protein (*LEA4*)"; two genes, *BnaA05g10200D* and *BnaA05g10920D*, were located 405 and 33 kb from the marker Bn-A05-p6415983 (associated with SL under S5) and encoded orthologs to *A. thaliana* genes "mitogen-activated protein kinase kinase kinase 17 (MAPKKK17)" and "Caleosin-related family protein (*RD20*)," respectively; *BnaA05g05760D* was located 495 kb from marker Bn-scaff_15908_1-p140050, and encoded for ortholog to *A. thaliana* gene "delta1-pyrroline-5-carboxylate synthase 1 (*P5CS1*)"; *BnaA07g03740D* and *BnaA07g04670D* were 760 and 928 kb from SNP marker Bn-A10-p3063220, encoded for orthologs to *A. thaliana* genes "cold regulated 413 plasma membrane 1(*COR413PM1*)" and "RING/U-box superfamily protein (*XERICO*)," respectively;

TABLE 2 | Pearson's phenotypic correlation coefficients between seedling traits measured under normal (S0) and water stress (S5) conditions.

Trait[a]	RL (cm)	SL (cm)	R/S	FW (mg/seedling)	DSI (%)[b]
RL (cm)	**0.48****[c]	−0.12 ns	0.73**	0.56**	−0.13
SL (cm)	0.00 ns	**0.65****	−0.69**	0.16 ns	−0.18**
R/S	0.72**	−0.57**	**0.56****	0.23**	0.05
FW (mg/seedling)	0.68**	0.28**	0.33**	**0.49****	−0.28**
DSI (%)	0.46**	0.14 ns	0.26**	0.72**	**NA**

Data from water stress (S5, lower left) and normal (S0, upper right) conditions of 140 canola inbred lines and the cross correlation coefficients between S5 and S0 (diagonal in bold). Phenotypic correlation coefficients were calculated between the traits using means from five replications.

[a] *RL, root length; SL, shoot length; R/S, root to shoot length ratio; FW, fresh weight; DSI, drought stress index.*

[b]

$$DSI = \frac{FW\ under\ S5}{FW\ under\ S0} \times 100\%.$$

[c] ** and ** indicated the correlation coefficient was significant at the 5% and 1% level, respectively; ns indicated the correlation coefficient was not significant.*

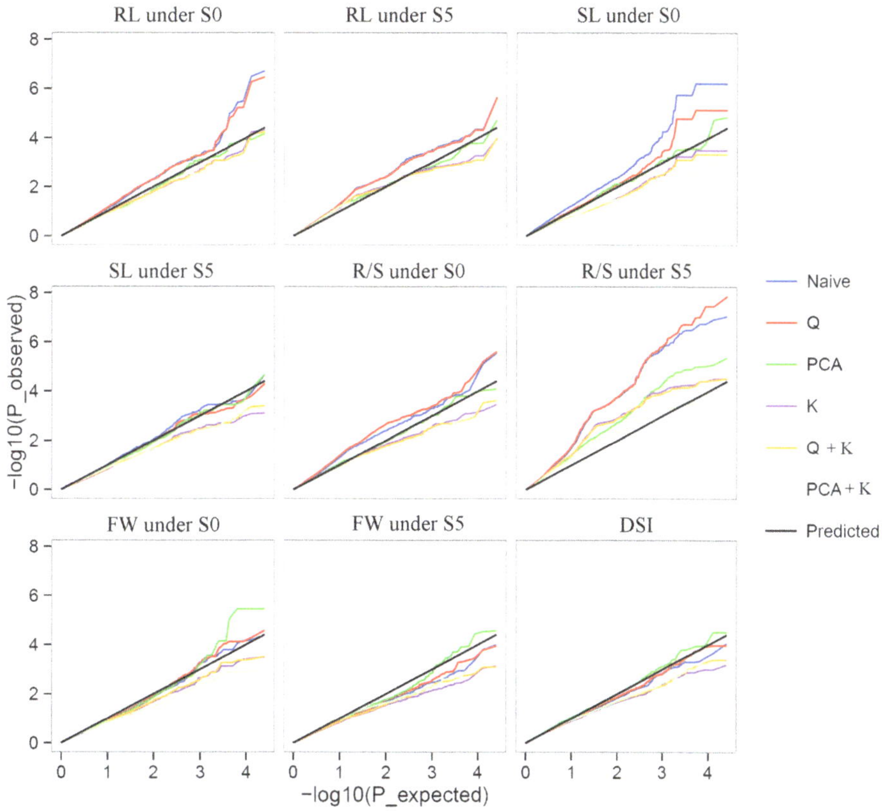

FIGURE 2 | Quantile-quantile plots of estimated $-\log_{10}(p)$ from association analysis using six models for drought stress index (DSI) and four traits under normal (S0) and water stress (S5) conditions. Root length (RL) under S0 and S5; Shoot length (SL) under S0 and S5; Root to shoot length ratio (R/S) under S0 and S5; Fresh weight (FW) under S0 and S5; Drought stress index (DSI).

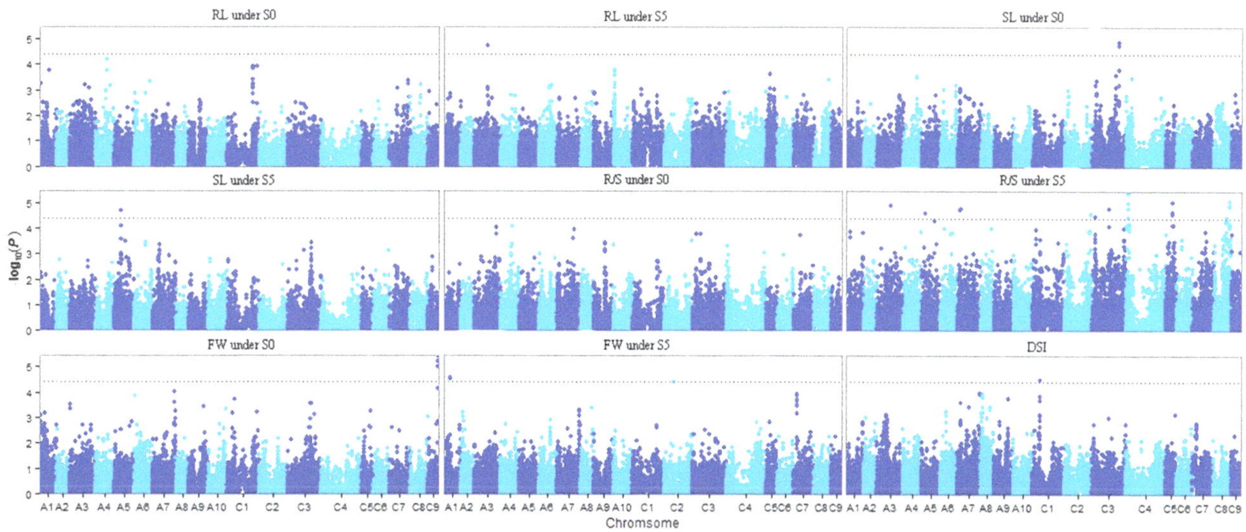

FIGURE 3 | Manhattan plots resulting from GWAS results for drought stress index and four traits under normal (S0) and water stress (S5) conditions. Root length (RL) under S0 and S5; Shoot length (SL) under S0 and S5; Root to shoot length ratio (R/S) under S0 and S5; Fresh weight (FW) under S0 and S5; Drought stress index (DSI). The dashed horizontal line depicts the significance threshold ($P < 3.92 \times 10^{-5}$).

TABLE 3 | Leading genome-wide association SNPs for seedling traits under normal (S0) and water stress (S5) conditions.

Trait[a,b]	SNP	Allele	Linkage group	Position	$-\log_{10}(P)$	R^2 (%)	Published drought-related genes or QTL	Position of published genes or QTL
SL under S0	Bn-scaff_16394_2-p1222322	[A/G]	C3	50058482	4.85	26.66		
SL under S5	Bn-A05-p6415983	[A/G]	A5	5964030	4.68	25.77	BnPIP1 (Yu et al., 2005; Kagale et al., 2007)	A5:6368730
RL under S5, R/S under S5	Bn-A03-p15141821	[T/C]	A3	14250181	4.73, 4.90	29.72	GU189589 (Chen et al., 2009)	A3:13907353
R/S under S5	Bn-scaff_15908_1-p140050	[T/C]	A5	3535320	4.58	26.88	BNECP63 (Singh et al., 2009)	A5:4193606
	Bn-A10-p11466727	[A/C]	A7	2036064	4.73	29.49	BnaCPK18 (Zhang et al., 2014)	A7:2037903
	Bn-A10-p3063220	[T/C]	A7	4149628	4.79	33.46		
	Bn-scaff_17623_1-p657238	[T/C]	C2	42770184	4.56	26.80	BnPIP1 (Yu et al., 2005; Kagale et al., 2007)	C2:42871610
	Bn-scaff_22728_1-p1185330	[A/G]	C3	5479710	4.45	26.77	BNLAS (Yang et al., 2010)	C03:4993284
	Bn-scaff_17521_1-p1398928	[T/G]	C3	20851748	4.77	27.81	GU189589 (Chen et al., 2009)	C3:20465796
	Bn-scaff_19248_1-p73640	[T/C]	C4	6627161	5.37	31.29	GU189586 (Chen et al., 2009)	C4:6662462
	Bn-scaff_16268_1-p444284	[A/G]	C5	12805031	5.02	30.31	BnPIP 1(Yu et al., 2005; Kagale et al., 2007)	C5:12258275
	Bn-scaff_16445_1-p1834116	[A/G]	C8	34965909	5.08	34.32	GU189582 (Chen et al., 2009)	C8:34681768
FW under S0	Bn-scaff_16362_1-p404058	[T/C]	C9	43721409	5.45	33.17	BN28a (Lee et al., 2012)	C9:43714965
FW under S5	Bn-A01-p7558345	[A/G]	A1	6898964	4.57	31.09	*rsdw1.4* (Li et al., 2014b)	A1: 6971413
	Bn-scaff_20221_1-p84872	[A/G]	C2	24189052	4.44	35.77	BnaCPK17 (Zhang et al., 2014)	C2:23752130
DSI	Bn-scaff_17369_1-p625180	[T/C]	C1	11746556	4.52	28.93	BnPIP1 (Yu et al., 2005; Kagale et al., 2007)	C1:11562792

[a]*RL, root length; SL, shoot length; R/S, root to shoot length ratio; FW, fresh weight; DSI, drought stress index.*

[b]

$$DSI = \frac{FW\ under\ S5}{FW\ under\ S0} \times 100\ \%.$$

TABLE 4 | Differentially-expressed genes (DEGs) between seedlings under control (S0) and water stress (S10) conditions.

Lines	Treatments	Total clean reads	Mapped reads	Uniquely mapped reads	Number of water stress-induced genes				
					Total	Up-regulated		Down-regulated	
						Known[a]	Unknown[b]	Known	Unknown
L72	S0	12583612	11791063 (93.7%)	8175037 (65.0%)	2566	1218	169	1078	101
	S10	12331549	11531448 (93.5%)	8021295 (65.0%)					
L106	S0	11656594	10926990 (93.7%)	7585794 (65.1%)	2155	972	153	960	70
	S10	12484808	11575069 (92.7%)	8145137 (65.2%)					

[a]*Known genes had Gene Ontology annotation.*
[b]*Unknown genes had no Gene Ontology annotation.*
http://www.ncbi.nlm.nih.gov/protein/; http://www.ebi.ac.uk/interpro/.

BnaC03g12050D and *BnaC03g12400D* were located 362 and 513 kb from marker Bn-scaff_22728_1-p1185330, and encoded for orthologs of *A. thaliana* genes "lipid transfer protein 3 (*LTP3*)" and "phospholipase C1 (*PLC1*)," both induced by dehydration stress; and *BnaA05g08020D* was 878 kb from Bn_scaff_15908_1-p140050 (associated with R/S under S5) and *BnaC04g09030D* was 148 kb from Bn-scaff_19248_1-p73640 (associated with R/S under S5), which both encoded for orthologs of *A. thaliana* genes "Basic-leucine zipper (bZIP) transcription factor family protein (ABI5)" (Supplementary Table 4).

Comparisons of Candidate Drought-Response Genes Using qRT-PCR with RNA-seq

In order to validate the expression profiles from RNA-seq, qRT-PCR was performed with 18 genes which were orthologs to drought responsive genes in *A. thaliana*. The Pearson's correlation coefficient of genes expression level between the two platforms was calculated. It showed that the gene expression level using RNA-seq was significantly ($R^2 = 0.915$, $P < 0.01$) correlated with those using qRT-PCR (**Figure 6**). The validated

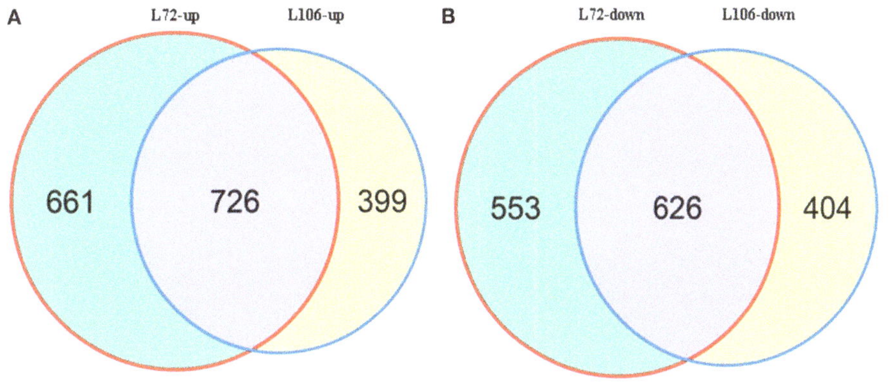

FIGURE 4 | Overlap between differentially-expressed genes (DEGs) in two *B. napus* accessions (L72: MUTU and L106: Alku). Venn diagrams display overlap between DEGs in L72 (green) and L106 (yellow) after exposure to −1.0 MPa (S10) water stress for 6 h. **(A)**, intersection of DEGs up-regulated in L72 with those up-regulated in L106; **(B)**, intersection of DEGs down-regulated in L72 with those down-regulated in L106.

genes included 8 down-regulated genes and 10 up-regulated genes, most of the genes showed consistent expression profiles with those from RNA-seq (**Figure 7**). The results suggested that combined GWAS with RNA-seq was an accurate and reliable way to screen for water stress tolerant candidate genes in canola.

DISCUSSION

Index of Seedling Water Stress Tolerance

Establishing an index of water stress tolerance under water stress conditions would facilitate population screening and breeding. In this study, it was noted that R/S was sensitive to water stress as reflected by the 91.0% increment, much higher variation level and relatively high broad-sense heritability (h_B^2) under water stress (**Table 1** and **Figure 1**). This result was in line with findings in wheat, where the authors suggested that selection on the basis of root to shoot length ratio could be beneficial (Dhanda et al., 2004). In addition, R/S showed much higher correlations with FW and DSI under drought conditions than those under control conditions (**Table 2**). It is possible that the plants avoided low water potential by balancing water uptake and water loss, while the continued root growth under water stress was important for water uptake and the shorter (smaller) plants resulted in less water loss (Dhanda et al., 2004; Ruta et al., 2010). Therefore, plants with relatively higher R/S may have relatively higher FW and DSI under water stress. More significant associations were observed for R/S under water stress conditions than in the normal. Therefore, R/S could be used as a reliable indicator for evaluating water stress tolerance at early seedling growth stages.

Genome-Wide Association of SNPs with Drought Tolerance

With the rapid development of new technologies in DNA sequencing and bioinformatics, and reduction in associated costs, GWAS are rapidly becoming a standard tool for detecting natural variation which accounts for complex quantitative phenotypes in plants (Li et al., 2013). Mixed linear models (MLM) are popular for detecting genotype–phenotype associations in plant GWAS, but are more stringent than GLMs and can result in type II errors (false negatives), while naïve GLMs can result in type I errors (false positives) (Pace et al., 2015). In this study, observed *P*-values fitted better to the expected *P*-values using the PCA model than other models, indicating that population structure correction in a PCA model was more efficient at removing false positives than the STRUCTURE algorithm (**Figure 2**) (Montilla-Bascón et al., 2015). The PCA model has been used to dissect nine agronomic traits in maize under well-watered and water-stressed conditions (Xue et al., 2013).

Although only 66 lines were used in the association panel in our study, these lines showed consistent water stress tolerant (16), moderate (34) and water stress sensitive (16) phenotypes. Of the 16 SNP loci associated with one or two phenotyped traits, some were located close to previously identified drought-responsive genes. For example, 10 drought-related QTLs or genes reported previously were close to 14 of the significantly associated SNPs we identified. Bn-A05-p6415983 associated with SL under S5 was 406 kb from *BnPIP1* (Yu et al., 2005; Kagale et al., 2007) and Bn-A03-P15141821 associated with both R/S under S5 and RL under S5 was 343 kb from drought-responsive genes-*GU189589* (Chen et al., 2009). Of the nine SNPs associated with R/S under S5, eight were 2–658 kb from published drought-related genes (Yu et al., 2005; Kagale et al., 2007; Chen et al., 2009; Singh et al., 2009; Yang et al., 2010; Zhang et al., 2014). Bn-scaff_16362_1-p404058 associated with FW under S0 was only 6 kb from *BN28a* (Lee et al., 2012); Bn-A01-p7558345 and Bn-scaff_20221_1-p84872 associated with FW under S5 were 72 and 437 kb from the reported drought-related QTL-*rsdw1.4* and gene-*BnaCPK17*, respectively (Li et al., 2014b; Zhang et al., 2014). Bn-scaff_17369_1-p625180 associated with DSI was 184 kb from *BnPIP1* (Yu et al., 2005; Kagale et al., 2007; **Table 3**). These results support the success of the association genetics approach used in this study.

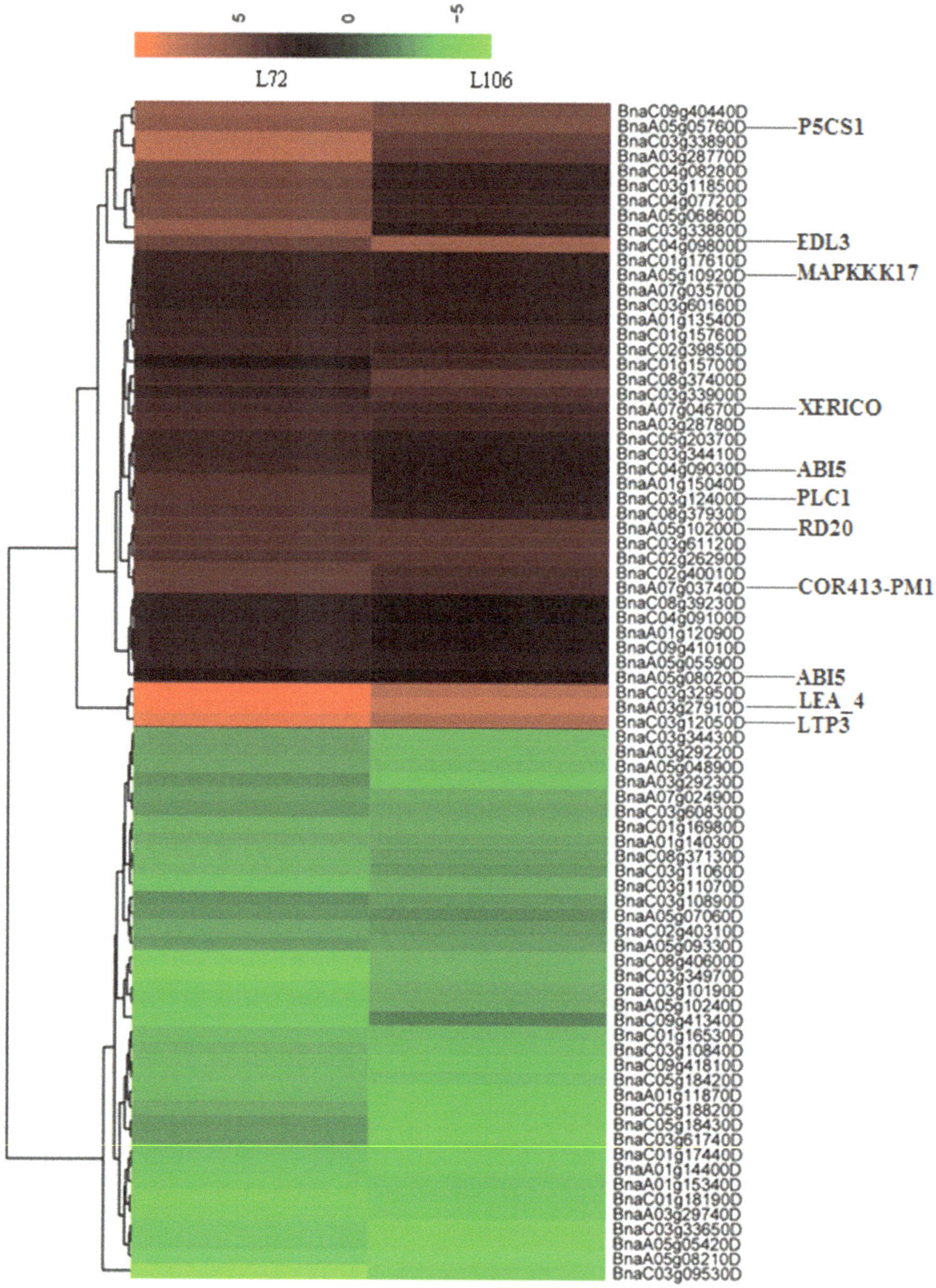

FIGURE 5 | Heatmap with hierarchical clustering showing differentially-expressed genes (DEGs) in two *B. napus* accessions (L72: MUTU and L106: Alku) within the vicinity of 1.0 Mb of 16 uniquely identified SNPs associated with water stress tolerance phenotypes. The genes which were orthologs to drought-related genes in *A. thaliana* are shown.

Details of Candidate Genes

For the DEGs in response to water stress, 40% of total genes in the two lines followed a similar pattern (**Figure 4**) while 60% of the total genes followed different patterns. This outcome may be due to the differed genetic backgroud of the two lines since they are originated from two different sources, one from Australia while the other from Finland. Their different responses to nutrients (the seedlings were exposed to ¼ strength Hoagland's solution supplemented with PEG 6000) may also result in DEGs (**Figure 4**). In order to exclude the above effects, the differentially expressed genes overlapped in both lines would be plausible to detect the drought responding genes. We therefore, combined GWAS with DEGs expressed in both tested lines to identify water stress tolerant candidate genes in *B. napus*.

Under drought, R/S and RL were strongly correlated with DSI, an indicator of drought tolerance (Kpoghomou et al., 1990). Thus, the SNP marker Bn-A03-p15141821, associated with both R/S and RL, is of particular interest. The gene, *BnaA03g27910D*, is located near SNP marker Bn-A03-p15141821 and encodes for an ortholog of *A. thaliana* gene *LEA4*. LEA proteins are hydrophilic, mostly intrinsically-disordered proteins, which play major roles in desiccation tolerance (Hundertmark and Hincha, 2008; Candat et al., 2014). Enhanced expression of LEA genes is thought to contribute to desiccation tolerance (Jakoby et al., 2002). Furthermore, transgenic *Arabidopsis* plants over-expressing *BnLEA4-1* showed enhanced tolerance to drought stress (Dalal et al., 2009). In this study, the expression level of gene-*BnaA03g27910D* was greatly up-regulated in both accessions tested. Hence, *BnaA03g27910D* is a plausible candidate gene for drought tolerance in *B. napus*.

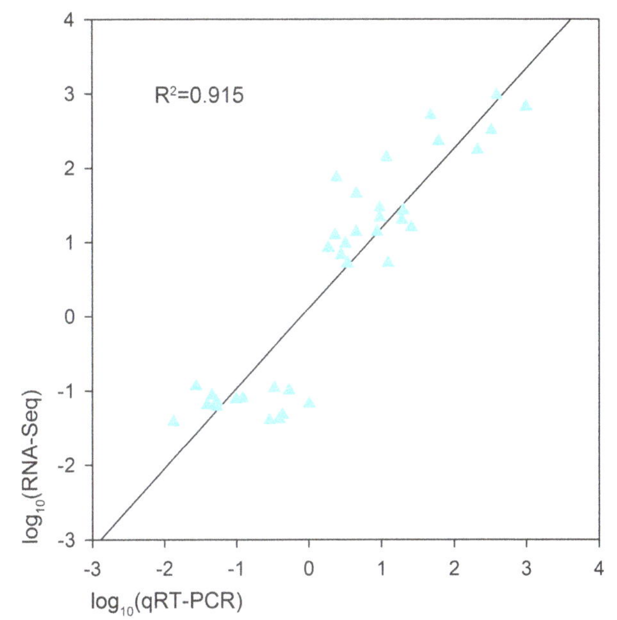

FIGURE 6 | Pearson's correlation between the expression profiles of 18 genes by qRT-PCR and RNA sequencing (RNA-seq) in two *B. napus* accessions (L72: MUTU and L106: Alku). A significant positive correlation is shown ($R^2 = 0.915$, $P < 0.01$).

This study suggests that R/S could be used as an index of drought tolerance under drought treatments, and R/S was involved in 10 of 16 identified marker-trait associations. Based on this approach, six particularly promising candidate genes located near four SNP markers associated with R/S were identified. The gene-*BnaA05g05760D* encoded for ortholog of *A. thaliana* gene *P5CS1*. *P5CS1* encodes the delta1-pyrroline-5-carboxylate synthase enzyme which catalyzes the rate-limiting step of proline biosynthesis (Székely et al., 2008). Enhanced expression of *P5CS1* was observed in both *B. napus* accessions under water stress, and may lead to augmented proline synthesis. Proline accumulation is believed to play a role in plant stress tolerance, and elevated proline levels may contribute to drought tolerance (Verbruggen and Hermans, 2008; Sharma and Verslues, 2010). *BnaA07g03740D* and *BnaA07g04670D* close to SNP marker Bn-A10-p3063220, encoded orthologs to *A. thaliana* genes *COR413PM1* and *XERICO*. *COR413PM1* is a stress-inducible

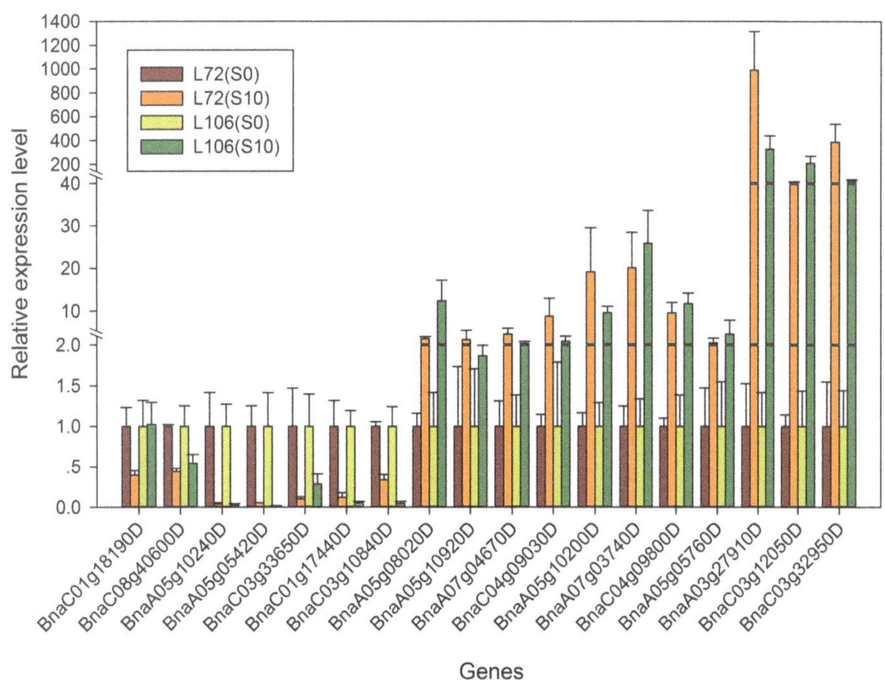

FIGURE 7 | The relative expression level of the 18 genes from real-time quantitative PCR (qRT-PCR) in two *B. napus* accessions (L72: MUTU and L106: Alku). S10, seedlings sampled after exposure to −1.0 MPa water stress for 6 h; S0, seedlings sampled in the parallel control condition.

gene, known to be up-regulated in response to cold and desiccation (Huang et al., 2008; Fang et al., 2015). Previous studies have indicated that up-regulation of *XERICO* confers drought tolerance in *A. thaliana* (Ko et al., 2006). In this study, the two genes were up-regulated under water stress, which was consistent with previous research. Another two genes-*BnaC03g12050D* and *BnaC03g12400D*, located near marker Bn-scaff_22728_1-p1185330, encode orthologs of *A. thaliana* genes *LTP3* and *PLC1* which were both induced by dehydration stress (Hirayama et al., 1995; Guo et al., 2013). The overexpression of *LTP3* and *PLC1* enhanced drought tolerance in *A. thaliana* and maize, respectively (Wang et al., 2008; Guo et al., 2013). In this study, *BnaC03g12050D* and *BnaC03g12400D* significantly enhanced expression under water stress. *BnaC04g09800D*, close to marker Bn-scaff_19248_1-p73640,encoded *EDL3* which is an F-box protein involved in the regulation of abscisic acid signaling, and strong and rapid induction of *EDL3* gene expression was observed under osmotic stress in *A. thaliana*, in line with the results in this study (Koops et al., 2011). In addition, enhanced expression of *BnaA05g10200D* under water stress was observed in this study. *BnaA05g10200D* is located 405 kb from the marker Bn-A05-p6415983 (associated with SL under S5), and encodes *RD20* which plays a role in drought tolerance through stomatal control under water deficit conditions (Aubert et al., 2010). Hence, *BnaA05g10200D* may be considered a possible candidate gene for drought tolerance.

Apart from the above eight genes which were orthologs to drought tolerant genes of *A. thaliana*. We found that *BnaC03g32950D* is located 684 kb from the marker Bn-scaff_17521_1-p1398928 (associated with R/S under S5) and clustered to the same category as *BnaA03g27910D and BnaC03g12050D* (**Figure 5**). Also, its expression level under water stress condition was significantly up-regulated in both tested lines. Therefore, we deduced that it may be a possible new candidate gene for water stress tolerance in *B. napus* not characterized in other plants as for water stress tolerance. Further investigation and validation is needed to confirm its function in water stress tolerance in *B. napus*.

In conclusion, this study supports the use of the root to shoot length ratio as a reliable indicator for drought tolerance at early seedling growth stages. To our knowledge, this is the first report for a genome-wide association of drought tolerance using the Illumina Brassica 60K SNP array in canola for the detection of 16 significantly associated loci. By combining the significantly associated loci and DEGs identified by RNA-seq, 79 candidate genes were identified. Of these genes, nine may be strongly associated with water stress tolerance. Further investigation and validation of these genes is underway to confirm their function in the tolerance of *B. napus* to water stress.

AUTHOR CONTRIBUTIONS

GY, JZ, and LH designed the study. JZ and XZ carried out the experiment work for phenotyping. AM performed the SNP genotyping and filtering. JZ performed the DNA and RNA extractions. JZ, JW, SL, and TL did the data analysis. BR contributed most of the seed materials. JZ wrote the manuscript. AM, GY, JB, BR, and LH critically revised the manuscript. All authors have read and approved the final version of the manuscript.

ACKNOWLEDGMENTS

This research was funded by the National Natural Science Foundation of China (31271664) and Agriculture Research System. The authors sincerely thank Dr. Xuanli Ma, Dr. Heping Han, and Dr. Hui Liu from The University of Western Australia for their help during the RNA and DNA extraction. JZ thanks the China Scholarship Council for providing a scholarship for her study in Australia.

REFERENCES

Altschul, S. F., Gish, W., Miller, W., Myers, E. W., and Lipman, D. J. (1990). Basic local alignment search tool. *J. Mol. Biol.* 215, 403–410. doi: 10.1016/S0022-2836(05)80360-2

An, X., Chen, J., Zhang, J., Liao, Y., Dai, L., Wang, B., et al. (2015). Transcriptome profiling and identification of transcription factors in ramie (*Boehmeria nivea* L. Gaud*)* in response to PEG treatment, using illumina paired-end sequencing technology. *Int. J. Mol. Sci.* 16, 3493–3511. doi: 10.3390/ijms16023493

Atwell, S., Huang, Y. S., Vilhjálmsson, B. J., Willems, G., Horton, M., Li, Y., et al. (2010). Genome-wide association study of 107 phenotypes in *Arabidopsis thaliana* inbred lines. *Nature* 465, 627–631. doi: 10.1038/nature08800

Aubert, Y., Vile, D., Pervent, M., Aldon, D., Ranty, B., Simonneau, T., et al. (2010). RD20, a stress-inducible caleosin, participates in stomatal control, transpiration and drought tolerance in *Arabidopsis thaliana*. *Plant Cell Physiol.* 51, 1975–1987. doi: 10.1093/pcp/pcq155

Bibi, A., Sadaqat, H. A., Tahir, M. H. N., and Akram, H. M. (2012). Screening of sorghum (*Sorghum bicolor Var Moench*) for drought tolerance at seedling stage in polyethylene glycol. *J. Anim. Plant Sci.* 22, 671–678.

Bouslama, M., and Schapaugh, W. T. Jr. (1984). Stress tolerance in soybeans. I. evaluation of three screening techniques for heat and drought tolerance. *Crop Sci.* 24, 933–937. doi: 10.2135/cropsci1984.0011183X002400050026x

Bradbury, P. J., Zhang, Z., Kroon, D. E., Casstevens, T. M., Ramdoss, Y., and Buckler, E. S. (2007). TASSEL: Software for association mapping of complex traits in diverse samples. *Bioinformatics* 23, 2633–2635. doi: 10.1093/bioinformatics/btm308

Burton, G. W., and Devanez, E. H. (1953). Estimating heritability in tall fescue (*Festuca Arundinacea*) from replicated clonal material. *Agron. J.* 45, 478–481. doi: 10.2134/agronj1953.00021962004500100005x

Cai, D., Xiao, Y., Yang, W., Ye, W., Wang, B., Younas, M., et al. (2014). Association mapping of six yieldrelated traits in rapeseed (*Brassica napus* L.). *Theor. Appl. Genet.* 127, 85–96. doi: 10.1007/s00122-013-2203-9

Candat, A., Paszkiewicz, G., Neveu, M., Gautier, R., Logan, D. C., Avelange-Macherel, M. H., et al. (2014). The ubiquitous distribution of late embryogenesis abundant proteins across cell compartments in *Arabidopsis* offers tailored protection against abiotic stress. *Plant Cell* 26, 3148–3166. doi: 10.1105/tpc.114.127316

Chalhoub, B., Denoeud, F., Liu, S., Parkin, I. A., Tang, H., Wang, X., et al. (2014). Plant genetics. Early allopolyploid evolution in the post-Neolithic *Brassica napus* oilseed genome. *Science* 345, 950–953. doi: 10.1126/science.1253435

Chen, L., Ren, F., Zhong, H., Jiang, W., and Li, X. (2009). Identification and expression analysis of genes in response to high-salinity and drought stresses in *Brassica napus*. *Acta Biochim. Biophys. Sin.* 42, 154–164. doi: 10.1093/abbs/gmp113

Chen, W., Gao, Y., Xie, W., Gong, L., Lu, K., Wang, W., et al. (2014). Genome-wide association analyses provide genetic and biochemical insights into natural variation in rice metabolism. *Nat. Genet.* 46, 714–721. doi: 10.1038/ng.3007

Conesa, A., Götz, S., García-Gómez, J. M., Terol, J., Talón, M., and Robles, M. (2005). Blast2GO: a universal tool for annotation, visualization and analysis in functional genomics research. *Bioinformatics* 21, 3674–3676. doi: 10.1093/bioinformatics/bti610

Cook, E. R., Seager, R., Cane, M. A., and Stahle, D. W. (2007). North American drought: reconstructions, causes, and consequences. *Earth Sci. Rev.* 81, 93–134. doi: 10.1016/j.earscirev.2006.12.002

Dalal, M., Tayal, D., Chinnusamy, V., and Bansal, K. C. (2009). Abiotic stress and ABA-inducible Group 4 *LEA* from *Brassica napus* plays a key role in salt and drought tolerance. *J. Biotechnol.* 139, 137–145. doi: 10.1016/j.jbiotec.2008.09.014

Dhanda, S. S., Sethi, G. S., and Behl, R. K. (2004). Indices of drought tolerance in wheat genotypes at early stages of plant growth. *J. Agron. Crop Sci.* 190, 6–12. doi: 10.1111/j.1439-037X.2004.00592.x

Fang, Z., Zhang, X., Gao, J., Wang, P., Xu, X., Liu, Z., et al. (2015). A buckwheat (*Fagopyrum esculentum*) DRE-binding transcription factor gene, *FeDREB1*, enhances freezing and drought tolerance of transgenic *Arabidopsis*. *Plant Mol. Biol. Rep.* 33, 1510–1525. doi: 10.1007/s11105-015-0851-4

Fletcher, R. S., Mullen, J. L., Heiliger, A., and Mckay, J. K. (2015). QTL analysis of root morphology, flowering time, and yield reveals trade-offs in response to drought in *Brassica napus*. *J. Exp. Bot.* 66, 245–256. doi: 10.1093/jxb/eru423

Fulton, T. M., Chunwongse, J., and Tanksley, S. D. (1995). Microprep protocol for extraction of DNA from tomato and other herbaceous plants. *Plant Mol. Biol. Rep.* 13, 207–209. doi: 10.1007/BF02670897

Guo, L., Yang, H., Zhang, X., and Yang, S. (2013). Lipid transfer protein 3 as a target of *MYB96* mediates freezing and drought stress in *Arabidopsis*. *J. Exp. Bot.* 64, 1755–1767. doi: 10.1093/jxb/ert040

Hao, Z., Li, X., Xie, C., Weng, J., Li, M., Zhang, D., et al. (2011). Identification of functional genetic variations underlying drought tolerance in maize using SNP markers. *J. Integr. Plant Biol.* 53, 641–652. doi: 10.1111/j.1744-7909.2011.01051.x

Hardy, O. J., and Vekemans, X. (2002). SPAGeDi: a versatile computer program to analyse spatial genetic structure at the individual or population levels. *Mol. Ecol. Notes* 2, 618–620. doi: 10.1046/j.1471-8286.2002.00305.x

Hatzig, S. V., Frisch, M., Breuer, F., Nesi, N., Ducournau, S., Wagner, M., et al. (2015). Genome-wide association mapping unravels the genetic control of seed germination and vigour in *Brassica napus*. *Front. Plant Sci.* 6:221. doi: 10.3389/fpls.2015.00221

Hirayama, T., Ohto, C., Mizoguchi, T., and Shinozaki, K. (1995). A gene encoding a phosphatidylinositol-specific phospholipase C is induced by dehydration and salt stress in *Arabidopsis thaliana*. *Proc. Natl. Acad. Sci. U.S.A.* 92, 3903–3907. doi: 10.1073/pnas.92.9.3903

Huang, D., Wu, W., Abrams, S. R., and Cutler, A. J. (2008). The relationship of drought-related gene expression in *Arabidopsis thaliana* to hormonal and environmental factors. *J. Exp. Bot.* 59, 2991–3007. doi: 10.1093/jxb/ern155

Huang, X., Wei, X., Sang, T., Zhao, Q., Feng, Q., Zhao, Y., et al. (2010). Genome-wide association studies of 14 agronomic traits in rice landraces. *Nat. Genet.* 42, 961–969. doi: 10.1038/ng.695

Huang, X., Zhao, Y., Wei, X., Li, C., Wang, A., Zhao, Q., et al. (2012). Genome-wide association study of flowering time and grain yield traits in a worldwide collection of rice germplasm. *Nat. Genet.* 44, 32–41. doi: 10.1038/ng.1018

Hundertmark, M., and Hincha, D. K. (2008). LEA (Late Embryogenesis Abundant) proteins and their encoding genes in *Arabidopsis thaliana*. *BMC Genomics* 9:118. doi: 10.1186/1471-2164-9-118

Jakoby, M., Weisshaar, B., Dröge-Laser, W., Vicente-Carbajosa, J., Tiedemann, J., Kroj, T., et al. (2002). bZIP transcription factors in *Arabidopsis*. *Trends Plant Sci.* 7, 106–111. doi: 10.1016/S1360-1385(01)02223-3

Jones, P., Binns, D., Chang, H.-Y., Fraser, M., Li, W., Mcanulla, C., et al. (2014). InterProScan 5: genome-scale protein function classification. *Bioinformatics* 30, 1236–1240. doi: 10.1093/bioinformatics/btu031

Kagale, S., Divi, U. K., Krochko, J. E., Keller, W. A., and Krishna, P. (2007). Brassinosteroid confers tolerance in *Arabidopsis thaliana* and *Brassica napus* to a range of abiotic stresses. *Planta* 225, 353–364. doi: 10.1007/s00425-006-0361-6

Ko, J. H., Yang, S. H., and Han, K. H. (2006). Upregulation of an *Arabidopsis* RING-H2 gene, *XERICO*, confers drought tolerance through increased abscisic acid biosynthesis. *Plant J.* 47, 343–355. doi: 10.1111/j.1365-313X.2006.02782.x

Koops, P., Pelser, S., Ignatz, M., Klose, C., Marrocco-Selden, K., and Kretsch, T. (2011). EDL3 is an F-box protein involved in the regulation of abscisic acid signalling in *Arabidopsis thaliana*. *J. Exp. Bot.* 62, 5547–5560. doi: 10.1093/jxb/err236

Kpoghomou, B. K., Sapra, V. T., and Beyl, C. A. (1990). Screening for drought tolerance: soybean germination and its relationship to seedling responses. *J. Agron. Crop Sci.* 164, 153–159. doi: 10.1111/j.1439-037X.1990.tb00801.x

Lee, S. C., Lim, M. H., Yu, J. G., Park, B. S., and Yang, T. J. (2012). Genome-wide characterization of the CBF/DREB1 gene family in *Brassica rapa*. *Plant Physiol. Biochem.* 61, 142–152. doi: 10.1016/j.plaphy.2012.09.016

Li, F., Chen, B., Xu, K., Wu, J., Song, W., Bancroft, I., et al. (2014a). Genome-wide association study dissects the genetic architecture of seed weight and seed quality in rapeseed (*Brassica napus* L.). *DNA Res.* 21, 355–367. doi: 10.1093/dnares/dsu002

Li, H., Peng, Z., Yang, X., Wang, W., Fu, J., Wang, J., et al. (2013). Genome-wide association study dissects the genetic architecture of oil biosynthesis in maize kernels. *Nat. Genet.* 45, 43–50. doi: 10.1038/ng.2484

Li, Z., Mei, S., Mei, Z., Liu, X., Fu, T., Zhou, G., et al. (2014b). Mapping of QTL associated with waterlogging tolerance and drought resistance during the seedling stage in oilseed rape (*Brassica napus*). *Euphytica* 197, 341–353. doi: 10.1007/s10681-014-1070-z

Liu, L., Qu, C., Wittkop, B., Yi, B., Xiao, Y., He, Y., et al. (2013). A high-density SNP map for accurate mapping of seed fibre QTL in *Brassica napus* L. *PLoS ONE* 8:e83052. doi: 10.1371/journal.pone.0083052

Lu, Y., Xu, J., Yuan, Z., Hao, Z., Xie, C., Li, X., et al. (2011). Comparative LD mapping using single SNPs and haplotypes identifies QTL for plant height and biomass as secondary traits of drought tolerance in maize. *Mol. Breed.* 30, 407–418. doi: 10.1007/s11032-011-9631-5

Mason, A. S., Zhang, J., Tollenaere, R., Vasquez Teuber, P., Dalton-Morgan, J., Hu, L., et al. (2015). High-throughput genotyping for species identification and diversity assessment in germplasm collections. *Mol. Ecol. Resour.* 15, 1091–1101. doi: 10.1111/1755-0998.12379

Matsuda, F., Nakabayashi, R., Yang, Z., Okazaki, Y., Yonemaru, J., Ebana, K., et al. (2015). Metabolome-genome-wide association study dissects genetic architecture for generating natural variation in rice secondary metabolism. *Plant J.* 81, 13–23. doi: 10.1111/tpj.12681

Michel, B. E. (1983). Evaluation of the water potentials of solutions of polyethylene glycol 8000 both in the absence and presence of other solutes. *Plant Physiol.* 72, 66–70. doi: 10.1104/pp.72.1.66

Michel, B. E., and Kaufmann, M. R. (1973). The osmotic potential of polyethylene glycol 6000. *Plant Physiol.* 51, 914–916. doi: 10.1104/pp.51.5.914

Montilla-Bascón, G., Rispail, N., Sánchez-Martín, J., Rubiales, D., Mur, L. A., Langdon, T., et al. (2015). Genome-wide association study for crown rust (*Puccinia coronata f. sp. avenae*) and powdery mildew (*Blumeria graminis f. sp. avenae*) resistance in an oat (*Avena sativa*) collection of commercial varieties and landraces. *Front. Plant Sci.* 6:103. doi: 10.3389/fpls.2015.00103

Ogura, T., and Busch, W. (2015). From phenotypes to causal sequences: using genome wide association studies to dissect the sequence basis for variation of plant development. *Curr. Opin. Plant Biol.* 23, 98–108. doi: 10.1016/j.pbi.2014.11.008

Pace, J., Gardner, C., Romay, C., Ganapathsybrumanian, B., and Lubberstedt, T. (2015). Genome-wide association analysis of seedling root development in maize (*Zea mays* L.). *BMC Genomics* 16:47. doi: 10.1186/s12864-015-1226-9

Patel, R. K., and Jain, M. (2012). NGS QC Toolkit: A toolkit for quality control of next generation sequencing data. *PLoS ONE* 7:e30619. doi: 10.1371/journal.pone.0030619

Pritchard, J. K., Stephens, M., and Donnelly, P. (2000). Inference of population structure using multilocus genotype data. *Genetics* 155, 945–959.

Raman, H., Dalton-Morgan, J., Diffey, S., Raman, R., Alamery, S., Edwards, D., et al. (2014). SNP markers-based map construction and genome-wide linkage analysis in *Brassica napus*. *Plant Biotechnol. J.* 12, 851–860. doi: 10.1111/pbi.12186

R Foundation for Statistical Computing (2012). *R: A Language and Environment for Statistical Computing*. Vienna.

Ruta, N., Liedgens, M., Fracheboud, Y., Stamp, P., and Hund, A. (2010). QTLs for the elongation of axile and lateral roots of maize in response to low water potential. *Theor. Appl. Genet.* 120, 621–631. doi: 10.1007/s00122-009-1180-5

Rychlik, W. (2007). OLIGO 7 primer analysis software. *Methods Mol. Biol.* 402, 35–60. doi: 10.1007/978-1-59745-528-2_2

Sharma, S., and Verslues, P. E. (2010). Mechanisms independent of abscisic acid (ABA) or proline feedback have a predominant role in transcriptional regulation of proline metabolism during low water potential and stress recovery. *Plant Cell Environ.* 33, 1838–1851. doi: 10.1111/j.1365-3040.2010.02188.x

Singh, J., Whitwill, S., Lacroix, G., Douglas, J., Dubuc, E., Allard, G., et al. (2009). The use of Group 3 LEA proteins as fusion partners in facilitating recombinant expression of recalcitrant proteins in *E. coli*. *Protein Expr. Purif.* 67, 15–22. doi: 10.1016/j.pep.2009.04.003

Székely, G., Abrahám, E., Cséplo, A., Rigó, G., Zsigmond, L., Csiszár, J., et al. (2008). Duplicated P5CS genes of *Arabidopsis* play distinct roles in stress regulation and developmental control of proline biosynthesis. *Plant J.* 53, 11–28. doi: 10.1111/j.1365-313X.2007.03318.x

Trapnell, C., Pachter, L., and Salzberg, S. L. (2009). TopHat: discovering splice junctions with RNA-Seq. *Bioinformatics* 25, 1105–1111. doi: 10.1093/bioinformatics/btp120

Untergasser, A., Cutcutache, I., Koressaar, T., Ye, J., Faircloth, B. C., Remm, M., et al. (2012). Primer3-new capabilities and interfaces. *Nucleic Acids Res.* 40, e115. doi: 10.1093/nar/gks596

Verbruggen, N., and Hermans, C. (2008). Proline accumulation in plants: a review. *Amino Acids* 35, 753–759. doi: 10.1007/s00726-008-0061-6

Wan, J., Griffiths, R., Ying, J., Mccourt, P., and Huang, Y. (2009). Development of drought-tolerant canola (*Brassica napus* L.) through genetic modulation of ABA-mediated stomatal responses. *Crop Sci.* 49, 1539–1554. doi: 10.2135/cropsci2008.09.0568

Wang, C. R., Yang, A. F., Yue, G. D., Gao, Q., Yin, H. Y., and Zhang, J. R. (2008). Enhanced expression of phospholipase C 1 (*ZmPLC1*) improves drought tolerance in transgenic maize. *Planta* 227, 1127–1140. doi: 10.1007/s00425-007-0686-9

Wang, M., Yan, J., Zhao, J., Song, W., Zhang, X., Xiao, Y., et al. (2012). Genome-wide association study (GWAS) of resistance to head smut in maize. *Plant Sci.* 196, 125–131. doi: 10.1016/j.plantsci.2012.08.004

Wen, W., Li, D., Li, X., Gao, Y., Li, W., Li, H., et al. (2014). Metabolome-based genome-wide association study of maize kernel leads to novel biochemical insights. *Nat. Commun.* 5, 3438. doi: 10.1038/ncomms4438

Xue, Y., Warburton, M. L., Sawkins, M., Zhang, X., Setter, T., Xu, Y., et al. (2013). Genome-wide association analysis for nine agronomic traits in maize under well-watered and water-stressed conditions. *Theor. Appl. Genet.* 126, 2587–2596. doi: 10.1007/s00122-013-2158-x

Yang, J., Lee, S. H., Goddard, M. E., and Visscher, P. M. (2011). GCTA: a tool for genome-wide complex trait analysis. *Am. J. Hum. Genet.* 88, 76–82. doi: 10.1016/j.ajhg.2010.11.011

Yang, M., Yang, Q., Fu, T., and Zhou, Y. (2010). Over expression of the *Brassica napus BnLAS* gene in *Arabidopsis* affects plant development and increases drought tolerance. *Plant Cell Rep.* 30, 373–388. doi: 10.1007/s00299-010-0940-7

Yu, Q., Hu, Y., Li, J., Wu, Q., and Lin, Z. (2005). Sense and antisense expression of plasma membrane aquaporin *BnPIP1* from *Brassica napus* in tobacco and its effects on plant drought resistance. *Plant Sci.* 169, 647–656. doi: 10.1016/j.plantsci.2005.04.013

Zhang, H., Liu, W.-Z, Zhang, Y., Deng, M., Niu, F., Yang, B., et al. (2014). Identification, expression and interaction analyses of calcium-dependent protein kinase (CPK) genes in canola (*Brassica napus* L.). *BMC Genomics* 15:211. doi: 10.1186/1471-2164-15-211

Zhao, K., Tung, C. W., Eizenga, G. C., Wright, M. H., Ali, M. L., Price, A. H., et al. (2011). Genome-wide association mapping reveals a rich genetic architecture of complex traits in *Oryza sativa*. *Nat. Commun.* 2:467. doi: 10.1038/ncomms1467

Conflict of Interest Statement: The authors declare that the research was conducted in the absence of any commercial or financial relationships that could be construed as a potential conflict of interest.

3

Genome-Wide Linkage Mapping of QTL for Yield Components, Plant Height and Yield-Related Physiological Traits in the Chinese Wheat Cross Zhou 8425B/Chinese Spring

Fengmei Gao[1,2,3], Weie Wen[2], Jindong Liu[2], Awais Rasheed[2,4], Guihong Yin[5], Xianchun Xia[2], Xiaoxia Wu[1*] and Zhonghu He[2,4*]

[1] Key Laboratory of Soybean Biology, Soybean Research Institute, Ministry of Education, Northeast Agricultural University, Harbin, China, [2] National Wheat Improvement Center, Institute of Crop Science, Chinese Academy of Agricultural Sciences, Beijing, China, [3] Keshan Sub-Academy, Heilongjiang Academy of Agricultural Sciences, Keshan, China, [4] International Maize and Wheat Improvement Center (CIMMYT) China Office, c/o Chinese Academy of Agricultural Sciences, Beijing, China, [5] Zhoukou Academy of Agricultural Sciences, Zhoukou, China

Edited by:
Soren K. Rasmussen,
University of Copenhagen, Denmark

Reviewed by:
Zhongyun Piao,
Shenyang Agricultural University, China
Wujun Ma,
Murdoch University, Australia

***Correspondence:**
Xiaoxia Wu
xxwu2012@126.com;
Zhonghu He
zhhecaas@163.com

Identification of genes for yield components, plant height (PH), and yield-related physiological traits and tightly linked molecular markers is of great importance in marker-assisted selection (MAS) in wheat breeding. In the present study, 246 F_8 RILs derived from the cross of Zhou 8425B/Chinese Spring were genotyped using the high-density Illumina iSelect 90K single nucleotide polymorphism (SNP) assay. Field trials were conducted at Zhengzhou and Zhoukou of Henan Province, during the 2012–2013 and 2013–2014 cropping season under irrigated conditions, providing data for four environments. Analysis of variance (ANOVA) of agronomic and physiological traits revealed significant differences ($P < 0.01$) among RILs, environments, and RILs × environments interactions. Broad-sense heritabilities of all traits including thousand kernel weight (TKW), PH, spike length (SL), kernel number per spike (KNS), spike number/m^2 (SN), normalized difference in vegetation index at anthesis (NDVI-A) and at 10 days post-anthesis (NDVI-10), SPAD value of chlorophyll content at anthesis (Chl-A) and at 10 days post-anthesis (Chl-10) ranged between 0.65 and 0.94. A linkage map spanning 3609.4 cM was constructed using 5636 polymorphic SNP markers, with an average chromosome length of 171.9 cM and marker density of 0.64 cM/marker. A total of 866 SNP markers were newly mapped to the hexaploid wheat linkage map. Eighty-six QTL for yield components, PH, and yield-related physiological traits were detected on 18 chromosomes except 1D, 5D, and 6D, explaining 2.3–33.2% of the phenotypic variance. Ten stable QTL were identified across four environments, viz. *QTKW.caas-6A.1*, *QTKW.caas-7AL*, *QKNS.caas-4AL*, *QSN.caas-1AL.1*, *QPH.caas-4BS.2*, *QPH.caas-4DS.1*, *QSL.caas-4AS*, *QSL.caas-4AL.1*, *QChl-A.caas-5AL,* and *QChl-10.caas-5BL*. Meanwhile, 10 QTL-rich regions were found on chromosome 1BS, 2AL (2), 3AL,

4AL (2), 4BS, 4DS, 5BL, and 7AL exhibiting pleiotropic effects. These QTL or QTL clusters are tightly linked to SNP markers, with genetic distances to the closest SNPs ranging from 0 to 1.5 cM, and could serve as target regions for fine mapping, candidate gene discovery, and MAS in wheat breeding.

Keywords: linkage analysis, molecular marker, QTL, *Triticum aestivum*, wheat 90K SNP array

INTRODUCTION

Wheat (*Triticum aestivum* L.) is the third most important cereal food crop after maize (*Zea mays* L.) and rice (*Oryza sativa* L.; Green et al., 2012; Edae et al., 2014). It accounts for about 19% of total grain production among the principal cereal crops, and provides 55% of the carbohydrate consumed by the human population in the world (Gupta et al., 1999; Bagge et al., 2007). Food security is becoming a serious concern for the future due to a rapidly increasing population, the gradual decrease in arable land area, shortages of water and other input resources, and predicted climate change impacts on crop yield. Thus, it is very important to increase the yields of all food crops to avert predicted food security crises (Yang et al., 2012).

Wheat GY is a complex quantitative trait with components such as spike number (SN), kernel number per spike (KNS), and thousand kernel weight (TKW). Potential yield is closely associated with plant photosynthesis (Reynolds et al., 2011). Genetic improvement of yield components and physiological traits can certainly increase grain yield (GY). Quantitative trait loci (QTL) mapping is a key approach for understanding the genetic architecture of yield components and physiological traits in wheat (Holland, 2007). Previously, QTL mapping using various segregating populations was conducted for plant height (PH), spike length (SL), SN, KNS, and TKW (Börner et al., 2002; Kumar et al., 2007; Cuthbert et al., 2008; Golabadi et al., 2011; Bennett et al., 2012). However, QTL were defined by relatively large genetic distances due to the limited numbers of markers. In addition, QTL for physiological traits were rarely reported, except a few association studies for SPAD value of chlorophyll content (Chl), normalized difference in vegetation index (NDVI), and canopy temperature (CT) in spring wheat (Edae et al., 2014; Pinto and Reynolds, 2015; Sukumaran et al., 2015).

The recently developed high-density single nucleotide polymorphism (SNP) gene-chip technology provides a superior approach for QTL mapping, because SNP markers have less errors in evaluation, higher accuracy and particularly higher numbers than SSR markers (Birkhead et al., 2010; Yu et al., 2011). In addition, SNPs can be employed to survey the structure and progressive history of populations, as a tool for association and linkage mapping to detect QTL and to build high-density linkage maps (Aranzana et al., 2005; Akhunov et al., 2009). During the past 5 years, high-density SNP data were increasingly used to identity QTL in bi-parental populations and genome-wide association studies (GWAS) in important crops and animals (Rafalski, 2002; Tian et al., 2011; Zhao et al., 2011; Cook et al., 2012; Jia et al., 2013) due to their high call frequency, locus specific, co-dominant inheritance, simple documentation, potential for analysis, and low error rates (Gupta et al., 1999; Schlotterer, 2004). Many QTL for agronomic and quality traits have been successfully identified in maize and rice using GWAS and high-throughput SNP genotyping (Huang et al., 2010, 2011; Li et al., 2012; Yang et al., 2013). QTL mapping using segregating populations and SNP chip technology has been reported in pea, potato, watermelon, and barley (Ariyadasa et al., 2014; Lambel et al., 2014; Prashar et al., 2014; Sindhu et al., 2014). During the last 2 years, several QTL and association mapping studies were conducted for disease resistance, pre-harvest sprouting, and yield related traits using 9K and 90K SNP chips in wheat (Cabral et al., 2014; Sela et al., 2014; Wang et al., 2014; Sukumaran et al., 2015).

Zhou 8425B, an elite Chinese wheat line developed by the Zhoukou Academy of Agricultural Sciences in 1984, has a semi-dwarf PH, large spike, high TKW and multiple disease resistance (Li et al., 2006; Zhao et al., 2008; Xiao et al., 2011). More than 100 cultivars derived from this line have been grown on an accumulated area of over 33 million ha in China during the past 20 years (Yin et al., 2009). Currently, more than half of the wheat cultivars in Henan province, the largest wheat production region in China, are derivatives of Zhou 8425B. Therefore, it should be interesting to dissect the genetic components of this elite line for more efficient use in breeding programs. The present study used a 90K Infinium iSelect SNP assay to screen 246 RILs from the cross of Zhou 8425B/Chinese Spring; 5636 polymorphic SNPs were used to generate a high-density chromosome linkage map. The objectives were to identify QTL for yield components, PH, and yield-related physiological traits, and their tightly linked SNP markers for marker-assisted selection (MAS) in wheat breeding.

MATERIALS AND METHODS

Plant Materials and Field Trials

A total of 246 F_8 RILs derived from the cross of Zhou 8425B/Chinese Spring were used in this study. Field trials were performed at Zhengzhou and Zhoukou of Henan Province, during the 2012–2013 and 2013–2014 cropping seasons, providing data for four environments. The RILs were planted in randomized complete blocks with three replicates at each location. Plots consisted of four 1.5 m rows with 20 cm between

Abbreviations: CT, canopy temperature; Chl-A, SPAD value of chlorophyll content at anthesis; Chl-10, SPAD value of chlorophyll content at 10 days post-anthesis; GY, grain yield; GWAS, genome-wide association study; KNS, kernel number per spike; MAS, marker-assisted selection; NDVI-A, normalized difference in vegetation index at anthesis; NDVI-10, normalized difference in vegetation index at 10 days post-anthesis; PH, plant height; QTL, quantitative trait locus/loci; RIL, recombinant inbred line; SN, spike number/m^2; SL, spike length; SNP, single nucleotide polymorphism; SSR, simple sequence repeat; TKW, thousand kernel weight.

rows. Approximately 50 seeds were sown evenly in each row. The field trials were managed following the local normal practice.

Phenotyping

The NDVI and Chl were measured at anthesis and 10 days post-anthesis in each plot. NDVI was measured by scanning plants with a portable spectroradiometer (GreenSeeker, Ntech Industries, Inc, Ukiah, CA). Chl was scored as the average of six flag leaves per plot using a chlorophyll meter SPAD-502 (Inolta, Japan). PH was measured from the ground to the tip of the spike excluding awns at the late grain-filling stage. SL was recorded as average values of five spikes per plot. SN was scored in a 1 m single row section and then transformed to SN per m^2. KNS was calculated from the mean of 30 randomly selected spikes in each plot. After harvest, TKW was measured by weighing duplicates of 500 kernels from each plot. GY was determined as the weight of grain harvested per unit area (kg/m^2).

Phenotypic Data Analysis

The phenotypic data analyses were conducted with SAS v. 9.2 software (SAS Institute Inc, Cary, NC). PROC GLM was used in ANOVA, where genotypes were considered as fixed effects, and environments and replicates nested in environments were considered as random effects. Correlation analysis between parameters was performed using the "PROC CORR" procedure. Broad-sense heritability was estimated in all environments aqs $h_B^2 = \sigma_g^2/(\sigma_g^2 + \sigma_{ge}^2/r + \sigma_\varepsilon^2/re)$, where the genetic variance $\sigma_g^2 = (MS_f - MS_{fe})/re$, genotype × environment interaction variance $\sigma_{ge}^2 = (MS_{fe} - MS_e)/r$, error variance $\sigma_\varepsilon^2 = MS_e$, MS_f = genotype mean square, MS_{fe} = genotype × environment interaction mean square, MS_e = error mean square, and r and e were the numbers of replicates and environments, respectively.

SNP Genotyping

The 246 RILs and their parents were genotyped with the 90K iSelect SNP array (Wang et al., 2014) from CapitalBio Corporation (Beijing, China; http://www.capitalbio.com). Genotypic clusters for every SNP were determined following the manual for Genome Studio version 1.9.4 with the polyploid clustering version 1.0.0 (Illumina; http://www.illumina.com), based on the data from all the genotypes. SNPs were filtered by excluding those with monomorphic or with poor quality data. SNP markers with missing parental genotype information, where the parental genotypes were inconsistent with progeny genotypic ratios were removed. SNPs with large numbers of missing values (20% or more) were not included in map construction. Molecular markers for dwarf genes *Rht-B1b* and *Rht-D1b* were used to confirm the association with PH.

Linkage Map Construction

SNP markers were grouped using IciMapping 4.0 software (http://www.isbreeding.net). Linkage analysis was performed using JoinMap 4.0 (Stam, 1993). Then, the linkage map was constructed using MapChart 2.2 (http://www.earthatlas.mapchart.com). Map distances between markers were calculated with the Kosambi mapping function. Each linkage group was oriented from the short (S) to long (L) chromosome arms, and the position and the order of the markers were compared with wheat 90K consensus SNP map (Wang et al., 2014).

QTL Analysis

QTL analysis was performed using inclusive composite interval mapping (ICIM) with IciMapping 4.0 software (Li et al., 2007a). Phenotypic values of all lines in each environment, and the averaged phenotypic values from the four environments, were used for QTL detection. Missing phenotypic data were deleted using the "Deletion" command. The walking speed chosen for all QTL was 1.0 cM, with $P = 0.001$ in stepwise regression. Based on 2000 permutations at a probability level of 0.01, the LOD scores to declare significant QTL for all traits ranged from 2.0 to 2.5 across four environments, thus a LOD threshold of 2.5 was chosen for declaration of putative QTL. Each QTL was represented by a 20 cM interval with the LOD maximum as center. The phenotypic variance explained (PVE) was estimated through stepwise regression (Li et al., 2007a).

RESULTS

Phenotypic Evaluation

ANOVA were conducted for TKW, KNS, SN, PH, SL, Chl-A, Chl-10, NDVI-A, and NDVI-10 across four environments. There were significant differences among the 246 RILs for all traits. The frequency distributions of TKW, KNS, SN, PH, SL, Chl-A, Chl-10, NDVI-A, and NDVI-10 for the RILs in each environment were continuous (**Figure S1**), indicating polygenic control. Based on data averaged across four environments, TKW ranged from 26.5 to 52.6 g with an average of 37.2 g, KNS ranged between 41 and 74 with an average of 53, and SN ranged from 318 to 671 with an average of 465. PH and SL ranged from 60.6 to 125.9 cm and 6.9 to 16.0 cm, with averages of 100.9 and 10.2 cm, respectively. Chl-A and Chl-10 ranged between 38.6 and 76.5 units and between 28.7 and 58.1 units, with an average of 46.5 and 47.8 units, respectively. Similarly, NDVI-A and NDVI-10 ranged from 0.51 to 0.81 and 0.40 to 0.71 units, with averages of 0.74 and 0.55, respectively (**Table S1**). TKW, PH, SL, and NDVI-10 showed higher heritabilities, ranging from 0.88 to 0.94, followed by Chl-10 (0.85), Chl-A (0.79), KNS (0.78), SN (0.74), and NDVI-A (0.65). ANOVA of the nine traits revealed significant differences ($P < 0.01$) among RILs, environments, and genotype × environment interactions (**Table 1**), confirming strong environmental influences on these traits.

Correlations Between Traits

Pearson's coefficients of correlation were calculated for all traits based on the data averaged over four environments (**Table 2**). KNS was positively correlated with SL ($r = 0.43$). The highest negative correlation was observed between TKW and SN ($r = -0.36$). KNS exhibited positive correlations with Chl-10 ($r = 0.45$) and NDVI-A ($r = 0.38$). SN was positively correlated with NDVI-A ($r = 0.41$) and NDVI-10 ($r = 0.37$). PH exhibited a negative correlation with Chl-A ($r = -0.44$). Chl-A was positively correlated with Chl-10 ($r = 0.65$). Chl-10 showed

TABLE 1 | Analysis of variance for yield components, plant height, and physiological traits in F_8RILs from Zhou 8425B/Chinese Spring across four environments.

Source of variance	*df*	Sum of squares								
		TKW	KNS	SN	PH	SL	Chl-A	Chl-10	NDVI-A	NDVI-10
Environments	3	30,765**	4676**	6,570,684**	163,966**	1806**	9428**	5974**	13.8**	9.09**
Lines	245	57,359**	89,731**	1,035,994**	605,189**	3971**	19,547**	46,644**	3.6**	9.03**
Replicates	2	29	170	4596	128	201**	113**	574**	0.01**	0.02**
Lines × Environments	735	90,428*	51,955**	717,092**	96,937**	1051**	10,527**	17,795**	2.5**	3.08**
Error	1966	6607	54,938	108,3116	103,483	1090	8235	22,346	1.2	2
Broad-sense heritability		0.94	0.78	0.74	0.94	0.90	0.79	0.85	0.65	0.88

**Significant at $P = 0.05$, **Significant at $P = 0.01$. TKW, thousand kernel weight; KNS, kernel number per spike; SN, spike number/m^2; PH, plant height; SL, spike length; Chl-A, SPAD value of chlorophyll content at anthesis; Chl-10, SPAD value of chlorophyll content at 10 days post-anthesis; NDVI-A, normalized difference in vegetation index at anthesis; NDVI-10, normalized difference in vegetation index at 10 days post-anthesis.*

TABLE 2 | Pearson's coefficient of correlation for average of yield components, plant height, and physiological traits across four environments.

Trait	TKW	KNS	SN	PH	SL	Chl-A	Chl-10	NDVI-A
KNS	−0.12							
SN	−0.36**	−0.25**						
PH	0.19*	−0.02	0.04					
SL	0.08	0.43**	−0.19**	−0.03				
Chl-A	0.24**	0.24**	−0.23**	−0.44**	0.09			
Chl-10	0.20**	0.45***	−0.13	−0.26**	0.21**	0.65**		
NDVI-A	−0.06	0.38**	0.41**	−0.06	0.24**	0.10	0.51**	
NDVI-10	−0.10	0.26**	0.37**	−0.11	0.14	0.01	0.51**	0.78**

**and ** represent significance at $P < 0.05$ and $P < 0.01$, respectively. TKW, thousand kernel weight; KNS, kernel number per spike; NS, number of spikes; PH, plant height; SL, spike length; Chl-A, SPAD value of chlorophyll content at anthesis; Chl-10, SPAD value of chlorophyll content at 10 days post-anthesis; NDVI-A, seedling normalized difference in vegetation index at anthesis; NDVI-10, seedling normalized difference in vegetation index at 10 days post-anthesis.*

positive correlation with NDVI-A ($r = 0.51$) and NDVI-10 ($r = 0.51$). The maximum positive correlation was between NDVI-A and NDVI-10 ($r = 0.78$).

SNP Genotyping

Among 81,587 SNPs used in screening the Zhou 8425B/Chinese Spring population, 7514 SNP markers (9.2%) were polymorphic between the two parental lines. Of those, 192 markers had more than 20% missing data points in the RILs, and 1686 were not anchored on the linkage map.

Linkage Map Construction

Twenty-one linkage groups corresponding to the 21 hexaploid wheat chromosomes were constructed from the 5636 high-quality polymorphic SNP markers (**Tables S2, S3**); 2457 (43.6%) were localized to the A genome with a total length of 1668.1 cM and average marker density of 0.68 cM, 2838 (50.4%) were mapped to the B genome with a total length of 1276.8 cM and average marker density of 0.45 cM, and 341 were mapped to the D genome with a total length of 664.5 cM and average marker density of 1.95 cM. Ninety-four percent of markers mapped to the A and B genomes, indicating that SNP markers on those genomes were much more polymorphic than those in the D genome. All linkage maps covered 3609.4 cM with an average chromosome length of 171.9 cM, ranging from 21.0 cM (6D) to 303.7 cM (7A). The number of SNP markers in each wheat chromosome ranged from 10 mapped on chromosome 3D to 599 on chromosome 5B. The SNP markers were well distributed throughout the genome, although chromosomes 3D, 4D, and 7D exhibited lower marker densities. The overall SNP density was 0.64 cM, with the highest density of 0.28 cM on chromosome 2B, and the lowest density of 9.10 cM on chromosome 3D.

QTL Analysis of Grain Yield and Related Traits

ICIM identified 86 QTL for yield components, PH, and yield-related physiological traits based on data from individual location and year, and data averaged across the four environments. These QTL were detected on 18 chromosomes, excluding 1D, 5D, and 6D (**Table 3**, **Figure 1**). QTL for individual traits are described below.

Thousand Kernel Weight

Thirteen QTL for TKW were identified on chromosomes 1AL, 2DL (2), 3DL, 4AL, 4BS, 5AL (2), 5AS, 5BL, 6A, 7AL, and 7BL, respectively (**Table 3**, **Figure 1**). Alleles that increased TKW at 11 loci were derived from the parent Zhou 8425B, however two positive alleles were contributed by Chinese Spring. *QTKW.caas-6A.1* and *QTKW.caas-7AL* were stably

TABLE 3 | QTL for yield components, plant height, and physiological traits in the Zhou8425B/Chinese Spring population.

Trait	Location and year	QTL	Position[a]	Marker interval	LOD[b]	PVE(%)[c]	Add[d]
TKW	Zhoukou2013	*QTKW.caas-3DL*	57	*IBV5136–Excalibur_c32309_395*	3.16	3.9	−0.88
		QTKW.caas-4AL	141	*wsnp_Ra_rep_c70233_67968353–RAC875_c29282_566*	2.62	3.3	−0.8
		QTKW.caas-4BS.1	24	*BobWhite_c162_145–Kukri_c66885_230*	2.69	3.3	0.92
		QTKW.caas-6A.1	73.5	*Ku_c32392_967–wsnp_RFL_Contig2523_2130662*	5.98	7.3	−1.14
		QTKW.caas-7AL	172	*Kukri_rep_c97425_164–RAC875_c18798_103*	3.66	5.4	−1.05
	Zhengzhou2013	*QTKW.caas-1AL.4*	125	*BS00036104_51–Ra_c5683_1762*	4.33	5.1	−0.96
		QTKW.caas-4AL	141	*wsnp_Ra_rep_c70233_67968353–RAC875_c29282_566*	4.59	5.5	−1.27
		QTKW.caas-5AL.1	220	*TA005992-0641–Tdurum_contig82476_184*	3.35	4.5	1.37
		QTKW.caas-6A.1	73.5	*Ku_c32392_967–wsnp_RFL_Contig2523_2130662*	7.71	4.8	−1.03
		QTKW.caas-7AL	172	*Kukri_rep_c97425_164 –RAC875_c18798_103*	2.86	6.5	−1.19
		QTKW.caas-7BL	131	*Tdurum_contig63207_82–Tdurum_contig15734_221*	3.71	4.3	−0.86
	Zhoukou2014	*QTKW.caas-2DL.1*	65	*D_GBB4FNX01D4DHE_47–RAC875_c79540_228*	3.38	4.5	−0.82
		QTKW.caas-5AS.1	44.5	*wsnp_Ex_rep_c71219_70023450–Kukri_c24642_426*	2.81	3.4	−0.9
		QTKW.caas-5BL	61	*wsnp_Ra_c5634_9952011–RAC875_c14882_275*	6.54	8.2	−1.1
		QTKW.caas-6A.1	73.5	*Ku_c32392_967–wsnp_RFL_Contig2523_2130662*	7.71	9.9	−1.31
		QTKW.caas-7AL	172	*Kukri_rep_c97425_164–RAC875_c18798_103*	2.86	3.5	−0.81
		QTKW.caas-7BL	131	*Tdurum_contig63207_82–Tdurum_contig15734_221*	3.55	4.3	−1.05
	Zhengzhou2014	*QTKW.caas-2DL.2*	1	*wsnp_Ku_c8712_14751858–Ku_c19185_1569*	4.44	5.3	−1.03
		QTKW.caas-4AL	141	*wsnp_Ra_rep_c70233_67968353–RAC875_c29282_566*	2.96	3.4	−1.07
		QTKW.caas-5AS.1	44.5	*wsnp_Ex_rep_c71219_70023450–Kukri_c24642_426*	2.9	3.5	−0.98
		QTKW.caas-5AL.2	183	*Kukri_rep_c102608_599–Tdurum_contig13810_485*	5.12	6.0	−1.25
		QTKW.caas-6A.1	73.5	*Ku_c32392_967–wsnp_RFL_Contig2523_2130662*	7.82	9.2	−1.31
		QTKW.caas-7AL	172	*Kukri_rep_c97425_164–RAC875_c18798_103*	4.56	5.4	−1.09
		QTKW.caas-7BL	131	*Tdurum_contig63207_82–Tdurum_contig15734_221*	3.53	4.0	−1.05
	Average	*QTKW.caas-3DL*	57	*IBV5136–Excalibur_c32309_395*	5.37	5.8	−1.01
		QTKW.caas-4AL	141	*wsnp_Ra_rep_c70233_67968353–RAC875_c29282_566*	5.83	6.6	−0.97
		QTKW.caas-5AL.2	183	*Kukri_rep_c102608_599–Tdurum_contig13810_485*	3.95	4.2	−1.16
		QTKW.caas-6A.1	73.5	*Ku_c32392_967–wsnp_RFL_Contig2523_2130662*	9.3	10.3	−0.85
		QTKW.caas-7AL	172	*Kukri_rep_c97425_164–RAC875_c18798_103*	4.52	4.8	−1.01
		QTKW.caas-7BL	131	*Tdurum_contig63207_82–Tdurum_contig15734_221*	5.22	5.5	−0.96
KNS	Zhoukou2013	*QKNS.caas-2B.1*	92	*Tdurum_contig10048_447–IAAV1381*	2.9	5.7	−1.79
		QKNS.caas-3AL	228.3	*RAC875_c61934_186–wsnp_Ex_c45877_51547406*	6.98	10.9	2.46
		QKNS.caas-4AL	139	*Kukri_rep_c106490_583–RAC875_c29282_566*	3.39	5.1	−1.68
		QKNS.caas-7BS	18	*BS00011652_51–BS00081132_51*	3.02	4.6	1.6
	Zhengzhou2013	*QKNS.caas-3AL*	228.3	*RAC875_c61934_186–wsnp_Ex_c45877_51547406*	3.18	5.1	1.67
		QKNS.caas-3B	166	*RAC875_c10909_1180–BobWhite_c22016_155*	3.42	5.6	−1.75
		QKNS.caas-4AL	139	*Kukri_rep_c106490_583–RAC875_c29282_566*	6.44	10.5	−2.4
	Zhoukou2014	*QKNS.caas-2B.2*	12	*Tdurum_contig98206_211–RFL_Contig1483_1765*	3.45	4.4	−1.54
		QKNS.caas-3AL	228.3	*RAC875_c61934_186–wsnp_Ex_c45877_51547406*	2.79	3.4	1.35
		QKNS.caas-3B	166	*RAC875_c10909_1180–BobWhite_c22016_155*	3.32	4.1	−1.48
		QKNS.caas-4AL	139	*Kukri_rep_c106490_583–RAC875_c29282_566*	7.23	9.1	−2.22
		QKNS.caas-4BL.1	100	*BobWhite_c8266_582–GENE-2826_154*	3.7	4.7	−1.59
		QKNS.caas-6BL	178	*RAC875_c28848_330–BS00065202_51*	3.84	5.0	1.7
	Zhengzhou2014	*QKNS.caas-1BS.1*	43	*Kukri_c1529_462–Kukri_c8390_547*	4.13	7.5	1.83
		QKNS.caas-2D	29	*Kukri_c14902_1112–RAC875_c77816_365*	2.62	5.3	−1.29
		QKNS.caas-4AL	139	*Kukri_rep_c106490_583–RAC875_c29282_566*	6.42	9.4	−1.59
		QKNS.caas-2AL.1	143	*BS00014251_51–IBV80*	5.11	5.4	−1.29
	Average	*QKNS.caas-2B.1*	92	*Tdurum_contig10048_447–IAAV1381*	4.88	6.1	−1.37
		QKNS.caas-3AL	228.3	*RAC875_c61934_186–wsnp_Ex_c45877_51547406*	7	7.2	1.48
		QKNS.caas-3B	166	*RAC875_c10909_1180–BobWhite_c22016_155*	5.21	8.8	−1.83
		QKNS.caas-4AL	139	*Kukri_rep_c106490_583–RAC875_c29282_566*	10.93	11.8	−1.89

(Continued)

TABLE 3 | Continued

Trait	Location and year	QTL	Position[a]	Marker interval	LOD[b]	PVE(%)[c]	Add[d]
SN	Zhoukou2013	*QSN.caas-1AL.1*	47.5	*IACX592–Jagger_c1403_60*	7.66	13.8	−17.49
		QSN.caas-3AL	141	*Ra_c14565_1056–Tdurum_contig64606_1104*	2.63	6.4	−21.19
		QSN.caas-7AL	170	*BS00023673_51–wsnp_JD_c18814_17164689*	3.39	6.4	−20.53
	Zhengzhou2013	*QSN.caas-1AL.1*	47.5	*IACX592–Jagger_c1403_60*	8.59	11.4	−34.46
		QSN.caas-1BL	122	*D_contig12192_450–Tdurum_contig8840_575*	2.62	3.4	−18.77
		QSN.caas-2BL	35	*Excalibur_c19260_105–IACX8096*	3.63	6.7	26.32
		QSN.caas-3AS	113	*wsnp_Ku_c40218_48484410–wsnp_Ex_rep_c106152_90334299*	2.95	3.8	−19.75
	Zhoukou2014	*QSN.caas-1AL.1*	47.5	*IACX592–Jagger_c1403_60*	5.51	8.0	−20.08
		QSN.caas-2AS.3	126	*JG_c883_445–Kukri_rep_c104727_91*	3.91	5.6	16.75
		QSN.caas-3AL	141	*Ra_c14565_1056–Tdurum_contig64606_1104*	3.93	7.9	−20.07
		QSN.caas-5BS.1	9	*BS00032003_51–BS00070871_51*	2.92	4.3	14.69
		QSN.caas-6AL.1	75	*RAC875_c7804_236–Excalibur_c36332_449*	3.66	5.3	−16.19
	Zhengzhou2014	*QSN.caas-1AL.1*	47.5	*IACX592–Jagger_c1403_60*	15.09	17.2	−34.6
		QSN.caas-1BL.1	66	*IAAV4702–wsnp_BG274294B_Ta_2_3*	3.03	2.3	−12.98
		QSN.caas-2AS.3	126	*JG_c883_445–Kukri_rep_c104727_91*	3.81	3.0	14.31
		QSN.caas-3AL	141	*Ra_c14565_1056–Tdurum_contig64606_1104*	9	11.7	−29.26
		QSN.caas-6AL.1	75	*RAC875_c7804_236–Excalibur_c36332_449*	7.12	5.5	−19.5
		QSN.caas-7AL	170	*BS00023673_51–wsnp_JD_c18814_17164689*	6.89	5.9	−20.09
	Average	*QSN.caas-1AL.1*	47.5	*IACX592–Jagger_c1403_60*	12.09	14.2	−24.72
		QSN.caas-2AS.3	128	*Kukri_c7914_99–wsnp_Ex_c36242_44232305*	4.17	4.9	14.57
		QSN.caas-3AL	141	*Ra_c14565_1056–Tdurum_contig64606_1104*	6.93	11.1	−22.1
		QSN.caas-6AL.1	75	*RAC875_c7804_236–Excalibur_c36332_449*	6.33	5.8	−17.92
PH	Zhoukou2013	*QPH.caas-4AL*	90	*IHX2890–RAC875_c35171_613*	4.37	3.9	−2.56
		QPH.caas-4BS.1	11	*RAC875_c86104_111–tplb0025f09_1853*	4.88	4.9	−2.87
		QPH.caas-4BS.2	55	*RAC875_c6749_954–BobWhite_c44691_648*	21.68	22.8	−6.35
		QPH.caas-4DS.1	63.8	*RAC875_c13945_597–BS00036421_51*	13.66	14.5	−4.99
		QPH.caas-5AS	50	*Kukri_c24642_426–RFL_Contig2251_434*	5.67	14.9	−2.89
		QPH.caas-7AL	170	*BS00023673_51–wsnp_JD_c18814_17164689*	2.99	2.6	2.1
	Zhengzhou2013	*QPH.caas-4BS.2*	55	*RAC875_c6749_954–BobWhite_c44691_648*	14.87	22.7	−6.67
		QPH.caas-4DS.1	63.8	*RAC875_c13945_597–BS00036421_51*	14.02	23.5	−6.71
		QPH.caas-5AS	50	*Kukri_c24642_426–RFL_Contig2251_434*	5.84	13.2	−2.61
	Zhoukou2014	*QPH.caas-2BL*	58	*Tdurum_contig47_148–RAC875_c40992_113*	2.62	2.3	2.4
		QPH.caas-4AL	90	*IHX2890–RAC875_c35171_613*	3.25	2.9	−2.69
		QPH.caas-4BS.2	55	*RAC875_c6749_954–BobWhite_c44691_648*	22.59	24.2	−7.98
		QPH.caas-4DS.1	63.8	*RAC875_c13945_597–BS00036421_51*	26.88	33.2	−9.2
		QPH.caas-5AS	50	*Kukri_c24642_426–RFL_Contig2251_434*	5.67	14.3	−2.71
	Zhengzhou2014	*QPH.caas-2BL*	58	*Tdurum_contig47_148–RAC875_c40992_113*	2.82	2.8	2.93
		QPH.caas-4BS.2	55	*RAC875_c6749_954–BobWhite_c44691_648*	23.1	29.3	−9.58
		QPH.caas-4DS.1	63.8	*RAC875_c13945_597–BS00036421_51*	21.21	29.6	−9.51
	Average	*QPH.caas-4AL*	90	*IHX2890–RAC875_c35171_613*	2.73	2.6	−2.26
		QPH.caas-4BS.2	55	*RAC875_c6749_954–BobWhite_c44691_648*	24.73	27.4	−7.53
		QPH.caas-4DS.1	63.8	*RAC875_c13945_597–BS00036421_51*	24.01	30.0	−7.78
		QPH.caas-5AS	50	*Kukri_c24642_426–RFL_Contig2251_434*	2.95	12.7	−2.31
SL	Zhoukou2013	*QSL.caas-1BL*	66	*IAAV4702–wsnp_BG274294B_Ta_2_3*	4.86	6.0	0.34
		QSL.caas-4AS	57.6	*Kukri_c46057_646–RAC875_rep_c77874_269*	3.68	4.5	0.28
		QSL.caas-4AL.1	146	*Kukri_c17417_571–BS00022076_51*	7.93	11.2	−0.44
		QSL.caas-5AL	157.5	*JD_c15758_288–BS00041911_51*	5.78	8.7	−0.39
		QSL.caas-7AS	168	*wsnp_Ex_c200_391493–Ex_c6870_1704*	4.1	4.9	0.29
		QSL.caas-7DS	60	*RAC875_c53629_483–Excalibur_c55782_55*	3.27	4.0	0.27

(Continued)

TABLE 3 | Continued

Trait	Location and year	QTL	Position[a]	Marker interval	LOD[b]	PVE(%)[c]	Add[d]
	Zhengzhou2013	QSL.caas-1BL	66	IAAV4702–wsnp_BG274294B_Ta_2_3	2.6	3.6	0.24
		QSL.caas-3BL	133	Ku_c12191_1202–Excalibur_c3556_520	2.77	3.9	0.24
		QSL.caas-4AS	57.6	Kukri_c46057_646–RAC875_rep_c77874_269	7.98	12.3	0.43
		QSL.caas-4AL.1	146	Kukri_c17417_571–BS00022076_51	4.67	6.8	−0.32
		QSL.caas-5AL	157.5	JD_c15758_288–BS00041911_51	4.24	6.4	−0.31
	Zhoukou2014	QSL.caas-2AL	212.5	BS00022265_51–wsnp_Ex_rep_c70299_69243835	3.54	4.3	0.26
		QSL.caas-4AS	57.6	Kukri_c46057_646–RAC875_rep_c77874_269	8.53	10.3	0.41
		QSL.caas-4AL.1	146	Kukri_c17417_571–BS00022076_51	9.52	11.8	−0.43
		QSL.caas-5AL.1	183.5	Kukri_rep_c102608_599–Kukri_c14187_243	7.09	8.7	−0.38
		QSL.caas-6BL	156	Ra_c2557_2531–BS00067417_51	3.16	3.6	0.26
		QSL.caas-7AS.1	124	Tdurum_contig82438_136–BS00034509_51	4.26	9.6	0.45
	Zhengzhou2014	QSL.caas-1BS	42	BS00070878_51–Kukri_c1529_462	4.79	5.2	0.36
		QSL.caas-2AL	212.5	BS00022265_51-wsnp_Ex_rep_c70299_69243835	3.65	4.0	0.28
		QSL.caas-4AS	57.6	Kukri_c46057_646-RAC875_rep_c77874_269	7.58	8.6	0.4
		QSL.caas-4AL.2	98	Ex_c6665_1067–D_GCE8AKX02GF3QZ_210	3.03	5.4	0.32
		QSL.caas-4AL.1	146	Kukri_c17417_571–BS00022076_51	9.81	11.9	−0.48
		QSL.caas-5AL.1	183.5	Kukri_rep_c102608_599–Kukri_c14187_243	5.37	6.0	−0.34
		QSL.caas-6BL	156	Ra_c2557_2531–BS00067417_51	2.91	3.1	0.26
		QSL.caas-7AS	168	wsnp_Ex_c200_391493–Ex_c6870_1704	3.58	5.4	0.33
	Average	QSL.caas-1BS	43	Kukri_c1529_462-Kukri_c8390_547	6.3	8.8	0.44
		QSL.caas-4AS	57.6	Kukri_c46057_646–RAC875_rep_c77874_269	11.24	13.3	0.42
		QSL.caas-4AL.1	146	Kukri_c17417_571–BS00022076_51	11.91	14.9	−0.44
		QSL.caas-4AL.2	98	Ex_c6665_1067–D_GCE8AKX02GF3QZ_210	2.61	2.7	0.19
		QSL.caas-5AL	159	JD_c15758_288–BS00041911_51	6.89	9.1	−0.35
		QSL.caas-7AS	168	wsnp_Ex_c200_391493–Ex_c6870_1704	4.55	4.8	0.25
Chl-A	Zhoukou2013	QChl-A.caas-2AS	45	wsnp_Ex_c322_624793–Tdurum_contig10785_103	2.59	4.5	0.51
		QChl-A.caas-3AS	111	wsnp_Ku_c11052_18135847–wsnp_Ra_c16278_24893033	3.46	5.6	0.56
		QChl-A.caas-5AL	68.5	BS00109052_51–wsnp_BE443187A_Ta_2_3	3.89	6.5	0.6
	Zhengzhou2013	QChl-A.caas-2AL.1	199	Kukri_c25901_348–Tdurum_contig11659_253	3.42	5.1	0.71
		QChl-A.caas-2DS	55	BS00081578_51–tplb0021c10_951	3.75	8.6	0.92
		QChl-A.caas-3AS	111	wsnp_Ku_c11052_18135847–wsnp_Ra_c16278_24893033	4.31	6.5	0.8
		QChl-A.caas-5AL	68.5	BS00109052_51–wsnp_BE443187A_Ta_2_3	4.33	16.3	0.72
	Zhoukou2014	QChl-A.caas-4AL	78	BobWhite_c15697_675–BobWhite_c2179_1476	3.56	7.1	0.92
		QChl-A.caas-4DS	67	RAC875_c13945_597–BS00036421_51	3.49	6.3	0.86
		QChl-A.caas-5AL	68.5	BS00109052_51–wsnp_BE443187A_Ta_2_3	5.55	8.8	1.01
	Zhengzhou2014	QChl-A.caas-2AL.2	140	Excalibur_c84687_162–BS00014251_51	3.94	5.7	−0.82
		QChl-A.caas-2DL.1	114	IBV8632–D_contig02226_528	2.75	3.9	0.68
		QChl-A.caas-5AL	68.5	BS00109052_51–wsnp_BE443187A_Ta_2_3	5.36	7.8	0.96
		QChl-A.caas-5AS.1	227	Tdurum_contig82476_184–Tdurum_contig30719_380	3.01	8.4	1.31
	Average	QChl-A.caas-2AL.2	140	Excalibur_c84687_162– BS00014251_51	4.27	6.0	−0.61
		QChl-A.caas-2AL.1	199	Kukri_c25901_348– Tdurum_contig11659_253	3.53	4.8	0.54
		QChl-A.caas-3AS	111	wsnp_Ku_c11052_18135847–wsnp_Ra_c16278_24893033	3.58	4.9	0.55
		QChl-A.caas-4DS	64	RAC875_c13945_597–BS00036421_51	4.04	6.2	0.63
		QChl-A.caas-5AL	68.5	BS00109052_51–wsnp_BE443187A_Ta_2_3	6.6	9.2	0.75
Chl-10	Zhoukou2013	QChl-10.caas-2D	29	Kukri_c14902_1112–RAC875_c77816_365	4.11	5.5	−1.41
		QChl-10.caas-5BL	60	wsnp_Ex_c12909_20457660–wsnp_Ra_c5634_9952011	4.83	8.2	1.78
	Zhengzhou2013	QChl-10.caas-2BS	19	RAC875_c21378_474–Tdurum_contig81323_291	3.78	6.3	−1.25
		QChl-10.caas-5BL	60	wsnp_Ex_c12909_20457660–wsnp_Ra_c5634_9952011	6.06	10.3	1.59

(Continued)

TABLE 3 | Continued

Trait	Location and year	QTL	Position[a]	Marker interval	LOD[b]	PVE(%)[c]	Add[d]
	Zhoukou2014	*QChl-10.caas-2D*	29	*Kukri_c14902_1112–RAC875_c77816_365*	4.04	9.2	−1.6
		QChl-10.caas-5BL	60	*wsnp_Ex_c12909_20457660– wsnp_Ra_c5634_9952011*	4.83	7.0	1.33
		QChl-10.caas-6AS	21	*BS00031062_51–RFL_Contig5170_1904*	3.07	4.3	−1
		QChl-10.caas-7A	177	*BS00023993_51–Ex_c52798_415*	3.98	7.9	1.38
	Zhengzhou2014	*QChl-10.caas-2AL*	140	*Excalibur_c84687_162–BS00014251_51*	5.44	7.0	−0.91
		QChl-10.caas-2AL.1	200	*Kukri_c25901_348–Tdurum_contig11659_253*	4.77	6.1	0.85
		QChl-10.caas-2BS	19	*RAC875_c21378_474–Tdurum_contig81323_291*	5.37	7.8	−0.96
		QChl-10.caas-5AL	72	*BS00067453_51–Excalibur_c24638_380*	4.89	6.8	0.9
		QChl-10.caas-5BL	60	*wsnp_Ex_c12909_20457660–wsnp_Ra_c5634_9952011*	8.09	10.6	1.13
		QChl-10.caas-7A	177	*BS00023993_51–Ex_c52798_415*	2.53	4.2	0.72
	Average	*QChl-10.caas-2BS*	19	*RAC875_c21378_474–Tdurum_contig81323_291*	5.31	8.3	−1.14
		QChl-10.caas-5AL	77	*RFL_Contig727_736–wsnp_JD_c3867_4934646*	2.71	3.5	0.73
		QChl-10.caas-5BL	60	*wsnp_Ex_c12909_20457660–wsnp_Ra_c5634_9952011*	10.19	14.2	1.49
		QChl-10.caas-7A	177	*BS00023993_51–Ex_c52798_415*	3.14	5.7	0.96
NDVI-A	Zhengzhou2013	*QNDVI-A.caas-4AL*	93	*Ex_c6665_1067–D_GCE8AKX02GF3QZ_210*	3.27	6.5	0.01
		QNDVI-A.caas-5BL	128	*Tdurum_contig23273_426–BS00065128_51*	3.74	6.5	−0.01
	Zhoukou2014	*QNDVI-A.caas-3AL*	230	*Tdurum_contig31235_99–wsnp_Ex_c45877_51547406*	2.9	4.7	0.01
		QNDVI-A.caas-5BS.1	33	*Kukri_c10970_573–tplb0046h23_602*	3.46	5.7	0.01
	Zhengzhou2014	*QNDVI-A.caas-1BS*	41	*BS00070878_51–Kukri_c1529_462*	3.88	6.7	0.01
		QNDVI-A.caas-4BS	56	*RAC875_c6749_954–BobWhite_c44691_648*	2.73	4.0	0.01
		QNDVI-A.caas-4DS	69	*RAC875_c13945_597–BS00036421_51*	6.03	9.8	0.01
		QNDVI-A.caas-5AL	91	*BS00082002_51–wsnp_Ku_c14275_22535576*	4.15	6.4	0.01
	Average	*QNDVI-A.caas-3AL*	230	*Tdurum_contig31235_99–wsnp_Ex_c45877_51547406*	2.54	4.3	0.01
NDVI-10	Zhoukou2013	*QNDVI-10.caas-2DS*	53	*BS00081578_51–tplb0021c10_951*	2.67	4.4	−0.02
		QNDVI-10.caas-5AL	204	*BS00055102_51–BS00067351_51*	3.94	6.3	−0.02
		QNDVI-10.caas-5BL	61	*wsnp_Ra_c5634_9952011–RAC875_c14882_275*	5.2	8.5	0.02
	Zhengzhou2013	*QNDVI-10.caas-4BS*	53	*RAC875_c6749_954– BobWhite_c44691_648*	2.93	5.5	0.02
		QNDVI-10.caas-5BL	61	*wsnp_Ra_c5634_9952011–RAC875_c14882_275*	3.52	6.1	0.02
	Zhoukou2014	*QNDVI-10.caas-5BL*	61	*wsnp_Ra_c5634_9952011–RAC875_c14882_275*	3.69	6.0	0.01
		QNDVI-10.caas-6BL	109	*Kukri_c63314_962–BobWhite_c36416_56*	3.37	5.8	0.01
	Zhengzhou2014	*QNDVI-10.caas-4DS*	68	*RAC875_c13945_597–BS00036421_51*	3.48	7.0	0.01
		QNDVI-10.caas-6BL	109	*Kukri_c63314_962–BobWhite_c36416_56*	3.86	7.3	0.01
	Averages	*QNDVI-10.caas-5BL*	61	*wsnp_Ra_c5634_9952011–RAC875_c14882_275*	5.09	8.5	0.02
		QNDVI-10.caas-6BL	109	*Kukri_c63314_962–BobWhite_c36416_56*	3.34	6.3	0.01

[a]*Position of QTL located on chromosome: as cM distance from the top of each map.*
[b]*A LOD threshold of 2.5 was used for declaration of QTL, based on 2000 permutations at a significance level of 0.01.*
[c]*Phenotypic variance explained by QTL.*
[d]*Positive "additive effect" indicates an increasing effect from Chinese Spring; negative "additive effect" indicates an increasing effect from Zhou 8425B.*
TKW, thousand kernel weight; KNS, kernel number per spike; SN, spike number/m^2; PH, plant height; SL, spike length; Chl-A, SPAD value of chlorophyll content at anthesis; Chl-10, SPAD value of chlorophyll content at 10 days post-anthesis; NDVI-A, normalized difference in vegetation index at anthesis; NDVI-10, normalized difference in vegetation index at 10 days post-anthesis.

identified across all environments, and explained 4.8–10.3% and 3.5–6.5% of the phenotypic variances, respectively. Two other QTL, *QTKW.caas-4AL,* and *QTKW.caas-7BL,* were detected in three environments, explaining 3.3–6.6% and 4.0–5.5% of the phenotypic variances, respectively. *QTKW.caas-5AS.1* was detected at Zhoukou2014 and Zhengzhou2014, and explained from 3.4 to 3.5% of the phenotypic variance. The positive alleles at *QTKW.caas-1AL.4, QTKW.caas-2DL.1, QTKW.caas-2DL.2, QTKW.caas-3DL, QTKW.caas-4AL, QTKW.caas-5AL.2, QTKW.caas-5AS.1, QTKW.caas-5BL, QTKW.caas-6A.1, QTKW.caas-7AL,* and *QTKW.caas-7BL* loci were contributed by Zhou 8425B. Those for increasing TKW at *QTKW.caas-4BS.1* and *QTKW.caas-5AL.1* loci were derived from Chinese Spring.

Kernel Number per Spike

Eleven QTL for KNS were identified on chromosomes 1BS, 2AL, 2B (2), 2D, 3AL, 3B, 4AL, 4BL, 6BL, and 7BS (**Table 3, Figure 1**). Alleles increasing KNS at the loci on chromosomes 2AL, 2B (2), 2D, 3B, 4AL, and 4BL were contributed by Zhou

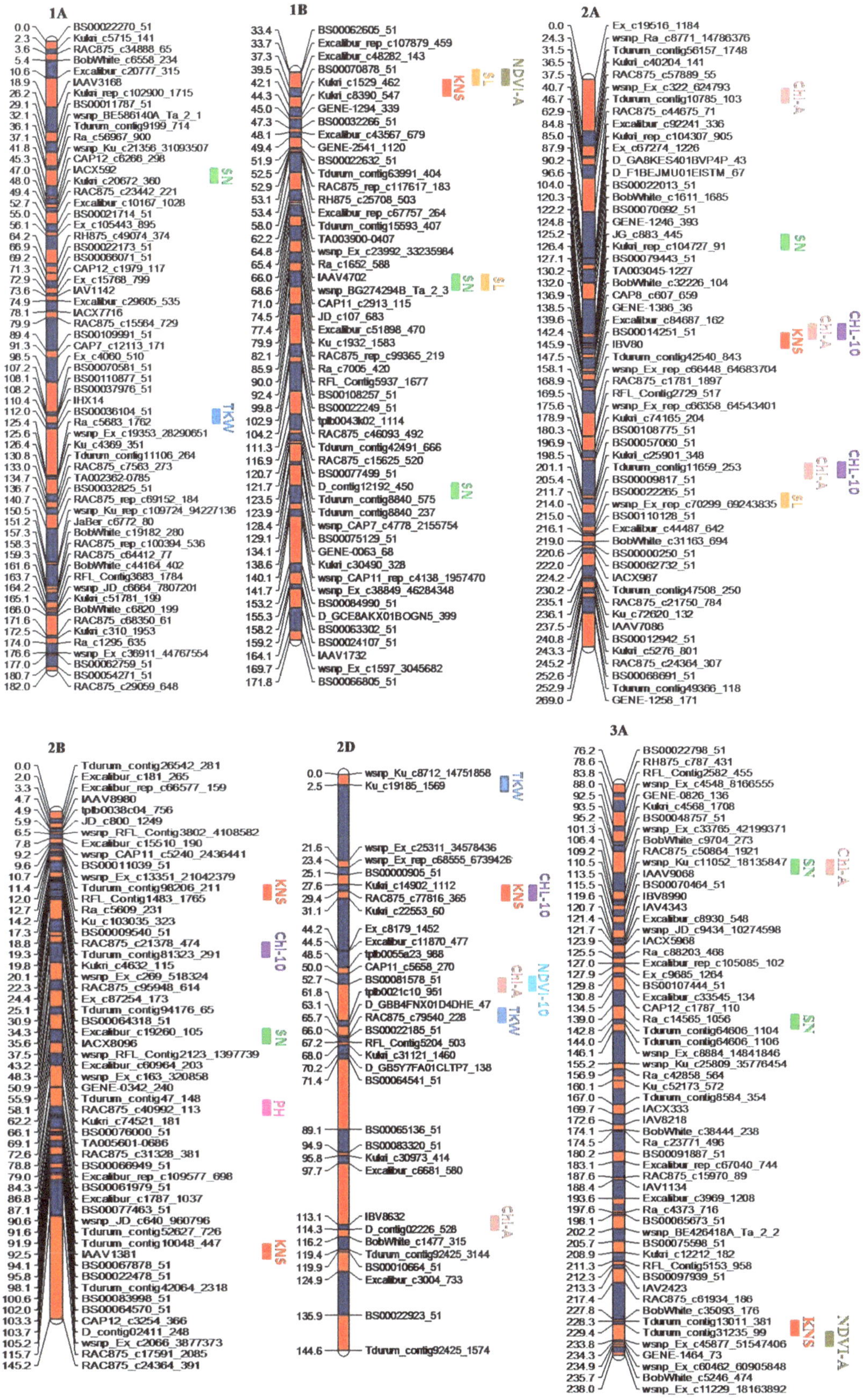

FIGURE 1 | Continued

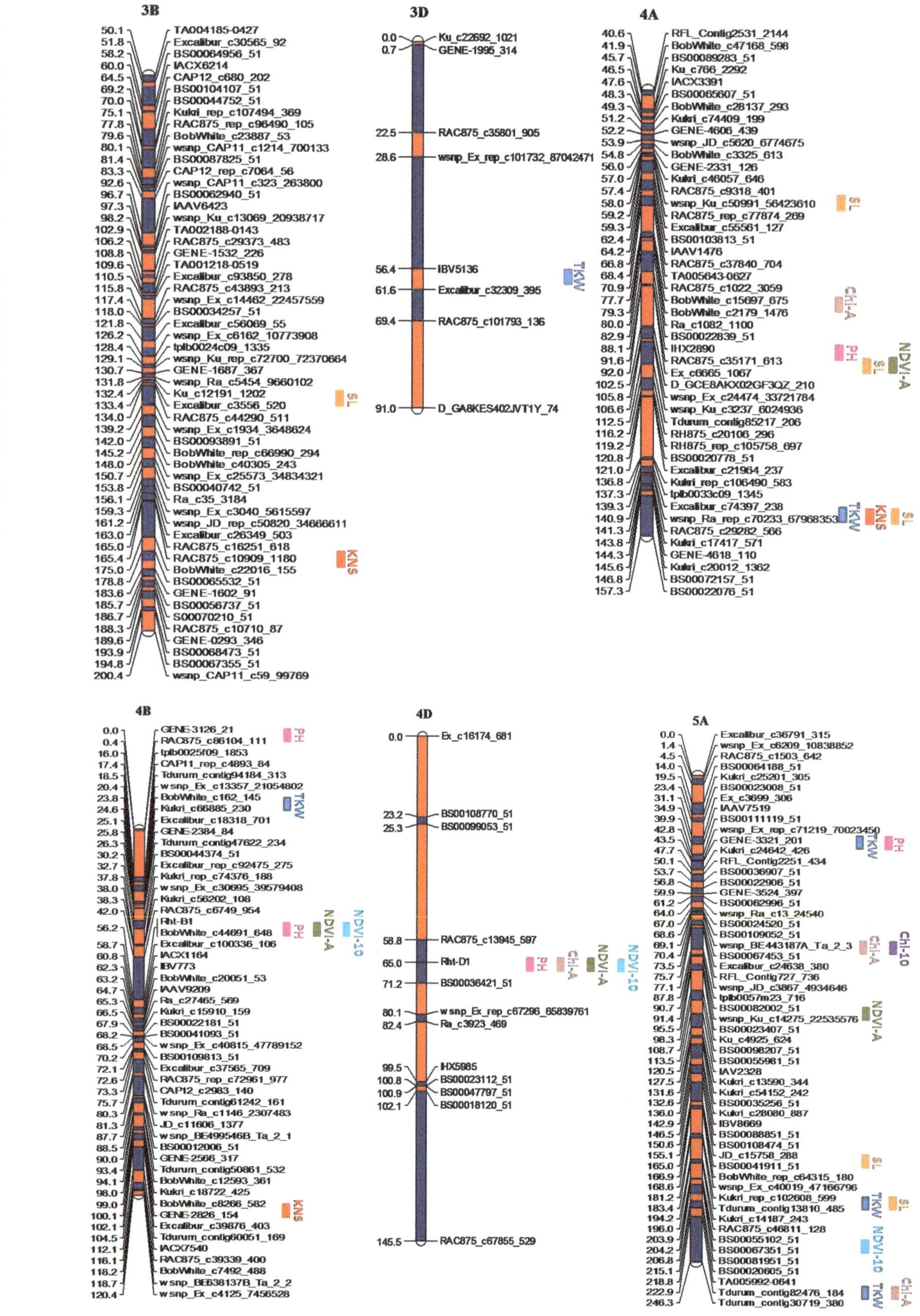

FIGURE 1 | Continued

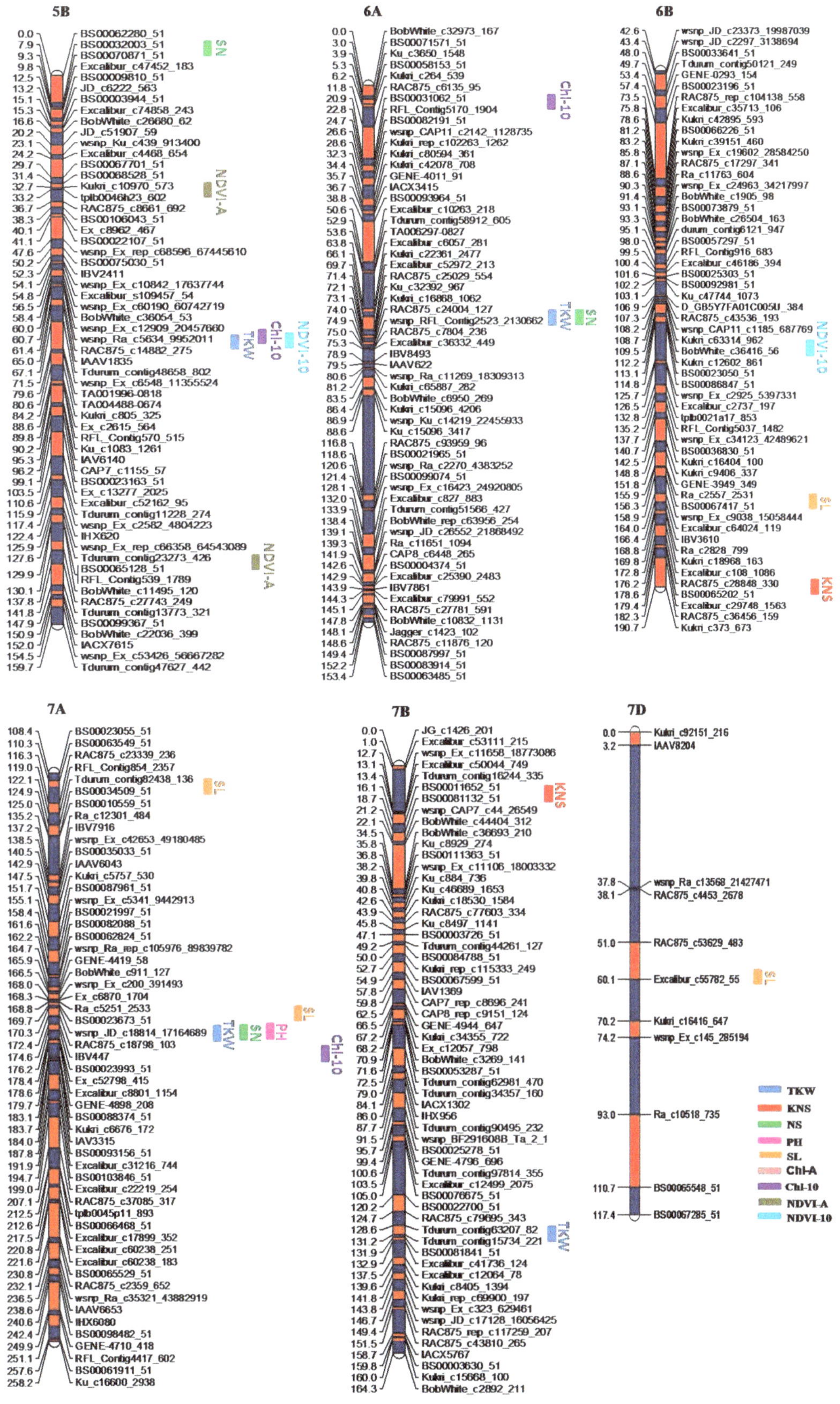

FIGURE 1 | Continued

FIGURE 1 | Genetic maps of chromosomes showing QTL for yield components, plant height, and yield-related physiological traits in the Zhou 8425B/Chinese Spring population. Traits are projected as solid bars with different colors for which the legend is given at the end of figure. TKW, thousand kernel weight; PH, plant height; SL, spike length; KNS, kernel number per spike; SN, spike number/m^2; Chl-A, SPAD value of chlorophyll content at anthesis; Chl-10, SPAD value of chlorophyll content at 10 days post-anthesis; NDVI-A, normalized difference in vegetation index at anthesis; NDVI-10, normalized difference in vegetation index at 10 days post-anthesis.

8425B, and those on chromosomes 1BS, 3AL, 6BL, and 7BS were come from Chinese Spring. *QKNS.caas-4AL* flanked by SNP markers *Kukri_rep_c106490_583* and *RAC875_c29282_566* was detected in all environments, explaining 5.1–10.5% of the phenotypic variance. The second QTL between SNP markers *RAC875_c61934_186* and *wsnp_Ex_c45877_51547406* on chromosome 3AL was found in three environments, and explained 3.4–10.9% of the phenotypic variance. *QKNS.caas-3B* detected at Zhengzhou2013 and Zhoukou2014 accounted for 4.1–5.6% of the phenotypic variance.

Spike Number

Ten QTL for SN were identified on chromosomes 1AL, 1BL (2), 2AS, 2BL, 3AL, 3AS, 5BS, 6AL, and 7AL (**Table 3**, **Figure 1**). Alleles for increased SN present on chromosomes 1AL, 1BL, 3AL, 3AS, 6AL, and 7AL were contributed by Zhou 8425B, and positive alleles on 2AS, 2BL and 5BS came from Chinese Spring. *QSN.caas-1AL.1* between the SNP markers *IACX592* and *Jagger_c1403_60* was identified in all four environments, explained 8.0–17.2% of the variation in SN. Another QTL, *QSN.caas-3AL*, between markers *Ra_c14565_1056* and *Tdurum_contig64606_1104*, was identified in three environments and explained 6.4–11.7% of the phenotypic variance. *QSN.caas-2AS.3*, *QSN.caas-6AL.1* and *QSN.caas-7AL* were detected in two environments and explained 3.0–6.4% of the phenotypic variance.

Plant Height

Seven QTL for PH were detected on chromosomes 2BL, 4AL, 4BS (2), 4DS, 5AS, and 7AL, respectively (**Table 3**, **Figure 1**). The alleles reducing PH on 4AL, 4BS, 4DS, and 5AS came from the shorter parent Zhou 8425B and those on 2BL and 7AL were from Chinese Spring. *QPH.caas-4BS.2* flanked by markers *RAC875_c6749_954* and *BobWhite_c44691_648* was identified in all four environments, and explained 22.7–29.3% of the phenotypic variance. *QPH.caas-4DS.1* between markers *RAC875_c13945_597* and *BS00036421_51* was also detected across all environments, accounting for 14.5–33.2% of the PH variance. These two major QTL are *Rht-B1b* and *Rht-D1b*, respectively, based on the gene-specific markers. *QPH.caas-5AS* was found in three environments and explained 12.7–14.9% of the phenotypic variance. *QPH.caas-2BL* and *QPH.caas-4AL* were consistently observed in two environments and explained 2.3–3.9% of the phenotypic variance.

Spike Length

Thirteen QTL for SL were mapped on chromosomes 1BL, 1BS, 2AL, 3BL, 4AS, 4AL (2), 5AL (2), 6BL, 7AS (2), and 7DS (**Table 3**, **Figure 1**). Alleles for increased SL at the loci on chromosomes 4AL (1) and 5AL were contributed by Zhou 8425B, and those at loci on chromosomes 1B, 2AL, 3BL, 4AS, 4AL (1), 6BL, 7AS (2), and 7DS were from Chinese Spring. *QSL.caas-4AS* between markers *Kukri_c46057_646* and *RAC875_rep_c77874_269* was detected in all environments, explaining 4.5–12.3% of the phenotypic variance. *QSL.caas-4AL.1* flanked by SNP markers *Kukri_c17417_571* and *BS00022076_51* was also found in four environments, accounting for 6.8–11.9% of the phenotypic variance. *QSL.caas-1BL*, *QSL.caas-2AL*, *QSL.caas-5AL*, *QSL.caas-5AL.1*, *QSL.caas-6BL*, and *QSL.caas-7AS* detected in two environments explained 3.1–8.7% of the phenotypic variance.

SPAD Value of Chlorophyll Content at Anthesis

Ten QTL for Chl-A were detected on chromosomes 2AS, 2AL (2), 2DS, 2DL, 3AS, 4AL, 4DS, 5AS, and 5AL (**Table 3**, **Figure 1**). Alleles increasing Chl-A at all loci except *QChl-A.caas-2AL.2* were contributed by Chinese Spring. *QChl-A.caas-5AL* flanked by SNP markers *BS00109052_51* and *wsnp_BE443187A_Ta_2_3* was detected in all environments, explaining 6.5–16.3% of the phenotypic variance. *QChl-A.caas-3AS* between markers *wsnp_Ku_c11052_18135847* and *wsnp_Ra_c16278_24893033* was identified in two environments, explaining 5.6–6.5% of the phenotypic variance.

SPAD Value of Chlorophyll Content at 10 Days Post-Anthesis

Eight putative QTL for Chl-10 were detected on chromosomes 2AL (2), 2BS, 2D, 5AL, 5BL, 6AS, and 7A (**Table 3**, **Figure 1**). *QChl-10.caas-5BL* between SNP markers *wsnp_Ex_c12909_20457660* and *wsnp_Ra_c5634_9952011* was detected in all four environments, explaining 7.0–10.6% of the phenotypic variance; the additive effect was for the Chinese Spring allele. *QChl-10.caas-2BS*, *QChl-10.caas-2D* and *QChl-10.caas-7A* were significant in two environments and explained 4.2–9.2% of the phenotypic variance. Alleles increasing Chl-10 at the *QChl-10.caas-2BS* and *QChl-10.caas-2D* loci came from Zhou 8425B, whereas that increasing Chl-10 at *QChl-10.caas-7A* locus was derived from Chinese Spring.

Normalized Difference in Vegetation Index at Anthesis

Eight QTL for NDVI-A were detected on chromosomes 1BS, 3AL, 4AL, 4BS, 4DS, 5AL, 5BL, and 5BS (**Table 3**, **Figure 1**), explaining 4.0–9.8% of the phenotypic variances. The alleles increasing NDVI-A at all loci except *QNDVI-A.caas-5BL* were from Chinese Spring.

Normalized Difference in Vegetation Index at 10 Days Post-Anthesis

Six QTL for NDVI-10 were identified on chromosomes 2DS, 4BS, 4DS, 5AL, 5BL, and 6BL (**Table 3**, **Figure 1**). Alleles for increasing NDVI-10 at the loci on chromosomes 2DS and 5AL were contributed by Zhou 8425B, and those at other loci were from Chinese Spring. *QNDVI-10.caas-5BL* flanked by markers *wsnp_Ra_c5634_9952011* and *RAC875_c14882_275* was significant in three environments and explained 6.0–8.5% of the phenotypic variance; the positive allele came from Chinese Spring. *QNDVI-10.caas-6BL* was found in two environments and explained 5.8–7.3% of the phenotypic variance.

DISCUSSION

SNP Discovery and Linkage Map Construction

SNP markers enable construction of high-density linkage maps and identification of QTL for complex agronomic traits in crop plants (Song et al., 2013). In the current study, we used 5636 polymorphic SNP markers from a 90K SNP assay (Wang et al., 2014), and constructed a high-density genetic map for a RIL population derived from the cross Zhou 8425B/Chinese Spring. Of these markers, 4770 (84.6%) were mapped by Wang et al. (2014) and 866 are newly mapped (**Table S4**). The order of SNP markers in the linkage map is generally consistent with Wang et al. (2014). The total length of the linkage map was 3609.4 cM, similar to previously reported maps in hexaploid wheat (Blanco et al., 1998; Marone et al., 2012). The average density of the map was 0.64 cM/marker, representing a considerable improvement over previously reported maps based on SSR, STS and DArT (Nachit et al., 2001; Marone et al., 2012). Markers for the A (43.6%) and B (50.4%) genomes were more abundant than those for the D genome (6%), again consistent with previous studies (Shiaoman et al., 2009; Wang et al., 2014), and this is attributed to the low level of polymorphism in D genome of hexaploid wheat. Although the average density is high, there were still some gaps, for instance, on chromosomes 1B and 4D.

The average number of mapped markers per chromosome was 268.4, ranging from 10 on chromosome 3D to 599 on chromosome 5B. However, 65.6% of the SNPs mapped displayed redundancy and only 6.9% of SNPs were used for linkage map construction in the present study, in agreement with previous reports (Barker and Edwards, 2009; Colasuonno et al., 2014). The low polymorphisms of SNPs in high-density assays identified in these studies may reflect an overall narrow range of genetic diversity in wheat. Because many SNP markers were co-located at the same genetic loci (Colasuonno et al., 2014), the BIN-Mapping function was employed in selecting markers for QTL mapping. BIN helps with automatic deletion of the high "nearest neighbor" markers in generating the input file that can be used for more efficient genetic map construction. The 5636 polymorphic SNP markers were optimized by the BIN-Mapping function and a linkage map based on 1938 skeleton SNPs were used to perform QTL analysis. Indeed, this generated a simplified genetic map for QTL mapping.

QTL Mapping

The green-revolution genes, *Rht1* and *Rht2* on chromosomes 4B and 4D, respectively, have been deployed worldwide (Ellis et al., 2005; Peng et al., 2011). In the present study, the most interesting QTL associated with PH were also detected on chromosomes 4B and 4D across all environments. *QPH.caas-4BS.2* and *QPH.caas-4DS.1* carrying positive alleles from the short parent Zhou 8425B explained the highest phenotypic variance, and represent polymorphisms was associated with *Rht-B1* and *Rht-D1*. Another stable QTL for PH, *QPH.caas-5AS*, was positioned at 50 cM. A QTL for PH reported previously on chromosome 5A in a spring wheat population derived from a Seri/Babax cross based on SSR marker *Xgwm617a* (Lopes et al., 2013) is likely the same gene. *QPH.caas-5AS* was detected in four environments and the reducing height allele was from Zhou 8425B. The gene could be used in MAS in wheat breeding. A minor PH QTL, *QPH.caas-4AL*, positioned at 90 cM, is different from a major QTL for PH at position 13.6cM on chromosome 4A in the Seri/Babax RIL population (Lopes et al., 2013). *QPH.caas-4AL*was detected in two environments and it is likely to be a new PH QTL.

An important QTL for TKW, *QTKW.caas-6A.1*, tightly linked to the SNP marker Ku_c32392_967 at a genetic distance of 1.4 cM was mapped at a similar locus to a QTL reported by Sukumaran et al. (2015) based on wheat 90K_consensus_map (Wang et al., 2014). *QTKW.caas-7BL*, positioned at 131cM across all four environments, is likely to be new. A minor QTL for TKW reported by Liu et al. (2014) in marker interval *Xcau30-7B-Xgwm66c-7B* positioned at 2cM in a mapping population derived from a cross of common wheat line ND3331 and Tibetan semi-wild wheat accession Zang1817 and is clearly different.

QKNS.caas-4AL mapped at 141 cM on chromosome 4AL in the present study, whereas Lopes et al. (2013) detected a different QTL for KNS at 5.55 cM on chromosome 4AS across 12 environments based on linkage with marker *C14p6*. A previously reported minor QTL for KNS (Liu et al., 2014) on chromosome 4AS, positioned at 0 cM, was also found in two environments but with very low contributions to phenotypic variation. Therefore, *QKNS.caas-4AL*, detected across all environments with higher phenotypic variation explained, is probably a new QTL. A new QTL *QKNS.caas-3AL* positioned at 228.3 cM in the current study differed from a major QTL at 56 cM reported by Ali et al. (2011) using a Cheyenne (CNN) × [CNN (Wichita 3A)] recombinant inbred chromosome line (RICL) population consisting of 223 CNN (RICLs3A) and seven check cultivars.

QSN.caas-1AL.1, at 47.5cM on chromosome 1AL across all environments explained more than 10% of the phenotypic variance, was different from a minor QTL for SN reported in a Chuang 35,050/Shannong 483 RIL population using SSR markers (Li et al., 2007b); the latter was found in single environment with lower explained phenotypic variation (5.7%) and it is likely to be a new QTL. Another *QSN.caas-3AL* at about 141 cM and detected in three environments is different from a minor QTL for SN on chromosome 3AL based on the linked markers *Xgwm-720* and *Xgwm-1063* positioned at 44.5cM (Kumar et al., 2007). Therefore, *QSN.caas-3AL* is also new.

Only a few QTL for physiological traits have been detected in wheat (Rebetzke et al., 2008a,b; Reynolds and Tuberosam,

2008). A QTL for Chl-10 was mapped 60 cM on chromosome 5BL in this study, whereas a major QTL for Chl was previously identified between *Xgwm639.1-5B* and *Xwmc388.4-5B* (Xu et al., 2012). A minor QTL, *QChl-10.caas-6AS,* tightly linked to the SNP marker *RFL_Contig5170_1904* at a genetic distance of 1.8 cM was different from that reported by Sukumaran et al. (2015) due to the large genetic distance between the two linked SNP markers (Wang et al., 2014). *QChl-10.caas-5AL* located on the long arm of chromosome 5A across all environments has not been reported previously, therefore, it is a new QTL.

A significant QTL, QNDVI-10.caas-*5BL*, was tightly linked to the SNP marker *RAC875_c14882_275,* at a genetic distance of 0.4 cM, which may be different from that reported by Sukumaran et al. (2015) based on wheat 90K_consensus_map (Wang et al., 2014).

Co-localization of QTL for Yield and Related Traits

Co-localization of QTL or QTL clusters for yield and related traits have been reported in previous studies (McCartney et al., 2005; Quarrie et al., 2006). In the current study, 10 QTL clusters on chromosomes 1BS, 2AL (2), 3AL, 4AL (2), 4BS, 4DS, 5BL, and 7AL were detected with each for more than two traits (**Table 4**, **Figure 1**) and QTL clusters associated with GY were detected on chromosomes 3AL, 4DS, and 5BL.

The interval 217.4–233.8 cM on chromosome 3AL is a pleiotropic locus impacting GY, KNS and NDVI-A. No similar pleiotropic region on chromosome 3A was reported previously. GY showed a significantly positive correlation with KNS ($r = 0.44$) and NDVI-A ($r = 0.6$), indicating that the increased GY at *QGY.caas-3AL* resulted from increased KNS and NDVI-A.

Another pleiotropic locus for GY, PH, Chl-A, NDVI-A, and NDVI-10 was identified at position 65 cM on chromosome 4DS. Co-localized QTL for GY and PH on chromosome 4DS detected in the present study and also reported by Li et al. (2015) are was associated with *Rht-D1b*. However, the co-localization of QTL for GY and Chl-A, NDVI-A, and NDVI-10 in same or similar region on chromosome 4DS has not been reported before.

QTL for GY, TKW, Chl-10, and NDVI-10 were associated in an interval of 54.1–61.4 cM on chromosome 5BL. A similar previously reported multiple trait region for GY and TKW on 5B (Edae et al., 2014) was positioned at 67.7–76.4 cM based on DArT markers in CIMMYT spring wheat lines. However, yield-related QTL for Chl-10 and NDVI-10 in a similar region on chromosome 5BL were not reported previously. Physiological traits Chl and NDVI were significantly correlated with GY in this study, with correlation coefficients of 0.5 and 0.34, respectively, implying that they may be related to the transfer of photosynthetic products in the grain filling (Lupton, 1966).

A QTL cluster for TKW, KNS and SL between markers *Kukri_rep_c106490_583* and *tplb0033c09_1345* and positioned in the interval 136.8–157.3 cM on chromosome 4AL, is likely the same or similar to a QTL cluster for KNS and SL reported by Liu et al. (2014) based on the interval *Xwmc491–Xwmc96.* Another QTL-rich region for PH, SL and NDVI-A, in interval 88.1–102.5 cM on chromosome 4AL, is new. *QTKW.caas-7AL* associated with SN, PH, SL, Chl-10, in interval 136.8–157.3 cM, was not reported previously.

In this study, a QTL-rich region for PH, NDVI-A and NDVI-10 on chromosome 4BS identified in interval 42.0–56.2 cM was different from the QTL for TKW and PH reported previously (Huang et al., 2004; McCartney et al., 2005). The 4BS QTL had a strong effect on PH and this QTL-rich region was associated with *Rht-B1b*.

Potential Application of QTL for MAS in Wheat Breeding

GY is highly affected by environments, and it is difficult to select high-yielding lines in smaller plots at the early stage of a breeding program. In contrast, environments have much less influence on yield components, PH and physiological traits, and some more stable QTL for these traits have been found, in agreement with previous reports (Lopes et al., 2013; Edae et al., 2014; Liu et al., 2014). Furthermore, yield was significantly and positively correlated with TKW, KNS, Chl-A, Chl-10, NDVI-A, and NDVI-10. Consequently, it is

TABLE 4 | Summary of pleiotropic QTL detected in the Zhou 8425B/Chinese Spring population.

Chromosome	Marker interval	Position (cM)	Trait
1BS	*BS00070878_51~Kukri_c8390_547*	39.5~44.3	KNS, SL, NDVI-A
2AL	*Excalibur_c84687_162~IBV80*	139.6~145.9	KNS, Chl-A, Chl-10
2AL	*Kukri_c25901_348~wsnp_Ex_rep_c70299_69243835*	198.5~214.0	SL, Chl-A, Chl-10
3AL	*RAC875_c61934_186~wsnp_Ex_c45877_51547406*	217.4~233.8	GY, KNS, NDVI-A
4AL	*IHX2890~D_GCE8AKX02GF3QZ_210*	88.1~102.5	PH, SL, NDVI-A
4AL	*Kukri_rep_c106490_583~tplb0033c09_1345*	136.8~157.3	TKW, KNS, SL
4BS	*RAC875_c6749_954~BobWhite_c44691_648*	42.0~56.2	PH, NDVI-A, NDVI-10
4DS	*RAC875_c13945_597~BS00036421_51*	58.8~71.2	GY, PH, Chl-A, NDVI-A, NDVI-10
5BL	*wsnp_Ex_c10842_17637744~RAC875_c14882_275*	54.1~61.4	GY, TKW, Chl-10, NDVI-10
7AL	*wsnp_Ex_c200_391493~Ex_c52798_415*	168.0~178.4	TKW, SN, PH, SL, Chl-10

GY, grain yield; TKW, thousand kernel weight; PH, plant height; SL, spike length; KNS, kernel number per spike; SN, spike number/m^2; Chl-A, SPAD value of chlorophyll content at anthesis; Chl-10, SPAD value of chlorophyll content at 10 days post-anthesis; NDVI-A, normalized difference in vegetation index at anthesis; NDVI-10, normalized difference in vegetation index at 10 days post-anthesis.

feasible to improve GY by selecting these yield-related traits in breeding programs because of the more accurate measurement and repeatability across environments in comparison with yield. Stable QTL such as *QTKW.caas-6A.1*, *QTKW.caas-7AL*, *QPH.caas-4BS.2*, *QPH.caas-4DS.1*, *QKNS.caas-3AL*, *QKNS.caas-4AL*, *QChl-A.caas-5AL*, *QChl-10.caas-5BL*, and *QNDVI-10.caas-5BL* could be used in breeding. Due to the availability of high-density SNP markers, it is more likely that these QTL represent actual candidate genes for the various traits, as previously identified for pre-harvest sprouting and yellow pigment content (Cabral et al., 2014; Colasuonno et al., 2014). If so they are potential candidates for fine mapping and ultimate candidate gene discovery.

CONCLUSION

A high-density linkage map was constructed in the Zhou 8425B/Chinese Spring population using the 90K SNP array; it proved powerful for mapping QTL for yield components, PH and yield-related physiological traits in wheat. Ten pleiotropic QTL clusters for yield related traits and eight novel QTL for TKW, PH, KNS (2), SN (2), and Chl-10 (2) were identified, with genetic distances of 0–1.5 cM from the closest linked SNP markers; therefore, these QTL could serve as target regions for fine mapping, candidate gene discovery, and MAS in wheat breeding.

AUTHOR CONTRIBUTIONS

FG carried out the experiment and wrote the paper. WW, JL, and AR performed SNP genotyping and data analysis. GY participated in field trials. XX, XW, and ZH designed the experiment and wrote the paper. All authors read and approved the final manuscript.

ACKNOWLEDGMENTS

The authors are grateful to Prof. R. A. McIntosh, Plant Breeding Institute, University of Sydney, for review of this manuscript. This study was supported by the National Natural Science Foundation of China (31461143021), Beijing Municipal Science and Technology Project (D151100004415003), International Science and Technology Cooperation Program of China (2013DFG30530, 2014DFG31690), and China Agriculture Research System (CARS-3-1-3).

REFERENCES

Akhunov, E., Nicolet, C., and Dvorak, J. (2009). Single nucleotide polymorphism genotyping in polyploid wheat with the Illumina Golden-Gate assay. *Theor. Appl. Genet.* 119, 507–517. doi: 10.1007/s00122-009-1059-5

Ali, M. L., Baenziger, P. S., Ajlouni, Z. A., Campbell, B. T., Gill, K. S., Eskridge, K. M., et al. (2011). Mapping QTL for agronomic traits on wheat chromosome 3A and a comparison of recombinant inbred chromosome line populations. *Crop Sci.* 51, 553–566. doi: 10.2135/cropsci2010.06.0359

Aranzana, M. J., Kim, S., Zhao, K. Y., Bakker, E., Horton, M., Jakob, K., et al. (2005). Genome-wide association mapping in Arabidopsis identifies previously known flowering time and pathogen resistance genes. *PLoS Genet.* 1:e60. doi: 10.1371/journal.pgen.0010060

Ariyadasa, R., Mascherm, M., Nussbaumer, T., Schulte, D., Frenkel, Z., Poursarebani, N., et al. (2014). A sequence-ready physical map of barley anchored genetically by two million single-nucleotide polymorphisms. *Plant Physiol.* 164, 412–423. doi: 10.1104/pp.113.228213

Bagge, M., Xia, X. C., and Lübberstedt, T. (2007). Functional markers in wheat. *Curr. Opin. Plant Biol.* 10, 211–216. doi: 10.1016/jpbi2007 01009

Birkhead, T. R., Ball, A., Stapleym, J., Dawson, D., Burke, T., and Slate, J. (2010). A comparison of SNPs and microsatellites as linkage mapping markers: lessons from the zebra finch (*Taeniopygia guttata*). *BMC Genomics* 11:218. doi: 10.1186/1471-2164-11-218

Barker, G. L. A., and Edwards, K. J. (2009). A genome-wide analysis of single nucleotide polymorphism diversity in the world's major cereal crops. *Plant Biotechnol. J.* 7, 318–325. doi: 10.1111/j.1467-7652.2009.00412.x

Bennett, D., Reynolds, M., Mullan, D., Izanloo, A., Kuchel, H., Langridge, P., et al. (2012). Detection of two major grain yield QTL in bread wheat (*Triticum aestivum* L.) under heat, drought and high yield potential environments. *Theor. Appl. Genet.* 125, 1473–1485. doi: 10.1007/s00122-012-1927-2

Blanco, A., Bellomo, M. P., Cenci, A., De Giovanni, C., D'Ovidio, R., Iacono, E., et al. (1998). A genetic linkage map of durum wheat. *Theor. Appl. Genet.* 97, 721–728. doi: 10.1007/s001220050948

Börner, A., Schumann, E., Fürste, A., Cöster, H., Leithold, B., Röder, M. S., et al. (2002). Mapping of quantitative trait loci determining agronomic important characters in hexapioid wheat (*Triticum aestivum* L.). *Theor. Appl. Genet.* 105, 921–936. doi: 10.1007/s00122-002-0994-1

Cabral, A. L., Jordan, M. C., McCartney, C. A., You, F. M., Humphreys, D. G., MacLachlan, R., et al. (2014). Identification of candidate genes, regions and markers for pre-harvest sprouting resistance in wheat (*Triticum aestivum* L.). *BMC Plant Biol.* 14:340. doi: 10.1186/s12870-014-0340-1

Colasuonno, P., Gadaleta, A., Giancaspro, A., Nigro, D., Giove, S., Incerti, O., et al. (2014). Development of a high-density SNP-based linkage map and detection of yellow pigment content QTLs in durum wheat. *Mol. Breeding* 34, 1563–1578. doi: 10.1007/s11032-014-0183-3

Cook, J. P., McMullen, M. D., Holland, J. B., Tian, F., Bradbury, P., Ross-Ibarra, J., et al. (2012). Genetic architecture of maize kernel composition in the nested association mapping and inbred association panels. *Plant Physiol.* 158, 824–834. doi: 10.1104/pp.111.185033

Cuthbert, J. L., Somers, D. J., Brûlé-Babel, A. L., Brown, P. D., and Crow, G. H. (2008). Molecular mapping of quantitative trait loci for yield and yield components in spring wheat (*Triticum aestivum* L.). *Theor. Appl. Genet.* 117, 595–608. doi: 10.1007/s00122-008-0804-5

Edae, E. A., Byrne, P. F., Haley, S. D., Lopes, M. S., and Reynolds, M. P. (2014). Genome-wide association mapping of yield and yield components of spring wheat under contrasting moisture regimes. *Theor. Appl. Genet.* 127, 791–807. doi: 10.1007/s00122-013-2257-8

Ellis, M. H., Rebetzke, G. J., Azanza, F., Richards, R. A., and Spielmeyer, W. (2005). Molecular mapping of gibberellin-responsive dwarfing genes in bread wheat. *Theor. Appl. Genet.* 111, 423–430. doi: 10.1007/s00122-005-2008-6

Golabadi, M., Arzani, A., Mirmohammadi Maibody, S. A. M., Tabatabaei, B. E. S., and Mohammadi, S. A. (2011). Identification of microsatellite markers linked with yield components under drought stress at terminal growth stages in durum wheat. *Euphytica* 177, 207–221. doi: 10.1007/s10681-010-0242-8

Green, A. J., Berger, G., Griffey, C. A., Pitman, R., Thomason, W., Balota, M., et al. (2012). Genetic yield improvement in soft red winter wheat in the Eastern United States from 1919 to 2009. *Crop Sci.* 52, 2097–2108. doi: 10.2135/cropsci2012010026

Gupta, P. K., Varshney, R. K., Sharma, P. C., and Ramesh, B. (1999). Molecular markers and their applications in wheat breeding. *Plant Breeding* 118, 369–390. doi: 10.1046/j1439-0523199900401x

Holland, J. B. (2007). Genetic architecture of complex traits in plants. *Curr. Opin. Plant Biol.* 10, 156–161. doi: 10.1016/jpbi200701003

Huang, X., Wei, X., Sang, T., Zhao, Q., Feng, Q., Zhao, Y., et al. (2010). Genome-wide association studies of 14 agronomic traits in rice landraces. *Nat. Genet.* 42, 961–967. doi: 10.1038/ng.695

Huang, X., Zhao, Y., Wei, X., Li, C., Wang, A., Zhao, Q., et al. (2011). Genome-wide association study of flowering time and grain yield traits in a worldwide collection of rice germplasm. *Nat. Genet.* 44, 32–39. doi: 10.1038/ng.1018

Huang, X. Q., Kempf, H., Ganal, M. W., and Röder, M. S. (2004). Advanced backcross QTL analysis in progenies derived from a cross between a German elite winter wheat variety and a synthetic wheat (*Triticum aestivum* L.). *Theor. Appl. Genet.* 109, 933–943. doi: 10.1007/s00122-004-1708-7

Jia, G., Huang, X., Zhi, H., Zhao, Y., Zhao, Q., Li, W., et al. (2013). A haplotype map of genomic variations and genome-wide association studies of agronomic traits in foxtail millet (*Setaria italica*). *Nat. Genet.* 45, 957–961. doi: 10.1038/ng.2673

Kumar, N., Kulwal, P. L., Balyan, H. S., and Gupta, P. K. (2007). QTL mapping for yield and yield contributing traits in two mapping populations of bread wheat. *Mol. Breeding* 19, 163–177. doi: 10.1007/s11032-006-9056-8

Lambel, S., Lanini, B., Vivoda, E., Fauve, J., Wechter, W. P., Harris-Shultz, K. R., et al. (2014). A major QTL associated with *Fusarium oxysporum* race resistance identified in genetic populations derived from closely related watermelon lines using selective genotyping and genotyping-by-sequencing for SNP discovery. *Theor. Appl. Genet.* 127, 2105–2115. doi: 10.1007/s00122-014-2363-2

Li, H., Peng, Z., Yang, X., Wang, W., Fu, J., Wang, J., et al. (2012). Genome-wide association study dissects the genetic architecture of oil biosynthesis in maize kernels. *Nat. Genet.* 45, 43–50. doi: 10.1038/ng.2484

Li, H. H., Ye, G. Y., and Wang, J. K. (2007a). A modified algorithm for the improvement of composite interval mapping. *Genetics* 175, 361–374. doi: 10.1534/genetics.106.06681

Li, S. S., Jia, J. Z., Wei, X. Y., Zhang, X. C., Li, L. Z., Chen, H. M., et al. (2007b). An intervarietal genetic map and QTL analysis for yield traits in wheat. *Mol. Breeding* 20, 167–178. doi: 10.1007/s11032-007-9080-3

Li, X. M., Xia, X. C., Xiao, Y. G., He, Z. H., Wang, D. S., Trethowan, R., et al. (2015). QTL mapping for plant height and yield components in common wheat under water limited and full irrigation environments. *Crop Pasture Sci.* 67, 660–670. doi: 10.1071/CP14236

Li, Z. F., Zheng, T. C., He, Z. H., Li, G. Q., Xu, S. C., Li, X. P., et al. (2006). Molecular tagging of stripe rust resistance gene *YrZH84* in Chinese wheat line Zhou 8425B. *Theor. Appl. Genet.* 112, 1098–1103. doi: 10.1007/s00122-006-0211-8

Liu, G., Li, J. J., Lu, L. H., Qin, D. D., Zhang, J. P., Guan, P. F., et al. (2014). Mapping QTLs of yield-related traits using RIL population derived from common wheat and Tibetan semi-wild wheat. *Theor. Appl. Genet.* 127, 2415–2432. doi: 10.1007/s00122-014-2387-7

Lopes, M. S., Reynolds, M. P., McIntyre, L., Mathews, K. L., Jalal Kamali, M. R., Mossad, M., et al. (2013). QTL for yield and associated traits in the Seri/Babax population grown across several environments in Mexico, in the West Asia, North Africa, and South Asia regions. *Theor. Appl. Genet.* 126, 971–984. doi: 10.1007/s00122-012-2030-4

Lupton, F. H. (1966). Translocation of photosynthetic assimilates in wheat. *Ann. Appl. Biol.* 57, 355–364. doi: 10.1111/j.1744-7348.1966.tb03829.x

Marone, D., Laidò, G., Gadaleta, A., Colasuonno, P., Ficco, D. B. M., Giancaspro, A., et al. (2012). A high-density consensus map of A and B wheat genomes. *Theor. Appl. Genet.* 125, 1619–1638. doi: 10.1007/s00122-012-1939-y

McCartney, C. A., Somers, D. J., Humphreys, D. J., and Lukow, O. (2005). Mapping quantitative trait loci controlling agronomic traits in the spring wheat cross RL 4452 ×AC 'Domain'. *Genome* 48, 870–883. doi: 10.1139/g05-055

Nachit, M. M., Elouafi, I., Pagnotta, M. A., Ei, S. A., Iacono, E., Labhilili, M., et al. (2001). Molecular linkage map for an intraspecific recombinant inbred population of durum wheat (*Triticum turgidum* L. var. *durum*). *Theor. Appl. Genet.* 102, 177–186. doi: 10.1007/s001220051633

Peng, Z. S., Li, X., Yang, Z. J., and Liao, M. L. (2011). A new reduced height gene found in the tetraploid semi-dwarf wheat landrace Aiganfanmai. *Genet. Mol. Res.* 10, 2349–2357. doi: 10.4238/2011.October.5.5

Pinto, R. S., and Reynolds, M. P. (2015). Common genetic basis for canopy temperature depression under heat and drought stress associated with optimized root distribution in bread wheat. *Theor. Appl. Genet.* 128, 575–585. doi: 10.1007/s00122-015-2453-9

Prashar, A., Hornyik, C., Young, V., McLean, K., Sharma, S. K., Dale, M. F. B., et al. (2014). Construction of a dense SNP map of a highly heterozygous diploid potato population and QTL analysis of tuber shape and eye depth. *Theor. Appl. Genet.* 127, 2159–2171. doi: 10.1007/s00122-014-2369-9

Quarrie, S. A., Quarrie, S. P., Radosevic, R., Rancic, D., Kaminska, A., Barnes, J. D., et al. (2006). Dissecting a wheat QTL for yield present in a range of environments: from the QTL to candidate genes. *J. Exp. Bot.* 57, 2627–2637. doi: 10.1093/jxb/erl026

Rafalski, A. (2002). Applications of single nucleotide polymorphisms in crop genetics. *Curr. Opin. Plant Biol.* 5, 94–100. doi: 10.1016/S1369-5266(02)00240-6

Rebetzke, G. J., Condon, A. G., Farquhar, G. D., Appels, R., and Richards, R. A. (2008a). Quantitative trait loci for carbon isotope discrimination are repeatable across environments and wheat mapping populations. *Theor. Appl. Genet.* 118, 123–137. doi: 10.1007/s00122-008-0882-4

Rebetzke, G. J., van Herwaarden, A. F., Jenkins, C., Weiss, M., Lewis, D., Ruuska, S., et al. (2008b). Quantitative trait loci for water-soluble carbohydrates and associations with agronomic traits in wheat. *Aust. J. Agric. Res.* 59, 891–905. doi: 10.1071/AR08067

Reynolds, M., Bonnett, D., Chapman, S. C., Furbank, R. T., Manes, Y., Mather, D. E., et al. (2011). Raising yield potential of wheat. I. Overview of a consortium approach and breeding strategies. *J. Exp. Bot.* 62, 439–452. doi: 10.1093/jxb/erq311

Reynolds, M., and Tuberosam, R. (2008). Translational research impacting on crop productivity in drought-prone environments. *Curr. Opin. Plant Biol.* 11, 171–179. doi: 10.1016/j.pbi.2008.02.005

Schlotterer, C. (2004). The evolution of molecular markers - just a matter of fashion. *Nat. Rev. Genet.* 5, 63–69. doi: 10.1038/nrg1249

Sela, H., Ezratim, S., Ben-Yehuda, P., Manisterski, J., Akhunov, E., Dvorak, J., et al. (2014). Linkage disequilibrium and association analysis of stripe rust resistance in wild emmer wheat (*Triticum turgidum* ssp. *dicoccoides)* population in Israel. *Theor. Appl. Genet.* 127, 2453–2463. doi: 10.1007/s00122-014-2389-5

Shiaoman, C., Zhang, W. J., Akhunov, E., Sherman, J., Ma, Y. Q., Luo, M. C., et al. (2009). Analysis of gene-derived SNP marker polymorphism in US wheat cultivars. *Mol. Breeding* 23, 23–33. doi: 10.1007/s11032-008-9210-6

Sindhu, A., Ramsay, L., Sanderson, L. A., Stonehouse, R., Li, R., Condie, J., et al. (2014). Gene-based SNP discovery and genetic mapping in pea. *Theor. Appl. Genet.* 127, 2225–2241. doi: 10.1007/s00122-014-2375-y

Song, Q., Hyten, D. L., Jia, G., Quigley, C. V., Fickus, E. W., Nelson, R. L., et al. (2013). Development and evaluation of Soy SNP 50K, a high-density genotyping array for soybean. *PLoS ONE* 8:e54985. doi: 10.1371/journal.pone.0054985

Stam, P. (1993). Construction of integrated genetic linkage maps by means of a new computer package: JoinMap. *Plant J.* 3, 739–744. doi: 10.1111/j.1365-313X.1993.00739.x

Sukumaran, S., Dreisigacker, S., Lopes, M., Chavez, P., and Reynolds, M. P. (2015). Genome-wide association study for grain yield and related traits in an elite spring wheat population grown in temperate irrigated environments. *Theor. Appl. Genet.* 128, 353–363. doi: 10.1007/s00122-014-2435-3

Tian, F., Bradbury, P. J., Brown, P. J., Hung, H., Sun, Q., Flint-Garcia, S., et al. (2011). Genome-wide association study of leaf architecture in the maize nested association mapping population. *Nat. Genet.* 43, 159–162. doi: 10.1038/ng.746

Wang, S. C., Wong, D., Forrest, K., Allen, A., Chao, S. M., Huang, B. E., et al. (2014). Characterization of polyploid wheat genomic diversity using a high-density 90000 single nucleotide polymorphism array. *Plant Biotechnol. J.* 12, 87–96. doi: 10.1111/pbi.12183

Xiao, Y. G., Yin, G. H., Li, H. H., Xia, X. C., Yan, J., Zheng, T. C., et al. (2011). Genetic diversity and genome-wide association analysis of stripe rust resistance among the core wheat parent Zhou 8425B and its derivatives. *Sci. Agric. Sin.* 44, 3919–3929. doi: 10.3864/j.issn.0578-1752.2011.19.001

Xu, Y. F., An, D. G., Liu, D. C., Zhang, A. M., Xu, H. X., and Li, B. (2012). Mapping QTLs with epistatic effects and QTL×treatment interactions for salt tolerance at stage of wheat. *Euphytica* 186, 233–245. doi: 10.1007/s10681-012-0647-7

Yang, Q., Li, Z., Li, W. Q., Ku, L. X., Wang, C., Ye, J. R., et al. (2013). CACTA-like transposable element in *ZmCCT* attenuated photoperiod sensitivity and accelerated the postdomestication spread of maize. *Proc. Natl. Acad. Sci. U.S.A.* 110, 16969–16974. doi: 10.1073/pnas.1310949110

Yang, Z. B., Bai, Z. Y., Li, X. L., Wang, P., Wu, Q. X., Yang, L., et al. (2012). SNP identification and allelic-specific PCR markers development for *TaGW2*, a gene linked to wheat kernel weight. *Theor. Appl. Genet.* 125, 1057–1068. doi: 10.1007/s00122-012-1895-6

Yin, G. H., Wang, J. W., Wen, W. E., He, Z. H., Li, Z. F., Wang, H., et al. (2009). Mapping of wheat stripe rust resistance gene *YrZH84* with RGAP markers and its application. *Acta Agron. Sin.* 35, 1274–1281. doi: 10.3724/SP.J.1006.2009.01274

Yu, H., Xie, W., Wang, J., Xing, Y., Xu, C., Li, X., et al. (2011). Gains in QTL detection using an ultra-high density SNP map based on population sequencing relative to traditional RFLP/SSR markers. *PLoS ONE* 6:e17595. doi: 10.1371/journal. pone.0017595

Zhao, K., Tung, C. W., Eizenga, G. C., Wright, M. H., Ali, M. L., Price, A. H., et al. (2011). Genome-wide association mapping reveals a rich genetic architecture of complex traits in *Oryza sativa*. *Nat. Commun.* 2:467. doi: 10.1038/ncomms1467

Zhao, X. L., Zheng, T. C., Xia, X. C., He, Z. H., Liu, D. Q., Yang, W. X., et al. (2008). Molecular mapping of leaf rust resistance gene *LrZH84* in Chinese wheat line Zhou 8425B. *Theor. Appl. Genet.* 117, 1069–1075. doi: 10.1007/s00122-008-0845-9

Conflict of Interest Statement: The authors declare that the research was conducted in the absence of any commercial or financial relationships that could be construed as a potential conflict of interest.

4

Hydrogen Sulfide Regulates Salt Tolerance in Rice by Maintaining Na^+/K^+ Balance, Mineral Homeostasis and Oxidative Metabolism Under Excessive Salt Stress

Mohammad G. Mostofa[1], Daisuke Saegusa[2], Masayuki Fujita[2] and Lam-Son Phan Tran[3,4]**

[1] Department of Biochemistry and Molecular Biology, Bangabandhu Sheikh Mujibur Rahman Agricultural University, Gazipur, Bangladesh, [2] Laboratory of Plant Stress Responses, Department of Applied Biological Science, Faculty of Agriculture, Kagawa University, Miki, Japan, [3] Plant Abiotic Stress Research Group & Faculty of Applied Sciences, Ton Duc Thang University, Ho Chi Minh City, Vietnam, [4] Signaling Pathway Research Unit, RIKEN Center for Sustainable Resource Science, Yokohama, Japan

Edited by:
Shabir Hussain Wani, Sher-e-Kashmir University of Agricultural Sciences and Technology of Kashmir, India

Reviewed by:
Hong-Bo Shao, Qingdao University of Science and Technology, China
Abdul Hameed, University of Karachi, Pakistan

***Correspondence:**
Mohammad G. Mostofa
mostofa@bsmrau.edu.bd;
Lam-Son Phan Tran
sontran@tdt.edu.vn,
son.tran@riken.jp

Being a salt sensitive crop, rice growth and development are frequently affected by soil salinity. Hydrogen sulfide (H_2S) has been recently explored as an important priming agent regulating diverse physiological processes of plant growth and development. Despite its enormous prospects in plant systems, the role of H_2S in plant stress tolerance is still elusive. Here, a combined pharmacological, physiological and biochemical approach was executed aiming to examine the possible mechanism of H_2S in enhancement of rice salt stress tolerance. We showed that pretreating rice plants with H_2S donor sodium bisulfide (NaHS) clearly improved, but application of H_2S scavenger hypotaurine with NaHS decreased growth and biomass-related parameters under salt stress. NaHS-pretreated salt-stressed plants exhibited increased chlorophyll, carotenoid and soluble protein contents, as well as suppressed accumulation of reactive oxygen species (ROS), contributing to oxidative damage protection. The protective mechanism of H_2S against oxidative stress was correlated with the elevated levels of ascorbic acid, glutathione, redox states, and the enhanced activities of ROS- and methylglyoxal-detoxifying enzymes. Notably, the ability to decrease the uptake of Na^+ and the Na^+/K^+ ratio, as well as to balance mineral contents indicated a role of H_2S in ion homeostasis under salt stress. Altogether, our results highlight that modulation of the level of endogenous H_2S genetically or exogenously could be employed to attain better growth and development of rice, and perhaps other crops, under salt stress. Furthermore, our study reveals the importance of the implication of gasotransmitters like H_2S for the management of salt stress, thus assisting rice plants to adapt to adverse environmental changes.

Keywords: global salinity, gasotransmitter, hydrogen sulfide, ionic stress, physiological and biochemical mechanisms, reactive oxygen species, rice, salt tolerance

INTRODUCTION

Plants, being sessile in nature, are constantly exposed to an array of abiotic stresses throughout their life cycles. Salinity is one of the most brutal environmental constraints, affecting plant growth, development and productivity, especially in the arid and semi-arid regions (Parihar et al., 2015). High soil salinity has arisen as an important global concern, which hampers sustainable rice production in many agrarian countries like Bangladesh. Moreover, water shortage, hot and dry climate, and rising sea level due to global warming are aggravating the existing salinity problems, and thus worsening rice production in inland as well as coastal areas (Rabbani et al., 2013; Islam et al., 2015). Efforts to improve rice tolerance to salt stress through classical breeding have produced limited success and most of the existing varieties are unable to endure high salinity (Rabbani et al., 2013). Therefore, understanding the salt tolerance mechanisms of rice is indispensable in developing high salt-tolerant rice varieties to assure sustainable rice production in salt-affected regions.

Excessive salinity can induce various negative effects in plant cells, including ionic stress by accumulating Na^+ and Cl^-, osmotic stress by provoking water deficit, and oxidative stress by overproducing reactive oxygen species (ROS) (Munns and Tester, 2008; Gupta and Huang, 2014). Moreover, salinity can cause nutritional imbalance, especially K^+ deficiency by inhibiting its uptake and triggering its leakage from the cells (Cuin et al., 2008; Munns and Tester, 2008; Parihar et al., 2015). Like other abiotic stresses, salinity also accelerates the generation of methylglyoxal (MG), a highly cytotoxic compound, which can crosslink with DNA, RNA, and proteins, thereby exacerbating oxidative stress (Yadav et al., 2005; Saito et al., 2011; Hossain et al., 2012).

Salt tolerance is a complex trait that includes a variety of defense mechanisms, including ionic, osmotic and ROS homeostases (Miller et al., 2010; Gupta and Huang, 2014). Plants respond to ionic imbalance by regulating ion fluxes and mineral homeostases. Several plant species have the potential to defend ionic and osmotic stresses by producing compatible solutes, such as proline and trehalose, to maintain water relations, stabilize enzyme and protein complexes and membranes under high salinity conditions (Mostofa et al., 2014b; Iqbal et al., 2015; Reddy et al., 2015). Plant protection against salt-induced ROS is closely related to the maintenance of cellular redox balance, which is mainly composed of several non-enzymatic antioxidants, such as ascorbic acid (AsA) and glutathione (GSH), and enzymatic antioxidants, such as superoxide dismutase (SOD), catalase (CAT), glutathione peroxidase (GPX), glutathione *S*-transferase (GST), ascorbate peroxidase (APX), monodehydroascorbate reductase (MDHAR), dehydroascorbate reductase (DHAR), and glutathione reductase (GR) (Shao et al., 2007; Gill and Tuteja, 2010; Miller et al., 2010). Likewise, plant cells also possess a GSH-based glyoxalase (Gly) system that detoxifies reactive MG by consecutive actions of Gly I and Gly II (Hossain et al., 2012). Transgenic studies revealed that the antioxidant defense and glyoxalase systems are intermingled, and efficient induction of these systems was correlated with enhanced tolerance to abiotic stresses in different crop varieties (Mustafiz et al., 2011; Upadhyaya et al., 2011; Alvarez Viveros et al., 2013).

Efforts are continuing to enhance plants' capacity to survive under environmental stresses. In this context, execution of the potential roles of small biological molecules, such as phytohormones and signal molecules could be considered as a powerful tool in modifying plants' adaptability against adverse effects resulted from ever-changing global conditions. Hydrogen sulfide (H_2S), a small bioactive gas with rotten egg disodour, has emerged as a third gasotransmitter after nitric oxide and carbon monoxide in animal systems, and is implicated in numerous health benefits (Mancardi et al., 2009; Toit, 2015). In plant systems, H_2S has recently come into light because of its association with adaptive responses against environmental stressors. Recent studies showed that priming of plants with H_2S donor sodium bisulfide (NaHS) offered significant tolerance to a number of abiotic stresses, including drought, heat, heavy metals and osmotic stresses (Christou et al., 2013; Shi et al., 2013; Ali et al., 2014; Mostofa et al., 2015d). The positive effects of H_2S in stress mitigation have been attributed to its association with multiple defense mechanisms, such as antioxidant activities and modulation of ROS detoxification system (Chen et al., 2011; Mostofa et al., 2015d). However, the underlying mechanisms H_2S uses to regulate salt tolerance, especially in major crops like rice, are still elusive, which requires in-depth analysis at the physiological and biochemical levels.

Thus, in the current study, we aimed to evaluate the potential roles of H_2S in addressing the problems associated with the detrimental consequences of global salinity in an economically important crop. For this purpose, we have examined the physiological and biochemical mechanisms associated with H_2S-induced salt tolerance in rice, with particular emphasis on ROS- and MG-detoxifying systems. At the same time, the role of H_2S in regulating Na^+/K^+ balance and mineral homeostases was also examined. Our results support the notion that H_2S could serve as an important priming agent in enhancing rice tolerance to salt stress, particularly by inhibiting Na^+ uptake, reestablishing redox and mineral homeostases, as well as reducing oxidative stress by stimulating ROS and MG detoxifications.

MATERIALS AND METHODS

Plant Materials, Growth Conditions and Treatments

The germination and cultivation of rice (*Oryza sativa* L. cv. BRRI dhan52) in a hydroponic condition were carried out according to the methods described by Mostofa et al. (2014b, 2015a,b). After 12 days, healthy plants were pretreated with H_2S for 48 h by supplying 0, 25, 50, 100, and 200 μM NaHS in the nutrient solution. Hypotaurine (HT), a scavenger of H_2S (Christou et al., 2013), was applied only with 50 μM NaHS at a concentration of 500 μM to justify the H_2S-dependent role of NaHS in salt tolerance. Pretreated and non-pretreated plants were exposed to nutrient solution containing 150 mM NaCl and further grown for 4 days in the specified conditions. This environmentally relevant NaCl concentration, which exhibited visible toxic symptoms on rice plants after 4 days of treatment, was selected based on our previous studies (Mostofa et al., 2014a, 2015a). The second leaf of rice seedlings was harvested to determine various physiological and biochemical responses altered by the application of NaHS, HT and NaCl. Each treatment was replicated three times under the same experimental conditions.

Assessment of Growth Parameters

The growth of rice seedlings was assessed by measuring plant height, fresh weight (FW) and dry weight (DW). Plant height was determined by measuring the length from the bottom of the main stem to the end of the emerging third leaf. To determine FW and DW, 10 seedlings from each treatment were weighed (FW) and oven-dried at 80°C for 48 h (DW), then expressed as g seedling^{-1}.

Determination of H_2S Content

H_2S content in rice leaves was determined following the method described by Christou et al. (2013) with modifications. Rice leaves (0.25 g) were homogenized in 1 mL of 100 mM K-P buffer (pH 7.0) containing 10 mM EDTA, then centrifuged at 11,200 × *g* for 15 min. The supernatant (100 μL) was mixed with 1,880 μL extraction buffer and 20 μL of 20 mM 5,5′-dithiobis (2-nitrobenzoic acid) and incubated at 25°C for 5 min. The absorbance was read at 412 nm, and H_2S was quantified using a standard curve developed with known concentrations of NaHS.

Measurement of Na, K and Mineral Nutrient Contents in Roots and Leaves

To determine Na, K and mineral nutrient contents (Ca, Mg, Fe, Zn, and Mn), the root and leaf samples were harvested separately and oven-dried at 80°C for 48 h. Dried samples (0.1 g) were ground and digested with a HNO_3:$HClO_4$ (5:1 v/v) mixture at 80°C until the yellow color disappeared. The Na, K, Ca, Mg, Fe, Zn, and Mn contents were measured by using flame atomic absorption spectrophotometry.

Relative Water Content and Proline Content

Relative water content (RWC) was determined as described by Mostofa et al. (2015a). Proline (Pro) content was determined according to the method of Bates et al. (1973) with minor modifications as reported in Mostofa and Fujita (2013). Freshly collected leaves (0.5 g) were homogenized with 5 mL of 3% aqueous sulfosalicylic acid. After centrifugation at 11,500 × *g* for 15 min, the supernatant (1 mL) was mixed with 2 mL solution containing glacial acetic acid and acid ninhydrin (1:1, v/v). Subsequently, the mixture was boiled at 100°C for 1 h and then transferred to ice to stop the reaction. Toluene (2 mL) was used to extract the developed color of the chromophore and the absorption was measured at 520 nm. Pro content was determined using a calibration curve developed by Pro standard.

Photosynthetic Pigments

After extracting 0.5 g leaves in 80% chilled acetone, the absorbance of the supernatants was recorded at 463, 445, and 470 nm for determining the contents of chlorophylls (Chl) and carotenoids according to the formulas suggested by Arnon (1949) and Lichtenthaler and Wellburn (1983), respectively.

Lipid Peroxidation and H_2O_2 Content

Lipid peroxidation of the leaves was measured by estimating malondialdehyde (MDA) according to the method of Heath and Packer (1968). H_2O_2 was extracted and determined after reaction with 0.1% $TiCl_4$ in 20% H_2SO_4 following the method of Mostofa and Fujita (2013).

Histochemical Analyses

For observing the accumulation of ROS, namely $O_2^{\cdot-}$ and H_2O_2, in leaves, histochemical analyses were carried out in rice leaves following the method of Mostofa and Fujita (2013). Briefly, 0.1% nitroblue tetrazolium (NBT) or 1% 3, 3-diaminobenzidine (DAB) solution was used to stain the cut leaves under dark and light, respectively. After incubation for 24 h, incubated leaves were decolorized by immersing in boiling ethanol to visualize the blue insoluble formazan for $O_2^{\cdot-}$ or deep brown polymerization product for H_2O_2. After cooling, photographs of the stained leaves were taken by placing between two glass plates.

Extraction and Estimation of Non-enzymatic Antioxidants

Fresh leaves (0.5 g) was homogenized in 3 mL of ice-cold 5% meta-phosphoric acid containing 1 mM EDTA and centrifuged at 11,500 × *g* for 15 min. Reduced and total AsA contents were assayed at 265 nm following the method of Mostofa et al. (2015b). Oxidized ascorbate (DHA; dehydroascorbic acid) content was determined by subtracting reduced AsA value from total AsA content. Total GSH and oxidized glutathione (GSSG) contents were calculated as described by Griffiths (1980). Reduced GSH content was measured after deducting the GSSG amount from total GSH level.

Extraction and Assay of Enzymes

Extraction of enzymes was carried out following the method of Mostofa et al. (2015a,b). Activities of antioxidant and glyoxalase enzymes were determined by the standard methods reported in Doderer et al. (1992) for lipoxygenase (LOX, EC 1.13.11.12), Mostofa and Fujita (2013) for SOD (EC 1.15.1.1) and CAT (EC 1.11.1.6), Nakano and Asada (1981) for APX (EC 1.11.1.11) and DHAR (EC 1.8.5.1), Hossain et al. (1984) for MDHAR (EC 1.6.5.4), Mostofa and Fujita (2013) for GR (EC 1.6.4.2), GST (EC 2.5.1.18) and GPX (EC: 1.11.1.9), Hossain et al. (2009) for Gly I (EC 4.4.1.5) and Mostofa and Fujita (2013) for Gly II (EC 3.1.2.6), respectively. Protein content was determined following the method of Bradford (1976) using bovine serum albumin (BSA) as a standard.

Estimation of Methylglyoxal (MG) Content

The MG content in rice leaves was determined following the method of Wild et al. (2012) as essentially described in Mostofa et al. (2015b) with exception that 0.5 g of leaves was used instead of 0.25 g.

Statistical Analysis

The data were subjected to one-way analysis of variance (ANOVA) and different letters indicate significant differences among treatments at $p < 0.05$, according to Duncan's multiple range test (DMRT) using IRRISTAT version 3 (IRRI, Biometrics Unit, Manila, Philippines). Data represented in the Tables and Figures are means ± standard deviations (SDs) of three replicates for each treatment.

RESULTS

Effects of Exogenous NaHS on Endogenous Level of H_2S and Selection of Effective Dose of NaHS

The protective role of H_2S was assessed by examining the effects of H_2S donor NaHS on the level of endogenous H_2S, and the levels of H_2O_2 and MDA in the leaves of rice plants under salt stress. Under non-stressed conditions, the pretreatment of rice plants with 25 (T1), 50 (T2) and 100 μM NaHS (T3) for 48 h increased the levels of H_2S by 28, 44, and 47%, respectively, as compared with the non-pretreated control plants (**Figure 1A**). We also observed that the pretreated rice plants maintained a higher level of H_2S when subjected to 150 mM NaCl stress for 4 days (14, 23 and 28%, respectively), as compared with salt-treated only plants. To clarify the role of H_2S released from NaHS, we applied 500 μM HT with 50 μM NaHS (T4) under both normal and stress conditions and observed that HT addition reversed NaHS-induced increase in H_2S level (**Figure 1A**). This result indicated the contribution of exogenous NaHS in elevating the level of endogenous H_2S. To further confirm the effective dose of NaHS, oxidative stress indicators, such as H_2O_2 and MDA, were estimated in the plants treated with varying concentrations of NaHS prior to salt stress. The pretreatment with 50 μM NaHS was shown to be most effective in decreasing the levels of H_2O_2 and MDA in salt-stressed seedlings (**Figure 1B**). Interestingly, a further increase in the concentration of NaHS aggravated the levels of these stress indicators. Taken together, 50 μM NaHS and 500 μM HT were used to test the protective role of H_2S against high salinity (150 mM) in the subsequent experiments.

H_2S Maintains Na^+ and K^+ Homeostases

Maintaining Na^+ and K^+ homeostases is crucial for plant survival under salt stress conditions. Salt treatment resulted in a noticeable increase in Na^+ content (15,069 and 5,515% in roots and leaves, respectively) and a significant decrease in K^+ content (46 and 38% in roots and leaves, respectively) in salt-stressed plants compared with untreated control (**Figures 2A,B**). Accordingly, Na^+/K^+ ratio significantly increased in roots and leaves of salt-treated plants in comparison with untreated control plants. On the other hand, pretreatment with NaHS maintained Na^+ and K^+ homeostases, as Na^+ content decreased by 34 and 25%, and K^+ content increased by 58 and 47% in roots and leaves of salt-stressed plants, respectively, compared with salt-stressed only plants (**Figures 2A,B**). Consequently, Na^+/K^+ ratio was significantly lower in NaHS-pretreated salt-stressed plants when compared with the plants treated with salt alone (**Figure 2C**). However, adding both HT and NaHS disturbed Na^+ and K^+ homeostases, resulting in similar accumulation of Na^+ and K^+ in the salt-stressed plants pretreated with and without both HT and NaHS (**Figures 2A–C**), which indicated that scavenging of H_2S by HT predisposed the salt-stressed plants to suffering from NaCl toxicity. Thus, H_2S eases the life of rice plants under salt stress, notably by inhibiting excessive Na^+ uptake and by maintaining K^+ homeostasis at the cellular level.

H_2S Regulates Plant Growth and Biomass

To evaluate the effects of H_2S on growth and biomass of rice plants under NaCl stress for 4 days, we determined the height, FW and DW of rice plants. Salt stress decreased the plant height by 39% in salt-stressed only seedlings compared

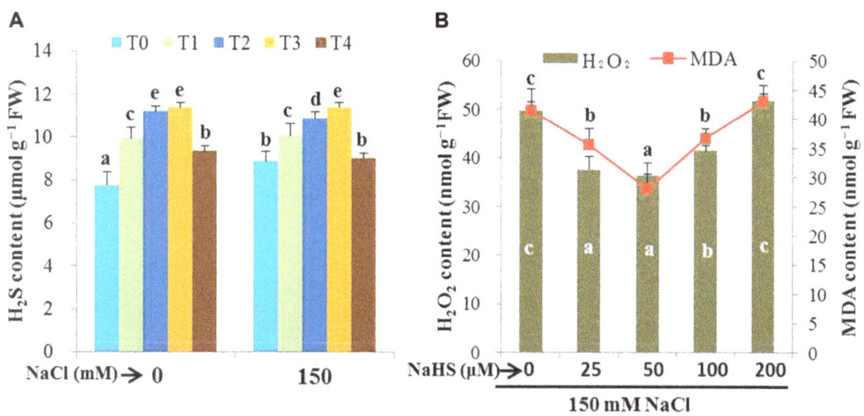

FIGURE 1 | Evaluation of effective dose of NaHS. Effects of exogenous NaHS on the levels of endogenous H_2S **(A)** and oxidative stress indicators, H_2O_2 and MDA **(B)** in the leaves of salt-stressed rice plants pretreated with or without varying doses of NaHS. T0, T1, T2, T3, and T4 indicate 0 μM NaHS, 25 μM NaHS, 50 μM NaHS, 100 μM NaHS and 50 μM NaHS + 500 μM HT, respectively. Bars represent standard deviation (SD) of the mean ($n = 3$). Different letters (a, b, c, d, and e) indicate significant differences among the treatments at $P < 0.05$, according to Duncan's multiple range test. HT, hypotaurine.

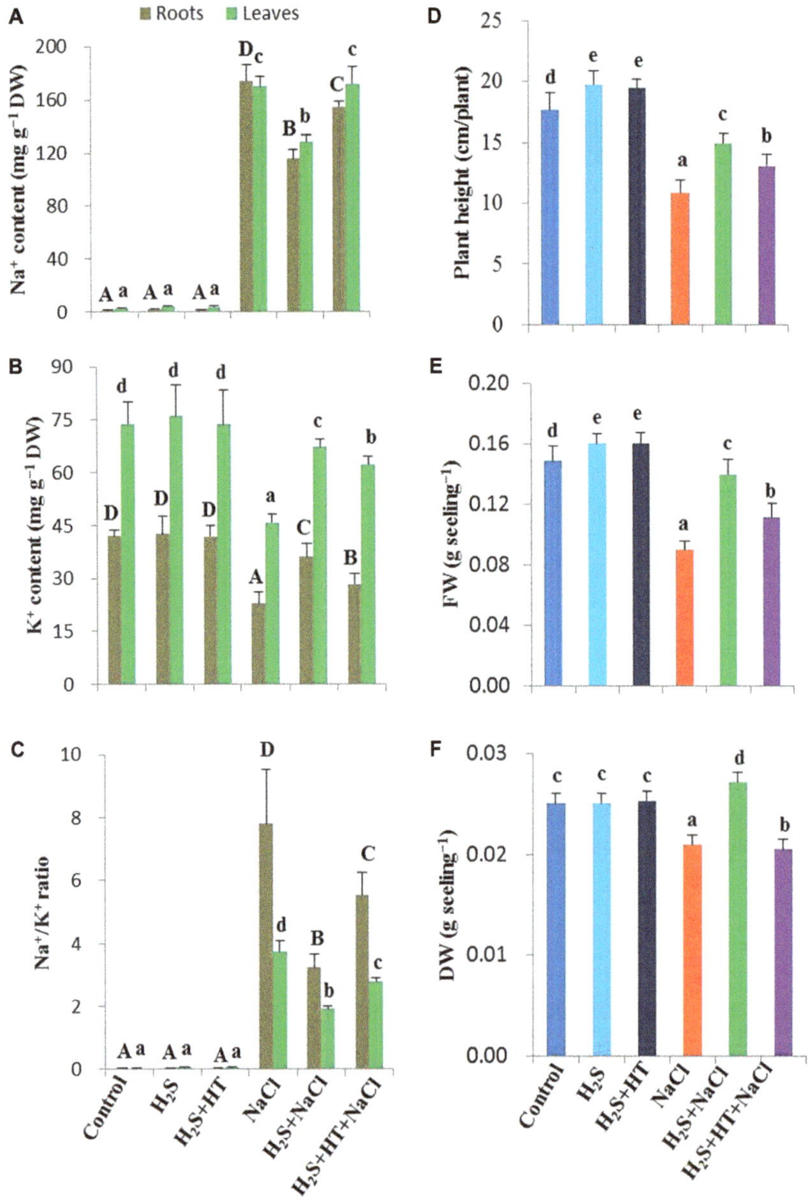

FIGURE 2 | Effects of H_2S on Na^+ and K^+ status, growth and biomass of salt-stressed rice plants pretreated with or without NaHS and HT. (A) Na^+ content, **(B)** K^+ content, **(C)** Na^+/K^+ ratio, **(D)** plant height, **(E)** fresh weight and **(F)** dry weight. Control, H_2S, H_2S + HT, NaCl, H_2S + NaCl, H_2S + HT + NaCl correspond to the group of seedlings subjected to only nutrients, 50 μM NaHS, 50 μM NaHS + 500 μM HT, 150 mM NaCl, 50 μM NaHS + 150 mM NaCl and 50 μM NaHS + 500 μM HT + 150 mM NaCl, respectively. Bars represent standard deviation (SD) of the mean ($n = 3$). Different letters (A, B, C, and D for roots and a, b, c, d, and e for leaves) indicate significant differences among the treatments at $P < 0.05$, according to Duncan's multiple range test. DW, dry weight; HT, hypotaurine.

with untreated control (**Figure 2D**). FW and DW also reduced by 40 and 16%, respectively, in salt-stressed seedlings when compared with control (**Figures 2E,F**). The pretreatment with NaHS; however, significantly increased the plant height, FW and DW in salt-stressed plants by 38, 78, and 29%, respectively, in comparison with salt-stressed only plants (**Figures 2D–F**). On the other hand, adding HT with NaHS during pretreatment negatively affected the plant growth and biomass recovery in salt-stressed plants. Under non-stress conditions, plant height and FW significantly increased, whereas DW remained constant in NaHS– and NaHS+HT-pretreated plants relative to control plants (**Figures 2D–F**).

H_2S Reestablishes Essential Mineral Homeostases

To provide an insight into the role of H_2S in regulating mineral homeostases under salt stress, we estimated the levels of Ca, Mg, Fe, Zn, and Mn in both roots and leaves of rice plants (**Table 1**). In

TABLE 1 | Effects of H_2S on the levels of essential minerals in the roots and leaves of rice plants with or without NaCl stress.

Treatment	Ca (mg g^{-1}DW)		Mg (mg g^{-1}DW)		Fe (mg g^{-1}DW)		Zn (mg g^{-1}DW)		Mn (mg g^{-1}DW)	
	Roots	Leaves	Roots	Leaves	Roots	Leaves	Roots	Leaves	Roots	Leaves
Control	3.96 ± 0.10^{c}	9.15 ± 1.12^{c}	32.47 ± 5.18^{c}	53.52 ± 5.26^{c}	8.68 ± 0.66^{b}	0.68 ± 0.03^{a}	0.32 ± 0.03^{b}	0.20 ± 0.02^{a}	0.14 ± 0.02^{c}	0.32 ± 0.04^{a}
H_2S	4.26 ± 0.05^{d}	9.33 ± 0.95^{c}	34.61 ± 0.42^{d}	59.49 ± 3.45^{e}	9.48 ± 1.81^{c}	1.13 ± 0.21^{c}	0.39 ± 0.03^{c}	0.26 ± 0.02^{b}	0.21 ± 0.03^{e}	0.67 ± 0.05^{c}
H_2S + HT	3.80 ± 0.18^{b}	8.18 ± 0.55^{b}	35.43 ± 2.01^{d}	57.08 ± 2.26^{d}	9.92 ± 1.24^{c}	1.19 ± 0.11^{c}	0.35 ± 0.04^{b}	0.26 ± 0.03^{b}	0.17 ± 0.03^{d}	0.79 ± 0.05^{e}
NaCl	2.62 ± 0.16^{a}	6.94 ± 0.47^{a}	23.07 ± 1.14^{a}	39.94 ± 1.61^{a}	7.07 ± 0.52^{a}	1.02 ± 0.10^{b}	0.25 ± 0.03^{a}	0.34 ± 0.04^{d}	0.12 ± 0.02^{b}	0.52 ± 0.06^{b}
H_2S + NaCl	3.73 ± 0.12^{b}	8.51 ± 0.53^{b}	34.50 ± 4.01^{d}	56.50 ± 2.01^{d}	9.79 ± 1.65^{c}	1.32 ± 0.06^{d}	0.44 ± 0.06^{d}	0.42 ± 0.02^{e}	0.21 ± 0.02^{e}	0.93 ± 0.03^{f}
H_2S + HT + NaCl	2.55 ± 0.22^{a}	6.78 ± 0.47^{a}	26.29 ± 3.14^{b}	49.75 ± 3.35^{b}	8.36 ± 0.57^{b}	1.16 ± 0.11^{c}	0.33 ± 0.02^{b}	0.31 ± 0.07^{c}	0.09 ± 0.07^{a}	0.72 ± 0.09^{d}

Control, H_2S, H_2S + HT, NaCl, H_2S + NaCl, H_2S + HT + NaCl correspond to the group of seedlings exposed to only nutrients, 50 μM NaHS, 50 μM NaHS + 500 μM hypotaurine, 150 mM NaCl, 50 μM NaHS + 150 mM NaCl and 50 μM NaHS + 500 μM hypotaurine + 150 mM NaCl, respectively. FW, fresh weight. Values are means ± SD of three independent replications ($n = 3$). Different superscripted letters (a, b, c, d, e, and f) within the column indicate statistically significant differences among the treatments according to Duncan's multiple range test ($P < 0.05$).

the roots of salt-stressed plants, Ca, Mg, Fe, Zn, and Mn contents decreased by 34, 29, 19, 22, and 14%, respectively, relative to the control plants. As for leaves, salt stress caused a decrease in the contents of Ca and Mg by 24 and 25%, respectively, whereas it resulted in an increase in Fe, Zn, and Mn contents by 50, 70, and 63%, respectively, over the control values. On the other hand, NaHS treatment prior to salt exposure significantly decreased the negative effects of high salinity on the contents of these minerals (**Table 1**). A lower decrease of Ca content in roots (6%) and leaves (7%) of NaHS-pretreated salt-stressed plants was observed in relation to control plants. In comparison with salt-stressed only plants, Mg, Fe, Zn, and Mn contents remained 50, 38, 76, and 75% higher in roots and 41, 29, 24, and 79% higher in leaves of NaHS-pretreated salt-stressed plants, respectively. Conversely, co-application of NaHS and HT prior to salt stress treatment disturbed the mineral homeostases in both roots and leaves, in which Ca, Mg, Fe, Zn, and Mn contents remained at the levels similar to those of salt-stressed only plants. Under non-stress conditions, addition of NaHS significantly improved the contents of these minerals in both roots and leaves compared with control (**Table 1**). Collectively, our results implied that NaHS pretreatment played a significant role in maintaining mineral homeostases in rice plants under both normal and salt stress conditions.

H_2S Suppresses ROS Accumulation

To clarify the protective role of H_2S against NaCl-induced oxidative stress, we carried out histochemical detection of major ROS like $O_2^{\cdot -}$ and H_2O_2 in the leaves of rice plants with or without salt stress. As shown in **Figures 3A,B** the rice plants treated with NaCl alone showed extensive staining with NBT as dark blue spots (for $O_2^{\cdot -}$) and DAB as brown polymerization products (for H_2O_2), while those pretreated with NaHS had only slight staining. Pretreatment with both NaHS and HT exacerbated the accumulation of $O_2^{\cdot -}$ and H_2O_2, showing similar staining intensity to that of salt-stressed only seedlings (**Figures 3A,B**). This result suggested that enhanced H_2S level conferred ROS homeostasis, thereby protecting rice plants from NaCl-induced oxidative stress.

H_2S Reduces Oxidative Stress

Reactive oxygen species, such as H_2O_2, and lipid peroxidation products, such as MDA, are recognized as cellular indicators of oxidative stress (Miller et al., 2010). Salt stress caused a dramatic increase of H_2O_2 by 50% in comparison with untreated control (**Figure 3C**). LOX, as an oxidative enzyme, often participates in lipid peroxidation contributing to oxidative stress (Kumar et al., 2013). We observed that salt stress increased the activity of LOX by 26% compared with control (**Figure 3D**). As a consequence, MDA content increased by 74% in salt-stressed plants over control plants that coincided with the higher H_2O_2 level and increased LOX activity. However, treating rice plants with NaHS prior to salt treatment contributed to the reduction of H_2O_2 (27%), LOX activity (28%) and subsequent MDA level (34%) compared with salt-stressed only plants (**Figure 3E**). In contrast, co-application of HT and NaHS during pretreatment

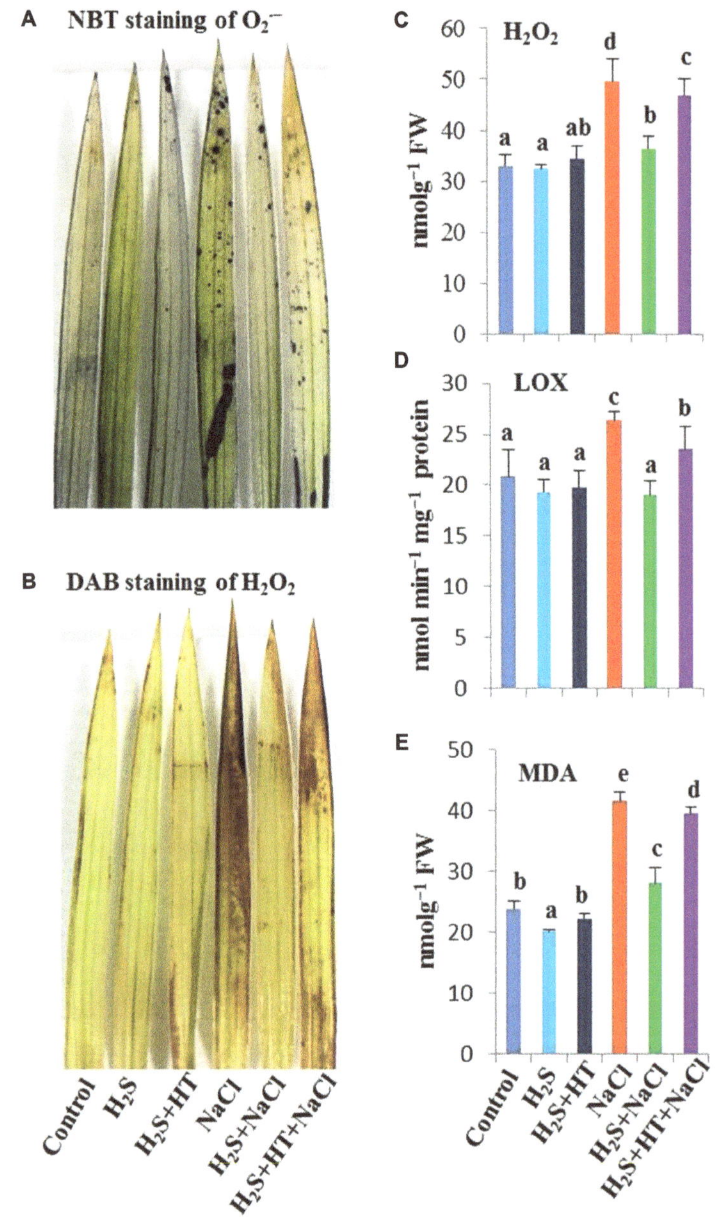

FIGURE 3 | Effect of H_2S on oxidative stress indices in the leaves of salt-stressed rice plants pretreated with or without NaHS and HT. (A) Dark blue spots indicated accumulation of superoxide ($O_2^{·-}$), **(B)** brown polymerization products indicated accumulation of hydrogen peroxide (H_2O_2), **(C)** H_2O_2 content, **(D)** lipoxygenase (LOX) activity and **(E)** malondialdehyde (MDA) content. Control, H_2S, H_2S + HT, NaCl, H_2S + NaCl, H_2S + HT + NaCl correspond to the group of seedlings subjected to only nutrients, 50 μM NaHS, 50 μM NaHS + 500 μM HT, 150 mM NaCl, 50 μM NaHS + 150 mM NaCl and 50 μM NaHS + 500 μM HT + 150 mM NaCl, respectively. Bars represent standard deviation (SD) of the mean ($n = 3$). Different letters (a, b, c, d, and e) indicate significant differences among the treatments at $P < 0.05$, according to Duncan's multiple range test. FW, fresh weight; HT, hypotaurine; NBT, nitroblue tetrazolium; DAB, diaminobenzidine.

inclined rice plants to suffer from higher oxidative load. In these plants, H_2O_2 content, LOX activity and MDA content remained significantly higher than that of NaHS-pretreated salt-exposed plants (**Figures 3C–E**), indicating that H_2S effectively protected rice plants from oxidative stress. Compared with untreated control, NaHS and NaHS + HT pretreatments alone did not significantly affect the levels of H_2O_2 and MDA as well as the LOX activity.

H_2S Protects Photosynthesis Pigments, Water Soluble Proteins and Water Status

In comparison with control, a sharp decline in the content of Chl a (18%), Chl b (31%), total Chl (21%), carotenoids (16%) and water soluble proteins (20%) was recorded in the leaves of salt-stressed rice plants (**Table 2**). Applying NaHS prior to salt stress treatment significantly improved the contents of these

TABLE 2 | Effects of H_2S on the levels of chlorophyll (Chl) a, b, total Chl, carotenoids, water soluble proteins, relative water content (RWC) and proline (Pro) in the leaves of rice plants with or without NaCl stress.

Treatment	Chl a (mg g^{-1} FW)	Chl b (mg g^{-1} FW)	Chl (a + b) (mg g^{-1} FW)	Carotenoids (mg g^{-1} FW)	Water soluble proteins (mg g^{-1} FW)	RWC (%)	Pro (μmol g^{-1} FW)
Control	2.68 ± 0.05^{c}	0.81 ± 0.06^{d}	3.49 ± 0.11^{c}	0.70 ± 0.02^{c}	17.49 ± 0.05^{c}	98.85 ± 0.89^{e}	0.15 ± 0.02^{a}
H_2S	2.70 ± 0.06^{c}	0.80 ± 0.07^{d}	3.50 ± 0.06^{c}	0.71 ± 0.01^{c}	17.53 ± 0.47^{c}	97.79 ± 1.53^{de}	0.15 ± 0.04^{a}
H_2S + HT	2.72 ± 0.06^{c}	0.74 ± 0.04^{c}	3.51 ± 0.06^{c}	0.68 ± 0.01^{c}	16.57 ± 0.24^{b}	96.93 ± 1.99^{d}	0.20 ± 0.02^{a}
NaCl	2.20 ± 0.10^{a}	0.56 ± 0.04^{a}	2.75 ± 0.07^{a}	0.58 ± 0.02^{a}	13.98 ± 0.74^{a}	74.18 ± 1.33^{a}	4.93 ± 0.20^{d}
H_2S + NaCl	2.83 ± 0.04^{d}	0.82 ± 0.06^{d}	3.65 ± 0.05^{d}	0.71 ± 0.02^{c}	18.57 ± 0.41^{d}	88.20 ± 1.69^{c}	3.10 ± 0.61^{b}
H_2S + HT + NaCl	2.39 ± 0.16^{b}	0.61 ± 0.05^{b}	3.00 ± 0.18^{b}	0.66 ± 0.05^{b}	16.35 ± 0.11^{b}	77.40 ± 1.14^{b}	3.98 ± 0.57^{c}

Control, H_2S, H_2S + HT, NaCl, H_2S + NaCl, H_2S + HT + NaCl correspond to the group of seedlings exposed to only nutrients, 50 μM NaHS, 50 μM NaHS + 500 μM hypotaurine, 150 mM NaCl, 50 μM NaHS + 150 mM NaCl and 50 μM NaHS + 500 μM hypotaurine + 150 mM NaCl, respectively. FW, fresh weight. Values are means ± SD of three independent replications (n = 3). Different superscripted letters (a, b, c, d, and e) within the column indicate statistically significant differences among the treatments according to Duncan's multiple range test ($P < 0.05$).

parameters. The levels of Chl a, Chl b, total Chl, carotenoids and soluble proteins were 22, 46, 33, 22, and 33%, respectively, higher in the leaves of NaHS-pretreated salt-stressed plants than in that of salt-stressed only plants (**Table 2**). Nevertheless, the H_2S-mediated recovery on the losses of photosynthetic pigments and soluble protein contents was significantly reduced due to simultaneous application of NaHS and HT prior to salt treatment, when compared with NaHS pretreatment alone (**Table 2**). Under non-stress conditions, NaHS did not alter the levels of photosynthetic pigments and water soluble proteins but co-pretreatment of NaHS and HT significantly reduced the levels of Chl b and water soluble proteins compared with control.

Leaf RWC decreased by 25% in salt-stressed only plants compared with the untreated control (**Table 2**). However, the observed 19% increase in leaf RWC in NaHS-pretreated salt-stressed plants relative to the plants stressed with salt alone indicated that exogenous NaHS significantly prevented the reduction of RWC in rice plants under salt stress. On the other hand, H_2S-mediated restoration of leaf RWC significantly reversed when NaHS and HT were applied together (**Table 2**). Salt stress also resulted in a significant increase in Pro content (3,187%), whereas NaHS pretreatment showed lower increase in Pro content (1,967%) in the leaves of salt-stressed plants in comparison with untreated control plants (**Table 2**). Conversely, adding HT with NaHS during pretreatment significantly increased Pro level by 2,553% in salt-stressed plants compared with control plants. NaHS did not affect the leaf RWC and Pro level under non-stress conditions. However, co-pretreatment of NaHS and HT increased Pro content by 33% compared with control (**Table 2**).

H_2S Enhances the Levels of AsA and GSH, and Modulation of Redox States

Ascorbic acid and GSH, two vital water soluble non-enzymatic antioxidants, can detoxify ROS either directly or indirectly by helping enzymes involved in H_2O_2 metabolism (Shao et al., 2008; Gill and Tuteja, 2010). In parallel with ROS, we investigated the H_2S-induced modulation of AsA, GSH and redox states in rice leaves under salt stress. With respect to the untreated control, salt stress led to a noticeable reduction of AsA content (39%) and increment of DHA content (17%), while a significant increase in GSH and GSSG contents by 43 and 189%, respectively (**Table 3**). This salt-induced modulation of AsA and GSH pools also led to decrease in AsA/DHA and GSH/GSSG ratios in the leaves of the plants treated with salt alone. However, NaHS treatment prior to salt stress significantly recovered salt-induced loss of AsA content as well as further enhanced the GSH content, leading to a maintenance in AsA/DHA and GSH/GSSG ratios (**Table 3**) that are important determinants of cell survival under stressful conditions. A contrasting change in AsA and GSH contents and the AsA/DHA and GSH/GSSG ratios was observed when HT was applied along with NaHS during pretreatment (**Table 3**). Thus, H_2S-mediated elevation of these antioxidants and redox states might have contributed to ROS detoxification under salt stress.

H_2S Stimulates the Activities of Various Antioxidant Enzymes

In our subsequent experiments, we evaluated the regulatory role of H_2S on the activities of various antioxidant enzymes involved in ROS metabolism. Experimental results revealed that NaCl treatment displayed an increase in SOD activity by 22% and a sharp decline in CAT activity by 20%, as compared with NaCl-free control (**Figures 4A,B**). NaHS pretreatment further enhanced the SOD activity, while rescuing CAT activity, but both of which were differentially inhibited when the rice plants were co-treated with NaHS and HT prior to salt stress treatment. H_2S-induced elevation of SOD and CAT activities corroborated with the levels of $O_2^{\cdot-}$ and H_2O_2 in the leaves of rice plants under salt stress (**Figures 3A,B** and **4A,B**). Interestingly, under NaCl-free conditions, NaHS pretreatment significantly enhanced the activities of SOD and CAT compared with control.

The synergistic actions of the enzymes of ascorbate-glutathione cycle eliminate H_2O_2 using AsA and GSH as co-factors (Gill and Tuteja, 2010). Our results showed a considerable increase in the activity of APX in salt-stressed alone seedlings, with 22.4% higher than the untreated control (**Figure 4C**). We also observed that salt stress alone significantly enhanced the activities of AsA- and GSH-regenerating enzymes, including MDHAR, DHAR and GR, by 44, 29, and 12%, respectively (**Figures 4D–F**). Although pretreatment with NaHS did not

TABLE 3 | Effects of H_2S on the levels of non-enzymatic antioxidants ascorbic acid (AsA) and glutathione (GSH), as well as their redox states (AsA/DHA and GSH/GSSG) in the leaves of rice plants with or without NaCl stress.

Treatment	AsA (nmol g^{-1} FW)	DHA (nmol g^{-1} FW)	AsA/DHA ratio	GSH (nmol g^{-1} FW)	GSSG (nmolg^{-1} FW)	GSH/GSSG ratio
Control	4194.19 ± 260.34^{d}	792.58 ± 51.61^{b}	5.32 ± 0.61^{cd}	465.57 ± 13.02^{a}	27.41 ± 2.78^{a}	17.13 ± 2.17^{c}
H_2S	4287.10 ± 161.55^{d}	773.23 ± 87.84^{b}	5.83 ± 1.75^{d}	506.60 ± 9.44^{b}	29.88 ± 4.59^{ab}	17.23 ± 2.71^{c}
H_2S+ HT	3827.42 ± 318.59^{c}	778.06 ± 49.37^{b}	5.06 ± 1.15^{c}	475.04 ± 16.52^{a}	39.10 ± 4.21^{c}	12.25 ± 1.48^{b}
NaCl	2555.81 ± 116.90^{a}	926.13 ± 90.56^{c}	2.78 ± 0.31^{a}	667.96 ± 64.52^{c}	79.25 ± 7.86^{d}	8.48 ± 1.09^{a}
H_2S+ NaCl	3541.94 ± 256.95^{b}	667.26 ± 65.89^{a}	5.35 ± 0.73^{cd}	787.98 ± 43.05^{d}	32.81 ± 5.83^{b}	24.42 ± 3.48^{d}
H_2S+ HT+ NaCl	2456.45 ± 172.61^{a}	625.76 ± 48.76^{a}	3.95 ± 0.48^{b}	687.81 ± 12.85^{c}	39.75 ± 4.26^{c}	17.43 ± 1.87^{c}

Control, H_2S, H_2S + HT, NaCl, H_2S + NaCl, H_2S + HT + NaCl correspond to the group of seedlings exposed to only nutrients, 50 μM NaHS, 50 μM NaHS + 500 μM hypotaurine, 150 mM NaCl, 50 μM NaHS + 150 mM NaCl and 50 μM NaHS + 500 μM hypotaurine + 150 mM NaCl, respectively. FW, fresh weight; DHA, dehydroascorbic acid; GSSG, oxidized glutathione. Values are means ± SD of three independent replications (n = 3). Different superscripted letters (a, b, c, and d) within the column indicate statistically significant differences among the treatments according to Duncan's multiple range test ($P < 0.05$).

show boosting effects on APX, MDHAR and DHAR activities under salt stress, their activity remained significantly higher than that of control. On the other hand, GR activity was further enhanced by 28% upon NaHS pretreatment of salt-stressed plants compared with that in salt-stressed only plants (**Figure 4F**). By contrast, co-application of NaHS and HT prior to salt exposure displayed a trend of these enzyme activities similar to salt stressed only seedlings. Notably, under NaCl-free conditions, NaHS and NaHS + HT pretreatments significantly increased the activities of these four enzymes, as compared with untreated control.

We further evaluated the effects of H_2S on the activities of GSH-metabolizing enzymes, such as GPX and GST. Our results showed differential responses of GPX and GST under salt stress, with GPX activity being increased by 25%, whereas GST activity being drastically decreased by 41% in comparison with untreated control (**Figures 4G,H**). The inducible tendency of GPX activity was further strengthened by NaHS pretreatment, but reversed to a lower level by NaHS and HT co-treatment. In case of GST, NaHS pretreatment significantly rescued the activity of GST, while NaHS + HT pretreatment further aggravated its decreasing tendency upon NaCl stress exposure. Following NaHS and NaHS + HT treatments, GPX activity showed significant enhancements but GST activity decreased by 11 and 20%, respectively, in unstressed plants as compared with control (**Figures 4G,H**).

H_2S Upregulates MG-detoxifying Gly System

The cytotoxic aldehyde MG is detoxified by the maintenance of GSH homeostasis via Gly enzymes (Yadav et al., 2005; Upadhyaya et al., 2011). NaCl treatment resulted in a noticeable increase of MG level by 58%, as compared with control (**Figure 4I**). A significant inhibition of MG production was observed in NaHS-pretreated salt-stressed plants when compared with the plants stressed with salt alone. Moreover, NaCl-triggered production of MG was sustained in NaHS + HT-pretreated salt-stressed plants. In the presence or absence of NaHS and HT, MG-detoxifying enzymes exhibited differential responses in salt-stressed seedlings (**Figures 4J,K**). Salt stress increased Gly I activity by 12%, but did not alter the Gly II activity as compared with the untreated control. On the other hand, NaHS and NaHS + HT pretreatments prior to salt stress kept the Gly I activity at the level of NaCl-stressed only plants. However, NaHS pretreatment showed significant increase in the activity of Gly II in salt-stressed plants when compared with salt-stressed only plants. By contrast, adding H_2S scavenger HT along with NaHS showed a decrease in Gly II activity in salt-stressed seedlings (**Figure 4K**). Thus, upregulation of Gly enzymes by administration of NaHS might contribute to the reduction of NaCl-induced MG toxicity in rice plants.

H_2S Improves Phenotypes of Rice Plants

As shown in **Figure 5** salt treatment of rice plants for 4 days exhibited toxicity symptoms, which were manifested as severe yellowing, chlorosis, burning and roiling of the leaves of rice plants (**Figure 5B**). Salt stress also induced stunted growth, as observed by reduction in plant height (**Figure 2D**). In comparison with salt-stressed only seedlings, NaHS-pretreated rice plants showed considerably less toxicity upon salt treatment. Among the doses of NaHS, 50 μM NaHS was best in terms of the physiological and morphological observations. Contrary, co-pretreatment with NaHS and HT further aggravated salt-induced toxicity symptoms. Under non-stress conditions, treatment of rice plants with NaHS or NaHS + HT for 48 h did not exert any toxicity symptoms (**Figure 5A**).

DISCUSSION

Plant tolerance to salinity is a genetic trait that differs greatly among species, as reflected in their growth responses under certain conditions. Of the various cereal crops, barley is the most tolerant and rice is the most sensitive to salt stress (Munns and Tester, 2008). Therefore, a slight accumulation of Na^+ and Cl^- in cells may adversely affect rice growth and development by impairing various physiological and biochemical processes, including membrane permeability, mineral homoeostases, photosynthesis, stomatal aperture, redox homeostasis, and ROS production and detoxification (**Figure 6**) (Munns and Tester, 2008; Gupta and Huang, 2014; Parihar et al., 2015). Consequently, plants have evolved sophisticated mechanisms to flexibly adapt themselves to overcome salinity-induced damage. However, under stressful conditions, plants' basal capacity overwhelms and plants need to boost the protective

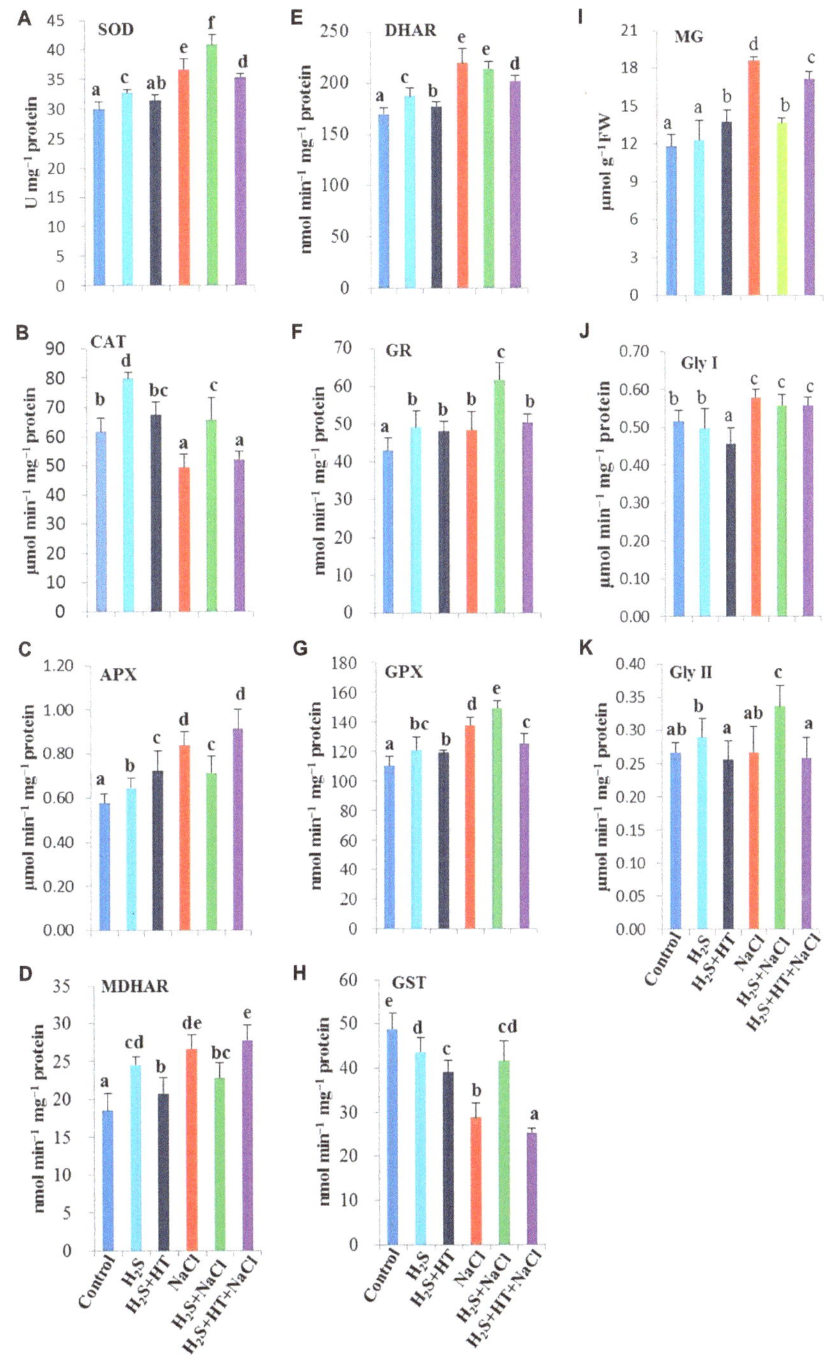

FIGURE 4 | Effect of H_2S on ROS and MG detoxification systems in the leaves of salt-stressed rice plants pretreated with or without NaHS and HT. Activities of ROS detoxifying enzymes **(A)** superoxide dismutase (SOD), **(B)** catalase (CAT), **(C)** ascorbate peroxidase (APX), **(D)** monodehydroascorbate reductase (MDHAR), **(E)** dehydroascorbate reductase (DHAR), **(F)** glutathione reductase (GR), **(G)** glutathione peroxidase (GPX) and **(H)** glutathione *S*-transferase (GST). **(I)** methylglyoxal (MG) content, **(J)** glyoxalase (Gly) I and **(K)** Gly II activities. Control, H_2S, H_2S + HT, NaCl, H_2S + NaCl, H_2S + HT + NaCl correspond to the group of seedlings subjected to only nutrients, 50 μM NaHS, 50 μM NaHS + 500 μM HT, 150 mM NaCl, 50 μM NaHS + 150 mM NaCl and 50 μM NaHS + 500 μM HT + 150 mM NaCl, respectively. Bars represent standard deviation (SD) of the mean ($n = 3$). Different letters (a, b, c, d, e, and f) indicate significant differences among the treatments at $P < 0.05$, according to Duncan's multiple range test. FW, fresh weight; HT, hypotaurine.

mechanisms in order to surmount the detrimental consequences. Our investigations clearly showed that the promotion of H_2S level alleviated, and reduction of its level aggravated salt-induced toxic effects, which were clearly observed in the levels of H_2O_2 and MDA as well as in the phenotypes of rice plants (**Figures 1A,B, 3A–E, 5** and **6**). The benefit of modulation of endogenous H_2S

FIGURE 5 | Effects of H_2S on phenotypes of rice plants in the absence (A) and presence (B) of salt (NaCl). Twelve-day-old plants were pretreated with 0, 25 μM H_2S, 50 μM H_2S, 100 μM H_2S and 50 μM H_2S + 500 μM HT for 48 h and then subjected to 150 mM NaCl stress for 4 days. Photographs of the plants were taken after 4 days of NaCl treatment. T0, T1, T2, T3, and T4 correspond to 0 μM NaHS, 25 μM NaHS, 50 μM NaHS, 100 μM NaHS and 50 μM NaHS + 500 μM HT, respectively. HT, hypotaurine.

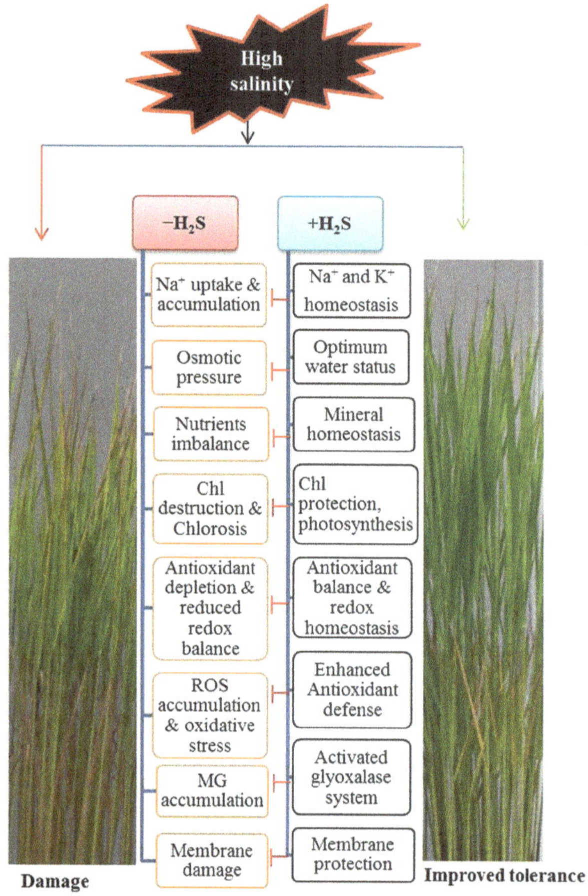

FIGURE 6 | A possible model showing the mechanism of H_2S-induced rice tolerance to salt stress. Exposure of rice plants to NaCl caused an increased in uptake and accumulation of Na^+ in plant cells. This increased level of toxic ions is responsible for inducing osmotic stress, chlorosis, reactive oxygen species (ROS) and methylglyoxal (MG) accumulation, membrane damage and ionic disorder, all of which significantly affect the plant health, leading to cell death and damage of plants. On the other hand, H_2S supplementation resulted in Na^+ and K^+ homeostasis, osmoregulation, increased photosynthesis, higher antioxidant capacity, enhanced MG detoxification, membrane stabilization and ion balance, all of which contributed to the alleviation of salt-induced damage, leading to improved tolerance to salt stress. Chl, chlorophyll; Blunted arrow indicates inhibitory effects.

has also been observed in other plant species like *Zea mays* under heat, *Arabidopsis thaliana* under drought and *Medicago sativa* under salt stress conditions (Jin et al., 2011; Li et al., 2013; Lai et al., 2014). Our study provided firm evidence that H_2S considerably prevented Na^+ uptake, whereas increasing the uptake of K^+, thereby restraining the increased Na^+/K^+ ratio (**Figures 2A–C** and **4**). This might be achieved through the maintenance of plasma membrane integrity, as was observed in *Arabidopsis* and alfalfa (Li et al., 2013; Lai et al., 2014). Furthermore, NaHS enhanced the contents of essential minerals under salt stress (**Table 1**) that are essential for activating a number of physiological and biochemical events associated with stress adaptations. One of the initial effects of high salinity is the reduction of plant growth, which is accompanied by salt-induced osmotic stress (Munns and Tester, 2008). Our results implied that H_2S improved overall growth and biomass of salt-stressed rice plants which could be attributed to its role in protecting Chl a, Chl b, carotenoids and proteins from salt-induced damage (**Table 2, Figures 2D–F** and **6**). Moreover, H_2S might have contributed to improving photosynthetic performance, and thus overall performance of rice plants by delivering Mg, Fe, Zn, and Mn ions as well as by promoting biogenesis of chloroplast and increasing the ability of CO_2 fixation, as observed in *Spinacea oleracea* (Chen et al., 2011). Under osmotic stress, accumulation of compatible solutes like Pro is a universal response plants adapted to retain water status (Hayat et al., 2012). However, Pro accumulation is not always correlated with stress tolerance, which rather depends on the turnover of Pro metabolism (Kishor and Sreenivasulu, 2014). Several studies have claimed that the level of accumulated Pro reflects the intensity of stressed symptoms when plants were challenged with different types of abiotic stresses (Di Martino et al., 2003; Metwally et al., 2003; Loutfy et al., 2012; Mostofa et al., 2015c) as also observed in this study. Similarly, Chang et al. (2014) recently reported that Pro accumulation in salt-stressed *Cathranthus roseus* negatively correlated with RWC, biomass and K^+ accumulation. We assumed that H_2S pretreatment reduced accumulation of Na^+, and as a consequence reduced water loss rate; thus, high level of Pro synthesis might be unnecessary (**Table 2**). Consistent with our results, Li-xu et al. (2013) observed that NaHS application reduced salt-induced accumulation of Pro in the hypocotyls and radicles of cucumber seedlings.

Salt stress-induced water deficit and -increased ionic and osmotic effects often lead to the formation of excessive ROS that can generate oxidative stress by oxidizing lipid, proteins and nucleic acids (Miller et al., 2010; Bose et al., 2014). As expected, high salinity-triggered overproduction of $O_2^{\cdot -}$ and H_2O_2, and increased LOX activity in leaf tissues were correlated with the significant increase in MDA level (**Figures 3A–E**). This detrimental effect was also coincided with the impaired ROS detoxifying capacity, as evident by dropped level of AsA and imbalance of the redox states in terms of reduced AsA/DHA and GSH/GSSG ratios (**Table 3**). Furthermore, unsynchronized activities of SOD and CAT (**Figures 4A,B**), which constitute the frontline enzymatic network by converting $O_2^{\cdot -}$ and H_2O_2 consecutively into H_2O, accelerated the accumulation of ROS in salt-stressed rice plants (**Figures 3A,B**). On the other hand, H_2S successfully alleviated salt-induced oxidative damage, as evident by reduced accumulation of $O_2^{\cdot -}$ and H_2O_2, as well as lipid peroxidation product MDA in rice leaves (**Figures 3A–E** and **6**). H_2S executed its antioxidant role, mainly by increasing the levels of AsA and GSH, and by reestablishing redox homeostasis (**Table 3**). Moreover, enhancement of SOD and CAT activities also indicated that H_2S kept a balance between ROS production and detoxification, thereby protecting cells from oxidative damage (**Figure 6**). The enzymes APX, MDHAR, DHAR and GR of the ascorbate-glutathione cycle act coordinately in eliminating H_2O_2 and regenerating AsA and GSH, which is crucial for promoting plant tolerance to abiotic stresses (Gill and Tuteja, 2010; Miller et al., 2010). In consistence with the findings of Mishra et al. (2013), this study observed an accumulation of H_2O_2 even after increasing the activities of these four enzymes, suggesting that production of H_2O_2 exceeded ROS-scavenging capacity in salt stressed plants (**Figures 3B,C** and **4C–F**). In contrast, H_2S regulates the ascorbate-glutathione cycle differentially by keeping APX, MDHAR and DHAR activities above the control level, while further intensifying GR activity (**Figures 4C–F**). We assumed that coordinating role of SOD and CAT exerted reduced oxidative load on ascorbate-glutathione cycle, resulting in less stimulation of the activities of these enzymes. Furthermore, H_2S-induced enhancement of GR activity contributed well to maintaining redox status, presumably by regenerating GSH from GSSG, which corroborated with the observed increase in GSH level and GSH/GSSG ratio (**Table 3**, **Figure 4F**). It is well known that GSH-dependent defense mechanisms play significant role in protecting plants against various stresses, including salt stress (Mittova et al., 2003; Anjum et al., 2014; Nazar et al., 2014). Besides scavenging ROS directly, GSH detoxifies a wide variety of lipid hydroperoxides, reactive aldehydes and endobiotic substrates with the help of GPX and GST (Mittova et al., 2003). A careful analysis of GSH-related defense mechanism under H_2S addition revealed that H_2S further enhanced GSH level, possibly by inducing its biosynthesis or by upregulating GR activity under salt stress, which then participates in redox homeostasis, as well as GPX and GST-mediated efficient scavenging of peroxides and other electrophiles (**Figures 4G,H**). Similar to our findings, Dawood et al. (2012) and Li-xu et al. (2013) also reported that H_2S-induced GPX and GST activities contributed to enhanced tolerance to salt and aluminum stresses in *Cucumis sativus* and *Hordeum vulgare*, respectively.

Additionally, we investigated the effects of H_2S on another GSH-dependent defense mechanism, namely Gly system, which is involved in MG detoxification and redox balance (Upadhyaya et al., 2011; Hossain et al., 2012; Alvarez Viveros et al., 2013). We noticed that salt stress resulted in a significant increase in MG content, which might be due to the inefficient and desynchronizing activities of Gly enzymes, as Gly I activity increased but Gly II activity remained at the control level (**Figures 4I–K**). Moreover, an inefficient Gly II activity might also trap GSH by accumulating *S*-lactoylglutathione (SLG), which is also mutagenic and cytotoxic (Hossain et al., 2012). However, H_2S addition exhibited an encouraging result on MG detoxification, as evident by a correlation between decreased MG content and increased activities of Gly enzymes, which ultimately protected the cells form MG and SLG toxicity. Furthermore, our result also assumed that enhanced Gly II activity efficiently recycled GSH into the system, which eventually facilitated GSH homeostasis and higher activities of GPX and GST (**Table 3**, **Figures 4G,H,J**) in preventing oxidative stress (**Figures 3A–E** and **6**). These results are consistent with previous reports (Mustafiz et al., 2011; Upadhyaya et al., 2011; Lakra et al., 2015; Mostofa et al., 2015a,b) and suggest that H_2S coordinates the biochemical actions of the antioxidant defense and Gly systems by maintaining GSH level to mitigate the salt-induced ROS and MG toxicity in rice plants.

In summary, we conclude (**Figure 6**) that modulating the level of endogenous H_2S could be an important strategy in improving performance of rice plants under salt stress for the following reasons. First, H_2S restricts the uptake of Na^+, thereby maintaining the Na^+/K^+ ratio. Second, H_2S maintains mineral homeostases, which smoothens minerals driven physiological and biochemical processes. Third, H_2S enhances photosynthetic capacity by protecting photosynthetic pigments. Fourth, H_2S reduces oxidative damage, principally by regulating the GSH-based antioxidant defense and MG detoxification systems. Thus, our findings provide a solid foundation for developing salt-tolerant rice and perhaps other crops by fine-tuning the level of endogenous H_2S through genetic engineering of H_2S biosynthesis.

AUTHOR CONTRIBUTIONS

MM conceived, designed, and conducted the experiments. DS helped in conducting experiments. MM and L-SPT analyzed the data and results. MM and L-SPT wrote the manuscript. MF monitored the experimental work and critically commented on the manuscript. All authors read and approved the final manuscript.

ACKNOWLEDGMENTS

MM thankfully acknowledges the financial support from the Ministry of Education, Culture, Sports, Science and Technology (MEXT), Japan during his Ph.D. study.

REFERENCES

Ali, B., Gill, R. A., Yang, S., Gill, M. B., Ali, S., Rafiq, M. T., et al. (2014). Hydrogen sulfide alleviates cadmium-induced morpho-physiological and ultrastructural changes in *Brassica napus*. *Ecotoxicol. Environ. Saf.* 110, 197–207. doi: 10.1016/j.ecoenv.2014.08.027

Alvarez Viveros, M. F., Inostroza-Blancheteau, C., Timmermann, T., González, M., and Arce-Johnson, P. (2013). Overexpression of Gly I and Gly II genes in transgenic tomato (*Solanum lycopersicum* Mill.) plants confers salt tolerance by decreasing oxidative stress. *Mol. Biol. Rep.* 40, 3281–3290. doi: 10.1007/s11033-012-2403-4

Anjum, N. A., Aref, I. M., Duarte, A. C., Pereira, E., Ahmad, I., and Iqbal, M. (2014). Glutathione and proline can coordinately make plants withstand the joint attack of metal(loid) and salinity stresses. *Front. Plant Sci.* 5:662. doi: 10.3389/fpls.2014.00662

Arnon, D. T. (1949). Copper enzymes in isolated chloroplast polyphenol oxidase in Beta Vulgaris. *Plant Physiol.* 24, 1–15. doi: 10.1104/pp.24.1.1

Bates, L. S., Waldren, R. P., and Teare, I. D. (1973). Rapid determination of free proline for water-stress studies. *Plant Soil* 39, 205–207. doi: 10.1016/j.dental.2010.07.006

Bose, J., Rodrigo-Moreno, A., and Shabala, S. (2014). ROS homeostasis in halophytes in the context of salinity stress tolerance. *J. Exp. Bot.* 65, 1241–1257. doi: 10.1093/jxb/ert430

Bradford, M. M. (1976). A rapid and sensitive method for the quantitation of microgram quantities of protein utilizing the principle of protein-dye binding. *Anal. Biochem.* 72, 248–254. doi: 10.1016/0003-2697(76)90527-3

Chang, B., Yang, L., Cong, W., Zu, Y., and Tang, Z. (2014). The improved resistance to high salinity induced by trehalose is associated with ionic regulation and osmotic adjustment in *Catharanthus roseus*. *Plant Physiol. Biochem.* 14, 140–148. doi: 10.1016/j.plaphy.2014.02.001

Chen, J., Wu, F. H., Wang, W. H., Zheng, C. J., Lin, G. H., Dong, X. J., et al. (2011). Hydrogen sulfide enhances photosynthesis through promoting chloroplast biogenesis, photosynthetic enzyme expression, and thiol redox modification in *Spinacia oleracea* seedlings. *J. Exp. Bot.* 62, 4481–4493. doi: 10.1093/jxb/err145

Christou, A., Manganaris, G. A., Papadopoulos, I., and Fotopoulos, V. (2013). Hydrogen sulfide induces systemic tolerance to salinity and non-ionic osmotic stress in strawberry plants through modification of reactive species biosynthesis and transcriptional regulation of multiple defence pathways. *J. Exp. Bot.* 64, 1953–1966. doi: 10.1093/jxb/ert055

Cuin, T. A., Betts, S. A., Chalmandrier, R., and Shabala, S. (2008). A root's ability to retain K+ correlates with salt tolerance in wheat. *J. Exp. Bot.* 59, 2697–2706. doi: 10.1093/jxb/ern128

Dawood, M., Cao, F., Jahangir, M. M., Zhang, G. P., and Wu, F. (2012). Alleviation of aluminum toxicity by hydrogen sulfide is related to elevated ATPase, and suppressed aluminum uptake and oxidative stress in barley. *J. Hazard. Mater.* 210, 121–128. doi: 10.1016/j.jhazmat.2011.12.076

Di Martino, C., Delfine, S., Pizzuto, R., Loreto, F., and Fuggi, A. (2003). Free amino acids and glycine betaine in leaf osmoregulation of spinach responding to increasing salt stress. *New Phytol.* 158, 455–463. doi: 10.1046/j.1469-8137.2003.00770.x

Doderer, A., Kokkelink, I., van der Veen, S., Valk, B. E., Schram, A. W., and Douma, A. C. (1992). Purification and characterization of two lipoxygenase isoenzymes from germinating barley. *Biochim. Biophys. Acta* 112, 97–104. doi: 10.1016/0167-4838(92)90429-H

Gill, S. S., and Tuteja, N. (2010). Reactive oxygen species and antioxidant machinery in abiotic stress tolerance in crop plants. *Plant Physiol. Biochem.* 48, 909–930. doi: 10.1016/j.plaphy.2010.08.016

Griffiths, O. W. (1980). Determination of glutathione and glutathione disulfide using glutathione reductase and 2-vinylpyridine. *Anal. Biochem.* 106, 207–212. doi: 10.1016/0003-2697(80)90139-6

Gupta, B., and Huang, B. (2014). Mechanism of salinity tolerance in plants: physiological, biochemical, and molecular characterization. *Int. J. Genomics* 2014, 1–18. doi: 10.1155/2014/701596

Hayat, S., Hayat, Q., Alyemeni, M. N., Wani, A. S., Pichtel, J., and Ahmad, A. (2012). Role of proline under changing environments. *Plant Signal. Behav.* 7, 1456–1466. doi: 10.4161/psb.21949

Heath, R. L., and Packer, L. (1968). Photoperoxidation in isolated chloroplasts: I. Kinetics and stochiometry of fatty acid peroxidation. *Arch. Biochem. Biophys.* 125, 189–198. doi: 10.1016/0003-9861(68)90654-1

Hossain, M. A., Hossain, M. Z., and Fujita, M. (2009). Stress induced changes of methylglyoxal level and Glyoxalase I activity in pumpkin seedlings and cDNA cloning of glyoxalase I gene. *Aust. J. Crop Sci.* 3, 53–64.

Hossain, M. A., Nakano, Y., and Asada, K. (1984). Monodehydroascorbate reductase in spinach chloroplasts and its participation in the regeneration of ascorbate for scavenging hydrogen peroxide. *Plant Cell Physiol.* 25, 385–395.

Hossain, M. A., Piyatida, P., Teixeira da Silva, J. A., and Fujita, M. (2012). Molecular mechanism of heavy metal toxicity and tolerance in plants: central role of glutathione in detoxification of reactive oxygen species and methylglyoxal and in heavy metal chelation. *J. Bot.* 2012, 1–37. doi: 10.1155/2012/872875

Iqbal, N., Shahid, U., and Khan, N. A. (2015). Nitrogen availability regulates proline and ethylene production and alleviates salinity stress in mustard (*Brassica juncea*). *J. Plant Physiol.* 178, 84–91. doi: 10.1016/j.jplph.2015.02.006

Islam, M. A., Shitangsu, P. K., and Hassan, M. Z. (2015). Agricultural vulnerability in Bangladesh to climate change induced sea level rise and options for adaptation: a study of a coastal Upazila. *J. Agric. Environ. Int. Dev.* 109, 19–39.

Jin, Z., Shen, J., Qiao, Z., Yang, G., Wang, R., and Pei, Y. (2011). Hydrogen sulfide improves drought resistance in *Arabidopsis thaliana*. *Biochem. Biophys. Res. Commun.* 414, 481–486. doi: 10.1016/j.bbrc.2011.09.090

Kishor, P. B. K., and Sreenivasulu, N. (2014). Is proline accumulation per se correlated with stress tolerance or is proline homeostasis a more critical issue? *Plant Cell Environ.* 37, 300–311. doi: 10.1111/pce.12157

Kumar, A., Prasad, M. N. V., Achary, V. M. M., and Panda, B. B. (2013). Elucidation of lead-induced oxidative stress in *Talinum triangulare* roots by analysis of antioxidant responses and DNA damage at cellular level. *Environ. Sci. Pollut. Res.* 20, 4551–4561. doi: 10.1007/s11356-012-1354-6

Lai, D., Mao, Y., Zhou, H., Li, F., Wu, M., Zhang, J., et al. (2014). Endogenous hydrogen sulfide enhances salt tolerance by coupling the reestablishment of redox homeostasis and preventing salt-induced K+ loss in seedlings of *Medicago sativa*. *Plant Sci.* 225, 117–129. doi: 10.1016/j.plantsci.2014.06.006

Lakra, N., Nutan, K., Das, P., Anwar, K., Singla-Pareek, S. L., and Pareek, A. (2015). A nuclear-localized histone-gene binding protein from rice (OsHBP1b) functions in salinity and drought stress tolerance by maintaining chlorophyll content and improving the antioxidant machinery. *J. Plant Physiol.* 176, 36–46. doi: 10.1016/j.jplph.2014.11.005

Li, Z. G., Yang, S. Z., Long, W. B., Yang, G. X., and Shen, Z. Z. (2013). Hydrogen sulphide may be a novel downstream signal molecule in nitric oxide-induced heat tolerance of maize (*Zea mays* L.) seedlings. *Plant Cell Environ.* 36, 1564–1572. doi: 10.1111/pce.12092

Lichtenthaler, H. K., and Wellburn, A. R. (1983). Determinations of total carotenoids and chlorophylls a and b of leaf extracts in different solvents. *Biochem. Soc. Trans.* 11, 591–592. doi: 10.1042/bst0110591

Li-xu, Y., Cun-jia, Z., Hong-qin, S., Xiu-feng, W., Min, W., Feng-juan, Y., et al. (2013). Exogenous hydrogen sulfide enhanced antioxidant capacity, amylase activities and salt tolerance of cucumber hypocotyls and radicles. *J. Integr. Agric.* 12, 445–456. doi: 10.1016/S2095-3119(13)60245-2

Loutfy, N., El-Tayeb, M. A., Hassanen, A. M., Moustafa, M. F., Sakuma, Y., and Inouhe, M. (2012). Changes in the water status and osmotic solute contents in response to drought and salicylic acid treatments in four different cultivars of wheat (*Triticum aestivum*). *J. Plant Res.* 125, 173–184. doi: 10.1007/s10265-011-0419-9

Mancardi, D., Penna, C., Merlino, A., Del Soldato, P., Wink, D. A., and Pagliaro, P. (2009). Physiological and pharmacological features of the novel gasotransmitter: hydrogen sulfide. *Biochim. Biophys. Acta* 1787, 864–872. doi: 10.1016/j.bbabio.2009.03.005

Metwally, A., Finkemeier, I., Georgi, M., and Dietz, K. J. (2003). Salicylic acid alleviates the cadmium toxicity in barley seedlings. *Plant Physiol.* 132, 272–281. doi: 10.1104/pp.102.018457

Miller, G., Suzuki, N., Ciftci-Yilmaz, S., and Mittler, R. (2010). Reactive oxygen species homeostasis and signalling during drought and salinity stresses. *Plant Cell Environ.* 33, 453–467. doi: 10.1111/j.1365-3040.2009.02041.x

Mishra, P., Bhoomika, K., and Dubey, R. S. (2013). Differential responses of antioxidative defense system to prolonged salinity stress in salt-tolerant and salt-sensitive Indica rice (*Oryza sativa* L.) seedlings. *Protoplasma* 250, 3–19. doi: 10.1007/s00709-011-0365-3

Mittova, V., Theodoulou, F. L., Kiddle, G., Gómez, L., Volokita, M., Tal, M., et al. (2003). Coordinate induction of glutathione biosynthesis and glutathione metabolizing enzymes is correlated with salt tolerance in tomato. *FEBS Lett.* 554, 417–421. doi: 10.1016/S0014-5793(03)01214-6

Mostofa, M. G., and Fujita, M. (2013). Salicylic acid alleviates copper toxicity in rice (*Oryza sativa* L.) seedlings by up-regulating antioxidative and glyoxalase systems. *Ecotoxicology* 22, 959–973. doi: 10.1007/s10646-013-1073-x

Mostofa, M. G., Fujita, M., and Tran, L. S. (2015a). Nitric oxide mediates hydrogen peroxide- and salicylic acid-induced salt tolerance in rice (*Oryza sativa* L.) seedlings. *Plant Growth Regul.* 77, 265–277. doi: 10.1007/s10725-015-0061-y

Mostofa, M. G., Hossain, M. A., Fujita, M., and Tran, L. S. (2015b). Physiological and biochemical mechanisms associated with trehalose-induced copper-stress tolerance in rice. *Sci. Rep.* 5, 11433. doi: 10.1038/srep11433

Mostofa, M. G., Seraj, Z. I., and Fujita, M. (2015c). Interactive effects of nitric oxide and glutathione in mitigating copper toxicity of rice (*Oryza sativa* L.) seedlings. *Plant Signal. Behav.* 10:e991570. doi: 10.4161/15592324.2014.991570

Mostofa, M. G., Rahman, A., Ansary, M. M. U., Watanabe, A., Fujita, M., and Tran, L. S. (2015d). Hydrogen sulfide modulates cadmium-induced physiological and biochemical responses to alleviate cadmium toxicity in rice. *Sci. Rep.* 5, 14078. doi: 10.1038/srep14078

Mostofa, M. G., Hossain, M. A., and Fujita, M. (2014a). Trehalose pretreatment induces salt tolerance in rice (*Oryza sativa* L.) seedlings: oxidative damage and co-induction of antioxidant defense and glyoxalase systems. *Protoplasma* 252, 461–475. doi: 10.1007/s00709-014-0691-3

Mostofa, M. G., Yoshida, N., and Fujita, M. (2014b). Spermidine pretreatment enhances heat tolerance in rice seedlings through modulating antioxidative and glyoxalase systems. *Plant Growth Regul.* 73, 31–44. doi: 10.1007/s10725-013-9865-9

Munns, R., and Tester, M. (2008). Mechanisms of salinity tolerance. *Annu. Rev. Plant Biol.* 59, 651–681. doi: 10.1146/annurev.arplant.59.032607.092911

Mustafiz, A., Singh, A. K., Pareek, A., Sopory, S. K., and Singla-Pareek, S. L. (2011). Genome-wide analysis of rice and *Arabidopsis* identifies two glyoxalase genes that are highly expressed in abiotic stress. *Funct. Integr. Genomics* 11, 293–305. doi: 10.1007/s10142-010-0203-2

Nakano, Y., and Asada, K. (1981). Hydrogen peroxide is scavenged by ascorbate-specific peroxidase in spinach chloroplasts. *Plant Cell Physiol.* 22, 867–880.

Nazar, R., Umar, S., and Khan, N. A. (2014). Involvement of salicylic acid in sulfur induced salinity tolerance: a role of glutathione. *Annu. Res. Rev. Biol.* 4, 3875–3893. doi: 10.9734/ARRB/2014/11773

Parihar, P., Singh, S., Singh, R., Singh, V. P., and Prasad, S. M. (2015). Effect of salinity stress on plants and its tolerance strategies: a review. *Environ. Sci. Pollut. Res.* 22, 4056–4075. doi: 10.1007/s11356-014-3739-1

Rabbani, G., Rahman, A., and Mainuddin, K. (2013). Salinity-induced loss and damage to farming households in coastal Bangladesh. *Int. J. Global Warming* 5, 400–415. doi: 10.1504/IJGW.2013.057284

Reddy, P. S., Jogeswar, G., Rasineni, G. K., Maheswari, M., Reddy, A. R., Varshney, R. K., et al. (2015). Proline over-accumulation alleviates salt stress and protects photosynthetic and antioxidant enzyme activities in transgenic sorghum [*Sorghum bicolor* (L.) Moench]. *Plant Physiol. Biochem.* 94, 104–113. doi: 10.1016/j.plaphy.2015.05.014

Saito, R., Yamamoto, H., Makino, A., Sugimoto, T., and Miyake, C. (2011). Methylglyoxal functions as Hill oxidant and stimulates the photoreduction of O2 at photosystem I: a symptom of plant diabetes. *Plant Cell Environ.* 34, 1454–1464. doi: 10.1111/j.1365-3040.2011.02344.x

Shao, H. B., Chu, L. Y., Lu, Z. H., and Kang, C. M. (2007). Primary antioxidant free radical scavenging and redox signaling pathways in higher plant cells. *Int. J. Biol. Sci.* 7, 8–14.

Shao, H. B., Chu, L. Y., Shao, M. A., Jaleel, C. A., and Mi, H. M. (2008). Higher plant antioxidants and redox signaling under environmental stresses. *C. R. Biol.* 331, 433–441. doi: 10.1016/j.crvi.2008.03.011

Shi, H., Ye, T., and Chan, Z. (2013). Exogenous application of hydrogen sulfide donor sodium hydrosulfide enhanced multiple abiotic stress tolerance in bermudagrass (*Cynodon dactylon* (L). Pers.). *Plant Physiol. Biochem.* 71, 226–234. doi: 10.1016/j.plaphy.2013.07.021

Toit, A. D. (2015). METABOLISM: the health benefits of hydrogen sulphide. *Nat. Rev. Mol. Cell Biol.* 16, 68. doi: 10.1038/nrm3946

Upadhyaya, C. P., Venkatesh, J., Gururani, M. A., Asnin, L., Sharma, K., Ajappala, H., et al. (2011). Transgenic potato overproducing L-ascorbic acid resisted an increase in methylglyoxal under salinity stress via maintaining higher reduced glutathione level and glyoxalase enzyme activity. *Biotechnol. Lett.* 33, 2297–2307. doi: 10.1007/s10529-011-0684-7

Wild, R., Ooi, L., Srikanth, V., and Münch, G. (2012). A quick, convenient and economical method for the reliable determination of methylglyoxal in millimolar concentrations: the N-acetyl-L-cysteine assay. *Anal. Bioanal. Chem.* 403, 2577–2581. doi: 10.1007/s00216-012-6086-4

Yadav, S. K., Singla-Pareek, S. L., Ray, M., Reddy, M. K., and Sopory, S. K. (2005). Methylglyoxal levels in plants under salinity stress are dependent on glyoxalase I and GSH. *Biochem. Biophys. Res. Commun.* 337, 61–67. doi: 10.1016/j.bbrc.2005.08.263

Conflict of Interest Statement: The authors declare that the research was conducted in the absence of any commercial or financial relationships that could be construed as a potential conflict of interest.

5

Ectopic Expression in *Arabidopsis thaliana* of an NB-ARC Encoding Putative Disease Resistance Gene from Wild Chinese *Vitis pseudoreticulata* Enhances Resistance to Phytopathogenic Fungi and Bacteria

*Zhifeng Wen[1,2], Liping Yao[3], Ran Wan[1,2], Zhi Li[1,2], Chonghuai Liu[4] and Xiping Wang[1,2]**

[1] State Key Laboratory of Crop Stress Biology in Arid Areas, College of Horticulture, Northwest A&F University, Yangling, China, [2] Key Laboratory of Horticultural Plant Biology and Germplasm Innovation in Northwest China, Ministry of Agriculture, Northwest A&F University, Yangling, China, [3] Key Laboratory of Stress Physiology and Molecular Biology for Tree Fruits of Beijing, Department of Pomology, College of Agriculture and Biotechnology, China Agricultural University, Beijing, China, [4] Zhengzhou Fruit Research Institute, Chinese Academy of Agricultural Sciences, Zhengzhou, China

Edited by:
Rex Brennan, James Hutton Institute, UK

Reviewed by:
Inger Martinussen, NIBIO – Norwegian Institute of Bioeconomy Research, Norway
Hao Peng, Washington State University, USA

***Correspondence:**
Xiping Wang
wangxiping@nwsuaf.edu.cn

Plant resistance proteins mediate pathogen recognition and activate innate immune responses to restrict pathogen proliferation. One common feature of these proteins is an NB-ARC domain. In this study, we characterized a gene encoding a protein with an NB-ARC domain from wild Chinese grapevine *Vitis pseudoreticulata* accession "Baihe-35-1," which was identified in a transcriptome analysis of the leaves following inoculation with *Erysiphe necator* (Schw.), a causal agent of powdery mildew. Transcript levels of this gene, designated *VpCN* (GenBank accession number KT265084), increased strongly after challenge of grapevine leaves with *E. necator.* The deduced amino acid sequence was predicted to contain an NB-ARC domain in the C-terminus and an RxCC-like domain similar to CC domain of Rx protein in the N-terminus. Ectopic expression of *VpCN* in *Arabidopsis thaliana* resulted in either a wild-type phenotype or a dwarf phenotype. The phenotypically normal transgenic *A. thaliana* showed enhance resistance to *A. thaliana* powdery mildew *Golovinomyces cichoracearum,* as well as to a virulent bacterial pathogen *Pseudomonas syringae* pv. tomato DC3000. Moreover, promoter::GUS (β-glucuronidase) analysis revealed that powdery mildew infection induced the promoter activity of *VpCN* in grapevine leaves. Finally, a promoter deletion analysis showed that TC rich repeat elements likely play an important role in the response to *E. necator* infection. Taken together, our results suggest that *VpCN* contribute to powdery mildew disease resistant in grapevine.

Keywords: wild Chinese *Vitis, VpCN*, disease resistance, powdery mildew, promoter analysis

INTRODUCTION

Plants have evolved multiple mechanisms to protect themselves against pathogens (Jones and Dangl, 2006). The first line of defense is microbe-associated molecular pattern (MAMP)-triggered immunity (MTI) following MAMP perception by membrane-resident pattern recognition receptors (Maekawa et al., 2011). MTI is thought to limit the growth of invasive pathogens. The second line of defense is plant innate immunity, which is activated by the specific recognition of pathogen-derived effectors by intracellular host resistance (R) proteins, and is termed effector-triggered immunity (ETI) (Chisholm et al., 2006). ETI typically leads to a hypersensitive response (HR) and gives rise to a faster and stronger defensive response than MTI-triggered immunity (Cesari et al., 2013). Understanding the function of R proteins, and the mechanisms by which they recognize pathogen effectors, can potentially lead to the development of a long-term strategy for the control and prevention of pathogen invasion.

Over the past few decades, numerous R genes have been cloned from model plants and important crops (Pan et al., 2000b; Collier and Moffett, 2009; Sekine et al., 2012). Most R proteins contain a nucleotide binding (NB) domain and a C-terminal leucine-rich repeat (LRR) domain, and belong to the so-called NB-LRR protein family (Ooijen et al., 2008). The most conserved domain in NB-LRR proteins is an NB domain that is found in proteins such as human Apaf-1, plant R proteins and *Caenorhabditis elegans* Ced-4 (ARC), and as such is referred to as the NB-ARC domain (Ooijen et al., 2008; van der Biezen and Jones, 1998). As a consequence of determining its three-dimensional structure, Albrecht and Takken (2006) proposed that the NB-ARC domain can be further divided into three sub-domains (NB, ARC1, and ARC2). Several conserved motifs have been identified thoughtout the NB-ARC domain in R proteins, such as Walker B, GxP, hhGRExE, Walker A or P-loop, MHD, and RNBS-A–D (Meyers et al., 1999; Pan et al., 2000a; Ooijen et al., 2008). Crystal structure analysis of the NB-ARC domain has led to the suggestion that it may function as a molecular switch to regulate signaling pathways through conformational changes (Riedl et al., 2005; Takken et al., 2006). It has also been shown that the nucleotide binding of the NB-ARC domain in the R proteins, I-2, and Mi-1, requires a P loop, since a P-loop mutant abolished the binding capacity (Tameling et al., 2010). Likewise, the oligomerization of an NB-ARC-LRR protein in the presence of its elicitor requires an intact P-loop in the NB-ARC domain (Mestre and Baulcombe, 2006).

Plant NB-LRR proteins can be divided into two distinct classes: the TNL and the CNL type, based on the domains present at their N terminus. Those that possess a Toll and human interleukin-1 receptor (TIR) domain are referred to as TIR-NB-ARC-LRR or TNL proteins, while those carrying a predicted coiled-coil (CC) domain are classified as CC-NB-ARC-LRR, or CNL proteins (Pan et al., 2000a; Lukasik-Shreepaathy et al., 2012). The potato (*Solanum tuberosum*) Rx protein is a typical CC-NB-ARC-LRR protein mediates resistance to potato virus X (PVX)(Kohm et al., 1993; Bendahmane et al., 1999), the CC domain of RX protein has a four bundle structure and forms a heterodimer with RanGAP2 WPP domain (Hao et al., 2013). The N-termini of the CC and TIR domains are thought to mediate downstream immune responses. It has been reported that in CNL proteins, the CC domain of NRG1 is capable of independently inducing defense responses (Collier et al., 2011), and in TIR proteins the TIR domain plays a crucial role in the cell death signaling pathway (Zhang et al., 2004; Weaver et al., 2006).

The identification and functional characterization of NB-ARC domain R proteins is of considerable interest in developing novel sources of disease resistance in crop plants that are threatened by phytopathogens. For example, *Erysiphe necator* is a fungus that causes powdery mildew (PM) disease in grapevine worldwide, resulting in serious losses in both grape yield and quality. The most economically important cultivated grapevine is *V. vinifera*, which is highly susceptible to PM (Gadoury et al., 2012). To combat the pathogen, fungicides are widely used, which causes environmental and financial pressure on grape growers and reduces wine quality. Thus, developing new grape cultivars with enhanced disease resistance mechanisms is of considerable interest. The wild Chinese *Vitis*, "Baihe-35-1," is an accession of wild Chinese *V. pseudoreticulata* W. T. Wang that possesses high resistance to multiple fungi, and particularly to *E. necator* (Wang et al., 1995; Lin et al., 2006; Yu et al., 2011). To elucidate the resistance mechanisms involved in the defense response to fungal infection in this species, we previously performed an RNA-seq based transcriptome analysis *V. pseudoreticulata* "Baihe-35-1" that had been inoculated with *E. necator* (Weng et al., 2014). Among the pathogen induced genes, one was predicted to encode an NB-ARC domain protein.

In this current study, we report the isolation of the full length cDNA of this gene, which we designated *VpCN*, and its functional characterization following ectopic expression in *Arabidopsis thaliana*. Conclusions regarding its role in conferring Chinese Wild *V. pseudoreticulata* "Baihe-35-1" with disease resistance to powdery mildew are presented.

MATERIALS AND METHODS

Plant Materials and Growth Conditions

Grapevines (Chinese wild *V. pseudoreticulata* accession Baihe-35-1 and *V. vinifera* cv. "Red globe") were maintained in the grape germplasm resources orchard, Northwest A&F University, Yangling Shaanxi, China. *A. thaliana* (ecotype type, Columbia-0) was grown in a growth chamber under the following conditions: 22°C, 50% humidity, a 16/8 h day/night intensity of 125 μmolm^{-2} s^{-1} provided by cool white fluorescent bulbs.

Cloning and Sequence Analysis

Total RNA was extracted from grapevine as previously described (Zhang et al., 2003). First strand cDNA was synthesized from 1 μg of total RNA with the PrimerScript™ II 1st Strand cDNA Synthesis kit (TaKaRa Bio Inc., Dalian, China), according to the manufacturer's instructions. LA *Taq* (Takara Bio. Inc.) was used to amplify the ORF sequence of *VpCN*. The PCR products were cloned into the T-easy vector (Promega, USA), sequenced (Beijing Genomics Institute, Beijing, China) and submitted to GenBank (accession number KT265084). The *VpCN* cDNA sequence was analyzed using BLAST (http://Ncbi.nlm.Nih.gov/blast) in the NCBI database. Grapevine DNA extraction was conducted as previously described (Yu et al., 2013), primers for amplify promoter sequence were designed according to acquired sequence from Grape Genome Database (12×; http://www.genoscope.cns.fr), after cloning into the T-easy vector and sequencing, the promoter sequence was analyzed using PlantCARE (http://bioinformatics.psb.ugent.be/webtools/plantcare/html/) (Lescot et al., 2002). The deduced amino acid sequence of *VpCN* was aligned with closely related proteins and a phylogenetic tree was generated using neighbor joining algorithm with 1000 bootstrapping with the ClustalW tool in the MegAlign program (Version 5.07, DNASTAR Inc.) (**Figure 1D**). A structural model of the NB-ARC domain of *VpCN* was constructed using the structure of PDB 4m9x.1.C (Huang et al., 2013) in SWWISS-MODEL (**Figure 1C**). Real time PCR was conducted using SYBR@ *Premix EX Taq*™II (Tli RNaseH Plus) (Takara Bio. Inc.) in a 20 μl volume reaction following the manufacturer's instructions using the CFX96TM real-time system (Bio-Rad, Hercules, CA, USA). The amplification cycles were as follows: initial denaturation at 94°C for 30 s, 40 cycles at 95°C 5 s, 60°C for 30 s. For melting curve analysis: 40 cycles at 95°C for 15 s followed by a constant increase from 60–95°C. The grapevine *Actin 1* (GenBank Accession number AY680701) was used as reference gene.

Construction of Vectors for Ectopic Expression and *A. thaliana* Transformation

To generate *35S:VpCN,* the open reading frame (ORF) region of *VpCN* was cloned into the binary vector, pCAMBIA 2300 (CAMBIA company), downstream of the CaMV 35S promoter. The construct was introduced into *Agrobacterium tumefaciens*, strain GV3101, via electroporation, and the transformed *A. tumefaciens* was used to transform *A. thaliana* using the floral dip method (Clough and Bent, 1998). Transgenic plants were screened on MS (Murshige and Skoog, 1962) medium containing 60 mg/mL kanamycin, PCR amplification was performed to identify transgenic plants with gene specific primers.

Construction of *VpCN* Promoter:: *GUS* Gene Fusion Vectors and *A. tumefaciens* Mediated Transient Expression Assays

To generate the *VpCN* promoter:*GUS* vector, the *VpCN* promoter was cloned into the T-easy vector, digested with *Bam*HI and *Pst*I, and finally cloned into the binary vector pC0380*GUS*. *35S:GUS* was used as a positive control (Xu et al., 2010). Four *pVpCN* promoter fragments with different 5′ deletions were amplified (**Supplement Table 1**). All the constructs were introduced into *A. tumefaciens* strain GV3101 via electroporation. The *A. tumefaciens* mediated transient expression assays were performed as previously described (Guan et al., 2011). *A. tumefaciens* GV3101 lines harboring the different constructs were grown in liquid Yeast Extract Phosphate (YEP) (Smith and Goodman, 1975) medium (supplemented with 100 μgml^{-1} kanamycin, 60 μgml^{-1} gentamycin, and 30 μgml^{-1} rifampicin) to an OD_{600} of 0.6, and harvested by centrifugation at 5000 ×g for 10 min, before being resuspended in filtration solution (10 mM 2-(N-morpholino) ethanesulfonic acid (MES), pH 5.7, 10 mM $MgCl_2$ and 15 μM acetosyringone) and adjusted to an OD_{600} of 0.6 for infiltration of young grapevine leaves using a vacuum infiltration method (Santos-Rosa et al., 2008). After infiltration, the leaves were kept in a chamber at 16/8 h day/night cycle at 23°C with 70% humidity for 48 h, before inoculation with *E. necator* (Guan et al., 2011; Yu et al., 2013).

Pathogen Inoculation Procedures

E. necator infected leaves were collected from a highly PM-susceptible wild Chinese wild *V. Adstricta*, Hance clone "Taishan-2." Leaves of the Chinese wild *V. pseudoreticulata* "Baihe-35-1" were inoculated by touching the adaxial epidermis of leaves with sporulating colonies on the surface of pathogen leaves, the inoculation were repeated three times (Guan et al., 2011). The samples were harvested 0, 6, 12, 24, 48, 72, 96, and 120 h after inoculation.

A. thaliana powdery mildew *G. cichoracearum* was maintained on highly susceptible *pad4 A. thaliana* mutant plants. The infection was conducted as previously described (Tang and Innes, 2002). The susceptibility or resistance phenotypes were scored 8 days after infection (Nie et al., 2011). Analyses of pathogenesis-related 1 (PR1) gene expression were performed using qRT-PCR using the same PCR program as for the *VpCN* analysis. The *A. thaliana* tubulin gene (GenBank Accession number NM_179953) was used as a reference. Rosett leaves from 4 week old *Arabidopsis* were harvested at 0, 12, 24, 36, and 48 h after inoculation.

P. st DC3000 cells grown in King's B medium (supplemented with 100 μgml^{-1} kanamycin and 30 μgml^{-1} rifampicin) to an OD_{600} of 0.6, harvested by centrifugation for 5000 × g for 10 min and re-suspended in 10 mM $MgSO_4$, adjusted to optical density at OD_{600} of 0.02. The bacterial suspension containing 0.025% Silwet-77, and the mixture were hand infiltrated into the abaxial side of the *A. thaliana* leaves using a needless 1 ml syringe (Fan et al., 2008). *P. st DC3000* bacterial growth were assessed 3 and 5 days after infection as described (Ahn et al., 2007).

Trypan Blue Staining

For trypan blue staining, *A. thaliana* leaves were collected 12 hpi (hours post-inoculation) and boiled in alcoholic lactophenol trypan blue solution (20 mL of ethanol, 10 mL of phenol, 10 mL of water, 10 mL of lactic acid [83%], and 30 mg of trypan blue). Stained leaves were cleared in chloral hydrate (2.5 g dissolved in 1 mL of water) for 3 h, before placing under a coverslip in 50% glycerol (Koch and Slusarenko, 1990; Frye and Innes, 1998).

Peroxide Assay

Peroxide (H_2O_2) was assayed using a hydrogen peroxide kit, according to the manufacturer's instructions (Nanjing Bio Ins., Nanjing, China). Quantification of dead cells was performed 12 hpi by staining leaf discs (0.5 mm in diameter) with 0.2% Evans blue (Sigma) for 30 min, followed by several washes with water to remove excess stain (Mino et al., 2002; Ahn et al., 2007). One

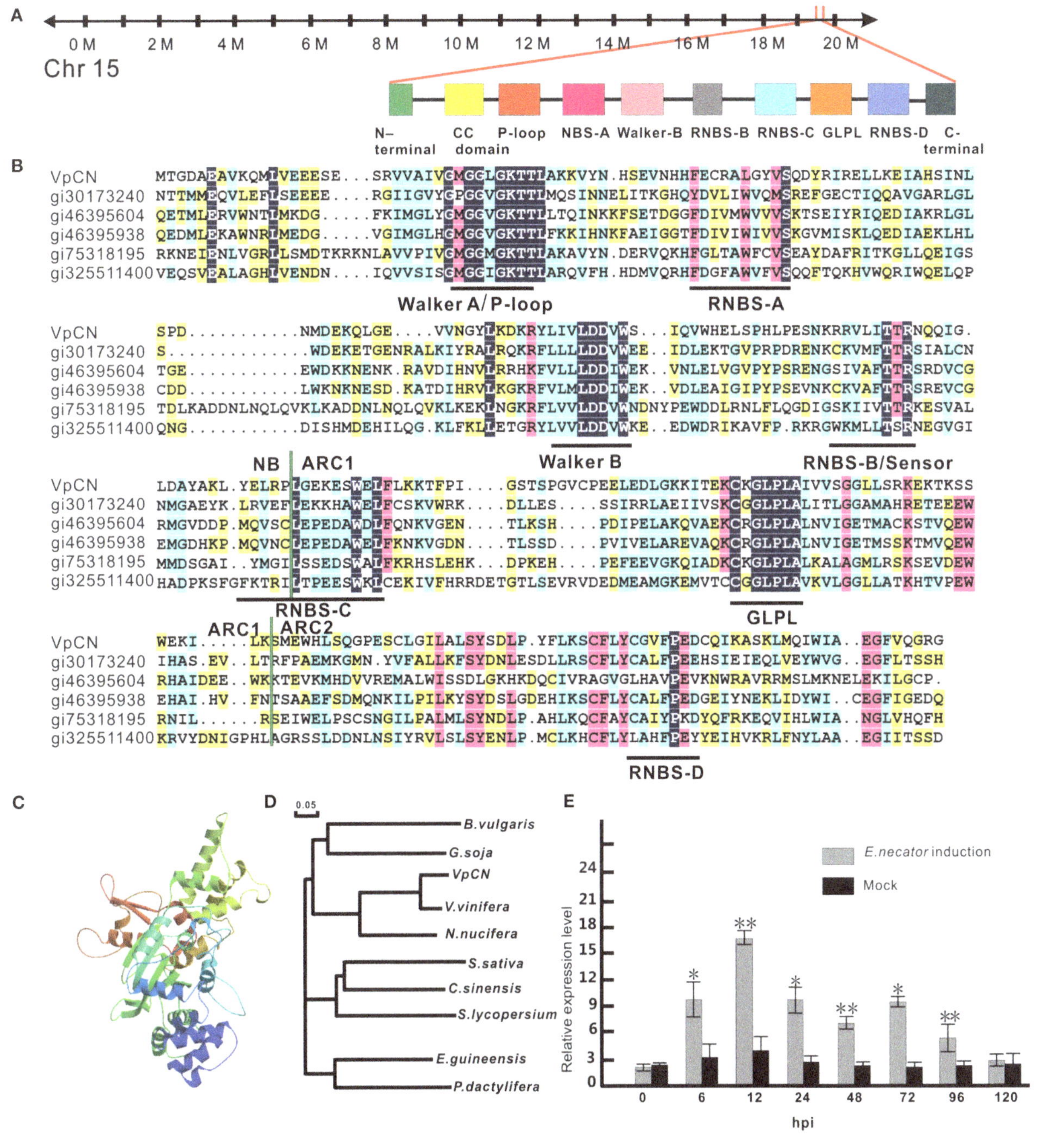

FIGURE 1 | Sequence analysis of *VpCN* and transcript level detection. (A) Schematic map of *VpCN* location and major motifs. **(B)** Multiple sequence alignment of the NB, ARC1 and ARC2 subdomains of NB-ARC in VpCN with closely related proteins. Domain borders are indicate as vertical green lines. Motifs are labeled by horizontal dark lines below the aligned sequences. gi30173240 (Bent et al., 1994), gi46395604 (Bevan et al., 1998), gi46395938 (Theologis et al., 2000), gi75318159 (Ori et al., 1997), gi325511400 (Theologis et al., 2000) **(C)** Structural model of the NB-ARC domain of *VpCN*. **(D)** Phylogenetic tree of *VpCN* and related proteins from other plant species. The tree was generated using the ClustalW function in the MegAlign program: *Vitis vinifera* (GenBank accession no.XP010661747), *Nelumbo nucifera* (GenBank accession no. XP0102588251), *Glycine soja* (GenBank accession no. KHN19144), *Elaeis guineensis* (GenBank accession no. XP010913221), *Solanum lycopersicum* (GenBank accession no. XP010319316), *Beta vulgaris subsp. Vulgaris* (GenBank accession no. XP010669409), *Phoenix dactylifera* (GenBank accession no. XP008791188), *Camelina sativa* (GenBank accession no. XP010426119), *Citrus sinensis* (GenBank accession no. XP006470644). The scale bar represents 0.05 substitutions per site. **(D)** Structure model of NB-ARC in VpCN. **(E)** Analysis of *VpCN* expression in response to *E. necator* inoculation. The third to fifth fully expanded young grapevine leaves beneath the apex were selected for samples. The experiment encompass three independent biological replicates, for each biological replicate three leaves haversted from three plant and three technical replicates were performed. Data represent means of three biological replicates ±SE, asterisksin indicate statistical significance in comparison with control (Student's*t*-test, significance levels of $^{*}P < 0.05$, $^{**}P < 0.01$ are indicated).

milliliter of 50% methanol supplemented with 1% SDS was added and the samples were incubated at 50°C for 1 h. Absorbance at OD_{600} was determined by ultraviolet spectrophotometry after a 10-fold dilution of the extracts (Ahn et al., 2007). The nitro blue terazolium (NBT) staining was performed as described (Kim et al., 2011).

Callose Accumulation

To observe callose accumulation, leaves (3 dpi) were immersed in destaining solution (10 ml phenol, 10 ml glycerin, 10 ml lactic acid, 10 ml H_2O, and 80 ml ethanol) and kept in an oven at 60°C for 1 h to remove chlorophyll. The samples were washed to remove the destaining solution, and stained with 0.1% aniline. The fluorescence of callose was detected using an epifluorescence microscope (E800, Nikon) with a V-2A filter (Reuber et al., 1998; Ahn et al., 2007). For quantitative determination of callose, *A. thaliana*, leaves (3 dpi) were immersed in ethanol for 2–3 days to remove the chlorophyll, before centrifugation at 5000 × g for 10 min. The supernatant was discarded and the pellet resuspended in 0.4 ml DMSO. One hundred microliter of the supernatant was supplemented with loading mixture [400 μl 0.1% (w/v) aniline blue, 590 mL 1 M glycine/NaOH (pH 9.5), 210 mL 1 M HCl] and 200 μl 1 M NaOH. The control samples were not supplemented with aniline. The samples were incubated in a water bath 50°C for 20 min and cooled to room temperature before detection with a fluorescence spectrophotometer (F-4600, Hitachi, Tokyo, Japan) under 393 nm excitation, 479 nm emission and a voltage of 400 v. The fluorescence of the samples was determined by subtracting the fluorescence value of the control from those of the samples (Kohler et al., 2000).

GUS Staining, Histochemical and Fluorometric Assays for Determining *GUS* Activity

A histochemical β-glucuronidase (GUS) assay of leaves was carried out as previously described (Jefferson, 1987). Briefly, leaves were immersed in GUS staining solution at 37°C for 24 h, before washing in 70% ethanol at 37°C and viewing macroscopically (Guan et al., 2011; Yu et al., 2013). GUS fluorescence was determined quantitatively according to Jefferson (1987). Protein concentrations in grapevine extracts was normalized by dilution with extraction buffer according to Bradford (1976). GUS activity was expressed as pmol 4MU (Sigma-Aldrich China, Shanghai, China) per minute per mg of protein. Sample fluorescence was detected with an infinite 200® PRO (Tecan Trading AG, Switzerland). Three independent experiments were performed.

RESULTS

VpCN Expression during Powdery Mildew Infection

To identify potential resistance mechanisms and resistance related genes in the response of wild Chinese *V. pseudoreticulata* to powdery mildew, we previously performed a transcriptome analysis of the "Baihe-35-1" using RNA-seq (Weng et al., 2014). We observed that the expression of *VpCN* (GenBank accession number KT265084) was strongly induced by inoculation with *E. necator*. To verify this, we performed quantitative real-time PCR (qPCR) analysis of *VpCN* expression in *V. pseudoreticulata* leaves that had been inoculated with *E. necator*, and observed 4.2-fold greater *VpCN* transcript levels than in leaves prior to inoculation. Subsequently, *VpCN* expression decreased but remained at a higher level than in mock inoculated plants (**Figure 1E**).

Cloning and Sequence Analysis of *VpCN*

To investigate the putative role of *VpCN* in providing resistance to pathogens, we first designed primers based on a cDNA sequence obtained from the Grape Genome Database (12×; http://www.genoscope.cns.fr), and isolated and designated the gene *VpCN* (GenBank accession number KT265084). The *VpCN* gene is located on chromosome 15 (**Figure 1A**), has an ORF of 1773 bp (**Supplement Figure 1**) and is predicted to encode a protein of 590 amino acids with a molecular mass of 67,390 Da and a theoretical pI value of 5.45. The amino acid sequence was further predicted to contain a RxCC-like domain in the N-terminus from residue 6–119, a Ran GTPase-acting protein 2 (RanGAP2) interaction site in the RxCC-like domain and an NB-ARC domain spanning residues 129–414. The NB-ARC sub-domains, NB, ARC1, and ARC2 were all present. Furthermore, several conserved motifs, such as a P-loop, RNBS A–D, and a GLPL (**Figure 1B**) were detected. In addition to a RxCC-like domain and an NB-ARC domain, we also found an AAA domain and a PLN03210 domain in the predicted amino acid sequence (picture not shown). A structure-based multiple amino acids sequence alignment was performed to compare the NB-ARC domain of *VpCN* with those of other closely related plant R proteins, including RPS2 (gi30173240) (Bent et al., 1994) and I-2 (gi75318159) (Ori et al., 1997). The amino acids sequence identity between the *VpCN* and the *A. thaliana* RPS2 NB-ARC domain was shown to be 33%, while the *VpCN* and I-2 NB-ARC domains had a 29%, sequence identity, concentrated on the conserved motifs of the NB-ARC subdomains (**Figure 1B**).

Ectopic Expression *VpCN* in *A. thaliana* Enhance Resistance to Powdery Mildew

We next transformed the *VpCN* in *A. thaliana* under the control of the constitutive 35S promoter (**Figure 2A**). A total of 42 independent transgenic T1 lines were obtained and the presence of the transgene confirmed by PCR using *VpCN* specific primers. The T2 progeny segregated so that 39 lines displayed wild type morphology while three lines exhibited a dwarfed phenotype and morphological abnormalities, such as small yellow leaves, stunted growth, and chlorotic tissue (**Figure 2B**). These dwarf lines eventually died. The lines with a wild type phenotype were challenged with *G. cichoracearum*, and three transgenic lines with higher resistance were chosen for the generation of homozygous T3 generation lines. The transgenic lines displayed few visible white powdery areas on their leaves at 8 dpi, whereas the wild-type (Col-0) exhibited abundant powdery mildew development (**Figures 2C,D**). To determine whether the enhanced resistance to *G. cichoracearum* in the transgenic lines was related to

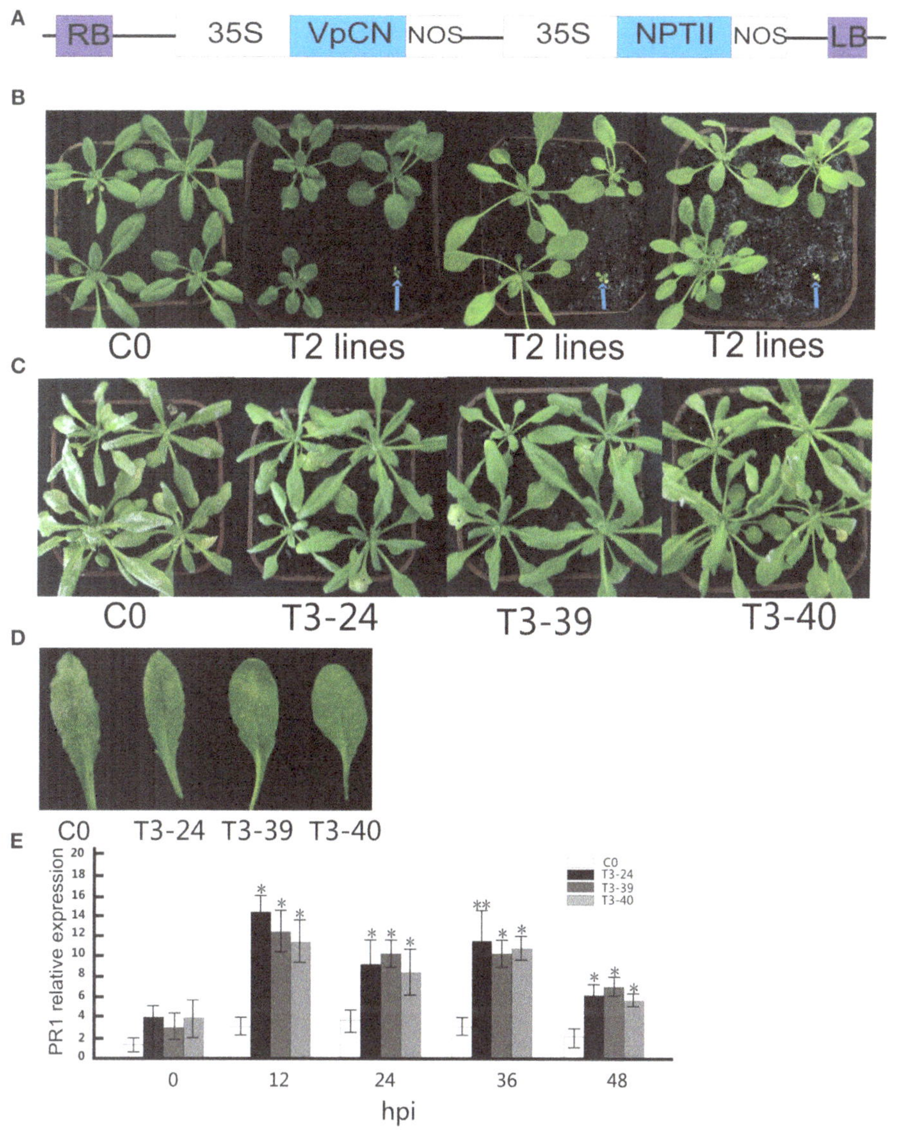

FIGURE 2 | Generation of CaMV 35S promoter-*VpCN* constructs used for transformation of*Arabidopsis thaliana*, morphology of wild type and transgenic *Arabidopsis thaliana* plants, with transgenic plants showing enhanced disease resistance to *G. cichoracearum* after ectopic expression of *VpCN*. (A) Structure of the CaMV 35S promoter-*VpCN* ectopic expression construct. LB, left border; RB, right border; 35S, CaMV 35S promoter; NOS, terminator; NPT II, aminoglycoside-3′- phosphotransferase. **(B)** Indicate T2 transgenic plants displayed either normal phenotypes or dwarfism. Blue arrows indicate the dwarf phenotype in 4 week old plants. **(C)** Transgenic *A. thaliana* leaves developed fungal spores 8 dpi with *G. cichoracearum*. **(D)** Disease symptoms developed on the leaves of transgenic lines and wild type plants 8 dpi with *G. cichoracearum*. **(E)** *A. thaliana* PR1 transcript levels in T3 lines and wild-type after inoculation with *G. cichoracearum*. Total RNA was extracted from *A. thaliana* leaves 0, 12, 24, 36, and 48 h post-inoculation (hpi) with *G. Cichoracearum*. The experiment encompass three independent biological replicates, for each biological replicate six rosette leaves were harvested from three plant and three technical replicates were performed. Data represent means of three biological replicates ±SE, asterisksin indicate statistical significance in comparison with WT (Student's *t*-test, significance levels of $^*P < 0.05$, $^{**}P < 0.01$ are indicated).

an increase in the expression of a known defense gene, we evaluated PR1 (Pathogenesis Related 1) (Friedrich et al., 1996) transcript levels at 0, 12, 24, 36, and 48 hpi. Three transgenic plants displayed higher PR1 transcript abundance after pathogen inoculation than wild type plants, reaching a maximum level at 12 hpi. The PR1 transcript levels of transgenic plants were ~4–5-fold higher after inoculation than in wild type at all time points (**Figure 2E**).

Ectopic Expression of *VpCN* Results In Enhanced Protection Against The Bacterial Pathogen, *P. st* DC3000

Since amino acid sequence of *VpCN* was predicted to contain a PLN03210 domain, which has been shown to be correlated with resistance to *Pseudomonas syringae pv. glycinea* race 6 (Kim et al., 2009), we hypothesized that it might function in providing resistance to bacterial infection. To test this, transgenic and control plants were challenged with the bacterial *P. st* DC3000 pathogen by leaf infiltration (**Figure 3A**). Most infiltrated wild type leaves exhibited water-soaking at 1 dpi, turned yellow and finally wilted at 5 dpi. In contrast, the transgenic plants infected with the pathogen showed fewer symptoms (**Figure 3B**), and when the growth of *P. st* DC3000 in the inoculated plants was quantified, it was found that the bacterial number in the transgenic plants was significantly lower than in the wild type plants (**Figure 3F**). To observe the effect of *VpCN* expression on cell death, trypan blue staining was performed of leaves and we observed that cell death was more widespread in the transgenic lines than the wild type plants (**Figure 3C**). Additionally, cell death quantification by Evans blue staining followed by spectrophotometric analysis, showed a 5-6 fold higher level cell death in the transgenic plants (**Figure 3H**). Nitroblue tetrazolium (NBT) staining for the superoxide anion also showed higher accumulation in the transgenic plants (**Figure 3D**), as did quantitative measurements of H_2O_2 (**Figure 3G**). Finally, the accumulation of the (1,3)-β-glucan polymer callose, which is known to be involved in plant defense responses (Brown et al., 1998), was visualized by aniline blue staining of wild type and transgenic plants after treated with *P. st* DC3000 (**Figure 3E**). Greater accumulation of callose was observed in the transgenic plants than in wild, and when callose levels were quantified, it was confirmed that the transgenic lines contained significantly ($P < 0.05$) more callose (**Figure 3I**).

Isolation and Analysis of the *VpCN* Promoter Sequence

A 1440 bp upstream sequence was cloned using wild Chinese *V. pseudoreticulata* "Baihe-35-1" genomic DNA by PCR, regulatory *cis*-acting elements predicted showed that several putative regulatory elements involved in the activation of defense-related genes, including 72 predicted TATA boxes, 32 CAAT boxes, and two TC-repeat elements, which are known to be involved in defense and stress responses, a TCA element, which is involved in salicylic acid (SA) responses, a TGACG motif, which is associate with methyl jasmonate-response, an HSE element, which is involved in heat stress responses and two TATC elements, which are related to gibberellin responses (**Figure 4A**). Additional predicted *cis*-regulatory elements included light response elements (TCCC-motif, MRE, I-box, GT1-motif, GAG-motif, GA-motif, G-box, CATT motif, Box-I, AT1-motif, and Box-4), as well as others *cis*-elements (5UTR Py-rich stretch, circadian element and, TATC box). Several of the predicted *cis*-elements are known to be involved in responses to environmental stresses, further suggesting that the *VpCN* promoter may play a role in defense responses.

Promoter::GUS (Glucuronidase) Assays

To test the activity of the *VpCN* promoter, the 1440-bp promoter fragment was fused to a reporter gene encoding β-glucuronidase (GUS), generating the construct pCVpCNGUS. As a positive control, a CaMV35S::GUS (PC35SGUS) construct was used and a construct with no promoter was used as a negative control (pC0380GUS) (Xu et al., 2010; **Figure 4B**). All the constructs were expressed transiently in grapevine leaves, which were subsequently subjected to GUS staining. Leaves transformed with the PC35SGUS construct showed strong GUS activity, while no activity was detected in wild type (WT) and very little in PC0380GUS. pCVpCNGUS transformed leaves showed GUS activity but at a lower level than leaves transformed with PC35SGUS (**Figure 4C**), and when leaves were infected with *E. necator* 2 dpi prior to GUS staining, the infected leaves exhibited stronger GUS activity than mock-inoculated control leaves. To further determine the location of the pathogen-responsive *cis*-regulatory region, we generated four promoter deletion fragments and fused them to *GUS* (−1360, −700, −400, and −240 bp) (**Figure 5A**). When the GUS activity was quantified fluorescently, the highest levels were measured in grapevines containing the −1440 bp fragment, where it was induced 1.57-fold after treatment with *E. necator* compared to mock controls. Leaves transformed with −1360, −700, and, −400 promoter fragments exhibited a relative low level of GUS activity; however, they showed increased GUS activity after being challenged with *E. necator* (**Figure 5B**). Since the leaves transformed with the −240 bp fragment showed no significant difference in GUS activity before and after treatment with *E. necator* (**Figure 5C**), the −400 bp promoter fragment was deduced to be the minimal promoter region required for the response to *E. necator* infection.

DISCUSSION

We previously reported the leaf transcriptome of wild Chinese grape (*V. Pseudoreticulata*, "Baihe-35-1") that had been inoculated with *E. necator*, and showed that expression of a unigene corresponding to *VpCN* was strongly induced by the infection (Weng et al., 2014). Here, we isolated the ORF sequence of *VpCN* and ectopically expressed it in *A. thaliana*. This resulted in enhanced disease resistance to the pathogens *G. cichoracearum* and *P. st* DC3000. The deduced amino acid sequence of the corresponding protein is predicted to contain an RxCC-like and an NB-ARC domain. Most currently known R proteins have a NB-ARC domain and the CC domain is thought to initiate signaling (Radirdan et al., 2008). Given the rapid and strong up-regulation of *VpCN* transcript accumulation in wild Chinese *Vitis* after treatment with *E. necator*, we suggest that *VpCN* may play a role in the early defense signaling pathways in pathogen recognition. In addition to these two domains, the deduced amino acid sequence also contained a PLN03210 domain, which is thought to contribute to the identification of resistance signaling components and to convey resistance to *P. syringae* (Kim et al., 2009), suggesting that *VpCN* may also be associated with bacterial disease resistance.

Several studies have already demonstrated that over-expression of an R-gene can cause growth retardation, spontaneous cell death, and constitutive defense activation (Tao et al., 2000; Bendahmane et al., 2002; Stokes et al., 2002; Mohr et al., 2010; Nandety et al., 2013) due an over activation of the ETI system. In this study, three independent transgenic lines exhibited dwarfism and stunted growth, as well as other

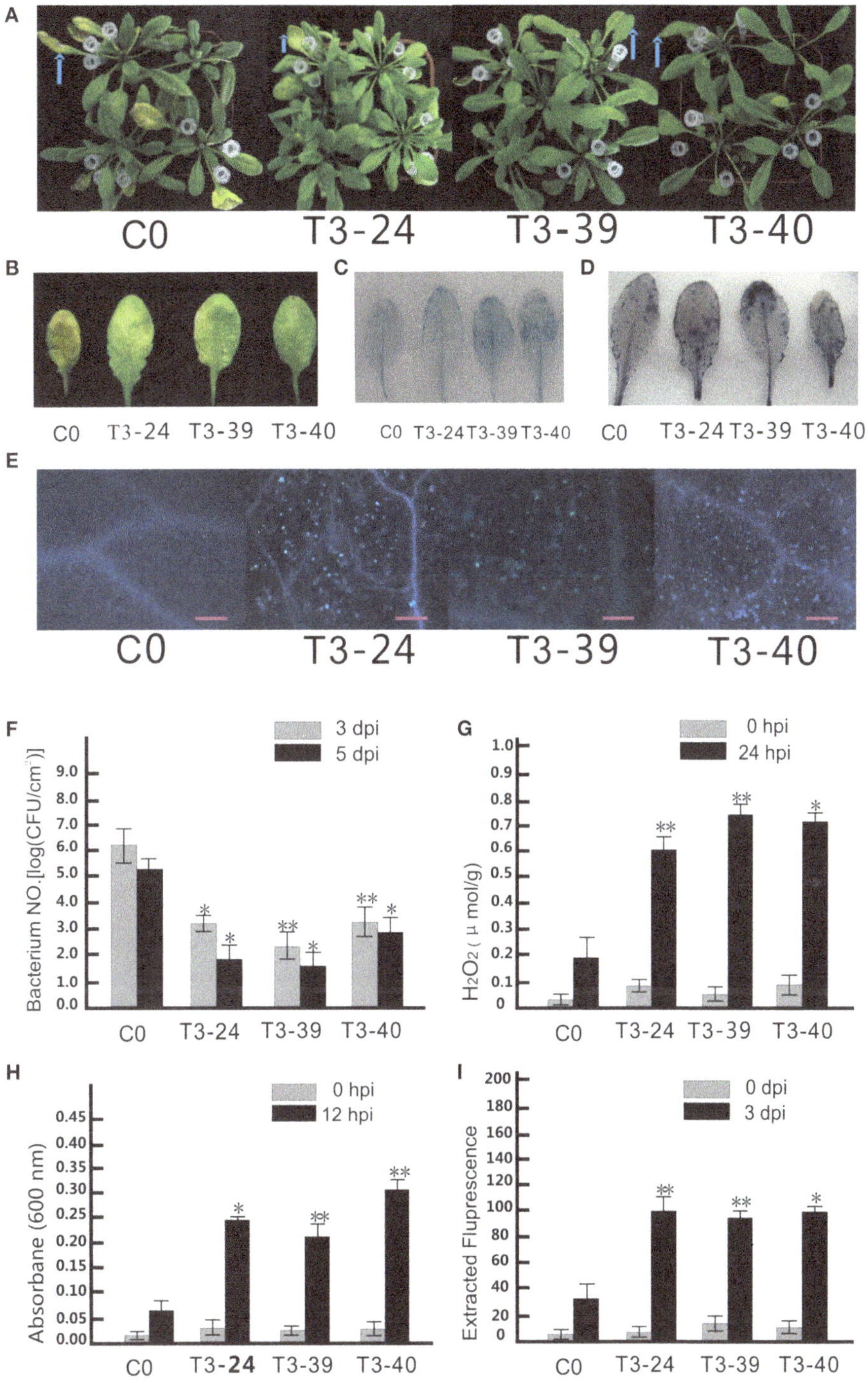

FIGURE 3 | Ectopic expression of *VpCN*in *Arabidopsis thaliana* enhanced disease resistance to *Pseudomonas syringae* pv. tomato DC3000. (A) *P. st* DC3000 was diluted to OD_{600} 0.02 and injected into the middle of a leaf with needleless syringes. The injected leaves were marked with white pipette tips, and pictures taken 3 dpi. **(B)** Disease symptoms developed on the leaves of transgenic lines and wild type plants 3 dpi with *P.st* DC3000. **(C)** Transgenic plants and wild type leaves were stained with trypan blue 12 hpi with *P.st* DC3000. **(D)** Transgenic plants and wild type leaves were stained with nitro blue terazolium (NBT). **(E)** Microscopic observation of callose deposition after 3 dpi. Bars = 50 μm. **(F)** The numbers of bacterial cells in the leaves were determined at 3 and 5 dpi. **(G)** Detection of H_2O_2 concentration in *Arabidopsis* leaf samples harvested at 24 hpi. **(H)** Quantification of dead cells at 12 hpi. **(I)** Quantification of callose from *A. thaliana* leaves at 3 dpi. The experiment encompass three independent biological replicates, for each biological replicate six rosette leaves were harvested from three plant and three technical replicates were performed. Data represent means of three biological replicates ±SE, asterisksin indicate statistical significance in comparison with WT (Student's *t*-test, significance levels of $^{*}P < 0.05$, $^{**}P < 0.01$ are indicated).

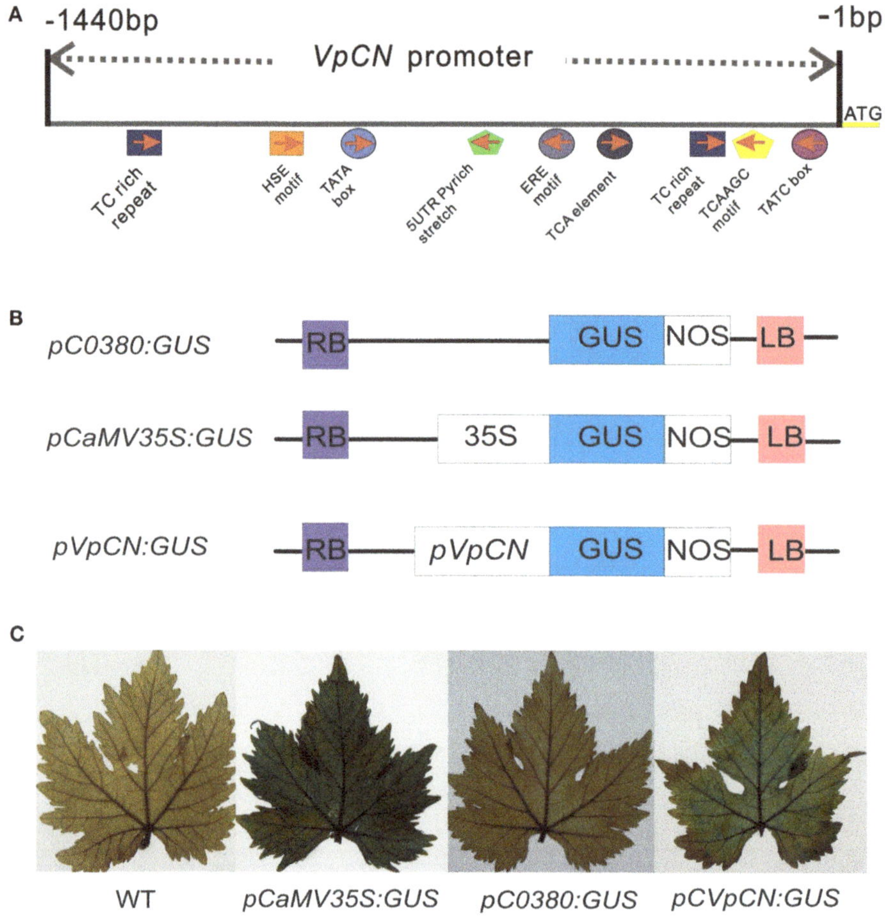

FIGURE 4 | The main predicted *cis*-acting elements in the *pVpCN* promoter sequence, structure of the *VpCN* promoter fused to the *GUS* reporter gene and GUS staining of the transient constructs in transformed grapevine leaves. (A) Schematic diagram of the main predicted *cis*-acting elements in the *VpCN* promoter sequence of Chinese wild *V. pseudoreticulata*. **(B)** The *pVpCN* promoter was fused to the *GUS* gene. The plasmid pCaMV35S:GUS was used as a positive control and pC0380:GUS was used as a negative control. **(C)** The fully expanded grapevine leaves of *V. vinifera* "Red globe" were collect from a grape germplasm resources orchard and used for agroinfiltration.

morphological defects, although since these plants eventually died, we were unable to investigate whether they also exhibited enhanced resistance to *G. Cichoracearum*. In agree with these results we suggest that *VpCN* ectopic expression may active ETI system and cause constitutive defense in three transgenic plants and cause growth retardation, spontaneous cell death. Further studies will investigate whether the three dwarf and lethal phenotypes is caused by toxic effects of high level of *VpCN* expression or the co-suppression between *VpCN* and *Arabidopsis* endogenous genes with VpCN-homologous sequences.

There have been several reports suggesting that over-expression of R genes enhances disease resistance due to constitutive SA accumulation, PR gene expression and active defense responses (Keller et al., 1999; Tang et al., 1999; Kim et al., 2001; Shirano et al., 2002; Stokes et al., 2002). In this study, ectopic expression of *VpCN* in *A. thaliana* enhanced disease resistance to *G. cichoracearum*, and when the PR1 transcript levels was assessed, a 4-5 fold increase in expression was observed in 12 hpi in transgenic plants compared to WT, and these levels remained higher over the time course. These results suggest that ectopic expression of *VpCN* in *A. thaliana* activate defense responses after pathogen inoculation.

The production of reactive oxygen species (ROS), mainly in the form of a superoxide burst and H_2O_2 accumulation, is thought to enhance plant defense responses and to be essential for the establishment of plant immunity (Alvarez et al., 1998; Grant and Loake, 2000; Punja, 2004; Choi and Hwang, 2011; Kim and Hwang, 2014). In agreement with these results, we found that higher levels of O_2^- anions and H_2O_2 in the transgenic plants than in WT after challenging with *P. st* DC3000. This suggests that ectopic expression of *VpCN* triggers an oxidative burst to induce plant immunity to *P. st* DC3000; however, further studies are needed to investigate how oxidative burst and H_2O_2 accumulation is mediated by *VpCN*. High concentrations of ROS can result in HR-like cell death (Kovtun et al., 2000;

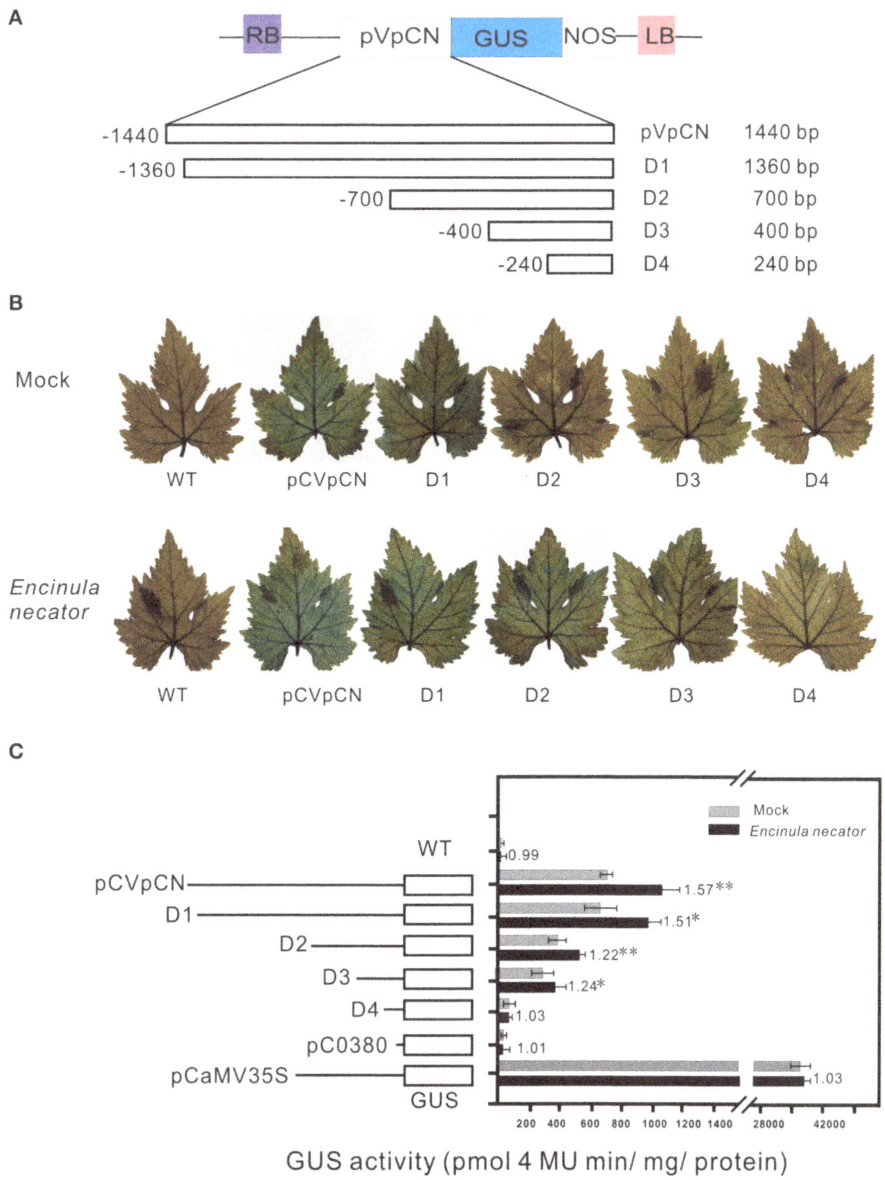

FIGURE 5 | Schematic map of the p*VpCN* promoter-*GUS* gene fusion deletion constructs, histochemical analysis of GUS expression in transiently transformed *V. vinifera* "Red globe" leaves after inoculation with *E.necator*, and fluorometric analysis of GUS activity in the transiently transformed grapevine leaves. (A) The *GUS* gene was driven by the *VpCN* promoter deletions, the exact locations of the promoter fragments are shown in **Supplement Figure 2**. The deletion size is indicated at the far right. **(B)** GUS staining was carried out 2 days after treatment with sterile water (upper panel) or *E. necator* (lower panel). **(C)** The various deletion fragments of the *VpCN* promoter fused to *GUS* and relative GUS activity driven in the transiently transformed grapevine leaves. The *dark bars* indicate the average *GUS* activity for deletion constructs in transiently transformed grapevine leaves treated with *E. necator*, the *gray bars* indicate the mock treatment (sterile water). Numbers adjacent to the bars indicate the fold difference in GUS activity leaves harboring the various constructs challenged with *E. necator* relative to the mock samples. The mean GUS activity (±SD) is averaged from three independent experiments ($n = 3$), the errors bars indicate the stand deviation. Significant difference between treatment and mock conditions was analyzed using one sided paired *t*-test (**and * meaning $P < 0.0.1$ or $P < 0.05$, respectively).

Wang et al., 2007; Zhang et al., 2012), and over-expression of a TIR-NB-LRR gene from wild north American grapevine in *V. vinifera* wine grape cultivars was reported to lead to HR-like cell death after inoculation with *E. necator* (Feechan et al., 2013). Moreover, over-expression of a RPP1A truncation in *A. thaliana* induced elicitor-independent HR-like cell death (Weaver et al., 2006). In this study, an increase in ROS (O_2^- and H_2O_2) accumulation followed by H_2O_2 induced HR-like cell death was observed after ectopic expression of *VpCN* in *A. thaliana*, when the transgenic plants were inoculated with *P. st* DC3000.

Callose-containing cell-wall appositions, called papillae, provide a physical barrier that slows pathogen invasion at the site of pathogen attack (Luna et al., 2011). Callose deposition is typically triggered by conserved pathogen-associated molecular patterns (PAMPs) and contributes to the innate immunity (Brown et al., 1998; Luna et al., 2011). Ellinger et al. (2013) demonstrated that over-expression of *PMR4* in transgenic plants promoted early callose accumulation at attempted fungal penetration sites, which provided complete resistance to *G. cichoracearum*, and the non-adapted PM agent, *B. graminis*. In this study, transgenic plants displayed more callose deposition than WT plants in response to treatment with *P. st* DC3000, suggesting that callose deposition may contribute to the enhanced disease resistance to the pathogen displayed by the transgenic plants.

To elucidate the molecular basis of *VpCN* transcript induction after inoculation with *E. necator*, the *VpCN* promoter was isolated and its activation investigated using *A. tumefaciens*-mediated transient expression of *VpCN* in *V. vinifera* leaves. Bioinformatic analysis of the promoter sequence revealed two TC-rich repeats ('5-ATTCTCTAAC-3'), which are thought to be involved in defense and stress responses (Diaz-De-Leon et al., 1993). We hypothesized that these might be involved in the response to *E. necator*, and generated four promoter deletion constructs to test this idea. Plants harboring a −1360, −700, or −400 bp region of the promoter sequence, all of which contain two or one TC rich repeat elements (**Supplement Figure 2**), showed increased GUS activity after challenge with *E. necator*. However, plants containing only a −240 bp region sequence, which has no TC-rich repeat elements (**Supplement Figure 2**), showed no significant change in GUS activity after inoculation with *E. necator*. Thus, we propose that the TC-rich repeat elements may play a role in the *VpCN* promoter activity in response to *E. necator* infection. This study suggests that *VpCN* is a disease resistance gene, and we will investigate that whether the *VpCN* is interact with AVR protein (effector) from *Erysiphe necator*. Further functional studies to the VpCN with other proteins and downstream defense signaling involved in the powdery mildew disease resistance will be helpful in understanding the molecular mechanisms of powdery mildew disease resistance in Chinese wild *V. pseudoreticulata*.

AUTHOR CONTRIBUTIONS

XW and ZW designed the experiments. ZW, LY, RW, and ZL performed the experiments. XW, ZW, and CL analyzed the results and wrote the manuscript. All authors read and approved the final manuscript.

ACKNOWLEDGMENTS

This work was supported by the National Natural Science Foundation of China (31272136) and the Program for Innovative Research Team of Grape Germplasm Resources and Breeding (2013KCT-25). We thank PlantScribe (http://www.plantscribe.com/) for editing the manuscript.

REFERENCES

Ahn, L. P., Kim, S., Lee, Y. H., and Suh, S. C. (2007). Vitamin B1-induced priming is dependent on hydrogen peroxide and the NPRl gene in *Arabidopsis*. *Plant Physiol.* 143, 838–848. doi: 10.1104/pp.106.092627

Albrecht, M., and Takken, F. L. W. (2006). Update on the domain architectures of NLRs and R proteins. *Biochem. Biophys. Res. Commun.* 339, 459–462. doi: 10.1016/j.bbrc.2005.10.074

Alvarez, M. E., Pennell, R. I., Meijer, P. J., Ishikawa, A., Dixon, R. A., and Lamb, C. (1998). Reactive oxygen intermediates mediate a systemic signal network in the establishment of plant immunity. *Cell* 92, 773–784. doi: 10.1016/S0092-8674(00)81405-1

Bendahmane, A., Farnham, G., Moffett, P., and Baulcombe, D. C. (2002). Constitutive gain-of-function mutants in a nucleotide binding site–leucine rich repeat protein encoded at the Rx locus of potato. *Plant J.* 32, 195–204. doi: 10.1046/j.1365-313X.2002.01413.x

Bendahmane, A., Kanyuka, K., and Baulcombe, D. C. (1999). The Rx gene from potato controls separate virus resistance and cell death responses. *Plant Cell* 11, 781–792. doi: 10.1105/tpc.11.5.781

Bent, A. F., Kunkel, B. N., Dahlbeck, D., Brown, K. L., Schmidt, R., Giraudat, J., et al. (1994). RPS2 of *Arabidopsis thaliana*: a leucine-rich repeat class of plant disease resistance genes. *Science* 265, 1856–1860. doi: 10.1126/science.8091210

Bevan, M., Bancroft, I., Bent, E., Love, K., Goodman, H., Dean, C., et al. (1998). Analysis of 1.9Mb of contiguous sequence from chromosome 4 of *Arabidopsis thaliana*. *Nature* 391, 485–493. doi: 10.1038/35140

Bradford, M. (1976). A rapid and sensitive method for the quantitation of microgram quantities of protein utilizing the principle of protein-dyebinding. *Anal. Biochem.* 72, 248–254. doi: 10.1016/0003-2697(76)90527-3

Brown, I., Trethowan, J., Kerry, M., Mansfiled, J., and Bolwell, G. P. (1998). Localization of components of the oxidative cross-linking of glycoproteins and of callose synthesis in papillae formed during the interaction between non-pathogenic strains of *Xanthomonas campestris* and French bean mesophyll cells. *Plant J.* 15, 333–343. doi: 10.1046/j.1365-313X.1998.00215.x

Cesari, S., Thilliez, G., Ribot, C., Chalvon, V., Michwl, C., Jauneau, A., et al. (2013). The rice resistance protein RGA4/RGA5 recognizes th *Magnaporthe oryzae* effectors AVR1-CO39 by direct binding. *Plant Cell* 25, 1463–1481. doi: 10.1105/tpc.112.107201

Chisholm, S. T., Coaker, G., Day, B., and Staskawicz, B. J. (2006). Host-microbe interactions: shaping the evolution of plant immune response. *Cell* 124, 803–814. doi: 10.1016/j.cell.2006.02.008

Choi, D. S., and Hwang, B. K. (2011). Proteomics and functional analyses of pepper abscisic acid-responsive 1 (ABR1), which is involved in cell death and defense signaling. *Plant Cell* 23, 823–842. doi: 10.1105/tpc.110.082081

Clough, S. J., and Bent, A. F. (1998). Floral dip: a simplified method for *Agrobacterium*-mediated transformation of *Arabidopsis thaliana*. *Plant J.* 16, 735–743. doi: 10.1046/j.1365-313x.1998.00343.x

Collier, S. M., Hamel, L. P., and Moffett, P. (2011). Cell death mediated by the N-terminal domains of a unique and highly conserved class of NB-LRR Protein. *Mol. Plant Microbe Interact.* 24, 918–931. doi: 10.1094/MPMI-03-11-0050

Collier, S. M., and Moffett, P. (2009). NB-LRRs works a "bait and switch" on pathogens. *Trends Plant Sci.* 14, 521–529. doi: 10.1016/j.tplants.2009.08.001

Diaz-De-Leon, F., Klotz, K. L., and Lagrimini, M. (1993). Nucleotide sequence of the tobacco (*Nicotiana tabacum*) anionic peroxides gene. *Plant Physiol.* 101, 1117–1118. doi: 10.1104/pp.101.3.1117

Ellinger, D., Naumann, M., Falter, C., Zwikowics, C., Jamrow, T., Manisseri, C., et al. (2013). Elevated early callose deposition results in complete penetration to powdery mildew in *Arabidopsis*. *Plant Physiol.* 161, 1433–1444. doi: 10.1104/pp.112.211011

Fan, J., Crooks, C., and Lamb, C. (2008). High-throughput quantitative luminescence assay of the grow thin planta of *Pseudomonas syringae* chromosomally tagged with *Photorhabdus luminescens* lux CDABE. *Plant J.* 53, 393–399. doi: 10.1111/j.1365-313X.2007.03303.x

Feechan, A., Anderson, C., Torregrosa, L., Jermakow, A., Mestre, P., Wiedemann, M. S., et al. (2013). Genetic dissection of a TIR-NB-LRR locus from the wild North American grapevine species *Muscadinia rotundifolia*identifies paralogous genes conferring resistance to major fungal and oomycete pathogens in cultivated grapevine. *Plant J.* 76, 661–674. doi: 10.1111/tpj.12327

Friedrich, L., Lawton, K., Ruess, W., Masnet, P., Specker, N., Rella, M. G., et al. (1996). A benzothiadiazole derivative induces systemic acquired resistance in tobacco. *Plant J.* 10, 61–70. doi: 10.1046/j.1365-313X.1996.10010061.x

Frye, C. A., and Innes, W. (1998). An *Arabidopsis* mutant with enhanced resistance to powdery mildew. *Plant Cell* 10, 947–956. doi: 10.1105/tpc.10.6.947

Gadoury, D. M., Cadle-Davidson, L., Wilcox, W. F., Dry, I. B., Seem, R. C., and Milgroom, M. G. (2012). Grapevine powdery mildew (*Erysiphe necator*): a fascinating system for the study of the biology, ecology and epidemiology of an obligate biotroph. *Mol. Plant Pathol.* 13, 1–16. doi: 10.1111/j.1364-3703.2011.00728.x

Grant, J. J., and Loake, G. J. (2000). The Role of reactive oxygen intermediates and cognate redox signaling in disease resistance. *Plant Physiol.* 124, 21–29. doi: 10.1104/pp.124.1.21

Guan, X., Zhao, H. Q., Xu, Y., and Wang, Y. J. (2011). Transient expression of glyoxal oxidase from the Chinese wild grape *Vitis pseudoreticulata* can suppress powdery mildew in a susceptible genotype. *Protoplasma* 248, 415–423. doi: 10.1007/s00709-010-0162-4

Hao, W., Collier, S. M., Moffett, P., and Chai, J. J. (2013). Structural basis for the interaction between the potato virus X resistance protein (Rx) and its cofactor Ran GTPase-activating protein 2 (RanGAP2). *J. Biol. Chem.* 288, 35868–35876. doi: 10.1074/jbc.M113.517417

Huang, W. J., Jiang, T. Y., Choi, W., Pang, Y. X., Hu, Q., Xu, Y. H., et al. (2013). Mechanistic insights into CED-4-mediated activation of CED-3. *Genes Dev.* 27, 2039–2048. doi: 10.1101/gad.224428.113

Jefferson, R. (1987). Assaying chimeric genes in plants: the GUS gene fusion system. *Plant Mol. Biol. Rep.* 5, 387–405. doi: 10.1007/BF02667740

Jones, J. D., and Dangl, K. L. (2006). The plant immune systems. *Nature* 444, 323–329. doi: 10.1038/nature05286

Keller, H., Pamboukdjian, N., Ponchet, M., Poupet, A., Delon, R., Verrier, J. L., et al. (1999). Pathogen-induced elicitin production in transgenic tobacco generates a hypersensitive response and nonspecific disease resistance. *Plant Cell* 11, 223–235. doi: 10.1105/tpc.11.2.223

Kim, D. S., and Hwang, B. K. (2014). An important role of the pepper phenylalanine ammonia-lyase gene (PAL1) in salicylic acid-dependent signaling of the defense response to microbial pathogens. *J. Exp. Bot.* 65, 2295–2306. doi: 10.1093/jxb/eru109

Kim, S., Ahn, I. P., Park, C., Park, S. G., Park, S. Y., Jwa, N. S., et al. (2001). Molecular characterization of the cDNA encoding an acidic isoform of PR-1 gene protein in rice. *Mol. Cells* 11, 115–121. Available online at: http://europepmc.org/abstract/MED/11266113

Kim, S. H., Kwon, S. I., Saha, D., Anyanwu, N. C., and Gassmann, W. (2009). Resistance to the *Pseudomonas syringae* effector HopA1 is governed by the TIR-NBS-LRR protein RPS6 and is enhanced by mutations in SRFR1. *Plant Physiol.* 150, 1723–1732. doi: 10.1104/pp.109.139238

Kim, S. H., Woo, D. H., Kim, J. M., Lee, S. Y., Chung, W. S., and Moon, Y. H. (2011). *Arabidopsis* MKK4 mediates osmotic-stress response via its regulation of MPK3 activity. *Biochem. Biophys. Res. Commun.* 412, 150–154. doi: 10.1016/j.bbrc.2011.07.064

Koch, E., and Slusarenko, A. (1990). *Arabidopsis* is susceptible to infection by a downy mildew fungus. *Plant Cell* 2, 437–445. doi: 10.1105/tpc.2.5.437

Kohler, A., Schwindling, S., and Conrath, U. (2000). Extraction and quantitative determination of callose from *Arabidopsis* leaves. *Biotechniques* 28, 1084–1086.

Kohm, B. A., Goulden, M. G., Gilbert, J. E., Kavanagh, T. A., and Baulcombe, D. C. (1993). A potato virus X resistance gene mediates an induced, nonspecific resistance in protoplasts. *Plant Cell* 5, 913–920. doi: 10.1105/tpc.5.8.913

Kovtun, Y., Chiu, W. L., Tena, G., and Sheen, J. (2000). Functional analysis of oxidative stress-activated mitogen-activated protein kinase cascade in plants. *Proc. Natl. Acad. Sci. U.S.A.* 97, 2940–2945. doi: 10.1073/pnas.97.6.2940

Lescot, M., Déhais, P., Thijs, G., Marchal, K., Moreau, Y., Van de Peer, Y., et al. (2002). PlantCARE, a database of plant *cis*-acting regulatory elements and a portal to tools for in silico analysis of promoter sequences. *Nucleic Acids Res.* 30:325. doi: 10.1093/nar/30.1.325

Lin, L., Wang, X. P., and Wang, Y. J. (2006). cDNA clone, fusion expression and purification of the novel gene related to ascorbate peroxidase from Chinese wild *Vitis pseudoreticulata* in *E.coli*. *Mol. Biol. Rep.* 33, 197–206. doi: 10.1007/s11033-006-0008-5

Lukasik-Shreepaathy, W., Slootweg, E., Richter, H., Goverse, A., Cornelissen, B. J. C., and Takken, F. L. W. (2012). Dual regulatory roles of the extended N terminus for activation of the tomato Mi-1.2 resistance protein. *Mol. Plant Microbe Interact.* 25, 1045–1057. doi: 10.1094/MPMI-11-11-0302

Luna, E., Pastor, V., Robert, J., Flors, V., Mauch-Mani, B., and Ton, J. (2011). Callose deposition: a multifaceted plant defense response. *Mol. Plant Microbe Interact.* 24, 183–193. doi: 10.1094/MPMI-07-10-0149

Maekawa, T., Cheng, W., Spiridon, L. N., Töller, A., Lukasik, E., and Saijo, Y. (2011). Coiled-coil domain-dependent homo-dimerization of intracellular barley immune receptors defines a minimal functional module for triggering cell death. *Cell Host Microbe* 9, 187–199. doi: 10.1016/j.chom.2011.02.008

Mestre, P., and Baulcombe, D. C. (2006). Elicitor-mediated oligomerization of the tobacco N disease resistance protein. *Plant Cell* 18, 491–501. doi: 10.1105/tpc.105.037234

Meyers, B. C., Dickerman, A. W., Michelmore, R. W., Sivaramakrishnan, S., Sobral, B. W., and Young, N. D. (1999). Plant disease resistance genes encode members of an ancient and diverse protein family within the nucleotide-binding superfamily. *Plant J.* 20, 317–333. doi: 10.1046/j.1365-313X.1999.t01-1-00606.x

Mino, M., Maekawa, K., Ogawa, K., Yamagishi, H., and Inoue, M. (2002). Cell death processes during expression of hybrid lethality in interspecific F1 hybrid between *Nicotiana gossei* domin and *Nicotiana tabacum*. *Plant Physiol.* 130, 1776–1787. doi: 10.1104/pp.006023

Mohr, T. J., Mammarella, N. D., Hoff, T., Woffenden, B. J., Jelesko, J. G., and McDowell, J. M. (2010). The *Arabidopsis* downy mildew resistance gene RPP8 is induced by pathogens and salicylic scid and is regulated by W Box *cis* elements. *Mol. Plant Microbe Interact.* 23, 1303–1315. doi: 10.1094/MPMI-01-10-0022

Murshige, T., and Skoog, F. (1962). A revised medium for rapid growth bioassays with tobacco tissue cultures. *Physiol. Plant.* 15, 473–497. doi: 10.1111/j.1399-3054.1962.tb08052.x

Nandety, R. S., Caplan, J. L., Cavanaugh, K., Perroud, B., Wroblewski, T., Michelmore, R. W., et al. (2013). The role of TIR-NBS and TIR-X proteins in plant basal defense response. *Plant Physiol.* 162, 1459–1472. doi: 10.1104/pp.113.219162

Nie, H. Z., Wu, X. Y., Yao, C. P., and Tang, D. Z. (2011). Suppression of edr2-mediated powdery mildew resistance, cell death and ethylene-induced senescenceby mutations in ALD1 in *Arabidopsis*. *J. Genet. Genomics* 38, 137–148. doi: 10.1016/j.jgg.2011.03.001

Ooijen, G. V., Mayr, G., Albrecht, M., Cornelissen, B. J. C., and Takken, F. L. W. (2008). Transcomplementation, but not physical association of the CC-NB-ARC and LRR domains of tomato R protein Mi-1.2 is altered by mutations in the ARC2 sub-domain. *Mol. Plant* 1, 401–410. doi: 10.1093/mp/ssn009

Ori, N., Eshed, Y., Paran, I., Presting, G., Aviv, D., Tanksley, S., et al. (1997). The I2C family from the wilt disease resistance locus I2 belongs to the nucleotide binding, leucine-rich repeat superfamily of plant resistance genes. *Plant Cell* 9, 521–532. doi: 10.1105/tpc.9.4.521

Pan, Q., Liu, Y. S., Budai-Hadrian, O., Sela, M., Carmel-Goren, L., Zamir, D., et al. (2000b). Comparative genetics of nucleotide binding site-leucine rich repeat resistance gene homologues in the genomes of two dicotyledons: tomato and *Arabidopsis*. *Genetics* 155, 309–322. Available online at: http://www.genetics.org/content/155/1/309.long

Pan, Q., Wendel, J., and Fluhr, R. (2000a). Divergent evolution of plant NBS-LRR resistance gene homologues in dicot and cereal genomes. *J. Mol. Evol.* 50, 203–213. doi: 10.1007/s002399910023

Punja, Z. K. (2004). *Fungal Disease Resistance in Plants*. New York, NY; London; Oxford: Food Products Press® An important of the Haworth Press, Inc.

Radirdan, G. J., Collier, S. M., Sacco, M. A., Baldwin, T. T., Boettrich, T., and Moffett, P. (2008). The coiled-coil and nucleotide binding domains of the potato Rx disease resistance protein function in pathogen recognition and signaling. *Plant Cell* 20, 739–751. doi: 10.1105/tpc.107.056036

Reuber, T. L., Plotnikova, J. M., Dewdney, J., Rogers, E. E., Wood, W., and Ausubel, F. M. (1998). Correlation of defense gene induction defects with powder mildew susceptibility in *Arabidopsis* enhanced disease susceptibility mutants. *Plant J.* 16, 473–485. doi: 10.1046/j.1365-313x.1998.00319.x

Riedl, S. J., Li, W., Chao, Y., Schwarzenbacher, R., and Shi, Y. (2005). Structure of the apoptotic protease-activating factor 1 bound to ADP. *Nature* 434, 926–933. doi: 10.1038/nature03465

Santos-Rosa, M., Poutaraud, A., Merdinoglu, D., and Mestre, P. (2008). Development of a transient expression system in grapevine viaagro-infiltration. *Plant Cell Rep.* 27, 1053–1063. doi: 10.1007/s00299-008-0531-z

Sekine, K. T., Tomita, R., Takeuchi, S., Atsumi, G., Saitoh, H., Miroyuki, H., et al. (2012). Functional differentiation in the leucine-rich repeat domains of closely related plant virus-resistance proteins that recognize common avr proteins. *Mol. Plant Microbe Interact.* 25, 1219–1229. doi: 10.1094/MPMI-11-11-0289

Shirano, Y., Kachroo, P., Shah, J., and Klessig, D. F. (2002). A gain-of-function mutation in an *Arabidopsis* Toll Interleukin1 receptor–nucleotide binding site–leucine-rich repeat type R gene triggers defense responses and results in enhanced disease resistance. *Plant Cell* 14, 3149–3162. doi: 10.1105/tpc.005348

Smith, L. D., and Goodman, N. L. (1975). Improved culture method for the isolation of *Histoplasma capsulatum* and *Blastomyces dermatitidis* from contamincated specimens. *Am. J. Clin. Pathol.* 63, 276–280. Available online at: http://europepmc.org/abstract/med/1115035

Stokes, T. L., Kunkel, B. L., and Richards, E. J. (2002). Epigenetic variation in *Arabidopsis* disease resistance. *Genes Dev.* 16, 171–182. doi: 10.1101/gad.952102

Takken, F. L. W., Albrecht, M., and Tameling, W. I. L. (2006). Resistance proteins: molecular switches of plant defence. *Curr. Opin. Plant Biol.* 9, 383–390. doi: 10.1016/j.pbi.2006.05.009

Tameling, W. I. L., Nooijen, C., Ludwig, N., Boter, E., Goverse, A., Shirasu, K., et al. (2010). RanGAP2 mediates nucleocytoplasmic partitioning of the NB-LRR immune receptor Rx in the *solanaceae*, There by dictating Rx function. *Plant Cell* 22, 4176–4194. doi: 10.1105/tpc.110.077461

Tang, D. Z., and Innes, R. W. (2002). Over-expression of a kinase-deficient form of the EDR1gene enhances powdery mildew resistance and ethylene-induced senescence in *Arabidopsis*. *Plant J.* 32, 975–983. doi: 10.1046/j.1365-313X.2002.01482.x

Tang, X. Y., Xie, M. T., Kim, Y. J., Zhou, J. M., Klessig, D. F., and Martin, G. B. (1999). Over expression of Pto activates defense responses and confers broad resistance. *Plant Cell* 11, 15–29.

Tao, Y., Yuan, F. H., Leister, R. T., Ausubel, F. M., and Katagiri, F. (2000). Mutational analysis of the *Arabidopsis* nucleotide binding site-leucine-rich repeat resistance gene RPS2. *Plant Cell* 12, 2541–2554.

Theologis, A., Ecker, J. R., Palm, C. J., Federspiel, N. A., Kaul, S., White, O., et al. (2000). Sequence and analysis of chromosome 1 of the plant *Arabidopsis thaliana*. *Nature* 408, 816–820. doi: 10.1038/35048500

van der Biezen, E. A., and Jones, J. D. G. (1998). The NB-ARC domain: a novel signaling motif shared by plant resistance gene products and regulators of cell death in animals. *Curr. Biol.* 8, 226–227. doi: 10.1016/S0960-9822(98)70145-9

Wang, W. M., Devoto, A., Turner, J. G., and Xiao, S. Y. (2007). Expression of the membrane-associated resistance protein RPW8 enhances basal defense against biotrophic pathogens. *Mol. Plant Microbe Interact.* 20, 966–976. doi: 10.1094/MPMI-20-8-0966

Wang, Y., Liu, Y., He, P., Chen, J., Lamikanra, O., and Lu, J. (1995). Evaluation of foliar resistance to *Uncinula necatorin* Chinese wild *Vitis* species. *Vitis* 3, 159–164.

Weaver, L. M., Swidersk, M. R., Li, Y., and Jone, J. D. G. (2006). The *Arabidopsis thaliana* TIR-NB-LRR R proteins, RPP1A; Protein localization and constitutive activation of defense by truncated alleles in tobacco and *Arabidopsis*. *Plant J.* 47, 829–840. doi: 10.1111/j.1365-313X.2006.02834.x

Weng, K., Li, Z. Q., Liu, Q. R., Wang, L., Wang, Y. J., and Xu, Y. (2014). Transcriptome of *Erysiphe necator*-infected *Vitis pseudoreticulata leaves* provides insight into grapevine resistance to powdery mildew. *Hortic. Res.* 1:14049. doi: 10.1038/hortres.2014.49

Xu, W. R., Yu, Y. H., Ding, J. H., Hua, Z. H., and Wang, Y. J. (2010). Characterization of a novel stilbene synthase promoter involved in pathogen- and stress-inducible expression from Chinese wild *Vitis pseudoreticulata*. *Planta* 231, 475–487. doi: 10.1007/s00425-009-1062-8

Yu, Y. H., Xu, R. W., Wang, J., Wang, L., Yao, W. K., Xu, Y., et al. (2013). A core functional region of the *RFP1* promoter from Chinese wild grapevine is activated by powdery mildew pathogen and heat stress. *Planta* 237, 293–303. doi: 10.1007/s00425-012-1769-9

Yu, Y. H., Xu, W. R., Wang, S. Y., Xu, Y., Li, H. E., Wang, Y. J., et al. (2011). *VpRFP1*, a novel C_4C_4-type RING finger protein gene from Chinese wild *Vitis pseudoreticulata*, functions as a transcriptional activator in defence response of grapevine. *J. Exp. Bot.* 62, 5671–5682. doi: 10.1093/jxb/err253

Zhang, J. J., Wang, Y. J., Wang, X. P., Yang, K. Q., and Yang, J. X. (2003). An improved method for rapidly extracting total RNA from *Vitis*. *J. Fruit Sci.* 53, 771–787. Available online at: http://www.gskk.cbpt.cnki.net/WKA/WebPublication/paperDigest.aspx?

Zhang, L., Li, Y. Z., Lu, W., Meng, F., Wu, C. A., and Guo, X. Q. (2012). Cotton GhMKK5 affects disease resistance, induces HR-like cell death, and reduces the tolerance to salt and drought stress in transgenic *Nicotiana benthamiana*. *J. Exp. Bot.* 63, 3935–3951. doi: 10.1093/jxb/ers086

Zhang, Y., Dorey, S., Swiderski, M., and Jones, J. D. (2004). Expression of RPS4 in tobacco induced an AvrRPS4-independent HR that requires EDS1, SGT1 and HSP90. *Plant J.* 40, 213–224. doi: 10.1111/j.1365-313X.2004.02201.x

Conflict of Interest Statement: The authors declare that the research was conducted in the absence of any commercial or financial relationships that could be construed as a potential conflict of interest.

6

High-Density SNP Map Construction and QTL Identification for the Apetalous Character in *Brassica napus* L.

Xiaodong Wang[1,2†], Kunjiang Yu[1†], Hongge Li[1†], Qi Peng[1], Feng Chen[1,3], Wei Zhang[1], Song Chen[1], Maolong Hu[1] and Jiefu Zhang[1,2]*

[1] Key Laboratory of Cotton and Rapeseed, Ministry of Agriculture, Institute of Industrial Crops, Jiangsu Academy of Agricultural Sciences, Nanjing, China, [2] Jiangsu Collaborative Innovation Center for Modern Crop Production, Nanjing, China, [3] Provincial Key Laboratory of Agrobiology, Jiangsu Academy of Agricultural Sciences, Nanjing, China

Edited by:
Juan Francisco Jimenez Bremont, Instituto Potosino de Investigacion Cientifica y Tecnologica, Mexico

Reviewed by:
Estefania Carrillo, Instituto Nacional de Investigaciones Agropecuarias, Ecuador
Javier Sanchez, University of Zürich, Switzerland

***Correspondence:**
Jiefu Zhang
jiefu_z@163.com

† These authors have contributed equally to this work.

Specialty section:
This article was submitted to Crop Science and Horticulture, a section of the journal Frontiers in Plant Science

The apetalous genotype is a morphological ideotype for increasing seed yield and should be of considerable agricultural use; however, only a few studies have focused on the genetic control of this trait in *Brassica napus*. In the present study, a recombinant inbred line, the AH population, containing 189 individuals was derived from a cross between an apetalous line 'APL01' and a normally petalled variety 'Holly'. The *Brassica* 60 K Infinium BeadChip Array harboring 52,157 single nucleotide polymorphism (SNP) markers was used to genotype the AH individuals. A high-density genetic linkage map was constructed based on 2,755 bins involving 11,458 SNPs and 57 simple sequence repeats, and was used to identify loci associated with petalous degree (PDgr). The linkage map covered 2,027.53 cM, with an average marker interval of 0.72 cM. The AH map had good collinearity with the *B. napus* reference genome, indicating its high quality and accuracy. After phenotypic analyses across five different experiments, a total of 19 identified quantitative trait loci (QTLs) distributed across chromosomes A3, A5, A6, A9 and C8 were obtained, and these QTLs were further integrated into nine consensus QTLs by a meta-analysis. Interestingly, the major QTL *qPD.C8-2* was consistently detected in all five experiments, and *qPD.A9-2* and *qPD.C8-3* were stably expressed in four experiments. Comparative mapping between the AH map and the *B. napus* reference genome suggested that there were 328 genes underlying the confidence intervals of the three steady QTLs. Based on the Gene Ontology assignments of 52 genes to the regulation of floral development in published studies, 146 genes were considered as potential candidate genes for PDgr. The current study carried out a QTL analysis for PDgr using a high-density SNP map in *B. napus*, providing novel targets for improving seed yield. These results advanced our understanding of the genetic control of PDgr regulation in *B. napus*.

Keywords: *Brassica napus* L., apetalous, single nucleotide polymorphism, high-density map, quantitative trait locus, recombinant inbred line

INTRODUCTION

Oilseed rape (*Brassica napus* L., AACC, 2n = 38) is a widely planted oil crop worldwide. Rapeseed oil is not only a desirable edible oil, but is also used as a biofuel in many parts of the world (Hoekman, 2007). Additionally, oil-extracted meal from *Brassica* seeds is an excellent protein source for animal feed (Linnemann and Dijkstra, 2002). As the global demand for rapeseed products is continuously increasing, developing a high-yield variety is a main goal of *B. napus* breeding programs. An effective approach is to seek the morphological ideotype (Virk et al., 2004), and apetalous genotypes are of particular interest in breeding programs (Habekotté, 1997; Jiang and Becker, 2003).

The apetalous trait was first reported in a naturally occurring mutant of turnip (*B. campestris* L.) in *Brassica* (Ramanujam, 1940), and Buzza (1983) first detected an apetalous mutant in a spring oilseed rape. Since then, other apetalous flowers in *Brassica* species have been discovered or bred. There are several advantages to apetalous rape in yield. First and foremost, photosynthesis in cultivars without petals is more efficient, with the thick and brightly colored flowers preventing *Brassica* oilseeds from efficiently using solar energy (Jiang, 2007). The petals at the top layer of the normal flower type were reported to reflect or absorb up to 60% of incoming radiation (Mendham et al., 1981). Additionally, the apetalous cultivars have higher yield potentials than the normal type. The petal is not a photosynthesizing organ, but it consumes considerable amounts of photosynthesized assimilates during its formation and respiration (Jiang, 2007). Mendham et al. (1991) revealed that the yield of apetalous lines was higher than normal petaled cultivars. Finally, the apetalous type of rapeseeds might have a lower rate of infection from diseases distributed by petals, such as *Sclerotinia sclerotiorum*. Deciduous petals can transmit the *Sclerotinia* pathogen to healthy tissue, whereas the ascospores that land directly on the leaf surface do not germinate (Jamaux and Spire, 1999). Compared with normal petaled controls, apetalous genotypes have a much lower incidence and severity of *Sclerotinia* infection (Lefol and Morrall, 1996; Zhao and Wang, 2004). Moreover, a multitude of other diseases, such as *Botrytis cinierea* and *Peronospora parasitica*, may be distributed by petals (Lefol and Morrall, 1996). In summary, genotypes with apetalous flowers are a component of the high-yielding ideotype.

In *B. napus*, various genetic models of the apetalous trait, with different origins, are documented in many literatures. One study found that petalous flower development was controlled by one gene locus that exhibited incomplete dominance over apetalous flower development (Zhao and Wang, 2004). Most of the other studies revealed that the apetalous character in *B. napus* was regulated by recessive genes, possibly by two to four loci (Buzza, 1983; Lu and Fu, 1990; Kelly et al., 1995; Chen et al., 2006a; Zhang et al., 2007b,c). Generally, these loci independently control flower morphology; however, epistatic interactions between recessive alleles were also identified (Fray et al., 1997). In addition, the apetalous character is governed by the interaction of cytoplasmic and nuclear genes (Jiang and Becker, 2003). Although much attention has been paid to the inheritance of the apetalous character in *B. napus*, questions concerning the genetic basis remain open. Plants having less than a 10% petalous degree (PDgr) were considered apetalous (Buzza, 1983), but the distribution of PDgr in segregation generation is consecutive and should be treated as a quantitative trait (Zhang et al., 2007a). Quantitative trait loci (QTLs) mapping is a preliminary step and an effective approach to unravel the genetic architecture of complex quantitative traits and to identify QTLs for knowledge-based breeding (Mauricio, 2001). Fray et al. (1997) reported that five restriction fragment length polymorphism markers were significantly associated with one of two *stamenoid petal* loci, one each on A4 and C4. One random amplified polymorphic DNA marker that was tightly linked to a petal-controlled gene in *B. napus* was identified by Tan et al. (2003). Using a bulked segregant analysis approach, one sequence-related amplified polymorphism and one amplified fragment length polymorphism marker mapped on A4 were found to be linked to the gene controlling the petal-loss trait (Chen et al., 2006b). Based on a genetic map containing 219 markers, Zhang et al. (2007a) identified four QTLs, which were located on chromosomes A5, A6, A8, and C5, associated with the apetalous phenotype.

High-density maps could increase the precision of QTL localization and the estimation of QTL effects in biparental populations (Stange et al., 2013). Since the first molecular linkage map in *B. napus* was reported by Landry et al. (1991), various types of populations have been constructed for mapping QTLs associated with seed oil content, seed fatty acid concentrations, flowering time, seed yield and yield-related traits (Qiu et al., 2006; Quijada et al., 2006; Udall et al., 2006; Long et al., 2007; Basunanda et al., 2010; Chen et al., 2010; Cai et al., 2012; Ding et al., 2012; Wang et al., 2013; Wang et al., 2015a). However, most of these genetic linkage maps were constructed based on PCR markers with low densities. Single nucleotide polymorphisms (SNPs) are the most frequent polymorphism in the genomes of crops (Vignal et al., 2002), and they have been widely used in rice (Huang et al., 2010), wheat (Mochida et al., 2003), and maize (Ching et al., 2002). In *B. napus*, SNPs were also used for high-density genetic map construction and the fine mapping of important genes. Delourme et al. (2013) developed an integrated genetic map, which was comprised of 5,764 SNPs and 1,603 PCR markers, with a genetic length of 2,250 cM. Based on the 6 K SNP array harboring 5,306 probes for *B. napus*, Raman et al. (2014) constructed a genetic linkage map covered 2,514.8 cM, including 613 SNPs and 228 non-SNPs, and Cai et al. (2014) constructed a genetic map containing 2,115 markers (1,667 SNPs and 448 SSRs), with a length of 2,477.4 cM. Chen et al. (2013) constructed a SNP bin map containing 8,780 SNP loci and a presence/absence variation map containing 12,423 dominant loci. In 2012, the *Brassica* 60 K SNP BeadChip Array comprised of 52,157 SNP loci was produced (Snowdon and Iniguez Luy, 2012; Edwards et al., 2013), which was developed by an international consortium using preferentially single-locus SNPs contributed from genomic and transcriptomic sequencing in genetically diverse *Brassica* germplasm (Liu et al., 2013). This paved the way for the high-throughput and cost-effective construction of a high-density genetic *B. napus* map. Using the 60 K SNP BeadChip Array,

Liu et al. (2013) constructed a linkage map containing 9,164 SNP markers covering 1,832.9 cM, and mapped the major QTL for seed color corresponding to a physical region of 620 kbp. Zhang et al. (2014) constructed a map covering a length of 2,139.5 cM with average distance of 1.6 cM between adjacent markers. Nevertheless, to the best of our knowledge, a QTL analysis for the apetalous trait using high-density SNP genetic linkage map in *B. napus* has not been performed.

The objectives of the present study were to: (1) construct a high-density genetic map using the *Brassica* 60 K Infinium SNP array and SSRs; and (2) investigate the QTLs for PDgr in *B. napus* across five experiments. The results will provide information useful for understanding the genetic control of apetalous in *B. napus*, and the major QTLs will lay a foundation for use in breeding programs to develop a variety with agronomic traits of interest for rapeseed production.

MATERIALS AND METHODS

Plant Materials

The *B. napus* segregating recombinant inbred line (RIL) population used in this study was derived from a cross between 'APL01' and 'Holly' using the single seed descent method. The parent 'APL01' is an apetalous line, developed at the Institute of Industrial Crops, Jiangsu Academy of Agricultural Sciences, Nanjing, China. 'APL01' was selected from F_6 generation of crosses between apetalous (Apeatlous No. 1) and normal petalous (Zhongshuang No. 4) rapeseed in 1998 (Zhang et al., 2002). 'Apeatlous No. 1' was bred from F_8 generation of crosses between China rapeseed cultivar with smaller petals (SP103) and *B. rapa* variety with lower petals (LP153). 'Zhongshuang No. 4' was developed at Oil Crops Research Institute of the Chinese Academy of Agricultural Sciences, Wuhan, China. Except for PDgr, other agronomic traits in 'APL01' are quite normal. At early flowering stage, 'APL01' is absolutely apetalous, however, there may be one cripple petal only in a few flowers at late flowering stage. The genotype 'Holly' is a normally and completely petaled variety. The two parents showed the similar flowering time, which recorded from the sowing day to the day when the first flower had opened on half of the plants in the plot. A total of 550 F_9 RIL lines were developed in 2014, and then a subset of 189 lines was randomly selected to compose the mapping population for the genetic linkage map construction. These were named the AH RIL population.

Field Trials and PDgr Measurements

The AH population, together with the two parents, was tested in five experiments. The materials were planted in a winter rapeseed area, Dali of Shaanxi Province (coded DL), in northwest China for one year (September–May of 2014–2015), and a semi-winter rapeseed area, Nanjing of Jiangsu Province (coded NJ), in eastern China for four years (September–May of 2011-2012, 2012-2013, 2013–2014, and 2014–2015). Year-location combinations were treated as experiments, for example, 14NJ indicates the experiment was conducted during 2014–2015 at the Nanjing location. The field experiments were conducted in a randomized complete block design with two replications in both NJ and DL. The experimental unit was a two-row plot with 20 plants per row and 40 cm between the rows. The field management followed the common agricultural practices.

At least five representational plants of each plot were selected to measure PDgr during the flowering stage. The percentage of petals of an individual plant was determined by counting the number of petals on the first 25 flowers to open (Buzza, 1983). The PDgr was calculated by the following formula as described by Buzza (1983):

$$\text{PDgr}\ (\%) = \left(\sum_{i=1}^{n} Pi/4n\right) \times 100\%,$$

in which *P* represents the number of petals for each flower that was counted, with a range of numerical value of 0–4, and *n* is the total number of flowers we investigated, $n \geq 50$.The average of the PDgr in each RIL line was used as raw data in the analysis.

Statistical Analysis

Basic statistical analyses of PDgr were performed using SPSS 18.0 software (SPSS Inc., Chicago, IL, USA). The software package SEA-G3DH, with the mixed major gene and polygene inheritance model, was used to analyze the inheritance of the PDgr character in the AH population (Cao et al., 2013). The best fitting model was selected from 39 different models that were included in the software package according to Wang et al. (2015b). Because the PDgr was expressed as a percentage and did not fit the normal distribution model, all data were subjected to an arcsine transformation for the genetic model analysis. Genetic parameters were estimated using the best fitting model with the default settings in the software.

SNP and SSR Marker Analysis

The genotypes of the AH RIL population and two parental lines were analyzed using the *Brassica* 60 K SNP BeadChip Array, which successfully assays 52,157 Infinium Type II SNP loci in *B. napus*. This array was developed by the international *Brassica* SNP consortium in cooperation with Illumina Inc. San Diego, CA, USA. DNA sample preparation, hybridization to the BeadChip, washing, primer extension and staining were strictly carried out according to the Infinium HD Assay Ultra manual. Imaging of the arrays was performed using an Illumina HiSCAN scanner. Allele calling for each locus was performed using the GenomeStudio genotyping software v2011 (Illumina, Inc.). SNP markers used the names that were assigned by GenomeStudio, such as "Bn-A01-p25032772". SSR primer pairs prefixed "CB" and "BRAS" were published by Piquemal et al. (2005); "Na", "Ol" and "Ra" were developed by Lowe et al. (2004); "MR" were published by Uzunova and Ecke (1999) and "BnGMS" were developed by Cheng et al. (2009).

Construction of the Genetic Linkage Map and Alignment of the *B. napus* Reference Genome

All SNPs that were polymorphic between 'APL01' and 'Holly', as well as having less than 5% missing data, were used for the genetic linkage map construction. Of the 52,157 SNPs in the array, 17,414 SNPs met the above requirements and were selected for further analysis. SNP marker pairs with no recombination were classified into one genetic bin (one bin corresponded to all of the markers having the same genotype scoring data) using a Perl script. Then, the selected 17,414 SNPs were grouped into 3,422 SNP-bins, containing 1 to 2,079 SNPs in each bin. Combined with 81 polymorphic SSR markers between the two parents, 3,503 loci (3,422 SNP-bins and 81 SSRs) were subsequently applied in map construction using JoinMap software Version 4.0 (Van Ooijen, 2006). Centimorgan (cM) distances were calculated by the Kosambi function for map distance (Kosambi, 1943). Markers with a mean chi-square value ≥ 3.0 were excluded in all genetic groups to ensure the high quality of the map (Wang et al., 2013).

In addition, the probe sequences of the SNPs that assigned the A and C sub-genomes of *B. napus*, were queried using the BLAST algorithm against the *B. napus* reference genome sequence to locate chromosomal positions with highly stringent parameters (*E* value $\leq$ 1e-10) (Chalhoub et al., 2014). Alignments between the SNP bin map and the *B. napus* reference genome were used to validate the quality of the genetic map. If a locus was mapped to multiple paralogous positions in the *B. napus* reference genome, only the location that corresponded to the particular linkage group of the locus was selected for the collinearity analysis.

QTL Detection and Meta-Analysis

The software Windows QTL Cartographer 2.5 with a composite interval mapping model was used to estimate putative QTLs with additive effects (Zeng, 1994; Wang et al., 2007). The walking speed was set to 2 cM, and a window size of 10 cM with five background cofactors was used. The LOD threshold (2.8–3.1) for detection of significant QTLs was set by a 1,000-permutation test based upon a 5% experiment-wise error rate, and these QTLs were termed 'identified QTLs'. Identified QTLs that were detected in different experiments with overlapping confidence intervals (CIs), may have been one single QTL. Then, identified QTLs were integrated into consensus QTLs using a meta-analysis method with the BioMercator V4.2 program (Arcade et al., 2004). If an identified QTL had no overlapping CIs with others, then it was also regarded as a consensus QTL (Wang et al., 2013).

The identified and consensus QTL nomenclature was based on the descriptions of McCouch et al. (1997) with minor modifications. For identified QTLs, a designation begins with the abbreviation "*iq*" (identified QTL), follow by the experiment and linkage group (A1–A10, C1–C9). If there was more than one identified QTL obtained in a linkage group, a serial number was added. For example, *iq12NJ.A9-2* indicates the second identified QTL for PDgr on A9 in the 12NJ experiment. For consensus QTLs, a designation begins with the abbreviation "*qPD*" (*q*, QTL; *PD*, petalous degree). For example, QTL *qPD.A9-2* indicates the second consensus QTL for PDgr on the A9 linkage group. In addition, candidate genes were identified by comparative mapping between the AH map and the *B. napus* reference genome based on the probe sequence of the SNPs. If the physical positions of aligned genes fell into the CI of a consensus QTL, then the orthologous candidate genes were assumed to be associated with the target QTL (Ding et al., 2012).

Results

Phenotypic Variation and Genetic Analysis for PDgr

'APL01' and 'Holly' are the two parental lines whose repeatability is very good in the five experiments. The mean values ± SE were 0.02% ± 0.03 and 99.99% ± 0.01 for 'APL01' and 'Holly', respectively. There was a wide range of variations and consecutive distributions in the AH population (**Figure 1**), suggesting that PDgr was governed by multiple genes. However, the phenotypic values did not fit the normal distribution in the majority of the RIL lines, mostly petalled, indicating that PDgr might be determined mainly by the major genes in *B. napus*.

The mixed major gene and polygene inheritance model has been used to analyze the two parents and AH population for PDgr in the 11NJ and 12NJ experiments (Li et al., 2014). The PDgr of the AH population was controlled by the two additive major genes and the additive polygene model (MX2-Additive-A model) (Li et al., 2014). Using the same method, the best fitness genetic models for PDgr in 13NJ, 14NJ, and 14DL were analyzed in the present study. The MX2-Additive-A model was also the best fitness genetic model in the 13NJ and 14NJ experiments. However, the genetic model MX2-EA-A, which mixed two equal additive major genes with the additive polygene model, was the most suitable model for the 14DL experiment. The heritabilities of major genes ranged from 68.52% to 88.01% in the

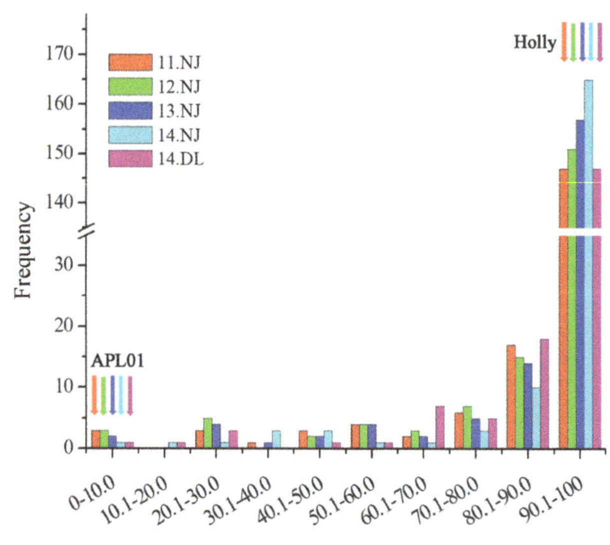

FIGURE 1 | Distribution of the petalous degree (PDgr) of the AH RIL population in five experiments. The units of the *x*-axis are the percentage of the PDgr, and the units of the *y*-axis are the number of lines. PDgr in different experiments was discriminated using different colored boxes (11NJ, red; 12NJ, green; 13NJ blue; 14NJ, cyan; 14DL, magenta). The color arrows represent the location of the parental lines.

five experiments (**Table 1**), significantly higher than that of the polygenes (11.99–31.48%), indicating that PDgr in *B. napus* was determined by the combination of major genes and polygenes, but mainly by the major genes.

High-Density SNP Map Construction

Among the 3,503 loci (3,422 SNP-bins and 81 SSRs) that were used for map construction, 2,755 SNP-bins and 57 SSRs were assigned to 19 linkage groups, including 1,686 loci in the A sub-genome (A1–A10) and 1,126 in C sub-genome (C1–C9; **Table 2**, Supplementary Table S1). The number of SNP markers varied considerably across the different bins, ranging from 1 to 317, and 11,458 SNPs involved in the 2,812 loci were assigned to the genetic map (**Table 2**; **Figure 2**). A8 has the lowest number of SNPs with only 139 SNPs spanning 76.77 cM, and C2 has the most SNPs with 1,788 SNPs spanning 112.08 cM (**Figure 2**).

The high-density map had a total length of 2,027.53 cM with an average marker interval of 0.72 cM, covering 1,033.31 and 994.22 cM of the A and C sub-genomes, respectively (**Table 2**). The average linkage group lengths of the A and C sub-genomes were similar at 103.3 and 110.5 cM, respectively. However, the lengths of each group showed great differences, ranging from 76.09 (A4) to 134.91 cM (A1) in the A sub-genome and 65.10 (C1) to 157.02 cM (C4) in the C sub-genome. In addition, no chromosome in the genetic map displayed gaps of more than 20 cM, while C5 showed the largest gap of 16.70 cM between Bn-scaff_16082_1-p33791 and Bn-scaff_15712_10-p52253 (Supplementary Table S1).

Alignment of the SNP Linkage Map to the *B. napus* Reference Genome

The probe sequences of all 2,755 SNP-bins that mapped to the 19 linkage groups were aligned to the *B. napus* reference genome to validate the genetic linkage maps (**Figure 3**). The results showed that 2,350 loci produced successful BLAST hits in the *B. napus* database, accounting for 85.30% of the 2,755 SNP-bins (85.63% for the A sub-genome and 84.80% for the C sub-genome).

Alignments indicated that the linkage map constructed in the present study had good collinearity with the *B. napus* reference genome sequence (**Figure 3**), suggesting the high quality of the AH RIL map. However, several inconsistencies on A1, A3, A7, and A8 were detected between the map and the *B. napus* reference genome. For example, a large inconsistency involving 80 SNP-bins and spanning a region from 100.82 to 133.75 cM (16.27 Mb of physical region) on A1 might be caused by the existence of paralogous sequences. The inconsistency from the 34.36 to 45.37 cM on the A3 chromosome, which includes 38 SNP-bins and corresponds to ~1.23 Mb of the physical interval, may have been caused by the presence of partial homologous sequences or fragment duplications. In addition, an inversion including 41 loci and spanning from 64.33 to 71.49 cM (2.49 Mb) was identified on the A7 chromosome, and an inversion with 20 loci from 4.09 to 12.40 cM (2.54 Mb) was also identified on A8.

The total genome size of *B. napus* is estimated to be 1,130 Mb, and ~645.4 Mb of the genome assembly was collectively comprised by the scaffolds (Chalhoub et al., 2014). The A sub-genome of the AH map showed good coverage of the reference *B. napus* genome, representing 93.72% of the genome assembly length, while the C sub-genome showed a much lower coverage of 76.91% (**Table 2**). The main reason was that three linkage groups, C1, C3, and C5, only represent 33.45, 43.70, and 34.17%, respectively, of the corresponding chromosomes (**Table 2**; **Figure 3**). The remaining six linkage groups of the C sub-genome account for 92.21to 99.71% of the genome assembly length.

QTL Detection and Meta-Analysis for PDgr

Phenotypic data of PDgr in the AH population were obtained from the five different environments. QTLs for PDgr were analyzed based on the high-density SNP map, and then identified QTLs were integrated into consensus QTLs. Detailed information on identified and consensus QTLs are summarized in **Table 3**.

A total of 19 identified QTLs distributed across A3 (1 QTL), A5 (1 QTL), A6 (1 QTL), A9 (6 QTLs), and C8 (10 QTLs) chromosomes were detected in the present study (**Table 3**). These QTLs have additive effects ranging from -6.09 to 6.63, and singly explaining 5.08%–11.29% of the estimated phenotypic variation (PV). The number of identified QTLs detected in different experiments was quite different, ranging from two (14NJ) to five (11NJ). Among them, up to 10 identified QTLs were distributed on C8, implying that the major genes for PDgr might exist on the C8 chromosome. Eighteen of the 19 identified QTLs had negative additive effects, suggesting that the normally petalled parent 'Holly' contributed favorable alleles for increasing PDgr. Only one identified QTL, *iq12NJ.A3*, had the positive additive effect of 6.63, indicating that the positive alleles for higher phenotypic values were inherited from the apetalous parent 'APL01'.

There were 13 identified QTLs with overlapping CIs, and these were further integrated into three consensus QTLs using a meta-analysis method (**Table 3**; **Figure 4**). As a result, the average CIs of these QTLs were reduced from 5.55 to 1.94 cM, which significantly increased the accuracy of the estimated positions of the meta-QTL. The other six non-overlapping QTLs were also considered as consensus QTLs. In total, nine consensus QTLs for PDgr were obtained in the present study. Among these consensus QTLs, one QTL (*qPD.C8-2*) with the closely linked marker Bn-scaff_18275_1-p1278049 of 0.15 cM, was consistently detected in all of the five experiments and had additive values in the range of −0.69 to −3.77 (**Table 3**). According to the description of Shi et al. (2009), if a consensus QTL presents at least once with PV $\geq$ 20% or at least twice with PV $\geq$ 10%, then the QTL can be regarded as a major QTL. The QTL *qPD.C8-2* with PV $\geq$ 10% in 11NJ, 12NJ and 13NJ (10.47, 11.29, and 10.24%, respectively) was the major QTL, which might harbor major genes responsible for PDgr. In addition, two QTLs, *qPD.A9-2* and *qPD.C8-3*, were expressed steadily in four experiments, with linked markers of Bn-A09-p29172005 and Bn-scaff_17227_1-p700248 and explained 6.05%–7.06% and 6.12%–8.40% of PV, respectively. The genes associated with these QTLs controlling PDgr may be less affected by environment,

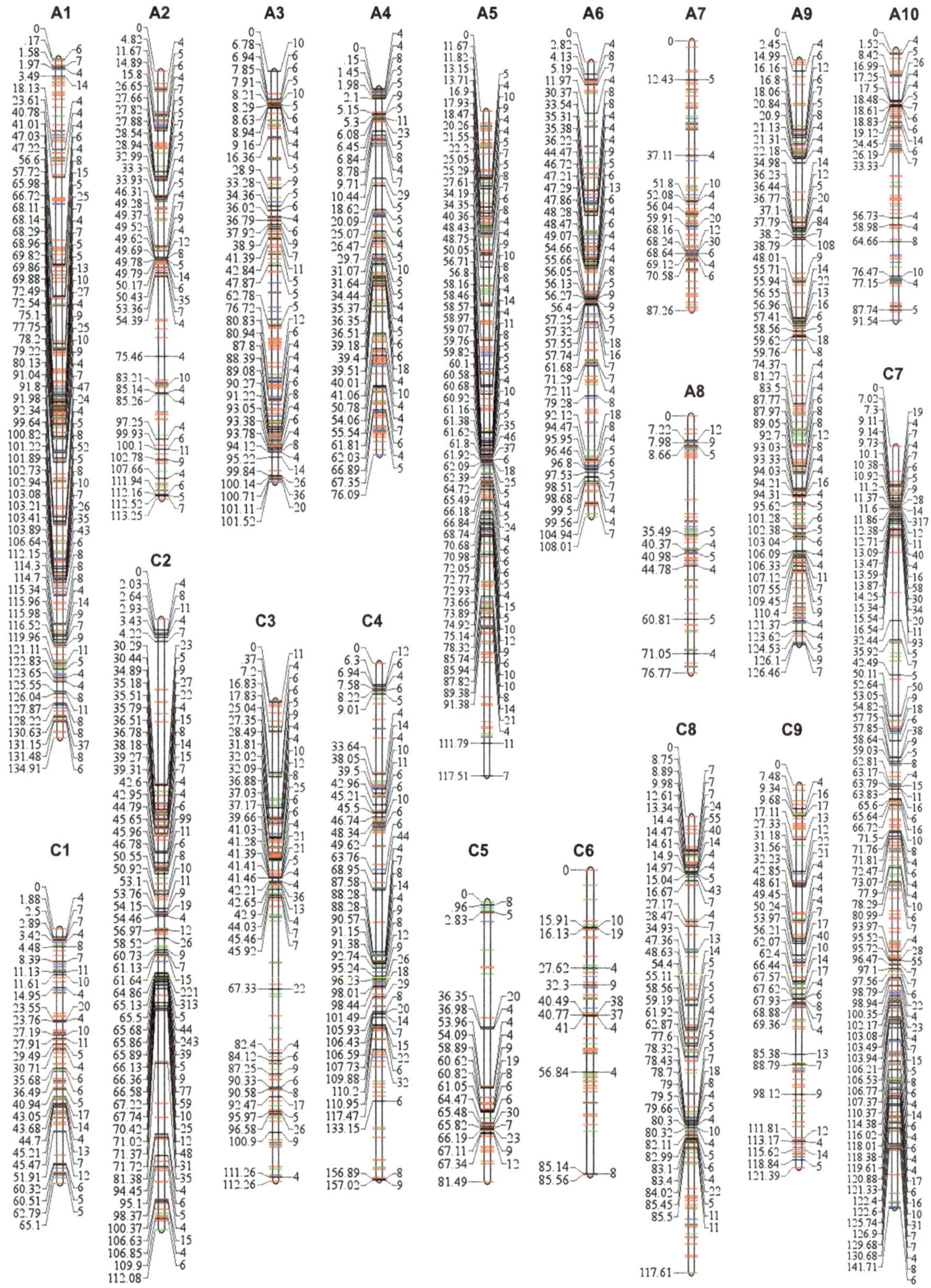

FIGURE 2 | SNPs distribution in each linkage group of the map constructed using 189 RIL individuals. The 19 linkage groups are represented by vertical bars, designated as A1–A10 in the A sub-genome and C1–C9 in the C sub-genome. The number of SNPs in each bin is listed on the right side of the linkage groups, while the positions of the bins are shown on the left side of the linkage groups. Bins with less than three SNPs were represented by colored lines (red lines, bins involve one SNP; green lines, two SNPs; and blue lines, three SNPs), for simplicity the numbers and positions are not shown. Full details are provided in Supplementary Table S1.

TABLE 1 | Genetic parameters estimated in the MX2-Additive-A or MX2-EA-A model in the AH population.

First order	Experiments[b]					Second order	Experiments				
Parameter[a]	11NJ	12NJ	13NJ	14NJ	14DL	parameter[c]	11NJ	12NJ	13NJ	14NJ	14DL
d(da)	−8.24	−9.31	−9.25	−10.85	−18.41	σ^2_p	332.53	235.52	333.76	286.01	185.5
Db					−18.41	σ^2_{mg}	287.11	198.41	290.57	251.65	127.1
I	−21.17	−20.9	−20.44	−20.35		σ^2_{pg}	44.87	36.56	43.18	34.29	58.38
[d]	−28.14	−26.03	−26.48	−23.3	0	σ^2_e	0.55	0.56	0.01	0.07	0.02
						h^2_{mg}(%)	86.41	84.34	87.06	88.01	68.52
						h^2_{pg}(%)	13.51	15.54	12.94	11.99	31.48

DL, Dali; NJ, Nanjing; 11, 12, 13 and 14 indicate the years 2011, 2012, 2013, and 2014, respectively. [a]d: The additive effect of a major gene; i: The epistasis effect of a major gene; [d]: The additive effect of polygene. [b]The best fitness genetic model in 11NJ, 12NJ, 13NJ and 14NJ experiments is the MX2-Additive-A model, while the MX2-EA-A model was the best fitness genetic model in the 14DL experiment. Genetic parameters estimated in 11NJ and 12NJ were reported by Li et al. (2014). [c]σ^2_p, Population variance; σ^2_{mg}, Major gene variance; σ^2_{pg}, Polygene variance; σ^2_e, Environmental variance; h^2_{mg} (%), Heritability of the major genes; h^2_{pg} (%), Heritability of the polygene.

which is consistent with the results of the genetic analysis, which indicated that PDgr in the AH population has a high heritability (**Table 1**). A striking finding was that, except for the three abovementioned QTLs, no other QTL was detected in multiple experiments, and the remaining six QTLs were environment-specific QTLs detected only in one environment (**Table 3**).

Identification of Candidate Genes Responsible for PDgr in the AH Population

As mentioned above, *qPD.A9-2*, *qPD.C8-2,* and *qPD.C8-3* were the three steady QTLs that showed significant effects in four, five, and four environments, respectively (**Table 3**). The three QTLs span regions of 75.97–76.75 cM, 15.69–19.7 cM, and 26.69–27.72 cM, with physical regions of 0.35, 2.65, and 0.26 Mb on chromosomes A9, C8, and C8, respectively. Based on the comparative mapping between the AH map and the *B. napus* reference genome, 58, 220, and 50 genes, respectively, were identified underlying the CIs of the three QTLs (Supplementary Table S2A). The other six consensus QTLs were only detected in certain geographical regions, suggesting that they were easily influenced by the environment, and genes underlying these QTLs were not analyzed in the present study.

Fifty-two genes regulating floral development were collected from 55 published studies that have been performed mainly in *Arabidopsis* (Supplementary Table S2B). These genes were further annotated with gene ontology (GO) terms, which were classified into three categories: biological process (BP, **Figure 5A**), cellular component (CC, **Figure 5B**), and molecular function (MF, **Figure 5C**). The 52 genes were classified into 92 functional groups, with the number of genes in each GO term ranging from 1 to 50 (Supplementary Table S2C). In the BP category, the most abundant GO terms were GO:0006355 (regulation of transcription, DNA-templated) and GO:0006351 (transcription, DNA-templated). In the CC category, GO:0005634 (nucleus) was the most abundant, followed by GO:0005737 (cytoplasm). In the MF category, GO:0005515 (protein binding) and GO:0003700 (sequence-specific DNA binding transcription factor activity) were the major GO terms.

Gene ontology assignments were also used to classify the functions of the 328 genes that underlie the three steady QTLs. The results showed that 146 genes were distributed under 38 of the 92 GO terms (**Figure 5**, Supplementary Table S2C), and thus are potential candidate genes for PDgr in *B. napus*. Among these, 34 (14 and 20, respectively), 25 (11 and 14, respectively), and 64 (57 and 7, respectively) genes were classified in the top two GO terms in the BP, CC and MF categories to which the published genes for petal development were assigned, respectively (**Figure 5**, Supplementary Table S2C). Some candidate genes were assigned to more than one category. For example, *BnaC08g10960D* and *BnaC08g14030D*, which underlie the CIs of *qPD.C8-2* and *qPD.C8-3*, respectively, were assigned to all three categories. In addition, seven, eight and five candidate genes were simultaneously assigned to BP and CC, BP and MF, CC and MF categories, respectively.

DISCUSSION

The apetalous genotype is a morphological ideotype for increasing seed yield compared with the fully petalled variety in *B. napus*, which improves light penetration through the floral canopy and lowers the incidence of *S. sclerotiorum* infection (Mendham et al., 1981, 1991; Jamaux and Spire, 1999; Jiang, 2007). Having suitable parental genotypes is critical for identifying the inheritance and the genetic bases of a character; however, the lack of completely apetalous genetic resources has been a limiting factor for understanding its genetic control in *B. napus*. Several apetalous flowers in *B. napus* were used to investigate the genetic models for the apetalous character (Buzza, 1983; Lu and Fu, 1990; Kelly et al., 1995; Zhao and Wang, 2004; Chen et al., 2006a; Zhang et al., 2007b,c). In the present study, an apetalous line 'APL01' and a normally petalled variety 'Holly' were used to construct the AH RIL population and perform a genetic analysis of the apetalous trait. The results showed that PDgr in *B. napus* was controlled by the combination of two major genes and polygenes, which was consistent with previous studies (Buzza, 1983; Kelly et al., 1995; Jiang and Becker, 2003; Chen et al., 2006a; Zhang et al., 2007b). Other genetic models

TABLE 2 | Summary of the high-density SNP map based on the AH RIL population.

Chromosome	SNP-bins	SNPs	SSRs	SNPs+SSRs	Bins+SSRs	Length (cM)	Density	No. of loci assigned to *B. napus* genome	Genome coverage (Mb)	Total length of *B. napus* genome (Mb)	% of genome coverage
A1	270	1008	7	1015	277	134.91	0.49	210	0.20–23.24 (23.04)	23.27	99.01
A2	151	430	0	430	151	113.25	0.75	126	0.06–24.79 (24.73)	24.79	99.76
A3	184	528	2	530	186	101.52	0.55	171	2.12–28.99 (26.87)	29.77	90.26
A4	141	413	1	414	142	76.09	0.54	119	0.74–19.08 (18.34)	19.15	95.77
A5	226	836	5	841	231	117.51	0.51	202	0.09–21.62 (21.53)	23.07	93.32
A6	175	493	4	497	179	108.01	0.60	148	1.21–23.84 (22.63)	24.40	92.75
A7	118	260	3	263	121	87.26	0.72	100	1.79–22.13 (20.34)	24.01	84.71
A8	68	139	4	143	72	76.77	1.07	65	0.73–18.86 (18.13)	18.96	95.62
A9	212	832	2	834	214	126.46	0.59	172	4.14–33.83 (29.69)	33.87	87.66
A10	111	269	2	271	113	91.54	0.81	105	0.26–17.37 (17.11)	17.40	98.33
C1	100	326	2	328	102	65.10	0.64	87	0.08–13.07 (12.99)	38.83	33.45
C2	146	1788	0	1788	146	112.08	0.77	116	0.52–45.79 (45.27)	46.22	97.94
C3	136	499	2	501	138	112.26	0.81	127	0.002–26.47 (26.47)	60.57	43.70
C4	124	585	5	590	129	157.02	1.22	104	0.002–48.79 (48.79)	48.93	99.71
C5	46	215	2	217	48	81.49	1.70	40	0.15–14.91 (14.76)	43.19	34.17
C6	63	212	1	213	64	85.56	1.34	50	2.86–37.19 (34.33)	37.23	92.21
C7	229	1612	8	1620	237	141.71	0.60	199	0.63–44.32 (43.69)	44.77	97.59
C8	153	587	2	589	155	117.61	0.76	136	0.80–37.98 (37.18)	38.48	96.62
C9	102	426	5	431	107	121.39	1.13	72	1.43–48.40 (46.97)	48.51	96.83
A subgenome	1656	5208	30	5238	1686	1033.31	0.61	1418	222.41	238.69	93.72
C subgenome	1099	6250	27	6277	1126	994.22	0.88	932	310.45	406.73	76.91
Whole Genome	2755	11458	57	11515	2812	2027.53	0.72	2350	532.86	645.42	85.76

The total length of the B. napus genome was reported by Chalhoub et al. (2014; http://www.genoscope.cns.fr/brassicanapus/).

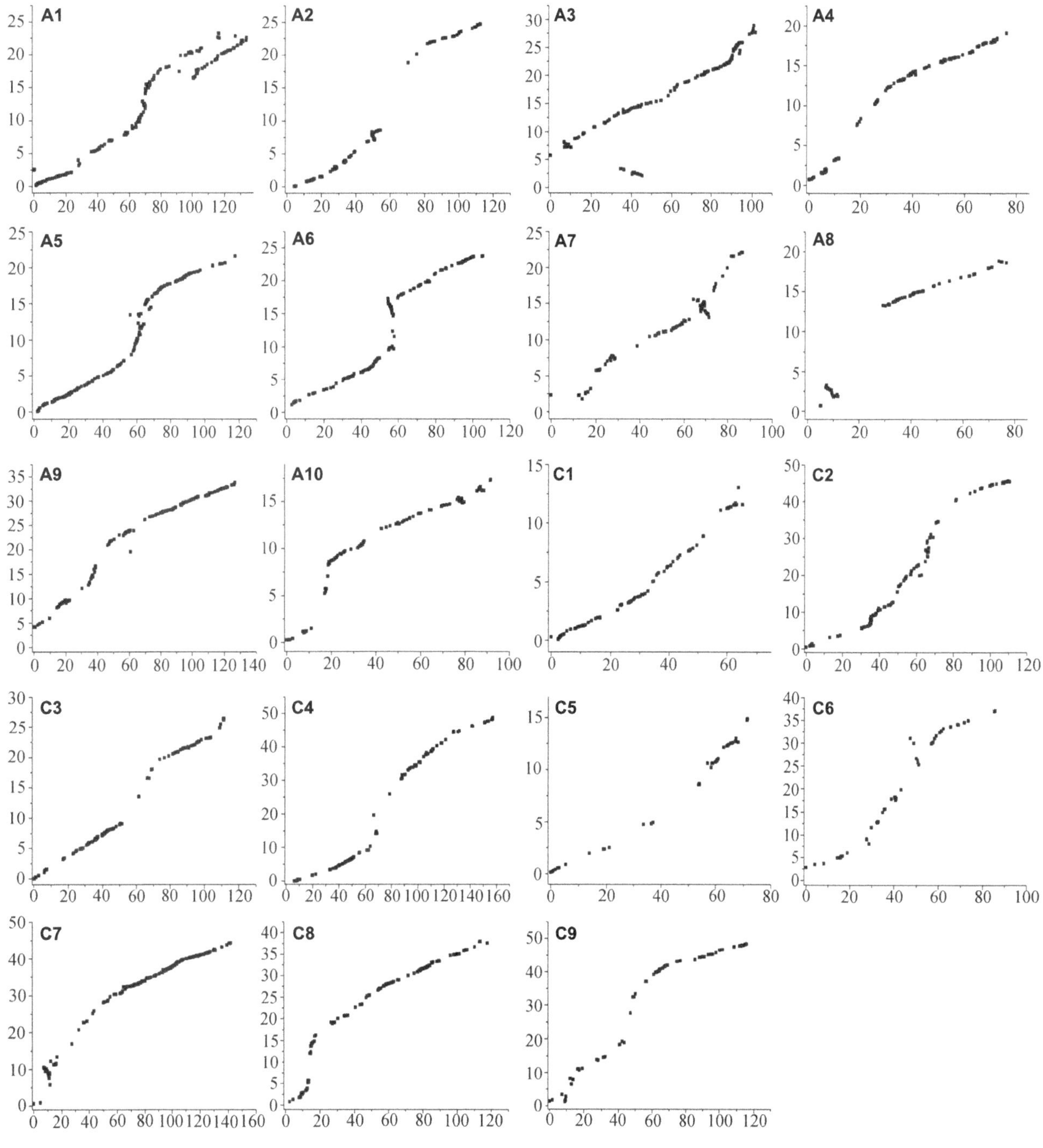

FIGURE 3 | Alignments between the SNP linkage map and the *Brassica napus* reference genome sequence. The *x*-axis represents the genetic distance in centimorgan (cM) of each linkage group in the AH map constructed in the present study. The units of the *y*-axis are the physical distances (Mb) based on the *B. napus* reference genome sequence.

for apetalous in *B. napus* have also been reported, such as the apetalous character being controlled by only one gene locus (Zhao and Wang, 2004), and four pairs of recessive genes (Lu and Fu, 1990; Zhang et al., 2007c). These studies demonstrated that PDgr in *B. napus* is a quantitative trait.

Quantitative trait loci mapping is an effective approach to dissect the genetic mechanisms of quantitative traits, while high-density map can increase the precision of QTL localization and effects, especially for small and medium sized QTLs (Almeida et al., 2013; Stange et al., 2013). However, linkage mapping studies in *B. napus* were mostly based on low-density genetic maps constructed using SSR markers, or higher-density maps based on anonymous markers, such as amplified fragment length polymorphisms or sequence-related amplified polymorphisms

TABLE 3 | A list of nine consensus QTLs for petalous degree obtained after the meta-analysis of 19 identified QTLs in five environments.

Consensus QTLs				Identified QTLs							
QTL	Peak	Marker[a]	CI	QTL	Chr.[b]	LOD	Peak	CI[c]	R^2(%)	Add[d]	Env.[e]
qPD.A3	47.82	Bn-A03-p15900189 (47.87)	46.9-49.5	*iq12NJ.A3*	A3	4.48	47.82	46.9–49.5	7.95	6.63	12NJ
qPD.A5	31.81	Bn-scaff_18338_1-p345782 (31.83)	27.6-34.0	*iq11NJ.A5*	A5	3.48	31.81	27.6–34.0	6.34	−4.51	11NJ
qPD.A6	54.8	Bn-A06-p10135140 (54.74)	49.5-55.5	*iq14DL.A6*	A6	2.82	54.8	49.5–55.5	5.08	−3.03	14DL
qPD.A9-1	64.06	Bn-A09-p25823689 (63.70)	59.66-74.36	*iq14NJ.A9*	A9	3.38	64.06	59.66–74.36	6.78	−3.69	14NJ
qPD.A9-2	76.36	Bn-A09-p29172005 (76.38)	75.97-76.75	*iq13NJ.A9*	A9	3.34	76.46	75.56–78.26	6.05	−3.95	13NJ
				iq11NJ.A9	A9	3.55	76.36	75.56–78.26	6.48	−4.52	11NJ
				iq12NJ.A9-1	A9	3.98	76.36	75.56–78.26	7.06	−4.83	12NJ
				iq14DL.A9	A9	3.26	76.36	75.56–76.46	6.12	−3.39	14DL
qPD.A9-3	81.36	Bn-A09-p30010889 (81.41)	79.86-82.96	*iq12NJ.A9-2*	A9	3.96	81.36	79.86–82.96	7.01	−4.81	12NJ
qPD.C8-1	11.8	Bn-scaff_17807_1-p364572 (11.79)	9.4-12.6	*iq11NJ.C8-1*	C8	4.63	11.8	9.4–12.6	8.77	−5.22	11NJ
qPD.C8-2	17.7	Bn-scaff_18275_1-p1278049 (17.55)	15.69-19.7	*iq11NJ.C8-2*	C8	5.59	17.6	16.7–22.8	10.47	−5.71	11NJ
				iq12NJ.C8-1	C8	6.21	17.6	14.7–25.8	11.29	−6.09	12NJ
				iq14NJ.C8	C8	3.65	17.6	15.6–26.5	7.11	−3.77	14NJ
				iq13NJ.C8-1	C8	5.48	17.9	16.7–26.5	10.24	−5.16	13NJ
				iq14DL.C8-1	C8	4.32	17.9	15.5–26.5	8.26	−3.83	14DL
qPD.C8-3	27.2	Bn-scaff_17227_1-p700248 (27.17)	26.69-27.72	*iq11NJ.C8-3*	C8	3.17	27.2	26.5–32.3	6.12	−4.42	11NJ
				iq12NJ.C8-2	C8	4.25	27.2	26.5–31.9	7.89	−5.12	12NJ
				iq13NJ.C8-2	C8	4.45	27.2	26.5–28.0	8.4	−4.7	13NJ
				iq14DL.C8-2	C8	3.69	27.2	26.5–28.0	7.06	−3.56	14DL

DL, Dali; NJ, Nanjing; 11, 12, 13 and 14 indicate the years 2011, 2012, 2013, and 2014, respectively. [a]The closest marker and the marker position in the AH map. [b]Chromosome. [c]The 2-LOD confidence interval of QTLs. [d]Additive effects. [e]The experiment in which the QTLs were detected.

(Sun et al., 2007). In the present study, a high-density SNP map was constructed with 2,812 loci involving 11,458 SNPs, covering a length of 2,027.53 cM and with an average marker interval of 0.72 cM, suggesting that it might be extremely useful in QTL detection. Compared with the previously published genetic maps constructed using SNPs in *B. napus* (Liu et al., 2013; Cai et al., 2014; Zhang et al., 2014), the AH map was one of the highest density maps with average distance between loci of less than 1.0 cM. In addition, the genome sequence of *B. napus* has been released (Chalhoub et al., 2014), which will facilitate the fine mapping of QTLs for quantitative traits and marker-assisted selection breeding in *B. napus*. A high-density SNP map is beneficial for the accurate alignment of the AH map to the physical chromosome segments in the assembled *B. napus* genomes. The results showed that the AH SNP map had good collinearity with the *B. napus* reference genome, indicating the high quality and accuracy of the map (**Figure 3**). Furthermore, the A sub-genome of the AH map showed a near-complete coverage of the *B. napus* genome (93.72%), considerably higher than the C sub-genome (76.91%; **Table 2**). One possible reason for the low coverage of the C sub-genome is that a form of distorted segregation in some chromosomes caused by rearrangements among homologous chromosomes in the parents occurred (Lu et al., 2013), resulting in the low coverage of C1 (33.45%), C3 (43.70%), and C5 (34.17%). The loci density in the A sub-genome (0.61 cM/marker) was also higher than that of the C sub-genome (0.88 cM/marker) in the AH map (**Table 2**), indicating the higher polymorphism rate in the A sub-genome. In agreement with these findings, Liu et al. (2013) constructed a map including 976 loci in the C genome and 1,819 loci in the A genome, with an average distance between markers of 0.53 cM in the A genome and 0.93 cM in the C genome.

A large population, a high-density genetic map and replicated experiments in multiple environments are three necessary factors for precise QTL detection (Yu et al., 2011; Wang et al., 2013). In the current study, a QTL analysis for PDgr was based on the high-density AH map and phenotypic data from five experiments, and nine consensus QTLs were identified on the A3, A5, A6, A9, and C8 chromosomes. Until now, only a few studies focused on the complex genetic mechanism of PDgr in *B. napus*. Several molecular markers associated with the petal-loss trait were located on A4 (Fray et al., 1997; Chen et al., 2006b) and C4 (Fray et al., 1997), and four QTLs for PDgr were located on A5, A6, A8, and C5 (Zhang et al., 2007a). These results suggest that the seven QTLs on A3, A9, and C8 in the present study were potential new QTLs, including the major QTL *qPD.C8-2* detected in all of the five experiments and two steadily expressed QTLs (*qPD.A9-2* and *qPD.C8-3*) identified in four experiments. The environment-specific QTLs *qPD.A5* on A5 and *qPD.A6* on A6 were not confirmed due to the lack of common markers between the different populations. The number of QTLs for PDgr in this study may be underestimated because the density of the AH genetic map was not saturated (**Table 2**), which may result in an incomplete resolution of QTLs. On the other hand, as AH population has been tested for 4 years in NJ, QTLs in DL might be underestimated based on the data of one year, and QTL detection could be improved in multiple experiments in DL location. These findings advanced our understanding of the genetic control of PDgr regulation in *B. napus*, and the

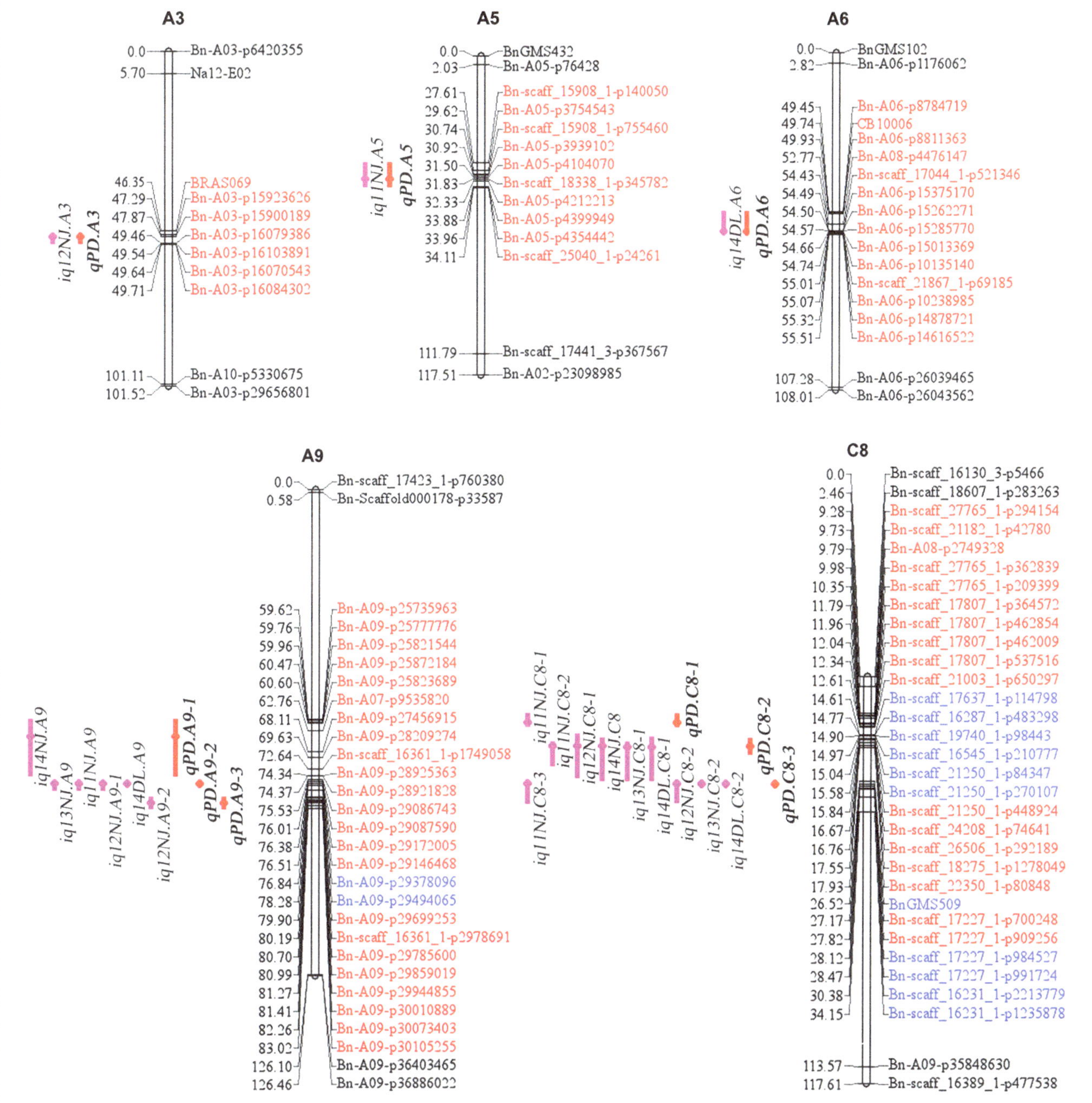

FIGURE 4 | The locations of identified and consensus QTLs associated with PDgr in the AH map. The linkage groups are represented by vertical bars. The loci names are listed on the right side of the linkage groups, while the loci positions are shown on the left side of the linkage groups. For simplicity, only the markers underlying the QTL CIs and the terminal two markers of each linkage group are shown. QTLs for PDgr are represented by bars with different backgrounds on the left side of each linkage group (purple bars, identified QTLs and red bars, consensus QTLs). Red loci are underlying the CIs of consensus QTLs and blue loci are underlying the CIs of identified QTLs but not consensus QTLs.

three stable QTLs are helpful for fine mapping and cloning of QTLs, and will enable a marker-accelerated backcrossing programs.

A typical flower consists of four different types of organs arranged in four whorls, with sepals, petals, stamens and carpels in the outermost to innermost whorls. According to the ABC model, combinatorial interactions between the three classes of floral homeotic genes are affected in the four floral organs, with 'A', 'A+B', 'B+C', and 'C' specifying sepals, petals, stamens and carpels, respectively (Coen and Meyerowitz, 1991; Honma and Goto, 2001; Theißen and Saedler, 2001). The isolation of novel floral mutants in *Arabidopsis* and other species has led to an expansion of the ABC model to include the 'D' and 'E' functions. Thus, the ABCE model was proposed, which established that the

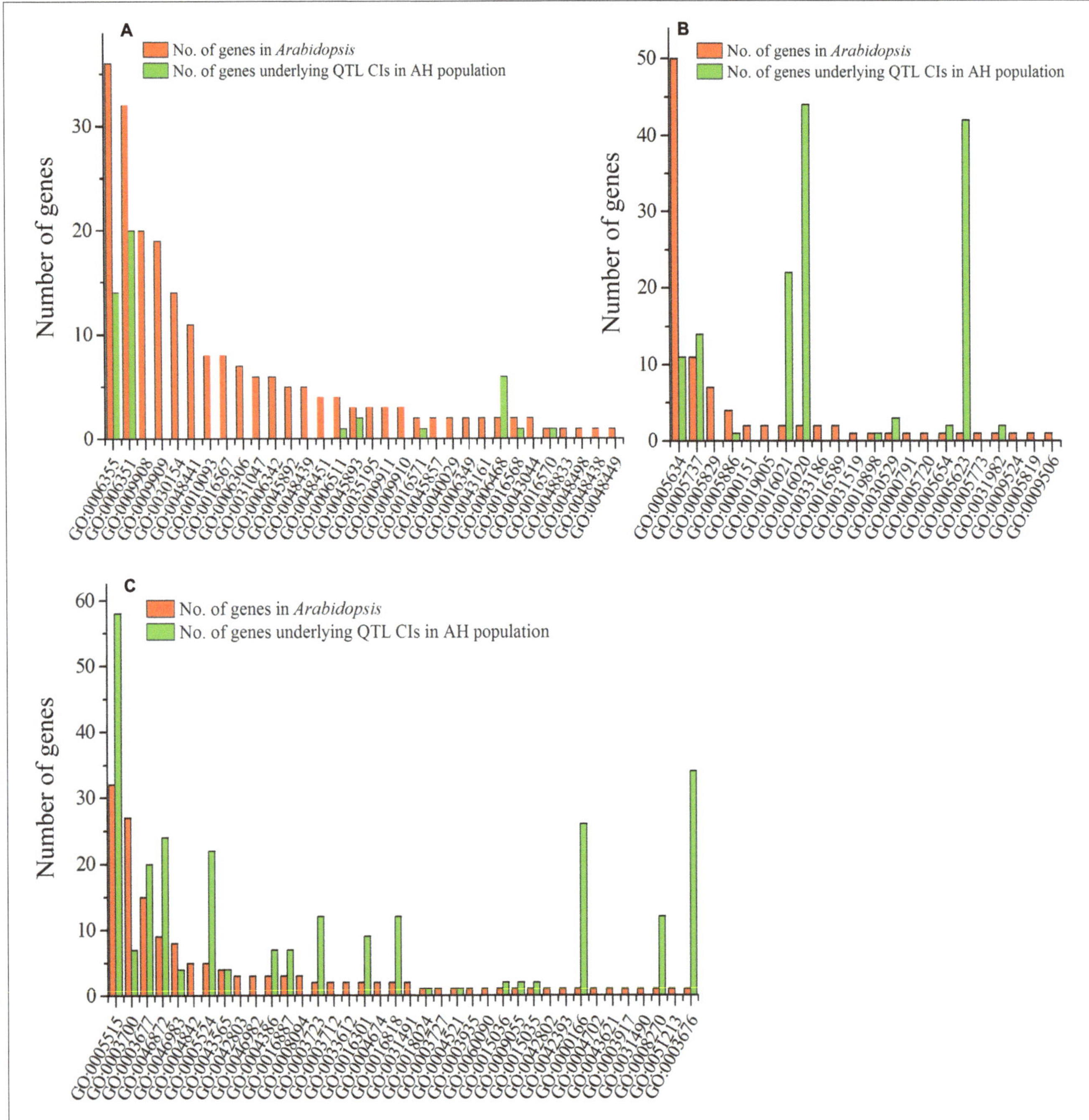

FIGURE 5 | Gene ontology (GO) annotation of collected genes regulating petal development and the candidate genes of three steady QTLs. The x-axis represents the functional groups (GO accessions), and the y-axis represents the number of genes in each category. **(A)** Biological process; **(B)** cellular component; and **(C)** molecular function.

organ-specific genes required the activity of *SEPALLATA* genes (Pinyopich et al., 2003; Krizek and Fletcher, 2005). These genes (termed class E genes), together with the class B and C genes, are required for the specification of organ identity in the petal ('A+B+E'), stamen ('B+C+E'), and carpel ('C+E') (Theißen and Saedler, 2001). Mutants with defects in the second and third whorls ('B' function) result in the homeotic conversion of petals to sepals and stamens to carpels (Causier et al., 2010). However, the apetalous line 'APL01' used in the present study had normal sepals, stamens and carpels, suggesting a more complex molecular mechanism of floral development in the allotetraploid species *B. napus*. In the present study, the genetic basis of floral development in *B. napus* was analyzed at the QTL level. Three QTLs (*qPD.A9-2*, *qPD.C8-2*, and *qPD.C8-3*) were stable across multiple environments, suggested that the genes controlling PDgr might underlying the CIs of the three QTLs.

B. napus has a common ancestor with *Arabidopsis*, and a high degree of sequence similarities and chromosomal colinearities are expected because of the progenitors diverged about 20 million years ago (Yang et al., 1999; Koch et al., 2000). Generally, the allotetraploid genomes *B. napus* may typically contain six distinct alleles for each gene present within *Arabidopsis* (Lysak et al., 2005), with the likelihood that genes carry out the core biological processes will be probable orthologs. Based on the GO assignments, 52 genes regulating of floral development mainly in *Arabidopsis* were classified into 92 functional groups, and 146 genes underlying the CIs of the three QTLs were distributed under 38 of the 92 GO terms. These genes were considered as potential candidate genes responsible for PDgr in the AH population. The B-class genes, represented in *Arabidopsis* by the MADS-box genes *APETALA3* (*AP3*) and *PISTILLATA* (*PI*), experienced gene duplication events (Kramer and Hall, 2005). In *B. napus*, two types of *AP3* genes, *B.AP3.a* and *B.AP3.b*, share a high similarity in amino acid sequences, except for an eight residue difference located at the C-terminus, and, the *B.AP3.a* specified petal and stamen development and *B.AP3.b* only specified stamen development (Zhang et al., 2011). Surprisingly, no A, B, C or E class genes from the ABCE model were identified underlying the three stable QTL CIs, indicating that novel genes for PDgr in *B. napus* might exist. To gain a better understanding of how the three QTLs control PDgr in rapeseed, it is necessary to isolate these loci through a map-based cloning strategy in the future study. This study provided useful information for understanding the genetic control of floral development in *B. napus*.

AUTHOR CONTRIBUTIONS

XW and KY carried out the QTL analysis and wrote the manuscript. HL, QP, FC, and WZ participated in the field experiment. SC and MH made helpful suggestions to the manuscript. JZ designed, led, and coordinated the overall study.

ACKNOWLEDGMENTS

The work was supported by National Natural Science Foundation of China (31371660), '948' Project of Ministry of Agriculture (2011-G23), the Industry Technology System of Rapeseed in China (CARS-13), Natural Science Foundation of Jiangsu Province (BK20151369), and Jiangsu Agriculture Science and Technology Innovation Fund (CX(14)5011).

REFERENCES

Almeida, G. D., Makumbi, D., Magorokosho, C., Nair, S., Borém, A., Ribaut, J., et al. (2013). QTL mapping in three tropical maize populations reveals a set of constitutive and adaptive genomic regions for drought tolerance. *Theor. Appl. Genet.* 126, 583–600. doi: 10.1007/s00122-012-2003-7

Arcade, A., Labourdette, A., Falque, M., Mangin, B., Chardon, F., Charcosset, A., et al. (2004). BioMercator: integrating genetic maps and QTL towards discovery of candidate genes. *Bioinformatics* 20, 2324–2326. doi: 10.1093/bioinformatics/bth230

Basunanda, P., Radoev, M., Ecke, W., Friedt, W., Becker, H. C., and Snowdon, R. J. (2010). Comparative mapping of quantitative trait loci involved in heterosis for seedling and yield traits in oilseed rape (*Brassica napus* L.). *Theor. Appl. Genet.* 120, 271–281. doi: 10.1007/s00122-009-1133-z

Buzza, G. (1983). The inheritance of an apetalous character in Canola (*Brassica napus*). *Cruciferae Newsletter* 8, 11–12.

Cai, G., Yang, Q., Yang, Q., Zhao, Z., Chen, H., Wu, J., et al. (2012). Identification of candidate genes of QTLs for seed weight in *Brassica napus* through comparative mapping among *Arabidopsis* and *Brassica* species. *BMC Genet.* 13:105. doi: 10.1186/1471-2156-13-105

Cai, G., Yang, Q., Yi, B., Fan, C., Edwards, D., Batley, J., et al. (2014). A complex recombination pattern in the genome of allotetraploid *Brassica napus* as revealed by a high-density genetic map. *PLoS ONE* 9:e109910. doi: 10.1371/journal.pone.0109910

Cao, X., Liu, B., and Zhang, Y. (2013). SEA: a software package of segregation analysis of quantitative traits in plants (in Chinese with an English abstract). *J. Nanjing Agric. Univ.* 36, 1–6.

Causier, B., Schwarz-Sommer, Z., and Davies, B. (2010). Floral organ identity: 20 years of ABCs. *Semin. Cell Dev. Biol.* 21, 73–79. doi: 10.1016/j.semcdb.2009.10.005

Chalhoub, B., Denoeud, F., Liu, S., Parkin, I. A., Tang, H., Wang, X., et al. (2014). Early allopolyploid evolution in the post-Neolithic *Brassica napus* oilseed genome. *Science* 345, 950–953. doi: 10.1126/science.1253435

Chen, B., Wu, X., Lu, G., Gao, G., Xu, K., and Li, X. (2006a). Inheritance for apetalous trait of two kinds of apetalous rapeseed mutants (in Chinese with an English abstract). *Chin. J. Oil Crop Sci.* 28, 263–267.

Chen, B., Wu, X., Lu, G., Gao, G., Xu, K., and Li, X. (2006b). Molecular mapping of the gene (s) controlling petal-loss trait in *Brassica napus*. *Hereditas(Beijing)* 28, 707–712.

Chen, G., Geng, J., Rahman, M., Liu, X., Tu, J., Fu, T., et al. (2010). Identification of QTL for oil content, seed yield, and flowering time in oilseed rape (*Brassica napus*). *Euphytica* 175, 161–174. doi: 10.1007/s10681-010-0144-9

Chen, X., Li, X., Zhang, B., Xu, J., Wu, Z., Wang, B., et al. (2013). Detection and genotyping of restriction fragment associated polymorphisms in polyploid crops with a pseudo-reference sequence: a case study in allotetraploid *Brassica napus*. *BMC Genomics* 14:346. doi: 10.1186/1471-2164-14-346

Cheng, X., Xu, J., Xia, S., Gu, J., Yang, Y., Fu, J., et al. (2009). Development and genetic mapping of microsatellite markers from genome survey sequences in *Brassica napus*. *Theor. Appl. Genet.* 118, 1121–1131. doi: 10.1007/s00122-009-0967-8

Ching, A., Caldwell, K. S., Jung, M., Dolan, M., Smith, O. S., Tingey, S., et al. (2002). SNP frequency, haplotype structure and linkage disequilibrium in elite maize inbred lines. *BMC Genet.* 3:19. doi: 10.1186/1471-2156-3-19

Coen, E. S., and Meyerowitz, E. M. (1991). The war of the whorls: genetic interactions controlling flower development. *Nature* 353, 31–37. doi: 10.1038/353031a0

Delourme, R., Falentin, C., Fomeju, B. F., Boillot, M., Lassalle, G., Andre, I., et al. (2013). High-density SNP-based genetic map development and linkage disequilibrium assessment in *Brassica napus* L. *BMC Genomics* 14:120. doi: 10.1186/1471-2164-14-120

Ding, G., Zhao, Z., Liao, Y., Hu, Y., Shi, L., Long, Y., et al. (2012). Quantitative trait loci for seed yield and yield-related traits, and their responses to reduced phosphorus supply in *Brassica napus*. *Ann. Bot.* 109, 747–759. doi: 10.1093/aob/mcr323

Edwards, D., Batley, J., and Snowdon, R. J. (2013). Accessing complex crop genomes with next-generation sequencing. *Theor. Appl. Genet.* 126, 1–11. doi: 10.1007/s00122-012-1964-x

Fray, M. J., Puangsomlee, P., Goodrich, J., Coupland, G., Evans, E. J., Arthur, A. E., et al. (1997). The genetics of stamenoid petal production in oilseed rape (*Brassica napus*) and equivalent variation in *Arabidopsis thaliana*. *Theor. Appl. Genet.* 94, 731–736. doi: 10.1007/s001220050472

Habekotté, B. (1997). Options for increasing seed yield of winter oilseed rape (*Brassica napus* L.) a simulation study. *Field Crop Res.* 54, 109–126. doi: 10.1016/S0378-4290(97)00041-5

Hoekman, B. (2007). *Global Economic Prospects 2008: Technology Diffusion in the Developing World.* Washington, DC: World Bank Publications, 40–41.

Honma, T., and Goto, K. (2001). Complexes of MADS-box proteins are sufficient to convert leaves into floral organs. *Nature* 409, 525–529. doi: 10.1038/35054083

Huang, X., Wei, X., Sang, T., Zhao, Q., Feng, Q., Zhao, Y., et al. (2010). Genome-wide association studies of 14 agronomic traits in rice landraces. *Nat. Genet.* 42, 961–967. doi: 10.1038/ng.695

Jamaux, D. I., and Spire, D. (1999). Comparison of responses of ascospores and mycelium by ELISA with anti-mycelium and anti-ascospore antisera for the development of a method to detect *Sclerotinia sclerotiorum* on petals of oilseed rape. *Ann. Appl. Biol.* 134, 171–179. doi: 10.1111/j.1744-7348.1999.tb05253.x

Jiang, L. (2007). Breeding for apetalous rape: inheritance and yield physiology. *Adv. Bot. Res.* 45, 217–231. doi: 10.1016/S0065-2296(07)45008-X

Jiang, L., and Becker, H. (2003). Inheritance of apetalous flowers in a mutant of oilseed rape. *Crop Sci.* 43:508. doi: 10.2135/cropsci2003.0508

Kelly, A., Fray, M., Arthur, E. A., Lydiate, D. J., and Evans, E. J. (1995). "The genetic control of petalless flowers and upright pods," in *Proceedings of the 9th International Rapeseed Congress*, Cambridge, 4–8.

Koch, M. A., Haubold, B., and Mitchell-Olds, T. (2000). Comparative evolutionary analysis of chalcone synthase and alcohol dehydrogenase loci in *Arabidopsis*, *Arabis*, and related genera (*Brassicaceae*). *Mol. Biol. Evol.* 17, 1483–1498. doi: 10.1093/oxfordjournals.molbev.a026248

Kosambi, D. D. (1943). The estimation of map distances from recombination values. *Ann. Eugen.* 12, 172–175. doi: 10.1111/j.1469-1809.1943.tb02321.x

Kramer, E. M., and Hall, J. C. (2005). Evolutionary dynamics of genes controlling floral development. *Curr. Opin. Plant. Biol.* 8, 13–18. doi: 10.1016/j.pbi.2004.09.019

Krizek, B. A., and Fletcher, J. C. (2005). Molecular mechanisms of flower development: an armchair guide. *Nat. Rev. Genet.* 6, 688–698. doi: 10.1038/nrg1675

Landry, B. S., Hubert, N., Etoh, T., Harada, J. J., and Lincoln, S. E. (1991). A genetic map for *Brassica napus* based on restriction fragment length polymorphisms detected with expressed DNA sequences. *Genome* 34, 543–552. doi: 10.1139/g91-084

Lefol, C., and Morrall, R. A. A. (1996). Immunofluorescent staining of *Sclerotinia* ascospores on canola petals. *Can. J. Plant Pathol.* 18, 237–241. doi: 10.1080/07060669609500618

Li, H., Yu, K., Guo, T., and Zhang, J. (2014). Genetic analysis of apetalous trait in *Brassica napus* using mixed model of major gene and polygene (in Chinese with an English abstract). *Jiangsu J. Agric. Sci.* 30, 253–258.

Linnemann, A. R., and Dijkstra, D. S. (2002). Toward sustainable production of protein-rich foods: appraisal of eight crops for Western Europe. Part I. Analysis of the primary links of the production chain. *Crit. Rev. Food Sci. Nutr.* 42, 377–401. doi: 10.1080/20024091054193

Liu, L., Qu, C., Wittkop, B., Yi, B., Xiao, Y., He, Y., et al. (2013). A high-density SNP map for accurate mapping of seed fibre QTL in *Brassica napus* L. *PLoS ONE* 8:e83052. doi: 10.1371/journal.pone.0083052

Long, Y., Shi, J., Qiu, D., Li, R., Zhang, C., Wang, J., et al. (2007). Flowering time quantitative trait loci analysis of oilseed *Brassica* in multiple environments and genomewide alignment with *Arabidopsis*. *Genetics* 177, 2433–2444. doi: 10.1534/genetics.107.080705

Lowe, A. J., Moule, C., Trick, M., and Edwards, K. J. (2004). Efficient large-scale development of microsatellites for marker and mapping applications in *Brassica* crop species. *Theor. Appl. Genet.* 108, 1103–1112. doi: 10.1007/s00122-003-1522-7

Lu, W., Liu, J., Xin, Q., Wan, L., Hong, D., and Yang, G. (2013). A triallelic genetic male sterility locus in *Brassica napus*: an integrative strategy for its physical mapping and possible local chromosome evolution around it. *Ann. Bot.* 111, 305–315. doi: 10.1093/aob/mcs260

Lu, Z., and Fu, S. (1990). Inheritance of apetalous character in rape (*Brassica napus* L.) and its implication in breeding (in Chinese with an English abstract). *Jiangsu J. Agric. Sci.* 6, 30–36.

Lysak, M. A., Koch, M. A., Pecinka, A., and Schubert, I. (2005). Chromosome triplication found across the tribe Brassiceae. *Genome Res.* 15, 516–525. doi: 10.1101/gr.3531105

Mauricio, R. (2001). Mapping quantitative trait loci in plants: uses and caveats for evolutionary biology. *Nat. Rev. Genet.* 2, 370–381. doi: 10.1038/35072085

McCouch, S. R., Cho, Y. G., Yano, M., Paul, E., Blinstrub, M., Morishima, H., et al. (1997). Report on QTL nomenclature. *Rice Genet. News* 14, 11–131. doi: 10.1007/s10142-013-0328-1

Mendham, N. J., Rao, M. S. S., and Buzza, G. C. (1991). "The apetalous flower character as a component of a high yielding ideotype," in *Proceedings of* the *8th International Rapeseed Congress*, Saskatoon, SK, 596–600.

Mendham, N. J., Shipway, P. A., and Scott, R. K. (1981). The effects of delayed sowing and weather on growth, development and yield of winter oil-seed rape (*Brassica napus*). *J. Agric. Sci.* 96, 389–416. doi: 10.1017/S0021859600 06617X

Mochida, K., Yamazaki, Y., and Ogihara, Y. (2003). Discrimination of homoeologous gene expression in hexaploid wheat by SNP analysis of contigs grouped from a large number of expressed sequence tags. *Mol. Genet. Genomics* 270, 371–377. doi: 10.1007/s00438-003-0939-7

Pinyopich, A., Ditta, G. S., Savidge, B., Liljegren, S. J., Baumann, E., Wisman, E., et al. (2003). Assessing the redundancy of MADS-box genes during carpel and ovule development. *Nature* 424, 85–88. doi: 10.1038/nature01741

Piquemal, J., Cinquin, E., Couton, F., Rondeau, C., Seignoret, E., Doucet, I., et al. (2005). Construction of an oilseed rape (*Brassica napus* L.) genetic map with SSR markers. *Theor. Appl. Genet.* 111, 1514–1523. doi: 10.1007/s00122-005-0080-6

Qiu, D., Morgan, C., Shi, J., Long, Y., Liu, J., Li, R., et al. (2006). A comparative linkage map of oilseed rape and its use for QTL analysis of seed oil and erucic acid content. *Theor. Appl. Genet.* 114, 67–80. doi: 10.1007/s00122-006-0411-2

Quijada, P. A., Udall, J. A., Lambert, B., and Osborn, T. C. (2006). Quantitative trait analysis of seed yield and other complex traits in hybrid spring rapeseed (*Brassica napus* L.): 1. Identification of genomic regions from winter germplasm. *Theor. Appl. Genet.* 113, 549–561. doi: 10.1007/s00122-006-0323-1

Raman, H., Dalton-Morgan, J., Diffey, S., Raman, R., Alamery, S., Edwards, D., et al. (2014). SNP markers-based map construction and genome-wide linkage analysis in *Brassica napus*. *Plant Biotechnol. J.* 12, 851–860. doi: 10.1111/pbi.12186

Ramanujam, S. (1940). An apetalous mutation in turnip (*Brassica* campestris L.). *Nature* 145, 552–553. doi: 10.1038/145552a0

Shi, J., Li, R., Qiu, D., Jiang, C., Long, Y., Morgan, C., et al. (2009). Unraveling the complex trait of crop yield with quantitative trait loci mapping in *Brassica napus*. *Genetics* 182, 851–861. doi: 10.1534/genetics.109.101642

Snowdon, R. J., and Iniguez Luy, F. L. (2012). Potential to improve oilseed rape and canola breeding in the genomics era. *Plant Breed.* 131, 351–360. doi: 10.1111/j.1439-0523.2012.01976.x

Stange, M., Utz, H. F., Schrag, T. A., Melchinger, A. E., and Würschum, T. (2013). High-density genotyping: an overkill for QTL mapping? Lessons learned from a case study in maize and simulations. *Theor. Appl. Genet.* 126, 2563–2574. doi: 10.1007/s00122-013-2155-0

Sun, Z., Wang, Z., Tu, J., Zhang, J., Yu, F., McVetty, P. B. E., et al. (2007). An ultradense genetic recombination map for *Brassica napus*, consisting of 13551 SRAP markers. *Theor. Appl. Genet.* 114, 1305–1317. doi: 10.1007/s00122-006-0483-z

Tan, X., Li, D., Wang, H., and Guo, A. (2003). Identification of a RAPD marker linked to a petal-controlled gene in *Brassica napus* (in Chinese with an English abstract). *Acta Bot. Boreal. Occident. Sin.* 23, 561–565.

Theißen, G., and Saedler, H. (2001). Plant biology: floral quartets. *Nature* 409, 469–471. doi: 10.1038/35054172

Udall, J. A., Quijada, P. A., Lambert, B., and Osborn, T. C. (2006). Quantitative trait analysis of seed yield and other complex traits in hybrid spring rapeseed (*Brassica napus* L.): 2. Identification of alleles from unadapted germplasm. *Theor. Appl. Genet.* 113, 597–609. doi: 10.1007/s00122-006-0324-0

Uzunova, M. I., and Ecke, W. (1999). Abundance, polymorphism and genetic mapping of microsatellites in oilseed rape (*Brassica napus* L.). *Plant Breed.* 118, 323–326. doi: 10.1046/j.1439-0523.1999.00371.x

Van Ooijen, J. W. (2006). *JoinMap 4. Software for the Calculation of Genetic Linkage Maps in Experimental Populations.* Kyazma BV.

Vignal, A., Milan, D., SanCristobal, M., and Eggen, A. (2002). A review on SNP and other types of molecular markers and their use in animal genetics. *Genet. Sel. Evol.* 34, 275–305. doi: 10.1051/gse:2002009

Virk, P. S., Khush, G. S., and Peng, S. (2004). Breeding to enhance yield potential of rice at IRRI: the ideotype approach. *Int. Rice Res. Notes* 29, 5–9.

Wang, S., Basten, C. J., and Zeng, Z. B. (2007). *Windows QTL Cartographer 2.5.* Raleigh, NC: North Carolina State University.

Wang, X., Long, Y., Yin, Y., Zhang, C., Gan, L., Liu, L., et al. (2015a). New insights into the genetic networks affecting seed fatty acid concentrations in *Brassica napus*. *BMC Plant Biol.* 15:91. doi: 10.1186/s12870-015-0475-8

Wang, X., Wang, H., Long, Y., Liu, L., Zhao, Y., Tian, J., et al. (2015b). Dynamic and comparative QTL analysis for plant height in different developmental stages of *Brassica napus* L. *Theor. Appl. Genet.* 128, 1175–1192. doi: 10.1007/s00122-015-2498-9

Wang, X., Wang, H., Long, Y., Li, D., Yin, Y., Tian, J., et al. (2013). Identification of QTLs associated with oil content in a high-oil *Brassica napus* cultivar and construction of a high-density consensus map for QTLs comparison in *B. napus*. *PLoS ONE* 8:e80569. doi: 10.1371/journal.pone.0080569

Yang, Y., Lai, K., Tai, P., and Li, W. (1999). Rates of nucleotide substitution in angiosperm mitochondrial DNA sequences and dates of divergence between *Brassica* and other angiosperm lineages. *J. Mol. Evol.* 48, 597–604. doi: 10.1007/PL00006502

Yu, H., Xie, W., Wang, J., Xing, Y., Xu, C., Li, X., et al. (2011). Gains in QTL detection using an ultra-high density SNP map based on population sequencing relative to traditional RFLP/SSR markers. *PLoS ONE* 6:e17595. doi: 10.1371/journal.pone.0017595

Zeng, Z. (1994). Precision mapping of quantitative trait loci. *Genetics* 136, 1457–1468.

Zhang, D., Hua, Y., Wang, X., Zhao, H., Shi, L., and Xu, F. (2014). A high-density genetic map identifies a novel major QTL for boron efficiency in oilseed rape (*Brassica napus* L.). *PLoS ONE* 9:e112089. doi: 10.1371/journal.pone.0112089

Zhang, J., Fu, S., Qi, C., Pu, H., and Wu, X. (2002). Preliminary report of breeding for near-isogenic lines of apetalous in rapeseed (*B. napus* L.) (in Chinese with an English abstract). *Chin. J. Oil Crop Sci.* 24, 14–16.

Zhang, J., Qi, C., Li, G., Pu, H., Chen, S., Chen, X., et al. (2007a). Genetic map construction and apetalousness QTLs identification in rapeseed (*Brassica napus* L.) (in Chinese with an English abstract). *Acta. Agron. Sin.* 33, 1246–1254.

Zhang, J., Qi, C., Pu, H., Chen, S., Chen, F., Chen, X., et al. (2007b). Genetic analysis of apetalous in *Brassica napus* L. using mixed model of major gene and polygene (in Chinese with an English abstract). *Chin. J. Oil Crop Sci.* 29, 227–232.

Zhang, R., Tang, X., Li, M., and Chen, D. (2007c). Study on genetics of apetalous line NF001 in rapeseed (*B. napus* L.) (in Chinese with an English abstract). *Seed* 26, 13–15.

Zhang, Y., Wang, X., Zhang, W., Yu, F., Tian, J., Li, D., et al. (2011). Functional analysis of the two *Brassica* AP3 genes involved in apetalous and stamen carpelloid phenotypes. *PLoS ONE* 6:e20930. doi: 10.1371/journal.pone.0020930

Zhao, Y., and Wang, M. (2004). Inheritance and agronomic performance of an apetalous flower mutant in *Brassica napus* L. *Euphytica* 137, 381–386. doi: 10.1023/B:EUPH.0000040522.37048.7a

Conflict of Interest Statement: The authors declare that the research was conducted in the absence of any commercial or financial relationships that could be construed as a potential conflict of interest.

7

Tensiometer-Based Irrigation Management of Subirrigated Soilless Tomato: Effects of Substrate Matric Potential Control on Crop Performance

Francesco F. Montesano[1], Francesco Serio[1*], Carlo Mininni[1], Angelo Signore[2], Angelo Parente[1] and Pietro Santamaria[2]

[1] Institute of Sciences of Food Production, National Research Council of Italy, Bari, Italy, [2] Department of Agricultural and Environmental Science, University of Bari Aldo Moro, Bari, Italy

Automatic irrigation scheduling based on real-time measurement of soilless substrate water status has been recognized as a promising approach for efficient greenhouse irrigation management. Identification of proper irrigation set points is crucial for optimal crop performance, both in terms of yield and quality, and optimal use of water resources. The objective of the present study was to determine the effects of irrigation management based on matric potential control on growth, plant–water relations, yield, fruit quality traits, and water-use efficiency of subirrigated (through bench system) soilless tomato. Tensiometers were used for automatic irrigation control. Two cultivars, "Kabiria" (cocktail type) and "Diana" (intermediate type), and substrate water potential set-points (−30 and −60 hPa, for "Diana," and −30, −60, and −90 hPa for "Kabiria"), were compared. Compared with −30 hPa, water stress (corresponding to a −60 hPa irrigation set-point) reduced water consumption (14%), leaf area (18%), specific leaf area (19%), total yield (10%), and mean fruit weight (13%), irrespective of the cultivars. At −60 hPa, leaf-water status of plants, irrespective of the cultivars, showed an osmotic adjustment corresponding to a 9% average osmotic potential decrease. Total yield, mean fruit weight, plant water, and osmotic potential decreased linearly when −30, −60, and −90 hPa irrigation set-points were used in "Kabiria." Unmarketable yield in "Diana" increased when water stress was imposed (187 vs. 349 g·plant^{-1}, respectively, at −30 and −60 hPa), whereas the opposite effect was observed in "Kabiria," where marketable yield loss decreased linearly [by 1.05 g·plant^{-1} per unit of substrate water potential (in the tested range from −30 to −90 hPa)]. In the second cluster, total soluble solids of the fruit and dry matter increased irrespective of the cultivars. In the seventh cluster, in "Diana," only a slight increase was observed from −30 vs. −60 hPa (3.3 and 1.3%, respectively, for TSS and dry matter), whereas in "Kabiria," the increase was more pronounced (8.7 and 12.0%, respectively, for TSS and dry matter), and further reduction in matric potential from −60 to −90 hPa confirmed the linear increase for both parameters. Both glucose and fructose concentrations increased linearly in "Kabiria" fruits on decreasing the substrate matric potential, whereas in "Diana," there was no increase. It is feasible

Edited by:
David Bryla,
U.S. Department of Agriculture, Agricultural Research Service, USA

Reviewed by:
Youssef Rouphael,
University of Naples Federico II, Italy
James S. Owen,
Virginia Tech, USA

*Correspondence:
Francesco Serio
francesco.serio@ispa.cnr.it

to act on matric potential irrigation set-points to control plant response in terms of fruit quality parameters. Precise control of substrate water status may offer the possibility to steer crop response by enhancing different crop-performance components, namely yield and fruit quality, in subirrigated tomato. Small-sized fruit varieties benefit more from controlled water stress in terms of reduced unmarketable yield loss and fruit quality improvements.

Keywords: irrigation set-point, water relations, through bench system, WUE

INTRODUCTION

Irrigation management directly affects crop performance and can lead to qualitative and quantitative improvements in vegetable production (Dukes et al., 2010). Under-irrigation generally results in reduced crop yield and quality, whereas over-irrigation may lead to increased crop vulnerability to diseases, energy costs for water pumping, water loss, and environmental pollution due to fertilizer runoff (Pardossi et al., 2009). Therefore, irrigation management needs to be efficient also in order to help reduce environmental impact and promote sustainable use of resources (Montesano et al., 2015).

Production of vegetable crops in greenhouses has expanded considerably over recent decades in Mediterranean region (Sonneveld and Voogt, 2009; FAO, 2013). Research efforts and the related introduction of technical innovations initially focused on high-quality, healthy products. However, concern over environmentally sustainable production has risen in the last decade, as industrial greenhouse crops are usually seen as entailing high environmental impact (Torrellas et al., 2012). On the other hand, there is also plenty of evidence that greenhouse vegetable production may decrease the environmental impact compared to field cultivation (Stanghellini, 2014). In this framework, providing the greenhouse sector with tools and skills for efficient irrigation management is a key factor.

Among the different approaches proposed to achieve better crop performance and efficient use of resources (water and fertilizers) in irrigated greenhouse agriculture, closed-cycle soilless cultivation, with recycling of nutrient solution (NS) aiming to minimize pollution (Vox et al., 2010), and smart sensor-based irrigation scheduling are promising and increasingly adopted strategies (van Iersel et al., 2013).

Visual assessment of plants and substrate and the use of timers to automate irrigation, still the most common approach in greenhouses, are generally inefficient (Nemali et al., 2007). Among the different possible approaches for improved irrigation management and automation, monitoring the water status of the soil/substrate through sensors, as well as making objective irrigation decisions based on real-time measurements, seems particularly suitable for greenhouse conditions (Jones, 2007; van Iersel et al., 2013). Sensors for measuring water status in the root zone are a dynamic and constantly developing area of technology for both technical and commercial reasons. Although technological developments in soil/soilless water sensors have generally focused on scientific applications, thereby aiming to study the effects of different soil moisture levels on plant physiological responses, interest in using sensors for practical irrigation management has grown in the last decade (Pardossi et al., 2009; Lichtenberg et al., 2013). Sensing soil/soilless substrate water status is becoming easier and more economically feasible, providing opportunities to integrate sensor networks into existing irrigation systems (van Iersel et al., 2013). The basic approach behind the automation of irrigation based on root-zone moisture measurements is simple: the moisture level in the growing media fluctuates according to evaporation and plant water use; sensors detect this change and automatically activate irrigation when the level reaches a set value predetermined by the operator, resulting in on-demand irrigation (van Iersel, 2015).

Two categories of sensors can be used to monitor soil/substrate water status based on measuring two basic inter-related properties: soil volumetric water content sensors measure how much water is present in the soil in terms of volume, whereas soil matric potential sensors measure how tightly water is bound to soil/substrate particles and, as a consequence, the amount of energy the plants need to exert to extract water from a soil or soilless substrate. The relation between volumetric water content and matric potential is characteristic of a specific porous medium and is described by the moisture release curve. Sensing matric potential has the advantage of providing direct information on whether and to what extent water in the growing medium is available to the plants (van Iersel et al., 2013).

While many technological innovations have been introduced into volumetric moisture sensors in the last decade, with a number of reliable and inexpensive sensors principally based on Frequency Domain Reflectometry (FDR) becoming available on the market, the water-filled tensiometer is still the most common sensor for matric potential measurement and has seen little improvement in comparison with the early models, apart from the use of improved pressure transducers and data logging systems (Whalley et al., 2013). Tensiometers are often preferred to other types of substrate moisture sensors due to their low cost, simplicity of use, high accuracy, and direct measurement of matric potential, and also, they are not influenced by temperature and soil osmotic potential; moreover, the possibility of electronic data acquisition through differential pressure transducers (Thalheimer, 2003; Sarkar et al., 2008a; Montesano et al., 2010) makes the tensiometer suitable for automated fertigation control. However, tensiometers must be operated carefully in order to avoid the formation of air bubbles in the shaft; they must be protected from frost and need regular maintenance, for instance, to refill the water in the tube. The risk of cavitation in very dry conditions is also a drawback. When tensiometers are used in pot cultures, some precautions are necessary in order to ensure good contact between the porous tip and the substrate, in particular in soilless conditions where high porosity is common, and to achieve correct sensor positioning, taking into account root distribution and the placing of nozzle(s) in the event of drip irrigation (Pardossi et al., 2009). Tensiometers specifically designed for soilless substrates are also available on the market, equipped with special ceramic tips, thereby allowing for faster equilibration with the surrounding substrate (van Iersel et al., 2013).

Soilless growing media generally hold easily available water (EAW) in a matric potential range from 0 to −100 hPa, with the majority of freely available water from 0 to −50 hPa (de Boodt and Verdonck, 1972; Argo, 1998). Other indications suggest that a matric potential of −200 to −300 hPa could be considered the lower limit for available water for plants, underlining the important role of hydraulic conductivity decrease in dry conditions in soilless substrates (van Iersel et al., 2013). Therefore, for practical purposes, water availability in the growing media can be sensed and measured on the base of matric potential; this has significant implications for the measurement and control of irrigation in soilless conditions, as there is a narrow range of available water for scheduling irrigation (Lea-Cox and Arguedas-Rodriguez, 2011). Matric potential control through tensiometers has also been used to impose controlled water stress in order to improve the quality response of crops in soilless conditions (Sarkar et al., 2008a).

Closed-cycle Zero Runoff Subirrigation (ZRS) is a practical and cost-effective method for achieving efficient water use in potted plants grown in greenhouses in soilless conditions. The basic principle of subirrigation systems is the same: the bottom of the pots is submerged in water (or NS) for a brief period (generally 5–20 min, according to the container size), which gives the growing medium enough time to absorb water by capillarity. An extensive overview on ZRS systems was recently published by Ferrarezi et al. (2015a), to which the readers are referred for a more in-depth review than space will allow herein.

Tomato is the most important greenhouse crop grown in soilless cultivation systems (Savvas et al., 2013). Several studies have been conducted on growing soilless tomato using subirrigation (Santamaria et al., 2003; Incrocci et al., 2006; Sarkar et al., 2008b; Montesano et al., 2010). Subirrigation systems have been successfully automated based on substrate water content using tensiometers (Rouphael et al., 2006; Montesano et al., 2010) as well as capacitance moisture sensors (Ferrarezi et al., 2015b). However, to the best of our knowledge, an evaluation of the overall performance of subirrigated tomato subjected to matric potential control using different irrigation set points has not been conducted.

Given the above considerations, the objective of the present study was to determine the effects of irrigation management based on matric potential control on growth, water relations, yield, quality, and water-use efficiency (WUE) of subirrigated tomato. We hypothesized that using matric potential control through a tensiometer with different irrigation set points would allow us to steer crop performance, with particular reference to WUE, yield, and fruit quality.

MATERIALS AND METHODS

Experiment Setup and Treatment Application

The experiment was carried out in a plastic (polymethacrylate) greenhouse at the Experimental Farm "La Noria" of the Institute of Sciences of Food Production (ISPA-CNR) in Mola di Bari, Southern Italy, during a winter-summer growing cycle (January–July, 2005).

Seedlings of two tomato cultivars [*Solanum lycopersicum* L., "Kabiria," a cocktail type, with small fruits (40–47 mm), and "Diana," an intermediate type, with large fruits (67–82 mm) (Schwarz et al., 2014)] were transplanted at the second true-leaf stage into 10 L plastic pots (one plant per pot) with bottom holes, filled with 8 L of a perlite [Agrilit 3, Perlite Italiana, Corsico (MI), Italy]: peat (Brill 3 Special, Brill Substrates, Georgsdorf, Germany) mixture (3:1, v:v).

Physical properties (total pore space—TPS, water-holding capacity—WC, air capacity—AC, available water—AW, easily available water—EAW, water-buffer capacity—WBC, and volumetric water content—VWC—at −10, −50, and −100 hPa) were determined according to the European Standard 13040 method (European Standard 13040, 1999) and are reported in **Table 1**. In brief, the material was equilibrated in water and then transferred into tubes made with two overlapping polyvinyl chloride rings (100 ± 1 mm diameter and 50 ± 1 mm height each). After filling, the double rings were saturated with water for 48 h and then transferred into a sandbox (Eijkelkamp Agrisearch Equipment, Giesbeck, The Netherlands) at −10 cm pressure head (−10 hPa) for 48 h. Thereafter, the double rings were removed from the sandbox and separated. The lower rings were weighted and dried at 105°C to constant mass. EAW and WBC were determined by increasing the values of suction pressure in the sandbox at −50 and −100 cm (−50 and −100 hPa, respectively).

The minimum (heating) temperature inside the greenhouse was set to 15°C (day) and 10°C (night), whereas above 20°C,

TABLE 1 | Physical properties of the growing medium used in the experiment (perlite:peat 3:1 v:v).

TPS	WC	AC	AW	EAW	WBC	VWC −10 hPa	VWC −50 hPa	VWC −100 hPa
% OF VOLUME								
94	52	42	13.4	11.1	2.3	52.0	40.9	38.6

TPS, total pore space; WC, water-holding capacity; AC, air capacity; AW, available water; EAW, easily-available water; WBC, water-buffer capacity; VWC, volumetric water content.

the greenhouse temperature was controlled by natural ventilation through the automatic ridge openings. Pots were placed 0.25 m apart at a density of 3.3 plants·m^{-2} on troughs measuring 6 m long and 26 cm wide with 6 cm high sides. The troughs were 1.30 m apart and were covered with polyethylene film and placed on a 2% slope. A 100-L reservoir tank containing NS was placed at the end of each trough [a layout of the subirrigated trough bench system used in the trial is reported by Santamaria et al. (2003)]. NS was recirculated (closed-cycle management) and never discharged at any time during the growing cycle.

The NS was supplied at the top end of each trough (with a flow rate of ≈2.0 L·min^{-1}). Excess NS was collected at the base of the troughs and reused for subsequent irrigation events. Reservoir tanks were refilled with fresh NS every 2 days, blending it with the recirculating NS still present in the container. Fertirrigation occurred when the substrate water potential reached the pre-fixed set points (see below); fertirrigation was automatically switched off when water potential increased to values higher than the set points.

The NS was prepared starting from well water with a pH of 7.6. Final macronutrient concentration of the NS was (Mm): N–NO_3 (10), K^+ (7.7), P–$H_2PO_4^-$ (1.9), Mg^{2+} (1.7), Ca^{2+} (3.7), S–SO_4^{2-} (3.4) resulting in an electrical conductivity (EC) of 2.0 dS·m^{-1}. Micronutrients were supplied according to Johnson et al. (1957). The pH of NS was adjusted to 5.5 when needed using 5 M H_3PO_4 or KOH.

During the first 2 weeks of the experiment, the fertirrigation schedule was controlled for all treatments by an operation timer; the timing varied from one to two fertirrigations (10 min each) per day.

The experimental treatments were differentiated 15 days after transplanting (DAT). Different irrigation set points based on substrate matric potential were compared: −30 and −60 hPa for "Diana," and −30, −60, and −90 hPa for "Kabiria."

Treatments were arranged in a randomized complete design with three replications. Each trough, containing 24 plants, represented an experimental unit. The first two plants at the upper and lower end of each trough were not taken into account for harvesting, sampling, and measurements. Two external rows served as guards.

Tensiometers (LT1 28 cm, Tensio-Technik, Geisenheim, Germany), placed with the porous cup at 7 cm from the bottom of the container (container eight, 26 cm), were used to measure the substrate water potential and to control irrigation. In each treatment, three tensiometers were installed (one per experimental unit/replication; Montesano et al., 2010). Tensiometers were connected to an electronic TensioSwitch (400C, Tensio-Technik) that controlled the beginning and the end of irrigation based on substrate water potential set point. A minimum duration of 3 min, corresponding to a supply of ∼200 mL of NS per pot per fertigation, was set.

Plants were trained vertically and topped at the eighth cluster (105 and 131 DAT, respectively, for "Kabiria" and "Diana"), and periodic binding, lateral-stem and basal-leaf pruning was carried out. Pollination was guaranteed by introducing a hive of bumblebees (*Bombus terrestris* L.) into the greenhouse beginning from the anthesis of the first cluster.

Harvests started on 100 DAT and ended on 148 and 156 DAT for "Kabiria" and "Diana," respectively. For "Kabiria," each harvest was carried out by picking the entire fruit cluster of a specific order when it had 80% red-ripened fruits on at least 80% of the plants at the experimental unit; for "Diana," single fruits were harvested at the "breaker" color stage.

Measurements

EC of the Substrate Aqueous Extract

At the end of the cycle, analysis of the growing media was conducted. Substrate was removed from the pot (one pot per experimental unit/replication) and the profile (18 cm from the bottom to the top) was divided horizontally into three equal layers (bottom, middle, and top). Approximately 100 mL of growing medium was sampled from the middle of each layer and an aqueous solution was extracted from the sample, using the 1:1.5 dilution method (Sonneveld et al., 1974). Substrate extract EC was measured using a handheld conductivity meter.

Growth Analysis

In order to assess the effects of treatments on plant growth, one plant from each experimental unit was sampled on 96 DAT. Plant height, total leaf area (Li-3100, Licor, NE, USA), total fresh and dry weight of leaves, stems, and fruits were determined. Plant material was dried to a constant weight in a forced draft oven at 65°C to determine dry weight. Specific Leaf Area (SLA) was calculated as the ratio between plant leaf area and total plant dry weight.

Yield and Fruit Quality Analysis

All harvested fruits were classified into marketable and unmarketable classes (i.e., those showing blossom-end rot or radial cracks on the epicarp). Representative samples of fruits harvested from the second and seventh cluster (105 and 140 DAT for "Kabiria" and 111 and 142 DAT for "Diana," respectively) were used to determine: (i) fruit total soluble solids (TSS), measured using a portable reflectometer (Brix-Stix BX 100 Hs; TechniQuip Corporation, Livermore, CA, USA) and expressed in °Brix at 20°C; (ii) fruit dry matter percentage (see above); (iii) glucose and fructose (only on the second cluster) concentrations were determined on fruits homogenized at 3000 rpm by ionic chromatography (Dionex model DX500; Dionex Corp., Sunnyvale, CA) with an amperometric detector using a Dionex CarboPac PA1 and isocratic elution with 50 mmol·L^{-1} NaOH (Caretto et al., 2008); K^+, Mg^{2+}, and Ca^{2+} concentrations (on fruits homogenized at 3000 rpm) were determined by ionic chromatography with a conductivity detector, using an IonPack CG12A pre-column and IonPack CS12A separation column (Di Gioia et al., 2013).

Plant–Water Relations

On 92 DAT, water (Ψ_i), osmotic (Ψ_o), and turgor (Ψ_t) *pre-dawn* potentials were measured; on 104 DAT, noon water potential on covered leaf (Ψ_x) was measured. For all parameters, measurements were replicated three times. Ψ_i and Ψ_x were measured using a Scholander–Hammel type pressure chamber (Scholander et al., 1965). Ψ_o was measured using a Roebling digital micro-osmometer (Gucci et al., 1991). Ψ_t was calculated as the difference between Ψ_i and Ψ_o.

Water-Use Efficiency

NS consumption was volumetrically measured by flow meters and recorded every 2 days, when the reservoir tanks were refilled. The WUE was calculated as the transformation efficiency of water through the cultivation system into yield, as the ratio of total fresh weight of fruits to the total volume of NS consumed by the system (Valenzano et al., 2008; De Pascale et al., 2011).

Statistical Analysis

All data were submitted to analysis of variance using the General Linear Model (GLM Proc; SAS 9.1; SAS Institute, Cary, NC, USA). For this purpose, all experimental factors were considered to be fixed.

According to the research objectives, means were compared using orthogonal contrasts with one degree of freedom (Steel and Torrie, 1988). Two polynomial contrasts were performed between the three levels of substrate water potential set points applied to "Kabiria" (linear and quadratic trend); three more contrasts were performed, depending on which variety or substrate water potential was studied: (1) "Diana" vs. "Kabiria"—on average, for −30 and −60 hPa; (2) −30 vs. −60 hPa; (3) the interaction ("Diana" vs. "Kabiria") × (−30 vs. −60 hPa).

RESULTS

Substrate Electrical Conductivity

Growing media of all treatments showed the salt stratification typical of subirrigated substrates. EC rose from the bottom to the top layer, with a higher increase in the trend when a −30-hPa matric potential was maintained in the growing media (**Figure 1**). The lower moisture content of substrates subjected to the lowest matric potential limited the upward salt flux, with a consequent higher EC in the bottom and middle layers of the substrate (3.2 and 3.6 vs. 2.6 and 3.1 dS/m, respectively, in the bottom and middle layers of −60 and −30 hPa substrates) and lower EC in the top layer (4.6 vs. 5.4, respectively, for −60 and −30 hPa substrates), resulting in a reduced EC gradient over the substrate profile.

Plant Growth and Yield

Water potential in the growing media affected tomato plant growth without significant interaction between the cultivar and water stress (**Table 2**). When comparing −30 with −60 hPa, a mean reduction of 18% for leaf area and 19% for SLA were observed, whereas plant height and total dry weight were unaffected. A further reduction in matric potential from −60 to −90 hPa in "Kabiria" had no effects on plant growth (**Table 2**). "Kabiria" showed, on average, higher plant height but lower leaf area and SLA than "Diana." For both cultivars, the lowering of the water potential in the growing media from −30 to −60 hPa led to a reduction in total yield (−10%, on average) and in mean fruit size (−13%, on average), with no interaction between cultivar and water stress; in the case of "Kabiria," where also a potential of −90 hPa was tested, the reduction of those two parameters followed a linear trend with the lowering of water availability in the growing media. "Diana" showed, on average, the highest total yield (3.1 vs. 2.7 kg·plant^{-1}, on average) and fruit mean weight (106 vs. 60 g) (**Table 2**).

Different water potential conditions in the growing media had opposite effects on the incidence of unmarketable yield for the two cultivars: while for "Diana" the unmarketable percentage rose from 5.7 to 11.6% of total yield, respectively, at −30 and −60 hPa, and for "Kabiria" the percentage decreased with the lowering of substrate water potential, following a linear trend (**Table 2**).

In both cultivars, the number of fruits per plant was not affected by water stress, with a mean value of 30 and 44, respectively, for "Diana" and "Kabiria" (data not reported).

Fruit Quality

In the second cluster, for both cultivars, the lowering of growing media water potential improved fruit quality in terms of TSS and

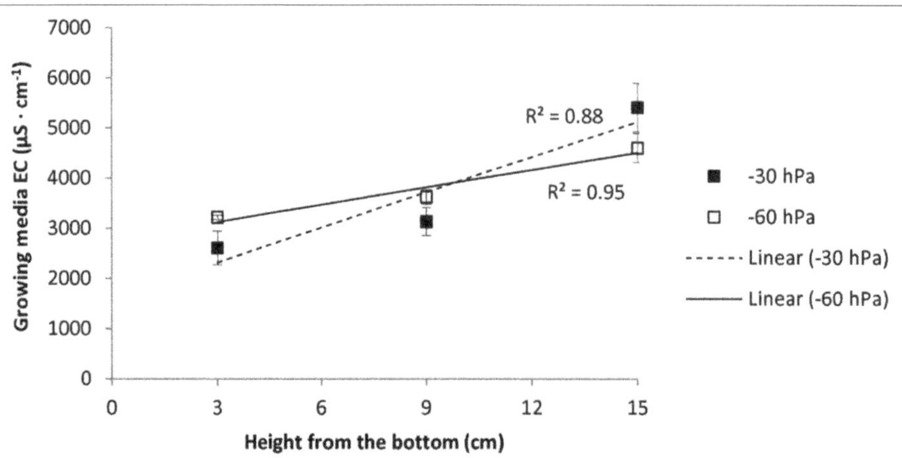

FIGURE 1 | Electrical conductivity (EC, 1:1.5 dilution method) at three layers of the vertical profile in subirrigated soilless substrates with constant −30 or −60 hPa matric potential irrigation set-point (vertical bars indicate ±S.E. of means).

TABLE 2 | Effect of cultivar (Diana—D and Kabiria—K) and growing-media matric potential (P) on plant height, leaf area, specific leaf area (SLA), total dry weight, total and unmarketable yield, and mean fruit weight of subirrigated soilless tomato under controlled water stress conditions.

Cultivar-Water potential (hPa)	Plant height (cm)	Leaf area (cm^2)	SLA ($cm^2.g^{-1}$)	Plant dry weight (g·$plant^{-1}$)	Total yield (g·$plant^{-1}$)	Unmarketable yield (g·$plant^{-1}$)	Mean fruit weight (g)
D-P30	164	10,989	67.1	164	3293	187	112
D-P60	161	8811	49.9	180	2997	349	99
K-P30	204	7271	41.9	174	2851	145	64
K-P60	204	6078	36.6	166	2532	108	55
K-P90	199	6137	38.0	163	2050	82	49
Significance of contrasts[a]							
30 vs. 60 hPa	ns	*	*	ns	**	**	**
D vs. K	***	**	**	ns	**	***	***
(D vs. K)*(30 vs. 60 hPa)	ns	ns	ns	ns	ns	***	ns
P (K) linear	ns	ns	ns	ns	***	*	**
P (K) quadratic	ns	ns	ns	ns	ns	ns	ns

[a]*F significance: ns, ***, **, and *, respectively, non-significant, $P \leq 0.001$, $P \leq 0.01$, and $P \leq 0.05$; fw, fresh weight.*

dry matter, with a significant increase in those two parameters (5 and 12.5% on average, respectively) with no interaction between cultivar and water stress (**Table 3**). In the case of "Kabiria," both TSS and dry matter increased linearly with decreasing growing media water potential. Both glucose and fructose concentrations increased linearly in "Kabiria" fruits with decreasing substrate matric potential, whereas the concentration was not affected by water stress in "Diana." The inorganic ion concentration in the fruit (K^+, Mg^{2+}, and Ca^{2+}) was not significantly affected by the matric potential irrigation set point, although a certain trend of increase in K^+ was observed (**Table 3**).

In the seventh cluster, the two cultivars reacted differently to water stress application in terms of TSS and dry matter. In "Diana" only a slight increase was observed from −30 to −60 hPa (3.3 and 1.3%, respectively, for TSS and dry matter), whereas in "Kabiria," the increase was more pronounced (8.7 and 12.0%, respectively, for TSS and dry matter), and further reduction in matric potential from −60 to −90 hPa in "Kabiria" confirmed the linear increase for both the parameters. The fruit concentration of Mg^{2+} tended to slightly increase in both cultivars under water stress conditions (**Table 3**).

Plant–Water Relations

No interaction was observed between cultivar and water stress with regard to plant–water relations. Regardless of the cultivar, water stress decreased Ψ_i and Ψ_x, with more remarkable differences in measurements taken at noon because of the greater influence of environmental conditions (**Table 4**). In the case of "Kabiria," Ψ_i and Ψ_x decreased following a linear trend with lowering of growing media water potential from −30 to −90 hPa. Plants of both cultivars showed an osmotic adjustment with a 9% decrease, on average, in Ψ_o-values from −30 to −60 hPa. No differences in Ψ_t were observed in plants (**Table 4**).

Water Consumption and Water-Use Efficiency

Water consumption was similar for both cultivars and decreased with decreasing growing media matric potential (104 and 89.5 L·$plant^{-1}$, on average, at −30 and −60 hPa, respectively—**Table 5**). The water consumption reduction in "Kabiria" followed a linear trend with the lowering of the growing media water potential (**Table 5**). WUE was unaffected by water potential but was 16% higher in "Diana."

DISCUSSION

In this study, we combined the tools for efficient water use and satisfactory crop performance in soilless greenhouse tomato cultivation. With this aim, a closed-cycle through bench subirrigation technique and on-demand sensor-based irrigation was used.

Subirrigation is an easy-management closed-cycle technique, which is suitable to be used also in low-tech greenhouse industry (Bouchaaba et al., 2015). As a closed-cycle system, the technique itself is an effective way of reducing water usage and nutrient runoff. However, proper irrigation management is required for the success of such a technique, and identification of proper irrigation set points is crucial for optimal crop performance, both in terms of yield and quality, and optimal use of water resources. In the present study, we demonstrated that sensor-based irrigation management can help implement a closed-cycle cultivation of soilless tomato using subirrigation. In addition, precise control of substrate water status may offer the possibility to steer crop response by enhancing two different components of crop performance, namely yield and fruit quality.

Maintaining the optimal water status of soilless growing media has been recognized as being critical, both because of its limited water-holding capacity and the difficulty in judging accurately when the plants require water (Fonteno et al., 1981; Karlovich and Fonteno, 1986). Substrate matric potential measurement and control through a tensiometer can be used with this aim, with the main advantage of directly determining whether water in the soil is available to the plants (van Iersel et al., 2013). In the present study, we used a tensiometer to completely automate the irrigation management of the crop and to impose controlled water stress corresponding to the different irrigation

TABLE 3 | Effect of cultivar (Diana—D and Kabiria—K) and growing-media matric potential (P) on total soluble solids (TSS), fruit dry matter, glucose, fructose, K, Mg, and Ca fruit concentration in the second and seventh cluster of subirrigated soilless tomato under controlled water stress conditions.

Cultivar-Water potential (hPa)	TSS (°Brix)	Dry matter (g·100 g^{-1} fw)	Glucose (mg·100 mL^{-1} fruit juice)	Fructose (mg·100 mL^{-1} fruit juice)	K^+ (mg·kg^{-1} fw)	Mg^{2+} (mg·kg^{-1} fw)	Ca^{2+} (mg·kg^{-1} fw)
SECOND CLUSTER							
D-P30	5.0	5.8	0.80	1.57	1823	44	52
D-P60	5.4	6.7	0.80	1.50	2098	49	45
K-P30	6.7	7.4	1.10	1.03	1387	35	46
K-P60	6.9	8.1	1.20	1.40	1426	31	31
K-P90	7.2	8.7	1.40	1.65	1681	35	31
Significance of contrasts[a]							
30 vs. 60 hPa	*	**	ns	ns	ns	ns	ns
D vs. K	***	***	***	**	*	*	ns
(D vs. K)*(30 vs. 60 hPa)	ns	ns	ns	*	ns	ns	ns
P (K) linear	*	**	**	***	ns	ns	ns
P (K) quadratic	ns	ns	ns	ns	ns	ns	ns
SEVENTH CLUSTER							
D-P30	6.1	7.8	–	–	2589	58	58
D-P60	6.3	7.9	–	–	2463	60	51
K-P30	6.9	8.3	–	–	2321	47	39
K-P60	7.5	9.3	–	–	2700	56	41
K-P90	8.3	10.5	–	–	2772	56	42
Significance of contrasts[a]			–	–			
30 vs. 60 hPa	*	**	–	–	ns	*	ns
D vs. K	***	***	–	–	ns	**	***
(D vs. K)*(30 vs. 60 hPa)	*	*	–	–	ns	ns	ns
P (K) linear	***	***	–	–	ns	ns	ns
P (K) quadratic	ns	ns	–	–	ns	ns	ns

[a]*F significance: ns, ***, **, and *, respectively, non-significant, $P \leq 0.001$, $P \leq 0.01$ and $P \leq 0.05$; fw, fresh weight.*

set points based on substrate matric potential. Although a generally accepted theory establishes the general matric potential value limits for EAW in soilless substrates, thus providing guidance for determining optimal irrigation set points, there are only a few specific studies that relate the cultivation technique and the response of the different crops and varieties to different matric potential values. The results of the present research outline the general negative effects of matric potential decrease on growth and total yield of subirrigated tomato, showing that an irrigation set point exceeding the conventional limit of −50 hPa leads to a reduction in growth and total yield (**Table 2**).

The differences in height, leaf area, and SLA between the two tested varieties are due to their different characters: "Diana" being characterized by lower plant height and higher leaf area than "Kabiria." Although "Diana" presented higher leaf area than "Kabiria," water use for the two cultivars was similar (**Table 5**). Our data seem to confirm that leaf area value is not an exhaustive parameter to explain differences in water use by plants of different cultivars, but it should be related to the plant's character. In a previous study, Santamaria et al. (2004) found that "Diana" and "Naomi," a cherry tomato variety, presented similar leaf area, but "Diana" consumed 8% less water than "Naomi." In the present study, we found the water consumption per unit of leaf area of 9.8 and 14.3 mL·cm^{-2}, respectively, for "Diana" and "Kabiria." The higher leaf area distributed in a shorter plant height may have resulted in higher shading of "Diana" leaves compared with "Kabiria," affecting the water consumption rate of the two varieties.

Although for most of the tested parameters the interaction between cultivar and water stress level was not significant, according to our findings, the specific response of the cultivar to controlled water stress, in terms of important crop-performance indicators, should be taken into account. In fact, while the unmarketable yield in "Diana" increased when water stress was imposed (**Table 2**) mainly due to blossom-end rot, an opposite effect was observed in "Kabiria," where marketable yield loss linearly decreased by 1.05 g·plant^{-1} per unit of substrate water potential (in the tested range from −30 to −90 hPa) mainly due to a reduced occurrence of fruit cracking, the latter variety being characterized by small-sized fruits and probably higher resistance to water stress. The reduced tolerance to water stress in "Diana," in terms of incidence of unmarketable fraction of the yield, confirmed the higher susceptibility to water stress of cultivars with large or intermediate fruits. Conversely, "Kabiria" showed higher adaptation to water stress conditions, typical of small-sized fruit genotypes (Santamaria et al., 2004; Serio et al., 2007).

Previous studies on drip-irrigated tomatoes cultivated in soilless peat–perlite mix-based (70:30, v:v) substrates showed a 6% yield reduction when using a −100-hPa compared to a −50-hPa irrigation trigger set-point (Xu et al., 1995), whereas no differences were observed when comparing −45 and −65 hPa matric potentials (Norrie et al., 1995). Sarkar et al. (2008a) found a progressive decrease in yield and growth parameters in tomato plants cultivated in an open-cycle drip-irrigated tomato on coir dust substrate. Beside the important effects on crop performance related to the adoption of proper matric potential irrigation set points, irrigation management using tensiometers reduced leaching and the amount of greenhouse effluents in a soilless tomato cultivation by 262 kg N·ha^{-1} when compared with the use of timer-based irrigation adjusted to solar radiation (Lemay et al., 2012).

The tensiometer has been previously used for the automated irrigation management of subirrigated containerized vegetables in some studies (Montesano et al., 2010; Bouchaaba et al., 2015), although only in rare cases specific comparisons between different water potential set points have been reported (Montesano et al., 2005a,b). In those studies, when comparing an irrigation set point of −40 hPa relative to −80 hPa, subirrigated tomato "Kabiria" grown on a perlite–peat mix similar to that used in the present research showed a reduction of 26% in yield and 16% in mean fruit weight, an increased dry matter percentage and TSS of fruits with the lowest matric potential, whereas WUE was not affected. Morever, in the present research, WUE was not affected by the matric potential control. In fact, both terms of WUE ratio (yield and water use) were similarly reduced when water stress (−60 hPa) was applied, without significant differences among varieties (**Table 5**).

Wang et al. (2007) claimed an increased tomato plant WUE and irrigation system WUE as soil matric potential decreased, with reduced irrigation amounts and no significant differences in tomato yield; however, the study was conducted on soil and a wide range of matric potential was tested (−10 to −50 kPa). WUE can be improved by modifying both terms of the ratio (De Pascale et al., 2011), but according to our findings, acting on matric potential control to impose controlled water stress is not a suitable approach for improving this parameter in subirrigated soilless tomato, thereby confirming that in soilless conditions, due to the low water-holding capacity of growing media, even slight decreases in soil matric potential could lead to severe effects on plant physiology and crop performance.

Tomato fruit quality, which is controlled by the interaction of genetic, environmental, and cultural factors, has become a major commercial concern in the highly competitive fresh fruit market, with increasing attention to setting up growing conditions and specific interventions capable of enhancing quality. In particular, there is an ever-increasing demand for high soluble solid content (Sarkar et al., 2008b). In the present study, a significant effect of controlled water stress on plants subjected to low water potentials lies in the improvement in fruit quality traits of commercial importance, namely TSS and dry matter content (**Table 3**). The two tested cultivars reacted differently in terms of benefit of water stress on fruit quality, as confirmed in particular by the quality traits analysis performed on the seventh cluster, where only slight quality improvements were observed in "Diana" (**Table 3**). A significant increase in TSS and fruit dry matter content was always observed in "Kabiria." In this cultivar, as a result of matric potential decrease from −30 to −60 hPa, the fructose/glucose ratio increased (from 0.94 to 1.17, respectively), whereas further decrease in substrate water potential (−90 hPa) did not result in further increases of this parameter. However, considering the sweetness index [sweetness index (SI) calculated by multiplying the sweetness coefficient for each individual sugar (glucose = 1, fructose = 2.3) as described by Keutgen and Pawelzik (2007)], we found that SI increased by 26 and 49%, respectively, at −60 and −90 hPa, when compared to −30 hPa. This indicates that the

TABLE 4 | Effect of cultivar (Diana—D and Kabiria—K) and growing-media matric potential (P) on water (Ψ_i), osmotic (Ψ_o), and turgor (Ψ_t) pre-dawn potential and noon water potential on covered leaf (Ψ_x) of subirrigated soilless tomato under controlled water stress conditions.

Cultivar-Water potential (hPa)	Ψ_i	Ψ_o	Ψ_t	Ψ_x
		(MPa)		
D-P30	−0.263	−0.953	0.690	−0.280
D-P60	−0.353	−1.023	0.669	−0.483
K-P30	−0.247	−0.922	0.675	−0.313
K-P60	−0.327	−1.021	0.694	−0.457
K-P90	−0.510	−1.093	0.583	−0.733
Significance of contrasts[a]				
30 vs. 60 hPa	*	**	ns	**
D vs. K	ns	ns	ns	ns
(D vs. K)*(30 vs. 60 hPa)	ns	ns	ns	ns
T (K) linear	**	**	ns	***
T (K) quadratic	ns	ns	ns	ns

[a]*F significance: ns, ***, **, and *, respectively, non-significant, $P \leq 0.001$, $P \leq 0.01$ and $P \leq 0.05$.*

TABLE 5 | Effect of cultivar (Diana—D and Kabiria—K) and growing-media matric potential (P) on total water consumption and water-use efficiency (WUE) of subirrigated soilless tomato under controlled water stress conditions.

Cultivar-Water potential (hPa)	Water consumption (L·plant^{-1})	WUE (g·L^{-1})
D-P30	107	31
D-P60	89	34
K-P30	101	28
K-P60	90	28
K-P90	73	28
Significance of contrasts[a]		
30 vs. 60 hPa	***	ns
D vs. K	Ns	**
(D vs. K)*(30 vs. 60 hPa)	Ns	ns
P (K) linear	***	ns
P (K) quadratic	Ns	ns

[a]*F significance: ns, ***, and **, respectively, non-significant, $P \leq 0.001$, and $P \leq 0.01$.*

effect of controlled water stress on tomato quality amelioration may involve effects on the ratio between the two principle reducing sugars present in fruits. Moreover, the higher benefit for small or medium-sized fruit tomato types seems to be confirmed in terms of quality amelioration as a response to controlled water stress (Elia et al., 2001; Di Gioia et al., 2013). For large fruit varieties, the commercial maturity of fruits corresponds to the "breaker" color stage, whereas smaller varieties are generally harvested when fruits are red-ripened. The fact that fruits of small-sized varieties are allowed to ripen on the plant before harvest may help maximize the effects of controlled water stress in terms of TSS and dry matter increase.

A common practice for fruit quality improvement in tomato is to apply controlled stress to plants by using saline water, high concentrated NS, or slight water stress in order to obtain fruits with higher dry matter and TSS concentration, although such practices generally lead to yield loss (Dorais et al., 2000; Guichard et al., 2001). In subirrigation conditions, since no leaching occurs in containers, salt accumulation, in particular at the substrate surface, is a major drawback (Elia et al., 2003; Santamaria et al., 2003; Rouphael and Colla, 2005). Therefore, many authors suggest reducing fertilizer NS concentration in subirrigation as compared to traditional drip irrigation systems (Cox, 2001; Mak and Yeh, 2001; Yeh et al., 2004; Montesano et al., 2010). However, an important effect of matric potential irrigation set point on substrate EC and salt stratification was found in the present study. In particular, lower water potential conditions impaired the typical stratification of unabsorbed salts in the substrate, leading to higher EC levels at the bottom and middle of the substrate profile (**Figure 1**). The accumulation of salts in the top surface layer of the substrate, and the consequent lower accumulation at the bottom where most of the roots are present, has often been reported as a natural counterbalance to the absence of leaching that occurs in subirrigation (Morvant et al., 1997; Bouchaaba et al., 2015). Therefore, it is important to take into account the possible effects of irrigation management conditions on the occurrence of such a phenomenon, in particular when saline water or unbalanced fertilizer solutions are used in subirrigation, with potential implications on salt stress for plants. Given such considerations, the common practice of using highly concentrated NS, as well as using NS containing NaCl, to obtain high-quality tomato fruits does not seem to lack possible negative implications when subirrigation techniques are adopted (Incrocci et al., 2006). However, according to the present study findings, it is feasible to act on matric potential irrigation set points to control the response of plants in terms of fruit quality parameters.

At the physiological level, as a consequence of the moisture decrease in the root zone, plants of both cultivars showed an osmotic adjustment (lowering of leaf osmotic potential—**Table 4**). Osmotic adjustment is an adaptive physiological mechanism involved in stress tolerance to drought and salinity, which permits the maintenance of turgor and cell functions under conditions of water deficit. It is reported that under salinity stress, this process is achieved mainly by uptake and accumulation of inorganic ions, whereas under drought stress, it is achieved by synthesis and accumulation of organic compatible solutes. However, it has been reported that drought tolerance in tomato varieties is not always correlated to tissue osmotic potentials, implying that osmotic adjustment is not the only process influencing tolerance (Alian et al., 2000).

A more pronounced tendency to overcome water stress by osmotic adjustment mechanisms was observed in "Kabiria," which showed a significant increase in glucose ($5\,\mu g \cdot 100\,mL^{-1}$ fruit juice per unit of substrate water potential in the tested range from -30 to -90 hPa) and fructose ($10.3\,\mu g \cdot 100\,mL^{-1}$ of fruit juice per unit of substrate water potential in the tested range) fruit concentrations under water stress conditions, whereas in "Diana," no significant changes in the concentration of those two sugars were observed as a result of water stress (**Table 3**). The role of accumulation of solutes in the cell (mostly compatible solutes, sugars or amino-acids in the cytoplasm, and inorganic solutes in the vacuole) for the osmotic adjustment has been demonstrated (Kramer and Boyer, 1995). However, while Veit-Kohler et al. (1999) claimed that a reduction in water supply led to an increase in sugars and titratable acids in tomato fruits, Nahar and Gretzmacher (2002) found similar responses only under extreme water deficit conditions, and Plaut et al. (2004) concluded that solutes contributing to the osmotic adjustment of fruits and leaves were only slightly affected by water stress.

CONCLUSIONS

Substrate matric potential control using a tensiometer can be used for automatic irrigation in soilless subirrigation systems. An irrigation set point exceeding the conventional limit of easily available water for soilless substrates leads to reduction of growth and total yield of subirrigated soilless tomato. Precise control of substrate water status may offer the possibility to steer crop response by enhancing different crop-performance components, namely yield and fruit quality. Small-sized fruit varieties benefit more from controlled water stress in terms of reduced unmarketable yield loss and fruit quality traits improvement.

REFERENCES

Alian, A., Altman, A., and Heuer, B. (2000). Genotypic difference in salinity and water stress tolerance of fresh market tomato cultivars. *Plant Sci.* 152, 59–65. doi: 10.1016/S0168-9452(99)00220-4

Argo, W. (1998). Root medium physical properties. *HortTechnology* 8, 481–485.

Bouchaaba, Z., Santamaria, P., Choukr-Allah, R., Lamaddalena, N., and Montesano, F. F. (2015). Open-cycle drip vs closed-cycle subirrigation: effects on growth and yield of greenhouse soilless green bean. *Sci. Hortic.* 182, 77–85. doi: 10.1016/j.scienta.2014.11.007

Caretto, S., Parente, A., Serio, F., and Santamaria, P. (2008). Influence of potassium and genotype on vitamin E content and reducing sugar of tomato fruits. *Hortscience* 43, 2048–2051.

Cox, D. A. (2001). Growth, nutrient content and growth medium electrical conductivity of poinsettia irrigated by subirrigation or from overhead. *J. Plant Nutr.* 24, 523–533. doi: 10.1081/PLN-100104977 de Boodt, M., and Verdonck, O. (1972). The physical properties of substrates in horticulture. *Acta Hort.* 26, 37–44. doi: 10.17660/ActaHortic.1972.26.5

De Pascale, S., Dalla Costa, L., Vallone, S., Barbieri, G., and Maggio, A. (2011). Increasing water use efficiency in vegetable crop production: from plant to irrigation system efficiency. *HortTechnology* 21, 301–308.

Di Gioia, F., Signore, A., Serio, F., and Santamaria, P. (2013). Grafting improves tomato salinity tolerance through sodium partitioning within the shoot. *Hortscience* 48, 855–862.

Dorais, M., Papadopoulos, A. P., Tucrotte, G., Hao, X., Ehret, D. L., and Gosselin, A. (2000). Control of tomato fruit quality and flavour by EC and water management. *Greenhouse Proc. Crops Res. Cent. Annu. Rep.* 18–21.

Dukes, M. D., Zotarelli, L., and Morgan, K. T. (2010). Use of irrigation technologies for vegetable crops in Florida. *HortTechnology* 20, 133–142.

Elia, A., Parente, A., Serio, F., Santamaria, P., and Ruiz Rodriguez, G. (2001). Electrical conductivity of nutrient solution, plant growth and fruit quality of soilless grown tomato. *Acta Hort.* 559, 503–508. doi: 10.17660/ActaHortic.2001.559.73

Elia, A., Santamaria, P., Parente, A., and Serio, F. (2003). Some aspects of the trough bench system and its performance in cherry tomato production. *Acta Hort.* 614, 161–166. doi: 10.17660/ActaHortic.2003.614.22

European Standard 13040 (1999). *Soil Improvers and Growing Media. Sample Preparation for Chemical and Physical Tests, Determination of Dry Matter Content, Moisture Content and Laboratory Compacted Bulk Density.* Brussels: European Committee for Standardization.

FAO (2013). Good Agricultural Practices for Greenhouse Vegetable Crops – Principles for Mediterranean Climate Areas. Rome: FAO.

Ferrarezi, R. S., van Iersel, M. W., and Testezlaf, R. (2015b). Monitoring and controlling Ebb-and-flow subirrigation with soil moisture sensors. *Hortscience* 50, 447–453.

Ferrarezi, R. S., Weaver, G. M., van Iersel, M. W., and Testezlaf, R. (2015a). Subirrigation: historical overview, challenges, and future prospects. *HortTechnology* 25, 262–276.

Fonteno, W. C., Cassel, D. K., and Larson, R. A. (1981). Physical properties of three container media and their effect on poinsettia growth. *J. Amer. Soc. Hort. Sci.* 106, 736–741.

Gucci, R., Xiloyannis, C., and Flore, J. A. (1991). Gas exchange parameters, water relations and carbohidrate partitioning in leaves of field-grown *Prunus domestica* following fruit removal. *Physiol. Plant.* 83, 497–505. doi: 10.1111/j.1399-3054.1991.tb00126.x

Guichard, S., Bertin, N., Leonardi, C., and Gary, C. (2001). Tomato fruit quality in relation to water and carbon fluxes. *Agronomie* 21, 385–392. doi: 10.1051/agro:2001131

Incrocci, L., Malorgio, F., Della Bartola, A., and Pardossi, A. (2006). The influence of drip irrigation or subirrigation on tomato grown in closed-loop substrate culture with saline water. *Sci. Hortic.* 107, 365–372. doi: 10.1016/j.scienta.2005.12.001

Johnson, C. M., Stout, P. R., Broyer, T. C., and Carlton, A. B. (1957). Comparative chlorine requirements of different plant species. *Plant Soil* 8, 337–353. doi: 10.1007/BF01666323

Jones, H. G. (2007). Monitoring plant and soil water status: established and novel methods revisited and their relevance to studies of drought tolerance. *J. Exp. Bot.* 58, 119–130. doi: 10.1093/jxb/erl118

Karlovich, P. T., and Fonteno, W. C. (1986). The effect of soil moisture tension and volume moisture on the growth of Chrysanthemum in three container media. *J. Amer. Soc. Hort. Sci.* 111, 191–195.

Keutgen, A., and Pawelzik, E. (2007). Modifications of taste-relevant compounds in strawberry fruit under NaCl salinity. *Food Chem.* 105, 1487–1494. doi: 10.1016/j.foodchem.2007.05.033

Kramer, P. J., and Boyer, J. S. (1995). *Water Relations of Plants.* San Diego, CA: Academic Press.

Lea-Cox, J. D., and Arguedas-Rodriguez, F. R. (2011). Relating real-time substrate matric potential measurements to plant water use for precision irrigation. *Acta Hort.* 891, 201–207. doi: 10.17660/ActaHortic.2011.891.23

Lemay, I., Caron, J., Dorais, M., and Pepin, S. (2012). Defining irrigation set points based on substrate properties for variable irrigation and constant matric potential devices in greenhouse tomato. *Hortscience* 47, 1141–1152.

Lichtenberg, E., Majsztrik, J., and Saavoss, M. (2013). Profitability of sensor-based irrigation in greenhouse and nursery crops. *HortTechnology* 23, 735–746.

Mak, A. T. Y., and Yeh, D. M. (2001). Nitrogen nutrition of Spathiphyllum 'Sensation' grown in sphagnum peat- and coir-based media with two irrigation methods. *Hortscience* 36, 645–649.

Montesano, F., Favuzzi, G., Parente, A., Serio, F., and Santamaria, P. (2005a). NFT vs subirrigation I: yield and water use efficiency of tomato grown in closed soilless systems under salinity or water stress. *Hortscience* 40, 1011.

Montesano, F., Ferulli, C., Parente, A., Serio, F., and Santamaria, P. (2005b). NFT vs subirrigation II: tomato fruit quality grown in closed soilless systems under salinity or water stress. *Hortscience* 40, 1011.

Montesano, F. F., Parente, A., Lamaddalena, N., Todorovic, M., and Trotta, L. (eds.). (2015). "Introduction," in *Modern Technologies, Strategies and Tools for Sustainable Irrigation Management and Governance in Mediterranean Agriculture (IrriMed 2015), Proceedings – Book of Abstracts* (Valenzano, IT: CIHEAM), 19–20. Available online at: http://www.irrigation-management.eu/images/Events/99IrriMed/IrriMED2015Abs.pdf

Montesano, F., Parente, A., and Santamaria, P. (2010). Closed cycle subirrigation with low concentration nutrient solution can be used for soilless tomato production in saline conditions. *Sci. Hortic.* 124, 338–344. doi: 10.1016/j.scienta.2010.01.017

Morvant, J. K., Dole, J. M., and Allen, E. (1997). Irrigation systems alter distribution of roots, soluble salts, nitrogen, and pH in the root medium. *HortTechnology* 7, 156–160.

Nahar, K., and Gretzmacher, R. (2002). Effect of water stress on nutrient uptake, yield and quality of tomato (*Lycopersicum esculentum* Mill.) under subtropical conditions. *Bodenkultur* 53, 45–51.

Nemali, K. S., Montesano, F., Dove, S. K., and van Iersel, M. W. (2007). Calibration and performance of moisture sensors in soilless substrates: ECH2O and theta probes. *Sci. Hortic.* 111, 227–334. doi: 10.1016/j.scienta.2006.12.013

Norrie, J., Charbonneau, J., Gosselin, A., and Graham, M. E. D. (1995). Impact of irrigation management of greenhouse tomato: yield, nutrition, and salinity of peat substrate. *Canadian J. Plant Sci.* 75, 497–503. doi: 10.4141/cjps95-087

Pardossi, A., Incrocci, L., Incrocci, G., Malorgio, F., Battista, P., Bacci, L., et al. (2009). Root zone sensors for irrigation management in intensive agriculture. *Sensors* 9, 2809–2835. doi: 10.3390/s90402809

Plaut, Z., Grava, A., Yehezkel, C., and Matan, E. (2004). How do salinity and water stress affect transport of water, assimilates and ions to tomato fruits? *Physiol. Plant.* 122, 429–442. doi: 10.1111/j.1399-3054.2004.00416.x

Rouphael, Y., Cardarelli, M., Rea, E., Battistelli, A., and Colla, G. (2006). Comparison of the subirrigation and drip-irrigation systems for greenhouse zucchini squash production using saline and non-saline nutrient solutions. *Agric. Water Manage.* 82, 99–117. doi: 10.1016/j.agwat.2005.07.018

Rouphael, Y., and Colla, G. (2005). Growth, yield, fruit quality and nutrient uptake of hydroponically cultivated zucchini squash as affected by irrigation systems and growing seasons. *Sci. Hortic.* 105, 177–195. doi: 10.1016/j.scienta.2005.01.025

Santamaria, P., Campanile, G., Parente, A., and Elia, A. (2003). Subirrigation vs drip-irrigation: effects on yield and quality of soilless grown cherry tomato. *J. Hort. Sci. Biotechnol.* 78, 290–296. doi: 10.1080/14620316.2003.11511620

Santamaria, P., Cantore, V., Conversa, G., and Serio, F. (2004). Effect of night salinity level on water use, physiological responses, yield, and quality of tomato. *J. Hort. Sci. Biotechnol.* 79, 59–66. doi: 10.1080/14620316.2004.11511737

Sarkar, S., Kiriiwa, Y., Endo, M., Kobayashi, T., and Nukaya, A. (2008b). Effect of fertigation management and the composition of nutrient solution on the yield and quality of high soluble solid content tomatoes. *J. Japan. Soc. Hort. Sci.* 77, 143–149. doi: 10.2503/jjshs1.77.143

Sarkar, S., Kiriiwa, Y., Endo, M., Uchino, S., and Nukaya, A. (2008a). Possibility of high soluble solid content tomato production under water stress conditions controlled by matric potential. *J. Japan. Soc. Hort. Sci.* 77, 251–258. doi: 10.2503/jjshs1.77.251

Savvas, D., Gianquinto, G., Tuzel, Y., and Gruda, N. (2013). "Soilless culture," in *Good Agricultural Practices for Greenhouse Vegetable Crops – Principles for Mediterranean Climate Areas,* eds W. Baudoin, R. Nono-Womdim, N. Lutaladio, A. Hodder, N. Castilla, C. Leonardi, S. De Pascale, and M. Qaryouti (Rome, IT: FAO), 303–354.

Scholander, P. F., Hammel, H. T., Bradstreet, E. D., and Hemmingsen, E. A. (1965). Sap pressure in vascular plants. *Science* 148, 339–346. doi: 10.1126/science.148.3668.339

Schwarz, D., Thompson, A. J., and Kläring, H. P. (2014). Guidelines to use tomato in experiments with a controlled environment. *Front. Plant Sci.* 5:625. doi: 10.3389/fpls.2014.00625

Serio, F., Parente, A., Leo, L., and Santamaria, P. (2007). Potassium nutrition increases the lycopene content of tomato fruit. *J. Hort. Sci. Biotechnol.* 82, 941–945. doi: 10.1080/14620316.2007.11512330

Sonneveld, C., van den Ende, J., and van Dijk, P. A. (1974). Analysis of growing media by means of 1:1.5 volume extract. *Commun. Soil Sci. Plan.* 5, 183–202. doi: 10.1080/00103627409366497

Sonneveld, C., and Voogt, W. (2009). *Plant Nutrition of Greenhouse Crops.* New York, NY: Springer.

Stanghellini, C. (2014). Horticultural production in greenhouses: efficient use of water. *Acta Hort.* 1034, 25–32. doi: 10.17660/ActaHortic.2014.1034.1

Steel, R. G., and Torrie, J. H. (1988). *Principles and Procedures of Statistics, 2nd Edn.* New York, NY, McGraw-Hill Co.

Thalheimer, M. (2003). Tensiometer modification for diminishing errors due to the fluctuating inner water column. *Soil Sci. Soc. Am. J.* 67, 737–739. doi: 10.2136/sssaj2003.0737

Torrellas, M., Antón, A., Ruijs, M., García Victoria, N., Stanghellini, C., and Montero, J. I. (2012). Environmental and economic assessment of protected crops in four European scenarios. *J. Cleaner Prod.* 28, 45–55. doi: 10.1016/j.jclepro.2011.11.012

Valenzano, V., Parente, A., Serio, F., and Santamaria, P. (2008). Effect of growing systems and cultivar on yield and water use efficiency of greenhouse-grown tomato. *J. Hort. Sci. Biotechnol.* 83, 71–75. doi: 10.1080/14620316.2008.11512349

van Iersel, M. W. (2015). "Modern strategies for irrigation management in protected environments," in *Modern Technologies, Strategies and Tools for Sustainable Irrigation Management and Governance in Mediterranean Agriculture (IrriMed 2015), Proceedings - Book of Abstracts*, eds F. F. Montesano, A. Parente, N. Lamaddalena, M. Todorovic, and L. Trotta (Valenzano, IT: CIHEAM), 127–130. Available online at: http://www.irrigation-management.eu/images/Events/99IrriMed/IrriMED2015Abs.pdf

van Iersel, M. W., Chappell, M., and Lea-Cox, J. D. (2013). Sensors for improved efficiency of irrigation in greenhouse and nursery production. *HortTechnology* 23, 735–746.

Veit-Kohler, U., Krumbein, A., and Kosegarten, H. (1999). Effect of different water supply on plant growth and fruit quality of *Lycopersicum esculentum. J Plant Nutr. Soil Sci.* 162, 583–588.

Vox, G., Teitel, M., Pardossi, A., Minuto, A., Tinivella, F., and Schettini, E. (2010). "Agriculture: technology, planning and management," in *Sustainable Greenhouse Systems*, eds A. Salazar, and I. Rios (New York, NY: Nova Science Publishers), 1–79.

Wang, D., Kang, Y., and Wan., S. (2007). Effect of soil matric potential on tomato yield and water use under drip irrigation condition. *Agric. Water Man.* 87, 180–186. doi: 10.1016/j.agwat.2006.06.021

Whalley, W. R., Ober, E. S., and Jenkins, M. (2013). Measurement of the matric potential of soil waterin the rhizosphere. *J. Exp. Bot.* 64, 3951–3396. doi: 10.1093/jxb/ert044

Xu, H. L., Gauthier, L., and Gosselin, A. (1995). Effects of fertigation management on growth and photosynthesis of tomato plants grown in peat, rockwool and NFT. *Sci. Hortic.* 63, 11–20. doi: 10.1016/0304-4238(95)00791-Q

Yeh, D. M., Hsu, P. H., and Atherton, J. G. (2004). Growth and flowering response of Canna × generalis to nitrogen supplied to the growing medium via top- or sub- irrigation. *J. Hort. Sci. Biotechnol.* 79, 511–514. doi: 10.1080/14620316.2004.11511797

Conflict of Interest Statement: The authors declare that the research was conducted in the absence of any commercial or financial relationships that could be construed as a potential conflict of interest.

8

Genome-Wide Analysis and Heavy Metal-Induced Expression Profiling of the HMA Gene Family in *Populus trichocarpa*

*Dandan Li[1], Xuemei Xu[2], Xiaoqing Hu[1], Quangang Liu[1], Zhanchao Wang[1], Haizhen Zhang[1], Han Wang[1], Ming Wei[1], Hanzeng Wang[1], Haimei Liu[1] and Chenghao Li[1]**

[1] State Key Laboratory of Tree Genetics and Breeding, Northeast Forestry University, Harbin, China, [2] Library of Northeast Forestry University, Harbin, China

Edited by:
Guo-Liang Jiang,
Virginia State University, USA

Reviewed by:
Li-Song Chen,
Fujian Agriculture and Forestry University, China
Hao Peng,
Washington State University, USA

***Correspondence:**
Chenghao Li
chli0@163.com

The heavy metal ATPase (HMA) family plays an important role in transition metal transport in plants. However, this gene family has not been extensively studied in *Populus trichocarpa*. We identified 17 HMA genes in *P. trichocarpa* (*PtHMA*s), of which *PtHMA1–PtHMA4* belonged to the zinc (Zn)/cobalt (Co)/cadmium (Cd)/lead (Pb) subgroup, and *PtHMA5–PtHMA8* were members of the copper (Cu)/silver (Ag) subgroup. Most of the genes were localized to chromosomes I and III. Gene structure, gene chromosomal location, and synteny analyses of *PtHMA*s indicated that tandem and segmental duplications likely contributed to the expansion and evolution of the *PtHMA*s. Most of the HMA genes contained abiotic stress-related *cis*-elements. Tissue-specific expression of *PtHMA* genes showed that *PtHMA1* and *PtHMA4* had relatively high expression levels in the leaves, whereas Cu/Ag subgroup (*PtHMA5.1- PtHMA8*) genes were upregulated in the roots. High concentrations of Cu, Ag, Zn, Cd, Co, Pb, and Mn differentially regulated the expression of *PtHMAs* in various tissues. The preliminary results of the present study generated basic information on the HMA family of *Populus* that may serve as foundation for future functional studies.

Keywords: *Populus*, P_{1B}-type ATPases, HMA gene, heavy metal stress, phytoremediation

INTRODUCTION

One of the negative effects of industrialization is heavy metal pollution, which is deleterious not only to the environment but also to human health. Heavy (transition) metals cause toxicity at the cellular level by binding to sulfhydryl groups in proteins and inhibiting enzyme activity or protein function. Exposure to heavy metals may also induce deficiencies in other essential ions by disrupting cellular transport processes and causing oxidative damage (Williams and Mills, 2005). Plants have evolved regulatory mechanisms on heavy metal ion toxicity tolerance and ensure an adequate supply of essential nutrients. Membrane transport proteins play an important role as scavengers of heavy metals. In recent years, progress in the development of genetic and molecular techniques has helped with in identification of various gene families that function in metal transportation (Williams and Mills, 2005).

The P_{1B}-type ATPase, also known as the heavy metal ATPases (HMAs), which belongs to the large P-type ATPase family, plays an important role in transition metal transport in plants (Mills et al., 2003; Gravot et al., 2004; Hussain et al., 2004). It can selectively absorb and transport essential metal ions (Cu^{2+}, Zn^{2+}, and Co^{2+}) for plant growth and development, as well as participate in the distribution of non-essential heavy metal ions (Cd^{2+} and Pb^{2+}). The typical P_{1B}-ATPase consists of approximately 6–8 transmembrane helices, a soluble nucleotide binding domain, phosphorylation domain, and a soluble actuator domain. The interactions of these three domains play an important part in the mechanism (Smith et al., 2014). In addition, both sides of the N-terminal and C-terminal metal binding sites comprise metal binding domains that interact with and bind specific metal ions (such as Cd^{2+}/Pb^{2+}). P_{1B}-type ATPase is also located in this site, which is a heavy metal-associated regulatory domain as well (Williams and Mills, 2005; Arguello et al., 2007).

HMAs provides metal resistance, absorption, and transport. Based on metal substrate specificity, HMAs can be clustered into two major phylogenetic subclasses, namely, the Cu/Ag P_{1B}-ATPase group and the Zn/Co/Cd/Pb P_{1B}-ATPase group (Axelsen and Palmgren, 1998). The HMA proteins have been studied at the genomic scale in *Arabidopsis thaliana* and rice (*Oryza sativa*). There are eight and nine members of P_{1B}-ATPase in *A. thaliana* and *O. sativa*, respectively (Williams and Mills, 2005). *AtHMA1-4* in *A. thaliana* and *OsHMA1-3* in *O. sativa* belong to the former group, whereas *AtHMA5-8* and *OsHMA4-9* belong to the latter group (Williams and Mills, 2005). The *Arabidopsis* disruption mutant of *AtHMA1* is sensitive to high concentrations of Zn (Moreno et al., 2008). Overexpression of *AtHMA3* induces Cd accumulation in plants, with a 2.5- and 2-fold higher in the roots and shoots, respectively (Morel et al., 2009). *AtHMA4* knockout plants are more sensitive to excess Zn(II) and Cd(II), and accumulate Zn less in shoots and more in roots than wild-type plants (Verret et al., 2004; Mills et al., 2005). High Cu concentrations elevate the expression of *OsHMA5* (Deng et al., 2013). Some members of P_{1B}-type ATPases were identified in other plant species, including barley, wheat, *Thlaspi caerulescens*, and *A. halleri* (Deng et al., 2013).

Early phytoremediation studies have shown that *Populus trichocarpa*, a fast-growing tree species, could be a suitable candidate for the treatment of heavy metal-polluted soils (Cunningham, 1996). Poplars transpire large amounts of water, thereby reducing contaminants from soil and water (Zacchini et al., 2009). Because of this feature, *Populus* spp. has attracted the attention of researchers working on remediation of heavy metal-contaminated soils. The completion of the *P. trichocarpa* genome sequence in 2006 presents an excellent opportunity to investigate the metal transporter families in this plant species (Tuskan et al., 2006). Although HMA genes have been extensively studied in various plants, including *Arabidopsis*, rice, and barley, investigations on the HMA gene family in *Populus* are limited. To compare the mechanisms on metal phytoremediation between woody and herbaceous plants that maintain different life cycles and considering the importance of the HMA gene family in plant responses to heavy metal stresses, we investigated *HMA* genes in *Populus*. In the present study, we performed a genome-wide analysis of the *P. trichocarpa* HMA gene family, its phylogenetic analysis, chromosomal distribution, and expressional analysis. We identified a total of 17 HMA genes in *P. trichocarpa*. Quantitative real-time RT-PCR (qRT-PCR) showed that the HMA genes in *Populus* were differentially regulated by excessive Cu, Ag, Zn, Cd, Co, Pb, and Mn stress. The results provide insights for future investigations into the roles of these candidate *HMA* genes in responds to metal stress in *Populus*.

MATERIALS AND METHODS

Identification of HMA Genes in Populus

We used Pfam (http://pfam.sanger.ac.uk/) to query HMA genes in the *P. trichocarpa* genome. All *Populus* HMA genome sequences could be directly downloaded from Phytozome (http://www.phytozome.net/) and NCBI (http://www.ncbi.nlm.nih.gov/). The downloaded HMA candidates sequences were analyzed manually using the SMART (http://smart.embl-heidelberg.de/) database to verify the presence of the HMA domain (Letunic et al., 2004). *A. thaliana* and *O. sativa* HMA subfamily sequences were downloaded from the *Arabidopsis* Information Resource (TAIR, http://www.arabidopsis.org/index.jsp, release 10.0) and Rice Information Resource (http://rice.plantbiology.msu.edu/index.shtml), respectively.

Phylogenetic Analysis

Multiple sequence alignment of the full-length protein sequences was performed using the ClustalX (version 1.83) program (Thompson et al., 1997) and manually adjusted using the BioEdit 7.1 software (Hall, 1999). A phylogenetic tree was constructed using the neighbor-joining (NJ) method, with a bootstrap test performed using 1000 iterations in MEGA5 (Tamura et al., 2007). Bootstrap analysis with 1000 replicates was used to evaluate the significance of the nodes. Gene clusters referred to the homologs within the three species (*P. trichocarpa*, *A. thaliana*, and *O. sativa*) that were identified based on the NCBI database (http://www.ncbi.nlm.nih.gov/).

Chromosomal Location and Subcellular Location of HMA Genes

The chromosomal localization of all genes was determined and plotted with PopGenIE (http://www.popgenie.org/). Genes separated by =5 gene loci within a range of 100-kb distance were considered tandem duplicates (Hu et al., 2010). Identification of homologous chromosome segments resulting from whole-genome duplication events was conducted as described by Tuskan et al. (2006) WoLF PSORT (http://wolfpsort.org/) was used to predict the subcellular localization of the *HMA* proteins (Horton et al., 2007).

Exon-Intron Structure and Motif Analysis

The online software Gene Structure Display Server (GSDS; Guo et al., 2007) was used to generate the exon/intron organization of HMA genes by comparing the cDNAs to its corresponding genomic DNA sequences in Phytozome (http://www.phytozome.net/search.php). The Multiple Expectation Maximization for Motif Elucidation (MEME) system (Version 4.9) was used to investigate conserved motifs for each HMA gene (Bailey et al., 2006).

Promoter *cis*-Element Identification and Microarray Analyses

Promoter sequences, located 2000 kb upstream of the translation start site, were searched using NCBI, and were analyzed in the Plant Care database (http://bioinformatics.psb.ugent.be/webtools/plantcare/html/) (Lescot et al., 2002). To obtain the expression of each gene in different tissues and organ, we searched the microarray dates from PopGenIE v2.0 (http://popgenie.org/gp) in the NCBI Gene Expression Omnibus (GEO) database using the series accession number GSE6422.

Plant Materials and Treatment

Clonally propagated *P. trichocarpa* genotype Nisqually-1 was cultured in half-strength Murashige and Skoog medium under long-day conditions (16 h light/8 h dark) at 23–25°C. Plants were exposed to the following: 100 μM $CdCl_2$, 100 μM $CuSO_4$,60 μM $MnSO_4$, 750 μM $Pb(NO_3)_2$, 200 μM $ZnSO_4$, and 1 mM $AgNO_3$, and 50 μM $CoCl_2$. Each treatment lasted for 0, 1, 3, and 10 h, and samples were collected at each time point. Three biological replicates were prepared for each stress treatment. All samples were immediately frozen in liquid nitrogen and stored at −80°C until analysis. Non-treated plants were used as control.

RNA Isolation and Real-time Quantitative PCR

Total RNA was extracted from the roots, stems, and leaves using the CTAB method (Jaakola et al., 2001). Prior to cDNA synthesis, RNA was treated with RQ1 RNase-free DNase (Promega, Madison, WI, USA), according to the manufacturer's instructions to rule out DNA contamination. First-strand cDNA synthesis was performed with the PrimeScript™ RT reagent kit (Perfect Real Time; Takara, Dalian, China). Primer Premier 5 was used to design gene-specific primers for all HMA genes based on its corresponding CDS. BLAST was used to check each primer to determine its specificity for the respective gene, and was further confirmed by melting curve analysis of real-time PCR reactions. All primer sequences are presented in **Table S1**.

SYBR Premix Ex Taq II (TaKaRa, Dalian, China) was used to perform qRT-PCR to determine the transcript levels of HMA genes in *P. trichocarpa* under different metal stresses. Reactions were prepared in a total volume of 20 μL that contained the following: 2 μL of template, 10 μL of 2× SYBR Premix, 6 μL of ddH_2O, and 1 μL of each specific primer, prepared to a final concentration of 10 μM. Three technical replicates were performed for each sample to ensure the accuracy of the results. The *P. trichocarpa actin* gene (GenBank Acc. No.: XM_002298674) was used as reference gene (An et al., 2011). A relative quantification method ($2^{-\Delta\Delta CT}$) was used to evaluate quantitative variation among replicates. The PCR conditions and relative gene expression calculations were as previously described (Zhang et al., 2013).

RESULTS

Identification of HMA Genes in *Populus*

We identified a total of 17 HMA genes (**Table 1**), of which *PtHMA4(1), PtHMA4(2), PtHMA4(3), PtHMA4(4), PtHMA4(5)* could be attributed to alternative splicing of mRNA with *PtHMA4*, which generates a variety of mature RNA transcripts that produce protein isoforms. The identified *PtHMA* genes encoded proteins that varied in length from 626 to 1228 amino acids (AA), and with an average of 928 AA. We used WoLF PSORT to predict the location of the proteins and noticed that most *PtHMA* genes were predicted as plasma membrane proteins, except *PtHMA1* and *PtHMA5.1*, which were located in the cytoplasm (**Table 1**).

Phylogenetic, Gene Structural, and Conserved Motif Analyses of *PtHMA* Genes

To examine the phylogenetic relationships among the *Populus* HMA domain proteins in *Populus*, an unrooted phylogenetic tree was constructed from alignments of the full-length HMA sequences (**Figure 1A**). We classified 12 HMA genes into two subgroups, namely, Zn/Co/Cd/Pb (*PtHMA1-4*) and Cu/Ag (*PtHMA5-8*). Near 8000 pairs of paralogous genes are distributed in the *Populus* genome (Guo et al., 2009). By phylogenetic analysis, four paralogous pairs among the 12 *Populus* HMA genes were identified (**Figure S1**).

A comparison of the exon/intron organization of the coding sequences of individual *PtHMA* genes showed that most closely related members shared similar exon/intron structures either according to the number of introns or exon length. These results were consistent with the characteristics defined in the above phylogenetic analysis (**Figure 1A**). The coding sequences of most of the HMA genes were disrupted by introns, with the number of introns ranging from 5 to 16. *PtHMA8* exhibited the largest number of introns (16), followed by *PtHMA6.1* and *PtHMA6.2* (15) and *PtHMA1* (12). Other genes had less than 10 introns (**Figure 1B**).

Table S2 presents the conserved motifs that were shared among related proteins within the family and 15 distinct motifs (**Table S2**). The number of HMA motifs in *Populus* HMA proteins and the spacing of amino acids between adjacent HMA motifs varied, which was similar to that observed in *Arabidopsis* and rice HMA proteins. Most of the closely related members in the phylogenetic tree shared common motifs, and all HMA genes contained common motifs, except for motifs 1, 4, 10, 12, and 15 (**Figure 1C**). All HMA family members contained motif 10, except *PtHMA1*, thereby indicating that at least one motif was lost in this gene during its divergence from a common ancestor. Moreover, differences in gene organization and motif composition among related members suggested that these genes may be functionally divergent.

Comparative Analysis of the HMA Genes in *Populus*, *Arabidopsis*, and Rice

We examined the phylogenetic relationship among the HMA domain proteins in *Populus* (12 genes), rice (9 genes), and *Arabidopsis* (8 genes) that were constructed from alignments of full-length HMA protein sequences (**Figure 2**). The tree topologies produced by the three algorithms were largely comparable, and there were only minor modifications at interior branches (data not shown). Therefore, we further analyzed only the NJ phylogenetic tree in the present study. The HMA gene

TABLE 1 | The HMA gene family in *Populus trichocarpa*.

Gene name	Accession number	NCBI Locus ID	Length(AA)	Protein subcellular Localization prediction
PtHMA1	POPTR_0007s10480	XM_002310080.2	833	cyto[a]
PtHMA4	POPTR_0006s07650	XM_002327518.1	1228	plas
PtHMA4(1)			873	
PtHMA4(2)			1126	
PtHMA4(3)			1169	
PtHMA4(4)			1170	
PtHMA4(5)			1193	
PtHMA5.1	POPTR_0003s12580	XM_002303545.1	983	cyto
PtHMA5.2	POPTR_0001s09210	XM_002299504.2	985	plas
PtHMA5.3	POPTR_0003s12570	XM_002303544.2	987	plas
PtHMA5.4	POPTR_0001s05650	XM_002299198.1	974	plas
PtHMA5.5	POPTR_0003s20380	XM_006385962.1	626	plas
PtHMA6.1	POPTR_0001s21290	XM_002299649.2	807	plas
PtHMA6.2	POPTR_0003s01850	XM_002304058.2	808	plas
PtHMA7.1	POPTR_0001s15900	XM_002326412.1	1010	plas
PtHMA7.2	POPTR_0003s07330	XM_002303313.1	1008	plas
PtHMA8	POPTR_0018s08380	XM_006371981.1	889	plas

[a] *plas, plasma membrane; cyto, cytoplasm.*

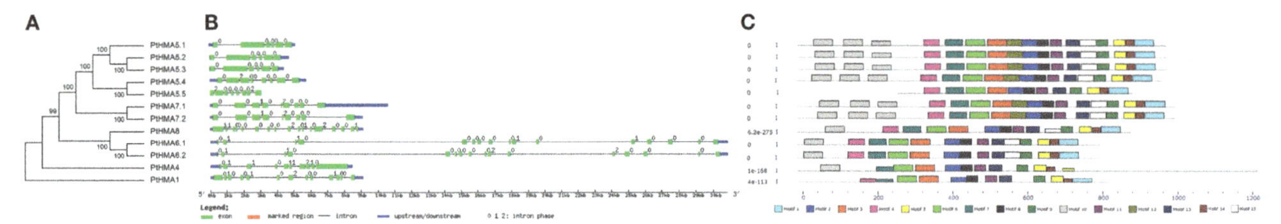

FIGURE 1 | Phylogenetic relationships, gene structure, and motif composition of HMA genes in *P. trichocarpa*. (A) Multiple alignments of 12 full-length amino acids of HMA genes were executed by ClustalX 1.83, and the phylogenetic tree was constructed using MEGA 5.0 using the neighbor-joining (NJ) method with 1000 bootstrap replicates. **(B)** Exon/intron structures. Exons and introns are represented by green boxes and black lines, respectively. **(C)** Schematic representation of the conserved motifs elucidated by MEME. Each motif is represented by a number in the colored box. The black lines represent the Non-conserved sequences.

family contains different numbers and types of HMAs. There are eight and nine members of P_{1B}-ATPase in *A. thaliana* and rice, respectively (Williams and Mills, 2005). These were then further divided into two groups: Zn/Cd/Co/Pb and Cu/silver (Ag) transporters (Axelsen and Palmgren, 1998). *AtHMA1–AtHMA4* and *OsHMA1–OsHMA3* belonged to the Zn/Co/Cd/Pb subgroup, whereas *AtHMA5–AtHMA8* and *OsHMA4–OsHMA9* comprised the Cu/Ag subgroup (Williams and Mills, 2005). *PtHMA1* and *PtHMA4* belonged to the Zn/Co/Cd/Pb subgroup, whereas the rest of the genes were classified into the Cu/Ag subgroup (**Figure 2**). The number of genes in the *Populus* (10 genes) Cu/Ag subgroup was significantly higher than that in rice (6 genes) and *Arabidopsis* (4 genes).

Chromosomal Location and Gene Duplication of *Populus* HMAs

In *silico* mapping of gene loci showed that all 12 *Populus* HMA genes were physically located on 19 linkage groups (LG). A previous study revealed that the *Populus* genome has undergone genome-wide duplications, followed by multiple segmental duplications, tandem duplications, and transposition events (Hu et al., 2012). *Populus* HMA genes were differentially expressed in various tissues (**Figure 4**). The *PtHMA* genes were located on chromosomes I, III, VI, VII, and XVII. Chromosome III harbored the highest number of HMA genes (five), followed by chromosome I (four). Only one HMA gene was detected on each of chromosomes VI, VII, and XVII (**Figure 3**).

The segmental duplication associated with the salicoid duplication event that occurred 65 million years ago largely contributed to the expansion of several multi-gene families (Kalluri et al., 2007; Barakat et al., 2009, 2011; Wilkins et al., 2009; Hu et al., 2010). We used the duplicated blocks established in a previous study (Tuskan et al., 2006) to confirm the possible relationship between HMA genes and potential segmental duplications. The distribution of HMA genes relative to the corresponding duplicate blocks is illustrated in **Figure 3**. In this study, of the 12 *Populus* HMA genes, 10 were located in duplicated regions and two were located outside duplicated

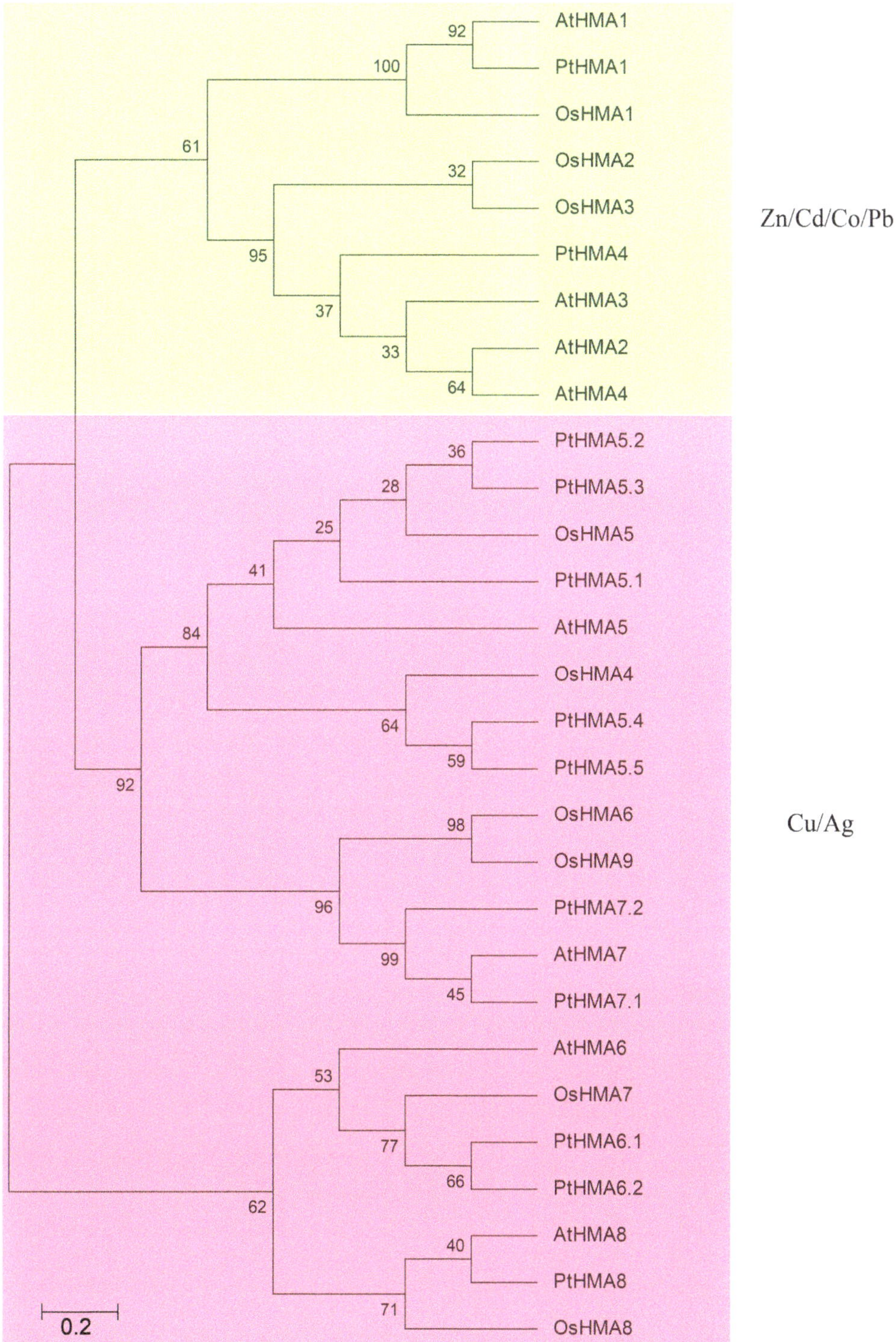

FIGURE 2 | All HMA proteins of *P. trichocarpa* (12), *A. thaliana* (8), and *O. sativa* (9) were divided into two distinct subfamilies. The phylogenetic tree was constructed using MEGA 5.0 using the neighbor-joining (NJ) method with 1000 bootstrap replicates. Green represents the Zn/Cd/Co/Pb subgroup; pink indicates the Cu/Ag subgroup.

blocks. Among these 10 genes, four were preferentially retained duplicates that were located in both duplicated regions, whereas the others were only in one of the blocks and lacked duplicates in the corresponding block. These results indicated that duplication events and tandem repeats contributed to the expansion of the HMA gene family in the *Populus* genome.

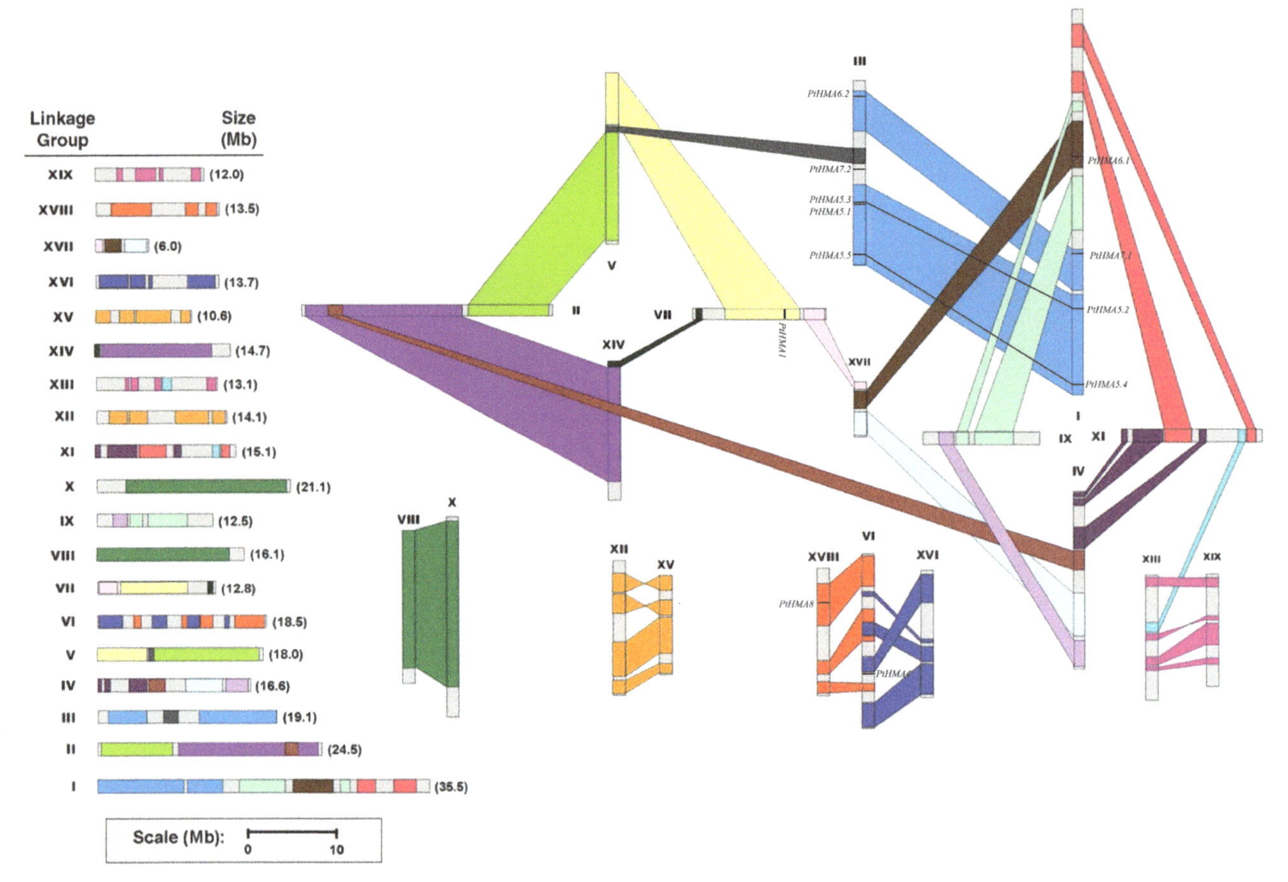

FIGURE 3 | Chromosomal location of *Populus* HMA genes. A total of 12 HMA genes are mapped to the 19 linkage groups (LG). A schematic view of chromosome reorganization by recent whole-genome duplication in *Populus* is shown. Segmentally duplicated homologous blocks are indicated with the same color. The *scale* represents megabases (Mb). The LG numbers are indicated at the top of each bar.

Promoter *cis*-Elements Analysis

Phytohormones such as salicylic acid (SA), jasmonic acid (JA), ethylene (ET), and abscisic acid (ABA) enable plants to adapt to abiotic stresses (Fujita et al., 2006; Santner and Estelle, 2009). We identified putative *cis*-acting regulatory DNA elements (**Table 2**) on the basis of the obtained promoter sequences (2000 bp). The *PtHMA* gene family promoter sequences showed several *cis*-elements that were related to phytohormones and environmental stress signal responsiveness. Most of the 12 HMA genes contained *cis*-elements such as ABA responsive (ABREs and AREs), heat stress responsive (HSEs), MYB binding site involved in drought-inducibility (MBS), and defense and stress responsive TC-rich repeat elements. *PtHMA1*, *PtHMA4*, *PtHMA5.1*, *PtHMA5.2*, *PtHMA6.2*, *PtHMA7.1*, and *PtHMA8* contained the *cis*-element TCA-element, which is involved in salicylic acid responsiveness. SA appears to be an effective therapeutic agent in plant responses to biotic and abiotic stresses (Rivas-San Vicente and Plasencia, 2011). It was also reported that SA induced adaptability to Cd, Hg, Ni, Pb, and Mn toxicity (Mostofa and Fujita, 2013). The CGTCA-motif or TGACG-motif involved in MeJA-responsiveness was identified in seven genes (**Table 2**). Moreover, *PtHMA1* and *5.4* contained the *cis*-acting regulatory element, as-2-box, and as1, which are involved in shoot-specific and root-specific expression, respectively. *PtHMA5.5* contained a CAT-box that is related to meristem expression.

Tissue-specific Expression Profile in PopGenIE v2.0

The expression profiles of 10 HMA genes from the Affymetrix (GSE6422) microarray data showed specific expression patterns in different tissues (**Figure 4**). Two HMA genes (*PtHMA6.1* and *PtHMA6.2*) did not have corresponding microarray sequences in the database, which suggested that the two genes were pseudogenes. Alternatively, the two genes showed a low level of transcript abundance or had special temporal and spatial expression patterns that were not examined in the libraries. The microarray data showed high expression levels of *PtHMA1* and *PtHMA4* in leaves. We also observed high expression levels of *PtHMA4* in the roots. Cu/Ag subgroup members (*PtHMA5.1*–*PtHMA8*) also showed higher expression levels in the roots than in other tissues. *PtHMA5.4* and *PtHMA5.5* had higher expression levels in young leaves than in other tissues. In addition, in promoters of *PtHMA5.4* and *PtHMA5.5*, the genes were related to meristem expression.

TABLE 2 | The *cis*-elements of HMA gene promoter in *P. trichocarpa* that are related to stress.

Gene name	cis-elements related to abiotic stress
PtHMA1	ABRE ARE HSE MBS CGTCA-motif TGACG-motif TCA-element as1 as-2-box
PtHMA4	ABRE ARE HSE TCA-element TC-rich repeats WUN-motif ERE
PtHMA5.1	ABRE ARE HSE MBS TGACG-motif TCA-element TC-rich repeats
PtHMA5.2	ABRE ARE HSE MBS ERE TCA-element CCAAT-box
PtHMA5.3	ABRE ARE HSE MBS ERE CGTCA-motif TGACG-motif TC-rich repeats
PtHMA5.4	ARE HSE MBS ERE CGTCA-motif TGACG-motif TC-rich repeats WUN-motif as1 as-2-box
PtHMA5.5	ABRE HSE MBS ERE CAT-box
PtHMA6.1	ABRE ARE HSE MBS CGTCA-motif TGACG-motif TC-rich repeats
PtHMA6.2	ABRE ARE HSE MBS CGTCA-motif TGACG-motif TCA-element TC-rich repeats
PtHMA7.1	ABRE ARE HSE TCA-element TC-rich repeats
PtHMA7.2	ABRE HSE MBS ERE CGTCA-motif TGACG-motif TC-rich repeats
PtHMA8	ARE HSE MBS ERE TCA-element TC-rich repeats

Expression of HMA Genes Under Different Heavy Metal Stresses

We studied the expression profiles of *PtHMA* genes in *P. trichocarpa* under heavy metal stress (**Figure 5**). The genes *PtHMA1* and *PtHMA4* and its five alternatively spliced genes (Zn/Co/Cd/Pb subgroup) were exposed to Zn, Cd, Co, and Pb stresses. Under Cd stress, the expression of *PtHMA1* gradually increased in the roots compared to the non-treated samples, the expression level increased in the stem after 1 h of treatment, and the expression level of *PtHMA1* in the leaves gradually increased with treatment time (**Figure 5B**). Therefore, *Populus HMA1* might not only be involved in Zn transport, but also in Cd transport from the roots to the leaves. The expression level of the *PtHMA4(2)* gene was higher compared to that of the other alternatively spliced genes [*PtHMA4(1)*, *PtHMA4(3)*, *PtHMA4(4)*, and *PtHMA4(5)*] under excess Zn, Cd, Co, and Pb stresses. In the 200 μM $ZnSO_4$ treatment, *PtHMA4(2)* was upregulated at 1 h in the roots and 10 h in the stems compared to that observed in the roots and leaves, and the expression in the roots peaked at 3 and 10 h of Zn treatment (**Figure 5A**). The transcription levels of *PtHMA4(2)* were higher in the roots and stems than that in the leaves with Cd treatment at all-time points. Moreover, *PtHMA4(2)* was upregulated in the roots, stems, and leaves after exposure to excessive Pb (**Figure 5D**), and

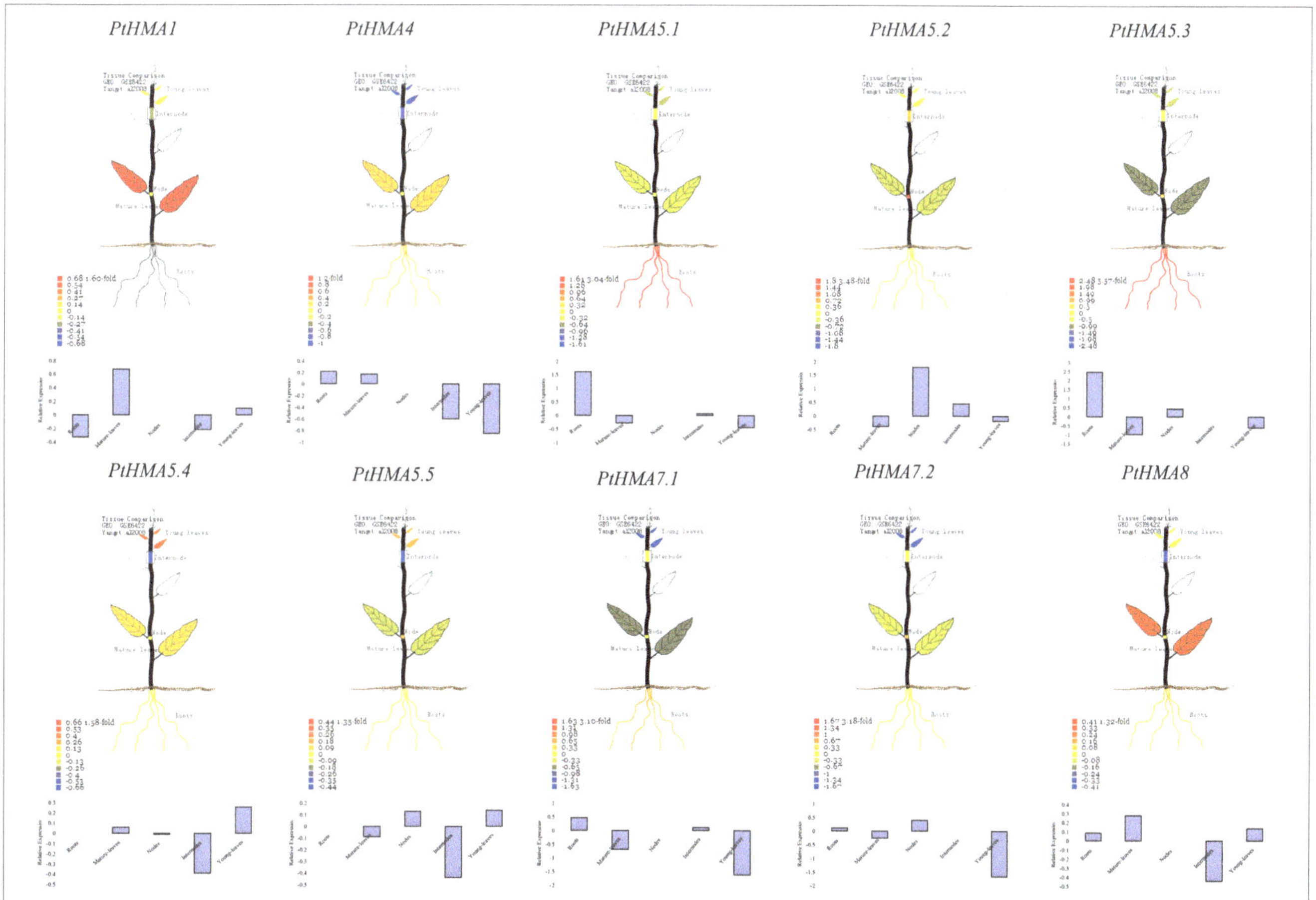

FIGURE 4 | Organ-specific expressions of *PtHMA* genes in mature-leaves, young-leaves, roots, nodes, and internodes. The visual images were searched from PopGenIE v2.0. Each gene is indicated in every graph.

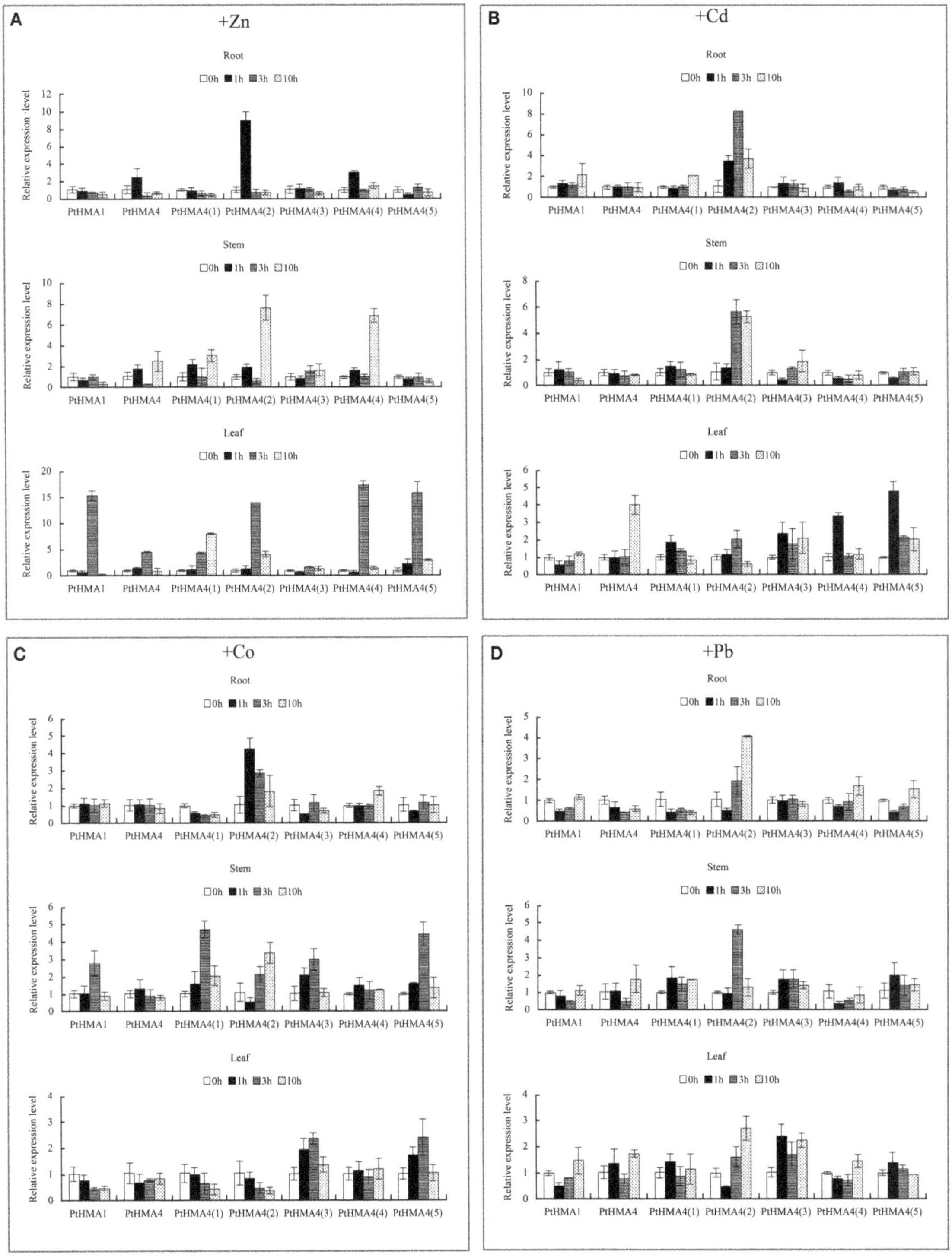

FIGURE 5 | Expression profiles of the *PtHMA* genes under heavy metal conditions. QRT-PCR was used to analyze the expression patterns of 7 *PtHMA* genes in the roots, stems, and leaves under excessive amounts of Zn **(A)**, Cd **(B)**, Co **(C)**, and Pb **(D)**. *X-axis* represents different genes. The *y-axis* represents relative gene expression levels normalized to the *Populus* reference gene, actin. Standard deviations were derived from three replicates of each experiment.

in the roots and leaves with exposure to high amounts of Co (**Figure 5C**).

We further evaluated the expression patterns of *PtHMA5*–*PtHMA8* of the Cu/Ag subfamily. The number of genes for *PtHMA5* (5 genes), *PtHMA6* (2 genes), *PtHMA7* (2 genes) was higher than those of its alternatively spliced *Arabidopsis* counterparts. We also examined the expression of nine candidate genes subjected to excessive Cu and Ag treatments (**Figure 6**).

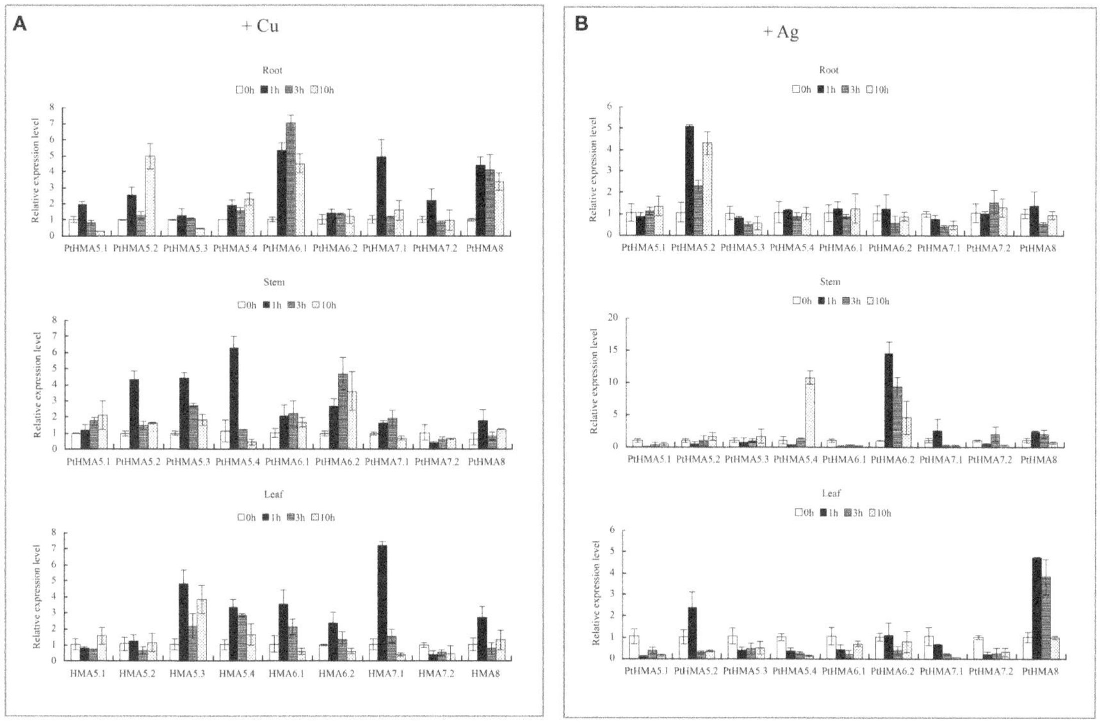

FIGURE 6 | qRT-PCR was performed to analyze the expression patterns of nine *PtHMA* genes in the roots, stems, and leaves under excessive amounts of Cu (A) and Ag (B). *X-axis* represents different genes. The *y-axis* represents relative gene expression levels normalized to the *Populus* reference gene, actin. Standard deviations were derived from three replicates of each experiment.

The genes *PtHMA5.2, 5.4, 6.1, 7.1,* and *8* were unregulated in response to Cu treatment, whereas *PtHMA5.1* and *5.3* were downregulated in the roots (**Figure 6A**). However, the expression of five genes (i.e., *PtHMA5.2, 5.4, 6.1, 7.1,* and *8*) in the roots remained unaffected with Cu treatment (**Figure 6A**). On the other hand, *PtHMA5.2, 5.3,* and *5.4* were considerably upregulated in poplar stems after exposure to Cu for 1 h, whereas *PtHMA 6.2* was upregulated at all time points, with the highest expression detected after 3 h. In leaves, *PtHMA5.3, 5.4, 6.1, 6.2,* and *7.1* showed considerably high expression levels after 1 h of exposure to excessive levels of Cu (**Figure 6A**). The gene expression patterns of the four *Populus PtHMA5* genes were divided into two groups based on the time point when the highest transcript levels were observed in different tissues and organs (**Figure 6A**). One group (*PtHMA5.1* and *PtHMA5.3*) accumulated the highest transcripts in the roots at 1 h after Cu treatment, whereas that of the other groups, *PtHMA5.2* and *PtHMA5.4*, exhibited two peaks, namely, at 1 and 10 h, of Cu treatment. The expression level of *PtHMA5.1* in the stems and leaves gradually increased with treatment time. The highest transcription levels for *PtHMA5.2, 5.3,* and *5.4* were observed in the roots at 3 and 10 h of excessive Cu treatment. *PtHMA6.1* and *PtHMA6.2* were highly expressed in the roots but were also expressed in stems and leaves. These results suggest that *PtHMA6.1* and *PtHMA6.2* play a role in Cu transport during plant growth and development. The expression of *PtHMA7.1, PtHMA7.2,* and *PtHMA8* was upregulated in the roots, stems, and leaves, and *PtHMA7.1* was considerably upregulated in the leaves (**Figure 6A**). The expression level of *PtHMA8* in the roots was higher than that in the stems and leaves. In the roots, *PtHMA5.1, 5.2,* and *7.2* were upregulated in response to Ag treatment, whereas the other genes remained unaffected at all-time points. In the stems, *PtHMA5.4, 6.2,* and *8* showed a significant increase in transcript levels at 10 h, and *6.2* presented a higher expression level at all-time points compared to that of the control (**Figure 6B**). Ag treatment increased the expression level of the *PtHMA5.2, PtHMA5.4, PtHMA6.2,* and *PtHMA8* transcripts compared to the five other HMA genes (*PtHMA5.1, PtHMA5.3, PtHMA6.1, PtHMA7.1,* and *PtHMA7.2).* The expression of *PtHMA5.4* and *PtHMA6.2* in the stems peaked at 10 and 1 h after excess Ag treatment, respectively. *PtHMA8* was expressed in the entire plantlet, whereas its expression level was higher in the leaves than in the stems and roots under excess Ag stress. In contrast, *PtHMA6.2* and *PtHMA8* had the highest levels of transcript abundance in the stems and leaves relative to those of the other HMA genes with Ag treatment.

Our qRT-PCR data further showed that *PtHMA5–PtHMA8* might have also been induced by Zn, Cd, Co, or Pb. Of these genes, two homologous genes (*PtHMA6.1* and *PtHMA6.2*) exhibited particularly high transcript accumulations in the roots, stems, and leaves after exposure to excessive amounts of Zn, Co and Pb, especially the roots (**Figure 7**). The expression level of *PtHMA5.2, PtHMA5.3, PtHMA5.4, PtHMA6.1, PtHMA7.1,* and *PtHMA7.2* increased in different tissues under excessive

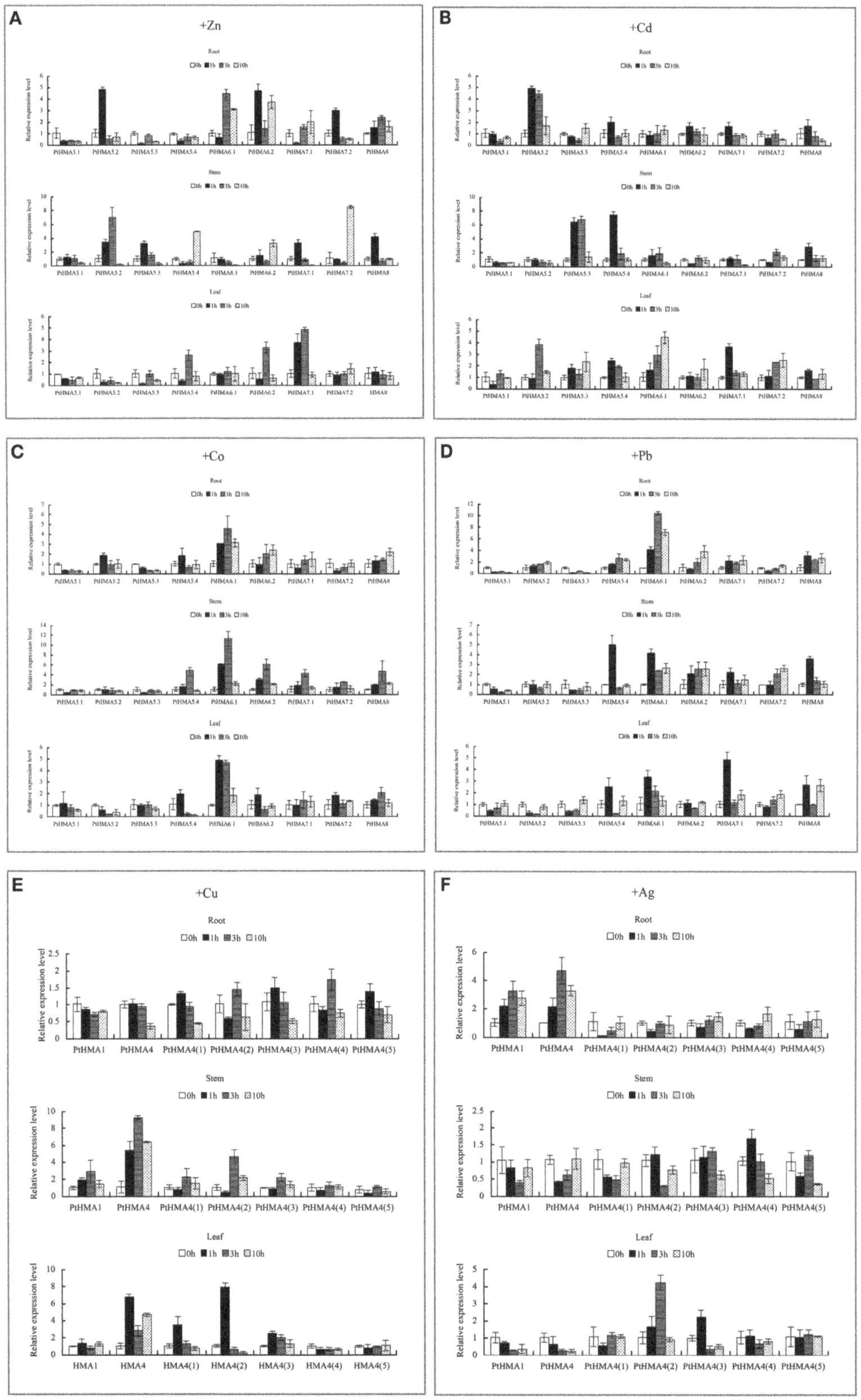

FIGURE 7 | qRT-PCR was used to analyze the expression patterns of nine *PtHMA* genes in the roots, stems, and leaves under excessive amounts of Zn (A), Cd (B), Co (C), and Pb (D), and seven *PtHMA* genes in the roots, stems, and leaves under excessive amounts of Cu (E) and Ag (F). *X-axis* represents different genes. The *y-axis* represents relative gene expression levels normalized to the *Populus* reference gene, actin. Standard deviations were derived from three replicates of each experiment.

Cd stress. The expression of *PtHMA6.1* was higher in the leaves than in the roots and stems (**Figure 7B**). The *PtHMA1–PtHMA4* phylogenetic cluster with the Zn/Cd/Co/Pb subclass of HMAs (**Figure 2**) was possibly induced by Cu or Ag. For example, *PtHMA4* and *PtHMA4(2)* were mainly expressed in the stems under excessive Cu stress (**Figure 7E**). Also, *PtHMA1* and *PtHMA4* were upregulated in the roots compared to that observed in the stems and leaves after excessive Ag stress (**Figure 7F**). Mn is an important micronutrient that is required in various stages of plant growth and development. Five genes (*PtHMA6.1*, *PtHMA6.2*, *PtHMA7.1*, *PtHMA7.2*, and *PtHMA8*) showed higher transcript levels in the roots, stems, and leaves under Mn stress, with peak expression at 1 and 10 h of excess Mn treatment (**Figure 8**).

DISCUSSION

The HMA Gene Family in *Populus*

In the present study, we identified 12 full-length HMA genes and five alternatively spliced genes of *PtHMA4* in the *P. trichocarpa* genome, each of which contains at least one HMA motif. The lengths of these sequences significantly varied, implying a high degree of complexity among the HMA genes. Furthermore, WoLF PSORT analyses allowed us to putatively localize the genes. However, experimental validation may be required for a more accurate localization. Preliminary analysis of the HMA gene family was performed in the plant models, *Arabidopsis* and rice (Williams and Mills, 2005; Takahashi et al., 2012). These studies revealed that *AtHMA1, AtHMA6,* and *AtHMA8* were located in the chloroplast; *AtHMA2, AtHMA4,* and *AtHMA5* were located in plasma membrane; and *AtHMA3* and *AtHMA7* were in the vacuole and Golgi, respectively (Williams and Mills, 2005). The present study showed that most of the HMA genes were predicted as plasma membrane proteins, except *PtHMA1* and *PtHMA5.1*. *PtHMA1* was located in the cytoplasm, though it was placed in the same group as *PtHMA4* (Zn/Cd/Co/Pb). *PtHMA5.1* was also located in the cytoplasm, which differed from the other members of *PtHMA5* (i.e., *PtHMA5.2–PtHMA5.5*). These results indicated that the same phylogenetic grouping based on sequence similarity did not necessarily correspond to the same subcellular localization. Therefore, homologous genes may show differences in gene function and signal transduction. Although our preliminary prediction showed that *PtHMA1* and *PtHMA5.1* were located in the cytoplasm, we could not determine the specific organelle which the genes were associated with. Further studies may help to determine the exact location of these genes. However, subcellular localization of these genes in *Populus* might change with varying metal supply, as seen in the mammalian Wilson's disease protein, ATP7B, which moves from the Golgi to the plasma membrane at higher metal concentrations (Petris et al., 1996). We speculate that the localization of HMA family proteins changes with differences in metal supply and utilize specific mechanisms with various metal stresses.

It has been reported that exon-intron increase or decrease can be caused by the integration and realignment of gene fragments (Xu et al., 2012). Therefore, structural gene variation plays a major role in the evolution of gene families (Xu et al., 2012). In the evolutionary history of *Populus*, members of the *HMA* gene family underwent rigorous selection. The structure of *Populus HMA* genes is well conserved, and these genes have different numbers of exons. Although *PtHMA5–8* belonged to the same phylogenetic group (group Zn/Cd/Co/Pb), their number of exons varied. *PtHMA6.1* and *PtHMA6.2* had more exons than the others, which in turn may be expressed in specific tissues.

Comparative Analysis of the HMA Genes in *Populus*, *Arabidopsis*, and Rice

In *Arabidopsis,* the Cu subgroup has three domains, E1-E2 ATPase domains (PF00122), a heavy metal-associated domain (PF00403), and a haloacid dehalogenase-like hydrolase (PF00702). The Zn subgroup has two domains, namely, the E1-E2 ATPase and haloacid dehalogenase-like hydrolase (**Table S3**). Unlike *Arabidopsis*, the HMAs of *P. trichocarpa* has similar domains. However, phylogenetic analysis showed that *PtHMA5.5* had only an E1-E2 ATPase and a haloacid dehalogenase-like hydrolase domain compared to the other members of its subgroup. Our analysis also revealed that the number of *Populus*, *Arabidopsis*, and rice HMA genes varied in most subfamilies. For example, *Arabidopsis* and rice just have one of each of the *HMA5, HMA6, HMA7* genes. However, *Populus* has five members of *HMA5* and two members of *HMA6* and *HMA7*. *Populus* lacks *HMA2* and *HMA3,* but alternative splicing of *PtHMA4* may occur and compensate for the two missing genes. Therefore, the present study provides a platform for future studies on the role of HMA genes in *Populus*.

Chromosomal Location and Gene Duplication

Tuskan et al. (2006) suggested that the *Populus* genome underwent at least three rounds of genome-wide duplication, followed by multiple segmental duplication, tandem duplication, and transposition events such as retroposition and replicative transposition. Gene duplication events, including tandem and segmental duplication, play an important role in genomic expansions and realignments (Vision et al., 2000; Kumar et al., 2011). In the present study, 5/6 of the *Populus* HMA genes were located in duplicated regions, whereas only two genes were located outside the duplicated blocks. These results suggest that dynamic rearrangements might have occurred following the segmental duplication, which led to the loss of some genes. These findings also showed that *PtHMA5.2* and *PtHMA5.3*, and *PtHMA5.4* and *PtHMA5.5* belonged to the same branch of the phylogenetic tree and were present in homologous regions of chromosomes I and III, respectively (**Figure 3**). This finding might explain the homology of the two genes. The tandem duplications might have impacted the expansion of the HMA gene family. Our results also showed that only one pair (*PtHMA5.1/5.3*) of the HMA genes underwent tandem duplication, as indicated by the >90% similarity in amino acid sequence. In comparison, two segmental duplications (*PtHMA5.2/5.3* and *PtHMA5.4/5.5*) were identified. These results corroborate the findings of previous studies involving other gene families (Kalluri et al., 2007).

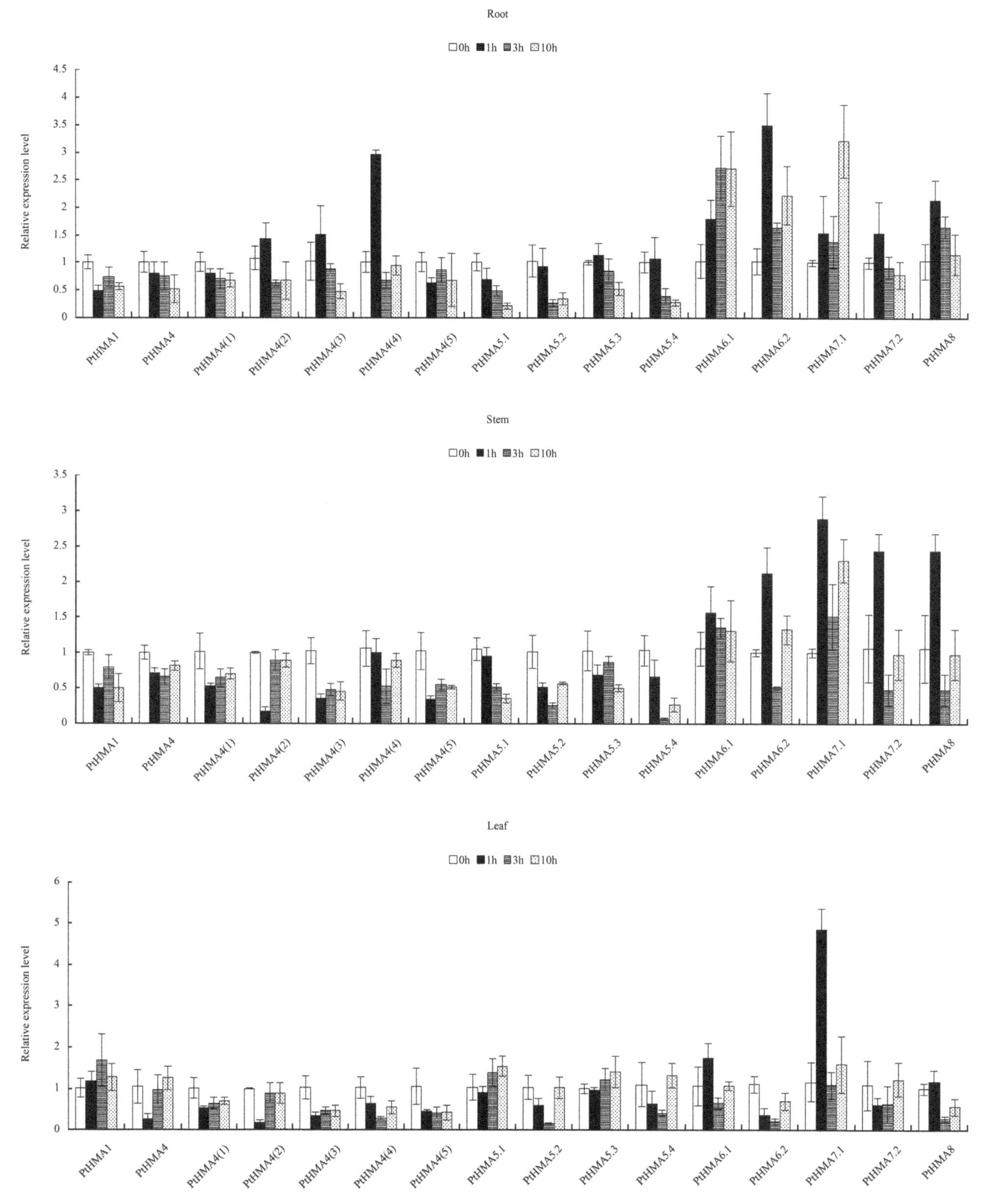

FIGURE 8 | qRT-PCR was performed to analyze the expression patterns of 16 *PtHMA* genes in the roots, stems, and leaves under excessive Mn stress. *X-axis* represents different genes. The *y-axis* represents relative gene expression levels normalized to the *Populus* reference gene, actin. Standard deviations were derived from three replicates of each experiment.

Transcript Profiles of HMA Genes in *Populus*

Plant metal homeostasis must be tightly regulated to ensure sufficient micronutrient (e.g., Zn, Cu, and Fe) supply to different organs, and to prevent non-essential metals (e.g., Cd and Pb) to reach toxic levels that may result in deleterious effects (Hall, 2002). Metal transporters play important roles in various aspects of essential and toxic metal distribution in plants. All members of HMAs in *A. thaliana* are functionally characterized (Deng et al., 2013). *AtHMA1* and *OsHMA1* are involved in Zn transport (Kim et al., 2009). *AtHMA1* transports Zn, Cu, and Ca (Williams and Mills, 2005; Seigneurin-Berny et al., 2006; Kim et al., 2009). However, the present study did not indicate transcript accumulation of *PtHMA1* under excessive Cu stress. Our data showed that the expression patterns of *PtHMA1* and *AtHMA1* were not similar. The *HMA1* in *Populus* not only transported Zn and Cu but also Cd from the roots to the leaves. *AtHMA4* is induced by Cd and Zn and is expressed in tissues surrounding the vascular vessels of the root vascular (Mills et al., 2003). In this study, in addition to Cd and Zn, *PtHMA4(2)* was also induced by Pb and Co. The transcription levels of *PtHMA4(2)* were almost unregulated in the roots, stems, and leaves (**Figure 5**). The above results indicated differential expression patterns for various *PtHMAs* in the roots, stems, and leaves, thus confirming tissue-specific expression. These findings may have significant implications in *Populus. AtHMA5* is involved in the Cu translocation from the roots to the shoots or Cu detoxification of roots (Kobayashi et al., 2008). *AtHMA6*, *AtHMA7*, and *AtHMA8* also play an important role in transport Cu (Woeste and Kieber, 2000; Shikanai et al., 2003; Abdel-Ghany et al., 2005; Catty et al., 2011). *OsHMA5* is involved in loading Cu to the xylem of the roots and other organs (Deng et al., 2013). The results of the present study suggested that the role of *PtHMA5* may differ from that of *AtHMA5* and *OsHMA5*. *AtHMA5* is involved in Cu detoxification, whereas *OsHMA5* is responsible for xylem loading of Cu. However, *PtHMA5* plays an important role in Ag detoxification in addition to Cu detoxification. *PtHMA6.1*, *PtHMA6.2*, *PtHMA7.1*, *PtHMA7.2*, and *PtHMA8* are similar to their homologs in *Arabidopsis* and function in transporting or detoxifying Cu. *PtHMA6.2* and *PtHMA8* had higher transcription levels under Ag stress. *PtHMA6.1*, *PtHMA6.2*, *PtHMA7.1*, *PtHMA7.2*, and *PtHMA8* were induced by Mn in the roots, stems, and leaves, suggesting that *PtHMA* genes help in scavenging excess Ag and Mn. The considerable differences in expression among *PtHMA* genes suggest that these genes perform varying physiological and biochemical functions to adapt to complicated challenges.

Metal transporters play important roles in various aspects of essential and toxic metal distribution in plants. Moreover, under different metal stress, the expression of *PtHMAs* varied in the roots, stems, and leaves, thereby confirming tissue-specific expression. In summary, the present study provides a theoretical basis to select metal-specific genes for related functional validation. Expression studies of *PtHMA* genes suggest their special roles during the plant growth and developmental stages. Their functions are similar to that in rice or *Arabidopsis*, although more complicated. The results of the present study led to a better understanding of the *Populus* phenotype and help to lay out the first step toward engineering a hyper accumulating *Populus* phenotype that may be utilized in phytoremediation. However, the specific functions of the most *PtHMA* genes remain unknown. Further experiments, such as Western blotting should be carried out to demonstrate that those HMA genes accounting for the responses to metal stress are real. Also, ectopic expression of *PtHMA* genes in *Arabidopsis* or other plant species is necessary to investigate heavy metal accumulation or tolerance phenotypes of the transgenic plants in response to heavy metal treatments. It would improve on the results of this study.

AUTHOR CONTRIBUTIONS

Conceived and designed the experiments: DL, CL. Performed the experiments: DL, HZ. Analyzed the data: DL, XH, XX, and CL. Contributed reagents/materials/analysis tools: DL, QL, ZW, HW, HL, MW, ZW, and CL. Wrote the paper: DL, CL. All authors approved the final manuscript.

ACKNOWLEDGMENTS

The Fundamental Research Funds for the Central Universities (Grant number 2572015EA05) and the Innovation Project of State Key Laboratory of Tree Genetics and Breeding (Northeast Forestry University; Grant number 2013A04) supported this study.

REFERENCES

Abdel-Ghany, S. E., Müller-Moulé, P., Niyogi, K. K., Pilon, M., and Shikanai, T. (2005). Two P-type ATPases are required for copper delivery in *Arabidopsis thaliana* chloroplasts. *Plant Cell* 17, 1233–1251. doi: 10.1105/tpc.104.030452

An, X. M., Ye, M. X., Wang, D. M., Wang, Z. L., Cao, G. L., Zheng, H. Q., et al. (2011). Ectopic expression of a poplar *APETALA3-like* gene in tobacco causes early flowering and fast growth. *Biotechnol. Lett.* 33, 1239–1247. doi: 10.1007/s10529-011-0545-4

Arguello, J. M., Eren, E., and Gonzalez-Guerrero, M. (2007). The structure and function of heavy metal transport P_{1B}-ATPases. *Biometals* 20, 233–248. doi: 10.1007/s10534-006-9055-6

Axelsen, K. B., and Palmgren, M. G. (1998). Evolution of substrate specificities in the P-type ATPase superfamily. *J. Mol. Evol.* 46, 84–101. doi: 10.1007/PL00006286

Bailey, T. L., Williams, N., Misleh, C., and Li, W. W. (2006). MEME: discovering and analyzing DNA and protein sequence motifs. *Nucleic Acids Res.* 34, W369–W373. doi: 10.1093/nar/gkl198

Barakat, A., Bagniewska-Zadworna, A., Choi, A., Plakkat, U., Diloreto, D. S., Yellanki, P., et al. (2009). The *cinnamyl alcohol dehydrogenase* gene family in *Populus*: phylogeny, organization, and expression. *BMC Plant Biol.* 9:26. doi: 10.1186/1471-2229-9-26

Barakat, A., Choi, A., Yassin, N. B., Park, J. S., Sun, Z., and Carlson, J. E. (2011). Comparative genomics and evolutionary analyses of the *O-methyltransferase* gene family in *Populus*. *Gene* 479, 37–46. doi: 10.1016/j.gene.2011.02.008

Catty, P., Boutigny, S., Miras, R., Joyard, J., Rolland, N., and Seigneurin-Berny, D. (2011). Biochemical characterization of AtHMA6/PAA1, a chloroplast envelope Cu(I)-ATPase. *J. Biol. Chem.* 286, 36188–36197. doi: 10.1074/jbc.M111.241034

Cunningham, S. D. (1996). Promises and prospects of phytoremediation. *Plant Physiol.* 110, 715–719.

Deng, F., Yamaji, N., Xia, J., and Ma, J. F. (2013). A member of the heavy metal P-type ATPase OsHMA5 is involved in xylem loading of copper in rice. *Plant Physiol.* 163, 1353–1362. doi: 10.1104/pp.113.226225

Fujita, M., Fujita, Y., Noutoshi, Y., Takahashi, F., Narusaka, Y., Yamaguchi-Shinozaki, K., et al. (2006). Crosstalk between abiotic and biotic stress responses: a current view from the points of convergence in the stress signaling networks. *Curr. Opin. Plant Biol.* 9, 436–442. doi: 10.1016/j.pbi.2006.05.014

Gravot, A., Lieutaud, A., Verret, F., Auroy, P., Vavasseur, A., and Richaud, P. (2004). AtHMA3, a plant P_{1B}-ATPase, functions as a Cd/Pb transporter in yeast. *FEBS Lett.* 561, 22–28. doi: 10.1016/S0014-5793(04)00072-9

Guo, A. Y., Zhu, Q. H., Chen, X., and Luo, J. C. (2007). GSDS: a gene structure display server. *Yi Chuan* 29, 1023–1026. doi: 10.1360/yc-007-1023

Guo, Y. H., Yu, Y. P., Wang, D., Wu, C. A., Yang, G. D., Huang, J. G., et al. (2009). GhZFP1, a novel CCCH-type zinc finger protein from cotton, enhances salt stress tolerance and fungal disease resistance in transgenic tobacco by interacting with GZIRD21A and GZIPR5. *New Phytol.* 183, 62–75. doi: 10.1111/j.1469-8137.2009.02838.x

Hall, J. L. (2002). Cellular mechanisms for heavy metal detoxification and tolerance. *J. Exp. Bot.* 53, 1–11. doi: 10.1093/jexbot/53.366.1

Hall, T. (1999). "BioEdit: a user-friendly biological sequence alignment editor and analysis program for Windows 95/98/NT," in *Nucleic Acids Symposium Series*, Vol. 41 (Oxford University Press), 95–98.

Horton, P., Park, K. J., Obayashi, T., Fujita, N., Harada, H., Adams-Collier, C. J., et al. (2007). WoLF PSORT: protein localization predictor. *Nucleic Acids Res.* 35, W585–W587. doi: 10.1093/nar/gkm259

Hu, R., Chi, X., Chai, G., Kong, Y., He, G., Wang, X., et al. (2012). Genome-wide identification, evolutionary expansion, and expression profile of homeodomain-leucine zipper gene family in poplar (*Populus trichocarpa*). *PLoS ONE* 7:e31149. doi: 10.1371/journal.pone.0031149

Hu, R., Qi, G., Kong, Y., Kong, D., Gao, Q., and Zhou, G. (2010). Comprehensive analysis of NAC domain transcription factor gene family in *Populus trichocarpa*. *BMC Plant Biol.* 10:145. doi: 10.1186/1471-2229-10-145

Hussain, D., Haydon, M. J., Wang, Y., Wong, E., Sherson, S. M., Young, J., et al. (2004). P-type ATPase heavy metal transporters with roles in essential zinc homeostasis in *Arabidopsis*. *Plant Cell* 16, 1327–1339. doi: 10.1105/tpc.020487

Jaakola, L., Pirttilä, A. M., Halonen, M., and Hohtola, A. (2001). Isolation of high quality RNA from bilberry (*Vaccinium myrtillus* L.) fruit. *Mol. Biotechnol.* 19, 201–203. doi: 10.1385/MB:19:2:201

Kalluri, U. C., Difazio, S. P., Brunner, A. M., and Tuskan, G. A. (2007). Genome-wide analysis of *Aux/IAA* and *ARF* gene families in *Populus trichocarpa*. *BMC Plant Biol.* 7:59. doi: 10.1186/1471-2229-7-59

Kim, Y. Y., Choi, H., Segami, S., Cho, H. T., Martinoia, E., Maeshima, M., et al. (2009). AtHMA1 contributes to the detoxification of excess Zn(II) in *Arabidopsis*. *Plant J.* 58, 737–753. doi: 10.1111/j.1365-313XX.2009.03818.x

Kobayashi, Y., Kuroda, K., Kimura, K., Southron-Francis, J. L., Furuzawa, A., Kimura, K., et al. (2008). Amino acid polymorphisms in strictly conserved domains of a P-type ATPase HMA5 are involved in the mechanism of copper tolerance variation in *Arabidopsis*. *Plant Physiol.* 148, 969–980. doi: 10.1104/pp.108.119933

Kumar, R., Tyagi, A. K., and Sharma, A. K. (2011). Genome-wide analysis of *auxin response factor* (*ARF*) gene family from tomato and analysis of their role in flower and fruit development. *Mol. Genet. Genom.* 285, 245–260. doi: 10.1007/s00438-011-0602-7

Lescot, M., Dehais, P., Thijs, G., Marchal, K., Moreau, Y., and Van De Peer, Y., et al. (2002). PlantCARE, a database of plant *cis*-acting regulatory elements and a portal to tools for *in silico* analysis of promoter sequences. *Nucleic Acids Res.* 30, 325–327. doi: 10.1093/nar/30.1.325

Letunic, I., Copley, R. R., Schmidt, S., Ciccarelli, F. D., Doerks, T., Schultz, J., et al. (2004). SMART 4.0: towards genomic data integration. *Nucleic Acids Res.* 32, D142–D144. doi: 10.1093/nar/gkh088

Mills, R. F., Francini, A., Ferreira Da Rocha, P. S., Baccarini, P. J., Aylett, M., Krijger, G. C., et al. (2005). The plant P_{1B}-type ATPase AtHMA4 transports Zn and Cd and plays a role in detoxification of transition metals supplied at elevated levels. *FEBS Lett.* 579, 783–791. doi: 10.1016/j.febslet.2004.12.040

Mills, R. F., Krijger, G. C., Baccarini, P. J., Hall, J. L., and Williams, L. E. (2003). Functional expression of AtHMA4, a P_{1B}-type ATPase of the Zn/Co/Cd/Pb subclass. *Plant J.* 35, 164–176. doi: 10.1046/j.1365-313XX.2003.01790.x

Morel, M., Crouzet, J., Gravot, A., Auroy, P., Leonhardt, N., Vavasseur, A., et al. (2009). AtHMA3, a P_{1B}-ATPase allowing Cd/Zn/Co/Pb vacuolar storage in *Arabidopsis*. *Plant Physiol.* 149, 894–904. doi: 10.1104/pp.108.130294

Moreno, I., Norambuena, L., Maturana, D., Toro, M., Vergara, C., Orellana, A., et al. (2008). AtHMA1 is a thapsigargin-sensitive Ca^{2+}/heavy metal pump. *J. Biol. Chem.* 283, 9633–9641. doi: 10.1074/jbc.M800736200

Mostofa, M. G., and Fujita, M. (2013). Salicylic acid alleviates copper toxicity in rice (Oryza sativa L.) seedlings by up-regulating antioxidative and glyoxalase systems. *Ecotoxicology* 22, 959–973. doi: 10.1007/s10646-013-1073-x

Petris, M. J., Mercer, J. F., Culvenor, J. G., Lockhart, P., Gleeson, P. A., and Camakaris, J. (1996). Ligand-regulated transport of the Menkes copper P-type ATPase efflux pump from the Golgi apparatus to the plasma membrane: a novel mechanism of regulated trafficking. *EMBO J.* 15, 6084–6095.

Rivas-San Vicente, M., and Plasencia, J. (2011). Salicylic acid beyond defence: its role in plant growth and development. *J. Exp. Bot.* 62, 3321–3338. doi: 10.1093/jxb/err031

Santner, A., and Estelle, M. (2009). Recent advances and emerging trends in plant hormone signalling. *Nature* 459, 1071–1078. doi: 10.1038/nature08122

Seigneurin-Berny, D., Gravot, A., Auroy, P., Mazard, C., Kraut, A., Finazzi, G., et al. (2006). HMA1, a new Cu-ATPase of the chloroplast envelope, is essential for growth under adverse light conditions. *J. Biol. Chem.* 281, 2882–2892. doi: 10.1074/jbc.M508333200

Shikanai, T., Müller-Moulé, P., Munekage, Y., Niyogi, K. K., and Pilon, M. (2003). PAA1, a P-type ATPase of *Arabidopsis*, functions in copper transport in chloroplasts. *Plant Cell* 15, 1333–1346. doi: 10.1105/tpc.011817

Smith, A. T., Smith, K. P., and Rosenzweig, A. C. (2014). Diversity of the metal-transporting P_{1B}-type ATPases. *J. Biol. Inorg. Chem.* 19, 947–960. doi: 10.1007/s00775-014-1129-2

Takahashi, R., Bashir, K., Ishimaru, Y., Nishizawa, N. K., and Nakanishi, H. (2012). The role of heavy-metal ATPases, HMAs, in zinc and cadmium transport in rice. *Plant Signal. Behav.* 7, 1605–1607. doi: 10.4161/psb.22454

Tamura, K., Dudley, J., Nei, M., and Kumar, S. (2007). MEGA4: molecular evolutionary genetics analysis (MEGA) software version 4.0. *Mol. Biol. Evol.* 24, 1596–1599. doi: 10.1093/molbev/msm092

Thompson, J. D., Gibson, T. J., Plewniak, F., Jeanmougin, F., and Higgins, D. G. (1997). The CLUSTAL_X windows interface: flexible strategies for multiple sequence alignment aided by quality analysis tools. *Nucleic Acids Res.* 25, 4876–4882. doi: 10.1093/nar/25.24.4876

Tuskan, G. A., Difazio, S., Jansson, S., Bohlmann, J., Grigoriev, I., Hellsten, U., et al. (2006). The genome of black cottonwood, *Populus trichocarpa* (Torr. & Gray). *Science* 313, 1596–1604. doi: 10.1126/science.1128691

Verret, F., Gravot, A., Auroy, P., Leonhardt, N., David, P., Nussaume, L., et al. (2004). Overexpression of *AtHMA4* enhances root-to-shoot translocation of zinc and cadmium and plant metal tolerance. *FEBS Lett.* 576, 306–312. doi: 10.1016/j.febslet.2004.09.023

Vision, T. J., Brown, D. G., and Tanksley, S. D. (2000). The origins of genomic duplications in *Arabidopsis*. *Science* 290, 2114–2117. doi: 10.1126/science.290.5499.2114

Wilkins, O., Nahal, H., Foong, J., Provart, N. J., and Campbell, M. M. (2009). Expansion and diversification of the *Populus* R2R3-MYB family of transcription factors. *Plant Physiol.* 149, 981–993. doi: 10.1104/pp.108.132795

Williams, L. E., and Mills, R. F. (2005). P_{1B}-ATPases-an ancient family of transition metal pumps with diverse functions in plants. *Trends Plant Sci.* 10, 491–502. doi: 10.1016/j.tplants.2005.08.008

Woeste, K. E., and Kieber, J. J. (2000). A strong loss-of-function mutation in *RAN1* results in constitutive activation of the ethylene response pathway as well as a rosette-lethal phenotype. *Plant Cell* 12, 443–455. doi: 10.1105/tpc.12.3.443

Xu, G., Guo, C., Shan, H., and Kong, H. (2012). Divergence of duplicate genes in exon-intron structure. *Proc. Natl. Acad. Sci. U.S.A.* 109, 1187–1192. doi: 10.1073/pnas.1109047109

Zacchini, M., Pietrini, F., Mugnozza, G. S., Iori, V., Pietrosanti, L., and Massacci, A. (2009). Metal tolerance, accumulation and translocation in poplar and willow clones treated with cadmium in hydroponics. *Water Air Soil Poll.* 197, 23–34. doi: 10.1007/s11270-008-9788-7

Zhang, H. Z., Yang, J. L., Chen, Y. L., Mao, X. L., Wang, Z. C., and Li, C. H. (2013). Identification and expression analysis of the *heat shock transcription factor* (*HSF*) gene family in *Populus trichocarpa*. *Plant Omics* 6, 415–424. doi: 10.1186/1471-2164-13-639

Conflict of Interest Statement: The authors declare that the research was conducted in the absence of any commercial or financial relationships that could be construed as a potential conflict of interest.

9

Chromosome Doubling of Microspore-Derived Plants from Cabbage (*Brassica oleracea* var. *capitata* L.) and Broccoli (*Brassica oleracea* var. *italica* L.)

Suxia Yuan †*, Yanbin Su* †*, Yumei Liu* **, Zhansheng Li, Zhiyuan Fang, Limei Yang, Mu Zhuang, Yangyong Zhang, Honghao Lv and Peitian Sun*

Institute of Vegetables and Flowers, Chinese Academy of Agricultural Sciences, Beijing, China

Edited by:
Naser A. Anjum, University of Aveiro, Portugal

Reviewed by:
Margherita Irene Beruto, Istituto Regionale per la Floricoltura, Italy
Fernando Martinez, University of Seville, Spain

***Correspondence:**
Yumei Liu
liuyumei@caas.cn

† *These authors have contributed equally to this work.*

Chromosome doubling of microspore-derived plants is an important factor in the practical application of microspore culture technology because breeding programs require a large number of genetically stable, homozygous doubled haploid plants with a high level of fertility. In the present paper, 29 populations of microspore-derived plantlets from cabbage (*Brassica oleracea* var. *capitata*) and broccoli (*Brassica oleracea* var. *italica*) were used to study the ploidy level and spontaneous chromosome doubling of these populations, the artificial chromosome doubling induced by colchicine, and the influence of tissue culture duration on the chromosomal ploidy of the microspore-derived regenerants. Spontaneous chromosome doubling occurred randomly and was genotype dependent. In the plant populations derived from microspores, there were haploids, diploids, and even a low frequency of polyploids and mixed-ploidy plantlets. The total spontaneous doubling in the 14 cabbage populations ranged from 0 to 76.9%, compared with 52.2 to 100% in the 15 broccoli populations. To improve the rate of chromosome doubling, an efficient and reliable artificial chromosome doubling protocol (i.e., the immersion of haploid plantlet roots in a colchicine solution) was developed for cabbage and broccoli microspore-derived haploids. The optimal chromosome doubling of the haploids was obtained with a solution of 0.2% colchicine for 9–12 h or 0.4% colchicine for 3–9 h for cabbage and 0.05% colchicine for 6–12 h for broccoli. This protocol produced chromosome doubling in over 50% of the haploid genotypes for most of the populations derived from cabbage and broccoli. Notably, after 1 or more years in tissue culture, the chromosomes of the haploids were doubled, and most of the haploids turned into doubled haploid or mixed-ploidy plants. This is the first report indicating that tissue culture duration can change the chromosomal ploidy of microspore-derived regenerants.

Keywords: cabbage (*Brassica oleracea* var. *capitata* L.), broccoli (*Brassica oleracea*. var. *italica* L.), microspore-derived plants, chromosome doubling, ploidy determination

INTRODUCTION

Microspore culture is an effective alternative technique for the production of doubled haploid (DH) parental lines to generate F_1 hybrids (Abercrombie et al., 2005). The development of DH lines accelerates the plant breeding process by saving time and labor (Ferrie and Caswell, 2011). In addition, DH lines can also be used for marker identification, gene mapping and various genetic manipulations (Forster et al., 2007; Ferrie and Möllers, 2011; Ferrie and Caswell, 2011). Consequently, this technique has been successfully used in cabbage and broccoli, and large-scale DH lines have been developed (Cao et al., 1990; Takahata and Keller, 1991; Duijs et al., 1992; Hansen, 1994; Pink, 1999; da Silva Dias, 2003; Yuan et al., 2009, 2011, 2012). Furthermore, some DH parental lines have even been introduced into breeding schemes (Hale et al., 2007; Lv et al., 2014). The procedure for DH production includes two major steps: haploid induction and chromosome doubling. Consequently, the chromosome doubling of haploids derived from microspores is an important step in the practical application of microspore culture technology. Microspore-derived haploids can spontaneously double their chromosomes during the very early stages of embryogenesis or can be induced to become DHs in the later stages of development (Palmer et al., 1996).

The ideal goal is to double the chromosome number of the original microspore and to then regenerate a plant from the resulting DH microspore. Theoretically, this would result in a stable, homozygous, completely fertile DH. Presently, the mechanism underlying this spontaneous chromosome doubling is unclear in many instances, with wide differences in the responses among and within species (Kasha, 2005). da Silva Dias (2003) and his group found 43–88% spontaneous diploids in broccoli and 7–91% spontaneous diploids in other coles. Because spontaneous doubling occurs randomly and is extremely genotype-dependent, it is important to ascertain the level of spontaneous diploids in the genotype used. da Silva Dias (2003) suggested that for genotypes with over 60% spontaneous doubling, it was not necessary to induce doubling. Obviously, for those genotypes with low spontaneous doubling rates, a successful chromosome-doubling process is essential for the production of homozygous plants after haploid plants are derived from the microspores of plants such as cabbage and broccoli. Various doubling agents have been studied, including caffeine (Thomas et al., 1997), nitrous oxide (Hansen et al., 1988), and the antimicrotubule herbicides trifluralin and amiprophos-methyl (APM; Hansen and Andersen, 1998b). However, the most commonly used chemical agent for chromosome doubling is colchicine (Niu et al., 2014), which disrupts mitosis by inhibiting the formation of spindle fibers and disturbing normal polar chromosomal migration, resulting in chromosome doubling (Jensen, 1974). da Silva Dias (2003) found that when the roots of plants derived from microspores were immersed in a 0.25% colchicine solution, a 53–71% doubling rate was achieved.

Previous studies examining the techniques of microspore culture in cole crops were focused strongly on improving embryogenesis and plant regeneration but ignored research on chromosome doubling. To date, successful microspore culture techniques have been established for cabbage and broccoli (Hansen, 2003; da Silva Dias, 2003; Yuan et al., 2009, 2011, 2012). Flower buds containing a large number of late uninucleate stage microspores and about 10–30% binucleate microspores were selected for microspore cultures in cabbage and broccoli (da Silva Dias, 2003). Heat shock pretreatment was shown to induce microspore embryogenesis in *Brassica oleracea* (Takahata and Keller, 1991; Duijs et al., 1992; da Silva Dias, 2001). In our previous research, the combination of cold pretreatment (4°C) for 1 or 2 days and heat shock (32.5°C) for 1 day significantly enhanced microspore embryogenesis in broccoli (Yuan et al., 2011), and 32.5°C for 1 or 2 days was optimal in cabbage (Yuan et al., 2012). NLN-13 medium and ½ NLN-13 medium were efficient on microspore embryogenesis in cabbage (Duijs et al., 1992) and broccoli (da Silva Dias, 2001), respectively. Furthermore, da Silva Dias (1999) reported that the addition of activated charcoal increased significantly embryo yields in nine genotypes of *B. oleracea*. Our previous study indicated that the combination of 10 mg l^{-1} gum arabic and 3 mM MES in NLN-13 at pH 6.4 significantly enhanced microspore embryogenesis in cabbage (Yuan et al., 2012). Based on the above-mentioned research, a large number of microspore-derived plant in cabbage and broccoli were obtained (Duijs et al., 1992; Yuan et al., 2009). For the useful application of microspore culture techniques in cabbage and broccoli breeding programs, the chromosome doubling of plants derived from microspores must be investigated and improved.

In the present paper, we used 29 populations of microspore-derived plantlets from cabbage (*Brassica oleracea* var. *capitata*) and broccoli (*Brassica oleracea* var. *italica*) to study the ploidy levels and spontaneous chromosome doubling in these populations. Additionally, we examined artificial chromosome doubling induced by colchicine and the influence of tissue culture duration on the chromosomal ploidy of the microspore-derived regenerants. Our objectives were to ascertain the spontaneous doubling of the populations, to assess the impact of tissue culture duration on chromosome doubling in the populations and to develop an efficient and reliable artificial chromosome doubling protocol for microspore-derived haploids of cabbage and broccoli.

MATERIALS AND METHODS

Plant Materials

From 2005 to 2013, the microspore culture of cabbage (*Brassica oleracea* var. *capitata*) and broccoli (*Brassica oleracea* var. *italica*) was undertaken from March to May at the Institute of Vegetables and Flowers, Chinese Academy of Agricultural Sciences, Beijing, China. The microspore isolation and culture procedures were performed as previously described by Yuan et al. (2011, 2012). Briefly, NLN-13 (Yuan et al., 2012) and ½ NLN-13 (da Silva Dias, 2001) were used as liquid microspore culture media for cabbage and broccoli, respectively. The microspores were incubated in the dark at 32.5 ± 1°C for 1 day and then maintained at 25 ± 1°C in the dark. Cotyledonary embryos were obtained from 24-day-cultured microspores. The embryo culture and plant regeneration were carried out according to the procedures described by Yuan et al. (2011). During subsequent tissue culture, the plantlets were subcultured using fresh solid MS-2 medium (Murashige and Skoog, 1962; 2% sucrose, 0.5% agar, 0.1 mg/l 1-naphthaleneacetic acid (NAA), 0.2 mg/l 6-benzylaminopurine (BAP), pH 5.8) every 2 months. A total of 29 populations of cabbage and broccoli microspore-derived plantlets were obtained (**Table 1**). One copy of the microspore-derived plants was continually grown in a tissue culture room, while another copy was transferred to a greenhouse after October and then planted in a field to allow flowering and set seed.

Chromosome Doubling

Spontaneous Doubling

To accurately determine the chromosomal ploidy of the regenerants, the ploidy level of each population mentioned above was identified using two methods: chloroplast counting and morphological identification. Chloroplast counting was used for early ploidy identification in tissue culture, and then to divide each population into haploid group, haploid group, polyploid group and ploid-mixed group in order to facilitate morphological identification in flowering.

Artificial Chromosome Doubling

The haploid genotype 04M1-93 derived from cabbage "Zhonggan No. 11" and the haploid genotype 05B743-49 derived from broccoli "TI-111" were tested in this experiment. To obtain sufficient plants for artificial chromosome doubling research, the two haploids were propagated in MS medium (Murashige and Skoog, 1962; 3% sucrose, 0.5% agar, 0.1 mg/l NAA, 1 mg/l BAP, pH 5.8). In the mid of September, 2007, all of the haploid plants were cut free from the hypocotyl tissue and were transferred to glass growth vessels containing solid ½ MS-2 medium (Murashige and Skoog, 1962; the concentration of the major salts was reduced to 50% compared with MS medium, 2% sucrose, 0.5% agar, 0.1 mg/l NAA, 0.1 mg/l indole-3-butytric acid (IBA), and pH 5.8), in which rooting took place.

The colchicine treatment was as follows: 2 weeks after rooting, the rooted plants were removed from the medium and washed completely with warm tap water; the roots were then trimmed to a length of 1–2 cm and immersed in a working solution of colchicine (supplemented with 2% DMSO). The plants were placed under intense light at 25°C. Next, the solution was poured off, and the roots were rinsed thoroughly in tap water. The treated plants were replanted in a soil mixture (peat soil: perlite: vermiculite, 8:1:1) in a pot and maintained for approximately 2 more weeks in a room at 24°C with a 16-h photoperiod, with low light intensity and high humidity. This was followed by gradual adaptation to greenhouse conditions. Approximately 6 weeks later, the plants were transferred to a cold frame for the duration of the winter. In April of the next year, the plants started flowering, and the ploidy level of the plants was determined using morphological identification.

In this experiment, four concentration levels of colchicine solution were tested: 0.05, 0.1, 0.2, and 0.4%. The plants were treated for 3, 6, 9, and 12 h, in each colchicine solution. Fifteen plants of every haploid genotype were used in each treatment. The same number of plants of each haploid genotype grown without colchicine served as the controls.

TABLE 1 | Microspore-derived cabbage and broccoli plantlet populations with different genotypes were obtained from 2005 to 2013.

Population genotype		Year	Population genotype		Year
Cabbage	Zhonggan No. 11	2005~2007, 2009	Broccoli	Lüxiu	2005~2007
	8398	2005~2007		Lüxing	2005~2007
	Xiwang	2005~2007		TI-111	2005~2007
	Zhonggan No. 18	2005~2007		TI-089	2005~2007
	08F8	2009		Lüyu	2005~2007
	08F9	2009		Yuguan	2005~2007
	08F15	2009		09B411	2010
	08F16	2009		09B524	2010
	08F17	2009		10B635	2011
	08F18	2009		10B644	2011
	08F24	2009		10B665	2011
	08F25	2009		10B908	2011
	Gen 165	2012		10B1092	2011
				12B520	2013
				12B1153	2013

Evaluation of Artificial Chromosome Doubling

Based on the experimental results of 2.2.2, artificial chromosome doubling induced by colchicine was applied to 576 haploid genotypes derived from six cabbage and six broccoli populations to test the chromosome doubling efficiency. In this experiment, 4–8 plants of each haploid genotype, i.e., a total of 3128 haploid plants, were treated with a colchicine solution.

Determination of the Ploidy Level

Chloroplast Number

In the tissue cultures, chloroplast counting (Yuan et al., 2009) was used to determine the initial ploidy level of the populations. In cole crops, haploid plants have at most 10 chloroplasts, diploid plants have 11–15, and polyploid plants have more than 15 chloroplasts.

Morphological Identification

In the mid of September each year, another copy of each of the above-mentioned microspore-derived plantlets obtained in the same year was cut free from the hypocotyl tissue and transferred to a glass growth vessel containing solid ½ MS-2 medium, in which rooting took place. After 2 weeks, the rooted plantlets were transferred to a soil-perlite mixture in a pot. This transfer was then followed by the gradual adaptation to greenhouse conditions. At the end of January of the following year, the plantlets were transferred to a cold frame for the winter. In April of that year, the regenerated plants began flowering, and the population ploidy level was determined using morphological identification.

The sizes of the plants, buds and flowers were observed in the field from 2006 to 2014. The presence or absence of pollen was an additional morphological feature that was determined. Normal diploid cabbage and broccoli plants were used as controls. The characteristics of the regenerated plants with different ploidy levels were described according to the following (**Figure 1**):

Haploid: The growth potential of the plant is weaker, and the plant size is smaller. The flower buds are smaller, flatter and without pollen. The stamens are missing, or the stamen development is not normal.

Diploid: The plant grows normally and has pollen. The stamens and pistils are normal.

Triploid: The flower buds are smaller, flatter and without pollen. The stamens are missing, or the stamen development is not normal.

Tetraploid: The plant has pollen and stronger growth vigor. The plant size and flower buds are larger. The stamens and pistil are normal.

FIGURE 1 | The characteristics of the inflorescences, buds and flowers from different chromosome ploidy plants derived from cabbage "Zhonggan No. 11" microspores. (A) Flowers from different chromosome ploidy plants; **(B)** Buds from different chromosome ploidy plants; **(C)** Inflorescences from different chromosome ploidy plants; (a) Tetraploid; (b) Triploid; (c) Diploid; (d) Haploid.

Data Analysis

The data were analyzed using Microsoft Excel 2003.

RESULTS

Ploidy Level and Spontaneous Chromosome Doubling of the Populations

In total, 1717 regenerants of cabbage and 622 of broccoli were derived and investigated for their ploidy level (**Table 2**). Spontaneous doubling occurred randomly and was extremely genotype dependent. In the microspore-derived populations, in addition to haploids and diploids, there was a low frequency of polyploids and mixed-ploidy plantlets (a plant having both haploid and diploid branches simultaneously; **Figure 2**). For the 14 cabbage genotypes, the spontaneous diploid rates for the populations were in the range of 0–76.9%, and the total spontaneous doubling (including diploids, polyploids, and mixed-ploidy plantlets) ranged from 0 to 84.6%. By contrast, in the 15 broccoli populations, the spontaneous diploid rate was 50.6–100%, and the total spontaneous doubling ranged from 52.2 to 100%. In this experiment, cabbage exhibited a larger variation in the rate of spontaneous doubling in the microspore-derived regenerated populations than did broccoli.

Artificial Chromosome Doubling Induced by Colchicine

The data presented in **Table 3** indicate that the plant survival rate gradually decreased with an increase in the concentration of the colchicine solution. Similarly, for the same concentration of colchicine, the plant survival rate decreased with an increase of the treatment time. Therefore, a higher concentration of colchicine or a longer duration of colchicine treatment had negative effects on the survival of the haploids; this phenomenon was more obvious in broccoli than in cabbage. In addition, colchicine not only induced the doubling of haploids but also induced a certain frequency of haploids to become mixed-ploidy plants (**Figure 2**).

The 0.2% colchicine solution produced negative effects on the survival of the 04M1-93 haploids derived from "Zhonggan No. 11" cabbage; however, the 0.05% colchicine solution produced negative effects on the survival of the 05B743-49 haploids derived from broccoli "TI-111."

When the cabbage haploids were treated with the 0.2% colchicine solution for 9 h, the DH frequency was the highest, up to 50%, followed by the 0.4% colchicine solution for 3 h. However, when the haploids were treated with the 0.4% colchicine solution for 6 h, the total chromosome doubling (including diploids and mixed-ploidy plantlets) was the highest, followed by the 0.4% colchicine for 9 h. Mixed-ploidy plants have diploid branches, the flowers of which exhibit normal fertility and can be pollinated; consequently, these plants have value for breeding. Considering the DH frequency, survival rate and total chromosome doubling, we concluded that a better chromosome doubling effect was obtained when the haploid plants were treated with 0.2% colchicine for 9–12 h or 0.4% colchicine for 3–9 h. The DH

TABLE 2 | Chromosomal ploidy levels of cabbage and broccoli populations derived from microspores.

Materials		Number of populations	Haploid		Diploid		Polyploid		Mixed-ploidy		Rate of chromosome doubling (%)
			Number of plants	Frequency (%)	Number of plants	Frequency (%)	Number of plants	Frequency (%)	Number of plants	Frequency (%)	
Cabbage	Zhonggan No. 11	210	35	16.7	158	75.2	2	1.0	15	7.1	83.3
	8398	381	205	53.8	125	32.8	5	1.3	46	12.1	46.2
	Xiwang	52	8	15.4	40	76.9	0	0	4	7.7	84.6
	Zhonggan No. 8	53	11	20.8	38	71.7	1	1.9	3	5.6	79.2
	Zhonggan No. 21	34	22	64.7	10	29.4	2	5.9	0	0.0	35.3
	08F8	106	52	49.1	49	46.2	2	1.9	3	2.8	50.9
	08F15	3	3	100.0	0	0.0	0	0.0	0	0.0	0.0
	08F16	2	2	100.0	0	0.0	0	0.0	0	0.0	0.0
	08F17	3	2	66.7	1	33.3	0	0.0	0	0.0	33.3
	08F9	116	95	81.9	12	10.3	4	3.5	5	4.3	18.1
	08F18	99	31	31.3	62	62.6	4	4.0	2	2.0	68.7
	08F24	380	171	45.0	192	50.5	3	0.8	14	3.7	55.0
	08F25	266	123	46.2	124	46.6	11	4.2	8	3.0	53.8
	gen 165	12	12	100.0	0	0.0	0	0.0	0	0.0	0.0
Broccoli	Lüxiu	23	11	47.8	12	52.2	0	0.0	0	0.0	52.2
	Lüxing	163	8	4.9	150	92.0	1	0.6	4	2.5	95.1
	TI-111	91	42	46.1	46	50.6	0	0.0	3	3.3	53.9
	TI-089	20	0	0.0	20	100.0	0	0.0	0	0.0	100.0
	Lüyu	152	4	2.6	144	94.7	1	0.7	3	2.0	97.4
	Yuguan	3	0	0.0	3	100.0	0	0	0	0.0	100.0
	09B411	32	6	18.8	23	71.9	1	3.1	2	6.2	81.2
	09B524	4	0	0.0	4	100.0	0	0.0	0	0.0	100.0
	10B908	36	16	44.4	19	52.8	0	0.0	1	2.8	55.6
	10B1092	36	14	38.9	21	58.3	0	0.0	1	2.8	61.1
	10B635	24	5	20.8	19	79.2	0	0.0	0	0.0	79.2
	10B644	21	5	23.8	15	71.4	1	4.8	0	0.0	76.2
	10B665	2	0	0.0	2	100.0	0	0.0	0	0.0	100.0
	12B520	12	3	25.0	9	75.0	0	0.0	0	0.0	75.0
	12B1153	3	0	0.0	3	100.0	0	0.0	0	0.0	100.0

frequency ranged from 23.1 to 50.0%, the total chromosome doubling ranged from 58.3 to 84.6%, and the survival rate ranged from 73.3 to 93.3%.

Similarly, for the 05B743-49 haploids derived from broccoli "TI-111," we concluded that a better chromosome doubling effect was obtained when the haploid plants were treated with 0.05% colchicine for 6–12 h; the DH frequency ranged from 16.7 to 37.5%, the total chromosome doubling ranged from 54.5 to 75.0%, and the survival rate ranged from 57.1 to 80.0%.

Evaluation of Chromosome Doubling in Cabbage and Broccoli Induced by Colchicine

According to the experimental results mentioned above, a 0.2% colchicine treatment for 9–12 h was used for cabbage, and a 0.05% colchicine treatment for 9–12 h was used for broccoli. The plant survival rate of the cabbage population "08F9" was low, at 61.9%, but the survival rates for the 11 other populations ranged from 77.3 to 87.0% (**Table 4**). Among the surviving plants, with the exception of the cabbage population "08F9" and the broccoli population "09B411," which had low total chromosome doubling rates of only 24.6 and 31.7%, respectively, the five other cabbage populations had total chromosome doubling rates ranging from 42.3 to 57.7%, and those for the five other broccoli populations ranged from 43.8 to 60.0%. It is worth mentioning that a higher frequency of chromosome-doubled genotypes was obtained for all of the haploid genotypes of each population compared with the total chromosome doubling efficiency of the surviving plants. For the cabbage population "08F9" and the broccoli population "09B411," the frequencies of the chromosome-doubled genotypes were 40.9 and 39.3%, respectively. For the five other cabbage and five other broccoli populations, the frequencies of the chromosome-doubled genotypes ranged from 49.2 to 69.6% and 64.3 to 72.7%, respectively.

FIGURE 2 | The morphological characteristics of mixed-ploidy plants (a plant having both haploid and diploid branches simultaneously) derived from cabbage "Zhonggan No. 11" microspores. (A) Flowers from a haploid branch; **(B)** Flowers from a diploid branch.

Effect of Tissue Culture Duration on the Chromosome Doubling of the Population

An experiment was performed to determine whether the tissue culture duration could change the chromosomal ploidy of the microspore-derived regenerants.

In this experiment, 9 populations of cabbage and 3 populations of broccoli were tested, and the ploidy variation of the plantlets was observed during the tissue culture. Notably, the chromosome doubling rate of the population gradually increased with an increase in the tissue culture time (**Table 5**). After 1 or more years of tissue culture, the number of chromosomes of the haploids doubled, and most of the haploids turned into DHs or mixed-ploidy plants (data not shown).

DISCUSSION

The chromosome doubling of microspore-derived haploids is an important factor in the practical application of microspore culture technology because breeding programs require a large number of genetically stable, homozygous DH plants with a high level of fertility. Theoretically, a microspore only carries half of the chromosomes of the somatic donor plant, and a plantlet derived from a microspore should be haploid. However, in practice, microspore-derived populations, in addition to haploids, also contain a certain proportion of diploids and even a small number of polyploids. Furthermore, in our experiments, in addition to haploids, diploids and polyploids, a low frequency of mixed-ploidy plantlets (a plant having both haploid and diploid branches simultaneously) was also obtained in cabbage and broccoli. The spontaneous doubling rate varies among crop species and genotypes. In *Brassica napus*, the spontaneous chromosome doubling rates of the microspores range from 10 to 26% (Chen et al., 1994), whereas higher spontaneous doubling rates (50–70%) were observed in Chinese cabbage (Zhang and Takahata, 2001). In broccoli, the diploid rate ranges from 43 to 88%, and in other coles, this rate ranges from 7 to 91% (da Silva Dias, 2003). In our research, the spontaneous diploid rate in cabbage ranged from 0 to 76.9%, and the total spontaneous doubling ranged from 0 to 84.6%. However, in broccoli, the spontaneous diploid rate ranged from 50.6 to 100%, and the total spontaneous doubling ranged from 52.2 to 100% (**Table 2**). Obviously, there was a larger range in the spontaneous chromosome doubling in the microspore-derived regenerated cabbage populations compared with those of broccoli. This is in good agreement with the results presented by da Silva Dias (2003).

A high proportion of spontaneous DH plants is particularly beneficial in breeding because there is no need to use colchicine to double the haploid chromosomes. Spontaneous doubling saves time and labor. To date, the mechanism of spontaneous chromosome doubling remains unclear, while in the mitotic divisions of microspores, the induction of androgenesis and chromosome doubling both appear to involve changes in microfilaments and microtubules. Microfilaments and microtubules may be responsible for the nuclear migration around the uni-nucleate microspore wall. If a pretreatment system for inducing embryogenesis disrupts the microtubules, this type of treatment might lead to chromosome doubling (Kasha, 2005). Colchicine and a number of other anti-microtubule agents have been used to improve chromosome doubling in microspore cultures of *Brassica* crops. Zhao et al. (1996) and Zhou et al. (2002) found that treatment with colchicine instead of heat shock improved the diploid frequency of microspore-derived *Brassica napus* plants. Li and Devaux (2003) found that cold and mannitol pretreatments of barley microspore cultures resulted in high rates of DHs, regardless of the explant (anther or spike) used. Kasha et al. (2001) concluded that the nuclear fusion of microspores was the main mechanism of spontaneous chromosome doubling in barley isolated microspore cultures following mannitol and cold pretreatments. In our previous study, the combination of cold pretreatment and heat shock resulted in a population with more spontaneous DHs compared with heat shock alone (Yuan et al., 2011).

Furthermore, the microspore culture stage could also influence spontaneous chromosome doubling. For example, culturing of early uni-nucleate stages produced predominantly haploid progeny, whereas culturing of bi-nucleate stages produced more DHs and polyploids (Kasha, 2005). Soriano et al. (2007) found that the spontaneous doubling rates (54–66%) of wheat microspores during the late uni-nucleate to early bi-nucleate stages were higher than those (33%) in the mid to late uni-nucleate stages. This may be related to the late uni-nucleate to early bi-nucleate stages being the best microspore stages for the highest embryo induction (Pechan and Keller, 1988).

Although, several factors can affect spontaneous chromosome doubling in microspores, the same microspore culture stage and type of pretreatment were used for the genotypes of the cabbage and broccoli microspore cultures in our study. Our

TABLE 3 | Efficiency of chromosome doubling of different haploids induced by colchicine.

Materials		Concentration of colchicine (%)	Treatment duration (h)	Number of plants	Number of surviving plants	Chromosomal ploidy level			Frequency of chromosome doubling (%)		Survival rate (%)
						Haploid	Diploid	Mixed-ploidy	Diploid frequency	Total chromosome doubling	
Cabbage	04M1-93	0 (CK)	0	15	15	15	0	0	0.0	0.0	100.0
		0.05	3	15	15	12	0	3	0.0	20.0	100.0
			6	15	15	12	0	3	0.0	20.0	100.0
			9	15	15	7	0	8	0.0	53.3	100.0
			12	15	15	7	1	7	6.7	53.3	100.0
		0.10	3	14	14	10	0	4	0.0	28.6	100.0
			6	15	15	9	0	6	0.0	40.0	100.0
			9	14	14	4	3	7	21.4	71.4	100.0
			12	15	15	8	4	3	26.7	46.7	100.0
		0.20	3	15	15	5	0	10	0.0	66.7	100.0
			6	15	13	7	3	3	23.1	46.2	86.7
			9	15	12	5	6	1	50.0	58.3	80.0
			12	15	12	3	3	6	25.0	75.0	80.0
		0.40	3	15	14	3	5	6	35.7	78.6	93.3
			6	15	13	2	3	8	23.1	84.6	86.7
			9	15	11	2	3	6	27.3	81.8	73.3
			12	15	6	3	0	3	0.0	50.0	40.0
Broccoli	05B743-49	0 (CK)	0	4	4	4	0	0	0.0	0.0	100.0
		0.05	3	15	14	8	4	2	28.6	42.9	93.3
			6	15	11	5	3	3	27.3	54.5	73.3
			9	14	8	2	3	3	37.5	75.0	57.1
			12	15	12	5	2	5	16.7	58.3	80.0
		0.10	3	15	13	7	1	5	7.7	46.2	86.7
			6	15	6	1	1	4	16.7	83.3	40.0
			9	15	6	1	1	4	16.7	83.3	40.0
			12	15	3	0	2	1	66.7	100.0	20.0
		0.20	3	15	7	2	1	3	14.3	57.1	46.7
			6	15	6	2	3	1	50.0	66.7	40.0
			9	15	2	0	0	1	0.0	50.0	13.3
			12	15	2	0	1	1	50.0	100.0	13.3
		0.40	3	15	3	1	1	1	33.3	66.7	20.0
			6	15	1	0	0	1	0.0	100.0	6.7
			9	15	0	0	0	0	0.0	0.0	0.0
			12	15	0	0	0	0	0.0	0.0	0.0

results indicate that the spontaneous chromosome doubling rate varies among *B. oleracea* crops and genotypes.

da Silva Dias (2003) suggested that is not necessary to induce doubling in genotypes with over 60% spontaneous doubling. In our research, the spontaneous doubling rate of 10 of the 14 microspore-derived populations (71.4%) of cabbage was less than 60% (**Table 2**). It is clear that the frequency of spontaneous doubling can be very low in cabbage. The spontaneous doubling rate exceeded 60% in most of the broccoli populations in the experiment (**Table 2**). However, a successful chromosome-doubling process is necessary for genotypes with low spontaneous doubling. Therefore, the development of efficient chromosome-doubling protocols is essential for the useful application of DH plants in cabbage and broccoli breeding programs.

Chromosome doubling can be induced in the early stages of gametic embryogenesis or during the developmental stage of haploid plantlets (Ferrie, 2003). Rudolf et al. (1999) found that *in vitro* treatment with anti-microtubule agents can enhance chromosome doubling, but this method can also reduce embryogenesis (Ferrie, 2003) and regenerant frequency (Hansen and Andersen, 1998a) in microspore cultures. Furthermore, this approach can increase the contamination of microspore cultures. In our study, haploid plantlet roots were immersed in a colchicine solution. This method is simple, convenient and targeted. This procedure produced double chromosomes in over 50% of the

TABLE 4 | Application of chromosome doubling in different cabbage and broccoli genotypes induced by colchicine.

Materials		Number of haploid genotypes	Number of plants	Number of surviving plants	Chromosomal ploidy level			Frequency of chromosome doubling (%)		Plant survival rate (%)	Total chromosome doubling	
					Haploid	Diploid	Mixed-ploidy	Diploid frequency	Total chromosome doubling		Number of genotypes	Frequency of genotypes (%)
Cabbage	08F8	72	360	288	146	124	18	43.1	49.3	80.0	39	54.2
	08F9	44	210	130	98	28	4	21.5	24.6	61.9	18	40.9
	08F18	44	189	156	66	84	6	53.8	57.7	82.5	30	68.2
	08F24	129	821	660	330	318	12	48.2	50.0	80.4	76	58.9
	08F25	118	682	532	307	211	14	39.7	42.3	78.0	58	49.2
	Zhonggan No. 11	23	92	80	44	40	6	50.0	57.5	87.0	16	69.6
Broccoli	09B411	56	286	243	166	55	22	22.6	31.7	85.0	22	39.3
	10B635	22	122	101	55	32	14	31.7	45.5	82.8	16	72.7
	10B644	17	88	68	36	24	8	35.3	47.1	77.3	11	64.7
	10B908	20	108	89	50	32	7	36.0	43.8	82.4	13	65.0
	10B1092	28	146	114	54	48	12	42.1	52.6	78.1	18	64.3
	12B520	3	24	20	8	11	1	55.0	60.0	83.3	2	66.7

TABLE 5 | Spontaneous doubling of populations during different tissue culture years.

Materials		Number of populations	Year			
			2010	2011	2012	2013
Cabbage	08F8	106	50.9	66.2	74.3	82.4
	08F9	116	18.1	24.0	38.5	47.8
	08F15	3	0.0	50.0	50.0	100.0
	08F16	2	0.0	25.0	50.0	100.0
	08F17	3	33.3	33.3	67.7	67.7
	08F18	99	68.7	78.6	82.4	96.0
	08F24	380	55.0	76.6	84.4	92.2
	08F25	266	53.8	75.0	82.4	90.0
	Zhonggan 11	145	65.5	77.4	85.2	100.0
Broccoli	09B411	32	–	81.3	93.6	100.0
	10B908	36	–	–	55.6	77.8
	10B1092	36	–	–	61.1	88.9

haploid genotypes for most of the populations (75%) derived from cabbage and broccoli (**Table 4**).

Many factors affect the chromosome-doubling process, such as the colchicine concentration, treatment duration, the addition of other synthetic compounds, and the developmental stage of the plants.

A higher colchicine concentration can increase the doubling rate, but a higher concentration also results in a low survival rate and increased cost. Similarly, the duration of colchicine exposure can have a considerable effect on the induction of chromosome doubling and the survival rate. The optimal colchicine treatment should result in a high plant survival rate and a high rate of chromosome doubling. In this study, 0.2% colchicine for 9–12 h or 0.4% colchicine for 3–9 h for cabbage and 0.05% colchicine for 6–12 h for broccoli produced optimal chromosome haploid doubling effects (**Table 3**).

In our study, the immersion of haploid plantlet roots in a colchicine solution was used to induce chromosome doubling; therefore, the haploid plants used for this purpose must have good root systems. The development of the root systems can affect the colchicine solution absorption efficiency, chromosome doubling and survival rates. Prior to the colchicine treatment, we selected haploid plants with good root systems and then trimmed the roots to better absorb the colchicine solution. Furthermore, 2% DMSO was added to the colchicine solution to increase the root absorption rate.

It is well known that two approaches can be used for chromosome doubling, i.e., spontaneous doubling and artificial chromosome doubling induced by colchicine or other anti-microtubule agents. In our study, which spanned a period of 4 years, we noted that after 1 or more years of tissue culture, the chromosome content of the haploids was doubled, and most of the haploid plants became DHs or mixed-ploidy plants. This phenomenon indicates that the chromosome number of haploids derived from cabbage and broccoli microspores is not stable and can easily be induced to change as a result of external conditions.

This is the first report suggesting that tissue culture duration can change the chromosomal ploidy of the microspore-derived regenerants.

AUTHOR CONTRIBUTIONS

SY performed the experiments, analyzed the data, wrote and revised the manuscript; YS performed the experiments, analyzed the data and revised the manuscript; YL designed the research and critically edited the manuscript; and ZL, ZF, LY, MZ, YZ, HL, and PS planted and managed the plants. All authors approved the final manuscript.

ACKNOWLEDGMENTS

This work was funded by the Chinese National Natural Science Foundation (30871708, 31372067), the Earmarked Fund for Modern Agro-industry Technology Research System (CARS-25-A), the Chinese National Key Technology R&D Program (2013BAD01B04), the National High Technology Research and Development Program (863 Program; 2012AA100104 and 2012AA100105) and the Key Laboratory of Biology and Genetic Improvement of Horticultural Crops, Ministry of Agriculture, P. R. China. We also acknowledge partial funding from the Science and Technology Innovation Program of the Chinese Academy of Agricultural Sciences (CAAS-ASTIP-IVFCAAS).

REFERENCES

Abercrombie, J. M., Farnham, M. W., and Rushing, J. W. (2005). Genetic combining ability of glucoraphanin level and other horticultural traits of broccoli. *Euphytica* 143, 145–151. doi: 10.1007/s10681-005-3059-0

Cao, M. Q., Charlot, F., and Doré, C. (1990). Embryogenesis and plant regeneration in sauerkraut cabbage (*Brassica oleracea* L. subsp. *Capitata*) by *in vitro* culture of isolated microspores. *C. R. Acad. Sci. Paris Ser.* 310, 203–209.

Chen, Z. Z., Snyder, S., Fan, Z. G., and Loh, W. H. (1994). Efficient production of doubled haploid plants through chromosome doubling of isolated microspores in *Brassica napus*. *Plant Breed.* 113, 217–221. doi: 10.1111/j.1439-0523.1994.tb00725.x

da Silva Dias, J. C. (1999). Effect of activated charcoal on *Brassica oleracea* microspore culture embryogenesis. *Euphytica* 108, 65–69.

da Silva Dias, J. C. (2001). Effect of incubation temperature regimes and culture medium on broccoli microspore culture embryogenesis. *Euphytica* 119, 389–394. doi: 10.1023/A:1017563915319

da Silva Dias, J. C. (2003). "Protocol for broccoli microspore culture," in *Doubled Haploid Production in Crop Plants*, eds M. Maluszynski, K. J. Kasha, B. P. Forster, and I. Szarejko (Dordrecht: Kluwer Academic Publishers), 195–204.

Duijs, J. G., Voorrips, R. E., Visser, D. L., and Custers, J. B. M. (1992). Microspore culture is successful in most crop types of *Brassica oleracea* L. *Euphytica* 60, 45–55.

Ferrie, A. (2003). "Microspore culture of Brassica species," in *Doubled Haploid Production in Crop Plants*, eds M. Maluszynski, K. J. Kasha, B. P. Forster, and I. Szarejko (Dordrecht: Kluwer Academic Publishers), 205–215.

Ferrie, A. M. R., and Caswell, K. L. (2011). Isolated microspore culture techniques and recent progress for haploid and doubled haploid plant production. *Plant Cell Tiss. Organ Cult.* 104, 301–309. doi: 10.1007/s11240-010-9800-y

Ferrie, A. M. R., and Möllers, C. (2011). Haploids and doubled haploids in Brassica spp. for genetic and genomic research. *Plant Cell Tiss. Organ Cult.* 104, 375–386. doi: 10.1007/s11240-010-9831-4

Forster, B. P., Heberle-Bors, E., Kasha, K. J., and Touraev, A. (2007). The resurgence of haploids in higher plants. *Trends Plant Sci.* 12, 368–375. doi: 10.1016/j.tplants.2007.06.007

Hale, A. L., Farnham, M. W., Nzaramba, M. N., and Kimbeng, C. A. (2007). Heterosis for horticultural traits in broccoli. *Theor. Appl. Genet.* 115, 351–360. doi: 10.1007/s00122-007-0569-2

Hansen, F. L., Andersen, S. B., Due, I. K., and Olesen, A. (1988). Nitrous oxide as a possible alternative agent for chromosome doubling of wheat haploids. *Plant Sci.* 54, 219–222. doi: 10.1016/0168-9452(88)90116-1

Hansen, M. (1994). "Gametic embryogenesis in *Brassica*: optimization of production and germination of embryos," in *Proceedings of the Int Colloquium on Impact of Plant Biotechnology on Agriculture*, eds B. Javornik, B. Bohanec, I. Kreft (Rogla: Centre for Plant Biotechnology and Breeding; University of Ljubjana), 15–18.

Hansen, M. (2003). "Protocol for microspore culture in Brassica," in *Doubled Haploid Production in Crop Plants*, eds M. Maluszynski, K. J. Kasha, B. P. Forster, and I. Szarejko (Dordrecht: Kluwer Academic Publishers), 195–204.

Hansen, N. J. P., and Andersen, S. B. (1998a). *In vitro* chromosome doubling with colchicine during microspore culture in wheat (*Triticum aestivum* L.). *Euphytica* 102, 101–108. doi: 10.1023/A:1018348816205

Hansen, N. J. P., and Andersen, S. B. (1998b). Efficient production of doubled haploid wheat plants by *in vitro* treatment of microspores with trifluralin or APM. *Plant Breed.* 117, 401–405. doi: 10.1111/j.1439-0523.1998.tb01963.x

Jensen, C. J. (1974). "Chromosome doubling techniques in haploids," in *Haploids in Higher Plants Advances and Potential*, ed K. L. Kasha (Guelph, ON: University of Guelph), 153–190.

Kasha, K. J. (2005). "Chromosome doubling and recovery of doubled haploid plants," in *Haploids in Crop Improvement?. Biotechnology in Agriculture and Forestry, Vol. 56*, eds C. E. Palmer, W. A. Keller, K. J. Kasha (Berlin: Springer), 123–152.

Kasha, K. J., Hu, T. C., Oro, R., Simion, E., and Shim, Y. S. (2001). Nuclear fusion leads to chromosome doubling during mannitol pretreatment of barley (*Hordeum vulgare* L.) microspores. *J. Exp. Bot.* 52, 1227–1238. doi: 10.1093/jexbot/52.359.1227

Li, H. C., and Devaux, P. (2003). High frequency regeneration of barley doubled haploid plants from isolated microspore culture. *Plant Sci.* 164, 379–386. doi: 10.1016/S0168-9452(02)00424-7

Lv, H., Wang, Q., Zhang, Y., Yang, L., Fang, Z., Wang, X., et al. (2014). Linkage map construction using InDel and SSR markers and QTL analysis of heading traits in *Brassica oleracea* var. *capitata* L. *Mol. Breed.* 34, 87–98. doi: 10.1007/s11032-014-0019-1

Murashige, T., and Skoog, F. (1962). A revised medium for rapid growth and bioassays with tobacco tissue cultures. *Plant Physiol.* 15, 473–497.

Niu, Z., Jiang, A., Abu Hammad, W., Oladzadabbasabadi, A., Xu, S. S., Mergoum, M., et al. (2014). Review of doubled haploid production in durum and common wheat through wheat × maize hybridization. *Plant Breed.* 133, 313–320. doi: 10.1111/pbr.12162

Palmer, C. E., Keller, W. A., and Arnison, P. G. (1996). "Utilization of Brassica haploids," in *In Vitro Haploid Production in Higher Plants*, eds S. M. Jain, S. K. Sopory, and R. E. Veilleux (Dordrecht: Kluwer Academic Publishers), 173–192.

Pechan, P. M., and Keller, W. A. (1988). Identification of potentially embryogenic microspores in *Brassica napus*. *Physiol. Plant* 74, 377–384.

Pink, D. (1999). "Application of doubled haploid technology and DNA Markers in breeding for Clubroot resistance in *Brassica oleracea*," in *COST-824 Gametic Embryogenesis Workshop, Book of Abstracts*, ed M. Hansen (Krakow: European Commission), 5–7.

Rudolf, K., Bohanec, B., and Hansen, M. (1999). Microspore culture of white cabbage, Brassica oleracea var. *capitata* L.: genetic improvement of non-responsive cultivars and effect of genome doubling agents. *Plant Breed.* 118, 237–241.

Soriano, M., Cistué, L., Vallés, M. P., and Castillo, A. M. (2007). Effects of colchicine on anther and microspore culture of bread wheat (*Triticum aestivum* L.). *Plant Cell Tiss. Organ Cult.* 91, 225–234. doi: 10.1007/s11240-007-9288-2

Takahata, Y., and Keller, W. A. (1991). High frequency embryogenesis and plant regeneration in isolated microspore culture of *Brassica oleracea* L. *Plant Sci.* 74, 235–242. doi: 10.1016/0168-9452(91)90051-9

Thomas, J., Chen, Q., and Howes, N. (1997). Chromosome doubling of haploids of common wheat with caffeine. *Genome* 40, 552–558. doi: 10.1139/g97-072

Yuan, S., Liu, Y., Fang, Z., Yang, L., Zhuang, M., Zhang, Y., et al. (2009). Study on the relationship between the ploidy level of microspore-derived plants and the number of chloroplast in stomatal guard cells in *Brassica oleracea*. *Agric. Sci. China* 8, 939–946. doi: 10.1016/S1671-2927(08)60298-9

Yuan, S., Su, Y., Liu, Y., Fang, Z., Yang, L., Zhuang, M., et al. (2012). Effects of pH, MES, arabinogalactan-proteins on microspore cultures in white cabbage. *Plant Cell Tiss. Organ Cult.* 110, 69–76. doi: 10.1007/s11240-012-0131-z

Yuan, S. X., Liu, Y. M., Fang, Z. Y., Yang, L. M., Zhuang, M., Zhang, Y. Y., et al. (2011). Effect of combined cold pretreatment and heat shock on microspore cultures in broccoli. *Plant Breed.* 130, 80–85. doi: 10.1111/j.1439-0523.2009.01754.x

Zhang, F. L., and Takahata, Y. (2001). Inheritance of microspore embryogenic ability in *Brassica* crops. *Theor. Appl. Genet.* 103, 254–258. doi: 10.1007/s001220100602

Zhao, J., Simmonds, D. H., and Newcomb, W. (1996). Induction of embryogenesis with colchicine instead of heat in microspores of *Brassica napus* L. cv. Topas. *Planta* 198, 433–439. doi: 10.1007/BF00620060

Zhou, W. J., Hagberg, P., and Tang, G. X. (2002). Increasing embryogenesis and doubleing efficiency by immediate colchicine treatment of isolated microspores in spring Brassica napus. *Euphytica* 128, 27–34. doi: 10.1023/A:1020687517888

Conflict of Interest Statement: The authors declare that the research was conducted in the absence of any commercial or financial relationships that could be construed as a potential conflict of interest.

10

Examination of the Abscission-Associated Transcriptomes for Soybean, Tomato, and Arabidopsis Highlights the Conserved Biosynthesis of an Extensible Extracellular Matrix and Boundary Layer

Joonyup Kim[1], Srivignesh Sundaresan[2,3], Sonia Philosoph-Hadas[2], Ronghui Yang[1], Shimon Meir[2] and Mark L. Tucker[1]*

[1] Soybean Genomics and Improvement Lab, Agricultural Research Service, United States Department of Agriculture, Beltsville, MD, USA, [2] Department of Postharvest Science of Fresh Produce, Agricultural Research Organization, The Volcani Center, Bet-Dagan, Israel, [3] The Robert H. Smith Faculty of Agriculture, Food and Environment, The Hebrew University of Jerusalem, Rehovot, Israel

Edited by:
Richard S. Winder, Natural Resources Canada, Canada

Reviewed by:
Panagiotis Kalaitzis, Mediterranean Agronomic Institute of Chania, Greece
Long Mao, Chinese Academy of Agricultural Sciences, China

***Correspondence:**
Mark L. Tucker
mark.tucker@ars.usda.gov

Abscission zone (AZ) development and the progression of abscission (detachment of plant organs) have been roughly separated into four stages: first, AZ differentiation; second, competence to respond to abscission signals; third, activation of abscission; and fourth, formation of a protective layer and post-abscission trans-differentiation. Stage three, activation of abscission, is when changes in the cell wall and extracellular matrix occur to support successful organ separation. Most abscission research has focused on gene expression for enzymes that disassemble the cell wall within the AZ and changes in phytohormones and other signaling events that regulate their expression. Here, transcriptome data for soybean, tomato and Arabidopsis were examined and compared with a focus not only on genes associated with disassembly of the cell wall but also on gene expression linked to the biosynthesis of a new extracellular matrix. AZ-specific up-regulation of genes associated with cell wall disassembly including cellulases (beta-1,4-endoglucanases, *CELs*), polygalacturonases (*PGs*), and expansins (*EXPs*) were much as expected; however, curiously, changes in expression of xyloglucan endotransglucosylase/hydrolases (*XTHs*) were not AZ-specific in soybean. Unexpectedly, we identified an early increase in the expression of genes underlying the synthesis of a waxy-like cuticle. Based on the expression data, we propose that the early up-regulation of an abundance of small pathogenesis-related (*PR*) genes is more closely linked to structural changes in the extracellular matrix of separating cells than an enzymatic role in pathogen resistance. Furthermore, these observations led us to propose that, in addition to cell wall loosening enzymes, abscission requires (or is enhanced by) biosynthesis and secretion of small proteins (15–25 kDa) and waxes that form an extensible extracellular matrix and boundary layer on the surface of separating cells. The synthesis of the boundary layer precedes what is typically associated with the post-abscission synthesis of a protective scar over the fracture plane. This modification in the abscission model is discussed in regard to how it influences our interpretation of the role of multiple abscission signals.

Keywords: abscission, Arabidopsis, ethylene, *IDA*, soybean, tomato, transcriptome, cuticle biosynthesis

INTRODUCTION

The process of abscission is roughly divided into four sequential stages with slight modifications by different authors (Patterson, 2001; Estornell et al., 2013): first, abscission zone (AZ) differentiation; second, competence to respond to abscission signals; third, activation of abscission; and fourth, formation of a protective layer and post-abscission trans-differentiation. It was recognized many years ago, based on light and electron microscopy, that during the activation of abscission (i.e., abscission stage 3) AZ cells expand at the fracture plane and the cells separate along the middle lamella (Hall and Sexton, 1974). It has been assumed that the swelling of these cells helps to create the forces necessary to break vascular connections (i.e., rigid xylem vessels), which then allows the organs to separate. Many genes and proteins linked to cell wall loosening and degradation of the middle lamella including cellulases (beta-1,4-endoglucanases, CELs), polygalacturonases (PGs), xyloglucan endotransglucosylase/hydrolases (XTHs), and expansins (EXPs) have been identified and characterized from AZ of several different plant species (Roberts et al., 2002). Also commonly observed at this stage in organ abscission is an up-regulation of defense genes, which includes pathogenesis-related (*PR*) genes (Del Campillo and Lewis, 1992; Meir et al., 2011; Gonzalez-Carranza et al., 2012; Estornell et al., 2013). The role of *PR* gene expression during abscission is often assumed to protect vulnerable abscising cells from opportunistic pathogen invasion (Del Campillo and Lewis, 1992).

In addition to cell wall loosening and *PR* genes, a great deal of attention has been directed toward the identification and characterization of signals that initiate abscission (Taylor and Whitelaw, 2001; Roberts et al., 2002; Liljegren, 2012). Phytohormones and other signals that have been shown to modulate abscission are many-fold: ethylene, auxin, abscisic acid, jasmonic acid, and IDA (a small secreted peptide named INFLORESCENCE DEFICIENT IN ABSCISSION) (Taylor and Whitelaw, 2001; Roberts et al., 2002; Liljegren, 2012; Kim et al., 2013). These abscission signals, however, may not all act in synchrony but sequentially or independently (Patterson and Bleecker, 2004). It is not our intention here to evaluate the role of a multitude of abscission signals, but rather focus on the cellular processes that these signals initiate and regulate. A more complete understanding of the processes and mechanisms utilized for successful organ separation will help to clarify the interactions and interdependencies of the various signals that regulate these processes.

Technological advances have made it possible to now obtain expression results for the entire transcriptome of any desired tissue or developmental program (Zhong et al., 2011; Grassi et al., 2013). In the present work we performed RNA-seq of soybean leaf AZs (LAZ) and the petioles after the AZs were removed (non-AZ petiole, NAZ-pet) harvested from explants exposed to ethylene for 0, 12, 24, 48, and 72 h, and compared the results to microarray data for RNA collected from tomato flower pedicel AZ (FAZ) and proximal non-AZ pedicel (NAZ) at 0, 4, 8, 12, 16, and 20 h after removal of the flowers. In addition, we compared the soybean and tomato results to RNA-seq results for the transcriptomes of Arabidopsis flower receptacles from wild type (WT) and the *hae-hsl2* double mutant (Niederhuth et al., 2013). If left untouched, the petals, sepals and stamens of the *hae-hsl2* mutant do not abscise (Niederhuth et al., 2013) The Arabidopsis data is not a time-course study of differential gene expression during floral organ abscission; nevertheless, the AZ tissues (receptacles) were collected at an early stage of abscission (stage 15) when changes in gene expression for cell separation were first evident in the WT AZ, and, therefore, useful for comparing to early stages of soybean and tomato abscission. Inclusion of the Arabidopsis results in comparison to the soybean and tomato results further highlights shared processes associated with abscission and also the independence and interdependence of regulatory pathways controlling abscission.

We focus here on genes associated with the disassembly and modification of the primary cell wall of the AZ cells, and also the synthesis of a new and different extracellular matrix on the AZ cells as abscission progressed. As expected, we found many homologous and orthologous genes annotated as cell wall loosening and PR proteins that were previously reported to be up-regulated during abscission in many species (Tucker et al., 1988; Kalaitzis et al., 1997; Roberts et al., 2002; Lashbrook and Cai, 2008; Meir et al., 2010). Of special interest here was the early and abundant expression in all three transcriptomes of genes encoding small (15–25 kDa) secreted proteins and genes associated with the deposition of a wax-like cuticle. The expression profiles for many of these genes were very similar to those for cell wall disassembly, i.e., cellulase, PGs, and EXPs. These observations led us to propose that secretion of small proteins and a wax-like cuticle might play an important role in restructuring the extracellular matrix to facilitate organ separation independent of the deposition of a protective scar after separation is complete. The need to synthesize a boundary layer during separation is discussed in regard to abscission signals and their regulation of gene expression required for successful organ separation.

METHODS

Plant Material

Soybean (*Glycine max*, cv. Williams82) plants were grown in the greenhouse and harvested when the primary leaves were fully expanded (19–24 days). Stem-petiole explants were prepared by cutting the stem approximately 4 cm below the leaf node and removing all but approximately a 5-mm triangular portion of the primary leaf blade (**Supplemental Figure S1**). Explants were placed in Erlenmeyer flasks with water and put into a darkened chamber wherein 25 μL/L ethylene in air saturated with water was passed through at a rate of 2 L/min at 25°C. In soybean, there is an AZ at the base of the petiole at the juncture with the stem (lower AZ) and another AZ at the distal end of the petiole approximately 1 mm below the leaf blade (upper AZ) (**Supplemental Figure S1**). In our system with ethylene treatment, the petiole separated from the stem at the lower AZ at approximately 48 h from just the weight of the petiole, but the distal portion of the upper AZ at the top of the petiole, which is rather small and light, did not sometimes fall away from the petiole even after 72 h of ethylene. To assess the extent of separation at the upper AZ, the distal part of the AZ was gently touched with forceps and, if it fell away from the petiole, the AZ was recorded as having fully abscised. Nonetheless, with high humidity, the petiole and AZ continue to senesce after separation from the parent plant, and did not dry out. Approximately 2 mm, which included approximately 1 mm of proximal and distal sides of the upper, leaf AZ (LAZ), was collected from 20 explants (2 upper AZs per explant or 40 AZs total) and flash frozen in liquid nitrogen. After excising the AZ, the petiole ends were again trimmed to avoid incidental collection of any AZ tissue and the petioles (NAZ-pet) were flash frozen. The lower AZ was not collected, because at this stage of growth the petioles are relatively small and we wanted to avoid incidental collection of a portion of the lateral bud that is very close to the lower AZ. Collection of a small part of a lateral bud with a meristem in it would compromise the interpretation of AZ-specific gene expression. AZs and petioles (NAZ-pet) were collected at 0 h (immediately prior to ethylene treatment) and at 12, 24, 48, and 72 h after the ethylene treatment had begun. Plants were grown, explants prepared and samples collected three separate times to give three independent experimental replicates.

Flower clusters (inflorescences) from tomato (*Solanum lycopersicum*) cv. New Yorker were harvested from 4-month-old greenhouse-grown plants between 08:00 and 10:00 a.m. Explants were prepared as previously described (Meir et al., 2010). Briefly, inflorescences bearing at least 2–4 freshly open flowers were brought to the laboratory under high humidity conditions. Closed young flower buds and senesced flowers including the entire pedicel and AZ joint were removed so that only freshly opened flowers and pedicels remained on the inflorescence. The stem ends were trimmed, and groups of 2–3 inflorescence explants were placed in vials containing double distilled water (DDW). Abscission was induced by cutting the flowers off the inflorescence at the base of the receptacle. Cutting off the flowers removes a source of auxin that inhibits abscission (Meir et al., 2010). Tissue samples for RNA extraction were taken from the flower AZ (FAZ) and proximal pedicel (NAZ) of 30 segments. The 2 mm FAZ collection included approximately 1 mm on either side of the AZ joint or fracture once it appeared. Samples were collected at 0 h (immediately before flower removal) and 4, 8, 12, 16, and 20 h after flower removal. Explants were harvested and tissue samples collected twice for two independent experimental replicates.

RNA Sequencing and Microarrays

RNA was isolated from soybean LAZ and NAZ-pet after 0, 12, 24, 48, and 72 h of exposure to ethylene using a Qiagen RNeasy Mini Kit following the standard protocol (Qiagen, Germantown, MD, USA). Each experiment produced 10 RNA samples, which resulted in 30 RNA samples for the 3 replicate experiments. Further RNA purification, cDNA synthesis and sequencing on an Illumina GAII sequencer were performed at Cornell University, Ithaca, NY, USA as previously described (Zhong et al., 2011; Grassi et al., 2013). The 30 RNA samples were processed, barcoded and run together on the GAII sequencer. The raw sequence files have been submitted to the NCBI SRA databases with the study accession SRP050050. On average, each RNA sample produced approximately 4 million reads (**Supplemental File S2**). Raw sequences were trimmed to remove ambiguous ends. Using Bowtie (Langmead et al., 2009), approximately 40,000 (1%) of the reads mapped to ribosomal RNA (rRNA) and were removed from the data set. Using TopHat (Trapnell et al., 2009), approximately 90% of the remaining RNA mapped to a predicted soybean transcriptome (cds) (*G. max* 189 genome assembly). Multiple versions (splice variants) were not taken into account. A single version (usually the last version) was used for alignment. A total of 54,175 transcripts were used for the alignment. The number of reads that aligned to a gene was normalized as Reads Per Kilobase per Million mapped reads (RPKM) (Mortazavi et al., 2008).

For a gene to be counted as expressed in the LAZ or NAZ-pet, we required that the mean RPKM for the three replicates be at least 1.0 or greater in at least one of the treatments. Using a cutoff of 1.0 RPKM resulted in the selection of 37,572 genes as being expressed in the soybean LAZ or NAZ-pet between 0 and 72 h of exposure to ethylene (**Supplemental File S2**). Then, to avoid ratios with a zero in the numerator or denominator, any RPKM of less than 0.1 was given the minimal value of 0.1. QPCR was performed as previously described (Tucker et al., 2007) on a few selected genes to confirm that the RNA-seq and RPKM normalization produced the expected expression profile (results not show).

The Meir lab at the Volcani Center, Bet-Dagan, Israel has prepared a tomato AZ-specific microarray chip in collaboration with Genotypic Technology Pvt. Ltd., Bangalore, India. Each chip includes 111,718 probe sets for more than 40,000 transcripts. Probe sets were designed using RNA-seq results for pooled RNA samples from non-induced and induced tomato flower pedicel and AZ tissue as described above. However, for the RNA sequencing, the tissues were collected from *Solanum lycopersicum* cv. "VF-36." Transcriptome libraries for sequencing were constructed according to the Illumina TruSeq RNA library protocol outlined in "TruSeq RNA Sample Preparation Guide"

(Part # 15008136; Rev. A, Illumina, USA). The DNA obtained from the prepared libraries was denatured and sequenced on the Illumina Genome Analyzer IIX, using the sequencing by synthesis method to read 72 bases per end. The raw sequencing data were then extracted from the server using the proprietary Illumina pipeline software to obtain a sequence data set in a Fastq format. Quality check of raw data was performed using SeqQC –V2.0 program. The raw sequence data and array information were submitted to the Gene Expression Omnibus (GEO) at the National Center for Biotechnology Information (NCBI) with GEO IDs GSE45355 and GSE45356, and array ID AMADID:043310. Expression was validated by qPCR (results not shown).

For the tomato microarray results, a gene was considered to be expressed, if the minimum mean signal for two replicates was >10 in at least one time point in either the NAZ or FAZ. This selection criterion resulted in the selection of 26,527 genes as being expressed in the FAZ or NAZ. Then, to avoid ratios with a zero in the numerator or denominator, any signal of less than 1.0 was given the minimal value of 1.0.

The RNA-seq results reported here for Arabidopsis WT and the *hae-hsl2* mutant are a reformulation of data from Additional file 2 in Niederhuth et al. (2013). Arabidopsis receptacles, which included the floral organ AZs, were collected as described (Niederhuth et al., 2013). To conform to the analysis of the soybean and tomato transcriptomes, the log2 fold change was reversed to WT over *hae-hsl2*. Similar to the other data sets, a minimum of >10 reads per gene was required to consider the gene as being expressed in either WT or the mutant, and any reads per gene of less than 1.0 was given a minimal value of 1.0 to avoid complications arising from having a zero in the numerator or denominator (**Supplemental File S4**).

RESULTS AND DISCUSSION

Differences in Experimental Design

The objective of this study was to identify processes directly linked to successful organ separation that are common to a variety of abscission systems. However, how abscission data is collected is not done the same by all researchers. Therefore, what first needs to be addressed and fully understood are the differences in experimental approaches used to study abscission in soybean, tomato and Arabidopsis. The explant system used to study abscission in soybean and tomato was established many years ago (Addicott, 1982; Sexton and Roberts, 1982). It is a versatile system in which a variety of hormones and chemicals can be exogenously applied. An important part of the explant system is the removal of the source of auxin, i.e., soybean leaves or tomato flowers. Moreover, in soybean and tomato, ethylene is essential for leaf and flower abscission, respectively (Lanahan et al., 1994; Meir et al., 2010; Tucker and Yang, 2012). In the soybean experiments, the explants were continuously exposed to a high concentration of ethylene (25 μL/L), which synchronizes and accelerates abscission, but also induces senescence and ethylene-regulated defense gene expression (Abeles et al., 1992). After a 72 h exposure to ethylene, both the petiole and AZ turned yellow. Thus, treating the entire explant with ethylene removes the variable of none uniform synthesis of ethylene between the AZ and petiole, and reduces differential gene expression linked solely to ethylene.

In the tomato experiments, the inflorescences were not treated with ethylene. However, because ethylene is essential for tomato flower abscission (Lanahan et al., 1994; Meir et al., 2010), it is likely that upon removal of the auxin source ethylene was synthesized in the inflorescence but the synthesis of ethylene may not have been equal in the FAZ and NAZ (discussed later). Nonetheless, the tomato system reflects a more natural process that would occur during flower abscission on the intact plant.

The leaves and flowers of Arabidopsis do not normally abscise and, therefore, Arabidopsis abscission scientists study the abscission of floral organs, petals, stamens, and sepals (Bleecker and Patterson, 1997; Patterson, 2001; Butenko et al., 2003). The Arabidopsis data we used here was collected and published by Niederhuth et al. (2013). They collected flower receptacles from WT and the double mutant *hae-hsl2* at developmental stage 15 flowers. HAE and HSL2 are redundant receptor-like kinases that bind the IDA (INFLORESCENCE DEFICIENT IN ABSCISSION) signaling peptide (Cho et al., 2008; Stenvik et al., 2008; Butenko et al., 2014). In the *ida* mutant, as in the *hae-hsl2* double mutant, the floral organs do not abscise if they are left untouched (Butenko et al., 2003). At flower stage 15 both WT and mutant flowers are fully open but the petals have not abscised (Niederhuth et al., 2013). Stage 15 flowers correspond to approximately the same developmental stage as flowers at positions 3 and 4 used in other studies of Arabidopsis floral organ abscission (Patterson and Bleecker, 2004; Cai and Lashbrook, 2008; Cho et al., 2008; Basu et al., 2013; Liu et al., 2013). In wild-type flowers at positions 3–4 abscission-associated gene expression has already begun and the break-strength of the petals has started to decline but the petals have not fallen off under their own weight (Kim and Patterson, 2006; Cai and Lashbrook, 2008; Niederhuth et al., 2013). The levels of auxin and ethylene in this system are unknown. Nonetheless, ethylene can accelerate abscission of Arabidopsis floral organs (Gonzalez-Carranza et al., 2007) but it does not appear to be essential because floral organ abscission is only delayed in ethylene-insensitive mutants (Patterson and Bleecker, 2004). Moreover, auxin also plays a role in Arabidopsis floral organ abscission (Basu et al., 2013). If auxin levels are genetically up-regulated or down-regulated in the AZ of Arabidopsis floral organs, abscission is delayed or accelerated, respectively (Basu et al., 2013). Thus, although the experimental approach to study Arabidopsis abscission is different from soybean and tomato and ethylene appears not to be essential to Arabidopsis floral organ abscission, comparison of differential gene expression in the transcriptomes of WT and the *hae-hsl2* mutant to soybean and tomato provides additional information and confirmation of conserved molecular and metabolic processes that contribute to organ separation.

Overview of Differential Gene Expression

We plotted the overall change in expression for those genes that changed markedly over time in the AZ and NAZ of soybean and tomato and between WT and *hae-hsl2* of Arabidopsis. For soybean and tomato we used an arbitrary threshold for a change in gene expression of >8-fold (log2>3 or <-3) (**Figure 1**). In soybean where the auxin source has been removed and the explants treated with ethylene, approximately 5% of all the expressed genes displayed an increase in expression of >8-fold in both the AZ and petiole; however, the expression of a much greater number of genes, 30%, decreased >8-fold (**Figure 1A**). Although 5% of all the genes increased and 30% decreased, only about 1% of the genes were expressed in an AZ-specific manner. Interestingly, at 72 h the 5% of the genes that increased in expression accounted for 50% of the transcripts in the transcriptome of both the AZ and petiole (**Figure 1A**). There was also a significant increase in transcripts that were AZ-specific but still much less than the total. In tomato, there was a similar but much smaller decline in gene expression that is common to both AZ and NAZ (**Figure 1B**). In tomato flowers, which abscise much more rapidly than the soybean leaves, approximately 2% of the genes changed expression >8-fold by 16 h, which accounted for almost 9% of the transcripts in the transcriptome; however, 90% of the gene expression was AZ-specific (**Figure 1B**). A possible explanation for why most of the differential expression >8-fold was AZ-specific in tomato may be that ethylene synthesis was greater in the AZ than the NAZ (discussed later).

In Arabidopsis, the magnitude of differential gene expression between WT and the mutant was relatively small. Less than 2% of the genes increased in expression more than 2-fold, log2>1, which accounted for only a 0.5% change in the total transcriptome (**Figure 1C**). There are at least two explanations for the small difference in gene expression between the WT and *hae-hsl2* receptacles. First, the receptacle collection was an early stage of abscission, and, second, the IDA signaling path is one of several signals that regulate gene expression in Arabidopsis abscission (discussed later).

The leaves and flowers were removed from the soybean and tomato explants, respectively, and, as expected, the removal of the auxin source is reflected in a general decline in auxin associated gene expression in both the AZ and NAZ (**Figures 2, 3**). Nonetheless, there were a few auxin-associated genes that increased in an AZ-specific manner in both soybean and tomato (**Figures 2, 3**) and many of these were linked to auxin movement (e.g., PIN) or auxin conjugation (e.g., GH3) (**Supplemental Files S2, S3**).

The soybean explants were treated with ethylene and because of this there was an expected non AZ-specific increase in gene expression for ethylene-associated genes linked to senescence and programmed cell death (PCD) (**Figures 2A, 3A**). There was also an expected increase in ethylene-induced defense gene expression (*PR* gene expression) (**Figures 2A, 3A**); nonetheless, several *PR* genes were expressed AZ-specifically (see below). In tomato, the increase in ethylene-associated genes (senescence, PCD and *PR* genes) was mostly AZ-specific (**Figures 2B, 3B**). In both soybean and tomato there was an AZ-specific increase in expression of genes for ethylene synthesis (i.e., *ACS* and *ACO*) (**Supplemental Files S2, S3**). This would suggest that in tomato ethylene synthesis was AZ specific whereas in soybean, because the explants were treated with a high concentration of ethylene, the AZ-specific synthesis of ethylene was inconsequential. Thus, in tomato, ethylene-induced gene expression was AZ-specific due to the synthesis of ethylene in the AZ tissue.

Changes in the Transcriptome Linked to Cell Wall Disassembly

Genes within the categories for cellulases (*CELs)*, pectinases (*PGs*) and pectin lyase-like (*PGs* and *PLs*), *EXPs*, *XTHs* were selected based primarily on their annotation, which is most often derived from the most similar gene sequence in Arabidopsis. The category labeled cellulases included any gene annotated as cellulase or glycoside hydrolase (GH) family 9, but genes in the subfamily GH9A1 were placed into the cellulose biosynthesis category, which is more appropriate for *KORRIGAN-like* cellulase genes (Doblin et al., 2002).

As expected, AZ-specific gene expression for cell wall disassembly proteins increased markedly in soybean, tomato and Arabidopsis (**Figures 1, 2**). Of interest in regard to cell wall disassembly was an unexpected result for *XTH* expression in soybean leaf abscission. Transmission electron micrographs indicate that cellulose microfibrils (striations) are untethered during abscission but not degraded (Hall and Sexton, 1974). It is generally accepted that a xyloglucan network plays an important role in tethering cellulose microfibrils and, along with pectins distributed within the primary cell wall, may influence the free movement of proteins and other large compounds across the primary cell wall out to the middle lamella (Carpita and Gibeaut, 1993; Cosgrove, 2001). It was surprising to find that in soybean no *XTHs* were up-regulated in an AZ-specific manner and most *XTHs* declined (**Figure 2**). However, in tomato and Arabidopsis there was an AZ-specific increase in expression of *XTH*. Ethylene treatment of soybean may be part of the reason for this difference. It is plausible that XTHs also play a role in in ethylene-induced senescence and this is why XTH was not LAZ-specific. However, there is another possibility. Cellulase (beta-1,4-endoglucanase) activity is assayed using carboxymethyl cellulose, which is a soluble cellulose derivative (Urbanowicz et al., 2007). *In vitro* assays with purified plant cellulases by themselves do not degrade crystalline cellulose microfibrils. The *in vivo* substrates for cellulases are not known. Therefore, it is possible that some cellulases may cleave chains of beta-1,4-glucans within xyloglucan polymers that fix the cellulose microfibrils in place (Hayashi and Kaida, 2011; Eklöf et al., 2012), or they cleave beta-1,4-glucans at the surface of the microfibrils that play a role in tethering and crosslinking the microfibril network. Transcription of *GmCEL01* increased earlier than other cell wall loosening enzymes during soybean abscission (**Figure 3A**). It is possible that the role of the GmCEL1 enzyme is to loosen the hemicellulosic fraction (xyloglucans, etc.) or their tethering to the cellulose microfibrils, which then opens up the cell wall for movement of protein and other compounds out to the middle lamella.

A heat map (**Figure 2**) is useful to display a general perspective on the expression patterns within each category of genes, but it does not take into account transcript abundance. For example, *GmCel01*, (Glyma11g02350, accession U34755) accounted for

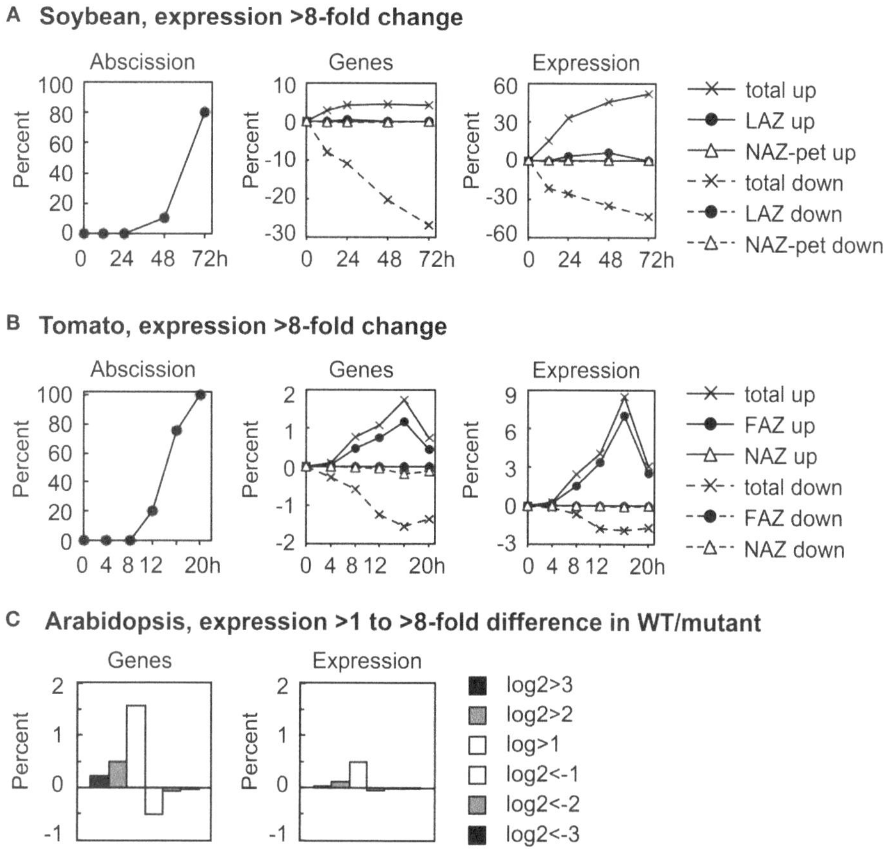

FIGURE 1 | Overall perspective on major changes in gene expression during abscission of soybean leaves, tomato flowers and Arabidopsis floral organs. (A) The percent of soybean pulvinar leaf abscission and the percent of genes with greater than 8-fold change (log2 >3 or < −3) in expression relative to the total number of genes expressed (37,572) and the differential expression of the same genes relative to the average total RPKM for each RNA sample (666,213 RPKM). **(B)** Similar to **(A)** except that tomato flower abscission had 26,572 expressed genes with a total average microarray signal of 64,696,620. **(C)** The percent of expressed genes in wild-type Arabidopsis receptacles relative to genes in the double mutant *hae-hsl2* and the differential expression of the same genes (Niederhuth et al., 2013). In the Arabidopsis data, 20,883 genes were counted as expressed with a total mean of 38,770,114 reads. Because differential expression was much less in the Arabidopsis data, 2-, 4-, and 8-fold differences are plotted as a bar graph.

approximately 75% of the RNA for the 27 *CELs* expressed in the soybean LAZ (**Figure 3A**). The same was true for tomato where *SlCEL1* accounted for 75% of the transcript for 17 *CELs* (**Figure 3B**). Similarly, one or just a few *PGs* accounted for most of the transcript in both soybean and tomato AZ (**Figures 3A,B**). This was not necessarily the case for other gene families (**Figure 3** and **Supplemental Files S2–S4**).

Expression of EXPs in abscission was more complex. In soybean, *EXPs* increased markedly in both the LAZ and petiole but only alpha *EXPs* were AZ-specific, but the AZ-specific expression of alpha *EXPs* was less than 10% of the overall expression for *EXPs* (**Figure 3A**). *EXP* expression in tomato had an AZ-specific component to it but the level of *EXP* gene expression was already relatively high at 0 h (**Figure 3B**). Based on our results, the role of EXPs in abscission is not easily interpreted.

Gene expression for cell wall disassembly in Arabidopsis abscission will be discussed in greater detail later. Nonetheless, the expression of *CELs*, *PGs*, *EXPs*, and *XTHs* were significantly higher in WT than the mutant (**Figure 3C** and **Supplemental File S4**). What was most surprising in contrast to soybean and tomato abscission was that the level of gene expression for all of the cell wall disassembly genes in both the WT and the *hae-hsl2* mutant started at fairly high levels (**Supplemental File S4**).

Expression of Cellulose Synthases and Typical Primary Cell Wall Proteins

We considered the possibility that the cell wall and middle lamella of fracture plane cells would be degraded but once separation was complete the same cells would synthesize a new cell wall much like the previous one that included new cellulose microfibrils, hemicelluloses and protein. The primary wall of a plant cell consists of approximately 20% protein (Carpita and Gibeaut, 1993). To assess whether or not a typical cell wall was synthesized we examined the expression of cellulose synthesis genes and typical primary cell wall proteins, extensins

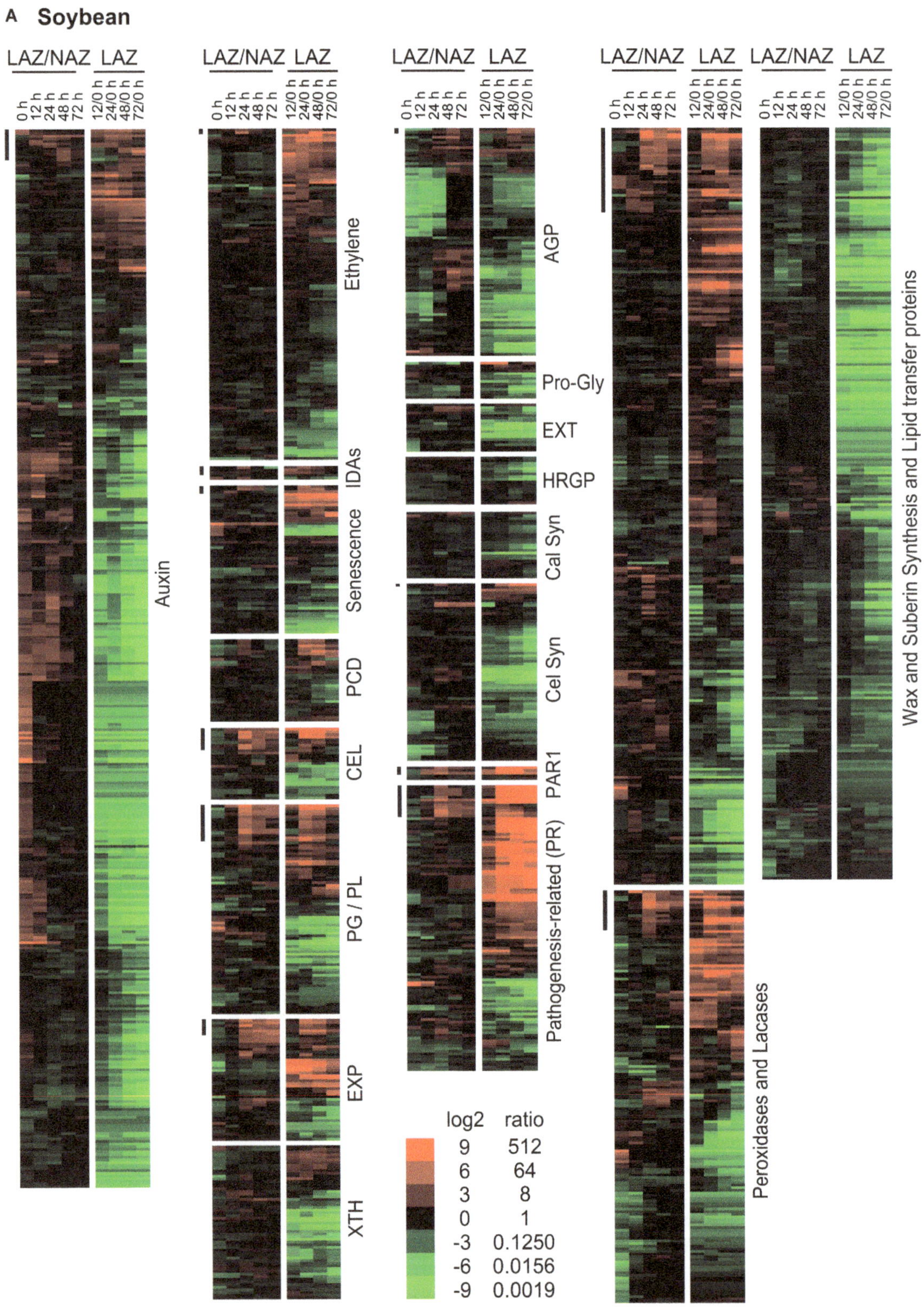

FIGURE 2 | Continued

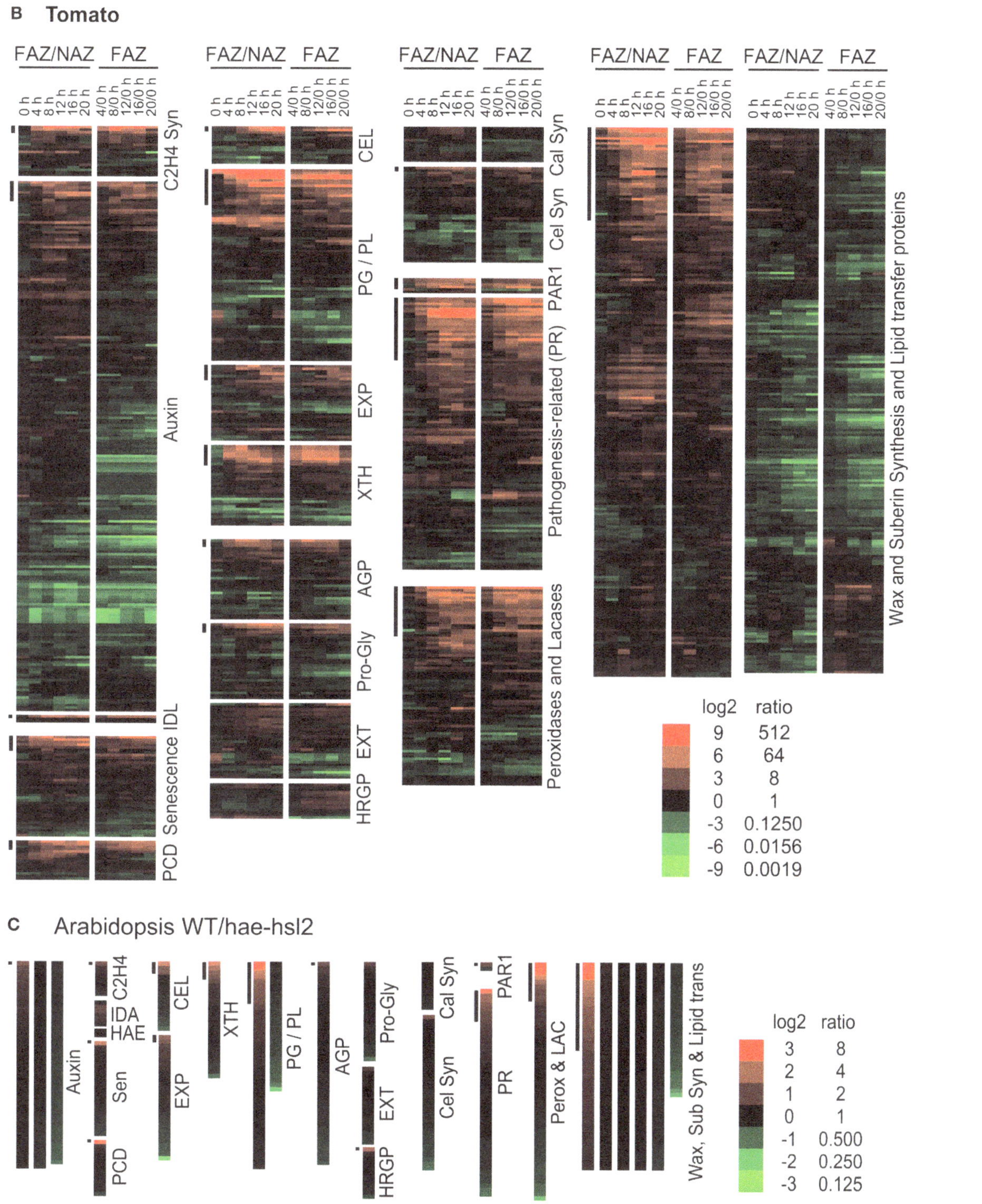

FIGURE 2 | Heat map display of expression profiles for genes in soybean leaf pulvinus AZs (LAZ) and petioles with the AZs removed (NAZ-pet), tomato flower AZs (FAZ) and proximal pedicel (NAZ) and Arabidopsis wild type (WT) and *hae-hsl2* mutant receptacle AZs. (A) Soybean, **(B)** tomato, and **(C)** Arabidopsis gene expression. The expression profiles are presented as ratios for AZ-specificity (AZ/NAZ), and as ratios for AZ expression at different time intervals relative to 0 h (start of ethylene treatment of soybean or removal of auxin source for tomato), e.g., 12/0, 24/0, 48/0, or 72/0 h. The bars at the top left of each category mark genes whose expression is >8-fold AZ-specific (AZ/NAZ) and >8-fold up-regulated (time interval/0 h). The scale at the bottom indicates the color representing the log2 ratio and ratio listed next to it. The color scale for Arabidopsis is different than soybean and tomato because differential expression was less in the Arabidopsis tissues. The genes in each category, their nomenclature, log2 ratios, and annotations are given in **Supplemental Files S2–S4** for soybean, tomato and Arabidopsis, respectively.

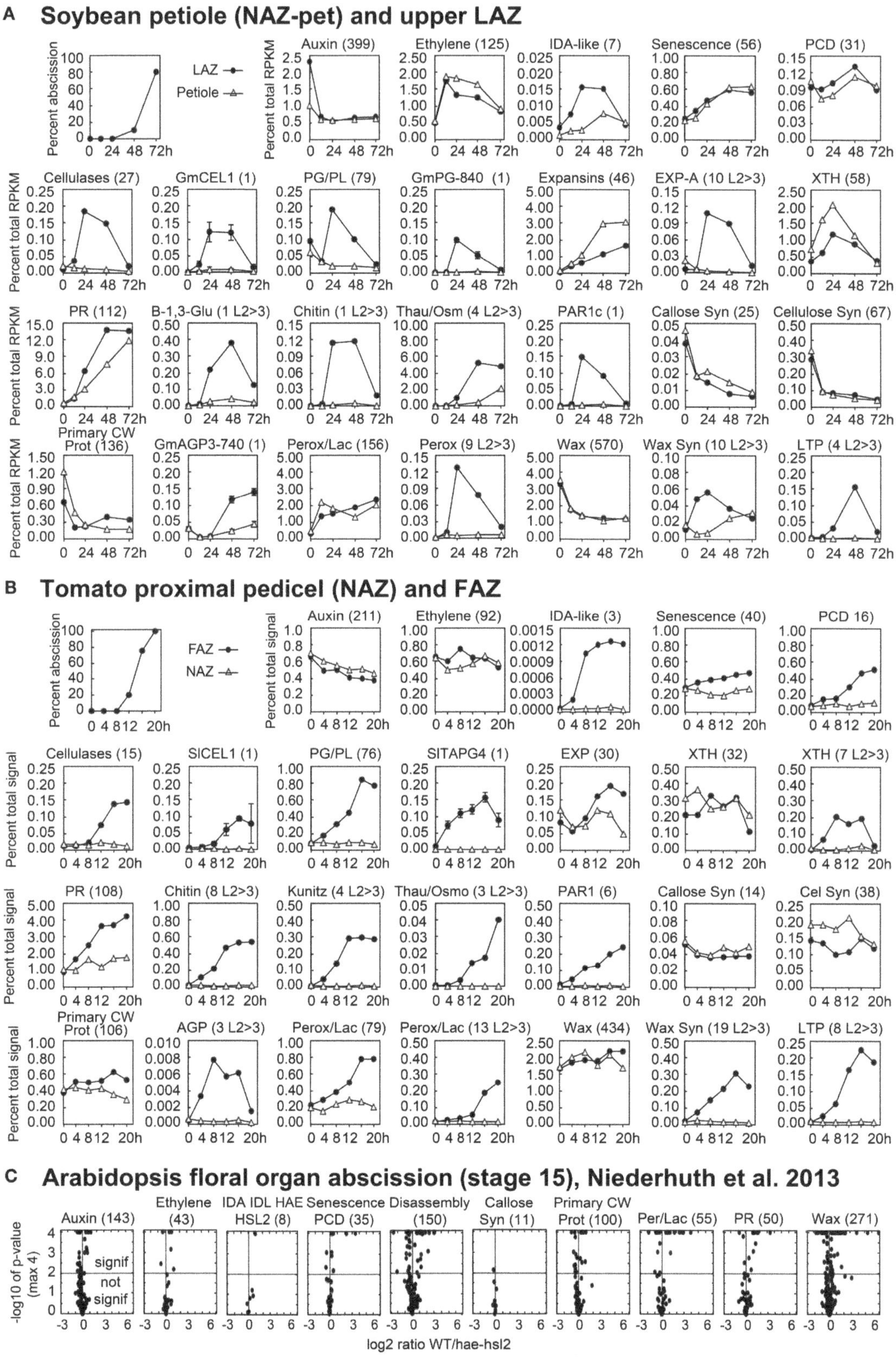

FIGURE 3 | Line graphs (soybean and tomato) and volcano plots (Arabidopsis) showing the change in gene expression for selected groupings of genes and some individual genes. The percent of abscission is indicated in the graph at the top left corner the sections for **(A)** soybean and **(B)** tomato as a reference. **(A)** The Y-axes indicate the percent expression relative to total RPKM for all 37,572 expressed soybean genes (transcriptome). The X-axes indicate the length of

(Continued)

FIGURE 3 | Continued
exposure to ethylene. Solid circles represent expression in the LAZ and the open triangles in petioles (NAZ-pet). **(B)** Similar to **(A)** but The Y-axes indicate the percent expression relative to total microarray signal for 26,527 expressed tomato genes. The X-axes indicate the hours after removal of the flowers, auxin source. Solid circles represent expression in the FAZ and the open triangles in NAZ. The number within the parenthesis to the right of each category or gene name is the number of genes included for that particular graph. Plots with an L2>3 after the number in parentheses show the cumulative expression for only those genes with a significant 8-fold increase in expression ($p < 0.015$ and log2>3 for any time relative to 0 h). Standard error bars cannot be calculated for plots of multiple genes. **(C)** Volcano plots of differential expression in *WT/hae-hsl2* receptacles at developmental stage 15. Y-axes are the −log10 of the *p*-values. A $p < 1\%$ (<0.015) is a *p*-value of −log10 >2, which is indicated by the horizontal line through the middle. Any expression with a *p*-value (significance) of less than 0.0001 (−log10>4) was set to a minimum of 0.0001, which is why many points align across the top. The X-axes are the log2 ratios for expression (reads) of *WT/hae-hsl2*. The genes in each category, their nomenclature, log2 ratios, and annotations are given in **Supplemental Files S2–S4** for soybean, tomato, and Arabidopsis, respectively.

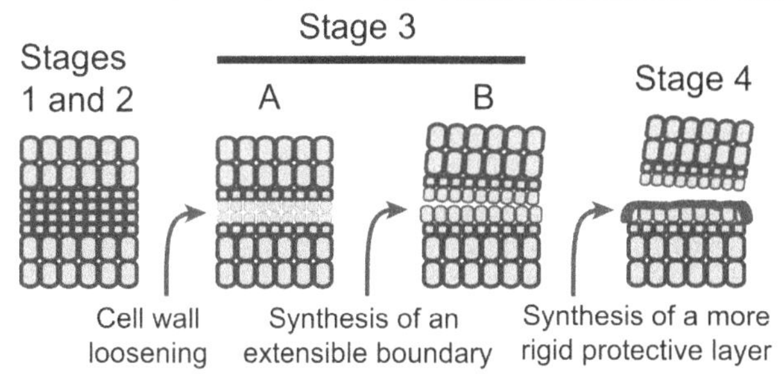

FIGURE 4 | A schematic model depicting the sequence of events in AZ formation and organ separation. Stages 1 and 2, AZ differentiation and competence to abscise are combined; Stage 3, activation of abscission, is divided into two temporally overlapping sub-stages, cell wall loosening and synthesis of a boundary layer; Stage 4, synthesis of a protective layer and post-abscission trans-differentiation.

(EXTs), arabinogalactan proteins (AGPs), proline and glycine-rich proteins, and hydroxyproline-rich glycoproteins (HRGPs). In soybean, the expression of almost all of these genes declined (**Figures 2A, 3A**); however, in tomato and Arabidopsis, this grouping of genes were mostly unchanged or increased slightly (**Figures 2B,C, 3B,C**). Based on these observations, we conclude that cellulose synthesis and synthesis of typical primary cell wall proteins are not a significant part of abscission or formation of a protective layer.

Synthesis of a More Proteinaceous Extracellular Matrix

Overall there was a strong up-regulation of *PR* genes. It might seem odd to include *PR* gene expression under the sub-heading of synthesis of a new and different extracellular matrix because the role of *PR* gene expression during abscission is customarily assumed to protect the vulnerable abscising cells from opportunistic pathogen invasion (Del Campillo and Lewis, 1992). However, the first and possibly the most important defense against pathogens is a structural barrier, which includes the cell wall and cuticle (Hamann, 2012). However, at the site of a microbial infection, the cell wall is further reinforced by the synthesis of papillae (Voigt, 2014). Callose, a 1-3 linked beta glucan polymer, is major component of papillae. Because gene expression for callose synthesis is either down-regulated or unchanged in all three abscission systems, we conclude that callose is not a primary component in the new extracellular matrix (**Figures 2, 3**). Nonetheless, what is important to successful organ separation is the creation of a flexible barrier to pathogens that both inhibits infection and allows cell expansion. We argue that in addition to an enzymatic role in the defense against pathogens, a remodeled extracellular matrix containing an abundance of small PR proteins is important to the actual separation process. An extracellular matrix made up of cross-linked protein rather than long chains of polysaccharides, i.e., callose, cellulose or hemicelluloses, might better allow for cell enlargement that creates the physical stress across the fracture plane, but still provide enough structure to prevent the fracture plane cells from rupturing.

What is the PR protein composition of the extracellular matrix? It does not appear to be the same in each of the three systems. In soybean, *thaumatin* begins to increase early and by 48 h expression of *thaumatin* accounted for 5% of the AZ transcriptome (**Figure 3A** and **Supplemental File S2**). Expression of *Chitinase* and *beta-1,3-glucanase* is also greatly up-regulated in an AZ-specific manner. In tomato, *chitinase* and *kunitz trypsin inhibitor* are abundantly expressed in the AZ but *thaumatin* is not as strongly expressed (**Figure 3B**). In Arabidopsis, *thaumatin*, *kunitz trypsin inhibitor*, and *chitinase* are all more highly expressed in WT than the mutant (**Supplemental File S4**).

PAR1 (*photoassimilate-responsive-1*) is also considered to be a *PR* gene (Herbers et al., 1995). We have separated *PAR1s* from the others because they exemplify our view that synthesis of a new protein-rich extracellular matrix is important to the separation process. *PAR1s* are expressed in an AZ-specific manner in all three systems with an expression profile much like the AZ-specific *CELs* and *PGs*. PAR1s are small proteins (approximately 18 kDa) with no known enzymatic function but have several cysteines that might support protein cross-linking in the cell wall. Also, *PAR1s* are highly conserved in both dicots and monocots (**Supplemental File S5**).

Synthesis of a Waxy Cuticle

Now, to detail and discuss our observation that was somewhat unexpected for us because it is not commonly discussed in the abscission literature as being a part of the separation process but rather a part of the synthesis of a protective layer after separation has occurred. We found an early AZ-specific increase in gene expression of several genes that are best linked to the synthesis and secretion of a waxy cuticle. Because both wax and suberin biosynthesis require protein for lipid modification and transport and phenylpropanoid metabolism, we have also included suberin-associated genes in this category. Genes included in the wax-suberin category were selected based on the cuticle synthesis genes listed by Suh et al. (2005), suberin biosynthesis genes listed by Soler et al. (2007) and genes whose gene ontology (GO) annotation included terms for lipid or fatty acid synthesis or modification. Peroxidases are also a part of wax and suberin synthesis. Gene expression for several peroxidases and laccases increased markedly during abscission and some quite early. Although peroxidases and laccases may be involved in the synthesis of waxes and suberin, they may also be tied to cross-linking of protein and pathogen defense (Matheis and Whitaker, 1984; Almagro et al., 2009). For this reason, we have put *peroxidase* and *laccase* genes into their own category. Nonetheless, collectively, we conclude that independent of the category they are included in peroxidases and laccases are a part of creating a new and different extracellular matrix.

In soybean, there was more than a 50% decline in the expression of the 570 genes in the wax-suberin category (**Figure 2A**). This is probably because cuticle synthesis is not generally required for senescing tissue. However, there was a subset of 10 wax synthesis genes that were significantly up-regulated in the AZ more than 8-fold within the first 12 h (**Figure 2A**). This was significantly earlier than the increase for *CELs* or *PGs*. Three of the 10 wax synthesis genes encoded GDSL-like lipases, which are important to cuticle synthesis (Yeats and Rose, 2013), and the most AZ-specific of the 10 was a *CER4*, Jojoba acyl CoA reductase, which is also important to the synthesis of a wax cuticle (Rowland et al., 2006) (**Supplemental File S2**). In addition to fatty acid and wax synthesis, there was a strong AZ-specific increase in four lipid transfer proteins (LTP) (**Figure 3A**). In tomato, we found 19 wax synthesis and 8 *LTP* genes that increased specifically in the AZ more than 8-fold (**Figure 3B**). Like soybean, the most AZ-specific wax synthesis gene in tomato was a *CER4*, Jojoba acyl CoA reductase (**Supplemental File S3**).

Gene expression for wax synthesis and lipid transfer is particularly interesting in the Arabidopsis data. This category of genes (wax-suberin) had the greatest difference between the WT and the *hae-hsl2* mutant (**Figure 3C** and **Supplemental File S4**). Moreover, if one sorts the original published Additional file 2 (fold-change calculated for the opposite ratio, i.e., mutant/WT) the gene showing the greatest difference between WT and the mutant was a *bifunctional inhibitor/lipid-transfer protein* (AT4G22485.1) and the third gene in the sorted file was a *GDSL-like lipase* (AT5G03810.1) (Niederhuth et al., 2013). Both of these cuticle-associated gene families are strongly up-regulated and AZ-specific in soybean and tomato (**Supplemental Files S2, S3**). When the original Arabidopsis Additional file 1 was sorted this way, the first cell wall disassembly gene, *PGAZAT* (AT2G41850), was not even in the top 50 differentially expressed genes. The large differential expression of putative cuticle synthesis genes in the Arabidopsis data may be significant to an interpretation of the role of IDA signaling in abscission (discussed below).

It should be noted that expression of *LTPs* in abscission has been previously reported (Agusti et al., 2009; Nakano et al., 2013). Here we put *LTPs* expression in context with additional observations and propose that synthesis of a waxy-like cuticle is important to successful organ separation.

Comparison to Separation During Organogenesis

Based on our abscission transcriptome results, it appears that an early up-regulation of gene expression for synthesis and secretion of a waxy cuticle-like substance during abscission is a common feature for abscission in many species. One possible explanation for this is that the separating cells need to be protected from water loss (Samuels et al., 2008). Also, synthesis of a cuticle could be part of a defense mechanism (Samuels et al., 2008). However, there is another possible role for synthesis of a waxy cuticle. Arabidopsis has proven to be an excellent model system for the study of cuticle synthesis (Samuels et al., 2008; Shi et al., 2011; Yeats and Rose, 2013). Interestingly, a common phenotype for knockout mutants for transcription factors and other genes related to the synthesis of the cuticle is that organs remain fused or display structural anomalies (Shi et al., 2011; Yeats and Rose, 2013). Fused floral organs were also observed in a wax-deficient mutant in tomato (Smirnova et al., 2013). The composition of the extracellular matrix around boundary cells that separate organs in the meristem is not well understood (Žádníková and Simon, 2014). Nevertheless, based on the genetic results, it seems likely that the synthesis of a cuticle like substance plays a role in the separation of organs in the meristem. It seems plausible that a similar process might be implemented during abscission of leaves, flowers, fruit, or floral organs. Moreover, expression of meristem-associated developmental genes in AZ seems to be consistent with this hypothesis (Wang et al., 2013). It should also be noted that modification of pectin structure was found to play a role in organogenesis (Peaucelle et al., 2011). Organogenesis appears to have many features in common with abscission.

Synthesis of a Boundary Layer and Hormonal and IDA-Like Signaling

If the synthesis of a waxy, cuticle boundary is important to successful organ separation, you might assume someone would have identified Arabidopsis mutants that were linked to this function. We suggest that the *ida*, *hae-hsl2* mutants are more closely linked to the synthesis of a cuticle-like boundary layer than expression of genes for cell wall disassembly. To explain, first, although we do not have a time course comparison of WT and *hae-hsl2*, differential expression of cuticle synthesis and peroxidases genes was greater in this data than for cell wall disassembly genes (**Figure 3C** and **Supplemental File S4**) (Niederhuth et al., 2013). Moreover, break-strength measurements for the *ida* and *hae-hsl2* mutants indicate that IDA-signaling affects only a part of the separation process (Cho et al., 2008; Liu et al., 2013). Break-strength is a measure of the force needed to pull the distal organ, e.g., petal, away from the proximal organ, e.g., receptacle (Craker and Abeles, 1969; Del Campillo and Bennett, 1996). In WT flowers, the petal break-strength begins to decline at flower position 3 soon after the flowers open (Patterson and Bleecker,

2004). Break-strength of WT petals continues to decline until positions 6–8 when the petals, stamens and sepals fall off from their own weight. In the *ida* mutant (Butenko et al., 2003), which has an abscission phenotype very similar to the *hae-hsl2* mutant (Cho et al., 2008), the petal break-strength also begins to decline at position 3 but declines at a slightly slower rate than WT; however, between flower positions 8–12 the petals begin to re-adhere to the receptacle so that by position 22 the petals are senescent but have regained a break-strength approximately equal to that of a flower at position 2 (Butenko et al., 2003). Scanning electron micrographs (SEM) of the fracture plane cells of WT petals on flowers at positions 10 show nicely rounded and enlarged cells. SEM micrographs of the petal fracture plane of *ida* at flower position 10, when break-strength was at its lowest, displayed partially enlarged and rounded cells (Butenko et al., 2003). However, at flower position 22 of the *ida* mutant, after the petals had re-adhered to the receptacle and were forcibly pulled off, the micrographs of the fracture plane displayed cells that were broken and torn (Butenko et al., 2003). Clearly, part of the abscission process is implemented in *ida* but the process is incomplete. The *nevershed* (*nev*) mutant, that has a defect in the cellular secretion process through the golgi apparatus, also displays a V-shaped break-strength profile; however, the cells in the fracture plane at the lowest break-strength appear to enlarge even more than in the WT cells (Liu et al., 2013). The phenotypes of *ida*, *hae-hsl2*, and *nev* are similar but crosses between them suggest that *NEVERSHED* expression is not solely dependent on IDA signaling (Liu et al., 2013). Nevertheless, the *nev* mutant may also affect secretion of components needed for synthesis of a waxy cuticle more than cell wall disassembly.

It is possible based on the results and interpretations we have presented here that IDA signaling in Arabidopsis plays a greater role in the synthesis of a new boundary layer than cell wall disassembly and that this boundary layer appears to aid in successful floral organ separation. Can we propose a similar role for IDA in soybean, tomato and other abscission processes? *IDA*-like genes are highly conserved in dicots and also found in some monocots (Vie et al., 2015). In soybean leaf abscission and tomato flower abscission there is an AZ-specific up-regulation of *IDA*-like gene expression (**Figures 2, 3** and **Supplemental Files S2, S3**). However, nobody to the best of our knowledge has demonstrated that IDA signaling is necessary for successful organ separation in any species other than Arabidopsis. It is worth mentioning here that we have used virus-induced gene silencing (VIGS) to suppress *IDA*-like expression in soybean and did not observe any effect on leaf abscission (results not shown). Regulation of abscission is a nexus of signaling events including ethylene, auxin, IDA, and others (Taylor and Whitelaw, 2001; Gonzalez-Carranza et al., 2012; Liljegren, 2012; Aalen et al., 2013; Kim, 2014; Tucker and Kim, 2015). It is possible that, even if IDA-signaling plays a more prominent role in regulating the synthesis of a boundary layer, a separation phenotype in *IDA*-like suppressed mutants will not be obvious in many abscission systems because disassembly of the cell wall and middle lamella is sufficient when the distal organ has a mass greater than that of a flower petal, stamen, or sepal. Nevertheless, we propose a modification to the abscission model to include synthesis of an extensible boundary layer early in the abscission process that is different from the deposition of a protective layer (**Figure 4**). However, the waxy cuticle synthesized early during separation may become a part of a more rigid protective layer deposited after separation. Moreover, as suggested much earlier by Patterson and Bleecker (2004), we reassert that multiple hormonal and peptide signals regulate the rate of abscission and successful organ separation and these different signals do not necessarily regulate the same set of genes. Of interest in this regard was that differential gene expression of ethylene (e.g., ERFs) and auxin-associated genes (e.g., SAUR, AUX.IAA) was not greatly changed between WT and *hae-hsl2* Arabidopsis receptacles at flower stage 15 (**Figures 2C, 3C**). This further supports a model where IDA signaling does not directly affect ethylene or auxin signaling in Arabidopsis floral organ abscission and that multiple signals influence abscission. What is necessary is discovering how these signals work together and independently to bring about a successful separation process.

CONCLUDING REMARKS

Fruit softening includes expression of cell wall and middle lamella degrading enzymes (Tucker, 2014). The fruit ripening genes that affect cell wall modifications are not necessarily the same genes as those expressed in abscission, but they likely have similar functions. However, cell separation is not the same in a ripe fruit as it is in the fracture plane. Evaluation of the transcriptome of the pericarp of a wild-type ripening tomato (Osorio et al., 2011) indicated strong up-regulation of *CELs*, *PGs*, *EXPs*, and *XTHs* but not a marked increase in wax synthesis genes or lipid transfer proteins (**Supplemental File S6**). Could it be that an important difference between fruit ripening and abscission is the synthesis of a cuticle-like boundary on the abscission cells? Our proposal that a waxy-like cuticle is important to the abscission process is further supported by observations for separation during organogenesis. It seems to make evolutionary sense that abscission of plant organs would be an adaptation of a primal process occurring in the meristem. Although we do not know the exact composition of the abscission boundary layer, our proposal can be tested genetically and biochemically.

SEQUENCE SUBMISSION

The raw soybean sequence files have been submitted to the National Center for Biotechnology Information (NCBI) SRA databases with the study accession SRP050050. The soybean results including RPKM and annotations for all 54,175 genes can be downloaded at http://sgil.ba.ars.usda.gov/mtucker/Public/Tucker.html. The raw tomato sequence data and array information were submitted to the Gene Expression Omnibus (GEO) at the NCBI with GEO IDs GSE45355 and GSE45356, and array ID AMADID:043310.

AUTHOR CONTRIBUTIONS

Conceived, designed experiments and interpreted data: JK and MT. Performed experiments and prepared and analyzed data: SS and RY. Analyzed and interpreted data: SH and SM.

FUNDING

This work was supported by a United States - Israel Binational Agricultural and Development Fund (BARD) US-4571-12C grant to MT, SM, and SP.

ACKNOWLEDGMENTS

We thank Jim Giovannoni and Zhangjun Fei of the Boyce Thompson Institute and Agricultural Research Service at Cornell University for performing the RNA sequencing of the soybean RNA.

REFERENCES

Aalen, R. B., Wildhagen, M., Stø, I. M., and Butenko, M. A. (2013). IDA: A peptide ligand regulating cell separation processes in *Arabidopsis*. *J. Exp. Bot.* 64, 5253–5261. doi: 10.1093/jxb/ert338

Abeles, F. B., Morgan, P. W., and Salveit, M. E. (1992). *Ethylene in Plant Biology.* New York, NY: Academic Press.

Addicott, F. T. (1982). *Abscission.* Berkely, CA: California University Press.

Agusti, J., Merelo, P., Cercós, M., Tadeo, F. R., and Talón, M. (2009). Comparative transcriptional survey between laser-microdissected cells from laminar abscission zone and petiolar cortical tissue during ethylene-promoted abscission in citrus leaves. *BMC Plant Biol.* 9:127. doi: 10.1186/1471-2229-9-127

Almagro, L., Gómez Ros, L. V., Belchi-Navarro, S., Bru, R., Ros Barceló, A., and Pedreño, M. A. (2009). Class III peroxidases in plant defence reactions. *J. Exp. Bot.* 60, 377–390. doi: 10.1093/jxb/ern277

Basu, M. M., González-Carranza, Z. H., Azam-Ali, S., Tang, S., Shahid, A. A., and Roberts, J. A. (2013). The manipulation of auxin in the abscission zone cells of *Arabidopsis* flowers reveals that indoleacetic acid signaling is a prerequisite for organ shedding. *Plant Physiol.* 162, 96–106. doi: 10.1104/pp.113.216234

Bleecker, A. B., and Patterson, S. E. (1997). Last exit: Senescence, abscission, and meristem arrest in Arabidopsis. *Plant Cell* 9, 1169–1179. doi: 10.1105/tpc.9.7.1169

Butenko, M. A., Patterson, S. E., Grini, P. E., Stenvik, G. E., Amundsen, S. S., Mandal, A., et al. (2003). *Inflorescence deficient in abscission* controls floral organ abscission in Arabidopsis and identifies a novel family of putative ligands in plants. *Plant Cell* 15, 2296–2307. doi: 10.1105/tpc.014365

Butenko, M. A., Wildhagen, M., Albert, M., Jehle, A., Kalbacher, H., Aalen, R. B., et al. (2014). Tools and strategies to match peptide-ligand receptor pairs. *Plant Cell* 26, 1838–1847. doi: 10.1105/tpc.113.120071

Cai, S., and Lashbrook, C. C. (2008). Stamen abscission zone transcriptome profiling reveals new candidates for abscission control: Enhanced retention of floral organs in transgenic plants overexpressing Arabidopsis ZINC FINGER PROTEIN2. *Plant Physiol.* 146, 1305–1321. doi: 10.1104/pp.107.110908

Carpita, N. C., and Gibeaut, D. M. (1993). Structural models of primary cell walls in flowering plants: Consistency of molecular structure with the physical properties of the walls during growth. *Plant J.* 3, 1–30. doi: 10.1111/j.1365-313X.1993.tb00007.x

Cho, S. K., Larue, C. T., Chevalier, D., Wang, H., Jinn, T. L., Zhang, S., et al. (2008). Regulation of floral organ abscission in *Arabidopsis thaliana*. *Proc. Natl. Acad. Sci. U.S.A.* 105, 15629–15634. doi: 10.1073/pnas.0805539105

Cosgrove, D. J. (2001). Wall structure and wall loosening. A look backwards and forwards. *Plant Physiol.* 125, 131–134. doi: 10.1104/pp.125.1.131

Craker, L. E., and Abeles, F. B. (1969). Abscission: Quantitative measurement with a recording abscissor. *Plant Physiol.* 44, 1139–1143. doi: 10.1104/pp.44.8.1139

Del Campillo, E., and Bennett, A. B. (1996). Pedicel breakstrength and cellulase gene expression during tomato flower abscission. *Plant Physiol.* 111, 813–820. doi: 10.1104/pp.111.3.813

Del Campillo, E., and Lewis, L. N. (1992). Identification and kinetics of accumulation of proteins induced by ethylene in bean abscission zones. *Plant Physiol.* 98, 955–961. doi: 10.1104/pp.98.3.955

Doblin, M. S., Kurek, I., Jacob-Wilk, D., and Delmer, D. P. (2002). Cellulose biosynthesis in plants: From genes to rosettes. *Plant Cell Physiol.* 43, 1407–1420. doi: 10.1093/pcp/pcf164

Eklöf, J. M., Ruda, M. C., and Brumer, H. (2012). Distinguishing xyloglucanase activity in endo-β(1→4)glucanases. *Methods Enzymol.* 510, 97–120. doi: 10.1016/B978-0-12-415931-0.00006-9

Estornell, L. H., Agustí, J., Merelo, P., Talón, M., and Tadeo, F. R. (2013). Elucidating mechanisms underlying organ abscission. *Plant Sci.* 199–200, 48–60. doi: 10.1016/j.plantsci.2012.10.008

González-Carranza, Z. H., Elliott, K. A., and Roberts, J. A. (2007). Expression of polygalacturonases and evidence to support their role during cell separation processes in *Arabidopsis thaliana*. *J. Exp. Bot.* 58, 3719–3730. doi: 10.1093/jxb/erm222

González-Carranza, Z. H., Shahid, A. A., Zhang, L., Liu, Y., Ninsuwan, U., and Roberts, J. A. (2012). A novel approach to dissect the abscission process in Arabidopsis. *Plant Physiol.* 160, 1342–1356. doi: 10.1104/pp.112.205955

Grassi, S., Piro, G., Lee, J. M., Zheng, Y., Fei, Z., Dalessandro, G., et al. (2013). Comparative genomics reveals candidate carotenoid pathway regulators of ripening watermelon fruit. *BMC Genomics* 14:781. doi: 10.1186/1471-2164-14-781

Hall, J. L., and Sexton, R. (1974). Fine structure and cytochemistry of the abscission zone cells of phaseolus leaves II. Localization of peroxidase and acid phosphatase in the separation zone cells. *Ann. Bot.* 38, 855–858.

Hamann, T. (2012). Plant cell wall integrity maintenance as an essential component of biotic stress response mechanisms. *Front. Plant Sci.* 3:77. doi: 10.3389/fpls.2012.00077

Hayashi, T., and Kaida, R. (2011). Functions of xyloglucan in plant cells. *Mol. Plant* 4, 17–24. doi: 10.1093/mp/ssq063

Herbers, K., Mönke, G., Badur, R., and Sonnewald, U. (1995). A simplified procedure for the subtractive cDNA cloning of photoassimilate-responding genes: Isolation of cDNAs encoding a new class of pathogenesis-related proteins. *Plant Mol. Biol.* 29, 1027–1038. doi: 10.1007/BF00014975

Kalaitzis, P., Solomos, T., and Tucker, M. L. (1997). Three different polygalacturonases are expressed in tomato leaf and flower abscission, each with a different temporal expression pattern. *Plant Physiol.* 113, 1303–1308. doi: 10.1104/pp.113.4.1303

Kim, J. (2014). Four shades of detachment: Regulation of floral organ abscission. *Plant Signal Behav.* 9:e976154. doi: 10.4161/15592324.2014.976154

Kim, J., Dotson, B., Rey, C., Lindsey, J., Bleecker, A. B., Binder, B. M., et al. (2013). New clothes for the jasmonic acid receptor *COI1*: Delayed abscission, meristem arrest and apical dominance. *PLoS ONE* 8:e60505. doi: 10.1371/journal.pone.0060505

Kim, J., and Patterson, S. E. (2006). Expression divergence and functional redundancy of polygalacturonases in floral organ abscission. *Plant Signal Behav.* 1, 281–283. doi: 10.4161/psb.1.6.3541

Lanahan, M. B., Yen, H. C., Giovannoni, J. J., and Klee, H. J. (1994). The *never ripe* mutation blocks ethylene perception in tomato. *Plant Cell* 6, 521–530. doi: 10.1105/tpc.6.4.521

Langmead, B., Trapnell, C., Pop, M., and Salzberg, S. L. (2009). Ultrafast and memory-efficient alignment of short DNA sequences to the human genome. *Genome Biol.* 10:R25. doi: 10.1186/gb-2009-10-3-r25

Lashbrook, C. C., and Cai, S. (2008). Cell wall remodeling in Arabidopsis stamen abscission zones: Temporal aspects of control inferred from transcriptional profiling. *Plant Signal Behav.* 3, 733–736. doi: 10.4161/psb.3.9.6489

Liljegren, S. J. (2012). Organ abscission: Exit strategies require signals and moving traffic. *Curr. Opin. Plant Biol.* 15, 670–676. doi: 10.1016/j.pbi.2012.09.012

Liu, B., Butenko, M. A., Shi, C. L., Bolivar, J. L., Winge, P., Stenvik, G. E., et al. (2013). NEVERSHED and INFLORESCENCE DEFICIENT IN ABSCISSION are differentially required for cell expansion and cell separation during floral organ abscission in *Arabidopsis thaliana*. *J. Exp. Bot.* 64, 5345–5357. doi: 10.1093/jxb/ert232

Matheis, G., and Whitaker, J. R. (1984). Peroxidase-catalyzed cross linking of proteins. *J. Protein Chem.* 3, 35–48. doi: 10.1007/BF01024835

Meir, S., Philosoph-Hadas, S., Sundaresan, S., Selvaraj, K. S., Burd, S., Ophir, R., et al. (2010). Microarray analysis of the abscission-related transcriptome in the tomato flower abscission zone in response to auxin depletion. *Plant Physiol.* 154, 1929–1956. doi: 10.1104/pp.110.160697

Meir, S., Philosoph-Hadas, S., Sundaresan, S., Selvaraj, K. S., Burd, S., Ophir, R., et al. (2011). Identification of defense-related genes newly-associated with tomato flower abscission. *Plant Signal Behav.* 6, 590–593. doi: 10.4161/psb.6.4.15043

Mortazavi, A., Williams, B. A., McCue, K., Schaeffer, L., and Wold, B. (2008). Mapping and quantifying mammalian transcriptomes by RNA-Seq. *Nat. Methods* 5, 621–628. doi: 10.1038/nmeth.1226

Nakano, T., Fujisawa, M., Shima, Y., and Ito, Y. (2013). Expression profiling of tomato pre-abscission pedicels provides insights into abscission zone properties including competence to respond to abscission signals. *BMC Plant Biol.* 13:40. doi: 10.1186/1471-2229-13-40

Niederhuth, C. E., Patharkar, O. R., and Walker, J. C. (2013). Transcriptional profiling of the Arabidopsis abscission mutant *hae hsl2* by RNA-seq. *BMC Genomics* 14:37. doi: 10.1186/1471-2164-14-37

Osorio, S., Alba, R., Damasceno, C. M., Lopez-Casado, G., Lohse, M., Zanor, M. I., et al. (2011). Systems biology of tomato fruit development: Combined transcript, protein, and metabolite analysis of tomato transcription factor (nor, rin) and ethylene receptor (Nr) mutants reveals novel regulatory interactions. *Plant Physiol.* 157, 405–425. doi: 10.1104/pp.111.175463

Patterson, S. E. (2001). Cutting loose. Abscission and dehiscence in Arabidopsis. *Plant Physiol.* 126, 494–500. doi: 10.1104/pp.126.2.494

Patterson, S. E., and Bleecker, A. B. (2004). Ethylene-dependent and -independent processes associated with floral organ abscission in Arabidopsis. *Plant Physiol.* 134, 194–203. doi: 10.1104/pp.103.028027

Peaucelle, A., Braybrook, S. A., Le Guillou, L., Bron, E., Kuhlemeier, C., and Höfte, H. (2011). Pectin-induced changes in cell wall mechanics underlie organ initiation in Arabidopsis. *Curr. Biol.* 21, 1720–1726. doi: 10.1016/j.cub.2011.08.057

Roberts, J. A., Elliott, K. A., and Gonzalez-Carranza, Z. H. (2002). Abscission, dehiscence, and other cell separation processes. *Annu. Rev. Plant Biol.* 53, 131–158. doi: 10.1146/annurev.arplant.53.092701.180236

Rowland, O., Zheng, H., Hepworth, S. R., Lam, P., Jetter, R., and Kunst, L. (2006). CER4 encodes an alcohol-forming fatty acyl-coenzyme A reductase involved in cuticular wax production in Arabidopsis. *Plant Physiol.* 142, 866–877. doi: 10.1104/pp.106.086785

Samuels, L., Kunst, L., and Jetter, R. (2008). Sealing plant surfaces: Cuticular wax formation by epidermal cells. *Annu. Rev. Plant Biol.* 59, 683–707. doi: 10.1146/annurev.arplant.59.103006.093219

Sexton, R., and Roberts, J. A. (1982). Cell biology of abscission. *Annu. Rev. Plant Physiol.* 33, 133–162. doi: 10.1146/annurev.pp.33.060182.001025

Shi, J. X., Malitsky, S., De Oliveira, S., Branigan, C., Franke, R. B., Schreiber, L., et al. (2011). SHINE transcription factors act redundantly to pattern the archetypal surface of Arabidopsis flower organs. *PLoS Genet.* 7:e1001388. doi: 10.1371/journal.pgen.1001388

Smirnova, A., Leide, J., and Riederer, M. (2013). Deficiency in a very-long-chain fatty acid beta-ketoacyl-coenzyme a synthase of tomato impairs microgametogenesis and causes floral organ fusion. *Plant Physiol.* 161, 196–209. doi: 10.1104/pp.112.206656

Soler, M., Serra, O., Molinas, M., Huguet, G., Fluch, S., and Figueras, M. (2007). A genomic approach to suberin biosynthesis and cork differentiation. *Plant Physiol.* 144, 419–431. doi: 10.1104/pp.106.094227

Stenvik, G. E., Butenko, M. A., and Aalen, R. B. (2008). Identification of a putative receptor-ligand pair controlling cell separation in plants. *Plant Signal Behav.* 3, 1109–1110. doi: 10.4161/psb.3.12.7009

Suh, M. C., Samuels, A. L., Jetter, R., Kunst, L., Pollard, M., Ohlrogge, J., et al. (2005). Cuticular lipid composition, surface structure, and gene expression in Arabidopsis stem epidermis. *Plant Physiol.* 139, 1649–1665. doi: 10.1104/pp.105.070805

Taylor, J. E., and Whitelaw, C. A. (2001). Signals in abscission. *New Phytol.* 151, 323–339. doi: 10.1046/j.0028-646x.2001.00194.x

Trapnell, C., Pachter, L., and Salzberg, S. L. (2009). TopHat: Discovering splice junctions with RNA-Seq. *Bioinformatics* 25, 1105–1111. doi: 10.1093/bioinformatics/btp120

Tucker, M. L. (2014). "Cell-wall metabolism and softening during ripening," in *Fruit Ripening: Physiology, Signalling and Genomics*, eds P. Nath, M. Bouzayen, A.K. Mattoo, and J. C. Pech (Wallingford, CT: CABI), 48–62.

Tucker, M. L., Burke, A., Murphy, C. A., Thai, V. K., and Ehrenfried, M. L. (2007). Gene expression profiles for cell wall-modifying proteins associated with soybean cyst nematode infection, petiole abscission, root tips, flowers, apical buds, and leaves. *J. Exp. Bot.* 58, 3395–3406. doi: 10.1093/jxb/erm188

Tucker, M. L., and Kim, J. (2015). Abscission research: What we know and what we still need to study. *Stewart Posthar. Rev.* 11, 1–7. doi: 10.2212/spr.2015.2.1

Tucker, M. L., Sexton, R., Del Campillo, E., and Lewis, L. N. (1988). Bean abscission cellulase: Characterization of a cDNA clone and regulation of gene expression by ethylene and auxin. *Plant Physiol.* 88, 1257–1262. doi: 10.1104/pp.88.4.1257

Tucker, M. L., and Yang, R. (2012). IDA-like gene expression in soybean and tomato leaf abscission and requirement for a diffusible stelar abscission signal. *AoB Plants* 2012:pls035. doi: 10.1093/aobpla/pls035

Urbanowicz, B. R., Bennett, A. B., Del Campillo, E., Catalá, C., Hayashi, T., Henrissat, B., et al. (2007). Structural organization and a standardized nomenclature for plant endo-1,4-β-glucanases (cellulases) of glycosyl hydrolase family 9. *Plant Physiol.* 144, 1693–1696. doi: 10.1104/pp.107.102574

Vie, A. K., Najafi, J., Liu, B., Winge, P., Butenko, M. A., Hornslien, K. S., et al. (2015). The *IDA/IDA-LIKE* and *PIP/PIP-LIKE* gene families in *Arabidopsis*: phylogenetic relationship, expression patterns, and transcriptional effect of the PIPL3 peptide. *J. Exp. Bot.* 66, 5351–5365. doi: 10.1093/jxb/erv285

Voigt, C. A. (2014). Callose-mediated resistance to pathogenic intruders in plant defense-related papillae. *Front. Plant Sci.* 5:168. doi: 10.3389/fpls.2014.00168

Wang, X., Liu, D., Li, A., Sun, X., Zhang, R., Wu, L., et al. (2013). Transcriptome analysis of tomato flower pedicel tissues reveals abscission zone-specific modulation of key meristem activity genes. *PLoS ONE* 8:e55238. doi: 10.1371/journal.pone.0055238

Yeats, T. H., and Rose, J. K. (2013). The formation and function of plant cuticles. *Plant Physiol.* 163, 5–20. doi: 10.1104/pp.113.222737

Žádníková, P., and Simon, R. (2014). How boundaries control plant development. *Curr. Opin. Plant Biol.* 17, 116–125. doi: 10.1016/j.pbi.2013.11.013

Zhong, S., Joung, J. G., Zheng, Y., Chen, Y. R., Liu, B., Shao, Y., et al. (2011). High-throughput illumina strand-specific RNA sequencing library preparation. *Cold Spring Harb. Protoc.* 2011, 940–949. doi: 10.1101/pdb.prot5652

Conflict of Interest Statement: The authors declare that the research was conducted in the absence of any commercial or financial relationships that could be construed as a potential conflict of interest.

11

Comparative Transcriptome Analysis of Resistant and Susceptible Tomato Lines in Response to Infection by *Xanthomonas perforans* Race T3

*Heshan Du, Yuqing Wang *, Jingjing Yang and Wencai Yang **

Beijing Key Laboratory of Growth and Developmental Regulation for Protected Vegetable Crops, Department of Vegetable Science, China Agricultural University, Beijing, China

Bacterial spot, incited by several *Xanthomonas* sp., is a serious disease in tomato (*Solanum lycopersicum* L.). Although genetics of resistance has been widely investigated, the interactions between the pathogen and tomato plants remain unclear. In this study, tanscriptomes of *X. perforans* race T3 infected tomato lines were compared to those of controls. An average of 7 million reads were generated with approximately 21,526 genes mapped in each sample post-inoculation at 6 h (6 HPI) and 6 days (6 DPI) using RNA-sequencing technology. Overall, the numbers of differentially expressed genes (DEGs) were higher in the resistant tomato line PI 114490 than in the susceptible line OH 88119, and the numbers of DEGs were higher at 6 DPI than at 6 HPI. Fewer genes (78 in PI 114490 and 15 in OH 88119) were up-regulated and most DEGs were down-regulated, suggesting that the inducible defense response might not be fully activated at 6 HPI. Accumulation expression levels of 326 co-up regulated genes in both tomato lines at 6 DPI might be involved in basal defense, while the specific and strongly induced genes at 6 DPI might be correlated with the resistance in PI 114490. Most DEGs were involved in plant hormone signal transduction, plant-pathogen interaction and phenylalanine metabolism, and the genes significantly up-regulated in PI 114490 at 6 DPI were associated with defense response pathways. DEGs containing NBS-LRR domain or defense-related WRKY transcription factors were also identified. The results will provide a valuable resource for understanding the interactions between *X. perforans* and tomato plants.

Edited by:
Vasileios Fotopoulos, Cyprus University of Technology, Cyprus

Reviewed by:
Jaime Auger, Universidad de Chile, Chile
Hao Peng, Washington State University, USA

***Correspondence:**
Yuqing Wang
wyq@cau.edu.cn;
Wencai Yang
yangwencai@cau.edu.cn

Keywords: ***Solanum lycopersicum*, *Xanthomonas perforans*, compatible and incompatible interactions, transcriptome analysis, RNA sequencing**

INTRODUCTION

Bacterial spot caused by at least four distinct species of *Xanthomonas* (*X. euvesicatoria*, *X. vesicatoria*, *X. perforans,* and *X. gardneri*), severely affects marketability of both fresh-market and processing tomato (Jones et al., 2000; Stall et al., 2009). Based on their virulence on a group of tomato genotypes, the causative agents of bacterial spot can be classified into five physiological races T1–T5 (Jones et al., 2005). It is difficult to manage the disease once the outbreak occurs (Yang et al., 2007; Stall et al., 2009). Identification and characterization of the host resistance as well as regulatory genes can provide insights into understanding the mechanisms of host resistance and the use of host resistance to develop new cultivars with durable resistance.

The resistance to *X. perforans* race T3 in tomato can be classified into two types, hypersensitive response (HR) and field resistance. HR is usually conditioned by single dominant genes, while field resistance is generally controlled by multiple genes (Yang, 2013). To date, four major dominant genes, *Rx4* in *Solanum pimpinellifolium* accession PI 128216 (Robbins et al., 2009; Pei et al., 2012), *Xv3* in an unimproved breeding line Hawaii 7981 (Wang et al., 2011), Rx_{LA1589} in *S. pimpinellifolium* accession LA 1589 (Sun et al., 2011a), and *RXopJ4* in *S. pennellii* accession LA 716 (Sharlach et al., 2013), conferring HR to *X. perforans* race T3 have been identified and mapped. A recent study suggests that *Rx4*, *Xv3*, and Rx_{LA1589} are possibly the same gene (Zhao et al., 2015). Although these genes also confer field resistance to race T3, they need modifiers and may be affected by genetic background (Scott et al., 2001; Robbins et al., 2009; Wang et al., 2011; Sharlach et al., 2013). Therefore, the tomato lines with HR usually show partial resistance to race T3 in the field. On the contrary, the *S. lycopersicum* var. *cerasiforme* accession PI 114490 shows no HR but high resistance to race T3 conditioned by at least five quantitative trait loci (QTL) in the field (Sun et al., 2011b, 2014; Scott et al., 2015). In addition to identification and genetic mapping of genes or QTLs, efforts have also been making on discovery of genes participating in a complex molecular network of regulation during the time-course of HR to race T3 in the unimproved tomato line Hawaii 7981 using microarray analysis approach (Gibly et al., 2004; Balaji et al., 2007), and identification of genes differentially expressed in the resistant line PI 114490 and a susceptible line OH 88119 during the time-course of the race T3 infection using cDNA-AFLP techniques (Du et al., 2014). Comparison of genes differentially expressed in PI 114490 and OH 88119 provides us some information to understand the mechanism of resistance to bacterial spot race T3 in tomato during the process of symptom development at the 3, 4, and 5 days post spray-inoculation stages (Du et al., 2014). However, both microarray and cDNA-AFLP can only identify part of the genes involved in resistance or defense processes due to their low throughput limitation. To fully unravel the mechanisms of field resistance to race T3 of bacterial spot, it is necessary to identify more genes differentially expressed during different infection times in tomato.

With the availability of a high-quality tomato genome sequence and second-generation sequencing, RNA-seq technology has rapidly become a popular tool for genome-wide expression profiling, providing the potential to better understanding the comprehensive host-pathogen interactions. Many studies have, therefore, tried to elucidate a nearly complete picture of inducible defense response pathways using RNA-seq analysis (Kim et al., 2011; Li et al., 2012; Gao et al., 2013; Tan et al., 2015). Through comparing the different RNA-seq data about hosts and pathogens interactions, similar or specific sets of genes activated in different plant pathosystems have been discovered. Meanwhile, using the expression pattern regulated by pathogen infection, it is also possible to identify the known defense gene family member involved in the tomato immune response. For example, some plant R genes, such as pepper *Bs3* and rice *Xa27* belonging to NBS-LRR gene family, are transcriptionally activated by corresponding transcription activator-like effector (TALE) protein (Gu et al., 2005; Röemer et al., 2007). Taking advantage of the TALE-induced expression pattern, a *Xanthomonas* TAL-effector activated resistance gene *Bs4C* has been identified using RNA-seq (Strauss et al., 2012). Therefore, RNA-seq technology can be used as a tool to isolate pathogen induced R gene, which could avoid laborious positional cloning and investigate the expression pattern of different defense signaling genes family members as well as identify novel defense related genes.

In the current study, it is first time using RNA-seq to investigate transcript dynamics of the field resistant line PI 114490 and the susceptible line OH 88119 in response to *X. perforans* race T3 at 6 h and 6 days post-infection. Large-scale expression profiling identified both common and specific differentially expressed genes in two tomato lines at different time points after inoculation. In addition, 8 induced genes were validated by quantitative real-time PCR (qRT-PCR) experiments. Our results may provide a subset of potential candidate defense-related genes in tomato-*X. perforans* interaction, which will help in better understanding the molecular basis of field resistance to bacterial spot in tomato.

MATERIALS AND METHODS

Plant Materials and *X. perforans* Race T3 Inoculation

Two tomato lines, a cherry tomato line PI 114490 with high level of field resistance and an elite processing breeding line OH 88119 with susceptibility to *X. perforans* race T3 (Sun et al., 2011b), were used in this study. For inoculation with *X. perforans* race T3, tomato seedlings were grown in a growth room at 28°C/25°C (light/dark), with a 14-h photoperiod. The race T3 strain *Xv829* was cultured on yeast, dextrose, calcium carbonate (YDC) agar medium (Lelliot and Stead, 1987) at 28°C for 48–72 h prior to inoculation. Bacterial cultures were diluted to a concentration of approximately 3×10^8 colony forming units per ml using sterile solution containing 10 mM $MgSO_4{\cdot}7H_2O$ and 0.025% (v/v) Silwet L77. Six-week-old plants were spray-inoculated with the bacterial suspension, and plants sprayed with the sterile solution containing 10 mM $MgSO_4{\cdot}7H_2O$ and 0.025% (v/v) Silwet L77 were used as controls (mock-treatment). The plants were misted with water twice a day (9:00 a.m. and 5:00 p.m.) from 1 day before inoculation to 6 days after inoculation to increase humidity and prolong leaf wetness for disease development.

RNA Isolation, RNA-seq Library Preparation, and Sequencing

Total RNA was isolated from leaf samples collected at started point (mock-treatment), 6 h-post-inoculation (HPI) and 6 day-post-inoculation (DPI) using the TRIzol reagent (Invitrogen, USA) following to the manufacturer's protocol. The concentration of total RNA was determined by NanoDrop 2000 Spectrophotometer (Thermo Fisher Scientific, USA), and the

RNA integrity value (RIN) was checked using RNA 6000 Pico LabChip of Agilent 2100 Bioanalyzer (Agilent, USA).

Three libraries (mock-treatment, race T3 treatment at 6 HPI and 6 DPI) for each tomato line were constructed for RNA-seq analysis. Two hundred micro-grams total RNA was incubated with 10 units DNase I (Ambion, USA) at 37°C for 1 h, and then nuclease-free water was added to bring the sample volume to 250 μl. The MicroPoly(A)Purist™ Kit (Ambion) was used to enrich mRNA according to the manufacture's protocol. The enriched mRNA was dissolved in 100 μl RNA storage solution and the final concentration was determined by NanoDrop 2000 Spectrophotometer. Double-stranded cDNA was synthesized from the mRNA according to the method described previously (Ng et al., 2005) with some modifications. First-strand cDNA was synthesized from 10 μg mRNA with GsuI-oligo dT primers using 1000 units of superscript II reverse transcriptase (Invitrogen). After incubation at 42°C for 1 h, the 5′-CAP structure of mRNA was oxidized by $NaIO_4$ (Sigma, USA) and ligated to biotin hydrazide, which was used to select complete mRNA/cDNA by binding Dynal M280 beads (Invitrogen). The second-strand cDNA synthesis was subsequently performed using Ex Taq polymerase (TaKaRa, Japan). The polyA and 5′ adaptor were removed by GsuI digestion. The double-strand cDNA was ultrasonically fractioned with a cDNA size fractionation column at a range of 300–500 bp, and then purified using Agencourt Ampure beads (Agencourt, USA). Sequencing libraries were generated using TruSeqTM DNA Sample Prep Kit—Set A (Illumina, USA). Libraries were clustered using TruSeq PE Cluster Kit (Illumina) according to manufacturer's instructions. After cluster generation, the libraries were sequenced with an Illumina HiSeq™ 2000 platform by Hanyu Bio-Tech Co., Ltd (Shanghai, China).

Analysis of Illumina Reads and Identification of Differentially Expressed Genes

Raw reads were firstly cleaned by removing adaptor sequences and unreliable low quality sequences (reads with unknown sequences "N"). The filtered clean reads were then mapped to *S. lycopersicum* reference genome sequence (SL2.40 version) and related annotation was obtained from the International Tomato Annotation Group (ITAG) Release 2.3 Predicted CDS (https://www.sgn.cornell.edu/) using Bowtie2 with default settings (Langmead et al., 2009). Clean reads mapped to reference genome sequence with multiple genes were filtered. Reads per kilo bases per million uniquely mapped reads (RPKM) values per sample for all genes were calculated based on both the total number of reads that mapped to the tomato reference CDS database and corresponding genes length (Mortazavi et al., 2008). Differentially expressed genes (DEG) were identified by DEG-seq package using the MA-plot-based method with Random Sampling model (MARS) (Wang et al., 2010). The *q*-value was adjusted using the method described in Benjamini and Hochberg (1995). "False Discovery Rate (FDR) ≤ 0.001 and the absolute value of Log2fold-change ≥ 2" were used as the threshold to determine the significance of gene expression difference.

Gene Ontology and KEGG Pathway Analysis

Gene ontology analysis for DEGs was performed using GoPipe (Chen et al., 2005) through BLASTP against Swiss-Prot and TrEMBL database. The metabolic pathway was constructed based on Kyoto Encyclopedia of Genes and Genomes (KEGG) database using BBH (bi-directional best hit) method (Kanehisa et al., 2010). The KO number was obtained for each protein and used for constructing metabolic pathways.

RT-PCR and qRT-PCR Validation

To verify the results of RNA-seq analysis, 8 differentially expressed genes (Receptor like kinase, Polyphenol oxidase, Harpin-induced protein-like, LRR receptor-like serine/threonine-protein kinase, 1-aminocyclopropane-1-carboxylate oxidase, Calmodulin 2, Universal stress protein family protein, and WRKY transcription factor 16) were selected for semi-quantitative RT-PCR and qRT-PCR analysis. First strand cDNA was synthesized by M-MLV reverse transcriptase (TaKaRa) with the same batch of RNA samples used in RNA-seq analysis. The specificity of primers for each gene (**Table S1**) was confirmed by analyzing PCR products on agarose gel and melting curve during real-time PCR. RT-PCR and qRT-PCR was conducted according to the previous protocols (Pei et al., 2012; Du et al., 2014).

RESULTS

Analysis of RNA-seq Libraries and Reads Alignment

Approximately 43 million Illumina raw reads were generated from six libraries with an average of 7 million 100 bp raw reads in each individual library. The raw reads were deposited in the National Center for Biotechnology Information Sequence Read Archive under the accession number SRP065597. In the six libraries, 97.9–99.3% of raw reads were clean. Of the clean reads, 46.5–77.3% could be mapped uniquely to one location within the tomato reference genome, 8.4–40.3% could be aligned to highly conserved domain sequences shared by different annotated genes, and 9.6–21.3% failed to map to the tomato genome sequence (**Table 1**). The uniquely mapped clean reads matched 21,168–21,993 with an average of 21,526 genes (approximately 62%) to the tomato annotated genes for the six libraries (**Table 1**). Approximately 20,000 genes were mapped at both mock- and T3-treatment samples (**Image S1**). The number of specific matched genes was higher in the PT6d library (890) than in other five libraries (**Image S1**).

The genes RPKM values were calculated only for those uniquely mapped reads. The number of genes with RPKM values of 0 varied from 12,734 to 13,559, with an average of 13,200 in the six libraries (**Table S2**). Of which 10,117 were common in all libraries, indicating a very low level or no expression of those genes in the six samples. The numbers of genes with RPKM value of 0 were slightly higher in the three libraries of the susceptible line OH 88119 than the corresponding samples in the resistant line PI 114490. Based on the expression analysis using RT-PCR

TABLE 1 | Summary of RNA-seq data.

Sample ID[a]	No. of raw reads	No. of clean reads				No. of matched genes
		Total	Mapped to unique genes	Mapped to multiple genes	Unmapped	
PM	10,937,043	10,783,924	5,402,226	4,347,202	1,034,496	21,776
PT6h	6,707,580	6,603,613	4,852,118	556,598	1,194,897	21,345
PT6d	7,148,378	7,096,195	5,421,494	548,184	1,126,517	21,993
OM	6,154,320	6,094,623	4,710,354	476,999	907,270	21,708
OT6h	5,381,872	5,316,213	3,927,090	399,894	989,229	21,168
OT6d	6,921,528	6,773,407	3,148,353	2,181,678	1,443,376	21,168
Average	7,208,454	7,111,329	4,576,939	1,418,426	1,115,964	21,526

[a]*PM and OM: PI 114490 and OH 88119 respectively mock-treatment with the sterile solution containing 10 mM $MgSO_4$. $7H_2O$ and 0.025%(v/v) Silwet L77. PT and OT: PI 114490 and OH 88119 respectively inoculated with bacterial spot race T3.*

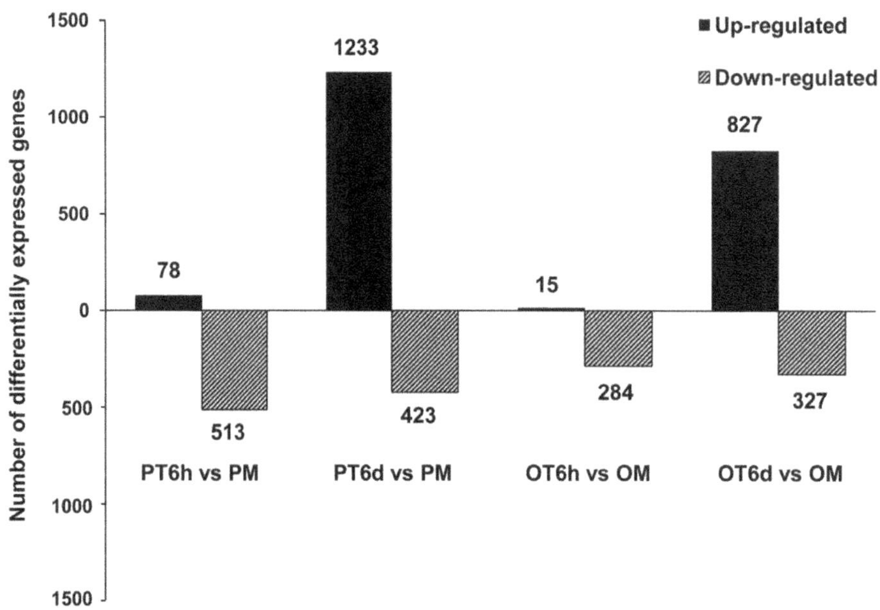

FIGURE 1 | Numbers of differentially expressed genes (DEGs) identified in tomato lines PI 114490 and OH 88119 at 6 h and 6 days post-inoculation of *Xanthomonas perforans* race T3. The DEGs were identified using FDR $\leq$ 0.001 and the absolute value of Log_2 fold-change $\geq$ 2 as the threshold for the significance of genes expression difference. PM and OM: PI 114490 and OH 88119 respectively mock-inoculated with the sterile solution containing 10 mM $MgSO_4 \cdot 7H_2O$ and 0.025% (v/v) Silwet L77. PT and OT: PI 114490 and OH 88119 respectively inoculated with T3.

and qRT-PCR, genes with RPKM value less than 2.28 needed more amount of cDNA and genes with RPLM value greater than 10.7 needed less amount of cDNA for obtaining clear bands (data not shown). Therefore, genes with RPKM values of 0–3 were considered as at low level of expression, while genes with RPKM values beyond 10 were considered to be expressed at high level. The number of highly expressed genes was less in PT6h library compared to OT6h, but more in PT6d than OT6d (**Table S2**).

Differentially Expressed Genes in Response to *X. perforans* Race T3

More stringent criteria with smaller FDR ($\leq$0.001) and higher fold-change value (absolute value of Log_2fold-change $\geq$ 2) were used to analyze the global transcriptome dynamics of both resistant line PI 114490 and susceptible line OH 88119 in response to race T3 during 6 h and 6 days post-inoculation. Overall, the number of differentially expressed genes (DEGs) was higher in PI 114490 than in OH 88119 at both infection stages, and the number of DEGs was considerably higher at 6 DPI than at 6 HPI in both tomato lines (**Figure 1**). At the 6 HPI, there were 591 and 299 genes were differentially expressed in PI 114490 and OH 88119, respectively. Most DEGs showed down-regulations in both tomato lines, whereas only 78 and 15 genes were up-regulated in PI 114490 and OH 88119, respectively (**Figure 1**). However, the numbers of DEGs dramatically increased to 1656 and 1154 at 6 DPI in PI 114490 and OH 88119, respectively, and the numbers of up-regulated genes were obviously greater than that of down-regulated genes (**Figure 1**). Consideration of

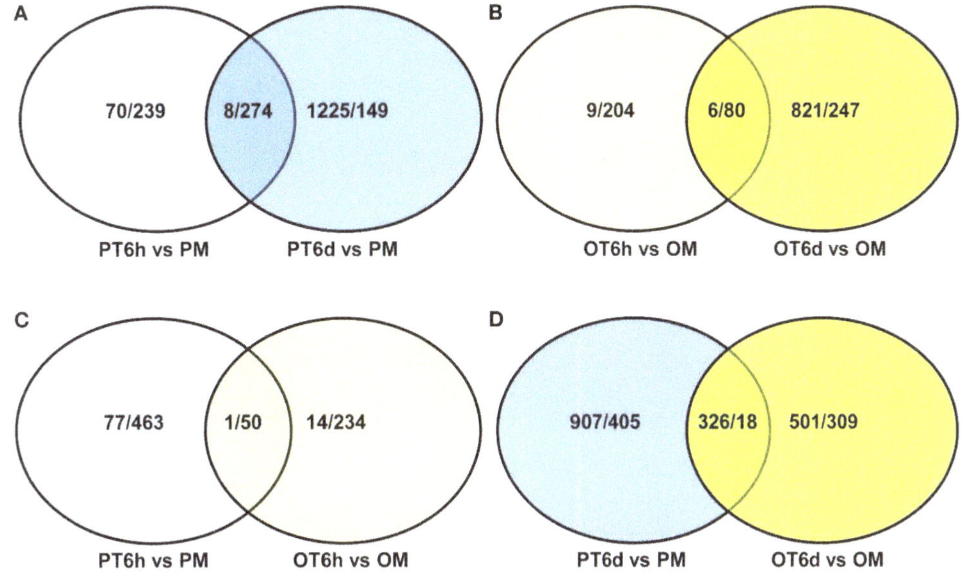

FIGURE 2 | Venn diagram showing the number of up/down-regulated genes between tomato lines PI 114490 and OH 88119 at 6 h and 6 days post-inoculation with *Xanthomonas perforans* race T3. PM and OM: PI 114490 and OH 88119 respectively mock-inoculated with the sterile solution containing 10 mM $MgSO_4{\cdot}7H_2O$ and 0.025% (v/v) Silwet L77. PT and OT: PI 114490 and OH 88119 respectively inoculated with race T3. **(A,B)** Comparison of co-DEGs between 6 HPI and 6 DPI in PI 114490 and OH 88119, respectively. **(C,D)** Comparison of co-DEGs between two tomato lines at 6 HPI and 6 DPI, respectively. Overlapping portion indicates co-regulated genes.

numbers and fold changes of these DEGs, resistant tomato line PI 114490 had a higher frequency and stronger fold changes of DEGs at 6 DPI, compared to PI 114490 at 6 HPI and OH 88119 at both time points (**Figure 1**).

Genes with Differentially Expressed Patterns in Two Tomato Lines during Infection with Race T3

In order to gain comprehensive insights into understanding molecular mechanisms of resistance induced by *X. perforans* race T3, the overlap differentially expressed patterns of DEGs between PI 114490 and OH 88119 at two infection stages were analyzed using a Venn diagram (**Figure 2**). For tomato line PI 114490, 282 genes (8 up-regulated and 274 down-regulated) were co-modulated in both 6 HPI and 6 DPI compared to mock-treatment (**Figure 2A**), while 86 genes (6 up-regulated and 80 down-regulated) were co-modulated in both 6 HPI and 6 DPI compared to mock-treatment in line OH 88119 (**Figure 2B**). Comparison between the two tomato lines discovered 51 and 344 DEGs co-regulated at 6 HPI and 6 DPI, respectively (**Figures 2C,D**). Among the 51 co-regulated DEGs at 6 HPI, 50 were co-down-regulated, and only 1 (Solyc03g114530.2.1) encoding Strictosidine synthase family protein was co-up-regulated. The expression of Strictosidine synthase (Str), the key gene of terpenoid indole alkaloid (TIA) biosynthetic pathway, has been found to be induced by fungal elicitor (Pasquali et al., 1992) and salinity stress in *Catharanthus roseus* (Dutta et al., 2013). Go classification analysis of the 50 down-regulated genes, showed that the dominant subcategories in biological process were response to stimulus (**Table S3**). Among the 344 DEGs co-regulated at 6 DPI, 326 genes were commonly up-regulated, and 18 genes were commonly down-regulated (**Figure 2D**). In the face of a large number of commonly up-regulated genes, particular emphasis was placed on the highly up-regulated genes resulting in 32 DEGs with the Log_2 fold-change ≥ 5 in both lines at 6 DPI (**Table 2**), of which a large proportion (40.6%) were assigned to Go term "response to stimulus." For example, marker genes of stress signaling such as pathogen-related genes (PR) and osmotin-like protein were intensely induced, and PR proteins are inducible by infection with various types of pathogens in many plant families (van Loon et al., 2006). In addition to these stimulus genes, other genes also have been clarified related to defense response.

Clustering analysis of DEGs among the four comparison groups resulted in 37 clusters (**Figure 3**). Despite there were substantial number of DEGs, none of these genes was commonly up-regulated at four comparison groups. Cluster 10 was the largest with 711 DEGs, followed by cluster 8 representing 223 common up-regulated genes in PI 114490 and OH 88119 at 6 DPI. Common up-regulated genes in cluster 8 indicated that both PI114490 and OH88119 used some same sets of genes to response to race T3 infection but the induced intensity of transcriptional levels were different between two genotypes. Cluster 10 and cluster 20 consisting of the DEGs were only up-regulated (compared to mock-inoculated) in PI 114490 and OH 88119 at 6 DPI response to race T3. These specific DEGs in PI 114490 from cluster 10 (**Table 3**) were promising candidates for revealing the resistance response mechanism. Cluster 32 comprised only one unknown function gene Solyc02g084850.2.1

TABLE 2 | RPKM for common highly up-regulated (Log_2fold-change ≥ 5) genes in tomato lines P I114490 and OH 88119 at 6 days post-inoculation (DPI).

Gene	Function description	PI 114490			OH 88119		
		Mock[b]	6 DPI[b]	Fold value[c]	Mock[b]	6 DPI[b]	Fold-change[c]
Solyc00g174330.2.1	Pathogenesis related protein[a]	6.37	11.43	5.06	2.88	9.42	6.54
Solyc09g007010.1.1	Pathogenesis related protein[a]	4.69	10.02	5.33	2.63	9.17	6.54
Solyc00g174340.1.1	Pathogenesis-related protein[a]	8.88	14.31	5.43	5.50	12.42	6.93
Solyc01g106620.2.1	Pathogenesis-related protein[a]	4.49	10.20	5.71	1.62	9.27	7.65
Solyc08g080620.1.1	Osmotin-like protein[a]	0	5.38	7.26	0	5.05	6.73
Solyc08g080650.1.1	Osmotin-like protein[a]	6.43	11.47	5.03	4.39	9.75	5.36
Solyc08g080670.1.1	Osmotin-like protein[a]	2.62	7.77	5.15	0	8.14	8.38
Solyc02g082960.2.1	Endochitinase[a]	0	3.30	6.67	0	4.37	7.54
Solyc02g086700.2.1	Beta-1 3-glucanase[a]	0	5.82	7.62	0	6.43	10.04
Solyc01g060020.2.1	Beta-glucanase[a]	4.12	9.30	5.18	0	7.27	7.70
Solyc07g056510.2.1	Glutathione S-transferase[a]	0	5.57	5.46	0	2.11	5.39
Solyc07g066330.2.1	NAC domain protein[a]	0	4.77	6.59	0	1.67	5.28
Solyc01g106630.2.1	Unknown Protein[a]	4.31	9.83	5.51	1.39	8.91	7.53
Solyc12g011150.1.1	Unknown Protein	0	5.52	7.02	0	3.87	5.17
Solyc03g098760.1.1	Kunitz-type protease inhibitor protein[a]	0	2.44	5.16	0	4.70	5.23
Solyc07g052150.2.1	Sesquiterpene synthase	0	6.84	8.65	0	4.83	6.44
Solyc02g093580.2.1	Pectate lyase	0	2.14	6.16	0	1.84	5.67
Solyc01g059990.2.1	Serine/threonine-protein phosphatase 7	0	5.81	5.06	0	3.99	6.36
Solyc02g090970.1.1	Serine/threonine-protein kinase 24	0	3.28	5.83	0	2.93	5.28
Solyc04g074000.2.1	Receptor like kinase, RLK	1.66	6.80	5.14	0	2.19	6.20
Solyc02g071470.2.1	1-aminocyclopropane-1-carboxylate oxidase 1	0	4.71	7.55	0	2.40	6.04
Solyc07g049530.2.1	1-aminocyclopropane-1-carboxylate oxidase	4.43	9.62	5.19	1.97	7.44	5.47
Solyc01g109140.2.1	Cytochrome P450	0	5.41	8.60	0	1.17	5.17
Solyc11g068940.1.1	U-box domain-containing 24	0	3.36	7.14	0	2.17	5.75
Solyc04g008100.1.1	U-box domain-containing	0	3.38	5.22	0	2.46	5.10
Solyc06g062330.1.1	UDP-glucosyltransferase	0	4.63	5.34	0	3.61	6.71
Solyc07g052120.2.1	(−)-germacrene D synthase	0	4.08	6.99	0	3.04	5.75
Solyc07g052140.2.1	(−)-germacrene D synthase	0	7.61	7.71	0	6.45	7.66
Solyc08g029000.2.1	Lipoxygenase	0	6.19	6.70	0	1.96	5.58
Solyc08g068870.2.1	Aspartic proteinase nepenthesin-1	2.28	7.35	5.07	0	5.31	5.38
Solyc12g044950.1.1	Omega-6 fatty acid desaturase	0	2.68	5.04	0	3.46	6.63
Solyc12g049030.1.1	Fatty acid desaturase	0	5.04	5.68	0	3.04	6.49

[a] *The representative genes are associated with "response to stimulus" according to Gene Ontology classification.*
[b] *the value of RPKM.*
[c] *the value of Log_2(fold-change).*

with strongly down-regulated in both PI 114490 and OH 88119 at two infection stages (**Table 3**). Cluster 27 contained 7 genes strongly down-regulated at 6 HPI but strongly up-regulated at 6 DPI (**Table 3**), with three genes annotated for pathogenesis-related (Solyc00g174330.2.1, Solyc00g174340.1.1, and Solyc06g009140.2.1). Cluster 13 exhibited that 8 genes were up-regulated in PI 114490 but down-regulated in OH 88119 (**Table 3**).

Gene Ontology Classification Analysis of DEGs

Go enrichment categorized for 52.5, 54.2, 50.5, and 51.2% of DEGs into functional groups from PT6h, OT6h, PT6d, and OT6d, respectively. More assigned function of DEGs was covered in biological process and molecular function categories than in cellular component categories (**Image S2**). The dominant subcategories of both PI 114490 and OH 88119 at 6 HPI and 6 DPI in each main category were cellular process, cell and catalytic activity, respectively (**Image S2**). Despite these enrichment subcategories were similar in both tomato lines, the individual genes contributing to the common enriched subcategories were substantial diversified between the two tomato lines.

The most significantly enriched GO terms in biological process in the OT6d library included "response to wounding (Go:0009611)," "response to external stimulus (GO:0009605)," "oxidation reduction (GO:0055114)," "response to stress (GO:0006950)," and "response to biotic stimulus (GO:0009607),"

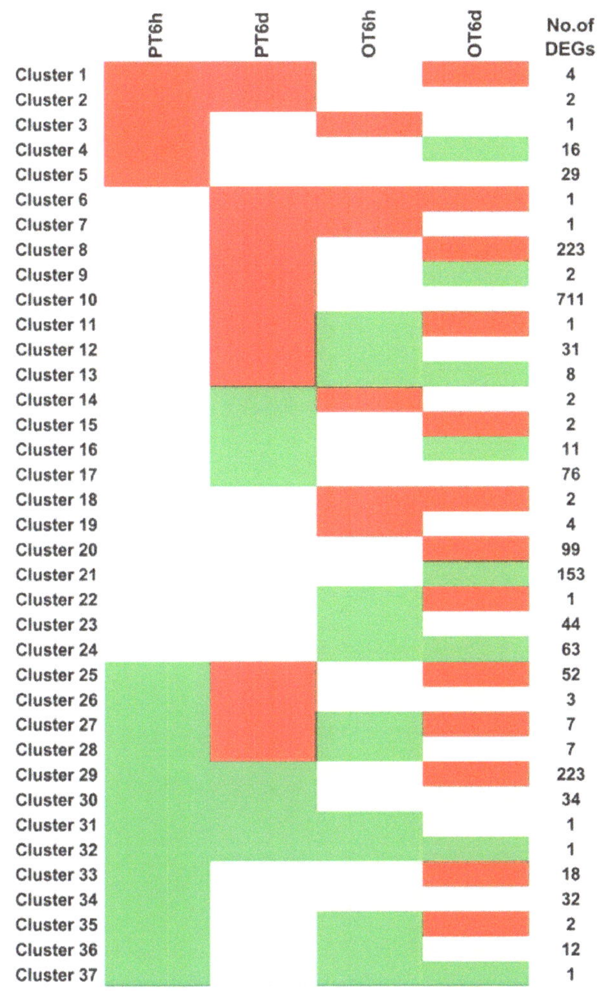

FIGURE 3 | Hierarchical clustering of differentially expressed genes (DEGs) among four groups. Up-regulation and down-regulation is represented by red shading and green shading, respectively. Gray shading indicates non-modulated. PT: PI 114490 inoculated with race T3. OT: OH 88119 inoculated with race T3.

For the library of PT6d, the most significantly enriched GO terms were "cell wall macromolecule metabolic process (GO:0044036)" and "response to water (GO:0009415)" (**Table S4**).

Metabolic Pathway by KEGG Analysis of DEGs

To characterize the pathway enrichment of the identified DEGs, gene classification was performed on the basis of KEGG analysis (Kanehisa et al., 2010). Only significant pathway categories among four comparisons were selected and listed in **Table 4**. Genes involved in photosynthesis and oxidative phosphorylation pathways were mainly up-regulated in OH 88119 at 6 DPI, but down-regulated in PI114490 at 6 HPI and 6 DPI (**Table 4**). It was obvious that genes associated with defense response pathways (labeled in bold in **Table 4**) were significantly up-regulated in PI114490 at 6 DPI. The defense-related pathways with most representation by DEGs were plant hormone signal transduction, plant-pathogen interaction and phenylalanine metabolism. A total of 16 defense response genes in KEGG pathway "plant-pathogen interaction" (KO: 04626) exhibited significant differences between PI 114490 and OH 88119 in response to T3 infection (**Table 5**). Of which, 5 and 3 were down-regulated at 6 HPI in PI 114490 and OH 88119, respectively. However, the numbers of DEGs increased in both tomato lines at 6 DPI. A total of 15 (14 up-regulated and 1 down-regulated) in PI 114490 and 8 (6 up-regulated and 2 down-regulated) gens in OH 88119 was detected. It was noteworthy that 5 DEGs (Solyc05g050350.1.1, Solyc10g079420.1.1, Solyc11g071740.1.1, Solyc09g014990.2.1, and Solyc12g009220.1.1) had the similar expression pattern between two tomato lines at both 6 HPI and 6 DPI (**Table 5**).

Validation of RNA-seq Data for Selected Genes by RT-PCR and qRT-PCR

To evaluate the validity of RNA-seq analysis, transcriptional levels of selected 8 DEGs representing a wide range of expression levels and patterns were determined by RT-PCR and qRT-PCR analysis. Based on the functional annotation in literatures, majority of these DEGs were associated with massive defense response processes including ethylene biosynthesis (Solyc07g049530.2.1), defense-related enzymes (Solyc02g072470.2.1, Solyc07g008600.1.1, Solyc01g057000.2.1, Solyc02g036480.1.1, Solyc04g072070.2.1 and Solyc10g081170.1.1), and second metabolite biosynthesis (Solyc08g074630.1.1). In general, the expression data provided by qRT-PCR were almost consistent with profiles detected by RNA-seq at all time-points (**Figure 4**), confirming the trends of up- or down-regulation of all the analyzed genes. The differences in the magnitudes of changes observed between the qRT-PCR and RNA-seq results might be caused by different algorithms.

It was validated by qRT-PCR that four genes Solyc02g072470.2.1, Solyc02g036480.1.1, Solyc08g074630.1.1, and Solyc07g008600.1.1 were steadily up-regulated during two time-points infection in both PI 114490 and OH 88119, differing in their induced strength. Solyc07g049530.2.1 and Solyc04g072070.2.1 initially displayed down-regulated at 6 HPI but significantly reinforced at 6DPI in both tomato lines, and the up-regulated fold-change was higher in PI 114490. The expression pattern of Solyc01g057000.2.1 was opposite between two tomato lines, which was up-regulated in the resistant line PI 114490 while down-regulated in the susceptible line OH 88119 (**Figure 4**).

DISSCUSSION

The *S. lycopersicum* var. *cerasiforme* accession PI 114490 has been considered as a durable source for resistance to bacterial spot due to its high level of quantitative field resistance to four races T1–T4 (Yang, 2013). There is no or few lesions on PI 114490, but the bacterial population of race T3 in its leaves is not significantly different from that in the susceptible line OH 88119 (Sun et al., 2014). So considering on the aspect of phenotype, PI 114490 shows higher tolerant response to avoid the

TABLE 3 | Part of significantly altered expression genes from Hierarchical clustering analysis.

Gene	Function description	Log$_2$Fold-change value[b]			
		PI 114490		OH 88119	
		6 HPI	6 DPI	6 HPI	6 DPI
CLUSTER 10 (SPECIFIC UP-REGULATED IN PT6d)[a]					
Solyc02g087070.2.1	Peroxidase family protein	No	7.6	No	No
Solyc03g020050.2.1	Proteinase inhibitor II	No	8.72	No	No
Solyc06g065060.1.1	FAD-binding domain protein	No	7.4	No	No
Solyc07g008600.1.1	LRR receptor-like serine/threonine-protein kinase	No	7.14	No	No
Solyc07g009500.1.1	Chitinase	No	7.33	No	No
Solyc08g079230.1.1	Cortical cell-delineating protein	No	7.32	No	No
Solyc08g080610.1.1	Osmotin-like protein	No	7.35	No	No
Solyc12g005720.1.1	Cysteine-rich receptor-like kinase	No	7.9	No	No
Solyc12g100250.1.1	Fatty acid desaturase	No	7.6	No	No
CLUSTER 13 (UP-REGULATED IN PT6d; DOWN-REGULATED IN OT6h AND OT6d)[a]					
Solyc01g057000.2.1	Universal stress protein	No	6.35	−6.87	−3.96
Solyc03g013440.2.1	Amino acid transporter	No	2.90	−2.91	−2.85
Solyc03g096670.2.1	Integrin-linked kinase-associated serine/threonine phosphatase 2C	No	2.46	−7.00	−4.22
Solyc04g063350.2.1	3-methyl-2-oxobutanoate dehydrogenase	No	2.16	−2.40	−2.33
Solyc06g007180.2.1	Asparagine synthase	No	3.77	−4.52	−2.56
Solyc07g006500.2.1	Alpha alpha-trehalose-phosphate	No	2.52	−2.21	−2.31
Solyc08g077900.2.1	Expansin-like protein	No	3.49	−5.73	−4.06
Solyc12g006230.1.1	RING-H2 finger protein	No	2.05	−4.02	−3.12
CLUSTER 27 (DOWN-REGULATED IN PT6h AND OT6h; UP-REGULATED IN PT6d AND OT6d)[a]					
Solyc00g174330.2.1	Pathogenesis related protein PR-1	−8.49	5.06	−5.7	6.54
Solyc00g174340.1.1	Pathogenesis-related protein 1b	−9.1	5.43	−5.41	6.93
Solyc03g025670.2.1	PAR-1c protein	−2.43	4.23	−2.5	2.46
Solyc04g077980.1.1	Zinc-finger protein	−3.75	2.06	−3.2	3.41
Solyc06g009140.2.1	Late embryogenesis abundant 3	−5.78	2.73	−4.74	2.38
Solyc07g049530.2.1	1-aminocyclopropane-1-carboxylate oxidase	−4.14	5.19	−4.54	5.47
Solyc10g085030.1.1	Soul heme-binding family protein	−3.19	2.29	−2.2	2.39
CLUSTER 32 (DOWN-REGULATED IN PT6h, PT6d, OT6h, AND OT6d)[a]					
Solyc02g084850.2.1	Unknown Protein	−4.12	−4.86	−12.62	−9.72

[a] *PT6h: 6 h after inoculation with bacterial spot race T3 in PI 114490. PT6d: 6 day after inoculation with bacterial spot race T3 in PI 114490. OH6h: 6 h after inoculation with bacterial spot race T3 in OH 88119. OH6d: 6 day after inoculation with bacterial spot race T3 in OH 88119.*

[b] *"NO" indicates that the expression level of the genes after infection with race T3 is not significant difference as relative to that of mock-treatment.*

development of symptoms. But the difference of global inducible defense response against *X. perforance* race T3 infection between PI 114490 and OH 88119 were not clear.

In this study, RNA-seq technique was adapted to detect the DEGs during the race T3 infection in PI 114490 and OH 88119. Our previous study using cDNA-AFLP techniques has already identified DEGs between the two tomato lines (Du et al., 2014). Therefore, we started with one biological replication for RNA-seq analysis. An average of 7 million sequence reads were obtained from RNA-seq and the clean reads could match approximately 62% of reference genes (**Table 1**), which was closed to previous tomato RNA-seq data with 20 million sequence reads (Tang et al., 2013), indicating that the sequencing depth was sufficient for the transcriptome coverage. The well correlation between qRT-PCR and RNA-seq data (**Figure 4**) also demonstrated the reliability of RNA-seq analysis. Of course, more replicates should provide more reliable information for understanding the genes involved in the process of resistance/defense to the pathogen of bacterial spot race T3 in tomato.

Inducible Defense Response to *X. perforans* Interfered at 6 HPI but Activated at 6 DPI in Both Resistant and Susceptible Tomato Lines

Despite PI 114490 and OH 88119 have different responses to race T3 infection (Sun et al., 2014), lots of common DEGs were regulated in both two tomato lines. At 6 HPI, 50 common down-regulated genes existed in both two tomato lines (**Figure 2C**) suggested that the inducible defense response pathways were not completely activated at 6 HPI. Meanwhile, down-regulation of the minority of pathogen-related genes in cluster 27 (**Figure 3**) indicated that at least some defense response had been interfered by pathogen infection as early as 6 HPI.

TABLE 4 | Significantly enriched KEGG pathways of differentially expressed genes by *Xanthomonas perforans* race T3 infection.

	Pathway	No. of up-regulated genes	No. of down-regulated genes	Pathway ID
PT6h vs. PM	Photosynthesis	0	23	Ko 00195
	Oxidative phosphorylation	0	27	Ko 00190
	Ribosome	0	13	Ko 03010
PT6d vs. PM	Photosynthesis	0	23	Ko 00195
	Oxidative phosphorylation	1	28	Ko 00190
	Phenylalanine metabolism	9	0	Ko 00360
	Glutathione metabolism	10	1	Ko 00480
	Phenylpropanoid biosynthesis	11	2	Ko 00940
	Ribosome	0	14	Ko 03010
	Plant hormone signal transduction	15	3	Ko 04075
	Plant-pathogen interaction	14	1	Ko 04626
OT6h vs. OM	Protein processing in endoplasmic reticulum	0	10	Ko 04141
	MAPK signaling pathway	0	4	Ko 04010
	Plant hormone signal transduction	0	9	Ko 04075
OT6d vs. OM	Oxidative phosphorylation	25	0	Ko 00190
	Photosynthesis	19	1	Ko 00195
	Ribosome	11	1	Ko 03010
	Phenylalanine metabolism	7	1	Ko 00360
	Metabolism of xenobiotics by cytochrome P450	9	0	Ko 00980

PM and OM: PI 114490 and OH 88119 respective mock-treatment with the sterile solution containing 10 mM $MgSO_4.7H_2O$ and 0.025%(v/v) Silwet L77. PT and OT: inoculation with bacterial spot race T3 in PI 114490 and OH 99119, respectively.

TABLE 5 | Differentially expressed genes in the common enriched KEGG pathway " Plant-pathogen interaction"(ko 04626) in tomato lines PI 114490 and OH 88119 at 6 h post-inoculation (HPI) and 6 days post-inoculation (DPI).

Gene	Homologous protein in KEGG	PI 11449					OH 88119				
		RPKM			Log_2Fold-change[a]		RPKM			Log_2Fold-change[a]	
		Mock	6 HPI	6 DPI	6 HPI vs. Mock	6 DPI vs. Mock	Mock	6 HPI	6 DPI	6 HPI vs. Mock	6 DPI vs. Mock
Solyc03g113390.2.1	CPK	7.25	7.21	38.92	−0.008	**2.42**	8.62	6.89	16.01	−0.32	0.89
Solyc06g065380.2.1	CPK	3.06	4.83	0.68	0.66	**−2.15**	3.17	5.97	1.86	0.91	−0.76
Solyc05g050350.1.1	CNGF	0.25	0	2.74	0	**3.41**	0.09	0.11	0.88	0.26	**3.16**
Solyc10g081170.1.1	CALM	336.89	336.1	1391.1	−0.003	**2.05**	189.18	336.69	387.5[a]	0.83	1.03
Solyc03g118810.1.1	CML	83.01	14.98	235.87	**−2.46**	1.51	14.15	8.74	96.89[a]	−0.69	**2.77**
Solyc06g068960.1.1	CML	31.84	30.13	161.84	−0.07	**2.35**	13.69	18.07	21.85	0.39	0.67
Solyc10g079420.1.1	CML	41.93	25.34	257.87	−0.72	**2.62**	19.58	23.48	131.57[a]	0.26	**2.75**
Solyc02g094000.1.1	CML	3.04	0	32.47	**−3.65**	**3.42**	0.49	0	11.18[a]	0	**4.49**
Solyc03g044900.2.1	CML	5.19	6.71	25.07	0.36	**2.27**	8.34	10.58	6.06	0.34	−0.46
Solyc06g073830.1.1	CML	0.66	0	19.27	0	**4.85**	1.147	0	1.71	0	0.58
Solyc11g071740.1.1	CML	5.19	0.34	74.57	**−3.93**	**3.84**	4.20	10.42	18.34	**−3.32**	**2.13**
Solyc11g071760.1.1	CML	0	0	8.34	0	**5.75**	1.42	0	2.66	0	0.90
Solyc09g014990.2.1	WRKY33	2.52	0.11	28.48	**−4.43**	**3.50**	0.84	0.14	12.26	**−2.54**	**3.86**
Solyc06g036290.2.1	HSP90	16.45	2.76	70.45	**−2.57**	**2.09**	32.76	2.02	18.11	**−4.0**	−0.85
Solyc07g042170.2.1	JAZ	15.32	10.98	69.14	−0.48	**2.17**	15.35	13.32	60.44	−0.20	1.98
Solyc12g009220.1.1	JAZ	3.41	7.33	75.82	1.10	**4.47**	6.43	7.71	33.47	0.26	**2.38**

[a]Significant difference (FDR $\leq$ 0.001 and the absolute value of Log_2 fold-change $\geq$ 2) were shown in boldface.

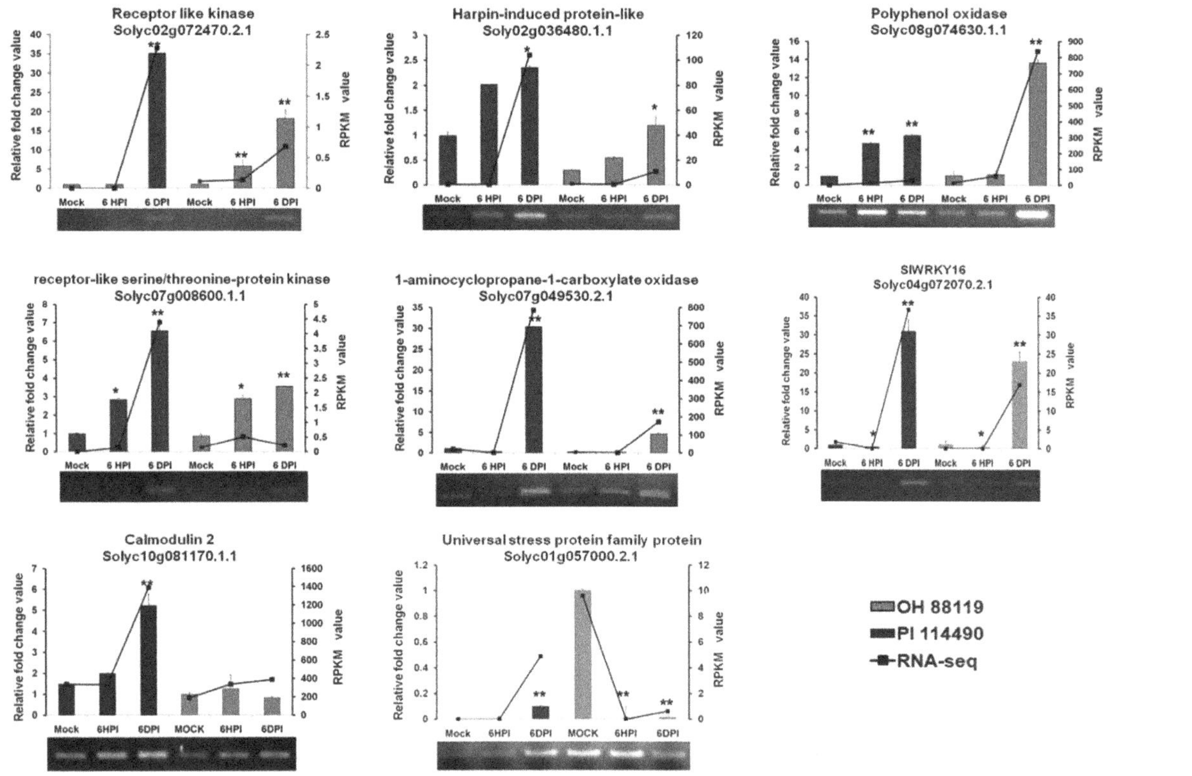

FIGURE 4 | RT-PCR and qRT-PCR validation of differentially expressed 8 genes related to defense response at 6 h post-inoculation (6 HPI) and 6 days post-inoculation (6 DPI). Expression levels of each sample were normalized on the basis of transcript amounts of *EF1*-α. Relative expression values for qRT-PCR analysis were determined against the average value of mock sample with PI 114490. Each experiment determined in three technical replicates and standard deviation ($n = 3$) is represented by error bars. The asterisk above the bars indicates statistically significant differences between the infected samples and corresponding mock samples.

Among the 344 common DEGs at 6 DPI, a majority of 326 genes were commonly up-regulated (**Figure 2D**) in both tomato lines. The dominant subcategories in biological process were cellular process, macromolecule metabolism and response to stimulus (data not shown). Six up-regulated genes in plant-pathogen interaction pathway were shared by the two tomato lines at 6 DPI (**Table 5**). Previous studies have proven that the function of these regulated genes plays important role in bacterial spot defense response (Cui et al., 2015). Therefore, defense response of tomato lines was activated 6 days after infection by race T3. Accumulation expression levels of those co-regulated genes were likely to increase basal defense response and improve the disease resistance to race T3 infection in both resistant and susceptible tomato. So the susceptible line OH 88119 also succeeded to properly activate the expression of many defense-related genes.

Resistant and Susceptible Tomato Lines Displayed Differentially Expressed Patterns in Photosynthesis Pathway

Plant-pathogen interaction response usually alters the expressional level of genes associated with photosynthesis (Major et al., 2010). Many photosynthesis-related genes including chlorophyll a/b binding proteins and photosynthetic reaction center proteins were strongly repressed in susceptible hybrid poplar leaves at 9 days inoculated with *Melampsora medusa* (Miranda et al., 2007). However, our RNA-seq data showed that race T3 pathogen caused opposite impact on photosynthesis between resistance and susceptible tomato lines. Genes involved in photosynthesis pathway were mainly up-regulated in the susceptible tomato line OH 88119 at 6 DPI, but down-regulated in the resistant line PI 114490 at 6 HPI and 6 DPI (**Table 4**). To the best of our knowledge, the reason of this divergent response between resistant and susceptible plants in photosynthesis pathway has not previously been known in plant-pathogen interactions.

Analysis of Differentially Expressed Genes Involved in Plant Immune Response Pathways

Plants possess different sophisticated defense strategies which are usually accompanied by transcriptional changes to protect themselves against pathogen attacks. The plant immune response pathway can be mainly divided into two branches (Jones and Dangl, 2006). The first branch and initial defense response is usually activated by pathogen-associated molecular patterns (PAMP) whose presence is recognized by plant plasma membrane-localized pattern recognition receptors (PRRs), which results in PAMP-triggered immunity (PTI). The second branch is activated by nucleotide-binding leucine-rich repeat (NB-LRR) resistance gene, which directly or indirectly interacts with virulence factor, and results in activation of effector-triggered immunity (ETI). Since the plant-pathogen interaction pathways (KEGG: 04626) summarized the genes involved in defense network of PTI and ETI, so we focused on the transcript dynamics of this KEGG pathways and enumerated the DEGs imparting the plant immune response (**Table 5**).

Several PAMPs have been identified from bacteria. Flagellin is one of important and well-characterized PAMPs that can be recognized by the LRR receptor-like serine/threonine-protein kinase (FLS2) (Gómez-Gómez and Boller, 2000). In our study, RLK protein Solyc02g070890.2.1 with the highest amino acid similarity (54%) to FLS2 protein from *Arabidopsis thaliana* than other tomato RLK genes was down-regulated at 6 HPI, and slightly up-regulated at 6 DPI in PI 114490, but this gene was not differentially expressed in OH 88119 upon race T3 infection. RLK protein Solyc02g072470.2.1 was the highest induced member of RLK family in our RNA-seq data. qRT-PCR validated that Solyc02g072470.2.1 were significantly induced in both tomato lines at 6 DPI, which displayed more than 35- and 18-fold-induction in PI 114490 and OH 88119, respectively (**Figure 4**). The highest induced member of LRR receptor-like serine/threonine-protein kinase was Solyc07g008600.1.1 in RNA-seq analysis, which was verified up-regulated in both two tomato lines (**Figure 4**).

Recognition of PAMPs initiates downstream signaling pathways involving WRKY transcription factors to confer defense response against bacterial, fungal pathogens and nematodes (Asai et al., 2002; Bhattarai et al., 2010). Based on KEGG analysis, WRKY transcription factor Solyc09g014990.2.1 homologous to WRKY33 was down-regulated at 6 HPI but highly up-regulated at 6 DPI in both genotypes (**Table 5**). Since members of WRKY families are thought to play important role in defense response, we further investigated the expression patterns of all 85 WRKY families genes annotated in the International Tomato Annotation Group (ITAG) Release 2.3 Predicted CDS. A total of 29 WRKY genes were differentially expressed. Among them, none was significantly up-regulated at 6 HPI but 17 (58.6%) were up-regulated in both tomato lines at 6 DPI (**Figure 5**). Previous study indicated that the expression levels of SlWRKY8 (Solyc02g093050.2.1), SlWRKY23 (Solyc01g079260.2.1), SlWRKY39 (Solyc03g116890.2.1), SlWRKY53 (Solyc08g008280.2.1), SlWRKY80 (Solyc03g095770.2.1), and SlWRKY81 (Solyc09g015770.2.1) were enhanced in *S. lycopersicum* cultivar Alisa Craig under the invasion of *Pseudomonas syringae* (Huang et al., 2012). In the current study, none of these WRKY genes were detected at 6 HPI, but SlWRKY39/53/80/81 were significantly induced in both tomato lines, and SlWRKY8/23 were specific up-regulated in PI 114490 at 6 DPI. SlWRKY72b are transcriptionally up-regulated during disease resistance mediated by the R gene *Mi-1*, and virus-induced gene silencing of this gene in tomato resulted in a clear reduction of *Mi-1*-mediated resistance as well as basal defense against root-knot nematodes and potato aphids (Bhattarai et al., 2010). WRKY gene Solyc06g070990.2.1, the highest identity with SlWRKY72b than other WRKY protein in tomato, were significantly induced at 6 DPI in PI 114490 (**Figure 5**).

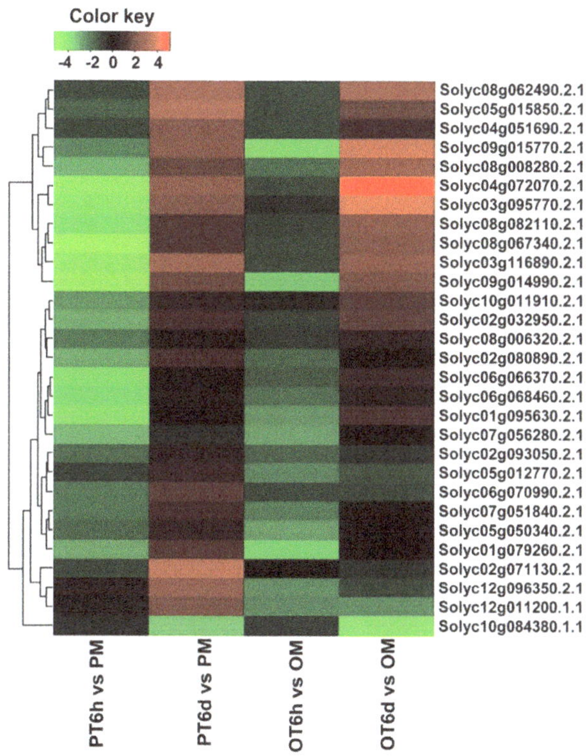

FIGURE 5 | Differentially expressed WRKY genes in tomato lines PI 1144490 and OH 88119 at 6 h and 6 days post-inoculation with *Xanthomonas perforans* race T3. PM and OM: PI 114490 and OH 88119 respectively mock-inoculated with the sterile solution containing 10 mM $MgSO_4 \cdot 7H_2O$ and 0.025% (v/v) Silwet L77. PT and OT: PI 114490 and OH 88119 respectively inoculated with race T3.

Intracellular Ca^{2+} influxes have also long been recognized as essential and early events downstream of multiple PAMP perception, leading to local and systemic acquired resistance (Lecourieux et al., 2005; Boudsocq et al., 2010). In the process of Ca^{2+} signaling, some members of Calcium-dependent protein kinase (CDPK) family was immediately induced by flg22 and Avr-Cf9 interaction (Romeis et al., 2000; Boudsocq et al., 2010). Unexpectedly, none of CDPK genes was induced as early as 6 HPI in both tomato lines. Calmodulin (CALM or CaM), a major Ca^{2+} sensor, play positive regulatory resistance roles in different host-pathogen interaction (Park et al., 2004; Choi et al., 2009). Only one CALM gene Solyc10g081170.1.1 was specifically up-regulated at 6 DPI in PI 114490 (**Table 5**), which was also validated by qRT-PCR analysis (**Figure 4**).

In the process of plant and pathogen interaction, bacterial type III effectors which are directly delivered into the host cells via the type III secretion system can bypass or disrupt host first line of PAMP-triggered immunity. But resistant plants contain

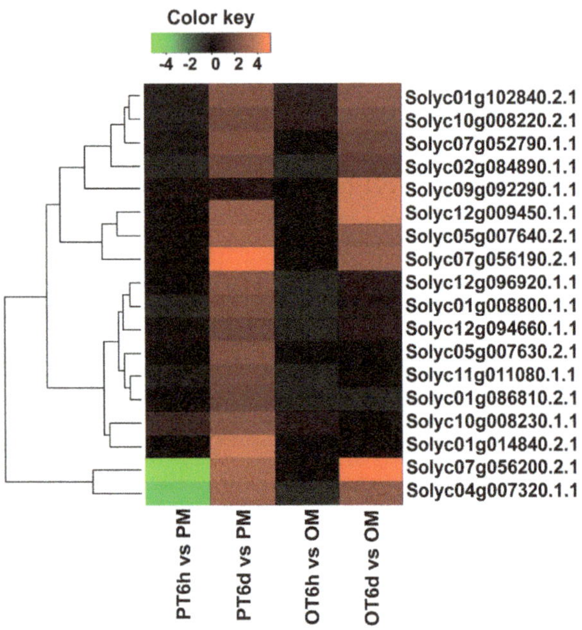

FIGURE 6 | Differentially expressed putative R genes containing NBS-LRR domain in tomato lines PI 1144490 and OH 88119 at 6 h and 6 days post-inoculation. PM and OM: PI 114490 and OH 88119 respectively mock-inoculated with the sterile solution containing 10 mM $MgSO_4 \cdot 7H_2O$ and 0.025% (v/v) Silwet L77. PT and OT: PI 114490 and OH 88119 respectively inoculated with race T3.

R proteins belonging to the nucleotide -binding site-leucine-rich repeat (NBS-LRR) family to directly or indirectly detect bacterial effectors, which will activate downstream signaling and lead to pathogen resistance (DeYoung and Innes, 2006). Previous papers have evidence that some R genes can be activated by the specific effectors, which are displayed by the accumulation of R genes expression levels. Therefore, we paid attention to those putative R genes containing NBS-LRR domain that could possibly be induced by the *X. perforans* infection. A total of 298 genes were annotated as NBS-LRR proteins for tomato in SGN database, and the RPKM values of most genes were low in the six sequenced samples, indicating the expression level of most putative R genes was low in tomato plant. The RNA-seq data showed that none of putative R genes were up-regulated in both tomato lines at 6 HPI. But the number of up-regulated putative R genes in PI 114490 at 6 DPI was more than that of in OH 881119 (**Figure 6**). Six putative R genes were commonly up-regulated in both tomato lines. Among these genes, Solyc07g056190.2.1 displayed the highest fold-change (**Figure 6**). In addition, there were 11 and 1 putative R genes were specifically up-regulated in P I114490 and OH88119, respectively (**Figure 6**).

Enumerate of Significant *X. perforans* Up-regulated Genes

Previously, we used cDNA-AFLP technique to identify differential expression genes in tomato lines PI 114490 and OH 88119 in response to race T3 infection at 3, 4, and 5 days. A total of 60 genes were commonly up-regulated at different levels in two tomato lines (Du et al., 2014). Among these genes, 14 genes were also up-regulated in this study (**Table S5**). Based on the functional annotation in literatures, majority of these genes were associated with massive defense response process including jasmonic acid biosynthesis pathways (Solyc03g122190.2.1 and Solyc04g079730.1.1), ethylene biosynthesis (Solyc07g049530.2.1), defense-related enzymes (Solyc10g055800.1.1, Solyc00g071180.2.1, Solyc03g007240.2.1, and Solyc04g015970.2.1), and cell death related (Solyc03g006700.2.1, Solyc01g105070.2.1, Solyc08g074680.2.1, and Solyc11g068940.1.1). The remaining (Solyc05g050120.2.1, Solyc08g068710.1.1, and Solyc04g010250.2.1) did not match any known defense response genes. Similar expression pattern in different expression profile techniques indicates that these well-validated and reproducible 14 up-regulated genes may play important role in tomato-*X. perforans* interaction.

CONCLUSION

In this study, high-throughput sequencing data was used to identify *X. perforans* responsive modules, which were enriched for differentially expressed genes between resistant line PI 114490 and susceptible line OH 88119 at different time point after spray-inoculation of race T3. The data provided a whole view on the variation tendency of genes involved in tomato-*X. perforans* interaction. A large number of candidate genes involved in tomato-*X. perforans* interaction were identified. Comparison of RNA-seq data with our previous cDNA-AFLP data revealed common up-regulated genes, and some of them were involved in jasmonic acid and ethylene biosynthesis.

AUTHOR CONTRIBUTIONS

Conceived and designed the experiments: YW, WY. Performed the experiments: HD. Analyzed the data: HD, JY, WY. Contributed reagents/materials/analysis tools: YW, JY, WY. Wrote the paper: HD, WY.

ACKNOWLEDGMENTS

The authors would like to thank Dr. David M. Francis in the Department of Horticulture and Crop Science at the Ohio State University for providing the seeds of tomato lines and Dr. Jeffrey B. Jones in the Department of Plant Pathology at the University of Florida for providing race T3 of the bacterial spot pathogen. The work was partly supported by the National Natural Science Foundation of China (31372073) and Beijing Modern Agro-industry Technology Research System (GCTDZJ2014033001).

REFERENCES

Asai, T., Tena, G., Plotnikova, J., Willmann, M. R., Chiu, W. L., Gomez-Gomez, L., et al. (2002). MAP kinase signalling cascade in Arabidopsis innate immunity. *Nature* 415, 977–983. doi: 10.1038/415977a

Balaji, V., Gibly, A., Debbie, P., and Sessa, G. (2007). Transcriptional analysis of the tomato resistance response triggered by recognition of the *Xanthomonas* type III effector AvrXv3. *Funct. Integr. Genomics* 7, 305–316. doi: 10.1007/s10142-007-0050-y

Benjamini, Y., and Hochberg, Y. (1995). Controlling the false discovery rate- a practical and powerful approach to multiple testing. *J. R. Statist. Soc. B* 57, 289–300.

Bhattarai, K. K., Atamian, H. S., Kaloshian, I., and Eulgem, T. (2010). WRKY72-type transcription factors contribute to basal immunity in tomato and Arabidopsis as well as gene-for-gene resistance mediated by the tomato R gene *Mi-1*. *Plant J.* 63, 229–240. doi: 10.1111/j.1365-313X.2010.04232.x

Boudsocq, M., Willmann, M. R., McCormack, M., Lee, H., Shan, L. B., He, P., et al. (2010). Differential innate immune signalling via Ca^{2+} sensor protein kinases. *Nature* 464, 418–422. doi: 10.1038/nature08794

Chen, Z. Z., Xue, C. H., Zhu, S., Zhou, F. F., Ling, X. F. B., Liu, G. P., et al. (2005). GoPipe: Streamlined Gene Ontology annotation for batch anonymous sequences with statistics. *Prog. Biochem. Biophys.* 32, 187–190.

Choi, H. W., Lee, D. H., and Hwang, B. K. (2009). The pepper calmodulin gene *CaCaM1* is involved in reactive oxygen species and nitric oxide generation required for cell death and the defense response. *Mol. Plant Microbe Interact.* 22, 1389–1400. doi: 10.1094/MPMI-22-11-1389

Cui, H., Tsuda, K., and Parker, J. E. (2015). Effector-Triggered immunity: from pathogen perception to robust defense. *Annu. Rev. Plant Biol.* 66, 487–511. doi: 10.1146/annurev-arplant-050213-040012

DeYoung, B. J., and Innes, R. W. (2006). Plant NBS-LRR proteins in pathogen sensing and host defense. *Nat. Immunol.* 7, 1243–1249. doi: 10.1038/ni1410

Du, H. S., Li, W. H., Wang, Y. Q., and Yang, W. C. (2014). Identification of genes differentially expressed between resistant and susceptible tomato lines during time-course interactions with *Xanthomonas perforans* race T3. *PLoS ONE* 9:e93476. doi: 10.1371/journal.pone.0093476

Dutta, A., Sen, J., and Deswal, R. (2013). New evidences about strictosidine synthase (Str) regulation by salinity, cold stress and nitric oxide in *Catharanthus roseus*. *J. Plant Biochem. Biotechnol.* 22, 124–131. doi: 10.1007/s13562-012-0118-1

Gao, L., Tu, Z. J., Millett, B. P., and Bradeen, J. M. (2013). Insights into organ-specific pathogen defense responses in plants: RNA-seq analysis of potato tuber-*Phytophthora infestans* interactions. *BMC Genomics* 14:340. doi: 10.1186/1471-2164-14-340

Gibly, A., Bonshtien, A., Balaji, V., Debbie, P., Martin, G. B., and Sessa, G. (2004). Identification and expression profiling of tomato genes differentially regulated during a resistance response to *Xanthomonas campestris* pv. *vesicatoria*. *Mol. Plant Microbe Interact.* 17, 1212–1222. doi: 10.1094/MPMI.2004.17.11.1212

Gómez-Gómez, L., and Boller, T. (2000). FLS2: an LRR receptor-like kinase involved in the perception of the bacterial elicitor flagellin in *Arabidopsis*. *Mol. Cell* 5, 1003–1011. doi: 10.1016/S1097-2765(00)80265-8

Gu, K. Y., Yang, B., Tian, D., Wu, L., Wang, D., Sreekala, C., et al. (2005). R gene expression induced by a type-III effector triggers disease resistance in rice. *Nature* 435, 1122–1125. doi: 10.1038/nature03630

Huang, S., Gao, Y., Liu, J., Peng, X., Niu, X., Fei, Z., et al. (2012). Genome-wide analysis of WRKY transcription factors in *Solanum lycopersicum*. *Mol. Genet. Genomics* 287, 495–513. doi: 10.1007/s00438-012-0696-6

Jones, J. B., Bouzar, H., Stall, R. E., Almira, E. C., Roberts, P. D., Bowen, B. W., et al. (2000). Systematic analysis of xanthomonads (*Xanthomonas* spp.) associated with pepper and tomato lesions. *Int. J. Syst. Evol. Microbiol.* 50, 1211–1219. doi: 10.1099/00207713-50-3-1211

Jones, J. B., Lacy, G. H., Bouzar, H., Minsavage, G. V., Stall, R. E., and Schaad, N. W. (2005). Bacterial spot—Worldwide distribution, importance and review. *Acta Hortic.* 695, 27–33. doi: 10.17660/ActaHortic.2005.695.1

Jones, J. D. G., and Dangl, J. L. (2006). The plant immune system. *Nature* 444, 323–329. doi: 10.1038/nature05286

Kanehisa, M., Goto, S., Furumichi, M., Tanabe, M., and Hirakawa, M. (2010). KEGG for representation and analysis of molecular networks involving diseases and drugs. *Nucleic Acids Res.* 38, D355–D360. doi: 10.1093/nar/gkp896

Kim, K. H., Kang, Y. J., Kim, D. H., Yoon, M. Y., Moon, J. K., Kim, M. Y., et al. (2011). RNA-Seq analysis of a soybean near-isogenic line carrying bacterial leaf pustule-resistant and -susceptible alleles. *DNA Res.* 18, 483–497. doi: 10.1093/dnares/dsr033

Langmead, B., Trapnell, C., Pop, M., and Salzberg, S. L. (2009). Ultrafast and memory-efficient alignment of short DNA sequences to the human genome. *Genome Biol.* 10:R25. doi: 10.1186/gb-2009-10-3-r25

Lecourieux, D., Lamotte, O., Bourque, S., Wendehenne, D., Mazars, C., Ranjeva, R., et al. (2005). Proteinaceous and oligosaccharidic elicitors induce different calcium signatures in the nucleus of tobacco cells. *Cell Calcium* 38, 527–538. doi: 10.1016/j.ceca.2005.06.036

Lelliot, R. A., and Stead, D. E. (1987). *Methods for the Diagnosis of Bacterial Diseases of Plants*. Oxford: Blackwell Scientific Publication Ltd.

Li, C. Y., Deng, G. M., Yang, J., Viljoen, A., Jin, Y., Kuang, R. B., et al. (2012). Transcriptome profiling of resistant and susceptible Cavendish banana roots following inoculation with *Fusarium oxysporum* f. sp *cubense* tropical race 4. *BMC Genomics* 13:374. doi: 10.1186/1471-2164-13-374

Major, I. T., Nicole, M. C., Duplessis, S., and Séguin, A. (2010). Photosynthetic and respiratory changes in leaves of poplar elicited by rust infection. *Photosynth. Res.* 104, 41–48. doi: 10.1007/s11120-009-9507-2

Miranda, M., Ralph, S. G., Mellway, R., White, R., Heath, M. C., Bohlmann, J., et al. (2007). The transcriptional response of hybrid poplar (*Populus trichocarpa x P-deltoides*) to infection by *Melampsora medusae* leaf rust involves induction of flavonoid pathway genes leading to the accumulation of proanthocyanidins. *Mol. Plant Microbe Interact.* 20, 816–831. doi: 10.1094/MPMI-20-7-0816

Mortazavi, A., Williams, B. A., McCue, K., Schaeffer, L., and Wold, B. (2008). Mapping and quantifying mammalian transcriptomes by RNA-Seq. *Nat. Methods* 5, 621–628. doi: 10.1038/nmeth.1226

Ng, P., Wei, C. L., Sung, W. K., Chiu, K. P., Lipovich, L., Ang, C. C., et al. (2005). Gene identification signature (GIS) analysis for transcriptome characterization and genome annotation. *Nat. Methods* 2, 105–111. doi: 10.1038/nmeth733

Park, C. Y., Heo, W. D., Yoo, J. H., Lee, J. H., Kim, M. C., Chun, H. J., et al. (2004). Pathogenesis-related gene expression by specific calmodulin Isoforms is dependent on NIM1, a key regulator of systemic acquired resistance. *Mol. Cells* 18, 207–213.

Pasquali, G., Goddijn, O. J., de Waal, A., Verpoorte, R., Schilperoort, R. A., Hoge, J. H., et al. (1992). Coordinated regulation of 2 indole alkaloid biosythetic genes from *Catharanthus-roseus* by auxin and elicitors. *Plant Mol. Biol.* 18, 1121–1131. doi: 10.1007/BF00047715

Pei, C. C., Wang, H., Zhang, J. Y., Wang, Y. Y., Francis, D. M., et al. (2012). Fine mapping and analysis of a candidate gene in tomato accession PI128216 conferring hypersensitive resistance to bacterial spot race T3. *Theor. Appl. Genet.* 124, 533–542. doi: 10.1007/s00122-011-1726-1

Robbins, M. D., Darrigues, A., Sim, S. C., Masud, M. A. T., and Francis, D. M. (2009). Characterization of hypersensitive resistance to bacterial spot race T3 (*Xanthomonas perforans*) from tomato accession PI 128216. *Phytopathology* 99, 1037–1044. doi: 10.1094/PHYTO-99-9-1037

Röemer, P., Hahn, S., Jordan, T., Strauss, T., Bonas, U., and Lahaye, T. (2007). Plant pathogen recognition mediated by promoter activation of the pepper *Bs3* resistance gene. *Science* 318, 645–648. doi: 10.1126/science.1144958

Romeis, T., Piedras, P., and Jones, J. D. G. (2000). Resistance gene-dependent activation of a calcium-dependent protein kinase in the plant defense response. *Plant Cell* 12, 803–815. doi: 10.1105/tpc.12.5.803

Scott, J. W., Hutton, S. F., Shekasteband, R., Sim, S. C., and Francis, D. M. (2015). Identification of tomato bacterial spot race T1, T2, T3, T4, and *Xanthomonas gardneri* resistance QTLs derived from PI 114490 populations selected for race T4. *Acta Hortic.* 1069, 53–58. doi: 10.17660/ActaHortic.2015.1069.6

Scott, J. W., Jones, J. B., and Somodi, G. C. (2001). Inheritance of resistance in tomato to race T3 of the bacterial spot pathogen. *J. Amer. Soc. Hort. Sci.* 126, 436–441.

Sharlach, M., Dahlbeck, D., Liu, L., Chiu, J., Jiménez-Gómez, J. M., Kimura, S., et al. (2013). Fine genetic mapping of *RXopJ4*, a bacterial spot disease resistance locus from *Solanum pennellii* LA716. *Theor. Appl. Genet.* 126, 601–609. doi: 10.1007/s00122-012-2004-6

Stall, R. E., Jones, J. B., and Minsavage, G. V. (2009). Durability of resistance in tomato and pepper to Xanthomonads causing bacterial spot. *Annu. Rev. Phytopathol.* 47, 265–284. doi: 10.1146/annurev-phyto-080508-081752

Strauss, T., van Poecke, R. M. P., Strauss, A., Röemer, P., Minsavage, G. V., Singh, S., et al. (2012). RNA-seq pinpoints a *Xanthomonas* TAL-effector activated resistance gene in a large-crop genome. *Proc. Natl. Acad. Sci. U.S.A.* 109, 19480–19485. doi: 10.1073/pnas.1212415109

Sun, H. J., Liu, X. X., Li, W. H., and Yang, W. C. (2011a). Preliminary mapping of a gene in tomato accession LA1589 conferring resistance to race T3 of bacterial spot. *J. Agric. Univ. Hebei* 34, 65–69.

Sun, H. J., Wei, J. L., Zhang, J. Y., and Yang, W. C. (2014). A comparison of disease severity measurements using image analysis and visual estimates using a category scale for genetic analysis of resistance to bacterial spot in tomato. *Eur. J. Plant Pathol.* 139, 125–136. doi: 10.1007/s10658-013-0371-8

Sun, H. J., Zhang, J. Y., Wang, Y. Y., Scott, J. W., Francis, D. M., and Yang, W. C. (2011b). QTL analysis of resistance to bacterial spot race T3 in tomato. *Acta Hortic. Sin.* 38, 2297–2308.

Tan, G. X., Liu, K., Kang, J. M., Xu, K. D., Zhang, Y., Hu, L. Z., et al. (2015). Transcriptome analysis of the compatible interaction of tomato with *Verticillium dahliae* using RNA-sequencing. *Front. Plant Sci.* 6:428. doi: 10.3389/fpls.2015.00428

Tang, X., Tang, Z., Huang, S., Liu, J., Liu, J., Shi, W., et al. (2013). Whole transcriptome sequencing reveals genes involved in plastid/chloroplast division and development are regulated by the HP1/DDB1 at an early stage of tomato fruit development. *Planta* 238, 923–936. doi: 10.1007/s00425-013-1942-9

van Loon, L. C., Rep, M., and Pieterse, C. M. J. (2006). Significance of inducible defense-related proteins in infected plants. *Annu. Rev. Phytopathol.* 44, 135–162. doi: 10.1146/annurev.phyto.44.070505.143425

Wang, H., Hutton, S. F., Robbins, M. D., Sim, S. C., Scott, J. W., Yang, W. C., et al. (2011). Molecular mapping of hypersensitive resistance from tomato 'Hawaii 7981' to *Xanthomonas perforans* race T3. *Phytopathology* 101, 1217–1223. doi: 10.1094/PHYTO-12-10-0345

Wang, L., Feng, Z., Wang, X., Wang, X., and Zhang, X. (2010). DEGseq: an R package for identifying differentially expressed genes from RNA-seq data. *Bioinformatics* 26, 136–138. doi: 10.1093/bioinformatics/btp612

Yang, W. C. (2013). Recent advances on genetics and mapping of resistance to bacterial spot in tomato. *Acta Hortic. Sin.* 40, 1731–1740.

Yang, W. C., Chen, J., Zhang, X. M., and Francis, D. M. (2007). Recent advances in classification of tomato bacterial spot pathogen, genetics of resistance, and marker-assisted selection. *Sci. Agr. Sin.* 40, 283–290.

Zhao, B. M., Cao, H. P., Duan, J. J., and Yang, W. C. (2015). Allelic tests and sequence analysis of three genes for resistance to *Xanthomonas perforans* race T3 in tomato. *Acta Hortic. Sin.* 42, 462–470.

Conflict of Interest Statement: The authors declare that the research was conducted in the absence of any commercial or financial relationships that could be construed as a potential conflict of interest.

12

Oviposition Preference of Pea Weevil, *Bruchus pisorum* L. Among Host and Non-host Plants and its Implication for Pest Management

Esayas Mendesil[1*†], *Birgitta Rämert*[1], *Salla Marttila*[1], *Ylva Hillbur*[1,2] *and Peter Anderson*[1]

[1] *Department of Plant Protection Biology, Swedish University of Agricultural Sciences, Alnarp, Sweden,* [2] *International Institute of Tropical Agriculture, Ibadan, Nigeria*

Edited by:
Maria Carlota Vaz Patto, Instituto de Tecnologia Quimica e Biologica/Universidade Nova de Lisboa, Portugal

Reviewed by:
Anne-Kristin Løes, Bioforsk Norwegian Institute of Agricultural and Environmental Research, Norway
Isabel Ribeiro Do Valle Teixeira, IFSULDEMINAS – Campus Poços de Caldas, Brazil

***Correspondence:**
Esayas Mendesil
esayas.mendesil@slu.se

†Present address:
Esayas Mendesil, Department of Horticulture and Plant Sciences, Jimma University, P.O. Box 307, Jimma, Ethiopia

The pea weevil, *Bruchus pisorum* L. is a major insect pest of field pea, *Pisum sativum* L. worldwide and current control practices mainly depend on the use of chemical insecticides that can cause adverse effects on environment and human health. Insecticides are also unaffordable by many small-scale farmers in developing countries, which highlights the need for investigating plant resistance traits and to develop alternative pest management strategies. The aim of this study was to determine oviposition preference of pea weevil among *P. sativum* genotypes with different level of resistance (*Adet*, *32410-1* and *235899-1*) and the non-host leguminous plants wild pea (*Pisum fulvum* Sibth. et Sm.) and grass pea (*Lathyrus sativus* L.), in no-choice and dual-choice tests. Pod thickness and micromorphological traits of the pods were also examined. In the no-choice tests significantly more eggs were laid on the susceptible genotype *Adet* than on the other genotypes. Very few eggs were laid on *P. fulvum* and *L. sativus*. In the dual-choice experiments *Adet* was preferred by the females for oviposition. Furthermore, combinations of *Adet* with either *235899-1* or non-host plants significantly reduced the total number of eggs laid by the weevil in the dual-choice tests. Female pea weevils were also found to discriminate between host and non-host plants during oviposition. The neoplasm (*Np*) formation on *235899-1* pods was negatively correlated with oviposition by pea weevil. Pod wall thickness and trichomes might have influenced oviposition preference of the weevils. These results on oviposition behavior of the weevils can be used in developing alternative pest management strategies such as trap cropping using highly attractive genotype and intercropping with the non-host plants.

Keywords: Bruchidae, host selection, insect behavior, legume, neoplasm, *Pisum sativum*, pea weevil

INTRODUCTION

Field pea, *Pisum sativum* L. is a cool season legume crop grown in tropical highlands and in many countries in temperate regions (Messiaen et al., 2006). It is an important crop both for human consumption and for animal feed mainly due to its high protein content, and thus nutritional value. Furthermore, it provides ecosystem services by improving soil fertility through symbiotic nitrogen fixation (French, 2004; Khan and Croser, 2004; Messiaen et al., 2006). Insect pests are one of the major constraints of field pea production (Clement et al., 2000), among which the pea weevil, *Bruchus pisorum* L. is an economically important pest of field pea worldwide. In Ethiopia, seed damage and weight loss up to 85 and 59%, respectively, has been reported after attack by the pea weevil (Teka, 2002; Seyoum et al., 2012). As a consequence, the damaged seeds have low marketable value, are less valuable for human consumption and animal feed and show poor germination rate (Brindley et al., 1956; Clement et al., 2000, 2002; Seyoum et al., 2012).

The pea weevil has one generation per year and it reproduces only on field pea. Upon emergence from hibernation sites adult weevils fly into the pea fields and start to search for mate and oviposition sites. Egg laying starts about 2–2.5 weeks after the arrival of the weevils. The female weevil lays its eggs on pods of peas and upon hatching the first instar larva bore directly to the seed. Larvae develop inside the seed by consuming the content of the seed, which results in damage to the crop (Brindley et al., 1956). This cryptic larval feeding habit within the seeds makes it difficult both to monitor the infestation and to control the pea weevil with chemical insecticides. Thus, the most suitable time to control the pest would be before females commence oviposition (Horne and Bailey, 1991; Baker, 1998; Clement et al., 2000). Due to the long infestation period of adult weevils it has been reported that repeated chemical spraying is required to be effective (Baker, 1998). Furthermore, fumigation of harvested peas in the store can prevent further damage by pea weevil (Baker, 1998; Clement et al., 2000). However, insecticides are often unaffordable for small-scale farmers in developing countries such as in Ethiopia. Furthermore, insecticides can have adverse effects on human health and the environment. For example, recent studies showed improper use of insecticides among field pea growers in Ethiopia can expose the farmers to pesticide risks (Mendesil et al., 2016). Thus, development of alternative pest management strategies is needed.

Understanding of oviposition preference behavior in relation to host and non-host plants may provide useful information for developing alternative pest management strategies such as intercropping and trap cropping strategy for insect pest management (Shelton and Badenes-Perez, 2006; Cook et al., 2007; Finch and Collier, 2012). Intercropping is a traditional agronomic practice in Africa which has been shown to reduce pest damage (Abate et al., 2000; Smith and McSorley, 2000) and increase productivity of farm land (Vandermeer, 1989). A study conducted in Ethiopia showed that intercropping of maize, *Zea mays* L. with Ethiopian mustard, *Brassica carinata* A. Braun and potato, *Solanum tuberosum* L. reduced infestation by the stem borers, *Busseola fusca* Füller and *Chilo partellus* (Swinhoe) (Wale et al., 2007). A plant species or variety which is attractive to insect pest can also be planted as a trap crop to protect the main crop (Shelton and Nault, 2004; Shelton and Badenes-Perez, 2006). Trap cropping has been developed for control of various insect pests (a review of Shelton and Badenes-Perez, 2006) and there is an increasing interest in the use of trap crops for pest management.

In many herbivorous insects, understanding female choice of oviposition site is important for evaluating plant resistance and interaction between plants. Among a variety of plants, insect herbivores often show higher preference for particular host plant species, crop varieties and/or crop stages for feeding and oviposition (Bernays and Chapman, 1994). Thus, there can be large differences in plant attractiveness and resistance between different host plants and varieties of the same crop (Smith, 2005). Furthermore, various studies have shown that non-host plants can influence insect herbivore behavior in different ways such as disturbing host finding, masking of host plants and as an oviposition repellent (Vandermeer, 1989; Finch and Collier, 2012; Ratnadass et al., 2012). Most insect herbivores rely on morphological and chemical cues in location of oviposition sites and both morphological traits and secondary chemical metabolites play a crucial role in plant resistance against insect pest attack (Bernays and Chapman, 1994). Plant traits such as different types of glandular structures, wax layers and tissue thickness have been shown to influence oviposition behavior of insect herbivores (Bernays and Chapman, 1994).

In previous field experiments, we found variation in the susceptibility to pea weevil attack between different field pea genotypes, of which *Adet* genotype is highly susceptible to the weevil, and *235899-1* and *32410-1* are moderately resistant based on mean percent seed damage (Teshome et al., 2015). There are also studies showing that non-host plants also can affect host plant choice behaviors of the pea weevil (Annis and O'Keeffe, 1984b). Furthermore, a specific morphological trait reported in peas is the growth of neoplasm on the pod surface, a 'postular-like outgrowth' that is controlled by a single dominant gene, *Np* (Nuttall and Lyall, 1964). Oviposition of pea weevil on peas with *Np* gene has been found to result in development of neoplasm (Berdnikov et al., 1992; Doss et al., 2000). Interestingly, neoplasms are also formed when peas with *Np* gene are grown in the greenhouse under reduced UV wavelengths (Nuttall and Lyall, 1964; Snoad and Matthews, 1969). However, there is little information about how oviposition behavior in the pea weevil reflects resistance among genotypes and how it is affected by non-host plants. There is also no information if it is possible to take advantage of neoplasm as a resistant trait against pea weevil.

Although intercropping and trap cropping pest management methods have been used for major insect pests in various cropping systems elsewhere, e.g., control of *B. fusca* and *C. partellus* in maize and sorghum, *Sorghum bicolor* (L.) Moench in Africa (Khan et al., 2014), there is no available information on such management methods for pea weevil. Identifying host plants that are preferred to female pea weevil and those that are less or non-preferred, may pave the way to develop intercropping and trap cropping as pest management strategies for the pea weevil. Therefore, the aim of this study was to determine oviposition preference of pea weevil to field pea genotypes with different level of pea weevil resistance and to non-host plants. We also wanted to determine the influence of pod morphological traits and neoplasm formation on oviposition preference by pea weevil.

MATERIALS AND METHODS

Plants

Three field pea, *P. sativum* L., genotypes were selected based on results obtained from field experiments conducted in northern and north-western Ethiopia during 2011–2012: Ebinat (12°10′ N 38° 05′ E), Liben (11° 50′N 37°10′ E), and Sekota (13° 00′ N 38° 50′E) (Teshome et al., 2015). The genotypes were: *Adet*, an improved variety which is highly susceptible to pea weevil, released by Adet Agricultural Research Center in Ethiopia in 1997, and *235899-1* and *32410-1* that both are gene bank accessions that were found moderately resistant (Teshome et al., 2015) obtained from the Ethiopian Institute of Biodiversity in Addis Ababa, Ethiopia. Furthermore, *235899-1* is a *Np* genotype with neoplasm formation in pods. Two non-host plants to pea weevil, wild pea, *Pisum fulvum* Sibth. et Sm. (*NGB 102148*) and grass pea, *Lathyrus sativus* L., were also included in the experiment. *P. fulvum* was obtained from the Nordic Genetic Resource Center (NordGen), Alnarp, Sweden whereas *L. sativus* was collected from Adet area, Ethiopia. Plants were grown to produce flowering branches and pods for the insect studies. All plants were grown in 2-L plastic pots with humus rich gardening soil (Weibulls, Sweden) in a biotron chamber (22°C, 75% RH, 12:12 h light/dark cycle) in 2014 at SLU, Alnarp.

Insects

Field pea seeds infested with *B. pisorum* were obtained from harvest of field experiments conducted in Liben, north-western Ethiopia during 2011–2012 and from farmers' seed stores in this area. Infested seeds were kept in a transparent plastic insect rearing cage (31 cm × 22.5 cm × 12 cm) at room temperature (20–24°C). Newly emerged adult pea weevils were used for the bioassays. The sex of the weevils was determined based on the presence of a small spine on the tibia of the middle leg of only male insects (Bousquet, 1990). Newly emerged weevils can be kept alive up to 1 year.

Oviposition Bioassay

No-Choice Tests

A no-choice oviposition assay was conducted following the methods of Hardie and Clement (2001) and Clement et al. (2002) with some modification. A pair of male and female *B. pisorum* was introduced into a plastic insect rearing cage (31 cm × 22.5 cm × 12 cm) that was placed in a climate chamber (24°C, 60% RH, 12:12 h light/dark cycle). The weevils were provided a branch of a field pea plant with four to six fresh flowers every 2 days for 10 days before each experiment. In addition, weevils were provided distilled water and sugar solution on a cotton swab, which was changed every week. At the start of the experiment two flat pea pods, which is the preferred stages for oviposition by pea weevil (Hardie and Clement, 2001), were provided for oviposition. Each pod was placed hanging from the top of the cage using paper clip without damaging the pod, then a staples magnet (10 mm, Staples, Inc. Amsterdam, The Netherlands) was placed on the outer surface of the cage in order to fix/attach the pod. Only pods of one genotype per cage (*Adet*, *235899-1*, *32410-1*, *P. fulvum* and *L. sativus*) were provided to each weevil. Pods were changed daily and the number of eggs laid on each pod was counted under stereo microscope. For each experimental setup ten replications were made in a completely randomized design. The weevils were allowed to oviposit for 10 days.

Dual-Choice Tests

A similar experimental procedure as described above was followed in the dual-choice experiments, except that one pod of the control (*Adet*) and one pod from four test genotypes (*235899-1*, *32410-1*, *P. fulvum* and *L. sativus*) were placed in each cage.

Degree of Neoplasm Formation on Pods of *235899-1*

In order to determine the association of degree of neoplasm formation and number of eggs laid by the weevil, pods of *235899-1*, which were selected for no-choice oviposition bioassay were first assessed for degree of neoplasm before oviposition assays. The degree of neoplasm formation was determined into four classes (1–4; **Figure 3A**), where 1 = <25% of the pod surface is covered with neoplasm, 2 = 25–50% of the pod surface is covered with neoplasm, 3 = 50–75% of the pod surface is covered with neoplasm and 4 = >75% of the pod surface is covered with neoplasm. Then the level of neoplasm formation was correlated with (related to) the number of eggs laid per pod. Twenty pods were sampled for each class and in total eighty pods were sampled for this study.

Morphological Traits of Test Genotypes

Pod Thickness

Thickness of pod wall was measured using an Absolute Digimatic Caliper (500-182-30, Mitutoyo, Japan). A measurement was done in the middle of both on the upper and bottom part of the pod. In total forty pods were measured per each test genotype (*Adet*, *235899-1*, *32410-1*, *P. fulvum* and *L. sativus*), where four fresh pods of flat stage were sampled from ten different plants.

Scanning Electron Microscopy

Scanning electron microscopy (SEM) was performed at SLU, Alnarp to examine if there were any differences on the pod anatomy of test genotypes. Fresh pods of flat stage were sampled from *Adet*, *32410-1*, *235899-1*, *P. fulvum* and *L. sativus*. Small pieces of the pods were fixed overnight at 4°C in a solution of 2.5% glutaraldehyde and 2% paraformaldehyde in 0,1 M Na-phosphate buffer, pH 7.2, dehydrated in a graded series of ethanol and critical-point dried (CPD 020, Balzers, Lichtenstein). Sample pieces were attached on stubs with double-sided tape external, internal or cros section surface of the pod upwards, and coated with a mixture of gold and palladium 3:2 in a sputter (JFC-1100, JOEL, Tokyo, Japan). Coated samples were examined with SEM (435VP, LEO Electron Microscopy Ltd., Cambridge, UK) with 10 kV.

Statistical Analysis

No-choice oviposition and genotype combination data, and pod wall thickness was analyzed using one-way analysis of variance

(ANOVA) using a generalized linear model. Mean oviposition were logarithmic transformed before analysis. For dual-choice test a Student's *t*-test was used to analyze differences in oviposition on control vs. test genotypes. Spearman's correlation analysis was used to determine the associations of number of eggs laid and level of neoplastic formation. All statistical analysis was done using MINITAB 16 statistical software.

RESULTS

Oviposition Preference Test

No-Choice Tests

There were clear differences in the number of eggs laid on the different genotypes ($F = 121.53$; $df = 4$; $P < 0.001$, **Figure 1**). Female *B. pisorum* laid significantly more eggs on *Adet* genotype (an average of 76 eggs per female) than on the other test plants. On the other field pea genotypes, intermediate amounts of eggs were deposited, with fewer eggs on *235899-1* (*Np* genotype; an average of 20 eggs) than on *32410-1* (an average of 45 eggs). The two non-host leguminous plants *P. fulvum* and *L. sativus* received on average only 2 and 1.5 eggs, respectively.

Dual-Choice Tests

Given dual-choice *B. pisorum* females consistently laid significantly more eggs on *Adet* (control plant) (49–66 eggs per female) over the other plants tested (**Figure 2A**). With the exception of *32410-1* (40 eggs), ($T = 4.66$; $df = 9$; $P < 0.005$), the weevil laid very few number of eggs on the other test plants, i.e., *235899-1* (three eggs; $T = 18.60$; $df = 9$; $P < 0.001$), *P. fulvum* (two eggs; $T = 16.55$; $df = 9$; $P < 0.001$) and no oviposition was recorded on *L. sativus* ($T = 7.68$; $df = 9$; $P < 0.001$).

The total number of eggs laid varied depending on plant combination ($F = 19.61$; $df = 3$; $P < 0.001$, **Figure 2B**). The weevils laid significantly higher number of eggs when *Adet* (control plant) was combined with *32410-1* (106 eggs per female) than when *Adet* was combined with either *235899-1* (*Np* genotype; 55 eggs) or non-host plants (49–57 eggs).

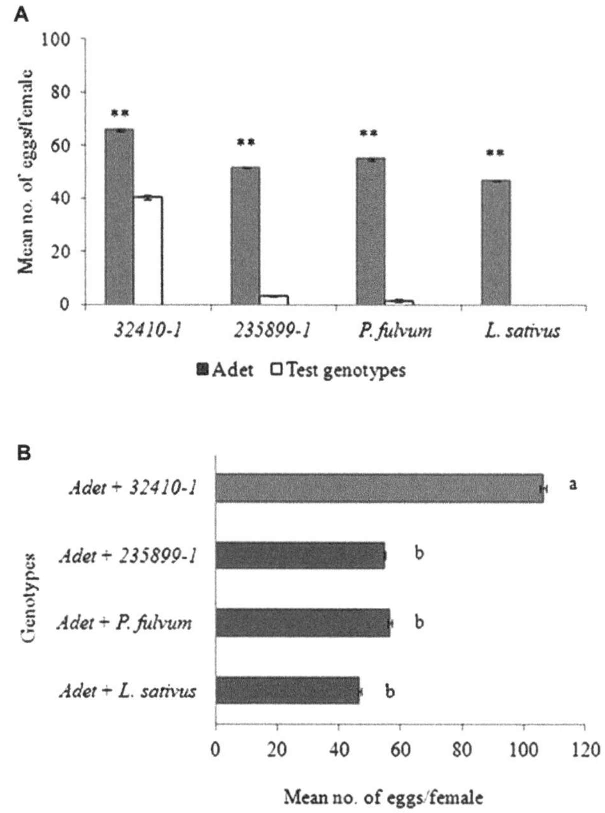

FIGURE 2 | (A) Distribution of eggs of pea weevil in dual-choice oviposition tests. **(B)** Total number of eggs laid by the pea weevil in the experiments where the susceptible genotype *Adet* was combined with other more resistant pea genotypes and non-host plants. Bars marked with different letters are significantly different (Tukey's test: $P < 0.001$; $^{**}P < 0.001$). Bars indicate mean ± standard error (SE).

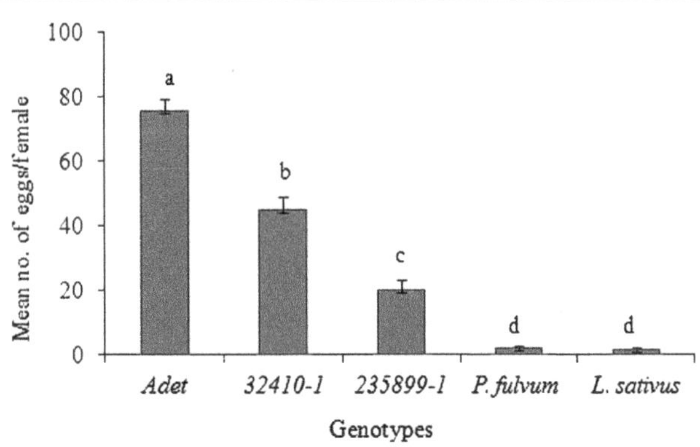

FIGURE 1 | Egg-laying activity (oviposition) of pea weevil females subjected to pods from *Pisum sativum*, *Pisum fulvum* and *Lathyrus sativus* genotypes in a no-choice test. Bars marked with different letters are significantly different (Tukey's test: $P < 0.001$). Bars indicate mean ± standard error (SE).

Degree of Neoplasm Formation on Pods of *235899-1* Compared to Oviposition

Bruchus pisorum females laid more eggs (11 eggs per pod) on the pods with <25% neoplasm formation followed by 25–50% (four eggs), while the weevil oviposited very few eggs (0.7–1) on pods with >51% neoplasm formation (**Figures 3A,B**). We found a negative correlation between the degree of neoplasm formation and number of eggs laid per pod ($r_s = -1.0$; $P < 0.01$).

Morphological Traits of Pod

Pod Thickness

The *Np* genotype *235899-1* and *P. fulvum* had significantly thicker pod wall with an average thickness of 1.36 and 1.31 mm, respectively, than *Adet* (1.04 mm) and *32410-1* (0.93 mm) ($F = 103.86$; $df = 4$; $P < 0.001$). *L. sativus* had an average pod wall thickness of 0.60 mm which was significantly lower than all other plants (**Figure 4**).

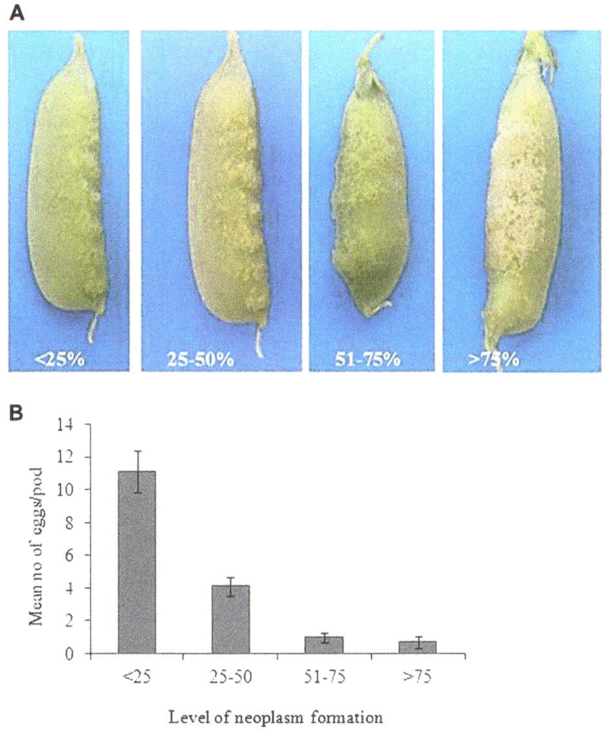

FIGURE 3 | (A) Illustration of the degree of neoplasm formation on pods of *235899-1*. **(B)** Mean number of eggs laid on pods of *Np* genotype with different level of neoplasm formation. Bars indicate mean ± standard error (SE).

Scanning Electron Microscopy

Scanning electron micrographs of pods are shown in **Figure 5**. The external surface of the pods of *Adet* and *32410-1* as well as non-host plants was covered by a thick wax layer (**Figures 5A,D,J,M**). Different formations of epicuticular wax were observed, but as pointed out by Butler (2002), surface wax is unstable when exposed to changes in the environment, e.g., temperature and humidity. Occasional hairy trichomes were seen at least on *32410-1* surface (**Figure 5E**). Large glandular trichomes were only found on the pod surface of *P. fulvum* and *L. sativum* (**Figures 5K,N**). The external surface of the *Np* genotype *235899-1* was largely covered by an intensive neoplastic outgrowth seen as a mass of trichome-like filaments (**Figures 5G–I**) corresponding to the original description by Nuttall and Lyall (1964). Cros section of the pod with neoplastic outgrowth showed no clear epidermal cell layer, giving an impression that the neoplasm proliferation was directly from the parenchymal tissue. The neoplastic layer was almost twice as thick as the pod wall without the outgrowth. The thickness measurements of the pods (**Figure 4**) were in agreement with the size of the cros sections. It was evident that the thin pod of *L. sativum* had less parenchymal cell layers than *P. fulvum*, *Adet* and *32410-1*. There was variation in the number of cell layers of *235899-1* in the pod wall (not shown). Internal surfaces of the pods did not show any particular differences (not shown).

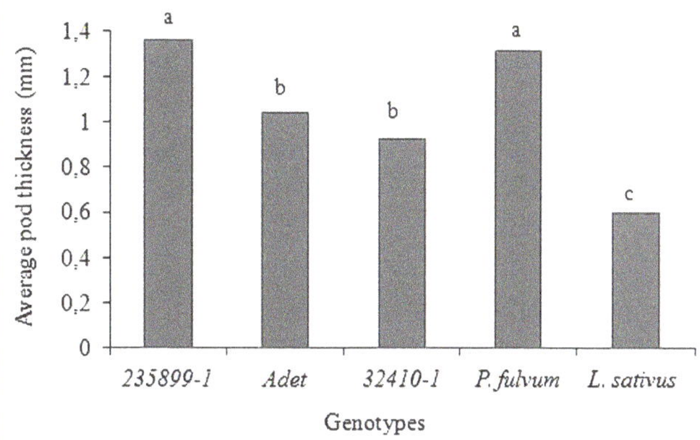

FIGURE 4 | Pod wall thickness of *P. sativum*, *P. fulvum* and *L. sativus* genotypes. Bars marked with different letters are significantly different (Tukey's test: $P < 0.001$).

DISCUSSION

Our results show that female pea weevils discriminate between the tested field pea genotypes during oviposition. Both under no-choice and dual-choice tests, the weevils laid more eggs on *Adet* than on the other plants, indicating that *Adet* is highly preferred for oviposition. These results are supported by a field experiment that showed that *Adet* was highly susceptible to pea weevils and had over 90% seed damage (Teshome et al., 2015). Furthermore, we found that the genotypes *32410-1* and *235899-1* received intermediate amount of eggs, which is also in agreement with field results where those accessions showed moderate level of resistance to pea weevil. This indicates that oviposition behavior in the pea weevil reflects very well the resistance of different genotypes of field pea found in a field situation.

A potential explanation for the higher resistance to egg laying females found for the genotypes *32410-1* and *235899-1* can be that they are gene bank accessions, while *Adet* is an improved variety. It has been shown that domesticated crops and improved varieties which have been developed for specific traits, such as high yield and seed quality, might have lost their inherent resistance ability (Keneni et al., 2011; Tamiru et al., 2011; Chen et al., 2015). Furthermore, studies have shown that genetically uniform varieties are more likely to be prone to insect pest damage as compared to genotypically diverse cultivar mixtures (Tooker and Frank, 2012). Thus, as other improved crop varieties, it is conceivable that *Adet* may have lost its resistant ability as a consequence of breeding for other traits.

The present study also demonstrates that the pea weevil laid very few eggs on the non-host plants (*L. sativus* and *P. fulvum*) compared to the field pea genotypes tested. Furthermore, the combination of *Adet* with non-host plants resulted in lower total number of eggs laid by the weevils suggesting that the presence of non-host plants reduces oviposition by female weevils. This result corroborates earlier findings of reduced oviposition rate on *L. sativus* and *L. tingitanus* pods compared to *P. sativum* (Annis and O'Keeffe, 1984a,b) and of resistance in *P. fulvum* against the pea weevil (Clement et al., 2002). A recent review of Chen et al.

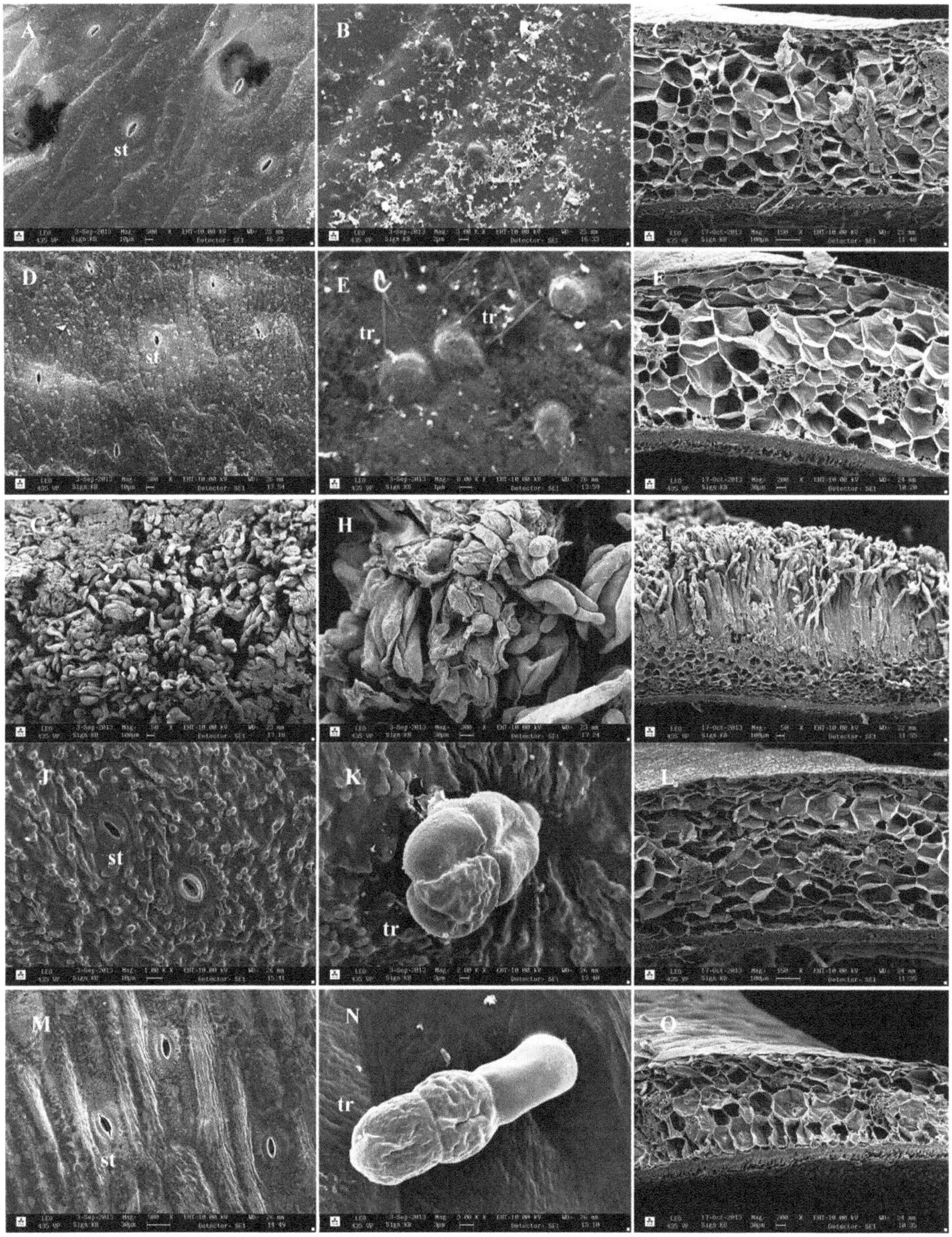

FIGURE 5 | Scanning electron micrographs of pod surface of field pea and non-host genotypes: (A–C; *Adet*), (D–F; *32410-1*), (G–I; *235899-1*), (J–L; *P. fulvum*), (M–O; *L. sativus*). The first and the second columns show external pod surface; the third column is cross section of the pod wall. st, stoma; tr, trichome.

(2015) also showed that gravid female insects in general prefer domesticated crops over their wild progenitors for oviposition. The reduced oviposition on the more resistant genotypes and on the non-host plants may depend on non-volatile chemical cues of the plants. It has been shown earlier that oviposition in pea weevil on *L. sativus* is probably attributed to deterrent compounds rather than absence of oviposition stimulants (Jermy and Szentesi, 1978; Annis and O'Keeffe, 1984b). Non-host plants commonly contain compounds that are deterrent for oviposition and feeding for most of insect herbivores (Bernays and Chapman, 1994). For example, in the Brassicaceae family glucosinolates act as a deterrent for generalist insects such as the green peach aphid,

Myzus persicae (Sulzer), while they are feeding and oviposition stimulants for specialist insect species such as cabbage root fly, *Delia radicum* (L.) (Hopkins et al., 2009).

Plant morphological traits such as surface wax, trichomes, and the toughness of plant tissues can also play a crucial or at least partial role in plant defense against insect herbivores as physical or chemical barriers (Bernays and Chapman, 1994). Scanning electron microscopy (SEM) depicts occasional simple trichomes on the pod surface of *32410-1* and these might have contributed to a reduced oviposition on this genotype as compared to *Adet* that is without trichomes. In addition, the pod surfaces of *P. fulvum* and *L. sativus* showed glandular trichomes, which have also been reported on pods of other legume crops (Butler, 2002) and that have been shown to confer resistance against insect herbivores (Levin, 1973; Bernays and Chapman, 1994). Furthermore, *P. fulvum* had a thicker pod wall than the susceptible host genotype, which may partly have contributed to a lower oviposition rate by female weevils on this genotype. A morphologically based resistance mechanism in peas has been reported for pea weevil and pea leaf weevil, *Sitona lineatus* L., both causing a higher damage on pea plants with a reduced wax-layer (White and Eigenbrode, 2000; Chang et al., 2006). Similarly, in pigeon pea, *Cajanus cajan* (L.) Millsp. pod wall thickness and trichome density served as resistant traits to the pod fly, *Melanagromyza obtusa* (Malloch) (Moudgal et al., 2008) and in sorghum, leaf glossiness (due to wax) and trichome density were reported as important resistant traits to the sorghum shoot fly, *Atherigona soccata* Rondani (Chamarthi et al., 2011).

The genotype *235899-1* pods showed a neoplasm formation on the surface of the pods, but with variations in the degree of neoplasm formation between different pods (**Figure 3**). Pod anatomical features of *235899-1* showed an intensive neoplastic outgrowth on the external pod surface that is likely to form a physical barrier to gravid weevils. Consequently, we found a reduced oviposition on this genotype when expressing the neoplasm and the number of eggs laid was correlated to the degree of neoplasm formation on the pods. We also observed that during oviposition gravid female weevils wander on neoplastic pod surface and spend longer time before commencing oviposition than on pods of preferred genotypes. Interestingly, the pod surface with neoplastic outgrowth did not have a proper epidermal cell layer. Possibly the neoplastic outgrowth represents uncontrolled proliferation and differentiation of cells directly from the parenchymal cell layer. It has been discussed before whether the proliferation comes from hypodermis or epidermis (Nuttall and Lyall, 1964; Dodds and Matthews, 1966; Snoad and Matthews, 1969). The latter has also described the hair-like filaments at the later stages of neoplasm formation as seen in light microscope.

Variation in neoplasm formation might be related with the position of the pod on the plant and the stage of the pods (Nuttall and Lyall, 1964). In addition, the level of neoplasm may vary between different field pea accessions (Berdnikov et al., 1992). Neoplasm growth on field pea pod has mainly been reported in greenhouse grown peas with *Np* gene (Nuttall and Lyall, 1964; Snoad and Matthews, 1969; Burgess and Fleming, 1973), but there are also some reports about its formation on field grown peas under shade (Nuttall and Lyall, 1964; Doss et al., 1995), indicating the possiblity to induce the neoplasm formation in the field through intercropping with crops which can able to provide shade to peas.

Neoplastic growth has earlier been reported as a resistant trait against the pea weevil, which is expressed at the site of egg deposition due to induced plant defense response (Berdnikov et al., 1992; Doss et al., 1995, 2000). It reduces larval entry into the pod and may expose the eggs to biotic and abiotic mortality factors and hence minimizes infestation caused by the weevils (Berdnikov et al., 1992; Doss et al., 1995, 2000).

It may be possible to exploit oviposition preference of the pea weevil for the development of pest management methods such as intercropping and trap cropping. Non-host plants (*L. sativus* and *P. fulvum*), which are the least preferred plants by female weevils could be used as an intercrop to control pea weevil, but field studies are needed to assess the potential of these plants in reducing pea weevil infestation. Various studies have demonstrated the role of intercropping in reducing insect pest damage in different cropping systems. For example, Ali et al. (2007) reported that intercropping of field pea with *B. carinata* reduced pea aphid (*Acyrthosiphon pisum*) infestation and gave higher grain yield as compared to field pea planted as a monoculture. In addition, barley, *Hordeum vulgare* L., cultivars sown in a mixture has reduced bird cherry-oat aphid, *Rhopalosiphum padi* L., infestation compared to barley sown in pure stands (Ninkovic et al., 2002). Furthermore, a highly preferred host genotype can serve as a trap crop. Winter pea, *P. sativum*, for instance, showed a promising result as a potential trap crop to control pea leaf weevil in peas (Cárcamo and Vankosky, 2011). *Adet* is highly attractive for oviposition of pea weevil, which is one of the criteria to be a candidate trap crop (Shelton and Badenes-Perez, 2006). However, it also considerably supports larval development (Teshome et al., 2015) and thus has to be treated with conventional insecticide to make it a 'dead-end-trap crop' (Shelton and Nault, 2004; Shelton and Badenes-Perez, 2006). It would be interesting to test *Adet* in the field as a border trap and/or early planting to determine its potential under field condition.

CONCLUSION

The present study demonstrates that female pea weevil discriminate between host genotypes and non-host plants during oviposition, *Adet* being highly attractive while the non-host plants were the least preferred by the females. The neoplasm formation attributed for a reduced oviposition rate on *235899-1*. Furthermore, thicknesses of pod wall and micromorphological traits appeared to influence oviposition behavior of gravid female weevils. Understanding oviposition behavior of the weevil among host and non-host plants may contribute to develop alternative pest management strategies such as trap cropping using a highly preferred genotype (*Adet*) and intercropping with the two non-host plants (*L. sativus* and *P. fulvum*). However, further studies are required under field conditions to test the preferred genotype for trap crop and the two non-host plants as an intercrops to reduce damage caused by the pea weevil.

AUTHOR CONTRIBUTIONS

EM, BR, YH, and PA conceived and designed the study. EM and SM collected the data and EM performed the analysis. BR and YH obtained the funding. EM wrote the manuscript with the assistance of PA, BR, YH, SM, and PA revised the manuscript.

ACKNOWLEDGMENTS

We would like to thank the Swedish International Development Agency (Sida) for funding this research project. We are also indebted to Ethiopian Institute of Biodiversity and NordGen, Sweden for providing us field pea accessions and wild pea genotype, respectively. We are also grateful to Mulatu Geleta and Abel Teshome for useful discussion during execution of this experiments as well as Kerstin Brismar for assistance with SEM. We would like to thank reviewers for improving the manuscript.

REFERENCES

Abate, T., van Huis, A., and Ampofo, J. K. O. (2000). Pest management strategies in traditional agriculture: an African perspective. *Annu. Rev. Entomol.* 45, 631–659. doi: 10.1146/annurev.ento.45.1.631

Ali, K., Louw, S. V. D., and Swart, W. J. (2007). Effect of mixed cropping of *Pisum sativum* L. on *Acyrthosiphon pisum* (Harris) infestation and ascochyta blight in Ethiopia. *Pest. Manage. J. Eth.* 11, 69–79.

Annis, B. A., and O'Keeffe, L. E. (1984a). Effect of pollen source on oogenesis in the pea weevil, *Bruchus pisorum* L. (Coleoptera: Bruchidae). *Prot. Ecol.* 6, 257–266.

Annis, B. A., and O'Keeffe, L. E. (1984b). Response of two *Lathyrus* species to infestation by the pea weevil, *Bruchus pisorum* L. (Coleoptera: Bruchidae). *Entomol. Exp. Appl.* 35, 83–87. doi: 10.1111/j.1570-7458.1984.tb03364.x

Baker, G. J. (1998). *Pea Weevil. Fact sheet.* Adelaide, SA: Primary Industries and Resources SA and the South Australian Research and Development Institute.

Berdnikov, V. A., Trusov, Y. A., Bogdanova, V. S., Kosterin, O. E., Rozov, S. M., Nedel'Kina, S. V., and Nikulina, Y. N. (1992). The neoplastic pod gene (Np) may be a factor for resistance to the pest *Bruchus pisorum* L. *Pisum Genet.* 24, 37–39.

Bernays, E. A., and Chapman, R. E. (1994). *Host-Plant Selection by Phytophagous Insects.* New York, NY: Chapman & Hall.

Bousquet, Y. (1990). *Beetles Associated with Stored Products in Canada: An Identification Guide, Publication 1837.* Ottawa: Agriculture Canada.

Brindley, T. A., Chamberlin, J. C., and Schopp, R. (1956). *The Pea Weevil and Methods for its Control. Farmers' Bulletin No. 1971.* Washington, DC: U.S. Department of Agriculture.

Burgess, J., and Fleming, E. N. (1973). The structure and development of a genetic tumour of the pea. *Protoplasma* 76, 315–325. doi: 10.1007/BF01279129

Butler, A. (2002). Investigations of pod characters in the Vicieae. *Veget. Hist. Archaeobot.* 11, 127–132. doi: 10.1007/s003340200014

Cárcamo, H., and Vankosky, M. (2011). Managing the pea leaf weevil in field peas. *Prairie Soils Crops* 4, 77–85.

Chamarthi, S. K., Sharma, H. C., Sahrawat, K. L., Narasu, L. M., and Dhillon, M. K. (2011). Physico-chemical mechanisms of resistance to shoot fly, *Atherigona soccata* in sorghum, *Sorghum bicolor*. *J. Appl. Entomol.* 135, 446–455. doi: 10.1111/j.1439-0418.2010.01564.x

Chang, G. C., Neufeld, J., and Eigenbrode, S. D. (2006). Leaf surface wax and plant morphology of peas influence insect density. *Entomol. Exp. Appl.* 119, 197–205. doi: 10.1111/j.1570-7458.2006.00410.x

Chen, Y. H., Gols, R., and Benrey, B. (2015). Crop domestication and its impact on naturally selected trophic interactions. *Annu. Rev. Entomol.* 60, 35–58. doi: 10.1146/annurev-ento-010814-020601

Clement, S. L., Hardie, D. C., and Elberson, L. R. (2002). Variation among accessions of *Pisum fulvum* for resistance to pea weevil. *Crop Sci.* 42, 2167–2173. doi: 10.2135/cropsci2002.2167

Clement, S. L., Wightman, J. A., Hardie, D. C., Bailey, P., Baker, G., and McDonald, G. (2000). "Opportunities for integrated management of insect pests of grain legumes," in *Linking Research and Marketing Opportunities for Pulses in the 21st Century*, ed. R. Knight (Dordrecht: Kluwer Academic), 467–480.

Cook, S. M., Khan, Z. R., and Pickett, J. A. (2007). The use of push-pull strategies in integrated pest management. *Annu. Rev. Entomol.* 52, 375–400. doi: 10.1146/annurev.ento.52.110405.091407

Dodds, K. S., and Matthews, P. (1966). Neoplastic pod in the pea. *J. Hered.* 57, 83–85.

Doss, R. P., Oliver, J. E., Proebsting, W. M., Potter, S. W., Kuy, S., Clement, S. L., et al. (2000). Bruchins: insect derived plant regulators that stimulate neoplasm formation. *Proc. Natl. Acad. Sci. U.S.A.* 97, 6218–6223. doi: 10.1073/pnas.110054697

Doss, R. P., Proebsting, W. M., Potter, S. W., and Clement, S. L. (1995). Response of the Np mutant of pea (*Pisum sativum* L.) to pea weevil (*Bruchus pisorum* L.) oviposition and extracts. *J. Chem. Ecol.* 21, 97–106. doi: 10.1007/BF02033665

Finch, S., and Collier, R. H. (2012). The influence of host and non-host companion plants on the behaviour of pest insects in field crops. *Entomol. Exp. Appl.* 142, 87–96. doi: 10.1111/j.1570-7458.2011.01191.x

French, R. J. (2004). "Pea agronomy," in *Encyclopaedia of Grain Science*, eds C. Wrigley, H. Corke, and C. Walker (Amsterdam: Elsevier ltd), 427–437.

Hardie, D. C., and Clement, S. L. (2001). Development of bioassays to evaluate wild pea germplasm for resistance to pea weevil (Coleoptera: Bruchidae). *Crop Prot.* 20, 517–522. doi: 10.1016/S0261-2194(01)00018-7

Hopkins, R. J., van Dam, N. M., and van Loon, J. J. A. (2009). Role of glucosinolates in insect-plant relationships and multitrophic interactions. *Annu. Rev. Entomol.* 54, 57–83. doi: 10.1146/annurev.ento.54.110807.090623

Horne, J., and Bailey, P. (1991). *Bruchus pisorum* L. (Coleoptera: Bruchidae) control by knockdown pyrethroid in field peas. *Crop Prot.* 10, 53–56. doi: 10.1016/0261-2194(91)90026-N

Jermy, T., and Szentesi, A. (1978). The role of inhibitory stimuli in the choice of oviposition site by phytophagous insects. *Entomol. Exp. Appl.* 24, 258–271. doi: 10.1111/j.1570-7458.1978.tb02806.x

Keneni, G., Bekele, E., Imtiaz, M., Getu, E., Dagne, K., and Assefa, F. (2011). Breeding chickpea (*Cicer arietinum* [Fabaceae]) for better seed quality inadvertently increased susceptibility to adzuki bean beetle (*Callosobruchus chinensis* [Coleoptera: Bruchidae]). *Int. J. Trop. Insect Sci.* 31, 249–261. doi: 10.1017/S1742758411000373

Khan, T. N., and Croser, J. S. (2004). "Pea: overview," in *Encyclopedia of Grain Science*, eds C. Wrigley, H. Corke, and C. Walker (Amsterdam: Elsevier ltd.), 287–295.

Khan, Z. R., Midega, C. A. O., Pittchar, J. O., Murage, A. W., Birkett, M. A., Bruce, T. J. A., et al. (2014). Achieving food security for one million sub-Saharan African poor through push-pull innovation by 2020. *Philos. Trans. R. Soc. B* 369:20120284. doi: 10.1098/rstb.2012.0284

Levin, A. D. (1973). The role of trichomes in plant defence. *Q. Rev. Biol.* 48, 3–15. doi: 10.1086/407484

Mendesil, E., Shumeta, Z., Anderson, P., and Rämert, B. (2016). Smallholder farmers' knowledge, perceptions and management of pea weevil in north and north-western Ethiopia. *Crop Prot.* 81, 30–37. doi: 10.1016/j.cropro.2015.12.001

Messiaen, C. M., Seif, A. A., Jarso, M., and Keneni, G. (2006). "*Pisum sativum* L," in *Record from PROTA4U*, eds M. Brink and G. Belay (Wageningen: PROTA (Plant Resources of Tropical Africa)).

Moudgal, R. K., Lakra, R. K., Dahiya, B., and Dhillon, M. K. (2008). Physico-chemical traits of *Cajanus cajan* (L.) Millsp. pod wall affecting *Melanagromyza obtusa* (Malloch) damage. *Euphytica* 161, 429–436. doi: 10.1007/s10681-007-9601-5

Ninkovic, V., Olsson, U., and Pettersson, J. (2002). Mixing barley cultivars affects aphid host plant acceptance in field experiments. *Entomol. Exp. Appl.* 102, 177–182. doi: 10.1046/j.1570-7458.2002.00937.x

Nuttall, V. W., and Lyall, L. H. (1964). Inheritance of neoplastic pod in the pea. *J. Hered.* 55, 184–186.

Ratnadass, A., Fernandes, P., Avelino, J., and Habib, R. (2012). Plant species diversity for sustainable management of crop pests and diseases in agroecosystems: a review. *Agron. Sustain. Dev.* 32, 273–303. doi: 10.1007/s13593-011-0022-4

Seyoum, E., Damte, T., Bejiga, G., and Tesfaye, A. (2012). "The status of pea weevil, *Bruchus pisorum* (Coleoptera: Chrysomelidae) in Ethiopia," in *Proceedings of the 17th Annual Conference. 26-27 November 2010 Invasive Plant Pests Threatening Ethiopian Agriculture, Ethiopiaed*, ed. B. Mulatu (Addis Ababa: Plant Protection Society of Ethiopia), 52–66.

Shelton, A. M., and Badenes-Perez, F. R. (2006). Concepts and applications of trap cropping in pest management. *Annu. Rev. Entomol.* 51, 285–308. doi: 10.1146/annurev.ento.51.110104.150959

Shelton, A. M., and Nault, B. A. (2004). Dead-end trap cropping: a technique to improve management of the diamondback moth, Plutella xylostella (Lepidoptera: Plutellidae). *Crop Prot.* 23, 497–503. doi: 10.1016/j.cropro.2003.10.005

Smith, C. M. (2005). *Plant Resistance to Arthropods. Molecular and Conventional Approaches.* Dordrecht: Springer.

Smith, H. A., and McSorley, R. (2000). Intercropping and pest management: a review of major concepts. *Am. Entomol.* 46, 154–161. doi: 10.1093/ae/46.3.154

Snoad, B., and Matthews, P. (1969). "Neoplasms of the pea pod," in *Chromosomes Today*, eds C. D. Darlington and K. R. Lewis (Edinburg, TX: Oliver & Boyd Ltd.), 126–131.

Tamiru, A., Bruce, T. J. A., Woodcock, C. M., Caulfield, J. C., Midega, C. A. O., Ogol, C. K. P. O., et al. (2011). Maize landraces recruit egg and larval parasitoids in response to egg deposition by a herbivore. *Ecol. Lett.* 14, 1075–1083. doi: 10.1111/j.1461-0248.2011.01674.x

Teka, W. (2002). "The importance and distribution of pea weevil (*Bruchus pisorum*) in the Amhara region," in *Proceedings of a National Workshop on the Management of Pea weevil, Bruchus pisorum, 25-27 November 2002*, Bahir Dar, 30–36.

Teshome, A., Mendesil, E., Geleta, M., Andargie, D., Anderson, P., Rämert, B., et al. (2015). Screening the primary gene pool of field pea (*Pisum sativum* L. subsp. sativum) in Ethiopia for resistance against pea weevil (*Bruchus pisorum* L.). *Genet. Resour. Crop Evol.* 62, 525–538. doi: 10.1007/s10722-014-0178-2

Tooker, J. F., and Frank, S. D. (2012). Genotypically diverse cultivar mixtures for insect pest management and increased crop yields. *J. Appl. Ecol.* 49, 974–985. doi: 10.1111/j.1365-2664.2012.02173.x

Vandermeer, J. H. (1989). *The Ecology of Intercropping Systems.* Cambridge: Cambridge Press.

Wale, M., Schulthess, F., Kairu, E. W., and Omwega, C. O. (2007). Effect of cropping systems on cereal stemborers in the cool-wet and semi-arid ecozones of the Amhara region of Ethiopia. *Agric. For. Entomol.* 9, 73–84. doi: 10.1111/j.1461-9563.2007.00324.x

White, C., and Eigenbrode, S. D. (2000). Effects of surface wax variation in *Pisum sativum* on herbivorous and entomophagous insects in the field. *Environ. Entomol.* 29, 773–780. doi: 10.1603/0046-225X-29.4.773

Conflict of Interest Statement: The authors declare that the research was conducted in the absence of any commercial or financial relationships that could be construed as a potential conflict of interest.

13

Four Tomato FLOWERING LOCUS T-Like Proteins Act Antagonistically to Regulate Floral Initiation

Kai Cao [1,2], Lirong Cui [1], Xiaoting Zhou [1], Lin Ye [1], Zhirong Zou [1] and Shulin Deng [2]**

[1] State Key Laboratory of Crop Stress Biology for Arid Areas, Horticulture College, Northwest A&F University, Yangling, China,
[2] Laboratory of Plant Molecular Biology, Rockefeller University, New York, NY, USA

The transition from vegetative growth to floral meristems in higher plants is regulated through the integration of internal cues and environmental signals. We were interested to examine the molecular mechanism of flowering in the day-neutral plant tomato (*Solanum lycopersicum* L.) and the effect of environmental conditions on tomato flowering. Analysis of the tomato genome uncovered 13 PEBP (phosphatidylethanolamine-binding protein) genes, and found six of them were *FT*-like genes which named as *SlSP3D*, *SlSP6A*, *SlSP5G*, *SlSP5G1*, *SlSP5G2,* and *SlSP5G3*. Six FT-like genes were analyzed to clarify their functional roles in flowering using transgenic and expression analyses. We found that SlSP5G, SlSP5G2, and SlSP5G3 proteins were floral inhibitors whereas only SlSP3D/SFT (*SINGLE FLOWER TRUSS*) was a floral inducer. *SlSP5G* was expressed at higher levels in long day (LD) conditions compared to short day (SD) conditions while *SlSP5G2* and *SlSP5G3* showed the opposite expression patterns. The silencing of *SlSP5G* by VIGS (Virus induced gene silencing) resulted in tomato plants that flowered early under LD conditions and the silencing of *SlSP5G2* and *SlSP5G3* led to early flowering under SD conditions. The higher expression levels of *SlSP5G* under LD conditions were not seen in *phyB1* mutants, and the expression levels of *SlSP5G2* and *SlSP5G3* were increased in *phyB1* mutants under both SD and LD conditions compared to wild type plants. These data suggest that *SlSP5G*, *SlSP5G2,* and *SlSP5G3* are controlled by photoperiod, and the different expression patterns of *FT*-like genes under different photoperiod may contribute to tomato being a day neutral plant. In addition, PHYB1 mediate the expression of *SlSP5G*, *SlSP5G2,* and *SlSP5G3* to regulate flowering in tomato.

Keywords: tomato, floral repressor, floral activator, PEBP protein, FT-like genes, phytochromes

Edited by:
John Doonan,
Aberystwyth University, UK

Reviewed by:
Richard Macknight,
University of Otago, New Zealand
Hao Peng,
Washington State University, USA

***Correspondence:**
Zhirong Zou
zouzhirong2005@hotmail.com;
Shulin Deng
sdeng@rockefeller.edu

INTRODUCTION

In flowering plant, the timing of the transition from vegetative to reproductive phase is a major event in the plant life cycle. Both physiological and genetic studies have revealed the complexity of mechanisms that tightly control switch from vegetative to reproductive growth in the apical meristem (Bernier et al., 1993; Shalit et al., 2009). The phosphatidylethanolamine-binding proteins (PEBPs), found in both angiosperms and gymnosperms, have evolved to become both activators and repressors of flowering and they can be classified into three clades (Gyllenstrand et al., 2007; Karlgren et al., 2011). An example of this functional diversification is seen in the six PEBP family members of Arabidopsis. FLOWERING LOCUS T (FT) and TWIN SISTER OF FT (TSF), which belong to the FT-like clade, function as flowering activators, TERMINAL FLOWER 1 (TFL1), BROTHER OF FT AND TFL1 (BFT), and ARABIDOPSIS THALIANA CENTRORADIALIS (ATC), which classify to the TFL1-like clade, are usually flowering repressors, and MOTHER OF FT AND TFL1(MFT), which defines the MFT-like clade, is predominantly a floral promoter (Karlgren et al., 2011).

In Arabidopsis, a long-day plant, FT is expressed in leaf phloem companion cells. This protein which triggers floral development in the shoot apical meristem (SAM) under long day (LD) conditions is a major output of the photoperiod pathway and controls floral transition in response to the changes in day length (Kardailsky et al., 1999; Kobayashi et al., 1999). CONSTANS (CO) encodes a zinc finger protein and promotes flowering under LD conditions (Putterill et al., 1995). In LD conditions, *FT* is activated by CO (Samach et al., 2000), and the FT protein then interacts with a novel endoplasmic reticulum membrane protein called FT-INTERACTING PROTEIN 1 (FTIP1; Liu et al., 2012). Following the interaction FT is transported from the companion cells to the sieve elements and entered the SAM by mass flow, where it associates with the basic leucine zipper domain (bZIP) transcription factor FD to activate downstream targets such as SUPPRESSOR OF OVEREXPRESSION OF CONSTANS 1 (SOC1) and the floral meristem identity gene APETALA 1 (AP1; Abe et al., 2005; Wigge et al., 2005; Corbesier et al., 2007). Also a PEBP family protein TSF probably acts in a similar way to FT (Yamaguchi et al., 2005). In SD conditions, Arabidopsis flowering is controlled by a gibberellin pathway, which promotes flowering through the activation of the flower meristem identity gene LEAFY (LFY) with no involvement of any PEBP family proteins (Moon et al., 2003). In SD plant rice, Hd3a, a FT homolog promotes flowering under SD conditions (Komiya et al., 2008, 2009). In the day-neutral plant tomato, the homolog of FT, SlSP3D/SFT (SINGLE-FLOWER TRSS), has been shown to encode the mobile florigen signal and promote tomato flowering (Molinero-Rosales et al., 2004).

Although almost all FT-like proteins act as floral activators an antagonistically functional switch has occurred in *Beta vulgaris* (sugar beet) and *Nicotiana tabacum* (tobacco) because of gene duplication event(s) generating other paralog(s). In sugar beet, BvFT1 protein acts as an inhibitor in floral development whereas another FT-like protein BvFT2 works as a promoter (Pin et al., 2010). Substitutions of specific amino acids can convert BvFT1 to a floral inducer and BvFT2 into a floral repressor (Pin et al., 2010). In tobacco, four *FT*-like proteins, NtFT1, NtFT2, and NtFT3 proteins are floral inhibitors whereas only NtFT4 is a floral inducer (Harig et al., 2012). These data suggest that some FT-like proteins, which are evolutionarily more related to FT than to TFL1/CEN, have evolved into flowering repressors.

Phytochromes are primary photosensory receptors that perceive red and far-red light of higher plants. These photochromic proteins exist in two photo-interconvertible isomeric forms: the red light absorbing form and the far-red light absorbing form (Hughes and Lamparter, 1999). Arabidopsis has five phytochrome genes, *PHYA* to *PHYE*, which encode the apoproteins of PHYA to PHYE, respectively (Quail et al., 1995). *PHYB* plays an inhibitory role in floral initiation in Arabidopsis; the *phyB* mutant flowered earlier than WT in both LD and SD conditions, but the early-flowering phenotype of the *phyB* mutant is more pronounced in SD than in LD conditions (Goto et al., 1991; Mockler et al., 1999). *phyB* mutations of the LD pea plant (Weller and Reid, 1993), SD plant sorghum (Childs et al., 1997), and rice (Izawa et al., 2002) showed early-flowering and decreased photoperiodic sensitivities. *PHYB* delays flowering by suppressing the expression of *FT* in Arabidopsis (Endo et al., 2005) and *Hd3a* in rice (Izawa et al., 2002). Tomato contains five phytochrome genes, named *PHYA, PHYB1, PHYB2, PHYE,* and *PHYF* (Hauser et al., 1997). The tomato *PHYB1* is mainly involved in the de-etiolation response of seedlings, unfolding of the hypocotyl hook, cotyledon expansion, hypocotyl elongation, and anthocyanin accumulation (Kerckhoffs et al., 1997; Weller et al., 2000). However, the function of phytochromes in tomato flowering have not yet been reported.

Tomato is a photoperiod-insensitive, perennial in its native habit. The flowering time of tomato is measured by the number of leaves in the initial segment, which is rather stable under various environmental conditions (Kinet, 1977). Here, we performed expression and transgenic studies to clarify the functional roles of four expressed *FT*-like genes in tomato. One of the *FT*-like genes has already been identified by Molinero-Rosales et al. (2004), whereas the other three genes have not been studied. Here, we demonstrate the functional differentiation between these genes in controlling flowering through overexpression in Arabidopsis and VIGS-mediated knocking down in tomato. Our data suggest that among four expressed *FT*-like proteins, three of them act as floral repressors and only one of them function as a floral promoter. We also showed the expression profiles of tomato *FT*-like genes under LD and SD conditions in tomato wild-type (WT) and *phy* mutants. The evolution of antagonistic *FT*-like paralogs may be a common strategy in *Solanaceous* plants to fine-tune floral development in response to internal and environmental cues.

MATERIALS AND METHODS

Plant Material and Growth Conditions

We used cv. MoneyMaker (*Solanum lycopersicum* L.) wild type (WT) as control in this study, and *phyA, phyB1, phyB2,* and *phyB1B2* mutants in the MoneyMaker background were provided by the Tomato Genetic Resource Center (Department of Vegetable Crops, University of California, Davis) and their TGR accession numbers were LA4356, LA4357, LA4358, and LA4364, respectively. Tomato seeds were soaked in 50% bleach for 30 min. After the treatment, seeds were rinsed thoroughly in running water, then sown directly on a germination paper and incubated at 25°C. After germination, seedings were sowed onto commercial substrate and grown in a growth chamber under LD (16 h of light/8 h of dark) conditions or SD (8 h of light/16 h of dark) at 300 μmol m^{-2} s^{-1} and 25°C (both day and night).

To study the spatial expression patterns of *FT*-like genes, we extracted total RNA from leaf, apex, stem, flower, and root tissues, pooled from three 7-week-old plants. For diurnal changes in the expression of *FT*-like genes, leaves were harvested every 4 h for

24 h (0, 4, 8, 12, 16, 20, and 24 h), pooled from 3 third leaves of 5-week old plants. To study the effect of photoperiod on the expression of these genes, 5-week old uniform plantlets were transferred from LD conditions to SD conditions and reversely. Three different leaves at the same level were harvested 1, 2, and 3 day after the transfer.

Phylogenetic Analysis

Tomato protein sequences of the PEBP family members were downloaded from https://solgenomics.net/, *Arabidopsis thaliana* PEBP family members were downloaded from https://www.arabidopsis.org/, tobacco FT-like proteins reported by Harig et al. (2012) were download from https://solgenomics.net/, and sugar beet FT-like proteins reported by Pin et al. (2010) were download from http://www.ncbi.nlm.nih.gov/. Protein sequences were aligned using the maximum-likelihood method implemented in ClustalW software (Thompson et al., 1994). An N-J tree was produced from the results of 1000 bootstrap replicates using the ClustalW program.

Gene Expression Studies

Total RNA was extracted using an RNeasy Plant Mini Kit (Qiagen) following the manufacturer's instructions. cDNA synthesis was performed by using the SuperscriptIII First strand synthesis system (Invitrogen) following the manufacturer's instructions. Real-time PCR was performed using SYBR Premix Ex Taq (TAKARA) in a Biorad CFX96 realtime PCR system. *ACTIN* was used as an internal control. The primers used were listed in **Supplementary Table S1**. Real-time quantitative PCR was repeated with three biological replicates, and each sample was assayed in triplicate by PCR.

Plasmid Constructs and Plant Transformation

The ORFs of *SlSP3D(Solyc03g063100)*, *SlSP5G(Solyc05g053850)*, *SlSP5G2(Solyc11g008640)*, *SlSP5G3(Solyc11g008650)* were amplified by PCR, cloned in *pENTR/3C* vector (Invitrogen) and then transferred into *pBCO-DC* by recombination (Jang et al., 2007) using LR Clonase enzyme (Invitrogen). The resultant plasmid was used to transform *A. thaliana* (Col-0) plants by the *Agrobacterium tumefaciens* strain GV3101-mediated floral dip method (cite the original ref as well; Zhang et al., 2006). Transformed plants were selected on 0.8% agar media containing Murashige and Skoog salts, 0.5 g/L MES, and 10 g/L Sucrose and containing 10 μg/L basta. Arabidopsis plants were grown in a growth chamber under LD conditions at a light intensity of 100 $\mu mol\ m^{-2}\ s^{-1}$ at 20°C (day and night).

Virus-Induced Gene Silencing (VIGS) in Tomato Plants

pTRV1 (*pYL192*) and *pTRV2* (*pYL156*) vectors had been described in Liu et al. (2002). The *pYL170 TRV2* vector was derived by cloning a *PstI*-blunt-*DraIII* fragment of *pYL156* into *EcoRI*-blunt-*DraIII*-cut *pCAMBIA3301*. This vector was identical to *pYL156*, except for a plant selection marker. To generate *pTRV2-SlPDS*, *pTRV2-SlSP5G*, *pTRV2SlSP5G2*, and *pTRV2SlSP5G3*, a cDNA fragment was PCR amplified using a tomato ecotype MoneyMaker cDNA library and primers were described in **Supplementary Table S1**. The resulting PCR products were cloned into *EcoRI-BamHI*-cut *pTRV2* (*PYL170*).

One-week-old tomato seedlings were used for the VIGS assay, *pTRV1* and *pTRV2* or its derivatives were introduced into *A. tumefaciens* strain GV3101 and the Agrobacterial strains mixed. A 5-mL culture was grown for 16 h at 28°C in 50 mg/L gentamycin and 50 mg/L kanamycin. The next day, the culture was inoculated into 30 mL of Luria-Bertani medium containing antibiotics, 10 mM MES, and 20 mM acetosyringone. The culture was grown 16 h in a 28°C shaker (200 r.p.m). *A. tumefaciens* cells were harvested and resuspended in infiltration media (10 mM $MgCl_2$, 10 mM MES, and 200 mM acetosyringone), adjusted to an OD600 of 1.5, and left at room temperature for 3–4 h. Agroinfiltration was performed with a needleless 1-mL syringe into two tomato cotyledons (Velásquez et al., 2009).

RESULTS

Identification and Phylogenetic Classification of Tomato FT-Like Genes

To identify FT-like proteins encoded by the tomato genome, the amino acid sequence of Arabidopsis FT protein was used to perform a BLAST survey against the tomato whole-genome database (https://solgenomics.net/). A total of 13 predicted PEBP genes were identified and annotated. In a previous study, the plant PEBP family could be classified into three main clades, described as FT-like, TFL1-like, and MFT-like (Chardon and Damerval, 2005). To evaluate the evolutionary relationships among the tomato, tobacco, sugar beet, and Arabidopsis FT-like proteins, specific and combined phylogenetic analysis based on their amino acid sequence were performed. We created a maximum-likelihood tree from an alignment of the 13 tomato PEBP proteins, the Arabidopsis 6 PEBP proteins, the tobacco FT-like (NtFT1-NtFT4) proteins and the sugar beet BvFT1 and BvFT2 proteins. **Figure 1** shows that there were six *FT*-like genes, five *TFL1*-like genes, and two *MFT*-like genes in the tomato PEBP family. PEBP family proteins contained two key motifs which are a putative ligand-binding pocket and an external loop. Protein sequence alignment also revealed a change of an amino acid residue from Tyr in tomato FT-like proteins to His in TFL1-like at the entrance of the binding pocket (**Supplementary Figure S1**), This amino acid residue in part determines the functional difference between FT and TFL1 in Arabidopsis (Hanzawa et al., 2005). Another amino acid residue was changed from Gln in tomato FT-like proteins to Asp in TFL1-like at the external loop encoded by the fourth exon (**Supplementary Figure S1**). This was another critical residue for the functional difference between FT and TFL1 in Arabidopsis (Ahn et al., 2006). These results suggest that *SlSP3D, SlSP6A, SlSP5G, SlSP5G1, SlSP5G2*, and *SlSP5G3* are *FT*-like genes.

Nucleotide sequence comparisons between genomic and predicted CDS allowed the identification of the exon-intron structures of tomato PEBP genes. Tomato PEBP genes showed conserved genomic organization and the exons were placed in

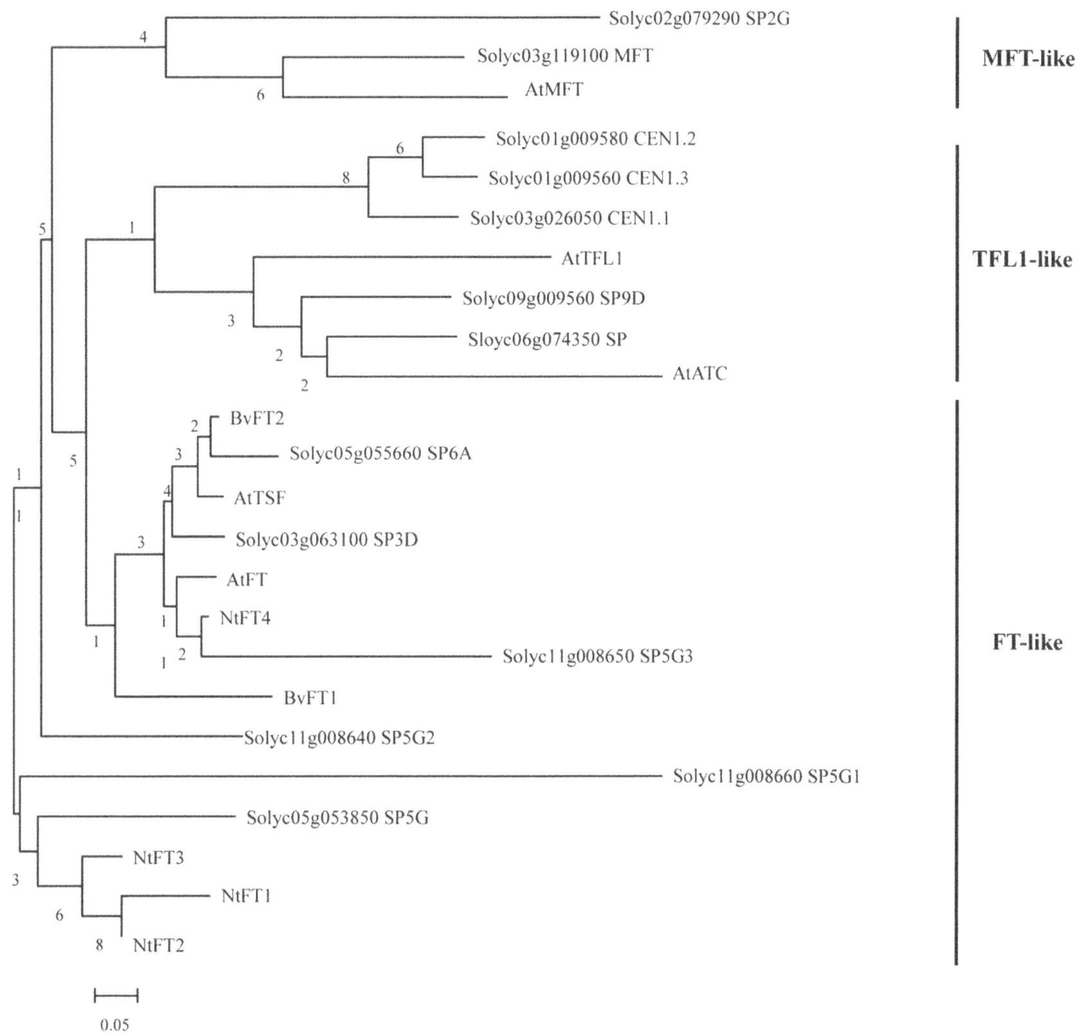

FIGURE 1 | Phylogenetic tree of PEBP proteins from tomato, sugar beet, tobacco, and Arabidopsis. Sequences were aligned with ClustalX and the results are displayed graphically using TreeView. The tomato PEBP genes are grouped into three major clades: the FT-like, TFL-like, and MFT-like clades. AtFT, *Arabidopsis thaliana* FLOWERING LOCUS T; AtTSF, *A. thaliana* TWIN SISTER OF FT; AtTFL1, *A. thaliana* TERMINAL FLOWER 1; AtATC, *A. thaliana* ARABIDOPSIS THALIANA CENTRORADIALIS; AtMFT, *A. thaliana* MOTHER OF FT AND TFL1; BvFT1-2, *Beta vulgaris* FLOWERING LOCUS T 1-2; NtFT1-4, *Nicotiana tabacum* FLOWERING LOCUS T 1-4.

identical positions relative to the amino acid sequence of the Arabidopsis PEBP genes family, except for *SlSP5G1* and *SlSP5G3* (**Supplementary Figure S2**). The length of exons was quite conserved compared among tomato FT-related genes themselves and with *Arabidopsis* FT-related genes, but the introns differed in length. For the FT-like genes exon-intron structures, *SlSP6A* and *SlSP5G1* were truncated by a premature stop codon in their last exon and there was only one 222 exon without intron for *SlSP5G3*. In sugar beet and tobacco, FT-like protein could be further divided into floral promoters and floral repressors.

Expression Pattern of FT-Like Genes in Different Organs Under LD and SD Conditions

To investigate the roles of the six tomato *FT*-like genes in flowering, we first monitored their expression levels in different organs. We isolated total RNA from the leaf, cotyledon, apex, stem, flower, and root tissues of 7-week-old tomato plants growth under LD and SD conditions. We compared the expression levels of *SlSP3D, SlSP6A, SlSP5G, SlSP5G1, SlSP5G2,* and *SlSP5G3* with those of the housekeeping gene *ACTIN* by qRT-PCR. No expression was detected for *SlSP6A* and *SlSP5G1* in all tissues. Considering there are premature stop codons in their last exons (Carmel-Goren et al., 2003; Consortium, 2012), these two genes probably do not encode functional proteins and are in fact pseudogenes. *SlSP3D, SlSP5G, SlSP5G2,* and *SlSP5G3* are mainly expressed in leaf and cotyledon under both LD and SD conditions (**Figures 2A,B**). A much higher expression level of *SlSP5G* was observed under LD conditions compared to SD conditions and *SlSP5G2* and *SlSP5G3* displayed an opposite expression pattern (**Figures 2A,B**). Under SD conditions, the number of leaves on the tomato main stem at flowering was eight on average, while this number increased to nine under LD conditions (**Figure 5A**).

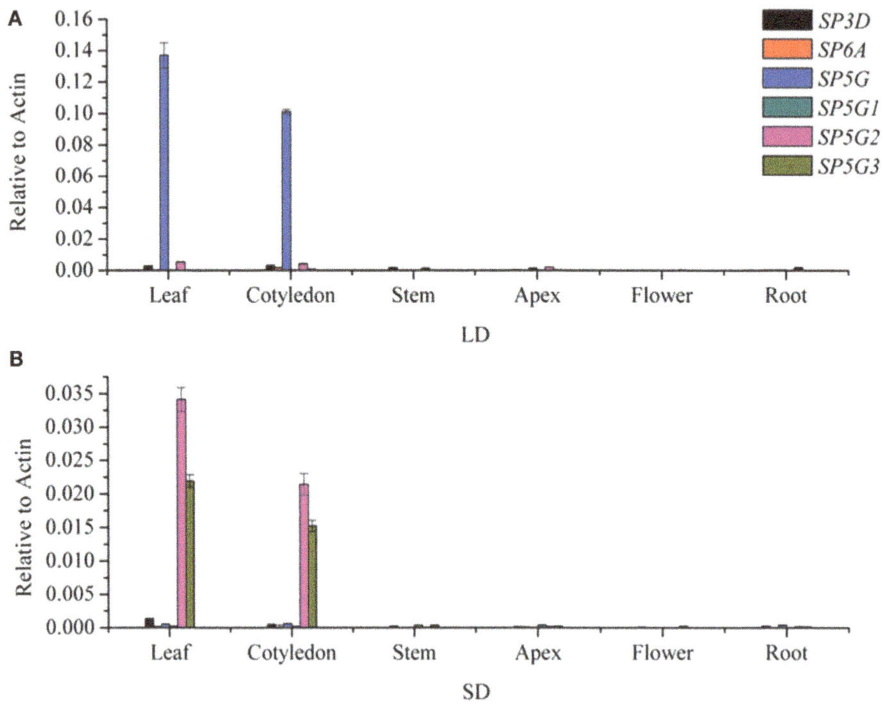

FIGURE 2 | Transcription analyses of tomato *FT*-like genes in different organs. *FT*-like genes expression analyzed by qRT-PCR using RNA extracted from leaf, cotyledon, stem, apex, flower, and root in 7-week old tomato plants grown under LD condition **(A)** and SD condition **(B)**. Expression of the tomato *ACTIN* gene was used as a reference. All data are showed as mean ±SE of three independent pools of extracts. Three technical replicates were performed for each extract.

Diurnal Rhythmic Expression Patterns of FT-Like Genes

To investigate the relationships among the four expressed FT-like genes, *SlSP3D, SlSP5G, SlSP5G2,* and *SlSP5G3*, we examined their diurnal expression patterns using the third leaves of 5-week-old seedlings. We performed qRT-PCR analyses using RNA from tomato plants grown in a LD diurnal cycle or a SD diurnal cycle. The expression of *SlSP3D* peaked at 4 h after dawn under LD conditions (**Figure 3A**) confirming previous results (Shalit et al., 2009). Our results also revealed that *SlSP5G* was transcribed at dawn and its expression peaked at the end of the day under LD conditions (**Figure 3B**). Under SD conditions, *SlSP5G* was constantly expressed at a lower level compared with its expression under LD conditions (**Figure 3B**). The expression pattern of *SlSP5G2* was different from that of *SlSP5G*, which showed a higher expression level under SD conditions, with expression peaking after 4 h of light under SD conditions (**Figure 3C**). On the other hand, *SlSP5G3* showed nearly the same diurnal oscillation pattern as *SlSP5G2*, and it peaked at 4 h after light under SD conditions (**Figure 3D**).

Tomato FT-Like Genes Have Antagonistic Functions in Floral Development in Transgenic Arabidopsis Plants

According to previous studies, *SlSP6A* and *SlSP5G1* were not expressed in tomato plants (Abelenda et al., 2014; **Figures 2A,B**) and consistent with this result we failed to clone *SlSP6A* and *SlSP5G1* from our tomato cDNA library. To investigate the functions of other *FT*-like genes in tomato flowering, *SlSP3D, SlSP5G, SlSP5G2,* and *SlSP5G3* were transferred into Arabidopsis plants under the control of a cauliflower mosaic virus (CaMV) *35S* promoter.

Overexpressing *SlSP3D* led to early flowering in transgenic Arabidopsis (**Figure 4B**) cofirming previous report that *SlSP3D* was a flowering promoter (Molinero-Rosales et al., 2004). Overexpression of *SlSP5G, SlSP5G2,* and *SlSP5G3* delayed flowering in transgenic Arabidopsis plants compared to wild-type controls (**Figures 4A,C–E**). The number of rosette leaves before flowering was seven in Col-0 under LD conditions. However, this number decreased to four in *SlSP3D* overexpressing plants (line 1), increased to 9.5 in *SlSP5G3* overexpressing plants (line 1), 12.5 in *SlSP5G* overexpressing plants (line 2) and 15.5 in *SlSP5G2* (line 2) overexpressing plants under LD conditions (**Figure 4F**). There were four overexpressing *SlSP3D, SlSP5G, SlSP5G2,* and *SlSP5G3* lines, respectively, and the number of rosette leaves before flowering in the other overexpressing *SlSP3D, SlSP5G, SlSP5G2,* and *SlSP5G3* lines were shown in **Supplementary Figure S3**. These results indicate that *SlSP3D* is a floral promoter, and *SlSP5G, SlSP5G2,* and *SlSP5G3* are floral repressors.

The Effect of Photoperiod on the Expression of SlSP3D, SlSP5G, SlSP5G2, and SlSP5G3 Genes

Figure 5B shows that *SlSP5G* expression increased under LD conditions, while *SlSP5G2* and *SlSP5G3* expression increased under SD conditions. These results suggested that *SlSP5G*,

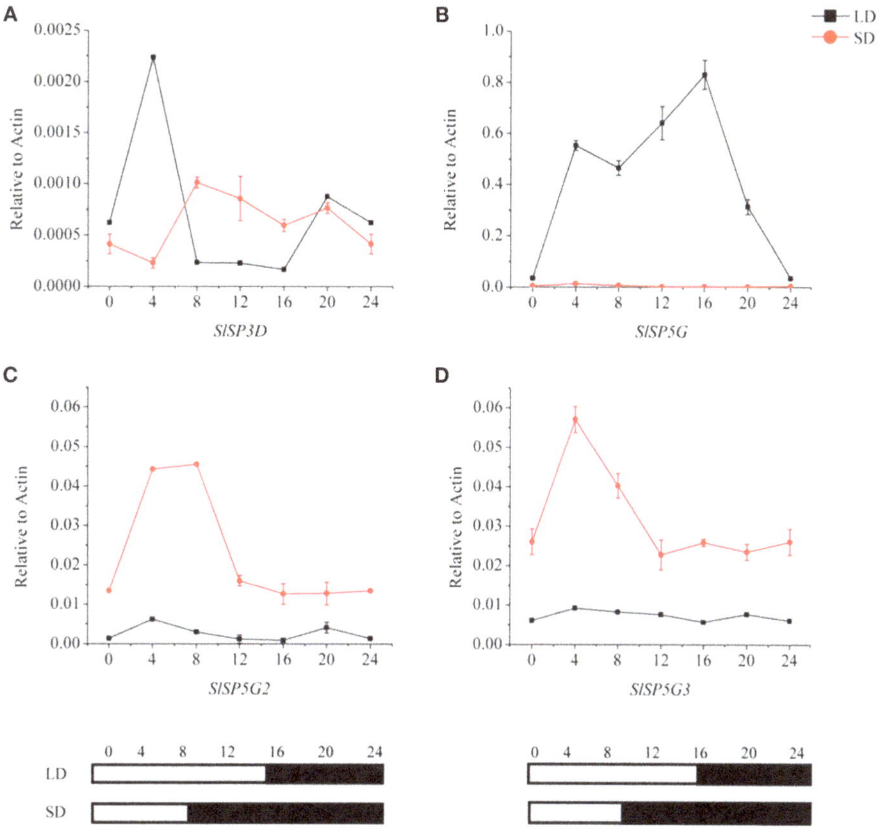

FIGURE 3 | Diurnal expression patterns of tomato *FT*-like genes, *SlSP3D* (A), *SlSP5G* (B), *SlSP5G2* (C), and *SlSP5G3* (D) under LD and SD conditions. The black line represents LD condition and the red line represents SD condition. Leaves were harvested from plants at 4-h intervals throughout a light cycle. The vertical axis shows relative mRNA levels of *FT*-like genes to *ACTIN* expression levels. All data are showed as mean ±SE of three independent pools of extracts. Three technical replicates were performed for each extract. White and black bars at the bottom indicate light and dark periods, respectively.

SlSP5G2, and *SlSP5G3* were targets of photoperiodic regulation. Therefore, we determined the expression levels of *SlSP3D, SlSP5G, SlSP5G2,* and *SlSP5G3* in tomato plants grown under LD conditions for 4 weeks and then transferred to SD conditions for 3 days, and vise-versa. There was no change of *SlSP3D* expression when tomato plants were transferred from LD conditions to SD conditions or from SD conditions to LD conditions (**Figure 5C**). Downregulation of *SlSP5G* and upregulation of *SlSP5G2* and *SlSP5G3* were apparent after tomato plants were transferred from LD conditions to SD conditions (**Figures 5D–F**). With tomato plants transferred from SD conditions to LD conditions, we found a directly increase of *SlSP5G* expression and a decrease of *SlSP5G2* and *SlSP5G3* expression after only one LD photoperiod (**Figures 5D–F**). These results indicated that *SlSP5G, SlSP5G2,* and *SlSP5G3* are directly regulated by day length.

Silencing of the Tomato SlSP5G, SlSP5G2, and SlSP5G3 Genes Using TRV-VIGS Vector

To study the function of *SlSP5G, SlSP5G2,* and *SlSP5G3* in tomato flowering under LD and SD conditions, we constructed a TRV-VIGS vector to suppress the expression of the endogenous *SlSP5G, SlSP5G2,* and *SlSP5G3*. A mixture of Agrobacterium cultures containing *pTRV1* and *pTRV2*, carrying tomato *SlSP5G* (*pTRV2-SlSP5G*), *SlSP5G2* (*pTRV2-SlSP5G2*), or *SlSP5G3* (*pTRV2-SlSP5G3*), were infiltrated into the cotyledon of 1-week-old tomato plants. We also used TRV-VIGS vector to suppress the expression of the endogenous phytoene desaturase gene (*PDS*) as a control. Tomato plants infected with *pTRV-SlPDS* developed a photo-bleached phenotype in the upper leaves 10 days post-agro-infiltration (**Supplementary Figure S4**). Under LD conditions, the number of leaves on tomato main stem upon flowering was nine on average. However, this number was reduced to seven when the tomato plants were infected with *pTRV1/pTRV2-SlSP5G* (**Figure 6A**). Sixteen out of twenty tomato plants showed early flowering after infiltration with *pTRV1/pTRV2-SlSP5G* compared with tomato plants infiltrated with *pTRV1/pTRV2*. We also extracted RNA from leaves of early flowering tomato plants to confirm that *SlSP5G* was indeed silenced by qRT-PCR. The primers that anneal to the *SlSP5G* gene outside the region targeted for silencing were used. In early flowering tomato plants infiltrated with *pTRV2-SlSP5G, SlSP5G* expression was reduced significantly compared with the *TRV* infected controls (**Figure 6B**). The results suggest that SlSP5G is a flowering repressor under LD conditions.

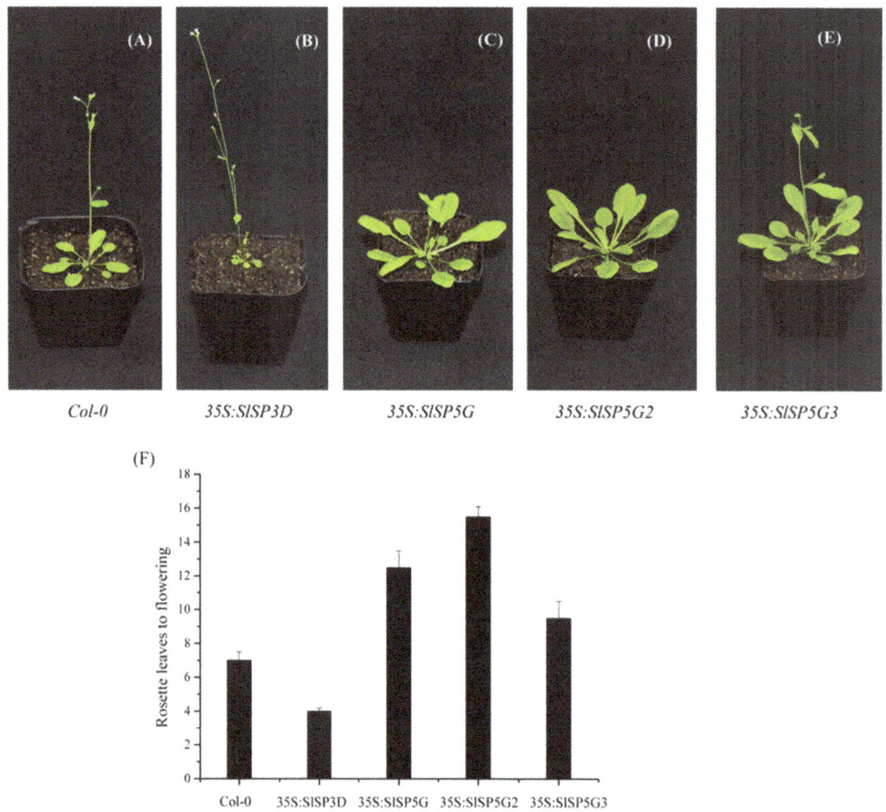

FIGURE 4 | Overexpression of *SlSP3D* in transgenic Arabidopsis promotes flowering, but overexpression of *SlSP5G*, *SlSP5G2,* and *SlSP5G3* in transgenic Arabidopsis causes a delay in flowering. (A–E) Wild type Col-0 **(A)** or representative transgenic Arabidopsis line overexpressing *SlSP3D* **(B)**, *SlSP5G* **(C)**, *SlSP5G2* **(D)**, and *SlSP5G3* **(E)** were grown under LD conditions. **(F)** The rosette leaves produced from SAM prior to flowering under LD conditions. The rosette leaves data are showed as mean ±SE of eight plants in each overexpression line. qRT-PCR data are showed as mean ±SE of three independent pools of extracts. Three technical replicates were performed for each extract.

Under SD conditions, the number of leaves on tomato main stem at flowering was eight on average; however, this number was reduced to 6.5 in tomato plants infected with *pTRV1/pTRV2-SlSP5G2* and *pTRV1/pTRV2-SLSP5G3* (**Figure 6C**). Fourteen out of twenty tomato plants showed slight early flowering after infiltration with *pTRV1/pTRV2-SlSP5G2* and *pTRV1/pTRV2-SLSP5G3*. RT-PCR also confirmed the decreased expression of *SlSP5G2* and *SlSP5G3* in infiltrated tomato plants (**Figure 6D**). These data suggest that *SlSP5G2* and *SlSP5G3* are factors that control tomato flowering under SD conditions.

Effects of Phytochrome B1 on the Expression of SlSP3D, SlSP5G, SlSP5G2, and SlSP5G3 Genes

As phytochromes are very important photoreceptors mediating flowering both in LD plants and SD plants (Izawa et al., 2002; Endo et al., 2005). We examined whether these photoreceptors have an effect on tomato flowering. We determined the number of leaves at flowering and the expression of the four expressed *FT*-like genes, *SlSP3D, SlSP5G, SlSP5G2,* and *SlSP5G3* in seedlings of *phyA*, *phyB1*, *phyB2,* and *phyB1B2* tomato mutants. We found that the number of leaves at flowering in *phyA* and *phyB2* mutants was the same as that in WT under both LD and SD conditions. However, the number of leaves at flowering in *phyB1* and *phyB1B2* mutants was 6 and 6.5, respectively, under LD conditions and this number was 8.5 on average under SD conditions (**Figures 7A–E**). In *phyB1* and *phyB1B2* mutants, there were a constant low expression level of *SlSP5G* under both LD and SD conditions (**Figures 7H,J**). However, there was a stable high expression level of *SlSP5G* in WT, *phyA* and *phyB2* mutants under LD conditions. Under SD conditions, there was a higher expression of *SlSP5G2* and *SlSP5G3* mRNA in *phyB1* and *phyB1B2* mutants compared to WT (**Figures 7F,H,J**). No difference was detected between *phyA, phyB2* mutants and WT on the expression levels of *SlSP5G, SlSP5G2,* and *SlSP5G3* under both LD and SD conditions (**Figures 7F,G,I**). Together, these results clearly demonstrate that PHYB1 has significant influence on the expression of *SlSP5G, SlSP5G2,* and *SlSP5G3*, and on the flowering time of tomato plants under both LD and SD conditions.

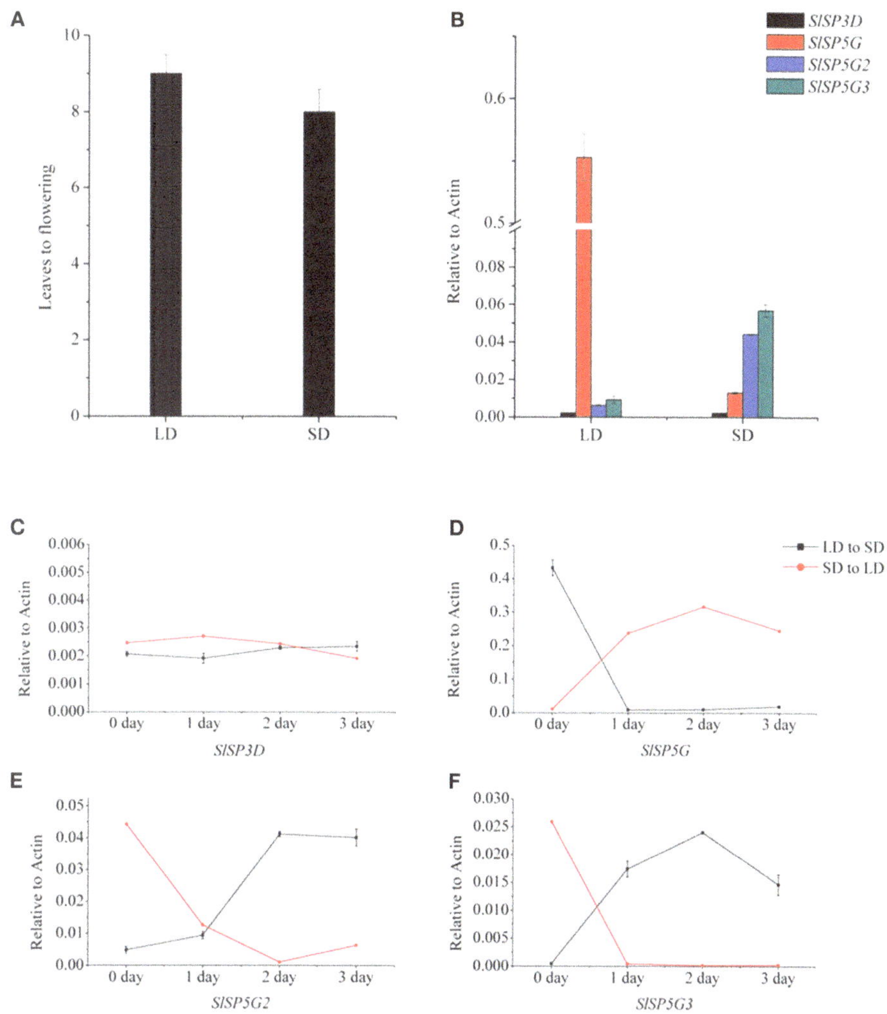

FIGURE 5 | The number of leaves at flowering of tomato plants under different photoperiod and transcription analysis of tomato *FT*-like genes by qRT-PCR in plants exposed to LD or SD conditions. (A) The number of leaves at flowering of tomato plants under LD or SD conditions. Data showed are mean ±SE (*n* = 10 plants). **(B)** The expression of *SlSP3D*, *SlSP5G*, *SlSP5G2*, and *SlSP5G3* under LD and SD conditions. **(C–F)** Expression patterns of *SlSP3D* **(C)**, *SlSP5G* **(D)**, *SlSP5G2* **(E)**, and *SlSP5G3* **(F)** of tomato plants transferred from SD to LD condition, and vise-versa. All data are expressed as means ±SE of three independent pools of extracts. Three technical replicates were performed for each extract.

DISCUSSION

FT-Like Genes Act Antagonistically to Regulate Floral Initiation in Tomato

Plant PEBP family proteins are divided into three major clades, with the FT-like and MFT-like clades primarily acting to promote and the TFL1-like clade primarily acting to repress floral development. In this study, we queried the complete tomato genome sequences and identified 13 PEBP genes, six of which belong to the FT-like clade, five are classified in the TFL-like clade and two are MFT-like clade. In the six FT-like clade, *SlSP3D, SlSP6A, SlSP5G, SlSP5G1, SlSP5G2,* and *SlSP5G3*, two, *SlSP6A* and *SlSP5G1* were not expressed in tomato plants. It has already been demonstrated that *SlSP3D/SFT*, the tomato ortholog of *FT*, induces flowering in day-neutral tomato and *sft* mutants show late flowering phenotype (Molinero-Rosales et al., 2004; Lifschitz et al., 2006). Here, we show that transgenic Arabidopsis plants possessing *SlSP3D* displayed much earlier flowering phenotype compared to control plants. FT-like proteins that promote flowering have been identified in many species such as *Populus* spp. (poplar; Böhlenius et al., 2006), *Malus domestica* (apple; Hättasch et al., 2008), *B. vulgaris* (sugar beet; Pin et al., 2010), *Solanum tuberosum* (potato; Navarro et al., 2011), *N. tabacum* (tobacco; Harig et al., 2012), and *Oryza sativa* (rice; Kojima et al., 2002).

Based on phylogenetic data *SlSP5G, SlSP5G2,* and *SlSP5G3* have been postulated to be orthologous to FT-like genes (Abelenda et al., 2014). However, overexpression of *SlSP5G, SlSP5G-2,* or *SlSP5G-3* in Arabidopsis resulted in late flowering phenotype compared control plants. In sugar beet and tobacco, FT-like genes can act as flowering promoters and repressors. The two sugar beet *FT*-like genes, *BvFT1* and *BvFT2* differ in

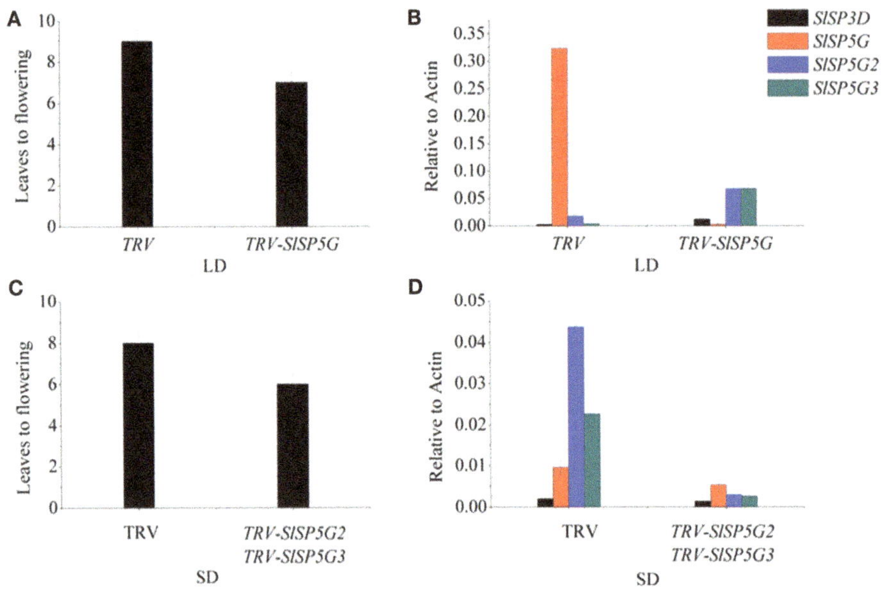

FIGURE 6 | Flowering time and transcript analysis of VIGS of SlSP5G or SlSP5G2 and *SlSP5G3* tomato plants. (A) The number of leaves at flowering of tomato plants when infected by *TRV-SlSP5G* vectors under LD conditions. Data are mean ±SE (*n* = 10 plants). **(B)** The expression of *SlSP3D*, *SlSP5G*, *SlSP5G2,* and *SlSP5G3* of tomato plants infected by *TRV-SP5G* vectors under LD conditions. **(C)** The number of leaves at flowering of tomato plants when infected by *TRV-SlSP5G2* and *SlSP5G3* vectors under SD conditions. **(D)** The expression of *SlSP3D*, *SlSP5G*, *SlSP5G2,* and *SlSP5G3* of tomato plants infected by *TRV-SP5G2* and *SlSP5G3* vectors under SD conditions. All data are expressed as means ±SE of three independent pools of extracts. Three technical replicates were performed for each extract.

three amino acid residues within the critical region encoded by the fourth exon (**Supplementary Figure S1**). Try (134), Gly (137), and Trp (138) are the most important three amino acids of the external loop for BvFT2 protein. Substitution of these three amino acid residues in BvFT2 was sufficient to convert it into a repressor (Pin et al., 2010). The change of BvFT1 Asn (138) into Try, Gln (141) into Gly and Gln (142) into Trp could completely revert its repressing function to promoting function in flowering. Four FT-like proteins have been reported in tobacco. NtFT4 is a flowering activator and the amino acid residues at the three conserved positions matched those of Arabidopsis FT and BvFT2 whereas the corresponding positions in the floral repressors NtFT1-3 were not conserved (Harig et al., 2012). Through protein sequence alignment, we also found that in the tomato SlSP3D the amino acid residues at the three critical positions were Tyr (133), Gly (136), and Trp (137) and these matched to those of the other floral activators, such as FT in Arabidopsis, BvFT2 in sugar beet and NtFT4 in tobacco (**Supplementary Figure S1**). However, the amino acid residues of SlSP5G in these three conserved positions were the same as those found in the floral repressors like NtFT1-3 in tobacco. The corresponding positions of SlSP5G2 and SlSP5G3 in these positions were not conserved compared with other floral activators and repressors (**Supplementary Figure S1**). These results suggest that SlSP5G, SlSP5G2, and SlSP5G3 were initially promoters of flowering but these mutations within the external loop converted its function to flowering repression. The three amino acids are critical for the activator vs. repressor function.

The Expression Profiles of *FT*-Like Genes is Influenced by Photoperiod

The expression of *FT*-like genes in many species is regulated in a photoperiod-dependent manner (Samach et al., 2000; Kojima et al., 2002). Termination and flowering in cultivated tomato are not sensitive to day length, but flower initiation occurs earlier and inflorescence development far better in SD conditions than in LD conditions (Kinet, 1977). All four tomato *FT*-like genes were expressed exclusively in leaf tissue (**Figure 2**), which was the same as tobacco *FT*-like genes (Harig et al., 2012). In tobacco, *NtFT1, NtFT2,* and *NtFT4* showed higher expression levels under SD conditions than under LD conditions (Harig et al., 2012). In sugar beet, the floral repressor *BvFT1* was expressed at high levels when plants were grown in SD or in non-vernalized biennials plants that were not competent to flower (Pin et al., 2010). We also found that *SlSP5G* mRNA expression was up-regulated under LD conditions, while *SlSP5G2* and *SlSP5G3* mRNA increased under SD conditions (**Figure 5B**). The expression of *SlSP3D* was similar under both LD and SD conditions (**Figure 5B**). Although tomato is day-neutral with respect to flowering, the expression of the *SlSP5G, SlSP5G2,* and *SlSP5G3* identified here seem to be photoperiod dependent. *SlSP5G* most likely controls tomato flowering under LD conditions while *SlSP5G2* and *SlSP5G3* seem to regulate flowering under SD conditions. Tomato plants have an adaptive mechanism to adjust flowering according to photoperiod using a combination of different *FT*-like genes.

In this study, the silencing of *SlSP5G* by *TRV-VIGS* vector under LD conditions resulted in early flowering of tomato

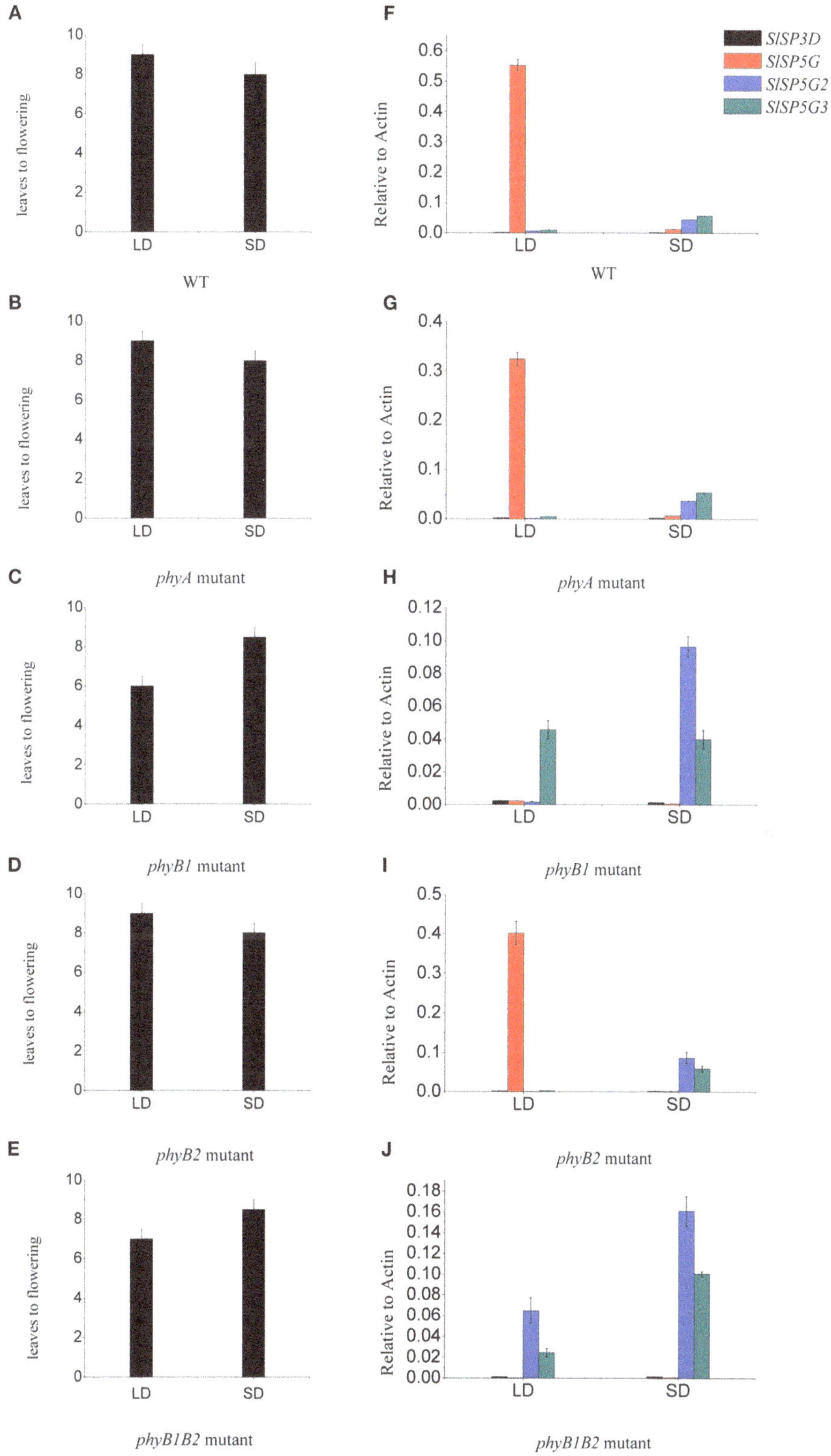

FIGURE 7 | Phytochrome B1 is responsible for the expression of *SlSP5G* and influences the flowering of tomato plants. (A–E) The number of leaves at flowering of tomato WT **(A)**, *phyA* **(B)**, *phyB1* **(C)**, *phyB2* **(D)**, and *phyB1B2* **(E)** mutants grown under LD and SD conditions. Data are mean ±SE ($n = 10$ plants). **(F–J)** Expression levels of *SlSP3D*, *SlSP5G*, *SlSP5G2*, and *SlSP5G3* of tomato WT **(F)**, *phyA* **(G)**, *phyB1* **(H)**, *phyB2* **(I)**, and *phyB1B2* **(J)** mutants that were grown in LD or SD conditions. Data are expressed as means ±SE of three independent pools of extracts. Three technical replicates were performed for each extract.

plants, and the silencing of *SlSP5G2* and *SlSP5G3* by *TRV-VIGS* vector under SD conditions led to early flowering of tomato plants. These results also showed that SlSP5G, SlSP5G2, and SlSP5G3 were floral repressors. SELF PRUNING (SP) is a homolog of TFL1-like gene and SP protein functions as an anti-terminator, maintaining vegetative growth (Pnueli et al., 1998). Mutant *sp* plants form progressively shorter sympodial units, until the shoots terminate in two successive inflorescences. In many species the ratio of floral activators and repressors, e.g., local ratios of SFT/SP3D (FT-like) and SP (TFL1-like), has been proposed to regulate local growth termination equilibria in all meristems of the tomato shoot system (Shalit et al., 2009; McGarry and Ayre, 2012). The three tomato FT-like floral repressors appear to have taken on the role usually played by TFL1 homologs in most other plants. Additional research is required to classify how FT-like floral activators and repressors and SP set the timing of the developmental switch from vegetative to reproductive growth. Both SFT/SP3D and SP of tomato bind to 14-3-3 and bZIP (SPGB, a homolog of FD) proteins in yeast, but each protein also has its own specific binding proteins (Pnueli et al., 2001). In Arabidopsis, FT protein is first transferred into the sieve elements and then subsequently transported by mass flow to the apex, where it interacts with FD to promote flowering (Abe et al., 2005; Wigge et al., 2005; Corbesier et al., 2007). In tomato, SlSP5G, SlSP5G2, and SlSP5G3 maybe like SlSP3D/SFT and they may interact with SPGB to control tomato flowering.

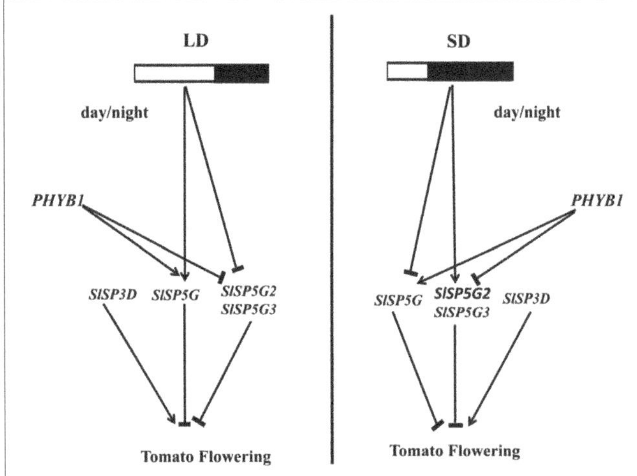

FIGURE 8 | Model of the photoperiod effect on flowering in tomato. The expression of FT-like genes was regulated by photoperiod and mediated by phytochrome B1. In LD conditions, the expression of *SlSP5G* was induced, and the expression of *SlSP5G2* and *SlSP5G3* were inhibited. In SD conditions, the expression of *SlSP5G* was inhibited, and the expression of *SlSP5G2* and *SlSP5G3* were induced. The different expression pattern of tomato FT-like genes under different photoperiod may contribute tomato being a day neutral plant. Phytochrome B1 could promote the expression of *SlSP5G*, and inhibit the expression of *SlSP5G2* and *SlSP5G3*.

Phytochrome B1 Regulates FT-Like Genes

Phytochromes are photochromic proteins that regulate light responses under different light conditions (quantity, quality, and timing). Our data showed that, in the tomato *phyB1* mutant, the expression of *SlSP5G* under both LD and SD conditions was very low. The expression of *SlSP5G2* and *SlSP5G3* was always in a fairly high level, compared with WT under both LD and SD conditions. Based on the results we obtained in tomato plants, we found that the PHYB1 could promote the expression of *SlSP5G* under LD conditions but suppress the expression of *SlSP5G2* and *SlSP5G3* under both LD and SD conditions. It has been shown that PHYB has a general inhibitory effect on flowering in both LD plants and SD Plants (Lin, 2000; Yanovsky and Kay, 2003). An inhibitory effect of PHYB on FT expression has been shown in Arabidopsis (Valverde et al., 2004; Endo et al., 2005). In rice, the *phyB* mutation abolishes the night break effect on flowering and *Hd3a* mRNA, and PHYB suppresses the expression of *Hd3a* (Izawa et al., 2002; Ishikawa et al., 2005).

Phytochromes need to interact with the circadian clock to regulate flowering time in different day-lengths, but the molecular details of such interactions remains unclear (Valverde et al., 2004; Song et al., 2012). *phyB* mutations of the SD plant sorghum and the LD plant Arabidopsis both caused an early flowering phenotype; tomato *phyB1* mutant also has an early flowering phenotype under LD conditions. One interpretation of this observation is that PHYB action may suppress floral initiation regardless of photoperiods, but the signal transduction or plant's responsiveness to PHYB signaling is gated by the action of the circadian clock, resulting in different day-length responses in the flowering time of different plants.

Based on the results we obtained in this study, we propose a model to explain the photoperiod effect on tomato flowering (**Figure 8**). This model is consistent with all of the results we obtained in our studies and suggests that the expression pattern of these *FT*-like genes is regulated by photoperiod and mediated by *PHYB1*. In addition, four tomato *FT*-like genes reveal they act antagonistically to regulate floral initiation. Understanding the molecular mechanism of flowering in a day-neutral plant has important implications for agriculture. Further studies are required to integrate the knowledge obtained from model species like LD plant Arabidopsis and SD plant rice to provide further insight on the mechanisms regulating flowering in day-neutral plant.

AUTHOR CONTRIBUTIONS

Conceived and designed the experiments: SD and ZZ. Performed the experiments: KC. Analyzed the data: LC and KC. Contributed reagents/materials/analysis tools: XZ. Amplify the seed: LY. Wrote the paper: KC.

ACKNOWLEDGMENTS

We thank Tomato Genetic Resource Center (Department of Vegetable Crops, University of California, Davis) for providing all the tomato mutants, Dr. Nam-Hai Chua for guiding this project and editing the manuscript, and Dr. Haixi Sun for his help in phylogenetic analysis.

REFERENCES

Abe, M., Kobayashi, Y., Yamamoto, S., Daimon, Y., Yamaguchi, A., Ikeda, Y., et al. (2005). FD, a bZIP protein mediating signals from the floral pathway integrator FT at the shoot apex. *Science* 309, 1052–1056. doi: 10.1126/science.1115983

Abelenda, J. A., Navarro, C., and Prat, S. (2014). Flowering and tuberization: a tale of two nightshades. *Trends Plant Sci.* 19, 115–122. doi: 10.1016/j.tplants.2013.09.010

Ahn, J. H., Miller, D., Winter, V. J., Banfield, M. J., Lee, J. H., Yoo, S. Y., et al. (2006). A divergent external loop confers antagonistic activity on floral regulators FT and TFL1. *EMBO J.* 25, 605–614. doi: 10.1038/sj.emboj.7600950

Böhlenius, H., Huang, T., Charbonnel-Campaa, L., Brunner, A. M., Jansson, S., Strauss, S. H., et al. (2006). CO/FT regulatory module controls timing of flowering and seasonal growth cessation in trees. *Science* 312, 1040–1043. doi: 10.1126/science.1126038

Bernier, G., Havelange, A. E., Houssa, C., Petitjean, A., and Lejeune, P. (1993). Physiological signals that induce flowering. *Plant Cell* 5, 1147.

Carmel-Goren, L., Liu, Y. S., Lifschitz, E., and Zamir, D. (2003). The SELF-PRUNING gene family in tomato. *Plant Mol. Biol.* 52, 1215–1222. doi: 10.1023/B:PLAN.0000004333.96451.11

Chardon, F., and Damerval, C. (2005). Phylogenomic analysis of the PEBP gene family in cereals. *J. Mol. Evol.* 61, 579–590. doi: 10.1007/s00239-004-0179-4

Childs, K. L., Miller, F. R., Cordonnier-Pratt, M.-M., Pratt, L. H., Morgan, P. W., and Mullet, J. E. (1997). The sorghum photoperiod sensitivity gene, Ma3, encodes a phytochrome B. *Plant Physiol.* 113, 611–619. doi: 10.1104/pp.113.2.611

Consortium, T. G. (2012). The tomato genome sequence provides insights into fleshy fruit evolution. *Nature* 485, 635–641. doi: 10.1038/nature11119

Corbesier, L., Vincent, C., Jang, S., Fornara, F., Fan, Q., Searle, I., et al. (2007). FT protein movement contributes to long-distance signaling in floral induction of Arabidopsis. *Science* 316, 1030–1033. doi: 10.1126/science.1141752

Endo, M., Nakamura, S., Araki, T., Mochizuki, N., and Nagatani, A. (2005). Phytochrome B in the mesophyll delays flowering by suppressing FLOWERING LOCUS T expression in Arabidopsis vascular bundles. *Plant Cell* 17, 1941–1952. doi: 10.1105/tpc.105.032342

Goto, N., Kumagai, T., and Koornneef, M. (1991). Flowering responses to light–breaks in photomorphogenic mutants of *Arabidopsis thaliana*, a long–day plant. *Physiol. Plant.* 83, 209–215. doi: 10.1111/j.1399-3054.1991.tb02144.x

Gyllenstrand, N., Clapham, D., Källman, T., and Lagercrantz, U. (2007). A Norway spruce FLOWERING LOCUS T homolog is implicated in control of growth rhythm in conifers. *Plant Physiol.* 144, 248–257. doi: 10.1104/pp.107.095802

Hanzawa, Y., Money, T., and Bradley, D. (2005). A single amino acid converts a repressor to an activator of flowering. *Proc. Natl. Acad. Sci. U.S.A.* 102, 7748–7753. doi: 10.1073/pnas.0500932102

Harig, L., Beinecke, F. A., Oltmanns, J., Muth, J., Müller, O., Rüping, B., et al. (2012). Proteins from the FLOWERING LOCUS T–like subclade of the PEBP family act antagonistically to regulate floral initiation in tobacco. *Plant J.* 72, 908–921. doi: 10.1111/j.1365-313X.2012.05125.x

Hättasch, C., Flachowsky, H., Kapturska, D., and Hanke, M.-V. (2008). Isolation of flowering genes and seasonal changes in their transcript levels related to flower induction and initiation in apple (*Malus domestica*). *Tree Physiol.* 28, 1459–1466. doi: 10.1093/treephys/28.10.1459

Hauser, B. A., Pratt, L. H., and Cordonnier-Pratt, M.-M. (1997). Absolute quantification of five phytochrome transcripts in seedlings and mature plants of tomato (*Solanum lycopersicum* L.). *Planta* 201, 379–387. doi: 10.1007/s004250050080

Hughes, J., and Lamparter, T. (1999). Prokaryotes and phytochrome. The connection to chromophores and signaling. *Plant Physiol.* 121, 1059–1068. doi: 10.1104/pp.121.4.1059

Ishikawa, R., Tamaki, S., Yokoi, S., Inagaki, N., Shinomura, T., Takano, M., et al. (2005). Suppression of the floral activator Hd3a is the principal cause of the night break effect in rice. *Plant Cell* 17, 3326–3336. doi: 10.1105/tpc.105.037028

Izawa, T., Oikawa, T., Sugiyama, N., Tanisaka, T., Yano, M., and Shimamoto, K. (2002). Phytochrome mediates the external light signal to repress FT orthologs in photoperiodic flowering of rice. *Genes Dev.* 16, 2006–2020. doi: 10.1101/gad.999202

Jang, I. C., Yang, S. W., Yang, J. Y., and Chua, N. H. (2007). Independent and interdependent functions of LAF1 and HFR1 in phytochrome A signaling. *Genes Dev.* 21, 2100–2111. doi: 10.1101/gad.1568207

Kardailsky, I., Shukla, V. K., Ahn, J. H., Dagenais, N., Christensen, S. K., Nguyen, J. T., et al. (1999). Activation tagging of the floral inducer FT. *Science* 286, 1962–1965. doi: 10.1126/science.286.5446.1962

Karlgren, A., Gyllenstrand, N., Källman, T., Sundström, J. F., Moore, D., Lascoux, M., et al. (2011). Evolution of the PEBP gene family in plants: functional diversification in seed plant evolution. *Plant Physiol.* 156, 1967–1977. doi: 10.1104/pp.111.176206

Kerckhoffs, L., Schreuder, M., Tuinen, A. V., Koornneef, M., and Kendrick, R. (1997). Phytochrome control of anthocyanin biosynthesis in tomato seedlings: analysis using photomorphogenic mutants. *Photochem. Photobiol.* 65, 374–381. doi: 10.1111/j.1751-1097.1997.tb08573.x

Kinet, J. (1977). Effect of light conditions on the development of the inflorescence in tomato. *Sci. Hortic.* 6, 15–26. doi: 10.1016/0304-4238(77)90074-7

Kobayashi, Y., Kaya, H., Goto, K., Iwabuchi, M., and Araki, T. (1999). A pair of related genes with antagonistic roles in mediating flowering signals. *Science* 286, 1960–1962. doi: 10.1126/science.286.5446.1960

Kojima, S., Takahashi, Y., Kobayashi, Y., Monna, L., Sasaki, T., Araki, T., et al. (2002). Hd3a, a rice ortholog of the Arabidopsis FT gene, promotes transition to flowering downstream of Hd1 under short-day conditions. *Plant Cell Physiol.* 43, 1096–1105. doi: 10.1093/pcp/pcf156

Komiya, R., Ikegami, A., Tamaki, S., Yokoi, S., and Shimamoto, K. (2008). Hd3a and RFT1 are essential for flowering in rice. *Development* 135, 767–774. doi: 10.1242/dev.008631

Komiya, R., Yokoi, S., and Shimamoto, K. (2009). A gene network for long-day flowering activates RFT1 encoding a mobile flowering signal in rice. *Development* 136, 3443–3450. doi: 10.1242/dev.040170

Lifschitz, E., Eviatar, T., Rozman, A., Shalit, A., Goldshmidt, A., Amsellem, Z., et al. (2006). The tomato FT ortholog triggers systemic signals that regulate growth and flowering and substitute for diverse environmental stimuli. *Proc. Natl. Acad. Sci. U.S.A.* 103, 6398–6403. doi: 10.1073/pnas.0601620103

Lin, C. (2000). Photoreceptors and regulation of flowering time. *Plant Physiol.* 163, 39–50. doi: 10.1104/pp.123.1.39

Liu, L., Liu, C., Hou, X., Xi, W., Shen, L., Tao, Z., et al. (2012). FTIP1 is an essential regulator required for florigen transport. *PLoS Biol.* 10:e1001313. doi: 10.1371/journal.pbio.1001313

Liu, Y., Schiff, M., Marathe, R., and Dinesh-Kumar, S. (2002). Tobacco Rar1, EDS1 and NPR1/NIM1 like genes are required for N-mediated resistance to tobacco mosaic virus. *Plant J.* 30, 415–429. doi: 10.1046/j.1365-313X.2002.01297.x

McGarry, R. C., and Ayre, B. G. (2012). Manipulating plant architecture with members of the CETS gene family. *Plant Sci.* 188–189, 71–81. doi: 10.1016/j.plantsci.2012.03.002

Mockler, T. C., Guo, H., Yang, H., Duong, H., and Lin, C. (1999). Antagonistic actions of Arabidopsis cryptochromes and phytochrome B in the regulation of floral induction. *Development* 126, 2073–2082.

Molinero-Rosales, N., Latorre, A., Jamilena, M., and Lozano, R. (2004). SINGLE FLOWER TRUSS regulates the transition and maintenance of flowering in tomato. *Planta* 218, 427–434. doi: 10.1007/s00425-003-1109-1

Moon, J., Suh, S. S., Lee, H., Choi, K. R., Hong, C. B., Paek, N. C., et al. (2003). The SOC1 MADS–box gene integrates vernalization and gibberellin signals for flowering in Arabidopsis. *Plant J.* 35, 613–623. doi: 10.1046/j.1365-313X.2003.01833.x

Navarro, C., Abelenda, J. A., Cruz-Oró, E., Cuéllar, C. A., Tamaki, S., Silva, J., et al. (2011). Control of flowering and storage organ formation in potato by FLOWERING LOCUS T. *Nature* 478, 119–122. doi: 10.1038/nature10431

Pin, P. A., Benlloch, R., Bonnet, D., Wremerth-Weich, E., Kraft, T., Gielen, J. J., et al. (2010). An antagonistic pair of FT homologs mediates the control of flowering time in sugar beet. *Science* 330, 1397–1400. doi: 10.1126/science.1197004

Pnueli, L., Carmel-Goren, L., Hareven, D., Gutfinger, T., Alvarez, J., Ganal, M., et al. (1998). The SELF-PRUNING gene of tomato regulates vegetative to reproductive switching of sympodial meristems and is the ortholog of CEN and TFL1. *Development* 125, 1979–1989.

Pnueli, L., Gutfinger, T., Hareven, D., Ben-Naim, O., Ron, N., Adir, N., et al. (2001). Tomato SP-interacting proteins define a conserved signaling system that regulates shoot architecture and flowering. *Plant Cell* 13, 2687–2702. doi: 10.1105/tpc.13.12.2687

Putterill, J., Robson, F., Lee, K., Simon, R., and Coupland, G. (1995). The CONSTANS gene of Arabidopsis promotes flowering and encodes a protein showing similarities to zinc finger transcription factors. *Cell* 80, 847–857. doi: 10.1016/0092-8674(95)90288-0

Quail, P. H., Boylan, M. T., Parks, B. M., Short, T. W., Xu, Y., and Wagner, D. (1995). Phytochromes: photosensory perception and signal transduction. *Science* 268, 675–680. doi: 10.1126/science.7732376

Samach, A., Onouchi, H., Gold, S. E., Ditta, G. S., Schwarz-Sommer, Z., Yanofsky, M. F., et al. (2000). Distinct roles of CONSTANS target genes in reproductive development of Arabidopsis. *Science* 288, 1613–1616. doi: 10.1126/science.288.5471.1613

Shalit, A., Rozman, A., Goldshmidt, A., Alvarez, J. P., Bowman, J. L., Eshed, Y., et al. (2009). The flowering hormone florigen functions as a general systemic regulator of growth and termination. *Proc. Natl. Acad. Sci. U.S.A.* 106, 8392–8397. doi: 10.1073/pnas.0810810106

Song, Y. H., Smith, R. W., To, B. J., Millar, A. J., and Imaizumi, T. (2012). FKF1 conveys timing information for CONSTANS stabilization in photoperiodic flowering. *Science* 336, 1045–1049. doi: 10.1126/science.1219644

Thompson, J. D., Higgins, D. G., and Gibson, T. J. (1994). CLUSTAL W: improving the sensitivity of progressive multiple sequence alignment through sequence weighting, position-specific gap penalties and weight matrix choice. *Nucleic Acids Res.* 22, 4673–4680. doi: 10.1093/nar/22.22.4673

Valverde, F., Mouradov, A., Soppe, W., Ravenscroft, D., Samach, A., and Coupland, G. (2004). Photoreceptor regulation of CONSTANS protein in photoperiodic flowering. *Science* 303, 1003–1006. doi: 10.1126/science.1091761

Velásquez, A. C., Chakravarthy, S., and Martin, G. B. (2009). Virus-induced gene silencing (VIGS) in *Nicotiana benthamiana* and tomato. *J. Vis. Exp.* 28:1292. doi: 10.3791/1292

Weller, J. L., and Reid, J. B. (1993). Photoperiodism and photocontrol of stem elongation in two photomorphogenic mutants of *Pisum sativum* L. *Planta* 189, 15–23. doi: 10.1007/BF00201338

Weller, J. L., Schreuder, M. E., Smith, H., Koornneef, M., and Kendrick, R. E. (2000). Physiological interactions of phytochromes A, B1 and B2 in the control of development in tomato. *Plant J.* 24, 345–356. doi: 10.1046/j.1365-313x.2000.00879.x

Wigge, P. A., Kim, M. C., Jaeger, K. E., Busch, W., Schmid, M., Lohmann, J. U., et al. (2005). Integration of spatial and temporal information during floral induction in Arabidopsis. *Science* 309, 1056–1059. doi: 10.1126/science.1114358

Yamaguchi, A., Kobayashi, Y., Goto, K., Abe, M., and Araki, T. (2005). TWIN SISTER OF FT (TSF) acts as a floral pathway integrator redundantly with FT. *Plant Cell Physiol.* 46, 1175–1189. doi: 10.1093/pcp/pci151

Yanovsky, M. J., and Kay, S. A. (2003). Living by the calendar: how plants know when to flower. *Nat. Rev. Mol. Cell Biol.* 4, 265–276. doi: 10.1038/nrm1077

Zhang, X., Henriques, R., Lin, S.-S., Niu, Q.-W., and Chua, N.-H. (2006). Agrobacterium-mediated transformation of *Arabidopsis thaliana* using the floral dip method. *Nat. Protoc.* 1, 641–646. doi: 10.1038/nprot.2006.97

Conflict of Interest Statement: The authors declare that the research was conducted in the absence of any commercial or financial relationships that could be construed as a potential conflict of interest.

14

A New Ethylene-Responsive Factor *CaPTI1* Gene of Pepper (*Capsicum annuum* L.) Involved in the Regulation of Defense Response to *Phytophthora capsici*

*Jing-Hao Jin, Huai-Xia Zhang, Jun-Yi Tan, Ming-Jia Yan, Da-Wei Li, Abid Khan and Zhen-Hui Gong**

College of Horticulture, Northwest A&F University, Yangling, China

Ethylene-responsive factors (*ERF*) are usually considered to play diverse roles in plant response to biotic and abiotic stresses. In this study, an *ERF* gene *CaPTI1* was isolated from pepper transcriptome database. *CaPTI1* contains an open reading frame (ORF) of 543 bp, which encodes a putative polypeptide of 180 amino acids with a theoretical molecular weight of 20.30 kDa. Results of expression profile showed that *CaPTI1* had a highest expression level in roots and this gene could not only response to the infection of *Phytophthora capsici* and the stresses of cold and drought, but also be induced by the signaling molecule (salicylic acid, Methyl Jasmonate, Ethephon, and hydogen peroxide). Furthermore, virus-induce gene silencing (VIGS) of *CaPTI1* in pepper weakened the defense response significantly by reducing the expression of defense related genes *CaPR1*, *CaDEF1* and *CaSAR82* and also the root activity. These results suggested that *CaPTI1* is involved in the regulation of defense response to *P. capsici* in pepper.

Keywords: pepper, ethylene-responsive factor (ERF), *Phytophtora capsici*, relative expression, virus-induce gene silencing (VIGS)

Edited by:
Elena Prats,
Consejo Superior de Investigaciones Cientificas, Spain

Reviewed by:
Chang-Jie Jiang,
National Institute of Agrobiological Sciences, Japan
Ky Young Park,
Sunchon National University, South Korea

****Correspondence:***
Zhen-Hui Gong
zhgong@nwsuaf.edu.cn

INTRODUCTION

Pepper (*Capsicum annuum* L.) is an agriculturally important vegetable crop worldwide. While the production of pepper is severely threatened by a variety of diseases, especially the soil borne disease *Phytophthora* blight which is caused by *Phytophthora capsici* (Hausbeck and Lamour, 2004; Granke et al., 2012). *P. capsici* could not only attack pepper plants, but also other crops such as cucurbits, tomato, eggplant, snap and lima beans (Lamour et al., 2012). *P. capsici* could affect the plant at any stage of development and cause damping-off, seedling blight, foliar blight and wilting follow by plant death. Infection on mature plants materializes as dark, rapidly expanding, water-soaked lesions (Baysal et al., 2005). Chemical and biological control are the most common strategies used to prevent the spreading of *P. capsici* (Oelke et al., 2003). Several fungicides have been reported to be effective against *P. capsici* on vegetable crops (Matheron and Porchas, 2000; Babadoost and Islam, 2003; Kousik et al., 2011). Although many studies have revealed that resistance to *P. capsici* is polygenic and could be controlled by quantitative trait loci (QTL) (Thabuis et al., 2003; Kim et al., 2008; Rehrig et al., 2014), little is known about pepper–*P. capsici* interactions at the molecular and genetic levels.

Plants have evolved many sophisticated defense mechanisms by activating multiple defense pathways in order to protect themselves from pathogen infection and cope with the pathogen invasion. salicylic acid (SA), Jasmonate (JA), and Ethylene (ET) are frequently induced in response to infection with various pathogens. The balance of these hormones is dependent on the recognized pathogen and plays a vital role in fine tuning appropriate defense responses (Dong, 1998; Dangl and Jones, 2001). SA typically mediates basal defense to biotrophic pathogens and is essential for the rapid activation of local and systemic resistance (Loake and Grant, 2007; Vlot et al., 2009). While JA controls defense reactions to necrotrophs. Together with ET, JA generally regulates induced systemic resistance (ISR) (Glazebrook, 2005; Alazem and Lin, 2015). ET may regulate pathogenesis-related (PR) genes expression through *ERFs*, and *ERFs* were supposed to act as either transcriptional activators or repressors of GCC-box mediated gene expression and be involved in the regulation of plant defense mechanisms (Fujimoto et al., 2000; Chakravarthy et al., 2003; Gutterson and Reuber, 2004; Yang et al., 2005; Alazem and Lin, 2015).

ERFs were firstly isolated in tobacco and found to regulate the expression of *PR* genes through binding specifically to the GCC-box (Ohmetakagi and Shinshi, 1995). Then the *ERFs* in other plants were constantly identified, such as *AtERFs*, *StERFs* and *LeERFs* (Zhou et al., 1997; Fujimoto et al., 2000; Wang et al., 2015). Fujimoto et al. (2000) isolated five *AtERFs* which respond to extracellular signals and modulate GCC box-mediated gene expression. Then Brown et al. (2003) found that *AtERF2* possibly play important role in regulating JA-dependent defense response via interaction with the GCC-box. The *ERFs* in tomato, *Pti4*, *Pti5*, and *Pti6* which were interacted with the resistance (R) gene *Pto* kinase were identified by yeast two-hybrid assays (Zhou et al., 1997). Further research found these *ERFs* could induced by *Pseudomonas* bacterium and also specifically recognize and bind to GCC-box of *PR* proteins (Thara et al., 1999). Moreover, *Pti4*, *Pti5*, and *Pti6* were suggested to indirectly regulate the SA response through interaction with other *TFs*, and activate the expression of *PR* genes (Gu et al., 2002; Chakravarthy et al., 2003). The over-expression of *Pti5* in tomato demonstrated the positive role of *Pti5* in defense genes regulation and disease resistance (He et al., 2001). Recently research revealed that *Pti5* contributes to potato aphid resistance in tomato independent of ethylene signaling (Wu et al., 2015). Besides, *NtTsi1* and *CaERFLP1* were also isolated and demonstrated to bind to the GCC-box (Park et al., 2001; Lee et al., 2004). Over-expression of *NtTsi1* in transgenic hot pepper plants induced constitutive expression of several *PR* genes in the absence of stress or pathogen treatment and enhanced the disease resistance to *Xanthomonas campestris pv. vesicatoria* and also the *P. capsici* (Shin et al., 2002).

Most of researches suggest that *ERFs* were involved in disease resistance responses as activators. There were also some *ERFs* whose protein contained an ERF-associated amphiphilic repression (EAR) motif in the C-terminal as repressors, such as *NtERF3*, *AtERF3*, *AtERF4* and *StERF3* (Ohta et al., 2001; McGrath et al., 2005; Tian et al., 2015). But recently Dong et al. (2015) found that the *GmERF5* which has an EAR motif could not only significantly induced by *P. sojae*, ETH, ABA and SA, but also bind to the GCC-box. Besides, transgenic soybean of *GmERF5* exhibited significant enhanced resistance to *P. sojae* and positively regulated the expression of the *PR* genes.

Although much progress has been made in elucidating the functional properties of *ERFs*, there was less research about *ERFs* in pepper. Previously, the *ERFs CaCBF1A*, *CaCBF1B* and *CaDREBLP1* who mainly response to environmental stresses (low temperature and drought) were isolated (Kim et al., 2004; Hong and Kim, 2005). In this study, an *ERF* gene, *CaPTI1* which was induced by pathogen was isolated from the transcriptome of interactions between pepper and *P. capsici*. The sequence character and structure were analyzed firstly. Then the expression profile of *CaPTI1* was analyzed by quantitative Real-Time-PCR (qRT-PCR). Furthermore, in past most of the virus-induced gene silencing (VIGS) concentrated in leaves but here we have performed in leaves as well as in roots of pepper to investigate the function of *CaPTI1* gene.

MATERIALS AND METHODS

Plant Materials and Growth Conditions

Pepper cultivar Y3 was used in this study, which was provided by pepper breeding group in Northwest A&F University, China. First presoaking of the seeds was done and when 80% of seeds germinated, they were transferred to plastic pots having growing media [grass charcoal/perlite (3/1, v/v)]. Pepper plants were grown in a growth chamber with a 16 h light and 8 h dark photoperiod at 25°C.

Pathogen Preparation and Inoculation

The virulent (HX-9) and avirulent strains (PC) of *P. capsici*, which was compatible and incompatible to cultivar Y3 respectively, were used in this research. Both of the strains were grown on potato dextrose agar (PDA) medium in the dark at 28°C for 7 days as described previously by Zhang et al. (2013). Then zoospores of *P. capsici* were induced by chilling at 4°C for 30 min and then incubated at room temperature for 30–60 min. The zoospores were collected by filtering through four layers of cheesecloth. After the zoospores were collected, a haemocytometer was used to adjust the concentration to 1×10^5 zoospores mL^{-1} with sterile water.

For the VIGS of *CaPTI1* gene, pepper plants were inoculated with 3 mL zoospore suspension of *P. capsici* virulent strain HX-9 (1×10^5 zoospores mL^{-1}) using root drenching method. While for inoculation of detached leaves, third to fifth leaf from top of the control (*pTRV2:00*) and *CaPTI1* silenced plants were taken and sterilized with 75% ethanol for 1 min, washed three times with sterile water, then 20 μL zoospore suspension (1×10^5 zoospores mL^{-1}) of HX-9 were injected with a needleless syringe. After inoculation the leaves were kept in petri dishes and sealed with parafilm immediately. All the plants inoculated with *P. capsici* were put in a growth chamber at 28°C with a photoperiod of 16 h light/8 h dark and the relative humidity was kept 60%.

Cloning and Sequence Analysis of *CaPTI1* Gene

Total RNA was extracted from the roots of Y3 plants which were inoculated with virulent strain of *P. capsici* and reverse transcribed into cDNA using a PrimeScript™ Kit (Takara, Bio Inc, China) following the manufacturer's protocols, and then the cDNA was used as a template to isolate the *CaPTI1* gene according to the transcriptome database of compatible and incompatible interactions of pepper (Y3) with *P. capsici*. After that *CaPTI1* gene was cloned into pMD19-T vector (Takara) and sequenced by Sangon Biotech (Shanghai) Co. Ltd.

The molecular weight (MW) and isoelectric point (pI) of *CaPTI1* were predicted by ExPASy[1]. Conserved domain of *CaPTI1* protein was identified in Conserved Domain of NCBI[2]. The full length of *CaPTI1* and other *ERF* proteins were used to construct the neighbor joining phylogenetic tree by MEGA 6.05 with 1000 bootstrap replicates. Multiple sequence alignments of the AP2/ERF domain in *CaPTI1* and other *ERFs* were performed by DNAMAN 5.0.

CaPTI1 Gene Expression Profile Analysis

To evaluate the expression levels of *CaPTI1* gene in different organs, the roots, stems, leaves, flowers, and immature fruits were collected from Y3 pepper plants, then frozen in liquid nitrogen and kept at −80°C for further RNA extraction and gene expression analysis.

Pepper plants at the six-true leaf stage were used for *P. capsici* inoculation and treatments of plant signaling molecule (ETH, SA, MeJA, and H_2O_2) and environmental stresses (drought and cold). For *P. capsici* inoculation treatment the root-drenching method was used as described previously (Zhang et al., 2013). Then leaves and roots from infected plants were collected at 0, 6, 12, 24, 48, 72, and 96 h after inoculation. All samples were collected and analyzed with three biological replicates.

In case of plant signaling molecule treatments, pepper plants were sprayed with 10 mM ETH, 5 mM SA, 50 μM MeJA or 10 mM H_2O_2 solutions, while in case of drought stress treatment, the roots of seedlings were watered with 0.4 M mannitol solutions. Pepper plants were placed at 4°C for cold stress treatment. Control plants were sprayed with sterile water. Pepper plants in above treatments were placed in a chamber at 25°C (except cold stress) under a 16 h light/8 h dark photoperiod cycle with 60% relative humidity. Leaves from treated plants were collected at 0, 3, 6, 9, 12, and 24 h after treatment, and frozen in liquid nitrogen immediately and kept at −80°C for RNA extraction. All samples were collected and analyzed with three biological replicates.

RNA Isolation and Quantitative RT-PCR for Gene Expression Analysis

Total RNA was extracted from collected sample after various stress treatments using Trizol (Invitrogen) method. Then the first strand cDNA was synthesized using of prime script TM kit (Takara) following the manufacturer's protocols. The NanoDrop instrument (Thermo Scientific NanoDrop 2000C, USA) was used to measure the concentration of cDNA. Later on qRT-PCR was carried out using SYBR® Premix Ex Taq™ II (TaKaRa) in an iCycler iQ™ Multicolor PCR Detection System (Bio-Rad, USA). The amplification cycling parameters of qRT-PCR were as follow: 95°C for 1 min and followed by 40 cycles at 95°C for 10 s, 56°C for 15 s, and 72°C for 15 s. All the primers used for qRT-PCR were shown in Supplementary Table S1. Relative expression of genes was calculated as described by Livak and Schmittgen (2001) and *CaUbi3* gene was used as the reference gene in this study (Wan et al., 2011).

VIGS of *CaPTI1* in Pepper

For VIGS assay, the 3′-untranslated region (UTR) of *CaPTI1* gene was selected to construct the combined vector *pTRV2:CaPTI1* and then this combine vector was used for subsequent VIGS experiment to ensure the specific silencing of *CaPTI1* (primers used for vector construction were shown in Supplementary Table S1). Then the *pTRV1*, *pTRV2* (negative control), *pTRV2:CaPDS* (positive control) (Wang et al., 2013a) and also the combined vector *pTRV2:CaPTI1* was transformed into *Agrobacterium tumefaciens* strain GV3101 by the freeze-thaw method. Agrodrench and leaf infiltration methods were used to silence *CaPTI1* in pepper (Ryu et al., 2006; Velasquez et al., 2009). The growth condition used for pepper seedling after inoculation was as described by Wang et al. (2013a). After inoculation for about 5 weeks, leaves and roots from *CaPTI1* silenced and control plants were collected for the analysis of silencing efficiency and expression level of relative defense genes. All experiments were performed and analyzed with three biological replicates.

Determination of Root Activity

The root activity was measured by a modified triphenyltetrazolium chloride (TTC) method (Ou et al., 2011). Before the TTC tests, the roots of silenced and control plants at different time points after inoculation with virulent strain of *P. capsici* were collected and washed with sterile water. Then the root surfaces were dried carefully with absorbent papers. After that root activity was measured by a modified TTC method as described by Wang et al. (2013b). The root samples were placed in small beaker and incubated with 10 ml 1:1 (v/v) mixture of 1% TTC solution and 0.1 M phosphate buffer (pH 7.0) at 37°C for 1 h in the dark. After the incubation, 2 ml of 1 M sulphuric acid was to added to inhibit the reaction. The root samples were rinsed twice with distilled water and ground in mortar with 3 ml ethyl acetate, in which the extracts of TTF (a derivative of TTC from the reduced reaction) were obtained and transferred to a 10 ml volumetric flask. The residues were rinsed with ethyl acetate and mixed with the earlier extracts, then the final volume was adjusted to10 ml. The absorption of the extraction was measured with a spectrophotometer at 485 nm. The control was similarly performed except the sulphuric acid was added first, the root sample second, and the mixture last. The reduced TTC amount was obtained from the standard curve and its intensity in the roots was calculated as follows: TTC reduction intensity [$mg\ g^{-1}\ h^{-1}$] = reduced TTC amount/FW h (FW-fresh root

[1] http://www.expasy.ch/tools/pi_tool.html

[2] http://www.ncbi.nlm.nih.gov/Structure/cdd/wrpsb.cgi

mass; h-the incubation time). Each treatment was conducted in three independent experiments and each measurement was repeated three times.

Statistical Analysis

All data were expressed as mean ± standard deviation (SD) of three independent replicates. Statistical Analysis System software (SAS 8.2, North Carolina State University, USA) was used for least significant difference (LSD) analysis and values of $p < 0.05$ was considered statistically significant. All experiments were performed and analyzed separately with three biological replicates.

RESULTS

Cloning and Sequence Analysis of *CaPTI1*

The *CaPTI1* gene (**Figure 1**) (GenBank No. KJ690096), which included an open reading frame (ORF) of 543 base pair (bp), was identified from pepper transcriptome database. The ORF of *CaPTI1* was predicted to encode a protein of 180 amino acids. The predicted molecular weight of the encoded protein was 20.30 kDa and *p*I 8.58. Sequence analysis found a single AP2/ERF domain in the *CaPTI1* protein.

Furthermore the evolutionary relationships and multiple sequence alignment of *CaPTI1* in comparison with *ERF* proteins of other plants were also investigated (**Figure 2**). The result showed that *CaPTI1* had a more closely relationship with *SlPti5*. Multiple sequence alignment showed the conservation of AP2/ERF domain in *ERFs* from different plants. There were three β-sheets and an α-helix in the AP2/ERF domain (Allen et al., 1998; Djemal and Khoudi, 2015). The amino acids difference at position 14th and 19th in AP2/ERF domain was also existed in these *ERFs* (Sakuma et al., 2002).

Tissue-Specific Expression of *CaPTI1* Gene in Pepper Plants

The expression levels of *CaPTI1* gene in different tissues of Y3 plants were investigated by qRT-PCR. As shown in **Figure 3**, significant difference of *CaPTI1* gene expression was observed in different tissues. The highest expression was detected in roots, which was almost 26-fold of that in leaves. The expression level of *CaPTI1* gene in stems and flowers were also higher than that in leaves, which was 3.50 and 2.45-fold of that in leaves respectively. While lower expression was detected in fruits as compared to leaves which was statistically non-significant.

Expression Analysis of the *CaPTI1* Gene in Response to *Phytophthora capsici* Infection

In virulent strain (HX-9) infection (**Figure 4**), the highest expression of *CaPTI1* in roots was detected at 24 h post inoculation (hpi) which was 10.24-fold of control, then decrease gradually and at 96 hpi the expression was 5.22-fold. The same trend of *CaPTI1* in roots was observed after inoculation with avirulent strain (PC), in which the expression of *CaPTI1* was increasing with the passage of time and reached to peak (4.62-fold) at 24 hpi and then decrease to the lowest (1.39-fold) at 96 hpi. Overall, the expression level of *CaPTI1* gene in the whole treatment with virulent strain was significantly higher than that with avirulent strain.

While in leaves, *CaPTI1* was up-regulated by virulent strain inoculation and the expression level was increased to 2.92-fold of control at 12 hpi, followed by a bit decrease, then reached to the peak (3.13-fold) at 48 hpi and at last decreased to 1.26-fold of control at 96 hpi. Less up regulation of *CaPTI1* was detected in the infection of avirulent strain except for a little increase at 6 hpi. Finally, the results showed that *CaPTI1* could be induced by *P. capsici*, especially in roots.

Expression Analysis of *CaPTI1* Gene Under Abiotic Stresses

The signaling molecules (ETH, SA, MeJA and H_2O_2) were used to investigate whether *CaPTI1* could be induced or not. As shown in **Figure 5**, *CaPTI1* in the leaves were strongly induced by foliar spraying of ETH, SA, and MeJA. As an *ERF*, *CaPTI1* in leaves showed a huge increase by ETH induction. The highest expression of *CaPTI1* was detected at 3 h post treatment (hpt), which was 102.57-fold of control. In the treatment of SA and MeJA, both the highest expression of *CaPTI1* was detected at 6 hpt, which were 114.53-fold and 108.80-fold of control respectively. While in the treatment with H_2O_2, the significant up regulation of *CaPTI1* was observed at 3 and 6 hpt, which was 6.26-fold and 5.80-fold of control respectively.

Moreover, in the stresses of cold and drought, the expression of *CaPTI1* in leaves also showed a significant increase and rise to the peak at 12 hpt, which were 11.32-fold and 6.82-fold of control respectively.

VIGS of *CaPTI1* Gene in Pepper Plants

To examine the effect of loss-of-function of the *CaPTI1* gene in pepper plants, VIGS was performed in pepper cultivar Y3. Empty vector (*pTRV2:00*) was used as a negative control. The *pTRV2:CaPDS* vector was used as a positive control, which will cause the silencing of *CaPDS* and induces a photo-bleaching phenotype in leaves.

Five weeks post inoculation, the phenotype of photo-bleaching was observed in leaves of positive control plants (**Figure 6A**). Then qRT-PCR analyses were performed to investigate the silencing efficiency of *CaPTI1* in leaves. *CaPTI1* in *CaPTI1* gene silenced plant exhibited half expression level of that in control (*pTRV2:00*) (**Figure 6B**). Then the third to fifth leaves from the top of control plants and *CaPTI1* silenced pepper plants (*pTRV2:CaPTI1*) were detached and inoculated by 20 μL zoospores suspension (10^5 zoospores mL^{-1}) of virulent strain of *P. capsici*. Disease symptoms were observed at 2 days post inoculation (dpi). The leaves of control plants exhibited less disease symptoms. While larger and more numerous lesions were observed in the leaves of *CaPTI1* silenced plants (**Figure 6C**). The

```
1      TCTGTAATTAGTTATGGTTCCAACTCACCAAAGTGATCTACCTCTTAATGAGAACGACTC
                      M  V  P  T  H  Q  S  D  L  P  L  N  E  N  D  S

61     ACAAGAGATGGTATTATACGAAGTCCTCAATGAAGCTAATAACCTCCACATCCCATATTT
        Q  E  M  V  L  Y  E  V  L  N  E  A  N  N  L  H  I  P  Y  L

121    GCCTCAAAGAAACCAATTGATACCTAGAAATAATCATATCCTTCGTCCATCACAAACCAT
        P  Q  R  N  Q  L  I  P  R  N  N  H  I  L  R  P  S  Q  T  I

181    AGGCAAGAAATACAGAGGCGTTCGACGTCGCCCGTGGGGAAATACGCAGCTGAAATTCG
        G  K  K  Y  R  G  V  R  R  R  P  W  G  K  Y  A  A  E  I  R

241    AGATTCCGCTCGACATGGTGCGAGAGTATGGCTAGGCACGTTTGAAACGGCCGAAGAAGC
        D  S  A  R  H  G  A  R  V  W  L  G  T  F  E  T  A  E  E  A

301    TGCTTTGGCTTATGATAGAGCGGCTTTTAGAATGCGTGGTGCTAAGGCACTACTTAATTT
        A  L  A  Y  D  R  A  A  F  R  M  R  G  A  K  A  L  L  N  F

361    CCCATCTGAAATAGTCACATCATCTGTTTCAGTAGATAAATTAAGTTTGTGCTCAAACAG
        P  S  E  I  V  T  S  S  V  S  V  D  K  L  S  L  C  S  N  S

421    TTACACTACTAATAGTTCAAATTCGAATTCAAATGAAGTTTCAAGTGGAGAGAATGTATT
        Y  T  T  N  S  S  N  S  N  S  N  E  V  S  S  G  E  N  V  F

481    TAAATCAAGAATTTCAAATCAGTTTTCTCAAGACGCTAAGACAGAATCGTGCATGGAGAA
        K  S  R  I  S  N  Q  F  S  Q  D  A  K  T  E  S  C  M  E  N

541    TTTCTTGGCGCTCTGAATAAATTGCTCGATCCATGTTAAATTATTGTTCTAGGAGGTTAC
        F  L  A  L
601    AGTGTGTCGACACATCAAGCATAATTTAGATGTGTTAGAATATATCTGTTAGATATATCA
661    ATCTATTCTGGGTTTTACTTTTACATTATATGCATGCTTGTATTTCCTTTTATTTTCTTC
721    TTGAATAGTTTGTAGCGATGCACGCTATAATCATGTGTAAAAAGATTGCTTAGGATAATT
781    AGCACGTTTATAAATTGTAATATACACATTCTATTTCTATTTTTTC
```

FIGURE 1 | The nucleotide sequence and deduced amino acid sequence of *CaPTI1*.

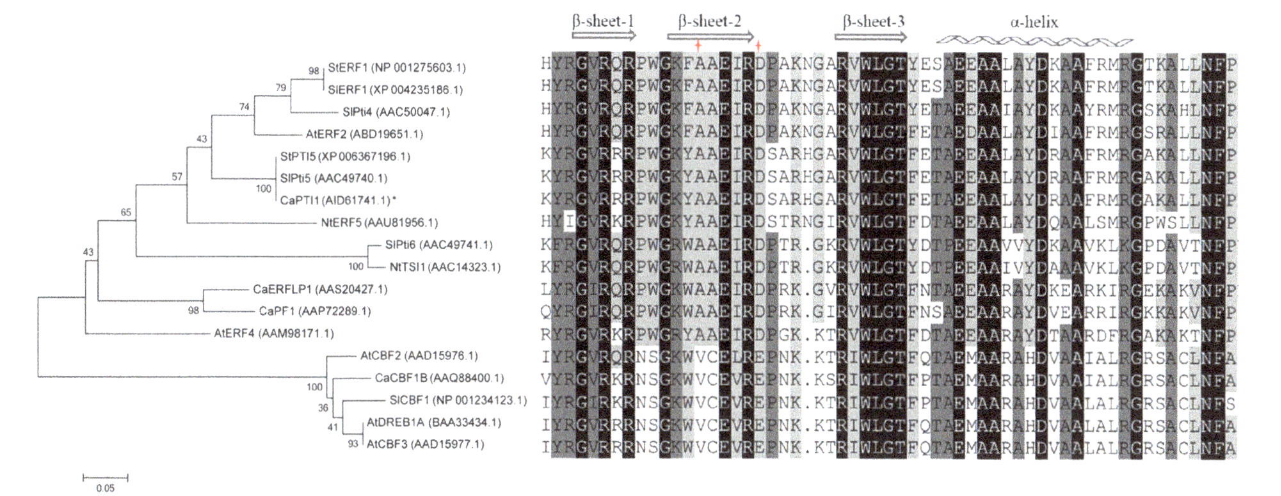

FIGURE 2 | Phylogenetic tree and multiple alignment of *CaPTI1* with other *ERFs*. Phylogenetic tree of *CaPTI1* with other Ethylene-responsive factors (ERF) proteins was generated by the MEGA 6.05 program with 1000 bootstrap replicates. GenBank No. of *ERFs* is in parentheses after each gene name. While the multiple alignment was conducted by DNAMAN5.0.

quantitative analysis of the lesion area in the detached leaves showed a significant increase in *CaPTI1* silenced plants (**Figure 6D**).

Besides, we also checked the expression of *CaPTI1* in roots of *CaPTI1* silenced plants and found it was only about 40% of the control (**Figure 7A**). After that we investigated whether

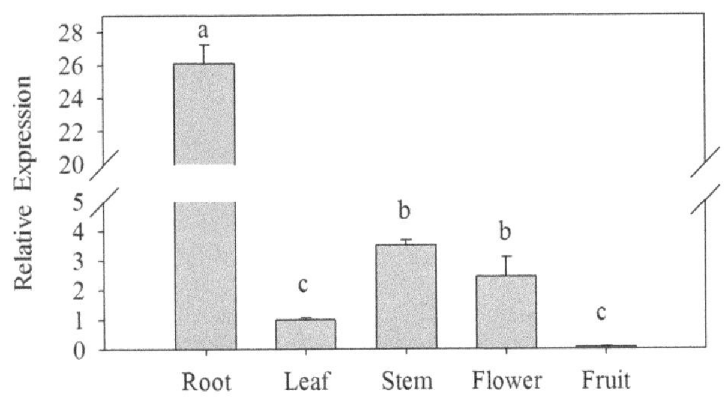

FIGURE 3 | Tissue specific expression of *CaPTI1* in pepper. Error bars represent standard deviation (SD) for three independent replicates. Bars with different lower case letters indicate significant differences using Duncan's multiple range test ($p < 0.05$).

silencing of *CaPTI1* altered the expression of defense related genes in roots during *P. capsici* infection. Significant decrease of *CaPR1* (a SA dependent gene) (Kim and Hwang, 2000), *CaDEF1* (a JA dependent gene) (Do et al., 2004) and *CaSAR82* (systemic acquired resistance gene) (Lee and Hwang, 2003) were detected in *CaPTI1* silenced plants. The time was expands to 4 dpi. Especially the *CaPR1* gene showed a really low expression (0.87-fold) in silenced plant, but a highest expression (29.23-fold) in control at 4 dpi. While at the same time, the expression of *CaPTI1* in silenced plant (2.02-fold) was still lower than that in control (6.88-fold).

In addition, the roots activity of silenced and control plants after inoculation with *P. capsici* were measured by TTC method. The *CaPTI1* silenced plants showed a lower root activity as compared to control, and a significant difference between silenced and control plants were detected at 4 dpi and enlarged to 7 dpi (**Figure 7B**). At 7 dpi disease symptoms of *Phytophthora* blight in pepper plants were observed. The *CaPTI1* silenced plants showed a serious disease symptom, such as the blacking and constricting of stem basement, the wilting and abscission of leaves and wilting of growing point of plant. While less and slightly disease symptoms were observed in the control plants (**Figure 7C**).

DISCUSSION

The AP2/ERF family is one of the largest transcription factor family and demonstrated to be involved in the regulation of plant development and response to biotic and abiotic stresses (Gutterson and Reuber, 2004; Wessler, 2005; Mizoi et al., 2012). In this study, we identified a new ERF in pepper, which named *CaPTI1* and belonged to the ERF family.

Sakuma et al. (2002) divided the ERF family into ERF subfamily and DREB subfamily according to the amino acids difference at position 14th and 19th in AP2/ERF domain. In addition, several ERFs which contain alanine and aspartic acid at 14th and 19th position of the AP2/ERF domains could bind to both the DRE and GCC boxes (Park et al., 2001; Wang et al., 2004). The 14th alanine and the 19th aspartic acid in the AP2/ERF domain of CaPTI1 suggested a member of ERF subfamily. And sequence character analysis showed that CaPTI1 does not contain an EAR motif, which maybe indicate the activation role of CaPTI1 in plant defense response.

Phytophthora capsici mainly invades the pepper plant by roots and *CaPTI1* has a highest expression in roots before inoculation. After infection of *P. capsici* the expression level of *CaPTI1* in roots got a significant increase. These possibly indicated that *CaPTI1* is involved in the interaction of pepper and *P. capsici*. *CaPTI1* could not only be induced by *P. capsici*, but also strongly induced by ETH, SA and MeJA. These hormones as signaling molecules had been showed to play crucial roles in defense response pathway (Dong, 1998; Reymond and Farmer, 1998; Pieterse and van Loon, 1999). The SA signaling transduction pathway usually acts as antagonism with ET/JA pathway. *ERF1* of *Arabidopsis* was found to be a key element in the integration of both ET and JA pathways for the regulation of defense response genes (Lorenzo et al., 2003). Besides, *AtERFs* may be involved in the cross-talk between SA and JA signaling pathways and important for plant defense to pathogen attack (Onate-Sanchez

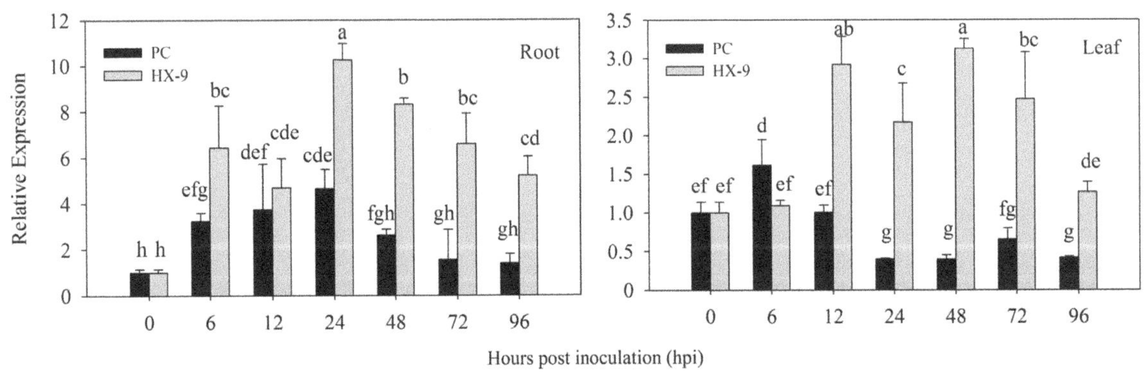

FIGURE 4 | Relative expression of *CaPTI1* in pepper plants exposed to *P. capsici* infection. Relative expression of *CaPTI1* in the roots and leaves of pepper cultivar Y3 after inoculation with virulent (HX-9) or avirulent (PC) strains of *P. capsici*. Error bars presented means ± SD from three replicates. Bars with different lower case letters indicate significant differences using Duncan's multiple range test ($p < 0.05$).

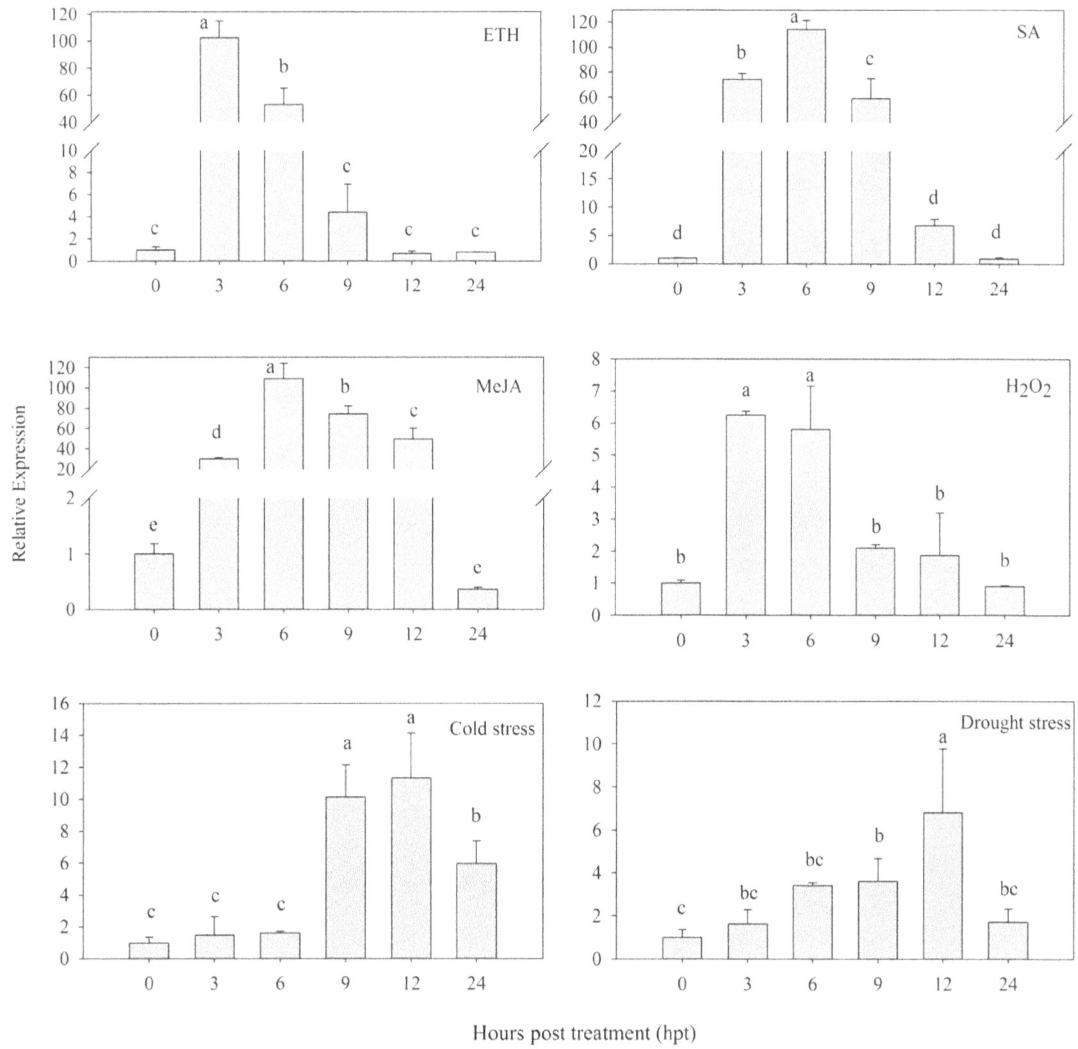

FIGURE 5 | Relative expression of *CaPTI1* responsed to various abiotic stresses. Relative expression of *CaPTI1* gene in leaves of pepper plants after different treatments [Ethephon (ETH), salicylic acid (SA), Methyl Jasmonate (MeJA), hydogen peroxide (H_2O_2), cold stress and drought stress]. Error bars represent the mean ± SD of three independent biological replicates. Bars with different lower case letters indicate significant differences using Duncan's multiple range test ($p < 0.05$).

and Singh, 2002). *ERF* transcription factor *ORA59* was supposed to be the integrator of JA and ET signaling pathways (Zarei et al., 2011). The *ERF* transcription factor in tomato, *Pti4* could be induced by SA, JA, and ETH. While *Pti5* and *Pti6* could not be induced by SA in tomato (Gu et al., 2000). Further studies found that *Pti4, Pti5,* and *Pti6* may be implicated indirectly in regulating the SA response through interaction with other *TFs*, and all of them could activate the expression of *PR* genes and play important roles in plant defense (Gu et al., 2002; Chakravarthy et al., 2003). All of these indicated that *CaPTI1* may be involved in the pathways which were mediated by the signaling molecules SA, MeJA and ETH.

We also detected the up regulation of *CaPTI1* by the spraying of H_2O_2. Reactive oxygen species (ROS) are produced constantly during normal plant growth and development and also perform essential roles as highly specific signaling molecules under stress conditions (Apel and Hirt, 2004). Plant defense hormones SA and JA could also modulate the plants ROS status (Mittler et al., 2011). *ERF6* was turned out to be a regulator of ROS in *Arabidopsis* and response to biotic and abiotic stresses (Sewelam et al., 2013). Moreover, *CaPTI1* could be induced by the cold and drought stresses which indicated *CaPTI1* response to abiotic stresses. Over-expression of the tobacco *Tsi1* gene also enhances resistance against pathogen attack and osmotic stress in tobacco (Park et al., 2001). *StERFs* also were shown to be regulated by drought stress (Wang et al., 2015). All these suggested the multiple role of *CaPTI1* in pepper plants response to diverse stresses.

We silenced the *CaPTI1* gene in leaves and roots of pepper plants successfully by VIGS. Detached leaves assay were frequently used to investigate the plant disease resistance to the

FIGURE 6 | Loss of function analysis of *CaPTI1* in leaves of pepper. (A) The photo-bleaching phenotype in leaves of positive control pepper plants (*pTRV2:CaPDS*); **(B)** The relative expression of *CaPTI1* in leaves of *CaPTI1* silenced and negative control plants; **(C)** Disease symptoms developed on the detached leaves of silenced and control plants at 2 day post inoculation (dpi); **(D)** Percentage of the lesion area of the leaves inoculated with *P. capsici*. Error bars represent the mean ± SD of three independent biological replicates. Bars with different lower case letters indicate significant differences using Duncan's multiple range test ($p < 0.05$).

invasion of pathogens (Wang et al., 2013c; Dong et al., 2015; Tian et al., 2015). After inoculation of *P. infestans* larger disease lesions were observed on the leaves of *StERF3* over-expressed potato lines as compared to the wild type, which indicated the repressor role of *StERF3* in defense response to *P. infestans* in potato. Less disease lesion were observed on the leaves of *GmERF5* transgenic soybean lines after infection with *P. sojae* (Dong et al., 2015). Detached leaves assay with the inoculation of *P. capsici* were also performed in this study after the silencing of *CaPTI1* in pepper, and then larger lesion area was observed in leaves of *CaPTI1* silenced plants as compared to control. This indicated that silencing of *CaPTI1* in pepper leaves make it more susceptible to *P. capsici* infection.

In addition, the roots of pepper were focused on because *Phytophthora* blight is a soil borne disease. Significant decrease of defense related genes *CaPR1*, *CaDEF1*, and *CaSAR82* was detected in roots of *CaPTI1* silenced plants when compared to the control plants, especially for the *CaPR1*, which is a SA dependent gene (Kim and Hwang, 2000). As mentioned before, *ERFs* could specially recognize and bind to the GCC-box. *Pti4* could regulate defense related genes expression via both GCC-box and non-GCC-box (Chakravarthy et al., 2003). Two GCC-box were found in the upstream promoter region (about 1500 bp) of *CaPR1* according the database of pepper genome (Kim et al., 2014). The silencing of *CaPTI1* maybe decrease the expression of *CaPR1* by affecting the binding of GCC-box which is in the promoter region of *CaPR1*. Moreover, Kim and Hwang (2000) found that treatment with SA in combination with ETH, caused a lower expression of the *CaPR1* in pepper leaves than the treatment with ETH alone, and pepper stems will synthesize ethylene at an exponential rate soon after infection with either virulent or avirulent strains of *P. capsici*. As *Pti4, Pti5,* and *Pti6* were supposed to be involved in the regulation of ethylene production in the plant singly or in combination (Gu et al., 2000). The silencing of *CaPTI1* may reduce the ethylene production of pepper plants. As a result, the silencing of *CaPTI1* may enhance the antagonism between SA and ET pathways by changing the ratio of SA to ET in *CaPTI1* silenced pepper plants and then decrease the expression of *CaPR1*. The enhanced antagonism may also affect the expression of *CaPTI1* in turn.

CaPTI1 silenced pepper plants showed a significant decrease in the expression of *CaDEF1* (a JA-dependent gene) compare to the control. JA and ETH usually appear to function synergistically to induce *PDF1.2* (a JA-dependent gene) in *Arabidopsis*, and

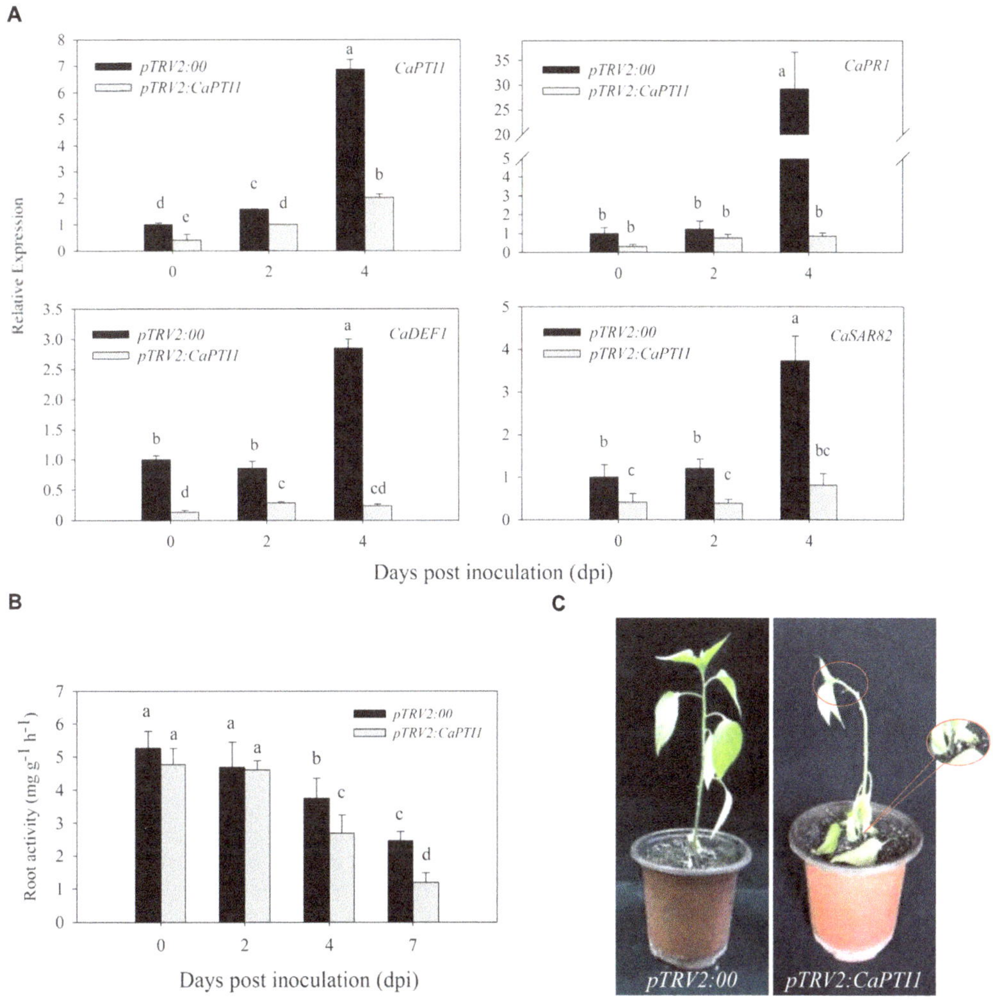

FIGURE 7 | The *CaPTI1* silenced pepper plants exhibit reduced resistance to *P. capsici*. (A) The expression of *CaPTI1*, *CaPR1*, *CaDEF1*, and *CaSAR82* in the root of *CaPTI1* silenced and control plants were analyzed; **(B)** Root activity in *CaPTI1* silenced and control plants after inoculation with *P. capsici*. Values are the means ± SD from three independent experiments. Small letters represent significant differences ($p < 0.05$); **(C)** Disease symptoms of control and silenced plants were observed at 7 day post inoculation (dpi).

osmotin and *PR1b* in tobacco (Xu et al., 1994; Penninckx et al., 1998; Norman-Setterblad et al., 2000). *CaDEF1* may act as the downstream of JA/ET pathway and be regulated by *CaPTI1* indirectly. The silencing of *CaPTI1* maybe also decreases the synergism between JA and ET by decreasing the ethylene production in pepper plants. *CaSAR82* serves as a molecular marker for the onset of ISR in pepper plants and could not only be induced by pathogens (*Xanthomonas campestris* pv. *vesicatoria*, *Colletotrichum coccodes*, and *Phytophthora capsici*), but also the signaling molecule (SA, JA, ETH and H_2O_2) and various stresses (Lee and Hwang, 2003). ISR was generally regulated by JA and ET. The significant decrease of *CaSAR82* in *CaPTI1* silenced pepper plants maybe indicated the systemic resistance mediated by JA/ET was weakened by the silencing of *CaPTI1*.

Besides a significant decrease of root activity was observed at 4 dpi in *CaPTI1* silenced plants as compared to control. The root activity which was measured by the TTC method was used as a cell vitality indicator (Chen et al., 2006). The cell vitality of root indicated the injury degree of pepper roots after the infection of *P. capsici.* Yeom et al. (2011) used this assay to measure TTC reductase activity in the roots of pepper cultivars CM334 (resistant to *P. capsici*) and Chilsungcho (susceptible to *P. capsici*) after inoculation with *P. capsici*. They found that significant differences were observed from 24 to 48 hpi, showing two or three times more activity in CM334 than in Chilsungcho. Moreover Wang et al. (2013b) also found that there is a significant difference in pepper plants after inoculation with the incompatible and compatible strains of *P. capsici*, and the root activity was much higher in the infection of incompatible strain than that of compatible strain. In this study, a significant decrease of root activity was observed at 4 dpi in *CaPTI1* silenced plants when compared to the control plants. This may indicate the serious injury to the roots of *CaPTI1* silenced pepper plants by *P. capsici*. Both of the significant decreased root activity and the reduced expression of defense related genes indicated the weakness of defense response in *CaPTI1* silenced pepper plants and suggested the significant role of *CaPTI1* in defense response to the infection of *P. capsici*.

Taken together, *CaPTI1* contains an AP2/ERF domain and belongs to the ERF family. *CaPTI1* could not only responses to the infection of *P. capsici* and the stresses of cold and drought, but also be induced by SA, MeJA, ETH and H_2O_2. VIGS of *CaPTI1* in pepper could weaken the defense response significantly by reducing the expression of defense related genes *CaPR1*, *CaDEF1* and *CaSAR82* and also the root activity. However, more detailed researches are still needed to reveal the exact role of *CaPTI1* in the regulation of defense response to *P. capsici* in pepper.

AUTHOR CONTRIBUTIONS

J-HJ, D-WL, and Z-HG conceived the research. J-HJ, H-XZ, J-YT, and M-JY performed the research. J-HJ and H-XZ performed statistical analyses. J-HJ wrote the paper. J-HJ, AK and Z-HG revised the paper. D-WL and Z-HG provided the materials and resources for the research. J-HJ, H-XZ, and Z-HG performed the integrity of the work. All authors read and approved the final manuscript.

ACKNOWLEDGMENTS

This work was supported by the National Natural Science Foundation of China (No. 31272163, 31401879), the Shaanxi Provincial Science and Technology Coordinating Innovative Engineering Project (2012KTCL02-09) and Northwest A&F University Basic Fund for Scientific Research (QN2013016).

REFERENCES

Alazem, M., and Lin, N. S. (2015). Roles of plant hormones in the regulation of host-virus interactions. *Mol. Plant Pathol.* 16, 529–540. doi: 10.1111/mpp.12204

Allen, M. D., Yamasaki, K., Ohme-Takagi, M., Tateno, M., and Suzuki, M. (1998). A novel mode of DNA recognition by a β-sheet revealed by the solution structure of the GCC-box binding domain in complex with DNA. *EMBO J.* 17, 5484–5496. doi: 10.1093/emboj/17.18.5484

Apel, K., and Hirt, H. (2004). Reactive oxygen species: metabolism, oxidative stress, and signal transduction. *Annu. Rev. Plant Biol.* 55, 373–399. doi: 10.1146/annurev.arplant.55.031903.141701

Babadoost, M., and Islam, S. Z. (2003). Fungicide seed treatment effects on seedling damping-off of pumpkin caused by *Phytophthora capsici*. *Plant Dis.* 87, 63–68. doi: 10.1094/Pdis.2003.87.1.63

Baysal, O., Turgut, C., and Mao, G. (2005). Acibenzolar-S-methyl induced resistance to *Phytophthora capsici* in pepper leaves. *Biol. Plant.* 49, 599–604. doi: 10.1007/s10535-005-0055-0

Brown, R. L., Kazan, K., Mcgrath, K. C., Maclean, D. J., and Manners, J. M. (2003). A role for the GCC-box in jasmonate-mediated activation of the PDF1.2 gene of *Arabidopsis*. *Plant Physiol.* 132, 1020–1032. doi: 10.1104/pp.102.017814

Chakravarthy, S., Tuori, R. P., D'Ascenzo, M. D., Fobert, P. R., Despres, C., and Martin, G. B. (2003). The tomato transcription factor Pti4 regulates defense-related gene expression via GCC box and non-GCC box cis elements. *Plant Cell* 15, 3033–3050. doi: 10.1105/Tpc.017574

Chen, C. W., Yang, Y. W., Lur, H. S., Tsai, Y. G., and Chang, M. C. (2006). A novel function of abscisic acid in the regulation of rice (*Oryza sativa* L.) root growth and development. *Plant Cell Physiol.* 47, 1–13. doi: 10.1093/pcp/pci216

Dangl, J. L., and Jones, J. D. G. (2001). Plant pathogens and integrated defence responses to infection. *Nature* 411, 826–833. doi: 10.1038/35081161

Djemal, R., and Khoudi, H. (2015). Isolation and molecular characterization of a novel WIN1/SHN1 ethylene-responsive transcription factor TdSHN1 from durum wheat (*Triticum turgidum*. L. subsp. durum). *Protoplasma* 252, 1461–1473. doi: 10.1007/s00709-015-0775-8

Do, H. M., Lee, S. C., Jung, H. W., Sohn, K. H., and Hwang, B. K. (2004). Differential expression and in situ localization of a pepper defensin (CADEF1) gene in response to pathogen infection, abiotic elicitors and environmental stresses in *Capsicum annuum*. *Plant Sci.* 166, 1297–1305. doi: 10.1016/j.plantsci.2004.01.008

Dong, L., Cheng, Y., Wu, J., Cheng, Q., Li, W., Fan, S., et al. (2015). Overexpression of GmERF5, a new member of the soybean EAR motif-containing ERF transcription factor, enhances resistance to *Phytophthora sojae* in soybean. *J. Exp. Bot.* 66, 2635–2647. doi: 10.1093/jxb/erv078

Dong, X. N. (1998). SA, JA, ethylene, and disease resistance in plants. *Curr. Opin. Plant Biol.* 1, 316–323. doi: 10.1016/1369-5266(88)80053-0

Fujimoto, S. Y., Ohta, M., Usui, A., Shinshi, H., and Ohme-Takagi, M. (2000). *Arabidopsis* ethylene-responsive element binding factors act as transcriptional activators or repressors of GCC box-mediated gene expression. *Plant Cell* 12, 393–404. doi: 10.1105/tpc.12.3.393

Glazebrook, J. (2005). Contrasting mechanisms of defense against biotrophic and necrotrophic pathogens. *Annu. Rev. Phytopathol.* 43, 205–227. doi: 10.1146/annurev.phyto.43.040204.135923

Granke, L. L., Quesada-Ocampo, L., Lamour, K., and Hausbeck, M. K. (2012). Advances in research on *Phytophthora capsici* on vegetable crops in the United States. *Plant Dis.* 96, 1588–1600. doi: 10.1094/Pdis-02-12-0211-Fe

Gu, Y. Q., Wildermuth, M. C., Chakravarthy, S., Loh, Y. T., Yang, C. M., He, X. H., et al. (2002). Tomato transcription factors Pti4, Pti5, and Pti6 activate defense responses when expressed in *Arabidopsis*. *Plant Cell* 14, 817–831. doi: 10.1105/Tpc.000794

Gu, Y. Q., Yang, C., Thara, V. K., Zhou, J., and Martin, G. B. (2000). Pti4 is induced by ethylene and salicylic acid, and its product is phosphorylated by the Pto kinase. *Plant Cell* 12, 771–785. doi: 10.1105/Tpc.12.5.771

Gutterson, N., and Reuber, T. L. (2004). Regulation of disease resistance pathways by AP2/ERF transcription factors. *Curr. Opin. Plant Biol.* 7, 465–471. doi: 10.1016/j.pbi.2004.04.007

Hausbeck, M. K., and Lamour, K. H. (2004). *Phytophthora capsici* on vegetable crops: research progress and management challenges. *Plant Dis.* 88, 1292–1303. doi: 10.1094/Pdis.2004.88.12.1292

He, P., Warren, R. F., Zhao, T., Shan, L., Zhu, L., Tang, X., et al. (2001). Overexpression of Pti5 in tomato potentiates pathogen-induced defense gene expression and enhances disease resistance to *Pseudomonas syringae* pv. tomato. *Mol. Plant Microbe. Interact.* 14, 1453–1457. doi: 10.1094/MPMI.2001.14.12.1453

Hong, J. P., and Kim, W. T. (2005). Isolation and functional characterization of the Ca-DREBLP1 gene encoding a dehydration-responsive element binding-factor-like protein 1 in hot pepper (*Capsicum annuum* L. cv. Pukang). *Planta* 220, 875–888. doi: 10.1007/s00425-004-1412-5

Kim, H. J., Nahm, S. H., Lee, H. R., Yoon, G. B., Kim, K. T., Kang, B. C., et al. (2008). BAC-derived markers converted from RFLP linked to *Phytophthora capsici* resistance in pepper (*Capsicum annuum* L.). *Theor. Appl. Gen.* 118, 15–27. doi: 10.1007/s00122-008-0873-5

Kim, S., An, C. S., Hong, Y. N., and Lee, K. W. (2004). Cold-inducible transcription factor, CaCBF, is associated with a homeodomain leucine zipper protein in hot pepper (*Capsicum annuum* L.). *Mol. Cells* 18, 300–308.

Kim, S., Park, M., Yeom, S. I., Kim, Y. M., Lee, J. M., Lee, H. A., et al. (2014). Genome sequence of the hot pepper provides insights into the evolution of pungency in *Capsicum* species. *Nat. Genet.* 46, 270–278. doi: 10.1038/Ng.2877

Kim, Y. J., and Hwang, B. K. (2000). Pepper gene encoding a basic pathogenesis-related 1 protein is pathogen and ethylene inducible. *Physiol. Plant.* 108, 51–60. doi: 10.1034/j.1399-3054.2000.108001051x./

Kousik, C. S., Adams, M. L., Jester, W. R., Hassell, R., Harrison, H. F., and Holmes, G. J. (2011). Effect of cultural practices and fungicides on *Phytophthora* fruit rot of watermelon in the Carolinas. *Crop Prot.* 30, 888–894. doi: 10.1016/j.cropro.2011.03.012

Lamour, K. H., Stam, R., Jupe, J., and Huitema, E. (2012). The oomycete broad-host-range pathogen *Phytophthora capsici*. *Mol. Plant Pathol.* 13, 329–337. doi: 10.1111/j.1364-3703.2011.00754.x

Lee, J. H., Hong, J. P., Oh, S. K., Lee, S., Choi, D., and Kim, W. T. (2004). The ethylene-responsive factor like protein 1 (CaERFLP1) of hot pepper (*Capsicum annuum* L.) interacts in vitro with both GCC and DRE/CRT sequences with different binding affinities: possible biological roles of CaERFLP1 in response to pathogen infection and high salinity conditions in transgenic tobacco plants. *Plant Mol. Biol.* 55, 61–81. doi: 10.1007/s11103-004-0417-6

Lee, S. C., and Hwang, B. K. (2003). Identification of the pepper SAR8.2 gene as a molecular marker for pathogen infection, abiotic elicitors and environmental stresses in *Capsicum annuum*. *Planta* 216, 387–396.

Livak, K. J., and Schmittgen, T. D. (2001). Analysis of relative gene expression data using real-time quantitative PCR and the 2-ΔΔCt method. *Methods* 25, 402–408. doi: 10.1006/meth.2001.1262

Loake, G., and Grant, M. (2007). Salicylic acid in plant defence-the players and protagonists. *Curr. Opin. Plant Biol.* 10, 466–472. doi: 10.1016/j.pbi.2007.08.008

Lorenzo, O., Piqueras, R., Sanchez-Serrano, J. J., and Solano, R. (2003). ETHYLENE RESPONSE FACTOR1 integrates signals from ethylene and jasmonate pathways in plant defense. *Plant Cell* 15, 165–178. doi: 10.1105/Tpc.007468

Matheron, M. E., and Porchas, H. (2000). Comparison of five fungicides on development of root, crown, and fruit rot of chile pepper and recovery of *Phytophthora capsici* from soil. *Plant Dis.* 84, 1038–1043. doi: 10.1094/Pdis.2000.84.9.1038

McGrath, K. C., Dombrecht, B., Manners, J. M., Schenk, P. M., Edgar, C. I., Maclean, D. J., et al. (2005). Repressor- and activator-type ethylene response factors functioning in jasmonate signaling and disease resistance identified via a genome-wide screen of *Arabidopsis* transcription factor gene expression. *Plant Physiol.* 139, 949–959. doi: 10.1104/pp.105.068544

Mittler, R., Vanderauwera, S., Suzuki, N., Miller, G., Tognetti, V. B., Vandepoele, K., et al. (2011). ROS signaling: the new wave? *Trends Plant Sci.* 16, 300–309. doi: 10.1016/j.tplants.2011.03.007

Mizoi, J., Shinozaki, K., and Yamaguchi-Shinozaki, K. (2012). AP2/ERF family transcription factors in plant abiotic stress responses. *Biochim. Et Biophys. Acta-Gene Regul. Mech.* 1819, 86–96. doi: 10.1016/j.bbagrm.2011.08.004

Norman-Setterblad, C., Vidal, S., and Palva, E. T. (2000). Interacting signal pathways control defense gene expression in *Arabidopsis* in response to cell wall-degrading enzymes from *Erwinia carotovora*. *Mol. Plant Microbe Interact.* 13, 430–438. doi: 10.1094/Mpmi.2000.13.4.430

Oelke, L. M., Bosland, P. W., and Steiner, R. (2003). Differentiation of race specific resistance to *Phytophthora* root rot and foliar blight in *Capsicum annuum*. *J. Am. Soc. Hortic. Sci.* 128, 213–218.

Ohmetakagi, M., and Shinshi, H. (1995). Ethylene-inducible DNA-binding proteins that interact with an ethylene-responsive element. *Plant Cell* 7, 173–182. doi: 10.2307/3869993

Ohta, M., Matsui, K., Hiratsu, K., Shinshi, H., and Ohme-Takagi, M. (2001). Repression domains of class II ERF transcriptional repressors share an essential motif for active repression. *Plant Cell* 13, 1959–1968. doi: 10.1105/tpc.13.8.1959

Onate-Sanchez, L., and Singh, K. B. (2002). Identification of *Arabidopsis* ethylene-responsive element binding factors with distinct induction kinetics after pathogen infection. *Plant Physiol.* 128, 1313–1322. doi: 10.1104/Pp.010862

Ou, L. J., Dai, X. Z., Zhang, Z. Q., and Zou, X. X. (2011). Responses of pepper to waterlogging stress. *Photosynthetica* 49, 339–345. doi: 10.1007/s11099-011-0043-x

Park, J. M., Park, C. J., Lee, S. B., Ham, B. K., Shin, R., and Paek, K. H. (2001). Overexpression of the tobacco Tsi1 gene encoding an EREBP/AP2-Type transcription factor enhances resistance against pathogen attack and osmotic stress in tobacco. *Plant Cell* 13, 1035–1046. doi: 10.1105/tpc.13.5.1035

Penninckx, I. A. M. A., Thomma, B. P. H. J., Buchala, A., Metraux, J. P., and Broekaert, W. F. (1998). Concomitant activation of jasmonate and ethylene response pathways is required for induction of a plant defensin gene in *Arabidopsis*. *Plant Cell* 10, 2103–2113. doi: 10.1105/tpc.10.12.2103

Pieterse, C. M. J., and van Loon, L. C. (1999). Salicylic acid-independent plant defence pathways. *Trends Plant Sci.* 4, 52–58. doi: 10.1016/S1360-1385(98)01364-8

Rehrig, W. Z., Ashrafi, H., Hill, T., Prince, J., and Van Deynze, A. (2014). *CaDMR1* Cosegregates with QTL *Pc5.1* for resistance to *Phytophthora capsici* in pepper (*Capsicum annuum*). *Plant Genome* 7, 1–12. doi: 10.3835/plantgenome2014.03.0011

Reymond, P., and Farmer, E. E. (1998). Jasmonate and salicylate as global signals for defense gene expression. *Curr. Opin. Plant Biol.* 1, 404–411. doi: 10.1016/S1369-5266(98)80264-1

Ryu, C. M., Anand, A., Kang, L., and Mysore, K. S. (2006). Agrodrench: a novel and effective agroinoculation method for virus-induced silencing in roots and diverse Solanaceous species. (vol 40, pg 322, 2004). *Plant J.* 45, 869–869.

Sakuma, Y., Liu, Q., Dubouzet, J. G., Abe, H., Shinozaki, K., and Yamaguchi-Shinozaki, K. (2002). DNA-binding specificity of the ERF/AP2 domain of *Arabidopsis* DREBs, transcription factors involved in dehydration- and cold-inducible gene expression. *Biochem. Biophys. Res. Commun.* 290, 998–1009. doi: 10.1006/bbrc.2001.6299

Sewelam, N., Kazan, K., Thomas-Hall, S. R., Kidd, B. N., Manners, J. M., and Schenk, P. M. (2013). Ethylene response factor 6 is a regulator of reactive oxygen species signaling in *Arabidopsis*. *PLoS ONE* 8:e70289. doi: 10.1371/journal.pone.0070289

Shin, R., Park, J. M., An, J. M., and Paek, K. H. (2002). Ectopic expression of Tsi1 in transgenic hot pepper plants enhances host resistance to viral, bacterial, and oomycete pathogens. *Mol. Plant Microbe Interact.* 15, 983–989. doi: 10.1094/Mpmi.2002.15.10.983

Thabuis, A., Palloix, A., Pflieger, S., Daubeze, A. M., Caranta, C., and Lefebvre, V. (2003). Comparative mapping of *Phytophthora* resistance loci in pepper germplasm: evidence for conserved resistance loci across Solanaceae and for a large genetic diversity. *Theor. Appl. Gen.* 106, 1473–1485.

Thara, V. K., Tang, X. Y., Gu, Y. Q., Martin, G. B., and Zhou, J. M. (1999). *Pseudomonas* syringae pv tomato induces the expression of tomato EREBP-like genes Pti4 and Pti5 independent of ethylene, salicylate and jasmonate. *Plant J.* 20, 475–483. doi: 10.1046/j.1365-313x.1999.00619.x

Tian, Z., He, Q., Wang, H., Liu, Y., Zhang, Y., Shao, F., et al. (2015). The potato ERF transcription factor StERF3 negatively regulates resistance to *Phytophthora* infestans and salt tolerance in potato. *Plant Cell Physiol.* 56, 992–1005. doi: 10.1093/pcp/pcv025

Velasquez, A. C., Chakravarthy, S., and Martin, G. B. (2009). Virus-induced gene silencing (VIGS) in *Nicotiana benthamiana* and tomato. *J. Vis. Exp.* 2009:1292. doi: 10.3791/1292

Vlot, A. C., Dempsey, D. A., and Klessig, D. F. (2009). Salicylic acid, a multifaceted hormone to combat disease. *Annu. Rev. Phytopathol.* 47, 177–206. doi: 10.1146/annurev.phyto.050908.135202

Wan, H. J., Yuan, W., Ruan, M. Y., Ye, Q. J., Wang, R. Q., Li, Z. M., et al. (2011). Identification of reference genes for reverse transcription quantitative real-time PCR normalization in pepper (*Capsicum annuum* L.). *Biochem. Biophys. Res. Commun.* 416, 24–30. doi: 10.1016/j.bbrc.2011.10.105

Wang, H., Huang, Z., Chen, Q., Zhang, Z., Zhang, H., Wu, Y., et al. (2004). Ectopic overexpression of tomato JERF3 in tobacco activates downstream gene expression and enhances salt tolerance. *Plant Mol. Biol.* 55, 183–192. doi: 10.1007/s11103-004-0113-6

Wang, J. E., Li, D. W., Gong, Z. H., and Zhang, Y. L. (2013a). Optimization of virus-induced gene silencing in pepper (*Capsicum annuum* L.). *Genet. Mol. Res.* 12, 2492–2506. doi: 10.4238/2013.July.24.4

Wang, J. E., Li, D. W., Zhang, Y. L., Zhao, Q., He, Y. M., and Gong, Z. H. (2013b). Defence responses of pepper (*Capsicum annuum* L.) infected with incompatible and compatible strains of *Phytophthora capsici*. *Eur. J. Plant Pathol.* 136, 625–638. doi: 10.1007/s10658-013-0193-8

Wang, J. E., Liu, K. K., Li, D. W., Zhang, Y. L., Zhao, Q., He, Y. M., et al. (2013c). A novel peroxidase CanPOD gene of pepper is involved in defense responses to phytophtora capsici infection as well as abiotic stress tolerance. *Int. J. Mol. Sci.* 14, 3158–3177. doi: 10.3390/ijms14023158

Wang, Z. M., Zhang, N., Zhou, X. Y., Fan, Q., Si, H. J., and Wang, D. (2015). Isolation and characterization of StERF transcription factor genes from potato (*Solanum tuberosum* L.). *C. R. Biol.* 338, 219–226. doi: 10.1016/j.crvi.2015.01.006

Wessler, S. R. (2005). Homing into the origin of the AP2 DNA binding domain. *Trends Plant Sci.* 10, 54–56. doi: 10.1016/j.tplants.2004.12.007

Wu, C. J., Avila, C. A., and Goggin, F. L. (2015). The ethylene response factor Pti5 contributes to potato aphid resistance in tomato independent of ethylene signalling. *J. Exp. Bot.* 66, 559–570. doi: 10.1093/Jxb/Eru472

Xu, Y., Chang, P. F. L., Liu, D., Narasimhan, M. L., Raghothama, K. G., Hasegawa, P. M., et al. (1994). Plant defense genes are synergistically induced by ethylene and methyl jasmonate. *Plant Cell* 6, 1077–1085. doi: 10.1105/Tpc.6.8.1077

Yang, Z., Tian, L. N., Latoszek-Green, M., Brown, D., and Wu, K. Q. (2005). *Arabidopsis* ERF4 is a transcriptional repressor capable of modulating ethylene and abscisic acid responses. *Plant Mol. Biol.* 58, 585–596. doi: 10.1007/s11103-005-7294-5

Yeom, S. I., Baek, H. K., Oh, S. K., Kang, W. H., Lee, S. J., Lee, J. M., et al. (2011). Use of a secretion trap screen in pepper following *Phytophthora capsici* infection reveals novel functions of secreted plant proteins in modulating cell death. *Mol. Plant-Microbe Interact.* 24, 671–684. doi: 10.1094/MPMI-08-10-0183

Zarei, A., Korbes, A. P., Younessi, P., Montiel, G., Champion, A., and Memelink, J. (2011). Two GCC boxes and AP2/ERF-domain transcription factor ORA59 in jasmonate/ethylene-mediated activation of the PDF1.2 promoter in *Arabidopsis. Plant Mol. Biol.* 75, 321–331. doi: 10.1007/s11103-010-9728-y

Zhang, Y. L., Jia, Q. L., Li, D. W., Wang, J. E., Yin, Y. X., and Gong, Z. H. (2013). Characteristic of the pepper CaRGA2 gene in defense responses against *Phytophthora capsici* leonian. *Int. J. Mol. Sci.* 14, 8985–9004. doi: 10.3390/ijms14058985

Zhou, J. M., Tang, X. Y., and Martin, G. B. (1997). The Pto kinase conferring resistance to tomato bacterial speck disease interacts with proteins that bind a cis-element of pathogenesis-related genes. *Embo J.* 16, 3207–3218. doi: 10.1093/emboj/16.11.3207

Conflict of Interest Statement: The authors declare that the research was conducted in the absence of any commercial or financial relationships that could be construed as a potential conflict of interest.

15

Transcriptome Changes in *Hirschfeldia incana* in Response to Lead Exposure

*Florence Auguy[1], Mouna Fahr[2], Patricia Moulin[3], Mohamed El Mzibri[2], Abdelaziz Smouni[4], Abdelkarim Filali-Maltouf[5], Gilles Béna[6] and Patrick Doumas[7]**

[1] Institut de Recherche pour le Développement, UMR DIADE, Equipe Rhizogenèse, Montpellier, France, [2] Centre National de l'Energie, des Sciences et des Techniques Nucléaires, Laboratoire de Biotechnologie des Plantes, UBRM-DSV, Rabat, Morocco, [3] Institut de Recherche pour le Développement, Laboratoire de Microbiologie et Biologie Moléculaire, Faculté des Sciences, Université Mohammed V-Rabat, Rabat, Morocco, [4] Laboratoire de Physiologie et Biotechnologie Végétale, Faculté des Sciences, Université Mohammed V-Rabat, Rabat, Morocco, [5] Laboratoire de Microbiologie et Biologie Moléculaire, Faculté des Sciences, Université Mohammed V-Rabat, Rabat, Morocco, [6] Institut de Recherche pour le Développement, UMR IPME, Equipe ABIP, Montpellier, France, [7] Institut National de la Recherche Agronomique, UMR Biochimie et Physiologie Moléculaire des Plantes, Montpellier, France

Edited by:
Shabir Hussain Wani, Sher-e-Kashmir Univeristy of Agricultural Sciences and Technology of Kashmir, India

Reviewed by:
Marc Hanikenne, University of Liège, Belgium
Anna Kozhevnikova, Timiryazev Institute of Plant Physiology, Russia

***Correspondence:**
Patrick Doumas
patrick.doumas@supagro.inra.fr

Hirschfeldia incana, a pseudometallophyte belonging to the *Brassicaceae* family and widespread in the Mediterranean region, was selected for its ability to grow on soils contaminated by lead (Pb). The global comparison of gene expression using microarrays between a plant susceptible to Pb (*Arabidopsis thaliana*) and a Pb tolerant plant (*H. incana*) enabled the identification of a set of specific genes expressed in response to lead exposure. Three groups of genes were particularly over-represented by the Pb exposure in the biological processes categorized as photosynthesis, cell wall, and metal handling. Each of these gene groups was shown to be directly involved in tolerance or in protection mechanisms to the phytotoxicity associated with Pb. Among these genes, we demonstrated that *MT2b*, a metallothionein gene, was involved in lead accumulation, confirming the important role of metallothioneins in the accumulation and the distribution of Pb in leaves. On the other hand, several genes involved in biosynthesis of ABA were shown to be up-regulated in the roots and shoots of *H. incana* treated with Pb, suggesting that ABA-mediated signaling is a possible mechanism in response to Pb treatment in *H. incana*. This latest finding is an important research direction for future studies.

Keywords: *Hirschfeldia incana*, *Arabidopsis thaliana*, Brassicaceae, functional genomics, lead tolerance, transcriptome

INTRODUCTION

Heavy metals including Cu, Mn, or Zn, play an essential role in many plant physiological processes but can be toxic if accumulated at high concentrations. Other metals including Cd, As, or Pb, have no biological functions and are extremely toxic even at low concentrations. Lead, a non-essential heavy metal, is widespread in the environment as a result of many human activities including mining and smelting, burning coal, cement manufacture and agricultural practices. It is widely accepted that lead is a serious pollutant, and is toxic not only for plant roots where Pb is taken up, but also when translocated to aboveground part where it can accumulate

(Sterckeman et al., 2006). Numerous previous studies have reported a wide range of negative effects following plant exposure to Pb; the visible symptoms of Pb toxicity are rapid inhibition of seed germination, reduced growth and the appearance of chlorosis. In addition, Pb can cause oxidative damage by stimulating the formation of free radicals and reactive oxygen species (ROS), resulting in oxidative stress and DNA damage (Seregin and Ivanov, 2001; Schutzendubel and Polle, 2002). However, some plant species tolerate the presence of Pb and, more interestingly, several species, including *Noccaea caerulescens, Agrostis tennuis,* or *Festuca ovina*, have developed the capacity to accumulate large amounts of lead in their root tissues with limited translocation to the aboveground parts (Vangronsveld et al., 2009). *Hirschfeldia incana*, commonly known as buchanweed, hoary mustard, or Mediterranean mustard, is a medium perennial shrub belonging to the *Brassicaceae* family. It has been identified as a potential hyperaccumulator of various toxic metals including lead (Auguy et al., 2013). Interestingly, *H. incana* was shown to be capable of accumulating more than 3% (dry weight) of lead in its shoots when grown in hydroponic solution containing 100 μM $Pb(NO_3)_2$ without any marked symptoms of toxicity. Such plant species have evolved physiological and molecular mechanisms to enable them to thrive under Pb metal stress. Many studies have already been published on the effects of lead on plants and on their cellular detoxification mechanisms and several molecular players in the Pb homeostatic network have been identified so far. For instance, the tobacco plasma membrane protein NtCBP4 and the *Arabidopsis* gene *CNGC1* have been shown to be components of a transport pathway responsible for Pb entry into plant cells (Sunkar et al., 2000). An *Arabidopsis* P-type ATPase, HMA3, has been reported to improve tolerance by sequestering Pb in vacuoles (Gravot et al., 2004; Morel et al., 2009). It has also been suggested that HMA4, another P-type ATPase, may play a role in Pb efflux in *N. caerulescens* and *H. incana* (Papoyan and Kochian, 2004; Auguy et al., 2013). In *Arabidopsis*, three members of the ABC (ATPase-binding cassette) transporter family AtATM3, AtPDR12, and AtPDR8 contribute to Pb resistance (Lee et al., 2005; Kim et al., 2006, 2007). ACBP1, an acyl-CoA-binding protein, has been reported to be involved in mediating Pb tolerance through accumulation in shoots (Xiao et al., 2008), and *AtMRP3* transcription has been shown to be strongly induced by Pb treatment in *Arabidopsis thaliana* (Zientara et al., 2009). Finally, chelation is an important mechanism controlling heavy metal tolerance involving small molecules such as metallothioneins, phytochelatins and glutathione (Cobbett and Goldsbrough, 2002).

Recent research has focused on identifying the mechanisms that allow organisms to adapt to or alleviate damage caused by metal stress, and in this context, microarray technology is a convenient tool for rapid analysis of plant gene expression patterns under a variety of environmental conditions. The use of cross-species hybridization (CSH), in which the target RNA and microarray probe come from different species, has increased in the last few years. CSH has been used to examine both closely and distantly related species. In the particular case of heavy metals, Becher et al. (2004) hybridized *A. halleri* RNA samples to *A. thaliana* Affymetrix microarrays and analyzed the effect of Zn treatments on gene expression. Similarly, using the same probe-transcript complex, Weber et al. (2006) analyzed the gene profiles after exposure to Cd. In the same line of thought, Hammond et al. (2006) compared Zn gene regulation by hybridizing *N. cearulescens* and *Thlaspi arvense* RNA to *A. thaliana* Affymetrix microarrays. Van de Mortel et al. (2006) examined the performance of CSH in cDNA microarrays of *A. thaliana* with RNA from *N. cearulescens* treated or not with excess Zn. *H. incana,* like *A. thaliana,* belongs to the *Brassicaceae* family. Based on eight cDNA sequences, from 400 to 1100 bp in length, cloned, and sequenced in a previous work, we estimated that *A. thaliana* and *H. incana* shared an average of 89% nucleotide identity within coding regions (Auguy et al., 2013). This level of identity is similar to those obtained between *A. thaliana* and *N. caerulescens* comparison, with on average 88.5% DNA identity in coding regions (Rigola et al., 2006) and 87% DNA identity in the intergenic transcribed spacer regions (Peer et al., 2003). Based on these data, we hypothesized that the use of *Arabidopsis* cDNA microarrays with RNA from *H. incana* should work and would provide information on changes in the transcriptome that occur during plant exposure to Pb. We performed transcriptome analysis of Pb responses in *A. thaliana* and *H. incana* roots and shoots by comparing the responses of a normal plant, *A. thaliana*, and a lead-hyperaccumulating metallophyte, *H. incana* (Auguy et al., 2013). Our aim was to identify any differential responses to Pb that would advance our understanding of metal-tolerance mechanisms. Based on the assumption that the metal responsiveness of putative signal-transduction components points to the functional involvement of the encoded proteins in mediating metal responses, we also hoped to identify candidate genes that could be analyzed further using reverse-genetic approaches.

MATERIALS AND METHODS

Plant Material and RNA Extraction

Four weeks old *H. incana* and *A. thaliana* seedlings were treated in hydroponic cultures with Pb for 3 days, by adding 100 or 40 μM $Pb(NO_3)_2$, respectively, in a fresh BD medium (Broughton and Dilworth, 1971) without phosphate. These two Pb concentrations correspond to approximately 50% of reduction of primary root growth for both species without any visible symptoms (data not shown). In control experiments, BD medium was replaced by fresh BD medium without phosphate.

RNA was extracted separately from the shoots and roots of the control and of the treated *H. incana* and *A. thaliana* seedlings. Total RNA extraction was performed with two different kits depending on the tissue concerned, RNeasy plant mini kit (Qiagen, USA) for roots and SV Total RNA isolation system (Promega, USA) for shoots. A post-treatment with Turbo DNA free (Ambion, USA) removed contaminated DNA. RNA quality was checked and confirmed using Nanodrop 2000 (Thermo Scientific, USA) analysis and non-denaturing gel electrophoresis. Three independent experiments were performed with 8–9 seedlings per treatment and per repetition and used for both microarray analysis and quantitative reverse transcriptase PCR (QRT-PCR) validation.

Amplification, Labeling, and Hybridization

Imaxio (France), a service provider accredited by Agilent Technologies (USA), performed amplification, labeling, hybridization, data capture, and primary analysis. The Agilent 44K *A. thaliana* microarrays were used for both species. The array, which comprised 43,803 probes, covered the entire *A. thaliana* transcriptome (27 235 cDNA). Two hundred ng of total RNA were used to synthesize cRNA, incorporating the Cy-3 for the control samples and the Cy-5 for the Pb treated samples. After amplification, 825 ng of fragmented cRNA were hybridized to each array at 60°C for 17 h following the manufacturer's instructions.

Microarray Data Processing and Analysis

Images were acquired with the Agilent G2505C scanner and treated with the Feature Extraction Agilent software (version 10.7.3). Two normalizations were applied: first, we applied quartile normalization to minimize variance between chips, and second, probe normalization using the median calculated for all the samples. A Gaussian distribution was obtained with the LOG transformation. Only probes for which all replicates for one condition had values higher than the background were retained. Finally, a Student's *t*-test between treatments was applied with the following parameters: asymptotic *p*-value, p-value < 0.05 and Benjamini–Hochberg correction. Microarray data from this article were submitted to the public NCBI Gene Expression Omnibus database (GEO) accession GSE65334 (Edgar et al., 2002). The microarray data is MIAME compliant.

Functional classifications of statistically significant general stress genes were obtained using the web-based Functional Classification SuperViewer (Provart and Zhu, 2003). This classification tool is based on functional information available from the Munich Information Center for Protein Sequences (MIPS) database (Schoof et al., 2004). Class scores were obtained to determine whether certain functional classes were over-represented among statistically significant general stress genes (Provart and Zhu, 2003). Class score means and standard errors were computed on the basis of 100 bootstrap samples of the input gene list, as described in Provart and Zhu (2003). It should be noted that with SuperViewer, genes can belong to more than one classification. Genevestigator software[1] was used to compare the expression profiles of selected genes with other expression profiles provided in the database.

QRT-PCR Validation

To validate the microarray results, the expression profile of a subset of differentially regulated genes was checked by qRT-PCR in *H. incana* and *A. thaliana*. Total RNA was extracted in both species in the same way as described for the microarray. Primer sets were designed with Primer 3 software[2] (Additional File 1 in Supplementary Material) from conserved gene sequences obtained from the *Arabidopsis* Information Resource database (TAIR[3]). QRT-PCR uses the method already described by Auguy et al. (2013). Dissociation (melting) curves for all the genes were checked for the presence of primer dimers or spurious and not specific products. The correlation coefficients between qRT-PCR and microarray values were calculated.

Identification of *Arabidopsis mt2b* T-DNA Insertion Mutants

Multiple alignments of nucleotide sequences revealed high sequence identity between *HiMT2b* and *AtMT2b*. In addition, phylogenic trees placed *HiMT2b* as a putative ortholog of *AtMT2b* (unpublished data). Consequently, *Arabidopsis* homozygote plants of T-DNA insertion lines Salk_144899 and Salk_037601 for the gene At5g02380 (*AtMT2b*) were identified by PCR using three different primers. In both T-DNA insertion lines, T-DNA specific primer (LBb1, 5′-GCGTGGACCGCTTGCTCAACT-3′) and two *AtMT2b* specific primers (AtMT2b-F, 5′-GATCCACAACCACAGCTTCC-3′ and AtMT2b-R, 5′-GGACAAAGATCGTTGACAGC-3′) were used. The genotype of the F3 individuals was checked by PCR using gene-specific primers and T-DNA primers. Individual homozygous mutants were backcrossed twice with *Arabidopsis* wild-type Col-0. Root growth and Pb content analysis were performed on *mt2b* mutants and Col-0 wild-type after 13 days culture on media with or without 40 μM $Pb(NO_3)_2$ as previously described (Auguy et al., 2013).

Lead Quantification

Roots and shoots of both *H. incana* and *A. thaliana* seedlings treated or not with Pb were washed at 4°C twice with 0.2 mM $CaSO_4$ and rinsed with cold distilled water. Samples were dried at 72°C for at least 7 days. The dried tissues were subjected to acid hydrolysis and the concentration of lead in the samples was determined by inductively coupled plasma-atomic emission (ICP AES Ultima 2JY, USA) according to the methods previously described (Temminghoff and Houba, 2004; Margui et al., 2007).

RESULTS AND DISCUSSION

Differences in Pb Content Between *H. incana* and *A. thaliana*

Optimal Pb concentrations for *H. incana* and *A. thaliana* were chosen based on the results we obtained in a previous experiment (Auguy et al., 2013). Pb content was measured in the roots and shoots of *H. incana* and *A. thaliana* cultivated in hydroponic conditions 3 days after treatment with, respectively, 100 and 40 μM Pb $(NO_3)_2$ (**Table 1**). The amount of Pb in roots was of the same order of magnitude in both the two species and showed very high accumulation [65.5 and 82.9 mg Pb/g DW (dry weight), respectively, in *H. incana* and *A. thaliana*]. In contrast, a marked

[1] https://genevestigator.com/gv/

[2] http://primer3.wi.mit.edu/

[3] http://www.arabidopsis.org/

TABLE 1 | Pb content in *Hirschfeldia incana* and *Arabidopsis thaliana* roots and shoots.

	Roots	Shoots
H. incana	65.5 ± 6.2	0.346 ± 0.034
A. thaliana	82.9 ± 38.6	0.006 ± 0.005

Pb content was measured 3 days after treatment with 100 and 40 μM de $Pb(NO_3)_2$, respectively. Data are expressed in Pb $mg.g^{-1}$ DW and are the average (±SE) of three independent measurements each made on 5–7 plants.

TABLE 2 | Number of regulated genes in roots and shoots of *A. thaliana* and *H. incana* treated with Pb relative to untreated plants.

	Roots		Shoots	
	Up-regulated	Down-regulated	Up-regulated	Down-regulated
H. incana	795	648	369	296
A. thaliana	6737	7095	407	561

difference in Pb content was measured in the shoots of the two species: 0.346 mg Pb/g DW in *H. incana* and 0.006 mg Pb/g DW in *A. thaliana*. As previously reported [5], Pb transfer from roots to shoots was observed in *H. incana* but not in *A. thaliana* in which almost all Pb only accumulated in the roots. This result confirms that *A. thaliana* has developed a metal exclusion strategy to cope with metal toxicity in the soil by limiting the amount of Pb translocated from the roots to the shoots (Baker, 1981). In contrast, *H. incana* was previously described to tolerate and accumulate Cu, Zn, Tl, and Pb (Poschenrieder et al., 2001; Gisbert et al., 2006; Madejon et al., 2007; Auguy et al., 2013), and indeed, in our experiment, *Hirschfeldia* plants accumulated significantly higher amounts of Pb in aboveground shoots than *A. thaliana* (**Table 1**). This significant difference between the two species confirmed the rightness of our choice of comparing their transcriptome in response to Pb treatment. The response of the non-tolerant and non-accumulating species *A. thaliana* to Pb was considered to be the standard plant response to Pb and was used as a filter to keep only the specific response to Pb in the tolerant Pb accumulating plant *H. incana*.

Overview of the Transcriptome Response to Pb Treatment

In order to understand the molecular events underlying the response of *H. incana* to Pb exposure, we measured gene expression in the shoots and roots of *H. incana* plants treated or not with 100 μM of $Pb(NO_3)_2$ for 3 days, using the Agilent 44k *Arabidopsis* array. In parallel, to evaluate differential genes expressed in a Pb tolerant plant versus a non-tolerant plant, we investigated the transcriptome response of *A. thaliana* treated or not with 40 μM of $Pb(NO_3)_2$ for 3 days also using the Agilent 44k *Arabidopsis* array.

In this study, the number of Pb-regulated genes was 2,108 for *H. incana* and 14,800 for *A. thaliana*, with a false detection rate (FDR) < 0.1. Additional Files 2–5 in Supplementary Material list the 20 most Pb-regulated genes in the roots and shoots of *H. incana* and *A. thaliana*. A total of 68.5% of Pb-regulated genes were distributed in the roots and 31.5% in the shoots of *H. incana*, and 93.5% in the roots and 6.5% in the shoots of *A. thaliana* (**Table 2**). The ratio of up- and down-regulated genes in each type of tissue was approximately the same and reached 50%. The marked difference between the number of Pb-regulated genes in *H. incana* and *A. thaliana* could be due to the CHS technique used. Even if *H. incana* and *A. thaliana* are closely related species, it cannot be excluded that some *H. incana* sequences diverged too much from *A. thaliana* sequences and consequently prevented hybridization on the *A. thaliana* arrays. On the other hand, the small number of Pb-regulated genes detected in *Arabidopsis* shoots could be related to the weak concentration of Pb measured in this tissue after 3 days of treatment with 40 μM of $Pb(NO_3)_2$ (**Table 1**).

To validate our microarray results, we performed qRT-PCR to determine the levels of expression of 33 *H. incana* and 6 *A. thaliana* genes randomly selected from the list of genes differentially expressed after Pb treatments. In *H. incana*, the qRT-PCR gene expression results are highly correlated with the microarray gene expression data (**Figure 1**). The coefficient of determination (r_2) between microarray and qRT-PCR data was 0.92, indicating good consistency between the two methods. The genes we tested varied in the same way in the microarray and the qRT-PCR results (Additional File 6 in Supplementary Material). This clearly validates the transcriptomic results obtained in *H. incana* with Agilent microarray. In *A. thaliana*, in a similar way but with fewer genes, qRT-PCR data were similar to those obtained with the microarray (Additional File 7 in Supplementary Material), thus validating the *A. thaliana* results obtained with the same microarray.

Genes Specifically Regulated by Pb Treatment in *H. incana*

In order to highlight genes that were specifically regulated by Pb exposure in *H. incana*, we compared its transcriptome with the transcriptome of the Pb-sensitive plant *A. thaliana*. We focused particularly on genes that were specifically Pb-regulated in *H. incana* roots and shoots, (Pb-specific genes), and genes that were regulated in both species (Pb-common genes). These comparisons are illustrated by Venn diagrams (**Figure 2**). The number of Pb-specific genes was higher in shoots than in roots, with 602 and 341 genes, respectively, suggesting a stronger acclimation response by the shoots. Conversely, the number of Pb-common genes was much higher in the roots, with 1,102 Pb-regulated genes in comparison to only 63 genes in the shoots (**Figure 2**). The distribution of up- and down-regulated genes was similar in shoots and roots (55% up-regulated genes). In the Pb-common gene group, we distinguished four categories: genes up-regulated in both roots and shoots, genes down-regulated in both roots and shoots, genes up-regulated in shoots and down-regulated in roots, and genes up-regulated in roots and down-regulated in shoots. The distribution of genes in these four categories was similar in the roots and shoots (**Figure 2**).

In the Pb-specific gene group, only regulated genes with a twofold change (FC) in gene expression were retained. They represented 121genes in roots and 331 in shoots (**Table 3**).

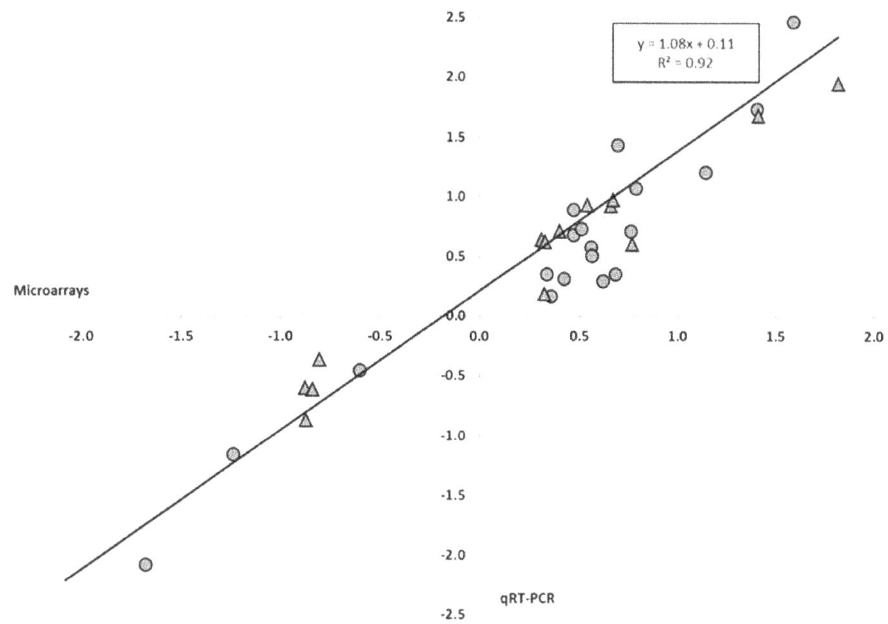

FIGURE 1 | Pearson's correlation between gene expression levels determined by quantitative reverse transcriptase PCR (qRT-PCR) and microarrays for *Hirschfeldia incana*. The diagram represents the correlation between gene expression obtained with the microarray and qRT-PCR for *H. incana* plantlets treated with 100 μM $Pb(NO_3)_2$ for 3 days. Values are Log10 of the fold change (FC) obtained for the same genes with the microarray (*x* axis) and qRT-PCR (*y* axis). FC values for roots (circles) and shoots (triangles) are reported in Additional File 6 in Supplementary Material.

For the Pb-common gene category, we introduced another selection step involving a comparison between *H. incana* and *A. thaliana*: only genes differentially expressed in *H. incana* with a FC *H. incana*/FC *A. thaliana* > 2.0 or < 0.5 were retained. This category represented 315 genes in roots and 28 in shoots (**Table 3**). Within this differentially expressed gene category, in the same way as Pb-specific genes, only regulated genes (those with a FC > 2.0 or FC < 0.5) were selected. After this second step, the Pb-common gene list included 227 genes in roots and 26 in shoots (**Table 3**). Finally, including specific and common genes, the selected gene lists comprised 348 genes in roots and 356 genes for roots and in shoots (**Table 3**, Additional Files 8 and 9 in Supplementary Material). Among these, only 42 genes were regulated both in roots and shoots (in bold in Additional Files 8 and 9 in Supplementary Material). As mentioned in the introduction, several genes were already known to be involved in plant tolerant response to Pb exposure including *ACBP1*, *AtATM3*, *NtCBP4*, *CNGC1*, *HMA3*, *HMA4*, *AtMRP3*, *AtPDR8*, and *AtPDR12*. Contrary to the data in the literature, no hybridization signal was measurable for *ACPB1*, *MRP3*, *PRD8*, and *PRD12* and a FC less than 2 or 0.5 was measurable for *ATM3*, *CBP4*, *CNGC1*, *HMA3*, and *HMA4*. This particular focus on these Pb-specific genes illustrates the limits of the DNA CSH technique.

Functional Classification of Pb-Responsive Genes

To evaluate the functional significance of the differentially expressed genes in response to Pb exposure in *H. incana*, the biological processes with over-represented regulated genes were identified in our analysis using Classification SuperViewer tools (Provart and Zhu, 2003) available at the Bio-Array Resource for Plant Functional Genomics (BAR) website with MapMan classification source. Several biological processes, categorized as photosynthesis, cell wall, and metal handling, were particularly enriched by Pb exposure in both roots and shoots (**Figure 3**).

Photosynthesis is considered to be one of the most sensitive metabolic processes to Pb toxicity (Singh et al., 1997). Lead toxicity has multifunctional adverse effects on photosynthetic CO_2 fixation and reduces the level of photosynthetic pigments, alters chloroplast ultra-structure and reduces the enzymatic activity of CO_2 assimilation (Parys et al., 1998; Islam et al., 2008; Belatik et al., 2013). In the present study, several genes directly involved in photosystem II encoding, respectively, a light-harvesting chlorophyll-a/b (Lhc) protein (At2g05070), a Mogl/PsbP/DUF1795-like protein (At3g56650), a subunit of the NAD(P)H dehydrogenase complex (At1g14150), a PsbP-like protein 1 (At3g55330) and a photosystem II reaction center subunit W (At2g30570) were all down-regulated in shoots (**Table 4**). Three genes involved in the Calvin cycle (At3g12780, At1g56190, and At1g73110) encoding phosphoglycerate kinase proteins were also down-regulated. Pb may play a role similar to that of Mg, which at high concentrations non-competitively inhibits phosphoglycerate kinase activity (Larsson-Raźnikiewicz, 1967). Interestingly, increased expression of photosynthetic genes was observed in roots, with the exception of At1g10960 encoding a ferredoxin

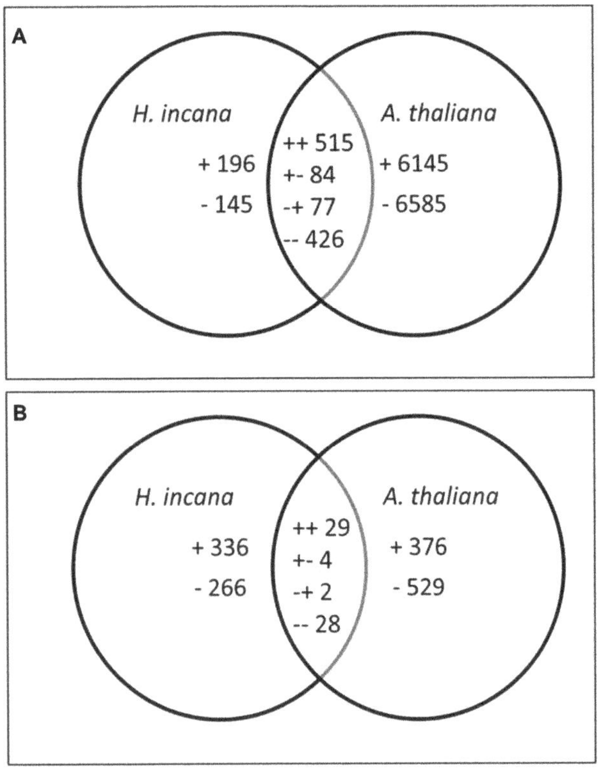

FIGURE 2 | Venn diagrams representing the distribution of regulated genes between *H. incana* and *Arabidopsis thaliana* in roots (A) and shoots (B) after Pb exposure. +, up-regulated; −, down-regulated; ++, up-regulated in both species, −−, down-regulated in both species, +−, up-regulated in *H. incana* and down-regulated in *A. thaliana* and −+, down-regulated in *H. incana* and up-regulated in *A. thaliana*.

1 for which the gene expression was down regulated by 0.4 (**Table 5**). Because roots are thought to be a non-photosynthetic organ, the biological relevance of this phenomenon has received little attention but the majority of the genes affected (8/11) are members of Lhc family of photosynthesis-associated nuclear genes (At1g29910, At1g29930, At2g34420, At2g34430, At3g47470, At3g54890, At5g28450, and At5g54270) and can be expressed in specialized territories, including the root tip (Sawchuk et al., 2008). The putative interactions between Lhc gene expression, root localization and Pb exposure are somewhat unclear but what is particularly interesting is that abscisic acid (ABA) was required for full expression of different Lhc members and physiologically high levels of ABA enhanced Lhc expression (Liu et al., 2013).

The second biological process group particularly enriched by Pb exposure contained genes encoding cell wall proteins in both shoots and roots (**Tables 4** and **5**). Recent studies in different plant species have shown that the plant cell wall could be modified in response to Pb (Krzeslowska, 2011). Modifications of the cell wall mainly concern an increase in the amount of polysaccharides, in particular pectins. In the present study, Pb treatment induced a reduction in xyloglucan endotransglucosylase gene expression (At2g06850, At2g23270, and At4g03210) and could play an important role in Pb-induced root growth inhibition by causing cell wall modification and reduction of cell elongation. Similar observations have been reported for Al toxicity in *Arabidopsis* (Yang et al., 2011).

The last biological group enriched by Pb exposure that emerged from this analysis encompasses the group of genes involved in metal handling. The frequency of up-regulated genes in roots was greater than 10, i.e., it was the most enriched category of the data (**Figure 3A**). Among the genes associated with this category, four belonged to the metallochaperone-like protein family (At4g39700, At5g17450, At1g22990, and At4g08570). Expression of these genes was increased by Pb-exposure with a fold-change ranging from 2.6 to 13.9 (**Table 5**). The function of these metalloproteins is unknown, but two of them (At1g22990 and At4g08570) are known to be involved in Cd tolerance (Tehseen et al., 2010). Another gene in the metal handling category, At2g28660, encodes putative soluble proteins with a heavy metal binding domain like Cys-x-x-Cys and a putative chloroplast targeting sequence is associated with the Cu chaperone family (Abdel-Ghanya et al., 2005). The expression of this gene was up regulated 6.2-fold in *H. incana* roots treated with Pb compared to the control. This protein possesses a heavy metal fixation domain and could act as a putative Pb fixation site. Only three genes up-regulated in shoots were identified in the metal handling category (**Table 4**) and two genes code for metallothioneins *MT2a* (At3g09390) and *MT2b* (At5g02380). Expression levels

TABLE 3 | Selection steps applied to highlight *H. incana* candidate genes regulated in response to Pb exposure in roots and shoots.

Category	Treatment	Roots		Shoots	
Specific genes	Total	341		601	
	FC > 2.0 or FC < 0.5	62 Up-reg	59 Down-reg	192 Up-reg	138 Down-reg
Common genes	Total	1102		63	
	FC Hi/FC At > 2.0 or < 0.5	315		28	
	FC > 2.0 orFC < 0.5	125 Up-reg	102 Down-reg	18 Up-reg	8 Down-reg
Candidate genes		187 Up-reg	161 Down-reg	210 Up-reg	146 Down-reg

H. incana specific genes were directly chosen on the basis of their fold change (FC) ratio (FC > 2.0 or FC < 0.5). H. incana common genes were first selected on the basis of their FC related to those from A. thaliana (FC Hi/FC At > 2.0 or < 0.5) and then FC on the basis of the FC ratio (FC > 2.0 or FC < 0.5). (Fold Change, FC; Hi, H. incana; At, A. thaliana; down-reg, down-regulated genes; up-reg, up-regulated genes).

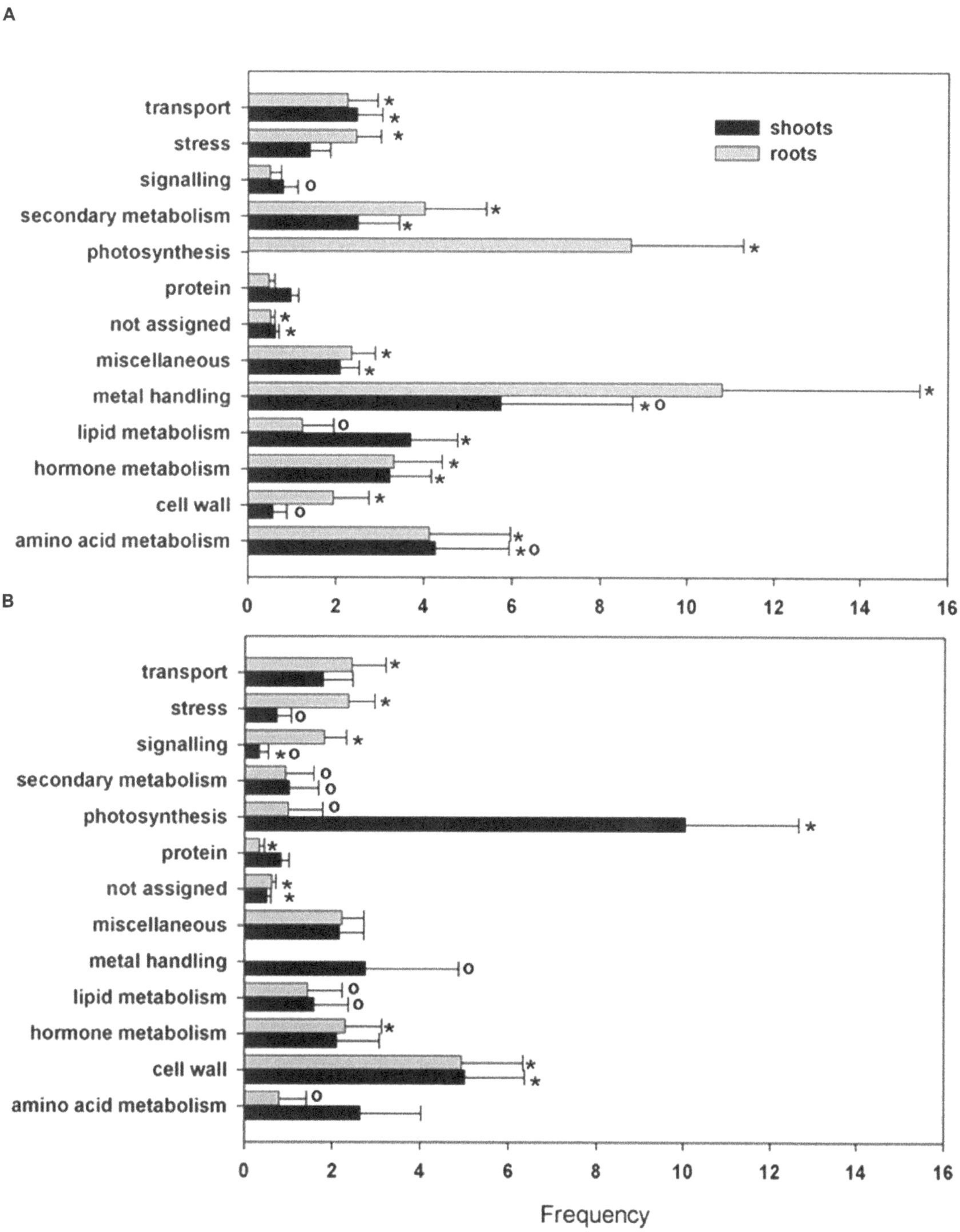

FIGURE 3 | Biological process classification of Pb responsive selected transcripts. The gene ontology was obtained using the Classification SuperViewer software with MapMan classification source for selected up- **(A)** and down-regulated **(B)** genes. Normed frequency was calculated as follows: (Number in class input set/Number in classified input set)/(Number in class reference set/Number in classified reference set). *Significant class ($p > 0.05$), o, size less than 5 genes.

of these two genes were, respectively 4.5 and 2.4 in shoots treated with Pb and in the control (**Table 4**). Although the protective role of metallothioneins against Cd is well known in mammals (Klaassen et al., 1999), their role in plant tolerance to heavy metals is less well known. MT2a is known to be localized the cytosol and could chelate heavy metals, but is not involved in vacuolar sequestration Lee et al., 2004). In a previous study, we demonstrated the role

TABLE 4 | GO enrichment analysis of differentially expressed genes in shoots.

Functional category	AGI	FC	Description
Photosynthesis	At1g56190	0.499	Phosphoglycerate kinase family protein
	At2g30570	0.427	Photosystem II reaction center W
	At3g55330	0.418	psbP-like protein 1
	At3g12780	0.413	Phosphoglycerate kinase 1
	At1g11860	0.372	Glycine cleavage T-protein family
	At1g14150	0.356	psbQ-like protein 2
	At1g73110	0.299	*p*-loop containing nucleoside triphosphate hydrolases
	At3g56650	0.266	Photosystem II reaction center PsbP family protein
	At2g05070	0.249	Light harvesting complex 2.2
Cell wall	At4g17030	16.664	Expansin-like B1
	At3g48530	2.564	Related protein kinase regulatory subunit gamma 1
	At5g16190	0.470	Cellulose synthase like A11
	At5g26670	0.387	Pectinacetylesterase family protein
	At1g70370	0.362	Polygalacturonase 2
	At1g04680	0.356	Pectin lyase-like superfamily protein
	At1g03870	0.302	Arabinogalactan 9
	At4g03210	0.301	Xyloglucan endotransglucosylase/hydrolase 9
	At1g23480	0.266	Cellulose synthase-like A3
	At3g10720	0.239	Plant invertase/pectin methylesterase inhibitor superfamily
	At3g23730	0.224	Xyloglucan endotransglucosylase/hydrolase 16
	At2g06850	0.212	Xyloglucan endotransglucosylase/hydrolase 4
	At5g04970	0.196	Plant invertase/pectin methylesterase inhibitor superfamily
	At4g16980	0.103	Arabinogalactan-protein family
Metal handling	At3g09390	4.577	Metallothionein 2A
	At5g01600	2.514	Ferretin 1
	At5g02380	2.415	Metallothionein 2B
	At3g24450	0.412	Heavy metal transport/detoxification superfamily protein

TABLE 5 | GO enrichment analysis of differentially expressed genes in roots.

Functional category	AGI	FC	Description
Photosynthesis	At5g54270	7.424	Chlorophyll B-binding protein 3
	At1g29910	7.103	Chlorophyll A/B binding protein 3
	At1g29930	6.547	Chlorophyll A/B binding protein 1
	At3g54890	5.506	Photosystem I light harvesting complex gene 1
	At5g28450	5.442	Chlorophyll A-B binding family protein
	At2g34430	5.218	Light harvesting complex gene B1B1
	At2g34420	5.141	Light harvesting complex gene B1B2
	At3g47470	5.022	Chlorophyll-protein complex I subunit A4
	At1g51400	4.379	Photosystem II 5 kD protein
	At2g30570	4.202	Photosystem II reaction center W
	At1g10960	0.414	Ferredoxin 1
Cell wall	At4g17030	25.232	Expansin B1
	At5g06860	4.580	Polygalacturonase inhibiting protein 1
	At5g17420	2.420	Cellulose synthase family protein
	At3g27400	2.197	Pectin lyase-like superfamily protein
	At1g67070	2.126	Mannose-6-phosphate isomerase, type I
	At2g37090	2.039	Nucleotide-diphospho-sugar transferase superfamily
	At1g26240	0.458	Proline-rich extensin-like family protein
	At2g28950	0.456	Expansin A6
	At1g57590	0.445	Pectinacetylesterase family protein
	At4g03210	0.423	Xyloglucan endotransglucosylase/hydrolase 9
	At3g15370	0.416	Expansin A12
	At4g08410	0.406	Proline-rich extensin-like family protein
	At5g53250	0.395	Arabinogalactan protein 22
	At5g45280	0.310	Pectinacetylesterase family protein
	At4g01220	0.250	Nucleotide-diphospho-sugar transferase family protein
	At4g01630	0.131	Expansin A17
	At4g30280	0.130	Xyloglucan endotransglucosylase/hydrolase 18
	At3g10720	0.120	Plant invertase/pectin methylesterase inhibitor superfamily
	At1g21310	0.090	Extensin 3
Metal handling	At4g08570	13.918	Heavy metal transport/detoxification superfamily protein
	At1g22990	5.758	Heavy metal transport/detoxification superfamily protein
	At2g28660	6.161	Chloroplast-targeted copper chaperone protein
	At4g39700	2.625	Heavy metal transport/detoxification superfamily protein
	At5g17450	2.765	Heavy metal transport/detoxification superfamily protein

of *MT2a* in Pb tolerance using *Arabidopsis* T-DNA insertional mutants (Auguy et al., 2013). Treatment with Pb highly significantly reduced primary root length in *mt2a* (48%) mutants. This reduction in primary root growth reflected increased sensitivity to Pb in *Arabidopsis* T-DNA insertion mutants and suggests that the *MT2a* gene is involved in lead tolerance. No effect of the mutation was observed on the Pb content in shoots and in roots of mt2a-mutant compared to wild-type plants (Auguy et al., 2013). Another metallothionein gene identified in the present study was *MT2b* (At5g02380). *MT2b* was previously described as being expressed in the phloem of all organs (Guo et al., 2008) and involved in Cu tolerance in *Silene vulgaris* or in Cd tolerance in Tobacco (Van Hoof et al., 2001; Grispen et al., 2011). In order to examine the functions of *MT2b* gene in Pb tolerance in *Arabidopsis*, we used a reverse-genetic approach with two independent *Arabidopsis* T-DNA insertion lines (Salk_144899 and Salk_037601) for the *MT2b* gene obtained from the Salk institute (**Figure 4A**). We measured the Pb content in roots and shoots of the 2 week-old wild-type and mt2b-mutants. Roots of both mutant lines contained 1.5-fold more Pb than the roots of wild-type plants (**Figure 4B**) and shoots of both

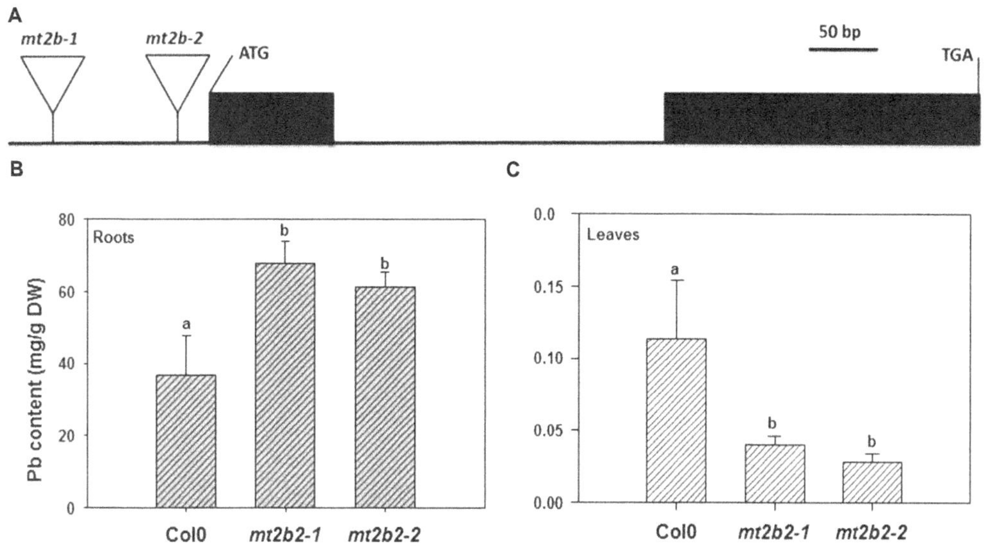

FIGURE 4 | Functional characterizations of the *Mt2b* gene. (A) Intron-exon organization of the *Arabidopsis Mt2b* gene (AT5G02380) and T-DNA locations. Solid black boxes and the solid line indicate coding regions and introns, respectively. The position of the 2 T-DNA insertions is indicated by a triangle. **(B)** and **(C)** Lead accumulation in roots and shoots of, respectively, wild type ecotype Columbia (Col0) and two mt2b mutant line seedlings of *A. thaliana*. Seedlings were collected after 13 days culture on media with or without 40 μM $Pb(NO_3)_2$. All results are the average value (±SE) of three independent replicates. The letters represent homogenous subgroups of roots and shoots, using LSD *post hoc* test at an $\alpha = 0.01$ significance level.

mutant lines contained 3.3-fold less than those of wild-type plants (**Figure 4C**). No effect of the mutations was observed on the primary root growth of *mt2b* mutants compared to the wild-type plants (data not shown). These results suggest a role of MT2b in Pb accumulation through a perturbation of root-to-shoot translocation by reducing Pb content in leaves and increasing in roots. Very recently, using a *MT* quadruple mutant (*mt1a/mt2a/mt2b/mt3*), Benatti et al. (2014) showed that *Arabidopsis* metallothioneins are important for the accumulation and distribution of Cu in leaves and seeds, confirming the role of MT in the accumulation of metals in plants.

TABLE 6 | Top 10 most similar profiles for *H. incana* roots from perturbation category (Genevestigator database).

Sample identification from Genevestigator	Relative similarity
Osmotic study 2 (late)/untreated root (late)	1.238
Salt study 5 (cpc-1 try-82)/mock treated primary root tip (cpc-1 try-82)	1.164
Salt study 5 (wer-1 myb23-1)/mock treated primary root tip (wer-1 myb23-1)	1.158
Salt study 5 (col-0)/mock treated primary root tip (col-0)	1.150
Osmotic study 2 (early)/untreated root (early)	1.149
Salt study 5 (scm-2)/mock treated primary root tip (scm-2)	1.148
Dicamba herbicide (10 h)/H2O treated seedling (10 h)	1.114
ABA study 6 (srk2cf)/untreated plant (srk2cf)	1.112
Salt study 2 (early)/untreated root (early)	1.111
ABA (3 h)/mock treated seedling (3 h)	1.111

Relative Similarity indicates the degree of their resemblance: the higher value the higher similarity relative to the average similarity.

Comparison of Profiles with Other Transcriptomic Experiments

Data were analyzed using Genevestigator software (Nebion AG, Switzerland) with the Genevestigator tool "signature[4]". Selected genes in both roots (348 genes) and shoots (356 genes) that responded specifically to Pb exposure in the Pb-tolerant *H. incana* were compared to the Genevestigator "Perturbation" database (**Tables 6** and **7**, respectively). A comparison of the complete profiles for roots and shoots is given in Additional Files 10 and 11 in Supplementary Material, respectively. In the Perturbation database, experiments with the most

[4]https://www.genevestigator.com/gv/doc/signature.jsp

TABLE 7 | Top 10 most similar profiles for *H. incana* shoots from perturbation category (Genevestigator database).

Sample identification from Genevestigator	Relative similarity
Osmotic (late)/untreated green tissue (late)	1.468
Drought study 2 (Trans.)/untreated leaf (Trans.)	1.414
Drought study 11 (hai1-2)/mock treated seedling (hai1-2)	1.412
Salt (late)/untreated green tissue (late)	1.411
Salt study 3 (atmy44)/H2O treated rosette leaf (atmy44)	1.399
ABA study (srk2cf)/untreated plant (srk2cf)	1.386
ABA study (col-0)/untreated plant (col-0)	1.366
Drought (wt)/untreated leaf (col)	1.362
Drought study 11 (col-0)/mock treated seedling (col-0)	1.354
Osmotic study 4 (shoot)/mock treated col-0 shoot	1.347

Relative Similarity indicates the degree of their resemblance: the higher value the higher similarity relative to the average similarity.

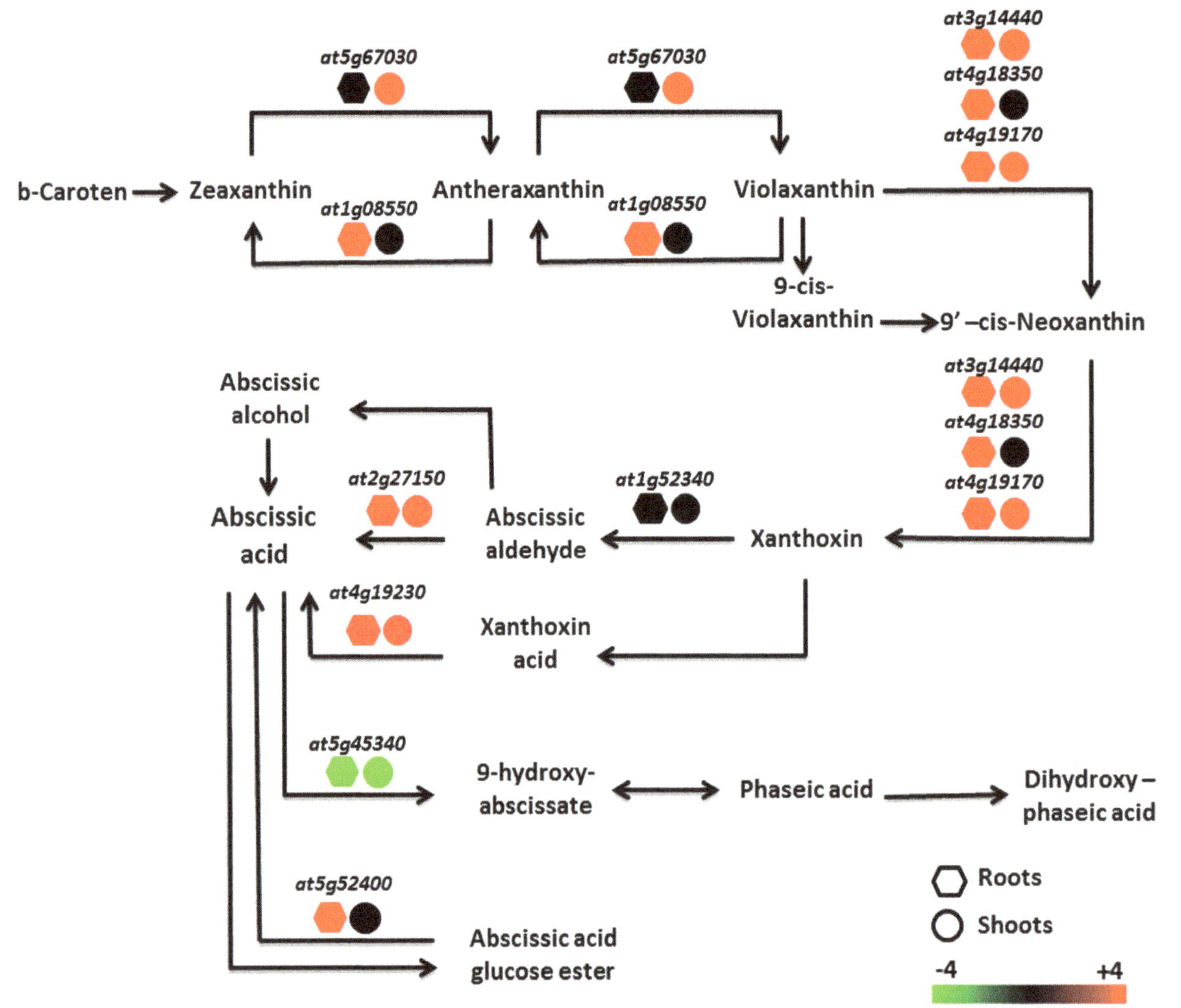

FIGURE 5 | Regulation of the abscisic acid metabolism pathway in *H. incana* roots and shoots treated with Pb. Red and green represent, respectively, an increase and a decrease in regulation, compared with untreated plants. Expression value scale is provided.

significant similarity to our Pb experiment include data from abiotic stress-related studies, mainly salt stress-related, osmotic-related and drought experiments as well as abscisic acid (ABA) study (**Tables 6** and **7**). Furthermore, it is important to note that, from the biological process classification of Pb responsive selected transcripts (**Figure 3**), the gene composition of the hormone metabolism class underline the over-representation of ABA subclass (Additional File 12 in Supplementary Material). To confirm this hypothesis, total expression data from our Pb experiment were re-analyzed. Several genes involved in the biosynthesis of ABA were shown to be up-regulated in *H. incana* roots and shoots treated with lead (**Figure 5**), suggesting induction of ABA biosynthesis when *H. incana* plants were exposed to Pb treatment. Interestingly, even the Atg52400 gene involved in the ABA catabolism pathway, via phaseic acid (PA), was largely down-regulated, also leading to potential ABA accumulation. Abscisic acid, as a stress signal, is known to enhance plant tolerance to several environmental stresses, including low temperature, salt, drought, and heavy metals (Bellaire et al., 2000; Verslues and Zhu, 2005; Hong et al., 2013; Li et al., 2014). The increase in tolerance is partly due to the enhancement of the antioxidant defense system, which prevents the accumulation of ROS (Bellaire et al., 2000; Jiang and Zhang, 2002). On the other hand, in plants in contact with Pb, transpiration is decreased (Barcelo and Poschenrieder, 1990). This decrease in transpiration could be due to a reduction in growth, or at least to a reduction in leaf surface area and/or to the fact that when Pb is fixed to the cell wall it reduces plasticity and eventually disturbs the osmotic balance. Accumulation of Pb in the cell leads to accumulation of ABA in the shoots, results in stomata closure and reduces loss of water through transpiration (Parys et al., 1998; Pourrut et al., 2011). Taken together, these data indicate an intricate relationship between plant tolerance to Pb and ABA pathways. Recently a similar conclusion was reached by Shukla et al. (2014), who suggested that ABA-mediated signaling could be a major mechanism in response to metal gold exposure in *Arabidopsis*.

CONCLUSION

The comparative transcriptomic analysis between the Pb-hyperaccumulating and Pb-tolerant plant *H. incana* and the Pb-sensible model plant *A. thaliana* identified a set of genes differentially expressed in response to Pb exposure. These genes were particularly over-represented in the photosynthesis, cell wall structure, and metal handling biological processes. In addition several genes involved in the ABA biosynthetic pathway were up regulated in response to lead exposure suggesting that ABA-mediated signaling is involved in plant response to lead. In addition *H. incana* could be considered as a good experimental model to identify genes involved in lead tolerance and accumulation in plants.

AVAILABILITY OF SUPPORTING DATA

The data discussed in this publication have been deposited in NCBI's GEO (Edgar et al., 2002) and are accessible through GEO Series accession number GSE65334 (http://www.ncbi.nlm.nih.gov/geo/query/acc.cgi?acc=GSE65334).

AUTHOR CONTRIBUTIONS

AF and PD conceived and designed the experiment. AF, MF, and PM collected the experimental data. ME, AF-M, and GB contributed reagents/materials/analysis tools. AF, AS, and PD analyzed data. AF and PD wrote the manuscript. GB assisted with the interpretation of the results and provided editorial support for the manuscript. All authors have read, edited, and approved the current version of the manuscript.

ACKNOWLEDGMENTS

The research reported here was financially supported by the *INSU EC2CO* program and by the *Laboratoire Mixte International* (LMI), *Biotechnologie Microbienne et Végétale*, (IRD/Mohammed V-Agdal Univ., Rabat, Morocco).

REFERENCES

Abdel-Ghanya, S. E., Burkheada, J. L., Gogolina, K. A., Andrés-Colásb, N., Bodeckera, J. R., Puigb, S., et al. (2005). AtCCS is a functional homolog of the yeast copper chaperone Ccs1/Lys7. *FEBS Lett.* 279, 2307–2312. doi: 10.1016/j.febslet.2005.03.025

Auguy, F., Fahr, M., Moulin, P., Brugel, A., Laplaze, L., El Mzibri, M., et al. (2013). Lead tolerance and accumulation in *Hirschfeldia incana*, a Mediterranean Brassicaceae from metalliferous mine spoils. *PLoS ONE* 8:e61932. doi: 10.1371/journal.pone.0061932

Baker, A. J. M. (1981). Accumulators and excluders-strategies in the response of plants to heavy metals. *J. Plant Nutr.* 3, 643–654. doi: 10.1080/01904168109362867

Barcelo, J., and Poschenrieder, C. (1990). Plant water relations as affected by heavy metal stress: a review. *J. Plant Nutr.* 13, 1–37. doi: 10.1080/01904169009364057

Becher, M., Talke, I. N., Krall, L., and Krämer, U. (2004). Cross-species microarray transcript profiling reveals high constitutive expression of metal homeostasis genes in shoots of the zinc hyperaccumulator *Arabidopsis halleri*. *Plant J.* 37, 251–268. doi: 10.1046/j.1365-313X.2003.01959.x

Belatik, A., Hotchandani, S., Tajmir-Riahi, H. A., and Carpentier, R. (2013). Alteration of the structure and function of photosystem I by $Pb2+$. *J. Photochem. Photobiol. B* 123, 41–47. doi: 10.1016/j.jphotobiol.2013.03.010

Bellaire, B. A., Carmody, J., Braud, J., Gossett, D. R., Banks, S. W., Lucas, M. C., et al. (2000). Involvement of abscissic acid-dependent and -independent pathways in the upregulation of antioxidant enzyme activity during NaCl stress in cotton callus tissue. *Free Radical Res.* 33, 531–545. doi: 10.1080/10715760000301071

Benatti, R. M., Yookongkaew, N., Meetam, M., Guo, J. W., Punyasuk, N., AbuQamar, S., et al. (2014). Metallothionein deficiency impacts copper accumulation and redistribution in leaves and seeds of *Arabidopsis*. *New Phytol.* 202, 940–951. doi: 10.1111/nph.12718

Broughton, W. J., and Dilworth, M. J. (1971). Control of leghaemoglobin synthesis in snake beans. *Biochem. J.* 125, 1075–1080. doi: 10.1042/bj1251075

Cobbett, C., and Goldsbrough, P. (2002). Phytochelatins and metallothioneins: roles in heavy metal detoxification and homeostasis. *Annu. Rev. Plant Biol.* 53, 159–182. doi: 10.1146/annurev.arplant.53.100301.135154

Edgar, R., Domrachev, M., and Lash, A. E. (2002). Gene expression omnibus: NCBI gene expression and hybridization array data repository. *Nucleic Acids Res.* 30, 207–210. doi: 10.1093/nar/30.1.207

Gisbert, C., Clemente, R., Navarro-Aviñó, J., Baixauli, C., Ginér, A., Serrano, R., et al. (2006). Tolerance and accumulation of heavy metals by Brassicaceae species grown in contaminated soils from Mediterranean regions of Spain. *Environ. Exp. Bot.* 56, 19–27. doi: 10.1016/j.envexpbot.2004.12.002

Gravot, A., Lieutaud, A., Verret, F., Auroy, P., Vavasseur, A., and Richaud, P. (2004). AtHMA3, a plant P1B-ATPase, functions as a Cd/Pb transporter in yeast. *FEBS Lett.* 561, 22–28. doi: 10.1016/S0014-5793(04)00072-9

Grispen, V. M. J., Hakvoort, H. W. J., Bliek, T., Verkleij, J. A. C., and Schat, H. (2011). Combined expression of the *Arabidopsis* metallothionein MT2b and the heavy metal transporting ATPase HMA4 enhances cadmium tolerance and the root to shoot translocation of cadmium and zinc in tobacco. *Environ. Exp. Bot.* 72, 71–76. doi: 10.1016/j.envexpbot.2010.01.005

Guo, W. J., Meetam, M., and Goldsbrough, P. B. (2008). Examining the specific contributions of individual *Arabidopsis* metallothioneins to copper distribution and metal tolerance. *Plant Physiol.* 146, 1697–1706. doi: 10.1104/pp.108.115782

Hammond, J. P., Bowen, H. C., White, P. J., Mills, V., Pyke, K. A., Baker, A. J., et al. (2006). A comparison of the *Thlaspi caerulescens* and *Thlaspi arvense* shoot transcriptomes. *New Phytol.* 170, 239–260. doi: 10.1111/j.1469-8137.2006.01662.x

Hong, J. H., Seah, S. W., and Xu, J. (2013). The root of ABA action in environmental stress response. *Plant Cell Rep.* 32, 971–983. doi: 10.1007/s00299-013-1439-9

Islam, E., Liu, D., Li, T., Yang, X., Jin, X., Mohmood, Q., et al. (2008). Effect of Pb toxicity on leaf growth, physiology and ultrastructure in the two ecotypes of *Elsholtzia argyi*. *J. Hazard. Mater.* 154, 914–926. doi: 10.1016/j.jhazmat.2007.10.121

Jiang, M., and Zhang, J. (2002). Water stress induced abscisic acid accumulation triggers the increased generation of reactive oxygen species and up regulates the activities of antioxidant enzymes in maize leaves. *J. Exp. Bot.* 53, 2401–2410. doi: 10.1093/jxb/erf090

Kim, D. Y., Bovet, L., Kushnir, S., Noh, E. W., Martinoia, E., and Lee, Y. (2006). AtATM3 is involved in heavy metal resistance in *Arabidopsis*. *Plant Physiol.* 140, 922–932. doi: 10.1104/pp.105.074146

Kim, D. Y., Bovet, L., Maeshima, M., Martinoia, E., and Lee, Y. (2007). The ABC transporter AtPDR8 is a cadmium extrusion pump conferring heavy metal resistance. *Plant J.* 50, 207–218. doi: 10.1111/j.1365-313X.2007.03044.x

Klaassen, C. D., Liu, J., and Choudhuri, S. (1999). Metallothionein: an intracellular protein to protect against cadmium toxicity. *Annu. Rev. Pharmacol. Toxicol.* 39, 267–294. doi: 10.1146/annurev.pharmtox.39.1.267

Krzeslowska, M. (2011). The cell wall in plant cell response to trace metals: polysaccharide remodeling and its role in defense strategy. *Acta Physiol. Plant.* 33, 35–42. doi: 10.1007/s11738-010-0581-z

Larsson-Raźnikiewicz, M. (1967). Kinetic studies on the reaction catalyzed by phosphoglycerate kinase. II - The kinetic relationships between 3-phosphoglycerate, MgATP2- and activating metal ion. *Biochim. Biophys. Acta* 132, 33–40. doi: 10.1016/0005-2744(67)90189-1

Lee, J., Shim, D., Song, W. Y., Hwang, I., and Lee, Y. (2004). *Arabidopsis* metallothioneins 2a and 3 enhance resistance to cadmium when expressed in *Vicia faba* guard cells. *Plant Mol. Biol.* 554, 805–815. doi: 10.1007/s11103-004-0190-6

Lee, M., Lee, K., Lee, J., Noh, E. W., and Lee, Y. (2005). AtPDR12 contributes to lead resistance in *Arabidopsis*. *Plant Physiol.* 138, 827–836. doi: 10.1104/pp.104.058107

Li, S. W., Leng, Y., Feng, L., and Zeng, X. Y. (2014). Involvement of abscisic acid in regulating antioxidative defense systems and IAA-oxidase activity and improving adventitious rooting in mung bean [*Vigna radiata* (L.) Wilczek] seedlings under cadmium stress. *Environ. Sci. Pollut. Res. Int.* 21, 525–537. doi: 10.1007/s11356-013-1942-0

Liu, R., Xu, Y. H., Jiang, S. C., Lu, K., Lu, Y. F., Feng, X. J., et al. (2013). Light-harvesting chlorophyll a/b-binding proteins, positively involved in abscisic acid signalling, require a transcription repressor, WRKY40, to balance their function. *J. Exp. Bot.* 64, 5443–5456. doi: 10.1093/jxb/ert307

Madejon, P., Murillo, J. M., Maranon, T., and Lepp, N. W. (2007). Factors affecting accumulation of thallium and other trace elements in two wild Brassicaceae spontaneously growing on soils contaminated by tailings dam waste. *Chemosphere* 67, 20–28. doi: 10.1016/j.chemosphere.2006.10.008

Margui, E., Queralt, I., Carvalho, M. L., and Hidalgo, M. (2007). Assessment of metal availability to vegetation (*Betula pendula*) in Pb-Zn ore concentrate residues with different features. *Env. Pollut.* 145, 179–184. doi: 10.1016/j.envpol.2006.03.028

Morel, M., Crouzet, J., Gravot, A., Auroy, P., Leonhardt, N., Vavasseur, A., et al. (2009). AtHMA3, a P1B-ATPase allowing Cd/Zn/Co/Pb vacuolar storage in *Arabidopsis*. *Plant Physiol.* 149, 894–904. doi: 10.1104/pp.108.130294

Papoyan, A., and Kochian, L. V. (2004). Identification of *Thlaspi caerulescens* genes that may be involved in heavy metal hyperaccumulation and tolerance, Characterization of a novel heavy metal transporting ATPase. *Plant Physiol.* 136, 3814–3823. doi: 10.1104/pp.104.044503

Parys, E., Romanowska, E., Siedlecka, M., and Poskuta, J. W. (1998). The effect of lead on photosynthesis and respiration in detached leaves and in mesophyll protoplasts of Pisum sativum. *Acta Physiol. Plant.* 20, 313–322. doi: 10.1007/s11738-998-0064-7

Peer, W. A., Mamoudian, M., Lahner, B. Z., Reeves, R. D., Murphy, A. S., and Salt, D. E. (2003). Identifying model metal hyperaccumulator plants: germplasm analysis of 20 Brassicaceae accessions from a wide geographical area. *New Phytol.* 159, 421–430. doi: 10.1046/j.1469-8137.2003.00822.x

Poschenrieder, P., Bech, J., Llugany, M., Pace, A., Fenés, E., and Barceló, J. (2001). Copper in plant species in a copper gradient in Catalonia (North East Spain) and their potential for phytoremediation. *Plant Soil* 2, 247–256. doi: 10.1023/A:1010374732486

Pourrut, B., Shahid, M., Dumat, C., Winterton, P., and Pinelli, E. (2011). Lead uptake, toxicity, and detoxification in plants. *Rev. Environ. Contam. Toxicol.* 213, 113–136.

Provart, N., and Zhu, T. (2003). A browser-based functional classification superviewer for *Arabidopsis* genomics. *C. C. Mol. Biol.* 2003, 271–272.

Rigola, D., Fiers, M., Vurro, E., and Aarts, M. G. M. (2006). The heavy metal hyperaccumulator *Thlaspi caerulescens* expresses many species-specific genes as identified by comparative EST analysis. *New Phytol.* 170, 753–766. doi: 10.1111/j.1469-8137.2006.01714.x

Sawchuk, M. G., Donner, T. J., Head, P., and Scarpella, E. (2008). Unique and overlapping expression patterns among members of photosynthesis-associated nuclear gene families in *Arabidopsis*. *Plant Physiol.* 148, 1908–1924. doi: 10.1104/pp.108.126946

Schoof, H., Ernst, R., Nazarov, V., Pfeifer, L., Mewes, H. W., and Mayer, K. F. (2004). MIPS *Arabidopsis thaliana* database (MAtDB): an integrated biological knowledge resource for plant genomics. *Nucleic Acids Res.* 32, 373–376. doi: 10.1093/nar/gkh068

Schutzendubel, A., and Polle, A. (2002). Plant responses to abiotic stresses: heavy metal-induced oxidative stress and protection by mycorrhization. *J. Exp. Bot.* 53, 1351–1365. doi: 10.1093/jexbot/53.372.1351

Seregin, I. V., and Ivanov, B. V. (2001). Physiological aspects of cadmium and lead toxic effects on higher plants. *Russ. J. Plant Physiol.* 48, 523–544. doi: 10.1023/A:1016719901147

Shukla, D., Krishnamurthy, S., and Sahi, S. V. (2014). Genome wide transcriptome analysis reveals ABA mediated response in *Arabidopsis* during gold (AuCl-4) treatment. *Front. Plant Sci.* 5:652. doi: 10.3389/fpls.2014.00652

Singh, R. P., Tripathi, R. D., Sinha, S. K., Maheshwari, R., and Srivastava, H. S. (1997). Response of higher plants to lead contaminated environment. *Chemosphere* 34, 2467–2493. doi: 10.1016/S0045-6535(97)00087-8

Sterckeman, T., Douay, F., Baize, D., Fourrier, H., Proix, N., and Schvartz, C. (2006). Trace elements in soils developed in sedimentary materials from Northern France. *Geoderma* 136, 912–929. doi: 10.1016/j.geoderma.2006.06.010

Sunkar, R., Kaplan, B., Bouche, N., Arazi, T., Dolev, D., Talke, I. N., et al. (2000). Expression of a truncated tobacco NtCBP4 channel in transgenic plants and disruption of the homologous *Arabidopsis* CNGC1 gene confer Pb2+ tolerance. *Plant J.* 24, 533–542. doi: 10.1111/j.1365-313X.2000.00901.x

Tehseen, M., Cairns, N., Sherson, S., and Cobbett, C. S. (2010). Metallochaperone-like genes in *Arabidopsis thaliana*. *Metallomics* 2, 556–564. doi: 10.1039/c003484c

Temminghoff, E. E., and Houba, V. J. (2004). *Plant Analysis Procedures*. Dordrecht: Kluwer Academic Publishers.

Van de Mortel, J. E., Villanueva, L. A., Schat, H., Kwekkeboom, J., Coughlan, S., Moerland, P. D., et al. (2006). Large expression differences in genes for iron and zinc homeostasis, stress response, and lignin biosynthesis distinguish roots of *Arabidopsis thaliana* and the related metal hyperaccumulator *Thlaspi caerulescens*. *Plant Physiol.* 142, 1127–1147. doi: 10.1104/pp.106.082073

Vangronsveld, J., Herzig, R., Weyens, N., Boulet, J., Adriaensen, K., Ruttens, A., et al. (2009). Phytoremediation of contaminated soils and groundwater: lessons from the field. *J. Environ. Sci. Pollut. Res.* 16, 765–794. doi: 10.1007/s11356-009-0213-6

Van Hoof, N., Hassinen, V. H., Ballintijn, K. F., Schat, H., Verkleij, J. A. C., Ernst, W., et al. (2001). Enhanced copper tolerance in *Silene vulgaris* (Moench) Garcke populations from copper mines is associated with increased transcript levels of a 2b-type metallothionein gene. *Plant Physiol.* 126, 1519–1526. doi: 10.1104/pp.126.4.1519

Verslues, P. E., and Zhu, J. K. (2005). Before and beyond ABA: upstream sensing and internal signals that determine ABA accumulation and response under abiotic stress. *Biochem. Soc. Texicol.* 33, 375–379. doi: 10.1042/BST0330375

Weber, M., Trampczynska, A., and Clemens, S. (2006). Comparative transcriptome analysis of toxic metal responses in *Arabidopsis thaliana* and the Cd2+-hypertolerant facultative metallophyte *Arabidopsis halleri*. *Plant Cell Environ.* 29, 950–963. doi: 10.1111/j.1365-3040.2005.01479.x

Xiao, S., Gao, W., Chen, Q. F., Ramalingam, S., and Chye, M. L. (2008). Overexpression of membrane-associated acyl-CoA-binding protein ACBP1 enhances lead tolerance in *Arabidopsis*. *Plant J.* 54, 141–151. doi: 10.1111/j.1365-313X.2008.03402.x

Yang, J., Zhu, X., Peng, Y., Zheng, C., Li, G., Liu, Y., et al. (2011). Cell wall hemicellulose contributes significantly to aluminum adsorption and root growth in *Arabidopsis*. *Plant Physiol.* 155, 1885–1892. doi: 10.1104/pp.111.172221

Zientara, K., Wawrzyńska, A., Lukomska, J., López-Moya, J. R., Liszewska, F., Assunção, A. G., et al. (2009). Activity of the AtMRP3 promoter in transgenic

Arabidopsis thaliana and *Nicotiana tabacum* plants is increased by cadmium, nickel, arsenic, cobalt and lead but not by zinc and iron. *J. Biotechnol.* 139, 258–263. doi: 10.1016/j.jbiotec.2008.12.001

Conflict of Interest Statement: The authors declare that the research was conducted in the absence of any commercial or financial relationships that could be construed as a potential conflict of interest.

16

Genome-Wide Association Mapping for Tomato Volatiles Positively Contributing to Tomato Flavor

Jing Zhang[1,2†], *Jiantao Zhao*[1,2†], *Yao Xu*[3], *Jing Liang*[4], *Peipei Chang*[1], *Fei Yan*[1,2], *Mingjun Li*[1], *Yan Liang*[1*] *and Zhirong Zou*[1,2*]

[1] State Key Laboratory of Crop Stress Biology for Arid Areas, College of Horticulture, Northwest A&F University, Yangling, China, [2] Key Laboratory of Protected Horticultural Engineering in Northwest, Ministry of Agriculture, Yangling, China, [3] College of Forestry, Northwest A&F University, Yangling, China, [4] Shaanxi Jinpeng Seed Industry Co. Ltd., Yangling, China

Edited by:
Jaime Prohens,
Universitat Politècnica de València, Spain

Reviewed by:
Antonio Granell,
Consejo Superior de Investigaciones Científicas, Spain
Amalia Barone,
Univeristy of Naples, Italy

***Correspondence:**
Yan Liang
liangyan@nwsuaf.edu.cn;
Zhirong Zou
zouzhirong2005@hotmail.com

†These authors have contributed equally to this work.

Tomato volatiles, mainly derived from essential nutrients and health-promoting precursors, affect tomato flavor. Taste volatiles present a major challenge for flavor improvement and quality breeding. In this study, we performed genome-wide association studies (GWAS) to investigate potential chromosome regions associated with the tomato flavor volatiles. We observed significant variation (1200x) among the selected 28 most important volatiles in tomato based on their concentration and odor threshold importance across our sampled accessions. Using 174 tomato accessions, GWAS identified 125 significant associations ($P < 0.005$) among 182 SSR markers and 28 volatiles (27 volatiles with at least one significant association). Several significant associations were co-localized in previously identified quantitative trait loci (QTL). This result provides new potential candidate loci affecting the metabolism of several volatiles.

Keywords: tomato, volatile, genome-wide association study, flavor, quantitative trait loci

INTRODUCTION

The perception of the tomato flavor, including aroma and taste, is a result of the interaction of sugars, acids, and many aromatic volatile compounds (Goff and Klee, 2006; Klee and Tieman, 2013). Plants produce a large variety of fruit flavor volatiles. These volatiles are defining elements of the distinct flavors of fruits and are mainly derived from essential compounds, including amino acids, fatty acids, carotenoids (Goff and Klee, 2006; Klee and Tieman, 2013). However, selecting for the quality of fruit aroma has never been a high priority for plant breeders (Goff and Klee, 2006; El Hadi et al., 2013). In fact, research that is focused on improving fruit size, yield, quality, and shelf life has often led to unintended, negative outcome for both aroma and, consequently, flavor (Kader et al., 1977; Ratanachinakorn et al., 1997).

The effect of a volatile on flavor perception is determined by its concentration and perceptible aroma, or odor threshold. As few as 20 of more than 400 volatiles in tomatoes have sufficient concentrations and odor thresholds to contribute to tomato flavor (Baldwin et al., 2000). However, the biosynthetic pathways and regulatory networks are known for volatiles of the greatest economic importance, such as the 20 volatiles aforementioned (Klee, 2010; Klee and Tieman, 2013). Important genes that regulate volatiles can be broadly subdivided into two classes: genes encoding enzymes responsible for synthesis of the end products and genes encoding factors that regulate the pathway (Klee, 2010). The regulation of the pathways of most volatiles is very poorly understood. In fact, even our knowledge on the 20 most important volatiles in tomato is limited.

The genetic and molecular bases of volatiles are also poorly understood. This is primarily due to the polygenic nature and biochemical complexity of flavor/aroma traits and our limited ability to quantify those volatiles (Klee and Tieman, 2013). Improvements in quantification techniques, as well as the capacity for high-throughput genotyping (Agarwal et al., 2008), provide the basis for quantitative trait locus (QTL) analysis of aroma components. Such QTL studies of aroma have already been performed with apple (Zini et al., 2005), grape (Doligez et al., 2006; Obando-Ulloa et al., 2010), and melon (Obando-Ulloa et al., 2010). In tomato, more than 100 quantitative trait loci (QTLs) affecting volatiles and their precursors have been identified (Saliba-Colombani et al., 2001; Causse et al., 2004; Schauer et al., 2006; Tieman et al., 2006; Mathieu et al., 2008; Zanor et al., 2009). Some QTLs specifically alter single volatiles while others can affect several related or even unrelated volatiles (Mathieu et al., 2008).

Genome-wide association studies (GWAS) is a method for mapping the loci responsible for natural variations in a target phenotype (Saidou et al., 2014; Matsuda et al., 2015; Pers et al., 2015). It is based on the identification of significantly associated genetic polymorphisms in a large population (Brachi et al., 2011). GWAS is a reliable, preliminary approach for identifying the locations of QTLs and has been conducted on major fruit quality traits in tomato (Mazzucato et al., 2008; Ranc et al., 2012; Xu et al., 2013; Ruggieri et al., 2014; Zhang et al., 2015). However, few association mapping studies have been performed on QTLs for the main flavor-enhancing volatiles in tomatoes or other crop plants (Kumar et al., 2015). In this study, we performed a GWAS to locate QTLs for flavor-affecting volatiles in tomato. In particular, we detected volatiles in a collection of 174 diverse tomato accessions using gas chromatography-mass spectrometry (GC/MS). We also detected a large number of loci to link their volatile composition with their genotypes. Our results confirmed some previous volatile QTLs, and we identified some chromosome regions that could be important in controlling volatile metabolism.

MATERIALS AND METHODS

Plant Materials

The tomato diversity panel consisted of 174 tomato accessions comprised of 123 cherry tomato accessions (*Solanum lycopersicum* var. *cerasiforme*) and 51 large-fruit cultivars (*S. lycopersicum*) (Table S1). All accessions were grown during the springs of 2013 and 2014 in a completely randomized block design with three replicates (10 plants per replicate), at the research greenhouse of the Tomato Research Group (34° 24_N, 108° 07_E) according to standard agronomic practices (Zhang et al., 2015).

Volatile Determinations

We performed analyses of fruit volatiles as described in Tikunov et al. (2005), with minor modifications. We combined all red, ripe fruit produced on the 10 accessions representing each horticultural type and immediately placed them in liquid nitrogen. We then kept them at −80°C. We transferred 15 g of the finely powdered tomato samples into a 40 mL Teflon cap vial (Thermo Fisher Scientific) with 5 g of NaCl. Ten microliters of 2-octanone (0.125 mg/mL in ethyl alcohol) were then added as an internal standard. We then sealed the vials using a silicone/PTFE septum and a magnetic cap. The closed vials were agitated (500 rpm) and sonicated for 10 min, and incubated at 50°C for 10 min prior to HS-SPME-GC-MS analysis. We performed three independent reactions for each sample. We extracted headspace volatiles by exposing each sample to a 75 μm carboxen-polydimethylsiloxane SPME fiber (Supelco, USA) for 30 min under continuous agitation (500 rpm) and heating at 40°C. The fiber was then inserted into an ISQ GC-MS (Thermo Scientific instruments, USA) injection port and the volatiles were desorbed for 3 min at 250°C. We performed chromatography on an HP-INNOWAX column (60 m × 0.25 mm × 0.25 μm) with helium as the carrier gas, at a constant flow of 1.0 mL min^{-1}. The temperature of both the GC interface and MS source was 230°C. The GC temperature program began at 40°C (2.5 min), and then was raised to 160°C (5°C min^{-1}) and to 230°C (10°C min^{-1}) before being held at 230°C for 5 min. The total run time, including oven cooling, was 40 min. Mass spectra were evaluated in the 35–450 m/z range at a scanning speed of 70 scans s^{-1} and an ionization energy of 70 eV. We processed the raw data obtained from GC-MS with Xcalibur and AMDIS software and identified the volatile compounds on the basis of the NIST/EPA/NIH Mass Spectral Library (NIST 2008) and Wiley Registry of Mass Spectral Data 8th edition. The retention index (RI) was calculated with a homologous series of n-alkanes (C7–C30, Sigma-Aldrich). The volatiles were semi-quantified according to the method of Baek and Cadwallader (1996) and Hopfer et al. (2012). We calculated the relative concentration of each volatile using the ratio of the areas of the target peak and the internal standard (2-octanone, 0.125 mg/mL, 10 μL) in a total ion current chromatogram with the following equation:

$$\text{Relative concentration} = \left[\frac{\text{Peak area of particular compound}}{\text{Peak area of internal standard (IS)}}\right] \times \text{Concentration of IS}$$

We used the relative concentrations to compare the differences in volatile profiles among the 174 accessions.

Genotyping

We extracted DNA from the 174 tomato accessions from fresh leaf tissue following the method of Fulton et al. (1995). All SSR markers were mainly selected from the SOL Genomics Network (http://solgenomics.wur.nl/) and the VegMarks database (http://vegmarks.nivot.affrc.go.jp/). We used the protocol of Sun et al. (2012) to amplify the markers. Only markers with minor allele frequency (MAF) > 0.05 were genotyped with the whole accessions (Zhang et al., 2015). Finally, we selected a set of 182 polymorphic simple sequence repeat (SSR) for further association mapping (Zhang et al., 2015).

Population Structure

We used the above set of 182 SSR markers to estimate the population structure of the 174 tomato accessions via

STRUCTURE 2.3.3 software (Pritchard et al., 2000). We set the number of hypothetical subpopulations (K) at 2–10 in order to evaluate the population structure with an admixture model. We performed 200,000 replicates of the Markov Chain Monte Carlo with a burn-in length of 100,000. We used Evanno transformation method to infer the optimal K of populations (Evanno et al., 2005). The kinship matrix was calculated via SPAGeDi software (Hardy and Vekemans, 2002). We set the diagonal of the matrix to two and negatives values to zero (Yu et al., 2005).

Association Mapping

Decay of LD and the corresponding significance level (*P*-value) were calculated using TASSEL 2.1 software (Bradbury et al., 2007). We also calculated associations between volatiles and SSR markers using TASSEL 2.1 software (Bradbury et al., 2007). Mixed linear model (MLM) was used in order to reduce false positive associations. We used $P < 0.005$ as the value to detect associations. We also took $P < 0.0003$ as the significant value to reduce false positive associations. The amount of phenotypic variation explained by each marker was estimated by R^2.

Statistics

We used either the SAS 8.1 program (SAS institute, Cary, NC) or the R statistical Software (http://www.r-project.org) version 3.0.2 for statistical analyses. We replaced the values of zero (undetectable) for all volatiles by the smallest non-zero value in the whole dataset (Mathieu et al., 2008). Then, we $\log_2$-transformed the volatile quantity values (ng g^{-1} fresh weight h^{-1}) before performing a Two-way ANOVA analysis for all traits. The resulting raw *P*-values were also corrected for multiple tests using the Benjamini and Hochberg FDR test (Benjamini and Hochberg, 1995). We estimated genetic variance, genetic by environment interaction variance, technical variance, heritability values according to the method of Xu et al. (2013). We developed correlation heat map via HemI 1.0., based on the analysis result among volatiles and accessions.

RESULTS

Variation in Volatiles

Of the more than 400 volatiles reported in tomato, fewer than 20 have been predicted to contribute to the unique tomato flavor based on the concentrations and odor thresholds (Buttery et al., 1987; Goff and Klee, 2006; Klee, 2010; Tieman et al., 2012). Here, we further investigated 28 volatiles that are in sufficient quantities to impact the tomato flavor. We found large variations of up to 1200 x in volatile contents across all of the sampled accessions (**Table 1**).

We calculated heritability values for all 28 volatiles based on 2 years of phenotypic characterization. Of these volatiles, nine had a value lower than 0.5 (**Table 1**). Therefore, volatile values in 2013 and 2014 were evaluated separately for further genome-wide association analyses.

Pearson correlation coefficients (*r*) among the 28 volatiles were relatively low, based on the mean value of 2 years data (the springs of 2013 and 2014) (lower than 0.5) (**Figure 1**). However, we observed significant coefficients among some volatiles (Table S2). For instance, beta-cyclocitral and (Z)-2-heptenal were positively correlated with 1-pentanol ($r = 0.464$ and 0.417, respectively). In addition, methyl salicylate was negatively

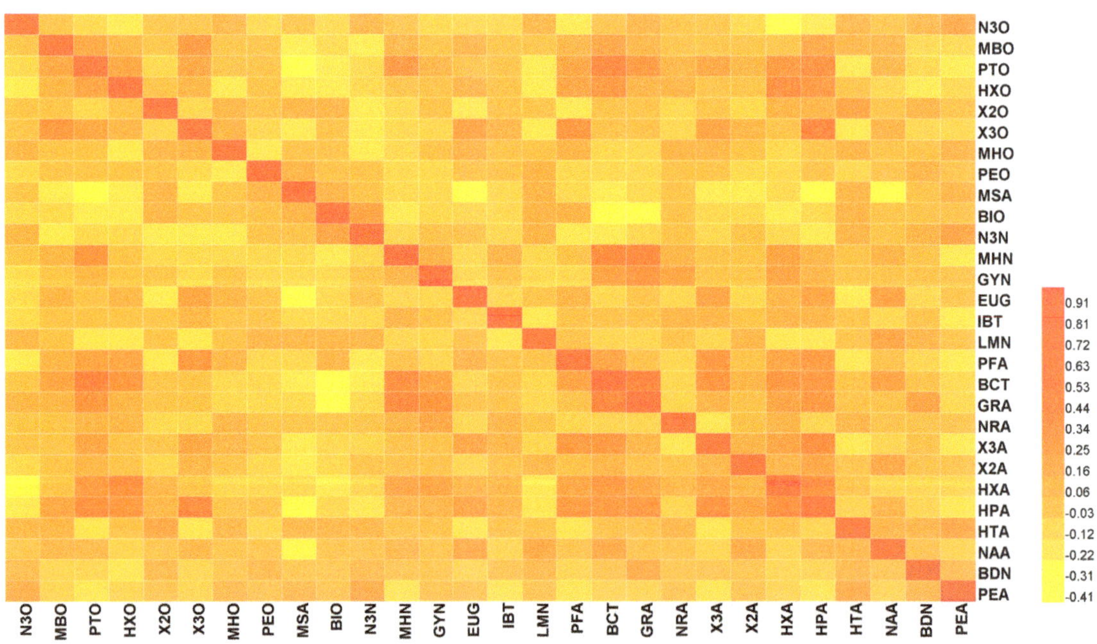

FIGURE 1 | Heat map showing the correlation analysis between 28 volatiles in the 174 tomato accessions. Regions in red and yellow indicate positive or negative correlations between traits, respectively (The complete data is available in Table S2).

TABLE 1 | Volatile variation within the 174 tomato accessions (ng/g fresh weight/hr).

Compound	Max	Min	Fold difference	Mean	SD	H^2
1-Penten-3-ol	33.73	1.37	25	7.35	6.66	0.71
3-Methylbutanol	224.42	1.13	199	13.17	25.30	0.58
1-Pentanol	139.68	1.29	109	11.22	15.14	0.54
1-Hexanol	937.23	1.18	797	164.77	187.10	0.76
(E)-2-Hexen-1-ol	85.68	1.16	74	14.53	14.97	0.66
(Z)-3-Hexen-1-ol	557.15	1.44	387	77.39	95.52	0.67
6-Methyl-5-hepten-2-ol	145.66	1.26	115	19.33	26.22	0.39
2-Phenylethanol	13.58	0.11	119	2.32	3.13	0.44
Methyl salicylate	2.16	0.12	18	0.52	0.45	0.30
Beta-ionone	0.98	0.11	9	0.43	0.25	0.41
1-Penten-3-one	5.44	0.12	44	1.03	1.16	0.56
6-Methyl-5-hepten-2-one	73.34	0.14	540	10.39	13.06	0.65
Geranylacetone	54.62	0.11	476	5.07	6.40	0.42
Eugenol	44.35	1.19	37	18.04	12.38	0.36
2-Isobutylthiazole	78.94	0.13	590	6.87	10.96	0.55
Limonene	43.12	1.55	28	11.25	10.55	0.47
2-Pentylfuran	91.84	1.13	81	24.56	19.94	0.40
Beta-cyclocitral	89.26	1.12	80	13.73	14.61	0.58
Geranial	355.69	0.29	1238	62.63	63.18	0.64
Neral	132.50	2.58	51	26.44	33.93	0.68
(Z)-3-Hexenal	524.47	1.16	451	60.15	96.12	0.75
(E)-2-Hexenal	131.58	1.44	92	12.38	16.80	0.71
Hexanal	666.31	1.44	463	79.14	114.64	0.69
(Z)-2-Heptenal	98.58	1.38	72	15.39	14.43	0.52
(E,E)-2,4-Heptadienal	75.54	1.20	63	10.62	12.87	0.60
(E,E)-2,4-Nonadienal	91.25	1.20	76	13.92	16.21	0.53
Beta-damascenone	13.15	1.22	11	4.98	3.35	0.32
(E)-2-Pentenal	30.56	1.69	18	9.71	8.67	0.54

Max, Maximum; Min, minimum; SD, standard deviation; H^2, heritability.

correlated with eugenol, (Z)-2-heptenal, and (f)-2,4-heptadienal ($r = -0.406$, -0.266, and -0.241, respectively).

We observed that volatiles derived from the same pathway had a tendency to cluster together (**Figure 2**). In addition, we observed large variation among the 28 volatiles in the cluster analysis for all of the accessions (**Figure 2**). The largest variations were mainly observed for 1-hexenol, hexanal, (Z)-3-hexen-1-ol, (Z)-3-hexenal. Additionally, these four volatiles were all linolenic acid-derived flavor molecules and were closely clustered. (Z)-3-hexen-1-ol can be derived from (Z)-3-hexenal by alcohol dehydrogenase (ADH). Similarly, 1-hexenol can also be derived from hexanal via ADH. Volatiles, structures, identification and their precursors of the selected 28 volatiles used in this research are presented in Table S3.

Molecular Polymorphism

The 174 sampled accessions were genotyped with 182 SSRs. We selected only markers with MAF > 5% for association mapping (Zhang et al., 2015). The distributions of the MAF were different among the three groups of accessions. The average MAF for all of the accessions, *S. l. cerasiforme* accessions and *S. lycopersicum* accessions has been described by Zhang et al. (2015).

LD decay was analyzed for all markers on all chromosomes for the 174 accessions. Pairwise r^2 was plotted according to the chromosome genetic distance between loci. Non-linear regression was fitted to the decay of LD over genetic distance. LD on the whole genome for all accessions extended on average over 8 cM for $r^2 = 0.2$ (**Figure 3**).

Population Structure

We assessed population structure of the 174 tomato accession using STRUCTURE 2.3.3 software with 182 SSR markers. We found an optimal $K = 2$ inferred according to Evanno method (Evanno et al., 2005). The inferred population was congruent with *S. l. cerasiforme* and *S. Lycopersicum* accessions, respectively (Zhang et al., 2015).

Genome-wide Association Analysis

In order to reduce false positive associations, we used the K+Q model (kinship matrix and genetic structure) to detect associations between the selected 28 volatiles and 182 SSR markers. We analyzed the phenotypic data in 2013 and 2014 separately. In total, we detected 125 marker-trait associations (MTAs) on 28 selected volatiles in 2013 and 2014 ($P < 0.005$)

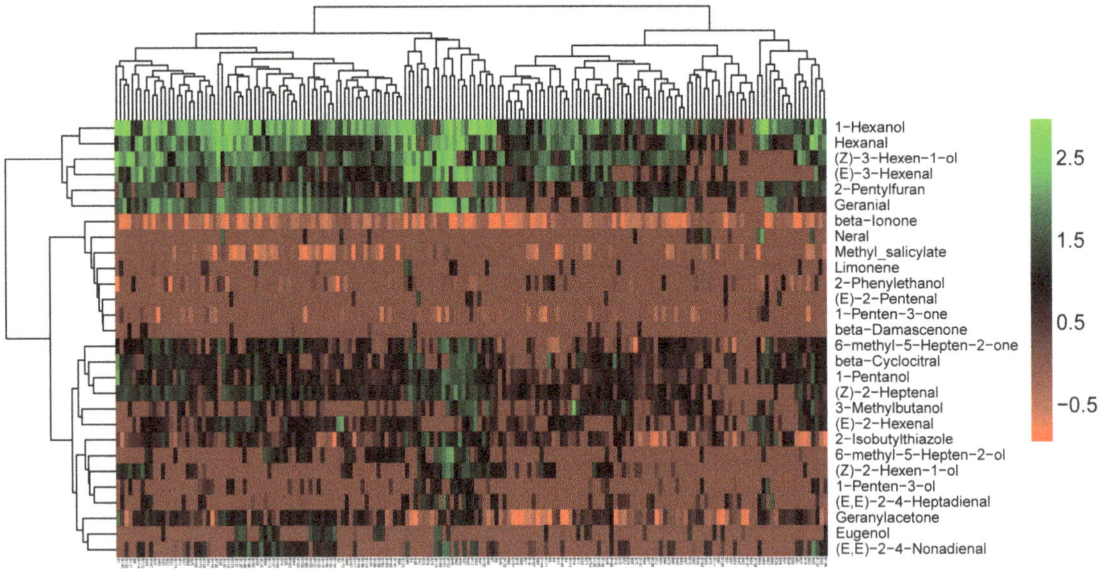

FIGURE 2 | Cluster analyses of the 28 selected volatiles among the whole accessions. The names of volatiles (right) and the accession codes (bottom) are shown (The accession codes can be seen in Table S1).

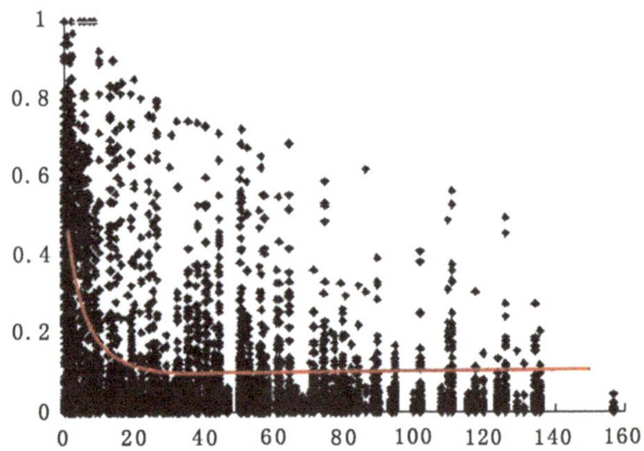

FIGURE 3 | Estimates of LD (r^2) over genetic distance on all chromosomes for all 174 tomato accessions. Only polymorphic sites with MAF >0.05 are indicated (see Materials and Methods). Plot of r^2 over genetic distance if fitted by non-linear regression (red curve).

(**Table 2**). Among these, 52 MTAs were detected in both years. Twenty-nine of them are significant associations ($P < 0.0003$). In 2013, 2014, we detected 82, 95 MTAs, respectively. We detected at least one MTA for each volatile. The only exception is for eugenol and we detected no MTAs for this volatile. The most significant MTA for all volatiles in both year was detected on 6-methyl-5-hepten-2-one. This MTA was detected on TES344 on chromosome 11 (Chr11), explaining 36.38, 33.25% of the phenotypic varation, in 2013 and 2014, respectively. The other most significant MTA was detected for (E)-2-hexen-1-ol. This MTA was detected on SSR 287 (Chr2), explaining 44.51, 42.13% of the phenotypic variation in 2013 and 2014, respectively.

Carotenoid-derived Volatiles

Of the 28 volatiles selected for association mapping, there are three important open chain carotenoid-derived volatiles, 6-methyl-5-hepten-2-one, 6-methyl-5-hepten-2-ol and geranylacetone. There are another three cyclic carotenoid-derived volatiles, including beta-ionone, beta-cyclocitral, and beta-damascenone. For 6-methyl-5-hepten-2-one, derived by oxidative cleavage of lycopene, five MTAs were detected. Among these, the associated marker TES344 (Chr11) showed the most significant association value ($P = 1.51E-26$, in 2013; $P = 1.84E-24$, in 2014) among all significant MTAs detected for 28 volatiles. For 6-methyl-5-hepten-2-ol, which is directly biochemically linked with 6-methyl-5-hepten-2-one via ADH, we detected six significant MTAs. Among these, three MTAs were detected in both years. The most significant associated marker was TOM166 (Chr3) in 2013, which explained about 19.69% of the total phenotypic variation. We also detected significant association on this marker in 2014. The significantly level is relatively lower, explaining 7.85% of the phenotypic variation. This marker was also significantly associated with 6-methyl-5-hepten-2-one, and explained approximately 9.79, 10.23% of the phenotypic variation, in 2013 and 2014, respectively. Marker TGS827 (Chr3) was also associated with 6-methyl-5-hepten-2-one and 6-methyl-5-hepten-2-ol, and explained 4.86 and 8.56% of the phenotypic variation in 2013, respectively. However, we observed no significant MTAs on both volatiles on this marker in 2014. For geranylacetone, we found 10 MTAs, and five of them were detected in both 2013 and 2014. The most significant of these was with marker SSR122 (Chr6), which explained 16.92, 15.65% of the phenotypic variation, respectively. The other most significant MTA of these was with SSR142 (Chr9). This MTA could explain 19.02, 17.65% of the phenotypic variation, in both years, respectively.

TABLE 2 | Marker-trait associations for 28 volatiles greatly affecting tomato flavor estimated with K+Q (MLM) model on 174 tomato accessions (only those where $P < 0.005$ are listed).

Compound	Locus	Chr[a]	Position[b]	2013		2014	
				Corrected *P*-value[c]	R^2	Corrected *P*-value	R^2
1-Penten-3-ol	TGS925	7	41.54	ns	–	0.0028	0.0562
	TES246	11	63.67	0.0044	0.043	ns	–
3-Methylbutanol	SSR92	1	0	4.07E-12	0.3069	6.87E-09	0.2171
	SSR26	2	77.5	0.0043	0.0487	1.12E-05	0.1213
	LEgata001	3	NG[d]	4.59E-04	0.1049	ns	–
	SSR320	3	107.3	ns	–	1.82E-04	0.1196
	SSR150	4	115	6.20E-06	0.147	3.18E-05	0.1078
	SSR325	5	18.5	ns	–	0.0026	0.0667
	SSR13	5	28	4.90E-12	0.3225	5.52E-10	0.2176
	TGS266	6	32.26	ns	–	0.0011	0.066
	TES1966	10	88.92	2.13E-04	0.0493	0.0011	0.1002
	TES1449	11	73.38	4.66E-10	0.191	1.12E-11	22.13
1-Pentanol	TES291	1	62.98	ns	–	0.0021	0.0539
	SSR598	2	78	0.0041	0.0572	ns	–
	SSR320	3	107.3	ns	–	0.0047	0.0657
	TES835	3	123.55	ns	–	0.0014	0.0891
	SSR133	4	30.6	8.73E-09	0.2672	7.45E-10	0.2534
	SSR345	12	72.5	6.20E-09	0.2284	3.78E-10	0.2215
1-Hexanol	SSR306	4	56	0.0034	0.0677	ns	–
	TES623	9	83.56	0.0016	0.0357	0.0035	0.0356
	TGS551	10	32.33	0.0043	0.0318	ns	–
(E)-2-Hexen-1-ol	SSR287	2	107	1.61E-24	0.4451	3.92E-24	0.4213
	TGS639	5	74.02	ns	–	0.0032	0.052
	TGS354	8	30.65	0.0016	0.0581	ns	–
	TGS607	8	37.89	0.0050	0.0413	ns	–
	TGS560	9	78.87	ns	–	0.0026	0.1222
	TGS738	10	1.6	0.0045	0.0424	ns	–
(Z)-3-Hexen-1-ol	SSR188	4	135.5	6.72E-04	0.0789	3.28E-04	0.0838
	TES99	5	125.75	0.0011	0.0255	ns	–
	TES628	6	2.23	1.88E-04	0.0789	4.28E-04	0.0759
	TES1427	8	16.59	0.0046	0.033	ns	–
	TGS607	8	37.89	0.0049	0.0307	0.0047	0.0356
	TES40	10	56.34	0.0044	0.0372	ns	–
6-Methyl-5-hepten-2-ol	SSR92	1	0	ns	–	1.21E-04	0.0079
	SSR20	2	58	3.53E-06	0.1458	6.35E-04	0.1011
	TES88	2	156.68	9.61E-04	0.0514	ns	–
	TGS827	3	4.42	1.89E-04	0.0856	ns	–
	SSR128	6	35	0.0021	0.1056	9.35E-04	0.0642
	TOM166	9	3.1	1.37E-07	0.1969	0.0049	0.0785
2-Phenylethanol	TES835	3	123.55	ns	–	0.0022	0.0844
	TES945	6	87.82	ns	–	9.35E-04	0.0642
	TES1521	7	79.06	3.84E-05	0.1059	7.87E-04	0.062
	TGS868	8	6.57	0.0018	0.1079	0.0016	0.0576
	TOM166	9	3.1	3.18E-04	0.0919	0.0044	0.0608

(Continued)

TABLE 2 | Continued

Compound	Locus	Chr[a]	Position[b]	2013		2014	
				Corrected *P*-value[c]	R^2	Corrected *P*-value	R^2
Methyl salicylate	TGS266	6	32.26	ns	–	2.53E-04	0.0657
	TGS1380	9	49.16	ns	–	0.0011	0.0511
	TES718	10	83.6	ns	–	0.0037	0.0467
Beta-ionone	SSR598	2	78	ns	–	5.28E-04	0.0701
	SSR45	7	75.5	ns	–	3.12E-05	0.0694
	TES1407	10	42.91	0.0050	0.0452	ns	–
	TES718	10	83.6	8.00E-04	0.0705	0.0018	0.033
	TES124	11	73.69	0.0049	0.0613	ns	–
1-Penten-3-one	TES363	5	1.54	ns	–	4.33E-04	0.0808
	TES628	6	2.23	ns	–	0.0030	0.0675
	TGS925	7	41.54	ns	–	9.03E-04	0.0734
	TES1427	8	16.59	0.0036	0.0513	0.0018	0.0536
6-Methyl-5-hepten-2-one	TGS827	3	4.42	0.0043	0.0486	ns	–
	SSR188	4	135.5	4.50E-07	0.1789	5.71E-06	0.1542
	TES36	9	0.73	0.0011	0.1202	ns	–
	TOM166	9	3.1	0.0010	0.0979	5.53E-04	0.1023
	TES344	11	51.42	1.51E-26	0.3638	1.84E-24	0.3325
Geranylacetone	SSR133	4	30.6	0.0016	0.1339	0.0012	0.143
	SSR325	5	18.5	ns	–	0.0032	0.0676
	SSR13	5	28	ns	–	8.13E-04	0.0423
	TES945	6	87.82	4.36E-04	0.0786	ns	–
	SSR122	6	101	1.19E-06	0.1692	1.69E-05	0.1565
	TGS2911	6	93.92	2.07E-04	0.0971	0.0037	0.054
	TES520	7	0.04	3.11E-04	0.0819	ns	–
	TGS2132	8	19.52	0.0018	0.0589	ns	–
	SSR142	9	16.5	2.27E-06	0.1902	1.09E-05	0.1765
	SSR110	9	55.7	2.72E-05	0.1439	4.67E-05	0.1322
2-Isobutylthiazole	SSR287	2	107	ns	–	3.21E-04	0.1052
	TES1179	6	40.18	ns	–	0.0013	0.1185
	SSR344	8	4	6.88E-04	0.1279	9.35E-04	0.0636
	TGS738	10	1.6	ns	–	0.0015	0.635
Limonene	TES816	6	67.53	1.11E-04	0.0955	0.0023	0.0741
	SSR142	9	16.5	ns	–	6.87E-04	0.0254
	TGS560	9	78.87	ns	–	5.28E-04	0.1529
2-Pentylfuran	TES1179	6	40.18	ns	–	0.0026	0.0782
Beta-cyclocitral	TGS1548	2	77.52	6.97E-06	0.143	5.51E-05	0.149
	TES1276	2	82.99	8.43E-06	0.1406	7.20E-05	0.1532
	TGS364	5	46.19	0.0043	0.0629	ns	–
	SSR128	6	35	6.75E-05	0.1561	8.14E-04	0.067
	TGS1973	6	80.07	0.0047	0.0632	2.41E-04	0.086
	TGS1629	9	6.41	ns	–	0.0014	0.0067
Geranial	SSR150	4	115	0.0041	0.0527	5.05E-05	0.0345
	TES1179	6	40.18	ns	–	9.86E-04	0.0453
	TES520	7	0.04	3.09E-04	0.0659	ns	–
	TGS2132	8	19.52	3.96E-04	0.0623	0.0021	0.0583
	SSR142	9	16.5	0.0028	0.075	0.0025	0.0702
	SSR110	9	55.7	0.0043	0.0597	0.0030	0.0583

(Continued)

TABLE 2 | Continued

Compound	Locus	Chr[a]	Position[b]	2013		2014	
				Corrected *P*-value[c]	R^2	Corrected *P*-value	R^2
Neral	SSR92	1	0	ns	–	2.72E-04	0.1422
	TGS420	3	21.07	0.0011	0.0807	8.49E-04	0.932
	SSR276	7	18	3.25E-04	0.0884	ns	–
	TGS153	12	28.96	0.0050	0.0398	0.0037	0.0796
(Z)-3-Hexenal	TGS1032	9	30.1	ns	–	0.0032	0.0143
(E)-2-Hexenal	TGS827	3	4.42	ns	–	0.0039	0.0322
	SSR342	6	19.1	0.0024	0.0728	ns	–
Hexanal	TES287	4	69.72	ns	–	0.0011	0.0697
	TES36	9	4.22	ns	–	0.0017	0.0621
	TES1028	9	83.73	ns	–	0.0019	0.0587
(Z)-2-Heptenal	TES358	5	60.96	0.0031	0.0348	ns	–
	TGS467	6	42.82	1.45E-05	0.1021	5.28E-04	0.0794
	TES1899	6	70.98	7.84E-04	0.0437	ns	–
	TGS1032	9	30.1	1.10E-08	0.1876	7.01E-08	0.1625
	TGS1380	9	49.16	3.28E-06	0.1017	1.39E-06	0.1519
	TES340	9	49.98	3.28E-06	0.1017	1.39E-06	0.1517
	TES1246	9	54.71	1.84E-05	0.0731	4.05E-04	0.0845
	TGS1713	9	56.86	3.63E-06	0.102	7.35E-06	0.1518
	SSR80	11	16.8	0.0021	0.042	ns	–
(E,E)-2,4-Heptadienal	SSR266	1	32.7	ns	–	0.0019	0.0779
	SSR20	2	58	0.0031	0.0687	ns	–
	TES748	3	88.04	ns	–	0.0016	0.0625
	TES618	12	15.07	ns	–	8.10E-04	0.0646
(E,E)-2,4-Nonadienal	TGS639	5	74.02	ns	–	1.14E-04	0.0984
	TES628	6	2.23	ns	–	5.28E-04	0.0969
	TES36	9	0.73	0.0041	0.0974	0.0011	0.1318
Beta-damascenone	SSR133	4	30.6	0.0014	0.1372	7.18E-06	0.1045
	TES816	6	67.53	3.69E-06	0.1355	1.69E-05	0.1364
	SSR122	6	101	6.56E-06	0.1286	5.98E-05	0.065
	SSR142	9	16.5	ns	–	0.0024	0.0751
	SSR110	9	55.7	ns	–	0.0036	0.0631
(E)-2-Pentenal	SSR320	3	107.3	0.0038	0.0762	ns	–
	SSR188	4	135.5	0.00	0.0619	ns	–
	TES363	5	1.54	6.19E-04	0.0707	0.0041	0.0467

[a]*Chromosome.*
[b]*Genetic distance of the marker was not found in EXPEN2000 reference map (http://www.solgenomics.net).*
[c]*P-values are corrected following the Benjamini and Hochberg (1995) procedure (see Materials and Methods).*
[d]*Genetic distance of the marker was not found in EXPEN2000 reference map (http://www.solgenomics.net).*
ns, no significant; –, not given.

Significant MTAs were also observed for three cyclic carotenoid-derived volatiles, with five, six and five MTAs, respectively. The most significant MTA for beta-cyclocitral was detected on marker TGS1548 (Chr2), both in 2013 and 2014. This MTA explained 14.3, 14.9% of the phenotypic variation, respectively. The most significant association for beta-damascenone was detected on TES816 (Chr6), explaining 13.55, 13.64% of the phenotypic variation, in 2013 and 2014, respectively.

Lipid-derived Volatiles

Of the 28 volatiles in sufficient quantities to impact the tomato flavor, we discovered MTAs for 11 lipid-derived volatiles (**Table 2**). For (E)-2-hexen-1-ol, we found six MTAs. The most significantly associated marker was SSR287 (Chr2). This MTA represented one of the most significant MTAs for all 28 volatiles, and explained 44.51, 42.13% of the total phenotypic variation, in 2013 and 2014, respectively. Only two MTAs were detected for (E)-2-hexenal, with one MTA in each year, respectively. For

1-penten-3-one, four MTAs were detected either in 2013 or 2014. For 1-penten-3-ol, which is directly biochemically linked with 6-methyl-5-hepten-2-one via ADH, only two MTAs were detected. For 1-pentanol, six MTAs were detected and two had a high association value. The two associated markers were SSR345 (Chr12) and SSR133 (Chr4). The associated marker SSR345 explained 22.84, 22.15% of the phenotypic variation, in 2013, 2014, respectively. Marker SSR133 accounted for 26.72, 25.34% of the phenotypic variation, respectively. For (Z)-2-heptenal, we discovered nine significant MTAs, and six of them were detected both in 2013 and 2014. Among the nine MTAs, five of which were located on Chr9 in the region from 30.1 to 56.86 cM. The most significant associated marker was TGS1032, located at about 30.1 cM on Chr9. This MTA which explained 18.76, 16.25% of the phenotypic variation, in 2013, 2014, respectively.

Amino Acid-derived Volatiles

For 3-methylbutanol, a leucine-derived flavor volatile, we found 10 MTAs. The two most significant MTAs involved with marker SSR92 (Chrl) and SSR13 (Chr5). These two MTAs responsible for 30.69 and 32.25% of the total phenotypic variation in 2013, respectively. In 2014, they accounted for 21.71, 21.76% of the total phenotypic variation. For 2-phenylethanol, a phenylalanine-derived volatile, we observed five MTAs in either 2013 and 2014. The most significant marker was TES1521 (Chr7), explaining 9.19, 6.2% of the phenotypic variation, respectively.

Terpenoid-derived Volatiles

The two most important terpenoid-derived volatiles in tomato are neral and geranial, which are primarily localized in tomato leaves and stems (Buttery and Ling, 1993). However, we still observed these two volatiles in fruits in many tomato accessions. We observed four MTAs for neral and six for geranial in either year. For neral, the most significant MTA was detected on SSR92 (Chrl) in 2014, explaining about 14% of the phenotypic variation. No significance was observed on this marker in 2013. For geranial, the most significant MTA was also detected in 2014. This MTA was detected on marker SSR150 (Chr4), accounting for 3.45% of the phenotypic variation.

DISCUSSION

Genome-wide association study (GWAS) is a useful tool to detect candidate loci responsible for the natural variations in a targeted phenotype. This tool can identify significant associations between polymorphic molecular markers and targeted traits in a large natural population (Weigel, 2012; Matsuda et al., 2015). However, many factors can impact the results of association mapping, including type and size of mapping population, targeted traits, number of environments and years for phenotypic evaluations and the type and genome coverage of molecular markers (Ruggieri et al., 2014). Thus, we used a large sample of cherry and large fruited tomato accessions and a MLM to reduce false positive associations in GWAS (Zhao et al., 2007; Bernardo, 2008). The population in our study composes 123 cherry tomato accessions and 51 large fruited accessions and we think the size of our collection was adequate for GWAS (Ranc et al., 2012; Xu et al., 2013; Ruggieri et al., 2014).

Phenotypic and Genetic Diversity

This whole studied population composes 123 cherry tomato and 51 large fruited accessions, representing a large diversity (**Table 1**). We found that this population has a large phenotyic diversity, such as fruit weight, soluble solid content, and lycopene content, etc. (Zhang et al., 2015). The large variations of the selected 28 crucial volatiles up to 1200x confirmed this (**Table 2**). The studied population could be mainly divided into two subgroups, cherry and large-fruited (Zhang et al., 2015). The higher MAF value among the studied population confirmed that *S. l. cerasiforme* (cherry tomato) is a mosaic of *S. lycopersicum* (large fruited tomato) and *S. pimpinellifolium* (wild species) (Frary et al., 2000; Zhao et al., 2007; Ranc et al., 2008; Xu et al., 2013; Zhang et al., 2015). The linkage disequilibrium of the whole genome decays at about 8 cM, which is consistent with previous studies (van Berloo et al., 2008; Xu et al., 2013; Zhang et al., 2015). Therefore, using a large collection of cherry tomato accessions together with cultivated tomato accessions is useful to overcome the high linkage disequilibrium value of tomato genome (Xu et al., 2013; Zhang et al., 2015).

Associations Confirmed Identified Volatile QTLs

The tomato is an excellent model for investigating the molecular basis of flavor using association mapping (Klee and Tieman, 2013). To date, few association mapping studies have focused on volatiles in major crops (Kumar et al., 2015). Here, we conducted GWAS between 28 most volatiles in tomato and SSR markers and found significant MTAs for most of the studied volatiles (**Table 2**). In tomato, over 50 QTLs affecting volatile levels have been identified, mainly using recombinant inbred lines (RIL) or introgression lines (IL) (Saliba-Colombani et al., 2001; Tieman et al., 2006; Mathieu et al., 2008; Zanor et al., 2009). However, the size of introgressed regions is large large (about 10–40 cM) (Zanor et al., 2009) and the results among prior studies differed. For example, prior studies have found different QTLs for the 6-emthyl-5-hepten-2-one, an important carotenoid-derived volatile. In particular, one major QTL *mhn4.1* impacting 6-emthyl-5-hepten-2-one was detected on chromosome 4 (Chr4) using an introgression line (Saliba-Colombani et al., 2001). However, two different QTLs were detected on 2A, 3C, and 12D in other IL populations (Tieman et al., 2006; Mathieu et al., 2008; **Figure 4**). In our research, we found six significant associations ($P < 0.005$) for 6-emthyl-5-hepten-2-one. Among them, the most significant associations were detected on Chr11 (TES344) and Chr4 (SSR188). These two associations were also found in the near region of two QTLs for 6-emthyl-5-hepten-2-one in a previous study by Tieman et al. (2006).

Twenty-five significant MTAs were detected for 15 volatiles on Chr4 (**Table 2**). The associations observed on Chr4 showed support for five previously-identified QTLs, including QTLs for3-methylbutanol and (E)-2-pentenal by Tieman et al. (2006); QTLs detected for 2-phenythanol and 1-penten-3-ol by Mathieu et al. (2008); and one QTL detected for eugenol by Zanor

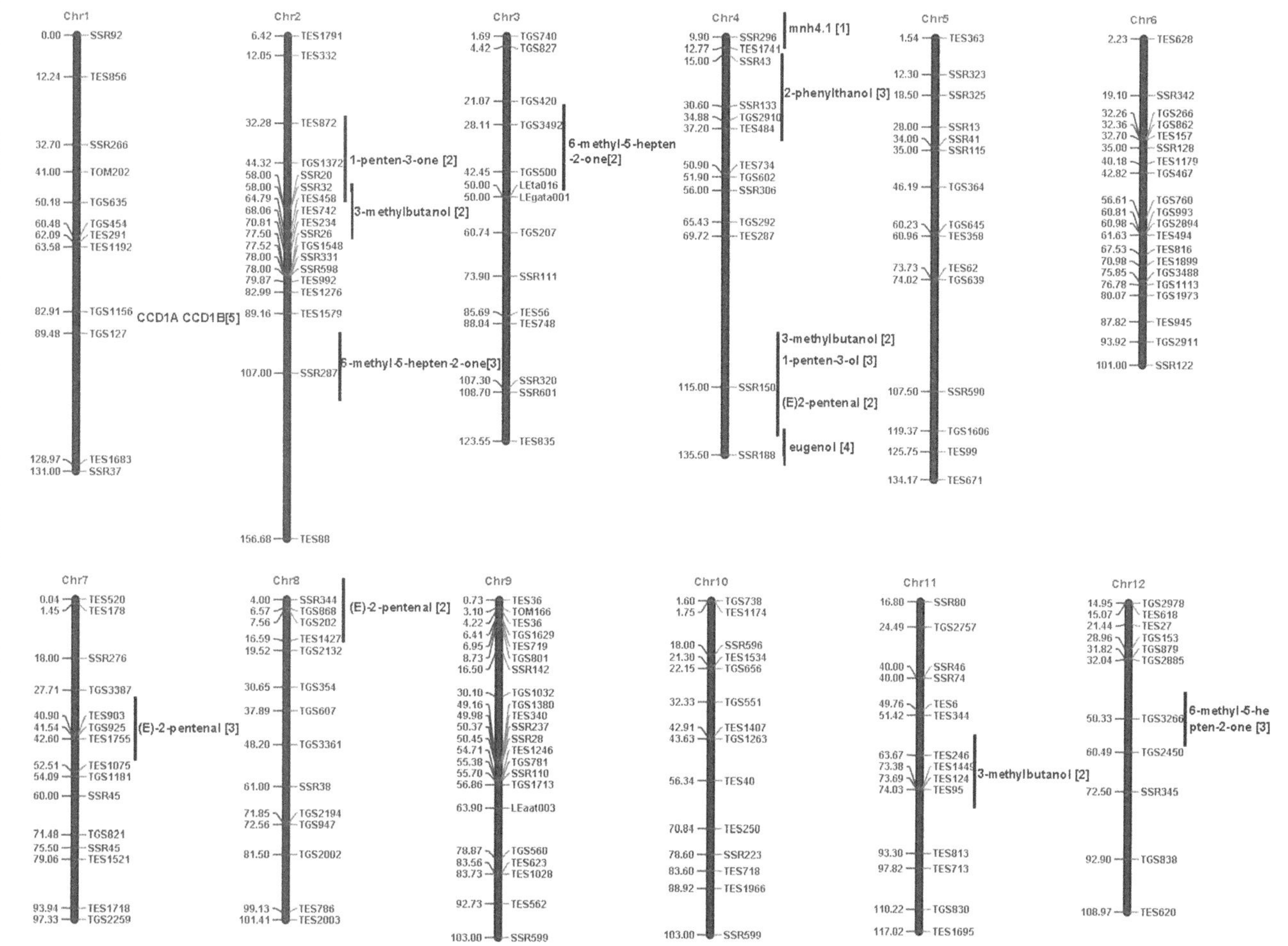

FIGURE 4 | Comparison of associations and QTLs identified by linkage mapping. Horizontal line corresponds to the genetic location of associated marker **(right)** and the associated volatiles **(left)**. Vertical line is the approximate regions of the identified QTLs. QTLs identified by Saliba-Colombani et al. (2001) were indicated by [1] to the end of volatiles; QTLs identified by Tieman et al. (2006) were indicated by adding [2] to the end of volatiles; QTLs identified by Mathieu et al. (2008) were indicated by [3] to the end of volatiles; QTLs identified by Zanor et al. (2009) were indicated by [4] to the end of volatiles. CCD, carotenoid cleavage dioxygenases. CCD1A and CCD1B are two genes from the tomato genome data and were indicated by [5] to the end of genes. In Simkin et al. (2004), these two genes were listed as *LeCCD1A* and *LeCCD1B*.

et al. (2009). However, only a few co-localized QTLs with the significant associations were observed on Chr4 (**Figure 4**). This could be mainly due to the limited volatiles sampled in previous studies and the limited molecular markers (Saliba-Colombani et al., 2001; Tieman et al., 2006; Mathieu et al., 2008; Zanor et al., 2009). In fact, among the28 volatiles that we sampled, only about 10 volatiles were mentioned in previous studies. Based on previous GWAS on tomato, the LD of tomato genome decayed at approximately 5–20 cM (Mazzucato et al., 2008; van Berloo et al., 2008; Ranc et al., 2012; Xu et al., 2013). Therefore, a 5–10 cM genome coverage should be enough to detect positive associations, especially by using SSRs. Our research revealed more polymorphic loci impacting tomato volatile profiles. However, among the detected 125 MTAs, only 52 of them were detected in both years. The overall significance value is still relatively low. A 5.2 cM genome coverage could detect positive associations. Still, more markers are needed to have a higher genome resolution to detected more candidate QTLs or genes. In addition, the tomato genome data has been available. In order to conduct more efficient association mapping, marker assisted selection and fine mapping of QTLs, high-throughput SNP chips via conducting re-sequencing on the core tomato accessions would greatly promote our further research.

Volatile Biosynthesis Pathways

Tomato volatiles are mainly derived from four pathways, including the fatty acid, amino acid, terpenoid, and carotenoid pathways. The metabolic pathways of the selected volatiles in this study are shown in **Figure 5**. All volatiles used in this study are directly or indirectly linked with the tricarboxlic acid cycle (TCA), indicating the fundamental significance of primary metabolism. However, our understanding of the biosynthesis pathways and regulatory networks is only known for a small

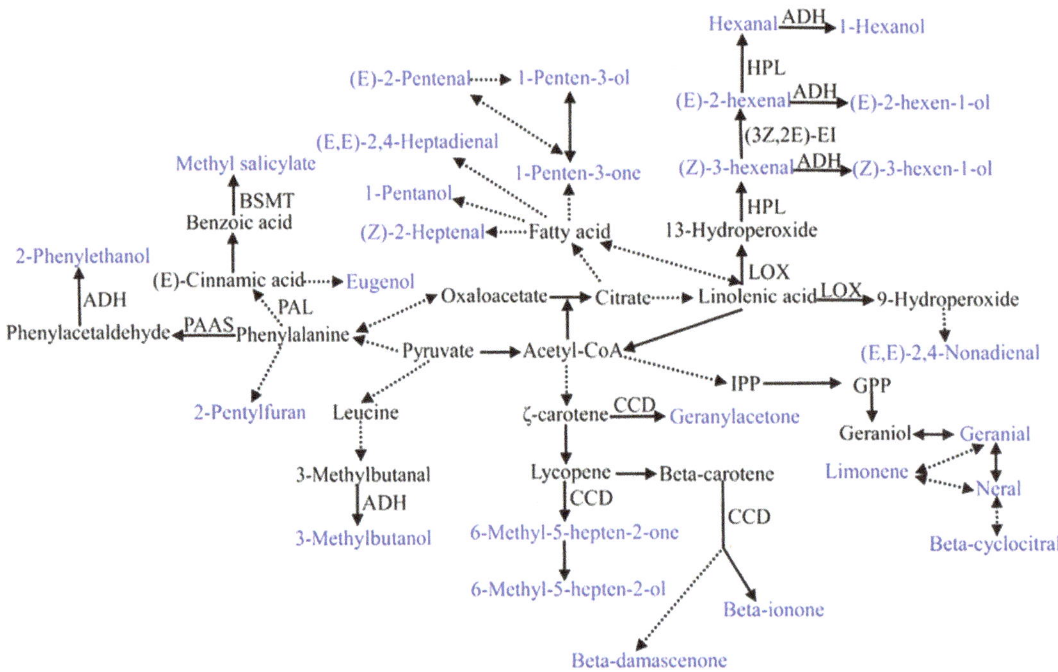

FIGURE 5 | Summary of metabolic pathways leading to the 28 important flavor-associated volatile synthesis. Volatiles used in this study are shown in blue. The precursor for 2-isobutylthiazole is not clear and is not listed in this summary. Dashed lines indicate multiple step reactions. Enzymes or genes involved in some volatile synthesis are listed. BSMT, benzoic acid and salicylic acid carboxyl methyltransferase; PAAS, phenylacetaldehyde synthase; PAL, phenylalanine ammonia-lyase; CCD, carotenoid cleavage dioxygenase; LOX, lipoxygenase; ADH, alcohol dehydrogenase; HPL, hydroperoxide lyaser; 3*Z*,2*E*-EI, 3*Z*,2*E*-enal isomerase; IPP, isopentenyl pyrophosphate; GPP, geranyl diphosphate.

portion of the most economically significant voaltiles (Klee, 2010). Even for some most important volatiles in tomato, the synthetic pathways have been recently established or remain unknown. For instance, the precursor and the corresponding regulation pathways is unknown for 2-isobutylthiazole, one important sulfur-containing compound in tomato (Iranshahi, 2012). Using significant correlations between traits could be used to build the network structure of the poorly known pathways (Carli et al., 2009).

Volatile Biosynthetic Genes

Knowledge of synthetic pathways and the regulatory networks can greatly facilitate the identification of genes encoding biosynthetic enzymes. This can be accomplished by exploiting the whole genomic or expressed sequence databases (Klee, 2010). For instance, Klee and Tieman (2013) reviewed several genes with validated functions in the metabolism of tomato volatiles, including PAR, phenylacetaldehyde reductase; loxC, 13-lipoxygenase; CCD1, carotenoid cleavage dioxygenase; and CXE1, carboxylesterase. LoxC catalyzes the first step in the metabolic pathway that converts 18:2 and 18:3 fatty acids to C6 volatiles, including (Z)-3-hexenal, hexanal, (Z)-3-hexen-1-ol, hexyl alcohol, and hexylacetate (Chen et al., 2004). Volatile terpenoid compounds, including neral, geranial, limonene and beta-cyclocitral, etc. could potentially be derived from carotenoids. These volatiles are all important component of flavor and aroma in tomato (Simkin et al., 2004). *LeCCD1A* (82,184,585–82,195,219 bp) and *LeCCD1B* (82,194,422–82,212,510 bp) are two closely related genes located on chromosome 1 potentially encoding carotenoid cleavage dioxygenases and LeCCD1B. *LeCCD1A* had a great impact on the concentration of beta-ionone and geranylacetone and *LeCCD1B* had a high expression level in ripening fruit (Simkin et al., 2004). In our search, we identified one significant MTA for 6-methyl-5-hepten-2-ol, one important volatile derived from lycopene and another significant MTA for geraylacetone. These two associations were both associated with marker TGS1156. However, the significant level was relatively low and not all carotenoid-derived volatiles were associated with this marker. This could due to the limited markers in this region or the weak marker polymorphic linkage with these two genes. At least five lipoxygenases (TomloxA, TomloxB, TomloxC, TomloxD, and TomloxE) in tomato have been identified. They can greatly impact the generation of C6 aldehyde and alcohol volatiles derived from fatty acids, such as n-hexanal, (Z)-3-hexenal, (E)-2-hexenal, and (Z)-3-hexenol, in both fruit and leaf tissues (Chen et al., 2004). However, researchers have not yet established the biosynthetic pathways for many of the most important volatiles. Here, we selected the 28 most important volatiles in tomato to perform genome-wide association mapping. Our research points to some chromosome regions that may play a significant role in tomato volatile metabolism. However, the marker coverage was relatively small. Combining with the availability of the tomato genome data, and higher density of SSR marker coverage (or SNPs), this research will promote the isolation of novel genes impacting volatiles.

CONCLUSIONS

Phenotypic evaluation on the 28 most important tomato volatiles detected by GC-MS revealed a broad phenotypic variability within diverse accessions across tomato. GWAS between the selected 28 volatiles and 182 SSR markers allowed detection of 125 significant MTAs ($P < 0.005$). We detected at least one MTA for 27 volatiles. Notably, some associations had a very high significant value. For instance, we found a highly significant association between 6-methyl-5-hepten-2-one. This MTA accounted for up to 30% of the phenotypic variation in both year. The other most significant association was detected between (E)-2-hexen-1-ol and SSR287 (Chr2). Some associations were co-localized with previously identified QTLs. We identified several chromosome regions that could greatly impact tomato volatile metabolism. Our results represent a step toward accelerating the rate of flavor related gene discovery.

AUTHOR CONTRIBUTIONS

JZ and ZZ designed the study. JZ and JTZ carried out the main GC-MS analysis and molecular mapping, analyzed the data, and drafted the manuscript. YL provided the tomato seeds and participated in its design. ML participated in its design and the GC-MS analysis. YX, JL, PC, and FY participated in the phenotype analysis and molecular mapping. All authors corrected and approved the final version.

FUNDING

This work was supported by the National Agricultural Science Foundation (No. 201203002), the Program for New Century Excellent Talents in University (No. NCET-12-0474) and National Natural Science Foundation of China (Grant No. 31301498).

ACKNOWLEDGMENTS

The authors thank Dr. Fengwang Ma for his encouragement and helpful advice; as well as Dr. Zheng Li, Xiaohui Hu, and Yanxu Yin for their technical support. We also thank Priscilla Licht for her help in revising our English composition. We thank Dr. Yanhong Hu for the helping the R analyses. We gratefully acknowledge the assistance of Xinli Huang, Xiaoting Zhou, and Lipan Hu in harvesting the fruits for this study.

REFERENCES

Agarwal, M., Shrivastava, N., and Padh, H. (2008). Advances in molecular marker techniques and their applications in plant sciences. *Plant Cell Rep.* 27, 617–631. doi: 10.1007/s00299-008-0507-z

Baek, H. H., and Cadwallader, K. R. (1996). Volatile compounds in flavor concentrates produced from crayfish-processing byproducts with and without protease treatment. *J. Agric. Food Chem.* 44, 3262–3267. doi: 10.1021/jf960023q

Baldwin, E. A., Scott, J. W., Shewmaker, C. K., and Schuch, W. (2000). Flavor trivia and tomato aroma: biochemistry and possible mechanisms for control of important aroma components. *HortScience* 35, 1013–1022.

Benjamini, Y., and Hochberg, Y. (1995). Controlling the false discovery rate: a practical and powerful approach to multiple testing. *J. R. Statist. Soc. B* 57, 289–300.

Bernardo, R. (2008). Molecular markers and selection for complex traits in plants: learning from the last 20 years. *Crop Sci.* 48, 1649. doi: 10.2135/cropsci2008.03.0131

Brachi, B., Morris, G. P., and Borevitz, J. O. (2011). Genome-wide association studies in plants: the missing heritability is in the field. *Genome Biol.* 12:232. doi: 10.1186/gb-2011-12-10-232

Bradbury, P. J., Zhang, Z., Kroon, D. E., Casstevens, T. M., Ramdoss, Y., and Buckler, E. S. (2007). TASSEL: software for association mapping of complex traits in diverse samples. *Bioinformatics* 23, 2633–2635. doi: 10.1093/bioinformatics/btm308

Buttery, R. G., and Ling, L. (1993). Volatile compounds of tomato fruit and plant parts: relationship and biogenesis. *ACS Symposium* 525, 23–24.

Buttery, R. G., Teranishi, R., and Ling, L. C. (1987). Fresh tomato aroma volatiles: a quantitative study. *J. Agric. Food Chem.* 35, 541–544. doi: 10.1021/jf00076a025

Carli, P., Arima, S., Fogliano, V., Tardella, L., Frusciante, L., and Ercolano, M. R. (2009). Use of network analysis to capture key traits affecting tomato organoleptic quality. *J. Exp. Bot.* 60, 3379–3386. doi: 10.1093/jxb/erp177

Causse, M., Duffe, P., Gomez, M. C., Buret, M., Damidaux, R., Zamir, D., et al. (2004). A genetic map of candidate genes and QTLs involved in tomato fruit size and composition. *J. Exp. Bot.* 55, 1671–1685. doi: 10.1093/jxb/erh207

Chen, G., Hackett, R., Walker, D., Taylor, A., Lin, Z., and Grierson, D. (2004). Identification of a specific isoform of tomato lipoxygenase (*TomloxC*) involved in the generation of fatty acid-derived flavor compounds. *Plant Physiol.* 136, 2641–2651. doi: 10.1104/pp.104.041608

Doligez, A., Audiot, E., Baumes, R., and This, P. (2006). QTLs for muscat flavor and monoterpenic odorant content in grapevine (*Vitis vinifera* L.). *Mol. Breed.* 18, 109–125. doi: 10.1007/s11032-006-9016-3

El Hadi, M., Zhang, F., Wu, F., Zhou, C., and Tao, J. (2013). Advances in fruit aroma volatile research. *Molecules* 18, 8200–8229. doi: 10.3390/molecules18078200

Evanno, G., Regnaut, S., and Goudet, J. (2005). Detecting the number of clusters of individuals using the software STRUCTURE: a simulation study. *Mol. Ecol.* 14, 2611–2620. doi: 10.1111/j.1365-294X.2005.02553.x

Frary, A., Nesbitt, T. C., Grandillo, S., Knaap, E., Cong, B., Liu, J., et al. (2000). *fw2.2*: a quantitative trait locus key to the evolution of tomato fruit size. *Science* 289, 85–88. doi: 10.1126/science.289.5476.85

Fulton, T. M., Chunzoongse, J., and Tanksley, S. D. (1995). Microprep protocol for extraction of DNA from tomato and other herbaceous plants. *Plant Mol. Biol. Rep.* 13, 207–209. doi: 10.1007/BF02670897

Goff, S. A., and Klee, H. J. (2006). Plant volatile compounds: sensory cues for health and nutritional value? *Science* 311, 815–819. doi: 10.1126/science.1112614

Hardy, O. J., and Vekemans, X. (2002). SPAGeDi: a versatile computer program to analyse spatial genetic structure at the individual or population levels. *Mol. Ecol. Notes* 2, 618–620. doi: 10.1046/j.1471-8286.2002.00305.x

Hopfer, H., Ebeler, S. E., and Heymann, H. (2012). The combined effects of storage temperature and packaging type on the sensory and chemical properties of chardonnay. *J. Agric. Food Chem.* 60, 10743–10754. doi: 10.1021/jf302910f

Iranshahi, M. (2012). A review of volatile sulfur-containing compounds from terrestrial plants: biosynthesis, distribution and analytical methods. *J. Essent. Oil Res.* 24, 393. doi: 10.1080/10412905.2012.692918

Kader, A. A., Stevens, M. A., Albright-Holton, M., Morris, L. L., and Algazi, M. (1977). Effect of fruit ripeness when picked on flavor and composition in fresh market tomatoes. *J. Am. Soc. Hortic. Sci.* 102, 724–731.

Klee, H. J. (2010). Improving the flavor of fresh fruits: genomics, biochemistry, and biotechnology. *New Phytol.* 187, 44–56. doi: 10.1111/j.1469-8137.2010.03281.x

Klee, H. J., and Tieman, D. M. (2013). Genetic challenges of flavor improvement in tomato. *Trends Genet.* 29, 257–262. doi: 10.1016/j.tig.2012.12.003

Kumar, S., Rowan, D., Hunt, M., Chagné, D., Whitworth, C., and Souleyre, E. (2015). Genome-wide scans reveal genetic architecture of apple flavour volatiles. *Mol. Breed.* 35, 118–134. doi: 10.1007/s11032-015-0312-7

Mathieu, S., Cin, V. D., Fei, Z., Li, H., Bliss, P., Taylor, M. G., et al. (2008). Flavour compounds in tomato fruits: identification of loci and potential pathways affecting volatile composition. *J. Exp. Bot.* 60, 325–337. doi: 10.1093/jxb/ern294

Matsuda, F., Nakabayashi, R., Yang, Z., Okazaki, Y., Yonemaru, J., Ebana, K., et al. (2015). Metabolome-genome-wide association study dissects genetic architecture for generating natural variation in rice secondary metabolism. *Plant J.* 81, 13–23. doi: 10.1111/tpj.12681

Mazzucato, A., Papa, R., Bitocchi, E., Mosconi, P., Nanni, L., Negri, V., et al. (2008). Genetic diversity, structure and marker-trait associations in a collection of Italian tomato (*Solanum lycopersicum* L.) landraces. *Theor. Appl. Genet.* 116, 657–669. doi: 10.1007/s00122-007-0699-6

Obando-Ulloa, J. M., Ruiz, J., Monforte, A. J., and Fernández-Trujillo, J. P. (2010). Aroma profile of a collection of near-isogenic lines of melon (*Cucumis melo* L.). *Food Chem.* 118, 815–822. doi: 10.1016/j.foodchem.2009.05.068

Pers, T. H., Karjalainen, J. M., Chan, Y., Westra, H., Wood, A. R., Yang, J., et al. (2015). Biological interpretation of genome-wide association studies using predicted gene functions. *Nat. Commun.* 6:5890. doi: 10.1038/ncomms6890

Pritchard, J. K., Stephens, M., and Donnelly, P. (2000). Inference of population structure using multilocus genotype data. *Genetics* 155, 945–959.

Ranc, N., Munos, S., Santoni, S., and Causse, M. (2008). A clarified position for *Solanum lycopersicum* var. cerasiforme in the evolutionary history of tomatoes (solanaceae). *BMC Plant Biol.* 8:130. doi: 10.1186/1471-2229-8-130

Ranc, N., Muños, S., Xu, J., Le Paslier, M. C., Chauveau, A., Bounon, R., et al. (2012). Genome-wide association mapping in tomato (*Solanum lycopersicum*) is possible using genome admixture of *Solanum lycopersicum* var. cerasiforme. *G3 (Bethesda)* 2, 853–64. doi: 10.1534/g3.112.002667

Ratanachinakorn, B., Klieber, A., and Simons, D. H. (1997). Effect of short-term controlled atmospheres and maturity on ripening and eating quality of tomatoes. *Postharvest Biol. Technol.* 11, 149–154. doi: 10.1016/S0925-5214(97)00021-5

Ruggieri, V., Francese, G., Sacco, A., D'Alessandro, A., Rigano, M. M., Parisi, M., et al. (2014). An association mapping approach to identify favourable alleles for tomato fruit quality breeding. *BMC Plant Biol.* 14:337. doi: 10.1186/s12870-014-0337-9

Saidou, A. A., Thuillet, A. C., Couderc, M., Mariac, C., and Vigouroux, Y. (2014). Association studies including genotype by environment interactions: prospects and limits. *BMC Genet.* 15:3. doi: 10.1186/1471-2156-15-3

Saliba-Colombani, V., Causse, M., Langlois, D., Philouze, J., and Buret, M. (2001). Genetic analysis of organoleptic quality in fresh market tomato. 1. Mapping QTLs for physical and chemical traits. *Theor. Appl. Genet.* 102, 259–272. doi: 10.1007/s001220051643

Schauer, N., Semel, Y., Roessner, U., Gur, A., Balbo, I., Carrari, F., et al. (2006). Comprehensive metabolic profiling and phenotyping of interspecific introgression lines for tomato improvement. *Nat. Biotechnol.* 24, 447–454. doi: 10.1038/nbt1192

Simkin, A. J., Schwartz, S. H., Auldridge, M., Taylor, M. G., and Klee, H. J. (2004). The tomato carotenoid cleavage dioxygenase 1 genes contribute to the formation of the flavor volatiles β-ionone, pseudoionone, and geranylacetone. *Plant J.* 40, 882–892. doi: 10.1111/j.1365-313X.2004.02263.x

Sun, Y. D., Liang, Y., Wu, J. M., Li, Y. Z., Cui, X., and Qin, L. (2012). Dynamic QTL analysis for fruit lycopene content and total soluble solid content in a *Solanum lycopersicum* x *S. Pimpinellifolium* cross. *Genet. Mol. Res.* 11, 3696–3710. doi: 10.4238/2012.August.17.8

Tieman, D., Bliss, P., McIntyre, L. M., Blandon-Ubeda, A., Bies, D., Odabasi, A. Z., et al. (2012). The Chemical interactions underlying tomato flavor preferences. *Curr. Biol.* 22, 1035–1039. doi: 10.1016/j.cub.2012.04.016

Tieman, D. M., Zeigler, M., Schmelz, E. A., Taylor, M. G., Bliss, P., Kirst, M., et al. (2006). Identification of loci affecting flavour volatile emissions in tomato fruits. *J. Exp. Bot.* 57, 887–896. doi: 10.1093/jxb/erj074

Tikunov, Y., Lommen, A., de Vos, C. H., Verhoeven, H. A., Bino, R. J., Hall, R. D., et al. (2005). A novel approach for nontargeted data analysis for metabolomics. Large-scale profiling of tomato fruit volatiles. *Plant Physiol.* 139, 1125–1137. doi: 10.1104/pp.105.068130

van Berloo, R., Zhu, A., Ursem, R., Verbakel, H., Gort, G., and van Eeuwijk, F. A. (2008). Diversity and linkage disequilibrium analysis within a selected set of cultivated tomatoes. *Theor. Appl. Genet.* 117, 89–101. doi: 10.1007/s00122-008-0755-x

Weigel, D. (2012). Natural variation in Arabidopsis: from molecular genetics to ecological genomics. *Plant Physiol.* 158, 2–22. doi: 10.1104/pp.111.189845

Xu, J., Ranc, N., Muños, S., Rolland, S., Bouchet, J., Desplat, N., et al. (2013). Phenotypic diversity and association mapping for fruit quality traits in cultivated tomato and related species. *Theor. Appl. Genet.* 126, 567–581. doi: 10.1007/s00122-012-2002-8

Yu, J., Pressoir, G., Briggs, W. H., Vroh Bi, I., Yamasaki, M., Doebley, J. F., et al. (2005). A unified mixed-model method for association mapping that accounts for multiple levels of relatedness. *Nat. Genet.* 38, 203–208. doi: 10.1038/ng1702

Zanor, M. I., Rambla, J. L., Chaïb, J., Steppa, A., Medina, A., Granell, A., et al. (2009). Metabolic characterization of loci affecting sensory attributes in tomato allows an assessment of the influence of the levels of primary metabolites and volatile organic contents. *J. Exp. Bot.* 60, 2139–2154. doi: 10.1093/jxb/erp086

Zhang, J., Zhao, J., Liang, Y., and Zou, Z. (2015). Genome-wide association-mapping for fruit quality traits in tomato. *Euphytica*. doi: 10.1007/s10681-015-1567-0. [Epub ahead of print].

Zhao, K., Aranzana, M. J., Kim, S., Lister, C., Shindo, C., Tang, C., et al. (2007). An Arabidopsis example of association mapping in structured samples. *PLoS Genet.* 3:e4. doi: 10.1371/journal.pgen.0030004

Zini, E., Biasioli, F., Gasperi, F., Mott, D., Aprea, E., Märk, T. D., et al. (2005). QTL mapping of volatile compounds in ripe apples detected by proton transfer reaction-mass spectrometry. *Euphytica* 145, 269–279. doi: 10.1007/s10681-005-1645-9

Conflict of Interest Statement: The authors declare that the research was conducted in the absence of any commercial or financial relationships that could be construed as a potential conflict of interest.

17

Isolation and Screening of Bacteria for Their Diazotrophic Potential and Their Influence on Growth Promotion of Maize Seedlings in Greenhouses

Medhin H. Kifle * *and Mark D. Laing*

Discipline of Plant Pathology, School of Agricultural, Earth and Environmental Sciences University of KwaZulu-Natal, Pietermaritzburg, South Africa

Edited by:
Humberto Rodríguez Fuentes, Universidad Autónoma de Nuevo León, Mexico

Reviewed by:
Teresa Orlikowska, Research Institute of Horticulture, Poland
Katarzyna Turnau, Jagiellonian University, Poland

***Correspondence:**
Medhin H. Kifle
medhin7@yahoo.com

Poor soil fertility is one of the major constraints for crop production. Nitrogen is the most limiting nutrient for increasing crop productivity. Therefore, there is a need to identify diazotrophic inoculants as an alternative or supplement to N-fertilizers for sustainable agriculture. In the current study, a number of free-living diazotrophic bacteria were isolated from soils collected from maize rhizosphere and from leaves and roots of maize within the KwaZulu-Natal Province, Republic of South Africa. Ninety-two isolates were selected for further screening because they were able to grow on N-free media containing different carbon sources. Isolates that were very slow to grow on N-free media were discarded. The isolates were screened *in vitro* for diazotrophic potential tests for ammonia production and acetylene reduction. Ethylene (C_2H_4) production was quantified and ranged from 4 to 73 nmoles of $C_2H_4h^{-1}$ culture^{-1}. The top 20 isolates were re-screened on maize seedlings, and eight isolates significantly ($P = 0.001$) enhanced some growth parameters of maize above the un-inoculated control. Isolates that showed significant effect on at least two growth parameters were identified at species or genera level. In conclusion, selected diazotrophic isolates may be potentially beneficial but they should be tested more in greenhouse and field conditions with maize to confirm their potential for application as biofertilizers.

Keywords: diazotrophs, nitrogen fixation, plant growth promotion, acetylene reduction assay

INTRODUCTION

Replacement of chemical fertilizers with biofertilizers is an attractive goal for sustainable agriculture. Nitrogen is the macro-nutrient that most frequently limits the growth and productivity of non-leguminous plants (Schepers et al., 1992) and is the most limiting factor in maize production (McCarty and Meisinger, 1997). A number of diazotrophic bacteria were previously found to interact with plants either in the rhizosphere or endophytes. Given the ability of diazotrophs to fix N, some strains may relieve N-deficiencies where there is inadequate application of N fertilizers. The genera *Bacillus*, *Burkholderia*, and *Enterobacter* are known to penetrate the roots of cereals and grow intercellularly as root endophytes, as well as in the rhizosphere (Reinhold-Hurek and Hurek, 1998; Wakelin and Ryder, 2004).

Some diazotrophic bacteria including those in the genera: *Rhanella*, *Pantoea*, *Rhizobium*, *Pseudomonas*, *Herbaspirillum*, *Enterobacter*, *Brevundimonas*, and *Burkholdera* were found to be associated to maize plants (Montañez et al., 2012). Studies confirmed that the positive effect of some diazotrophic bacterial species on non-leguminous plant yields may not only be due to nitrogen fixation but other mechanisms may also contribute to the growth responses observed in non-leguminous plants (Caballero-Mellado et al., 1992; Vessey, 2003). For example, Abiala et al. (2015) isolated a *Bacillus* strain with multiple growth promoting characterstics from maize rhizosphere. It is also believed that some endophytic diazotrophic bacteria contribute substantial amounts of N to certain graminaceous crops (Barraquio et al., 1997). When some plant seeds were inoculated with diazotrophic bacteria, enhanced seed germination, biomass, increased nitrogen, and chlorophyll content were observed in seedlings (Riggs et al., 2001; Meena et al., 2012). Egamberdiyeva (2007) reported that *Pseudomonas alcaligenes* inoculation stimulated the growth and nutrient uptake of maize. Inoculation of *Bacillus megaterium* and *Bacillus mucilaginous* increased the nutritional assimilation of maize plant (Wu et al., 2005).

In vitro screening techniques have been employed to select effective strains of bacteria with multiple plant growth-promotions (Ahmad et al., 2006, 2008; Çakmakçi et al., 2007). Although lack of consistency in correlation between the results obtained *in vitro* and *in vivo* have been reported (Khalid et al., 2004; Yobo et al., 2004), these techniques are simple and efficient for screening large numbers of isolates (Campbell, 1989). Combination of *in vitro* and *in vivo* screening may lead to identification of effective strains for sustainable agriculture. The current study was therefore aimed at isolating diazotrophic bacteria, their partial characterization and for preliminary screening of some growth parameters on maize seedlings in greenhouse experiment.

MATERIALS AND METHODS

Bacterial Isolation

Diazotrophic bacteria were isolated from rhizospheres, roots and leaves of maize and wheat plants collected from Cedara (Agricultural research collage, Hawick), Greytown, and Ukulinga (University of KwaZulu-Natal research farm, Pietermaritzburg), all within the KwaZulu-Natal province, Republic of South Africa. Roots and leaves were surface sterilized with 3.5% sodium hypochlorite for 5 min and subsequently rinsed three times with sterile distilled water, using a modified protocol of Kloepper et al. (1991). Roots and leaves were cut into pieces and grounded with 10 mL of distilled water. A modified protocol of Döbereiner (1988) (N-free semi-solid media) was used to isolate rhizospheric, rhizoplane, and endophytic diazotrophs. Pure cultures of diazotrophic bacteria were then isolated by serial dilution, and plated onto N-free (NF) media containing of either 20g $l^{-1}$of mannitol, sucrose, or malate as the carbon source; 0.2 g L^{-1} K_2HPO_4; 0.2 g L^{-1}NaCl; 0.2 g L^{-1}; $MgSO_4.7H_2O$; 0.1 g $L^{-1;}$ K_2SO_4; 5.0 g L^{-1} $CaCO_3$; 20 g L^{-1} agar (Merck) for solid agar media, or with 5 g of agar per liter for a semi-solid media. These bacteria were incubated at 30°C for 4 days.

Soil samples were collected from the rhizosphere of maize and wheat from different sites by uprooting the root system and placing them in plastic bags for transport to the laboratory. They were stored at 4°C for subsequent analysis. Excess soil was shaken off and the soil adhering to the plant roots was collected from each soil sample. Ten grams of each soil sample were transferred to a 250 mL-Erlenmeyer flask containing 90 mL sterile distilled water and shaken at 150 rpm in an orbital shaker incubator for 30 min. Plates with NF medium with mannitol as a carbon source for diazotrophic bacteria were inoculated with 0.1 mL of suspensions obtained from the above dilution procedure (three replicates per dilution). The pH was adjusted to 6.5 using 98% sulphuric acid, and 50% sodium hydroxide. After 5 days of incubation, colonies were transferred onto fresh N-free media, and after 2 days were streaked out onto tryptone soy agar (TSA; Merck) plates. Bacterial isolates were selected by size and shape of colony and by their ability to grow on N-free media. These colonies were sub-cultured onto TSA and incubated at 30°C, purified and stored in 15% glycerol at −80°C.

Ammonia Production Test

Ammonia production was analyzed using the qualitative method of Ahmad et al. (2008). Bacterial isolates were tested for the production of ammonia in peptone water. Freshly grown cultures were inoculated in 10 mL peptone water in each tube and incubated for 48–72 h at 28 ± 2°C. Nessler's reagent (0.5 mL) was added in each tube. Development of a brown to yellow color was a positive test for ammonia production.

Acetylene Reduction Assay (ARA)

The basis for the assay is the fact that nitrogenase, the enzyme complex in diazotrophic microorganisms that reduces nitrogen to ammonia, also reduces acetylene to ethylene. The diazotrophic nature of all the recovered isolates was determined by ARA. Ethylene was quantified by gas chromatography (GC) and the results were expressed in nano moles of C_2H_4 produced h^{-1} $culture^{-1}$. One milliliter of pure culture grown in tryptone soy broth (TSB) for 24 h were inoculated onto 10 mL of nitrogen-free semi-liquid medium, with 0.5% mannitol as a carbon source, solidified by 0.3% gellan gum in 20 mL serum bottles and closed with a red rubber septum (SIGMA-ALDRICH, William Freeman and Co., Ltd.) and incubated for 72 h at 28°C. Bottles that showed bacterial growth were assayed for acetylene reduction. Ten percent of the atmosphere in the bottles was replaced with acetylene (C_2H_2), whereas bottles without acetylene were used as the control. After 2 h, at 24°C, 0.25 mL gas samples were taken from each vial and analyzed for amount of ethylene formed using a gas chromatography. These processes were repeated three times from same vial and average were taken for the statistical analysis. A gas chromatograph (Hewlett-Packard 5830A) fitted with a 2–2.1 mm, 80–100 mesh, Poropak R column was used and the oven temperature was adjusted to 70°C. Injection and flame-ionization detector temperatures were adjusted to 150°C. Nitrogen carrier gas flow rate was adjusted to 50 mL min^{-1}.

Source of Seeds

Seeds of a white maize cultivar, Mac are Medium Pearl (an open pollinated variety), were purchased from McDonalds Seeds® (MacDonald's Seeds (Ltd).P.O. Box 40, Mkondeni, 3212,

Pietermaritzburg, Republic of South Africa) and were used throughout the experiment.

Inoculum Preparation

Bacterial isolates were grown in 100 mL Erlenmeyer flasks each containing a 25 mL tryptic soy broth (TSB; Merck) for 3 d at 28 ± 2°C in a shaker at 150 rpm. Flasks were inoculated with bacteria previously grown in TSA for 48 h. After 3 d bacteria were harvested by centrifugation using a Beckman J2-HS Centrifuge (Beckman Coulter Inc. 4300 N Harbor Boulevard, Box 3100, Fullerton, California, 92834-300, USA) at 9000 rpm for 15 min. The broth was decanted and bacterial pellets were re-suspended in sterile distilled water. Bacterial cells were then counted using a plate dilution technique on TSA plates, and adjusted to a concentration of 10^8colony forming unit (cfu) mL^{-1} of water.

Seed Treatment

Twenty out of 92 bacterial isolates which produced relatively high C_2H_4 levels were selected for further greenhouse screening on maize seedlings. Final cell pellets were diluted with sterile distilled water and cell numbers were adjusted to 2.4×10^8 cfu mL^{-1} for each of the bacterial isolates, as determined by serial dilution and plating. Two grams of a sticker, gum arabic, were dissolved in 100 mL of bacterial suspension, stirred, and allowed to stand for 1 h. This allowed the substance to dissolve and form a homogeneous suspension. Seed coating took place in a plastic bag. The bag was filled with 200 g seeds. The bacterial sticker suspension was added at a rate of 0.1 mL g^{-1} of seeds in order to increase the number of bacteria stuck onto each seed. The bag was closed in such a way to trap air as much as possible. The bag was shaken for 2 min, until all the seeds were uniformly wetted with the sticker suspension. The bag was opened and the seed spreads onto paper towels and air-dried overnight. Treated maize seeds were planted on pine bark artificial growing medium. Composition of the composted pine bark growing medium analyzed by KwaZulu-Natal Agriculture and Environmental Affairs (KZN Agriculture and Environmental Affairs, Private Bag X9059, Pietermaritzburg, 3200, Republic of South Africa) is as follows: 12.39 C (%); 0.11 S (%); 0.43 N (%); 0.61 Ca (%); 0.10 Mg(%); 0.28 K (%); 0.18 P (%); 25.07 Moisture (%); 405.0 Na (mg kg^{-1}); 73.0 Zn (mg kg^{-1}); 16.7 Cu (mg kg^{-1}); 802Mn (mg kg^{-1}); 12293 Fe (mg kg^{-1}); 7120 Al (mg kg^{-1}).

Measurements of Stomatal Conductance and Chlorophyll Content Index

Greenhouse measurements of stomatal conductance and leaf chlorophyll were made on 8 weeks old seedlings. Leaf stomatal conductance was measured with a portable porometer (SC-1 Leaf Porometer, Decagon Devices, Inc., 2365 NE Hopkins Court, Pullman, WA 99163, USA). Stomatal conductance readings were made between 9:00 am and 3:00 pm on sunny days on 10 leaves of each pot of nine plants per treatment. Chlorophyll was measured using a portable, handheld device called chlorophyll meter (CCM-200 Plus, Opti-Science Inc., 8 Winn Avenue, Hudson, NH, USA, 03051; that estimates the chlorophyll content of leaves). Measurements were made on eight leaves of each pot of nine plants per treatment.

DNA Extraction and 16S rRNA Sequence Analysis

One mL of 24 h bacterial culture was centrifuged at 14,000 relative centrifugal force (rcf) for 5 min. The pellet was suspended in 25 μL of (10 mM) Tris and 1 mL of buffer was added and incubated at 60°C for 1 h. One milliliter of CTAB buffer was added and gently mixed. Then the bacterial-buffer suspension were divided into two in 1.5 mL tubes, and 500 μL of chloroform-iso-amyl alcohol was added and mixed gently, and the resultant mixture was centrifuged at 14,000 rcf for 10 min. By avoiding the layer of impurities, 900 μl of clear supernatant was removed as the sample. To this, 600 μl of propan-2-ol was added and refrigerated at −20°C for 1 h. It was centrifuged for 15 min at 4000 rcf and the supernatant was discarded. The pellet were washed with 50 μl of 70% ethanol solution and dried in a laminar flow with lid of the tube being left open for 30 min. The pellet was then suspended in 50 μL of 10 mM Tris (pH 8) or 0.5 X TE buffer. At this point the DNA purity and quality were checked on the nano drop UV spectrophotometer equipment (nano drop 1000, Inqaba Biotech, P.O.Box 1435, Hatfield 0028, Pretoria, South Africa) and a 5 μL sample was run on 0.8% agarose gel [SeakemLE Agarose, Whitehead Scientific (Pty) Ltd www.whitesci.co.za] stained with SYBRSafe nucleic acid stain (Invitrogen) stained with SYBRSafe nucleic acid stain (Invitrogen), with a GeneRuler 1 kb DNA Ladder Plus molecular weight marker (Thermo Fisher Scientific Inc., 81 Wyman street, Waltham, MA 02454, US) to confirm the presence, size and quality of genomic DNA. Once the purity of the DNA was checked, it was sent to the Central Analytic Facility, Stellenbosch University, South Africa for sequencing and BLAST identification. The BLAST identifications were then confirmed by matrix assisted laser desorption ionization-time of flight (Maldi-TOF) classification [Brucker Daltonik Maldi-TOF Biotyper (www.bruker.com)].

Bruker Daltonik Maldi Biotyper Classification

Bacterial cultures were sub-cultured on 10% TSA for 24 h at 30°C. A single bacterial colony were taken and placed into a 2 mL eppendorf tube with 300 μL of ultra-pure water, and 900 μL of pure ethanol were added, mixed and the suspension was centrifuged at 14,000 rcf for 2 min. A small pellet of bacterial cells was visible at the bottom of the tube. The liquid was removed, and the pellet was briefly re-spun followed by the removal of residual ethanol. It was then re-suspended in 10 μL of 70% formic acid, and 10 μL of aceto-nitrile was added, and the sample was vortexed briefly. The mixture was centrifuged for 2 min at 14,000 rcf, and the supernatant transferred into a clean micro-tube. The sample to be analyzed was warmed to room temperature, and a 1 μL sample was spotted onto a steel target plate (Bruker Daltonics Inc., Billerica, MA, USA) and gently mixed with 2 μL of matrix solution.

TABLE 1 | Nitrogenase activity of the bacterial isolates measured by acetylene reduction assay (ARA).

Isolates	nmol of $C_2H_4h^{-1}$ culture^{-1}		Isolates	nmol of $C_2H_4h^{-1}$ culture^{-1}		Isolates	nmol of $C_2H_4h^{-1}$ culture^{-1}	
Broth	0.32	a	M12	21.86	cdefghijklmn	Bt12	29.01	ghijklmnop
Bt14	4.81	ab	StB12	21.91	cdefghijklmn	Bt7	29.46	ghijklmnop
M11	6.20	ab	V1	21.91	cdefghijklmn	StB8	29.87	ghijklmnop
Mr2	6.24	ab	Br2	21.96	cdefghijklmn	G3	31.78	hijklmnopq
SB1	6.46	ab	Bt1	21.96	cdefghijklmn	V3	31.99	ijklmnopq
Mr63	6.51	ab	RB1	21.96	cdefghijklmn	Mr27	32.27	ijklmnopq
Bt10	7.29	abc	Bt15	22.01	cdefghijklmn	V14	32.39	ijklmnopqr
Bt3	7.32	abc	RB6	22.01	cdefghijklmn	Mr57	32.77	jklmnopqr
Mr141	7.61	abc	V12	22.01	cdefghijklmn	Mr34	33.04	klmnopqr
Mr20	7.90	abc	Mr148	22.02	cdefghijklmn	Bt4	33.26	klmnopqr
Mr25	9.10	abcd	RB2	22.06	cdefghijklmn	Bt5	33.49	lmnopqr
Mr8	9.76	abcde	StB7	22.06	cdefghijklmn	V20	33.58	lmnopqr
StB3	10.46	abcde	LB7	22.11	cdefghijklmn	Bt11	33.69	lmnopqr
V15	10.96	abcde	Bt9	22.15	cdefghijklmn	Mr21	33.94	mnopqrs
Mr53	12.26	abcdef	Bt6	22.20	cdefghijklmn	StB13	35.09	nopqrst
Mr35	12.29	abcdef	Bt8	22.20	cdefghijklmn	V10	35.33	nopqrst
E9	12.66	abcdef	V2	22.20	cdefghijklmn	Mr150	35.54	nopqrstu
Mr38	15.40	bcdefg	StB17	22.30	cdefghijklmn	Mr131	37.13	nopqrstu
Mr58	15.96	bcdefg	V6	22.40	cdefghijklmn	StB1	37.23	nopqrstu
Mr6	16.44	bcdefgh	M9	22.41	cdefghijklmn	Mr37	38.82	opqrstuv
V5	17.02	bcdefghi	V7	22.45	cdefghijklmn	a2	42.00	pqrstuv
Mr16	17.22	bcdefghi	LB9	22.55	cdefghijklmn	a3	45.10	qrstuv
V4	17.25	bcdefghij	X	22.80	cdefghijklmn	B1	45.84	qrstuv
Mr19	17.32	bcdefghij	D6	24.19	defghijklmno	a61	46.69	qrstuv
Mr13	17.38	bcdefghij	Mr17	24.19	defghijklmno	a5	47.28	rstuv
Mr7	17.53	bcdefghij	Mr22	24.73	efghijklmno	LB5	48.36	stuv
Mr121	17.76	bcdefghijk	Mr9	25.09	efghijklmno	L1	48.75	tuv
Mr55	18.22	bcdefghijkl	V18	25.20	efghijklmno	a6	49.88	uv
LB2	18.75	bcdefghijklm	V17	27.48	fghijklmnop	StB5	52.37	v
V13	19.05	bcdefghijklm	V19	28.40	ghijklmnop	V16	65.15	w
Bt2	21.86	cdefghijklmn	V11	28.82	ghijklmnop	V9	73.20	w
CV%	30.4							
DMRT	12.035							
Sed	6.101							
F-test	9.662							
P-value	<0.001							

Means with the same letter are not significantly different at $P \leq 0.05$. Each treatment was replicated three times ($n = 3$). Each treatment vial was sampled three times.

Experimental Design

Maize seedlings treated with bacterial isolates were watered every day with 500 mL nutrient solution containing: 0.11 mL L^{-1} H_3PO_4; 0.13 g L^{-1} KOH; 0.14 g L^{-1} K_2SO_4; 0.74 g L^{-1} $CaCl_2.2H_2O$; 0.10 g L^{-1} $MgSO_4.7H_2O$; and 0.02 g L^{-1} of micronutrients (MICROPLEX®; Ocean Agriculture (Pty) Ltd, P.O. Box 741, Mulders drift, 1747, South Africa). There were two controls: one was untreated and supplemented with a soluble fertilizer NPK (3:1:3 [38] complete) which was parched from Ocean Agriculture (Pty) Ltd, P.O. Box 741, Mulders drift, 1747, South Africa at a rate of 1 gL^{-1} and the second was untreated control and supplemented with MICROPLEX® nutrient solution. Each treatment consisted of three pots with a top diameter of 200 mm that held 2 kg of composted pine bark. Each pot was seeded with five seeds. The seedlings were thinned to three plants per pot. Pots with each of the 20 isolates were watered every day with an equal amount of a nutrient solution containing phosphorus and potassium of levels adjusted to the full recommended amounts. The experiment was arranged in a randomized complete block design (RCBD), replicated three times. Two months after planting, chlorophyll and stomatal conductance were measured and plants were harvested and dry weight was taken after the biomass was dried in an oven for 72 h at 70°C.

Statistical Analysis

Experiments were repeated twice, unless otherwise stated. Data were analyzed using GenStat executable release 14th edition

TABLE 2 | Influence of maize seeds inoculation with diazotrophic bacteria on some growth parameters of 2 months seedlings in greenhouse experiment.

Bacterial isolates	Chlorophyll content index		%Over control	Dry weight(g)		%Over control	Stomatal conductance (m mol m^{-2} s^{-1})		%Over control
Control	4.45	a	–	2.77	a	–	54.2 (1.733)	a	–
Mr21	5.08	ab	14.16	3.63	ab	31.05	121.6 (2.085)	abc	124.35
Mr131	5.24	abc	17.75	3.62	ab	30.69	122.5 (2.086)	abc	126.01
Bt11	5.38	abcd	20.90	3.63	ab	31.05	122.5 (2.087)	abc	126.01
Mr150	5.59	abcde	25.62	4.13	abcd	49.10	142.0 (2.152)	abcd	161.99
StB1	5.83	abcde	31.01	5.17	bcd	86.64	137.7 (2.127)	abc	154.06
StB13	5.95	abcdef	33.71	3.79	abc	36.82	130.3 (2.115)	abc	140.41
Mr37	6.08	abcdef	36.63	3.72	abc	34.30	147.3 (2.165)	abcd	171.77
V20	6.11	abcdef	37.30	5.85	cd	111.19	147.6 (2.169)	abcd	172.32
V10	6.21	abcdefg	39.55	5.79	cd	109.03	134.6 (2.129)	abc	148.34
A61	6.28	bcdefg	41.12	6.23	d	124.91	206.5 (2.295)	cde	281.00
A2	6.37	bcdefg	43.15	5.84	cd	110.83	198.3 (2.274)	cde	265.87
B1	6.52	bcdefg	46.52	3.80	abc	37.18	244.1 (2.383)	de	350.37
A6	6.71	bcdefg	50.79	3.41	ab	23.10	218.1 (2.335)	cde	302.40
A5	7.03	cdefg	57.98	2.42	a	−12.64	188.8 (2.253)	bcde	248.34
A3	7.06	cdefg	58.65	5.26	bcd	89.89	219.3 (2.341)	cde	304.61
LB5	7.11	defg	59.78	5.51	bcd	98.92	203.4 (2.278)	cde	275.28
L1	7.26	efg	63.15	5.85	cd	111.19	206.2 (2.281)	cde	280.44
V16	7.70	fg	73.03	5.99	d	116.25	265.0 (2.423)	e	388.93
V9	7.72	fg	73.48	2.80	a	1.08	253.3 (2.404)	e	367.34
StB5	8.03	g	80.45	3.48	ab	25.63	244.7 (2.386)	de	351.48
NPK	13.62	h	206.07	9.90	e	257.40	517.5 (2.712)	f	854.80
CV%	14			23.9			23.20 (4.2)		
DMRT	1.544			1.833			90.41 (0.198)		
SED	0.765			0.908			43.00 (0.093)		
P	<0.001			<0.001			<0.001		

Means with the same letters in the same column are not significantly different at $P < 0.05$; values in parentheses are transformed data using log base 10 for the stomatal conductance. Means were calculated from $n = 3$ for dry weight, $n = 24$ for chlorophyll content index, and $n = 30$ for stomatal conductance.

statistical analysis software. Significant differences between treatments were determined using Duncan's multiple range test at 5% significant level.

RESULTS

Isolation and Preliminary Screening of Bacterial Diazotrophs

There were differences between diazotrophic isolates, in their ability to grow on semi-solid N-free medium, and using D-mannitol, D-malate or sucrose as carbon sources. Bacterial growth were considered slow when the mass doubling time was longer than 10–12 h and considered well when the mass doubling time was less than 10 h. All these isolates were able to grow well on the N-free semi-liquid medium when sucrose was used as the carbon source, and generated ammonia in peptone water. About 20% of the bacterial isolates grew well on N-free media with D-mannitol, sucrose, or malate as growth substrates. Approximately, 80% of the bacterial isolates showed slow growth on N-free medium with D-mannitol or malate.

Acetylene Reduction Assay (ARA)

All the isolates exhibited nitrogenase activity, but the level of activity varied with different isolates (**Table 1**). Approximately 17% of the isolates produced very little amount C_2H_4, no more than the control; 66% of the isolates produced significantly ($P < 0.001$) more C_2H_4 compared to the control (N-free semi-liquid medium). The 20 isolates which produced relatively high levels of C_2H_4 were further screened in the greenhouse.

Effect of Diazotrophic Inoculation on Chlorophyll Content Index, Dry Weight and Stomatal Conductance

Of the 20 diazotrophic isolates, about 50% showed significantly higher ($P < 0.001$) chlorophyll content index, stomatal conductance and dry weight, relative to the untreated control (**Table 2**). The rest of these isolates had no effect ($P < 0.05$) on the above parameters. Plants of the untreated control showed the lowest stomatal conductance, chlorophyll level and dry weight and plants of the 100% NPK fertilized (NPK control) the highest (**Table 2**). Isolate StB5 had the highest chlorophyll content index

TABLE 3 | Affiliation of the isolates in the GenBank and the identification of the closest type strain based on the 16S rRNA gene sequencing and Bruker Daltonik MALDI-TOF Biotyper classification.

Isolates	16S rRNA similarities (highest match)	BrukerMALDI biotype (highest score)
V16	*Bacillus megaterium* Strain As-30 (97%)	*Bacillus* sp. (1.722)
A5	*Burkholderia* sp. IBP-VNS127 (99%)	*Burkholderia* sp. (1.867)
V9	*Burkholderia ambifaria* (99%)	*Burkholderia ambifaria* (2.462)
L1	*Enterobacter cloacae* Strain G35-1(98%)	*Enterobacter cloacae (2.327)*
A2	*Klebsiella variicola* (99%)	*Klebsiella variicola* (2.243)
LB5	*Pantoea ananatis* (97%)	*Pantoea ananatis* (2.268)
A3	*Pseudomonas nitroreducens* Strain R5-791 (99%)	*Pseudomonas* sp. (1.901)
A6	*Pseudomonas nitroreducens* (99%)	*Pseudomonas* sp. (1.96)
B1	*Pseudomonas nitroreducens* (99%)	*Pseudomonas nitroreducens* (2.034)
StB5	*Pseudomonas nitroreducens* Strain R5-791 (99%)	*Pseudomonas* sp. (1.989)
A61	*Pseudomonas nitroreducens* (99%)	*Pseudomonas* sp. (1.882)

and stomatal conductance but its dry weight was not significantly higher than in the non-inoculated control. The highest dry weight was obtained in the seedlings inoculated with V16 and L1; their chlorophyll content index and stomatal conductance were significantly higher than in control plants.

Identification of Diazotrophic Isolates

Comparative analyses of nucleotide sequences of amplified 16S rRNA fragments, using a BLAST approach, revealed that some of the isolates with 99% sequence accuracy. Isolates StB5, A3, A6, B1, and A61 demonstrated this accuracy with *Pseudomonas nitroreducens*, A5 with *Burkholderia* sp., V9 with *Burkholderia ambifaria,* and A2 with *Klebsiella variicola*. Isolate L1 showed 98% similarity with *Enterobacter cloacae*; V16 (97%) with *B. megaterium* and LB5 (97%) with *Pantoea ananatis* (**Table 3**). With laser desorption ionization-time of flight (Maldi-TOF) classification system, a score of $\geq$2.000 indicates species level identification, a score of 1.700–1.999 indicates identification to the genus level, and a score of <1.700 is interpreted as no identification. Isolates StB5, A3, A6, and A61 were identified as *Pseudomonas* sp with MALDI-TOF scores of 1.98, 1.90, 1.96, and 1.88, respectively. Isolate B1 (2.03) as *P. nitroreducens*, V9 (2.46) as *Burkholderia ambiferia* and A5 (1.86) were identified as *Burkholderia* sp, isolate L1 (2.33) as *E. cloacae*, isolate V16 (1.72) as *Bacillus* sp, A2 (2.24) as *K. variicola* and isolate LB5 (2.27) as *Pantoea ananatis.* An independent identification of these isolates, based on 16S rRNA gene and Maldi-TOF biotyper, confirmed their identity to species and genera level.

DISCUSSION

Isolation and screening for potential diazotrophic bacteria are crucial steps in research on biofertilizers. In order to discover efficient nitrogen fixing bacteria, there is a need to develop simple, inexpensive and quick procedures with repeatable and reliable results (Döbereiner, 1988). Such as an *in vitro* screening procedure (growth on N-free semi-solid media, ARA, and the ammonia production test) and the combinations of which provides rapid and repeatable results. All the isolates proved to be nitrogen fixers. Though it is difficult to compare the nitrogenase activity of bacterial strains studied in this work with the results obtained by others, a study by Różycki et al. (1999) showed similar results of nitrogenase activity by some of diazotrophic isolates belonging to the genera *Pseudomonas* and *Bacillus.*

About 50% of the best isolates used in this study were identified as *Pseudomonas.* In another study, the genera of *Pseudomonas* and *Enterobacter* were reported to be dominant in maize cultivars (Rodríguez-Blanco et al., 2015). Moreover, Berge et al. (1991) reported that an *E. cloacae was* the most abundant diazotrophic bacterium in the rhizosphere of maize-growing soils in France. *Klebsiella* and *Burkholderia* were also dominant genera both in roots and rhizosphere of maize (Arruda et al., 2013). Estrada et al. (2002) isolated a strain of endophytic, N_2-fixing *Burkholderia* sp. associated with maize in Mexico. Similarly, Perin et al. (2006) isolated *Burkholderia* sp. from the rhizosphere of maize. The predominance of these genera both in the soil and in the root zone may be due to low nutritional requirements, its capacity to utilize numerous complex of organic substrates (Krotzky and Werner, 1987) and high tolerance to low pH (Eckford et al., 2002).

In this study, some of the bacterial isolates with high nitrogenase activity were found to improve chlorophyll content, stomatal conductance and dry weight in maize. All the tested isolates showed increases (124.4–351.5%) in stomatal conductance, nine isolates showed (86.6–24.9%) increases in dry weight, and 11 of the isolates (41.12–80.45%) increases in chlorophyll content index over the control. Diazotrophic microbes isolated either from soil or endophytes increased sweet corn biomass both in greenhouse and field trials (Mehnaz and Lazarovits, 2006; Mehnaz et al., 2010). In another study, interaction of diazotrophic plant growth promoting rhizobacteria with sugar cane, cotton, wheat, rice, and maize significantly increased the vegetative growth and grain yield (Kennedy et al., 2004). Likewise, maize plants inoculated with endophytic diazotrophs such as (*Herbaspirillum* sp, *Psuedomonas* sp., and *Burkholderia vietnamienis*) gained greater early biomass and higher rates of net CO_2 assimilation than the un-inoculated control (Knoth et al., 2013). Furthermore, dry matter of maize increased by 25–54% after inoculation with strains *Achromobacter* sp., *Burkholdria* sp. and *Arthrobacter* sp. (Arruda et al., 2013). In this study, inoculation of maize seeds with *Burkholderia ambifaria* (V9), *Bacillus* sp. (V16), *P. nitroreducens* (B1), Pseudomonas spp. (A3, A6, A61, StB5), *E. cloacae (L1), P. ananatis (LB5), and Klebsiella variicola (A2)* enhanced stomatal conductance and chlorophyll content index compared to the control. The increases could be a result of increased leaf N through nitrogen fixation by these diazotrophs. Nitrogen fixed by diazotrophic bacteria may not fully replace chemical fertilizers however; use of diazotrophic bacterial inoculants may be an important factor for sustainable agriculture.

CONCLUSION

It can be concluded that the selected diazotrophic isolates may be potentially beneficial biofertilizers and should be tested more in field conditions to confirm their potential to use as biofertilizers inoculants.

ACKNOWLEDGMENTS

We wish to acknowledge Matthew and Heather from microbiology department, UKZN, PMB campus, South Africa, for their assistance with the use of MALDI Biotyper and Mrs Susan Van der Merwe for her technical assistance with the greenhouse.

REFERENCES

Abiala, M., Odebode, A., Hsu, S., and Blackwood, C. (2015). Phytobeneficial properties of bacteria isolated from the rhizosphere of maize in southwestern nigerian soils. *Appl. Environ. Microbiol.* 81, 4736–4743. doi: 10.1128/AEM.00570-15

Ahmad, F., Ahmad, I., Aqil, F., Ahmed Wani, A., and Sousche, Y. S. (2006). Plant growth promoting potential of free–living diazotrophs and other rhizobacteria isolated from northern indian soil. *Biotechnol. J.* 1, 1112–1123. doi: 10.1002/biot.200600132

Ahmad, F., Ahmad, I., and Khan, M. (2008). Screening of free-living rhizospheric bacteria for their multiple plant growth promoting activities. *Microbiol. Res.* 163, 173–181. doi: 10.1016/j.micres.2006.04.001

Arruda, L., Beneduzi, A., Martins, A., Lisboa, B., Lopes, C., Bertolo, F., et al. (2013). Screening of rhizobacteria isolated from maize (*zea mays* L) in Rio grande do sul state (south Brazil) and analysis of their potential to improve plant growth. *Appl. Soil Ecol.* 63, 15–22. doi: 10.1016/j.apsoil.2012.09.001

Barraquio, W., Revilla, L., and Ladha, J. (1997). Isolation of endophytic diazotrophic bacteria from wetland rice. *Plant Soil* 194, 5–24. doi: 10.1023/a:1004246904803

Berge, O., Heulin, T., and Balandreau, J. (1991). Diversity of diazotroph populations in the rhizosphere of maize (*Zea mays* L) growing on different French soils. *Biol. Fertil. Soils* 11, 210–215. doi: 10.1007/bf00335769

Caballero-Mellado, J., Carcano-Montiel, M., and Mascarua-Esparza, M. (1992). Field inoculation of wheat (*Triticum aestivum*) with *Azospirillum brasilense* under temperate climate. *Symbiosis*13, 243–253.

Çakmakçi, R., Erat, M., Erdogan, U., and Dönmez, M. F. (2007). The influence of plant growth-promoting rhizobacteria on growth and enzyme activities in wheat and spinach plants. *J. Plant Nutr. Soil Sci.* 170, 288–295. doi: 10.1002/jpln.200625105

Campbell, R. (1989). *Biological Control of Microbial Plant Pathogens.* Cambridge, UK: Cambridge University Press.

Döbereiner, J. (1988). Isolation and identification of root associated diazotrophs. *Plant Soil* 110, 207–212.

Eckford, R., Cook, F. D., Saul, D., Aislabie, J., and Foght, J. (2002). Free-living heterotrophic nitrogen-fixing bacteria isolated from fuel-contaminated Antarctic soils. *Appl. Environ. Microbiol.* 68, 5181–5185. doi: 10.1128/AEM.68.10.5181-5185

Egamberdiyeva, D. (2007). The effect of plant growth promoting bacteria on growth and nutrient uptake of maize in two different soils. *Appl. Soil Ecol.* 36, 184–189. doi: 10.1016/j.apsoil.2007.02.005

Estrada, P., Mavingui, P., Cournoyer, B., Fontaine, F., Balandreau, J., and Caballero-Mellado, J. (2002). An N_2-fixing endophytic *Burkholderia* sp. associated with maize plants cultivated in Mexico. *Can. J. Microbiol.* 48, 285–294. doi: 10.1139/w02-023

Kennedy, I. R., Choudhury, A., and Kecskés, M. L. (2004). Non-symbiotic bacterial diazotrophs in crop-farming systems: can their potential for plant growth promotion be better exploited? *Soil Biol. Biochem.* 36, 1229–1244. doi: 10.1016/j.soilbio.2004.04.006

Khalid, A., Arshad, M., and Zahir, Z. (2004). Screening plant growth–promoting rhizobacteria for improving growth and yield of wheat. *J. Appl. Microbiol.* 96, 473–480. doi: 10.1046/j.1365-2672.2003.02161.x

Kloepper, J. W., Zablotowicz, R. M., Tipping, E. M., and Lifshitz, R. (1991). "Plant growth promotion mediated by bacterial rhizosphere colonizers," in *The Rhizosphere and Plant Growth*, eds L. Keister and P. B. Cregan (Dordrecht: Kluwer Academic Publishers), 315–326.

Knoth, J. L., Kim, S. H., Ettl, G. J., and Doty, S. L. (2013). Effects of cross host species inoculation of nitrogen-fixing endophytes on growth and leaf physiology of maize. *GCB Bioenergy* 5, 408–418. doi: 10.1111/gcbb.12006

Krotzky, A., and Werner, D. (1987). Nitrogen fixation in *Pseudomonas stutzeri. Arch. Microbiol.* 147, 48–57. doi: 10.1007/BF00492904

McCarty, G., and Meisinger, J. (1997). Effects of n fertilizer treatments on biologically active N pools in soils under plow and no tillage. *Biol. Fertil. Soils* 24, 406–412. doi: 10.1007/s003740050265

Meena, H., Sharma, G., Golada, S. L., and Jain, H. (2012). Yield and quality of sweet corn (*Zea mays* (L.) ssp. Saccharata) as affected by nitrogen levels, *Azotobacter* culture and nitrogen sources. *Res. Crops* 13, 486–492.

Mehnaz, S., Kowalik, T., Reynolds, B., and Lazarovits, G. (2010). Growth promoting effects of corn (*zea mays*) bacterial isolates under greenhouse and field conditions. *Soil Biol. Biochem.* 42, 1848–1856. doi: 10.1016/j.soilbio.2010.07.003

Mehnaz, S., and Lazarovits, G. (2006). Inoculation effects of *Pseudomonas putida*, *Gluconacetobacter azotocaptans*, and *Azospirillum lipoferum* on corn plant growth under greenhouse conditions. *Microb. Ecol.* 51, 326–335. doi: 10.1007/s00248-006-9039-7

Montañez, A., Blanco, A. R., Barlocco, C., Beracochea, M., and Sicardi, M. (2012). Characterization of cultivable putative endophytic plant growth promoting bacteria associated with maize cultivars (*zeamays* l.) and their inoculation effects *in vitro*. *Appl. Soil Ecol.* 58, 21–28. doi: 10.1016/j.apsoil.2012.02.009

Perin, L., Martínez-Aguilar, L., Paredes-Valdez, G., Baldani, J., Estrada-de Los Santos, P., Reis, V., et al. (2006). Burkholderia silvatlantica sp. nov., a diazotrophic bacterium associated with sugar cane and maize. *Int. J. Syst. Evol. Microbiol.* 56, 1931–1937. doi: 10.1099/ijs.0.64362-0

Reinhold-Hurek, B., and Hurek, T. (1998). Life in grasses: diazotrophic endophytes. *Trends Microbiol.* 6, 139–144. doi: 10.1016/S0966-842X(98)01229-3

Riggs, P. J., Chelius, M. K., Iniguez, A. L., Kaeppler, S. M., and Triplett, E. W. (2001). Enhanced maize productivity by inoculation with diazotrophic bacteria. *Funct. Plant Biol.* 28, 829–836. doi: 10.1071/PP01045

Rodríguez-Blanco, A., Sicardi, M., and Frioni, L. (2015). Plant genotype and nitrogen fertilization effects on abundance and diversity of diazotrophic bacteria associated with maize (*zea mays* l). *Biol. Fertil. Soils* 51, 391–402. doi: 10.1007/s00374-014-0986-8

Różycki, H., Dahm, H., Strzelczyk, E., and Li, C. Y. (1999). Diazotrophic bacteria in root-free soil and in the root zone of pine (*Pinus sylvestris* L) and oak (*Quercus robur* L). *Appl. Soil Ecol.* 12, 239–250. doi: 10.1016/S0929-1393(99)00008-6

Schepers, J., Francis, D., Vigil, M., and Below, F. (1992). Comparison of corn leaf nitrogen concentration and chlorophyll meter readings. *Commun. Soil Sci. Plant Anal.* 23, 2173–2187. doi: 10.1080/00103629209368733

Vessey, J. K. (2003). Plant growth promoting rhizobacteria as biofertilizers. *Plant Soil* 255, 571–586. doi: 10.1023/A:1026037216893

Wakelin, S. A., and Ryder, M. H. (2004). Plant growth-promoting inoculants in australian agriculture. *Crop Manag.* 3, 1–5. doi: 10.1094/cm-2004-0301-01-rv

Wu, S., Cao, Z., Li, Z., Cheung, K., and Wong, M. (2005). Effects of biofertilizer containing N-fixer, P and K solubilizers and AM fungi on maize growth: a greenhouse trial. *Geoderma* 125, 155–166. doi: 10.1016/j.geoderma.2004.07.003

Yobo, K., Laing, M., Hunter, C., and Morris, M. (2004). Biological control of Rhizoctonia solani by two Trichoderma species isolated from south african composted soil. *S. Afr. J. Plant Soil* 21, 139–144. doi: 10.1080/02571862.2004.10635039

Conflict of Interest Statement: The authors declare that the research was conducted in the absence of any commercial or financial relationships that could be construed as a potential conflict of interest.

18

The *qTSN* Positive Effect on Panicle and Flag Leaf Size of Rice is Associated with an Early Down-Regulation of Tillering

*Dewi E. Adriani[1,2], Tanguy Lafarge[1], Audrey Dardou[1], Aubrey Fabro[3], Anne Clément-Vidal[1], Sudirman Yahya[4], Michael Dingkuhn[1,3] and Delphine Luquet[1]**

[1] CIRAD, UMR AGAP, F-34398 Montpellier, France, [2] Faculty of Agriculture, University of Lambung Mangkurat, Banjarbaru, Indonesia, [3] Crop and Environment Science Division, International Rice Research Institute, Los Baños, Philippines, [4] Department of Agronomy and Horticulture, Bogor Agricultural University, Bogor, Indonesia

Edited by:
Alejandro Del Pozo,
Universidad de Talca, Chile

Reviewed by:
Maoteng Li,
Huazhong University of Science and Technology, China
Hao Peng,
Washington State University, USA

***Correspondence:**
Delphine Luquet
delphine.luquet@cirad.fr

The *qTSN4* was identified as rice QTL (Quantitative Traits Locus) increasing total spikelet number per panicle and flag leaf area but potentially reducing panicle number depending on the environment. So far, this trade-off was mainly observed at grain maturity and not specifically studied in details, limiting the apprehension of the agronomic interest of *qTSN4*. This study aimed to understand the effect of *qTSN4* and of the environment on panicle sizing, its trade-off with panicle number, and finally plant grain production. It compared two high yielding genotypes to their Near Isogenic Lines (NIL) carrying either QTL *qTSN4* or *qTSN12, two* distinct QTLs contributing to the enlarged panicle size, thereafter designated as *qTSN*. Traits describing C sink (organ appearance rate, size, biomass) and source (leaf area, photosynthesis, sugar availability) were dynamically characterized along plant and/or panicle development within two trials (greenhouse, field), each comparing two treatments contrasting for plant access to light (with or without shading, high or low planting densities). The positive effect of *qTSN* on panicle size and flag leaf area of the main tiller was confirmed. More precisely, it could be shown that *qTSN* increased leaf area and internode cross-section, and in some cases of the photosynthetic rate and starch reserves, of the top 3–4 phytomers of the main tiller. This was accompanied by an earlier tillering cessation, that coincided with the initiation of these phytomers, and an enhanced panicle size on the main tiller. Plant leaf area at flowering was not affected by *qTSN* but fertile tiller number was reduced to an extent that depended on the environment. Accordingly, plant grain production was enhanced by *qTSN* only under shading in the greenhouse experiment, where panicle number was not affected and photosynthesis and starch storage in internodes was enhanced. The effect of *qTSN* on rice phenotype was thus expressed before panicle initiation (PI). Whether early tillering reduction or organ oversizing at meristem level is affected first cannot be entirely unraveled. Further studies are needed to better understand any signal involved in this early regulation and the *qTSN* × Environment interactions underlying its agronomic interest.

Keywords: rice, *qTSN4*, down-regulation of tillering, panicle size, top leaf and internode size, main stem growth rate

INTRODUCTION

Grain yield elaboration in cereals depends on the establishment of panicle number per square meter and panicle size, grain filling rate, and individual grain size (Chen et al., 2008; Gaju et al., 2014). All these traits are known to compensate per unit area i.e., among plants (Zhang and Yamagishi, 2010) and within the plant where they compete for the same pool of C and N resources (Okawa et al., 2003; Hashida et al., 2013). Amongst these traits in rice, panicle number, and panicle size are those characterized with the highest plasticity under favorable growing conditions thus with the highest impact on yield elaboration. These two traits are determined along plant cycle, particularly based on tillering and green leaf area dynamics, internode reserve remobilization, and reproductive sink size and number. All these processes are depending on C assimilated and N availability, as reported by Lafarge and Bueno (2009) and Dingkuhn et al. (2015) in rice, and by Dreccer et al. (2008) in wheat. Regarding panicle size, many studies reported that panicle development, and the determination of spikelet number per panicle, is closely correlated to early plant vigor and underlying traits such as leaf appearance rate (Dong et al., 2004; Streck et al., 2009; Itoh and Shimizu, 2012; Rebolledo et al., 2012), tillering (Lafarge et al., 2002, 2010; Borràs-Gelonch et al., 2012), culm (Fujita and Yoshida, 1984; Wu et al., 2011), and peduncle (Liu et al., 2008) size, plant height (Wang et al., 2014; Chen et al., 2015), or even neck internode diameter (Zhang and Yamagishi, 2010).

Several genes were reported to control morphogenetic processes positively during both the vegetative and the reproductive phases. Amongst them, the genes *MOC1* (Monoculm 1, Li et al., 2003) and LAX (Komatsu et al., 2003) were shown to promote axillary meristem initiation and accordingly to regulate both shoot and panicle branching. The *Gn1* gene is involved in the control of plant height and grain number per panicle (Ashikari et al., 2005). The *DEP1* enhances shoot apical meristem activity and grain number per panicle (Huang et al., 2009). The *OsSPL14* promotes early tillering and grain number per panicle (Miura et al., 2010). The *APO1* increases grain number per plant and harvest index (Ikeda et al., 2007; Terao et al., 2010). The *OsEBS* enhances plant biomass and spikelet number per panicle (Dong et al., 2013). By contrast, other genes were shown to have trade-off effects on some morphogenetic traits expressed during the vegetative and the reproductive phases. This is the case of the rice mutant *this1* characterized by higher tiller number but lower plant height and fertile spikelet number compared to the wild type (Liu et al., 2013). The *Ghd7* increases spikelet number per panicle through panicle branching but decreases tillering in a density-dependent manner (Weng et al., 2014); the OsSPS1 gene inhibits plant height but enhances spikelet number per panicle (Hashida et al., 2013). The *Ltn* (low tiller number) and similarly *SPIKE* (*qTSN4*) induces higher total spikelet number per panicle but decreases final panicle number (Fujita et al., 2010, 2013).

These studies point out that a gene has unlikely a positive effect on both organ size and number and accordingly that molecular breeding efforts are frequently confronted to the issue of physiological trade-offs among traits, compromising the direct use of a candidate gene or QTL in crop improvement. The trade-off between tiller number and organ size is of major concern for rice, not only with respect to vegetative growth (Rebolledo et al., 2012) but also reproductive growth as it comes as a parameter regulating panicle size vs. number (Lafarge et al., 2010). The *qTSN4* was recently identified as a QTL enhancing flag leaf width and total spikelet number per panicle in IR64 background (Fujita et al., 2009, 2012, 2013). This QTL is known to co-locate with *Nal1* gene involved in the determination of leaf structure, veining pattern, and carboxylation (Qi et al., 2008), as well as other early traits like stem length and vascular bundle number (Fujita et al., 2012, 2013). It was also shown that its positive effect on panicle size was potentially depressive on panicle number (Fujita et al., 2012, 2013). Recently, Okami et al. (2015) confirmed that an isogenic line carrying *qTSN4* produced fewer but larger tillers than its parent (IR64).

No studies, however, were conducted to further understand at plant level the physiological and environmental determinisms regulating (i) the trade-off between panicle size and number in genotypes introgressed with *qTSN4* and (ii) accordingly, the positive effect of *qTSN4* on plant grain production. This is the aim of the present study that, analyzes the effect of *qTSN* on plant growth and development and thus on yield formation processes along plant cycle in contrasting situations of plant access to light. Two high yielding genotypes (IR64 and IRRI146) were compared to their Near Isogenic Lines (NIL) carrying QTL *qTSN4* (Fujita et al., 2013) and QTL *qTSN12*, a different QTL reported for having a similar function as *qTSN4*, i.e., enhanced panicle size (Ishimaru, personal communication). This was conducted within three trials under contrasted situations of plant access to light and thus to C assimilates in order to modulate the level of competition among sinks within the plant. Plant growth was then characterized in terms of morphogenesis and biomass accumulation per organ type along plant development.

MATERIALS AND METHODS

Plant Materials

Two pairs of parents/near-isogenic lines (P/NIL): IR64 vs. IR64 NIL and IRRI146 vs. IRRI146 NIL were studied in two experiments. In one of these experiments, an additional pair of P/NIL was considered asIR64 and NIL1 carrying *qTSN12*. This QTL has been observed to increase panicle size and reduce panicle number—as observed with *qTSN4*—and was detected in chromosome 12 from donor parent YP4. The NILs were developed by self-pollination of a plant selected from BC4F2 population (in IR64 background) and BC3F1 population (in IRRI146 background; Fujita et al., 2013). The IR64 NIL was identified by composite interval mapping carrying the high total spikelet number (TSN) QTL between Simple Sequence Repeat (SSR) markers RM3423 and RM17492 on the long arm of chromosome 4 (Fujita et al., 2012), whereas IRRI146 NIL was developed from recurrent backcrossing to IRRI146 and marker-assisted selection (MAS; Fujita et al., 2013). The details (cross combination, donor and category) of plant materials are available in **Table 1**. The recipient lines (parents) were chosen because

TABLE 1 | Details of plant materials.

Designation	Species	Cross combination	Donor	Category
IR64	Indica	IR64		Recurrent parent recipient
IR64 NIL	Isoline of IR64	(IR64/IR68522-10-2-2//3*IR64)/IR64	IR68522-10-2-2	IR64-qTSN4.4
IR64 NIL1	Isoline of IR64	(IR64/IR65564-2-2-3//3*IR64)/IR65	IR65564-2-2-3	IR64-qTSN12.2
IRRI146	Indica	NSIC Rc158		Recurrent parent recipient
IRRI146 NIL	Isoline of NSIC Rc158	NSIC Rc158/IR65564-2-2-3//3*NSIC Rc158	IR65564-2-2-3	NSIC Rc158-qTSN4.1

of the wide adaptability as of IR64 as a mega variety grown in many parts of the world (Khush, 1987) and IRRI146, a 2nd-generation New Plant Type (NPT) variety developed at IRRI and released in the Philippines in 2007 under the name NSIC Rc158, a high-yielding *indica* cultivar as well (Brennan and Malabayabas, 2011).

Experiments

GH-CNRS

Experiment I was conducted in the greenhouse from May to August 2013 at the National Center for Scientific Research (CNRS), Montpellier, France. This experiment tested *qTSN4* in IRRI146 and IR64 genetic backgrounds comparing one control treatment (C) and a shading (S) treatment with a 58% light attenuation from panicle initiation (PI) up to heading (H) by using a gray net all around the table bringing the plants.

The seeds were grown in a germination chamber at 29°C, then transplanted 4 days after germination in 3 l pots (three seeds per pot) when seedlings were about 3 cm tall. The thinning of plant population to one plant per pot (downsizing to 45.65 plants m^{-2}) was conducted at four-leaf stage. Pots contained about $^3/_4$ of their volume with EGOT 140 media (17N-10P-14K, pH of 5) and were placed side by side (corresponding to 14.8- cm spacing) on tables filled with 5 cm water depth. Basal fertilizer was applied using Basacot 6M+ at 2g l^{-1}, 11N-9P-19K +2Mg incorporated before transplanting. Pots were arranged in four aluminum tables and each table was containing 104 pots, including border plants, at the beginning of the experiment. The tables were moved every week from 2 weeks after transplanting until maturity to avoid any bias due to the green house structure.

Weather data were collected from the AWS (*Automatic Weather Station*) that was installed in the center of the tables measuring Photosynthetic Active Radiation (PAR), global radiation (Rg), air temperature (T), and relative humidity (RH). The average daily air temperature throughout the crop cycle was 27.3 ± 0.6°C. The average daily PAR for the whole crop cycle was 24.7 ± 7.1 mol m^{-2} d^{-1} under full light and 10.3 ± 3.4 mol m^{-2} d^{-1} under shading, and the average RH was 66.8 ± 7.7%.

Field-IRRI

Experiment II was performed in the field at International Rice Research Institute (IRRI) experiment station in Los Baños, Philippines (14°11′N, 121°15′E, 21 m altitude), from December 2013 to April 2014. This experiment tested *qTSN4* and *qTSN12* effect in IRRI146 and IR64 genetic backgrounds and at two planting densities, low (LD, 25 plants m^{-2}), and high density (HD, 100 plants m^{-2}).

The seeds were soaked for 24 h, drained and incubated for another 24 h, then sown in the seeding trays in the greenhouse on December 5, 2013. The 2-week old seedlings were transplanted in the field at one plant per hill in a 2 × 2.4 m^2 plots. The field was initially flooded to hold two puddlings and two harrowings, standing water level, 3–5 cm, was maintained as the IRRI guide field standard. Phosphorus (30 kg P ha^{-1}), potassium (40 kg K ha^{-1}), and zinc (5 kg Zn ha^{-1}) were applied and incorporated into all the plots 2 days before transplanting. 60 kg N ha^{-1} was applied 1 day before transplanting, then 40 kg N ha^{-1} and 60 kg N ha^{-1} were applied at mid-tillering and PI stage, respectively.

Weather data were collected from the IRRI meteorological station measuring radiation (MJ m^{-2}), daylight (h), rainfall (mm), evaporation (mm), average temperature (°C), vapor pressure (kPa), RH (%), and wind speed (m s^{-1}). Average daily air temperature throughout crop cycle was 25.6 ± 1.5°C. The average daily PAR for the whole crop cycle was 31.0 ± 11.3 mol m^{-2} d^{-1}, and the average RH was 84.2 ± 4.8%.

GH-IRRI

Experiment III was performed in an open-top green house during the wet season (August to November 2014) supporting the two experiments described above. This experiment adopted split plot design with three replications and was conducted at IRRI, Los Baños, Philippines (14°11′N, 121°15′E, 21 m altitude). The main factor was plant spacing: crowded density, 20 × 20 cm (Cr, 25 plants m^{-2}) and isolated density, 60 × 60 cm (Is, 2.78 plants m^{-2}) from PI up to flowering (FLO). The subsidiary factor was a pair of rice genotype: IRRI146 (NSIC Rc158) recipient line and its NIL (*qTSN4.1*-YP4).

The seeds were soaked for 24 h, drained and incubated for another 24 h, then sown in the 6 l pots (four seeds per pot). Pots contain about $^3/_4$ of its volume of Andaqueptic Haplaquoll with a topsoil of 39% clay, 46% silt, 14% sand, pH of 6.38. The thinning of plant population to one plant per pot was conducted 2 weeks after sowing. Water level in the pots was always maintained about 8 cm height. 4 g of Ammonium Sulfate $(NH_4)_2(SO_4)$, 2 g of Single Super Phosphate (SSP) and 2 g of muriate of Potash (KCl) were applied and incorporated into all the pots 2 days before sowing, and 2 g of Ammonium Sulfate was applied at PI stage.

Weather data were collected from IRRI meteorological station measuring radiation (MJ m^{-2}), daylight (h), rainfall (mm), evaporation (mm), average temperature (°C), vapor pressure (kPa), RH (%) and wind speed (m s^{-1}). Average daily air temperature throughout crop cycle was 27.7 ± 0.9°C. The average daily PAR for whole crop cycle was 29.8 ± 11.5 $molm^{-2}d^{-1}$, and RH was 85.7 ± 5.3%.

Plant Measurements

Plant Development and Biomass Accumulation

In both GH-CNRS and field-IRRI, leaf appearance on the main tiller and green tiller number were measured every week from 2 weeks after sowing (in GH-CNRS) and 2 weeks after transplanting (in the field) up to flag leaf stage on 3 and 4 sampled plants in GH-CNRS and field, respectively, in every replication. Thermal time was calculated by daily integration of air temperature minus a base temperature of 12°C (Rebolledo et al., 2012).

Dry weight (DW) of plant shoot organs (leaves, stems of the main tiller and the rest of the whole plant, and panicles) was measured after drying for 72 h in an oven at 70°C, during panicle development (PI + 3 weeks in GH-CNRS and GH IRRI; PI + 2 weeks in the field) and at heading (H; in GH-CNRS) or flowering (FLO; in the field). For each of this sampling main stem dry weight at H or FLO and MAT (MS DW H/FLO, MS DW MAT), shoot dry weight at FLO and MAT (Shoot DW H/FLO, Shoot DW MAT), main stem panicle dry weight at MAT (MS PDW MAT) and plant filled grain dry weight at MAT (Plant FGDW MAT) were measured.

At physiological maturity, the DW of all vegetative organs excluding root biomass was measured as well as MS PDW (after drying under the sun). The five plants harvested at maturity in GH were separated into panicles (after taken pictures for P-TRAP analysis), green leaf blades, senescent leaves, and productive stems (culms + sheaths). In the field, all the plants within a soil base area of 0.12 m^2 per plot were harvested, that is three plants under LD and 12 plants under HD. They were then separated into panicles, green leaf blades, dead tissues, and productive stems. The panicles were hand-threshed and then the filled spikelets were separated from the unfilled by a densitometric column (in GH) or submerging the spikelets in the water (in the field).

PI was determined by dissecting and observing the main tiller of randomized collected plants (border plants for field experiment) from each unit treatment every second day when PI was close. The occurrence of PI was considered when the first row of floral primordial was visible on the shoot apex. Flowering (FLO) was determined within each unit treatment when an average of 75% spikelets per panicle of the main tiller exerted their anthers. Plants were considered at physiological maturity when 75% of the grains of the panicles had turned yellow and the texture was in dough stage.

Leaf Area

In GH-CNRS, individual leaf area on the main stem was measured by using LI–3100C Area Meter (Lincoln, NE, USA). In the field and GH-IRRI, the length and maximum width of individual green leaf blades on the main stem was measured manually with a ruler, then leaf area was estimated as length × maximum width × 0.725 (Tivet et al., 2001). The measurement was done at PI + 3 weeks (four sampled plants in GH IRRI), and at FLO (four sampled plants in GH-CNRS and two sampled plants in the field) on the plants used for biomass measurement. The total area of all green leaves per plant was then measured by LI–3100C Area Meter in GH-CNRS and in the field.

Internode Profile and Anatomy

The length of each internode of the main tiller was measured at maturity in GH-CNRS and GH IRRI of sampled plants used for biomass measurement. In GH-CNRS, for anatomical observation purpose a middle part of 2 cm-long of peduncle and of internode-3 was sampled at FLO stage from the same plants used for biomass measurements. The samples were fixed in paraformaldehyde fixative solution and kept in desiccator overnight followed by dehydration with ethanol 70% for at least 24 h then conserved in the freezer prior to observation. The internodes were sliced into pieces of 60–80 μm with a Thermo Scientific Microm HM 650 V Vibration microtome. The pedunles were sliced into 75–90 μm after embedded in agarose 7% for 2 h with a Thermo Scientific Microm HM 650 V Vibration microtome.

In the field, internode profile was measured at FLO stage by sampling plants different from the plants used for biomass measurement. A 2 cm-long section of the peduncle and of the top internode was fixed in Formalin-Acetic-Alcohol (FAA) fixative solution until the date of sectioning. Prior to sectioning, the samples were dehydrate with ethanol 70% then sliced manually by razor blade.

The samples were then observed based on a high resolution imagery system (microscope Leica S8 APO equiped with camera QImaging MicroPublisher 3.3 in GH-CNRS and stereo microscope Olympus SZX7 equiped with camera Olympus DP71 in the field) to analyze the total and inner diameter as well as peripheric bundles number. In this study, peduncle is the uppermost internode between panicle neck node and node I, and top internode is the internode just below the peduncle, which is the same phytomer as flag leaf.

Carbon Assimilation Measurements

Actual assimilation rate (at homogenous level of light in the measurement chamber, i.e., 1500 or 1800 micromole m^{-2} s^{-1}), was measured by using Portable Photosynthesis System (GFS-3000 WALZ) during panicle development (2–3 weeks after PI), the same time as biomass measurement for all experiments. The last ligulated leaf of the main tiller from three tagged plants (in GH-CNRS) and two tagged plants (in the field and GH-IRRI) in every replication was chosen for this measurement.

Non Structural Carbohydrates (NSC) Analyses

NSC was analyzed during panicle development (2–3 weeks after PI) at the same time as C assimilation and biomass measurement in all experiments. In GH-CNRS, three plants per treatment (as three replications) were chosen homogenously for NSC analyses. In the field and GH-IRRI, two of sampled plants dedicated for biomass measurement were chosen for NSC analyses. In GH CNRS and GH IRRI trials, plants were dissected to sample the leaf blade of last ligulated leaf from the main stem. One base (only in GH CNRS) and one top internode (internode just below the peduncle) were sampled and immediately frozen in liquid nitrogen and store at −80°C until fine ground with a ball grinder (Mixer mill MM301, Retsch, Germany). In the field, all green leaves and internodes of the main tiller were sampled and dried for 72 h in an oven at 70°C before fine ground with a grinder (Thomas-Wiley Laboratory Mill Model 4, Thomas Scientific USA).

The method used for sugar content analysis in GH CNRS and GH IRRI was based on high performance liquid chromatography (HPLC). The sugars were extracted three times from 20 mg ground samples with 1 ml of 80% ethanol for 30 min at 75°C, and then centrifuged for 10 min at 10,000 rpm. Soluble sugars (sucrose, glucose and fructose) were contained in the supernatant and starch in the sediment. The supernatant was filtered in the presence of polyvinyl polypyrrolidone and activated carbon to eliminate pigments and polyphenols. After evaporation of solute with Speedvac (RC 1022 and RCT 90, Jouan SA, Saint Herblain, France), soluble sugars were quantified by high performance ionic chromatography (HPIC, standard Dionex) with pulsated amperometric detection (HPAE-PAD). The sediment was solubilized with 0.02 N sodas at 90°C for 1 h 30 min and then hydrolyzed with α-amyloglucosidase at 50°C, pH 4.2 for 1 h 30 min. Starch was quantified as described by Boehringer (1984) with 5 μL of hexokinase (glucose-6-phosphate dehydrogenase), followed by spectro-photometry of NADPH at 340 nm (spectrophotometer UV/VIS V-530, Jasco Corporation, Tokyo, Japan).

In the field IRRI, sugars were extracted two times from 200 mg ground samples with 7 ml of 80% ethanol for 10 min at 80°C, and then centrifuged for 5 min at 3000 rpm. The residue was washed with 5 ml of hot 80% ethanol three times and combined all washings with the supernatant. The residue was dried at 70°C for 24 h prior to starch assay. Total soluble sugars was determined through colorimetric by adding 5 ml anthrone to 0.5 ml aliquot (sugar extraction was diluted with 80% ethanol) then boiled for 10 min at 100°C. After vortex mix and cooling on ice bath for about 5 min, followed by spectro-photometry at 620 nm (DU 800 UV/Vis spectrophotometer, Beckman Coulter). Dried residue was dropped with absolute ethanol and added with 2 ml of acetate buffer then boiled for 3 h at 100°C. The tubes were cooled to 55°C and proceed to hydrolysis step by adding 1 ml acetate buffer and 1 ml amyloglucosidase, then vortex mix. After incubated for 24 h at 37°C, the hydrolysate (supernatant layer) was decanted and save combined with the residue that had been washed with 3 ml of distilled water. Starch was determined through colorimetric assay by adding enzyme Peroxidase Glucose Oxidase (PGO) to 0.6 ml aliquot (starch hydrolysis was diluted with distilled water), then followed by spectro-photometry at 450 nm (DU 800 UV/Vis spectrophotometer, Beckman Coulter) after incubated in the dark room for 30 min.

Physiological Maturity and Yield Component

In GH IRRI four plants were harvested per replication and separated into panicles, green leaf blades, dead tissues and productive tillers then processed as two other experiments described in Adriani et al. (unpublished data). We determined yield components as panicle number per plant and plant FGDW.

Response Rate

The rate of trait response to either *qTSN* introgression or to a reduced plant access to light was quantified as: (ref_value–mod_value)/ref_value; where ref_value is the trait value for the NIL (in the case of response to *qTSN* introgression) or for low plant access to light treatment (shading in GH-CNRS and HD in field-IRRI, in the case of response to low light quantification) and mod_value is the trait value for the parent or for full light treatment, i.e., control in GH-CNRS and LD in field-IRRI). Response rates are synthesized in **Table 3**.

Data Analyses

The graphs describing plant morphogenesis, individual leaf area, grain production, relative NIL-P, and tillering rate relation to growth were represented with mean values and standard error (standard deviation divided by square root of the number of samples). Data of **Tables 2**, **3** and **Figures 3**, **4** were analyzed by an ANOVA procedure and mean comparisons between parent vs. NIL and between treatments for each pair of genotype were analyzed by Duncan's multiple range test using Microsoft® Excel 2010/XLSTAT-PRO statistical software (version 2014, Addinsoft, Inc., Brooklyn, NY, USA). SigmaPlot® Version 11.2 software (for Windows XP and below, copyright 2009–2010), Systat Software Inc. (Chicago, IL, USA) was used for plotting data and nonlinear regressions.

RESULTS

QTL Effects on Plant Morphogenesis

Under Full Light Conditions

Tiller dynamic is delayed in the presence of *qTSN*. This effect was observed at early tillering stage, i.e., starting at 400°C days (accumulation of thermal time from sowing) when eight leaves had appeared on the main stem. This was true for both genetic backgrounds and experiments (**Figures 1A,D**, **2A,D**; **Figures S1B**, **S2B**) and resulted in a reduced tiller number at PI that was however only significant in the field ($P < 0.001$, see **Table 2** for ANOVA). In IR64 background, this reduction rate ranged from 8.5 to 16.7% in GH-CNRS and the highest reduction rate was observed for NIL1 in the field (between 26.3 and 32.7%; **Table 3A**). In IRRI146 background, the reduction rate ranged from 16 to 25.9% (GH CNRS) and up to 29.5% in the field (**Table 3A**). This reduction was associated with a smaller rate of tiller abortion until MAT for the NILs compared to the parents, resulting in a progressive convergence of tiller number of parents and NILs at MAT (**Figures 1A,D**, **2A,D**; **Figure S1B**; **Tables 2**, **3**). Nevertheless, fertile tiller number in IRRI146 background in the field kept smaller at MAT in the NIL compared to the parent (22.3 and 19% of reduction under LD and HD, respectively) as well as for the IR64 parent in HD treatment (21.3 and 24.4% less tillers in NIL and NIL1, respectively, compared to the parent).

The effect of *qTSN* on leaf appearance rate and final leaf number on the main stem was in general weak and not homogenous across treatments. The *qTSN4* had a depressive effect on final leaf number on the main tiller in the field, although less pronounced in IRRI146 background (**Figures 2B,E**; **Figure S2A**). This could be related to a slightly lower rate of

TABLE 2 | ANOVA of flag leaf area, tiller number at PI (Panicle Initiation), FLO (flowering; heading in GH-CNRS), MAT (grain physiological maturity), biomass related traits (DW, Dry Weight) at plant and main stem level at FLO and MAT for vegetative DW and at MAT only for panicle DW and FGDW (FG, Filled Grain).

Source	Genetic background (G)	QTL	Treatment (T)	Replication	G × QTL	G × T	QTL × T
FLAG LEAF AREA							
GH-CNRS	<0.0001	0.0003	0.2902	0.9291	0.8243	0.5827	0.1783
Field	0.652	0.001	0.059	0.245	0.626	0.937	0.212
TILLER NUMBER AT PI							
GH-CNRS	0.026	0.055	0.266	0.235	0.721	0.922	0.627
Field	0.097	0.0005	<0.0001	0.066	0.455	0.084	0.054
TILLER NUMBER AT FLO							
GH-CNRS	0.937	0.187	0.812	0.973	0.479	0.812	0.581
Field	0.099	0.079	<0.0001	0.489	0.980	0.056	0.367
TILLER NUMBER AT MAT							
GH-CNRS	0.959	0.720	0.574	0.924	0.878	0.878	0.959
Field	0.001	0.021	<0.0001	0.038	0.186	0.008	0.584
MS DW FLO							
GH-CNRS	0.005	0.003	<0.0001	0.154	0.224	0.035	0.502
Field	0.002	<0.0001	<0.0001	0.257	0.192	0.870	0.022
MS DW MAT							
GH-CNRS	0.0002	0.002	0.327	0.807	0.014	0.002	0.303
Field	0.126	0.165	0.003	0.949	0.657	0.452	0.475
SHOOT DW FLO							
GH-CNRS	0.492	0.224	<0.0001	0.332	0.197	0.821	0.444
Field	0.0002	0.126	<0.0001	0.868	0.729	0.013	0.112
SHOOT DW MAT							
GH-CNRS	0.080	0.180	0.122	0.565	0.478	0.095	0.502
Field	0.002	0.096	<0.0001	0.113	0.302	0.010	0.769
MS PDW MAT							
GH-CNRS	0.772	0.0001	<0.0001	0.619	0.249	0.678	0.209
Field	0.041	0.004	0.001	0.410	0.258	0.166	0.539
PLANT FGDW MAT							
GH-CNRS	0.089	<0.0001	<0.0001	0.946	0.073	0.863	0.922
Field	0.048	0.083	<0.0001	0.140	0.155	0.078	0.443

leaf appearance considering the similar duration of the vegetative phase. In contrast, in GH-CNRS, higher leaf number (one more in average) in the presence of *qTSN4* was observed in IR64 background only, which was appreciable at the end of panicle development (about 1300°C days, at time of appearance of leaf 14; **Figure 1B**). This can be related to the fact that the vegetative phase was slightly longer in the NIL (later PI), of approximately one phyllochron, i.e., duration between the appearance of two consecutive leaves (**Figures 1B,E; Figure S1A**).

Similarly to that observed with leaf number, *qTSN* effect on stem length (considered here as the successive internodes excluding the peduncle) also differed from GH-CNRS to field-IRRI trials as well as between genetic backgrounds. An appreciable decrease in final stem length was observed in IR64 background in the presence of *qTSN* in GH-CNRS (**Figure 1C**), whereas in the field no *qTSN* effect on stem length was observed (**Figure 2C; Figure S2C**). No clear difference was observed in IRRI146 background (**Figures 1F, 2F; Figure S1C**). The effect of *qTSN* on peduncle length differed with respect to the genetic background and the experiment. In GH-CNRS, the peduncle (the internode bearing the panicle) was significantly longer in the presence of *qTSN4* in IR64 background (not presented) but it was the opposite in IRRI146 background (**Figure S1E**), which was confirmed in GH-IRRI (data not presented). In the field, the peduncle was shorter in the presence of *qTSN12* in IR64 background (**Figure S2E**), whereas no effect was observed in IRRI146 background (not presented). In IRRI146 background in GH-CNRS, no significant difference in stem length was observed between parent and NIL until heading (**Figure S1C**), but thereafter *qTSN4* positively affected the length of the top three internodes located just below the peduncle. The peduncle was, however, shorter in the NIL compared to the parent (**Figure S1E**).

More stable effect of *qTSN* could be observed on peduncle anatomy and thickness for both genetic backgrounds and trials. The increase of peduncle thickness was 44% in IRRI146 background in GH-CNRS (**Figure S1F**), and 14% in IR64 background (not presented). In the field, the increase of peduncle

TABLE 3 | Response rate of traits to QTLs introgression in each pair of isoline in a given treatment (in the field-IRRI, LD is for Low Density, HD is for high density) and genetic background (IR64 and IRRI146) (A), to access to light in each trial (field-IRRI, GH-CNRS) (B), for each genotype (parent, NIL, NIL1).

A

Traits	IR64-QTL effect		IRRI146-QTL effect		IR64-QTL effect (NIL)		IR64-QTL effect (NIL1)		IRRI146-QTL effect	
	Control	Shading	Control	Shading	LD	HD	LD	HD	LD	HD
Flag leaf area	0.530	0.181	1.091	0.478	0.265	0.156	0.339	0.046	0.291	0.231
Tiller number at PI	−0.167	−0.085	−0.259	−0.160	−0.196	−0.183	−0.263	−0.327	−0.295	−0.161
Tiller number at FLO	−0.014	0.076	−0.147	−0.137	−0.124	−0.069	−0.292	−0.240	−0.072	−0.192
Tiller number at MAT	0.000	0.026	0.041	0.030	−0.052	−0.213	−0.069	−0.244	−0.223	−0.190
MS DW FLO	0.168	0.255	0.126	0.060	0.410	0.579	0.608	0.454	0.358	0.118
MS DW MAT	0.177	0.600	0.082	0.030	0.204	0.018	0.121	0.051	0.290	0.121
Shoot DW FLO	0.030	0.281	0.010	−0.020	0.037	0.068	0.128	0.103	0.133	−0.147
Shoot DW MAT	0.068	0.155	−0.004	0.080	−0.023	−0.179	0.025	−0.040	−0.141	−0.143
MS PDW MAT	0.346	1.158	0.275	0.326	0.058	−0.022	0.271	0.507	0.386	0.129
Plant FGDW MAT	0.310	0.948	0.195	0.198	−0.022	−0.097	0.131	0.075	−0.210	−0.232

B

Traits	IR64-light effect		IRRI146-light effect		IR64-density effect			IRRI146-density effect	
	Parent	NIL	Parent	NIL	Parent	NIL	NIL1	Parent	NIL
Flag leaf area	0.300	0.004	0.328	−0.061	0.024	−0.064	−0.200	−0.057	−0.101
Tiller number at PI					−0.635	−0.629	−0.667	−0.717	−0.664
Tiller number at FLO	−0.096	−0.014	0.176	0.190	−0.659	−0.637	−0.634	−0.540	−0.600
Tiller number at MAT	0.013	0.040	0.050	0.039	−0.588	−0.656	−0.665	−0.694	−0.681
MS DW FLO	−0.429	−0.386	−0.300	−0.338	−0.281	−0.195	−0.349	−0.157	−0.305
MS DW MAT	−0.326	−0.084	0.190	0.138	−0.091	−0.232	−0.148	−0.210	−0.313
Shoot DW FLO	−0.331	−0.168	−0.230	−0.250	−0.708	−0.699	−0.714	−0.674	−0.754
Shoot DW MAT	−0.184	−0.118	−0.030	0.046	−0.678	−0.729	−0.698	−0.735	−0.735
MS PDW MAT	−0.607	−0.369	−0.499	−0.479	−0.319	−0.370	−0.192	−0.048	−0.224
Plant FGDW MAT	−0.589	−0.388	−0.454	−0.453	−0.733	−0.754	−0.746	−0.772	−0.778

Gray columns indicate no treatment effect as shading in GH-CNRS just imposed at PI.

thickness in the presence of *qTSN* was 20% in IR64 background (**Figure S2F**) and 17% in IRRI146 background (not presented). The characteristics of the peduncle were associated with thicker top internode (in the third internode below the peduncle in GH-CNRS, **Figure S1F**; in top internode in the field, **Figure S2F**) and higher number of vascular bundles in the peduncle (data not presented) in the NILs compared to the parents.

A positive effect of *qTSN* on leaf area was observed for the flag leaf (FL) (significant in IR64 background for both trials) and the two to three leaves below the flag leaf (FL-3 or FL-2), but it was more pronounced for FL (**Figure 3**). In GH-CNRS, the increase was 53% for IR64 background (**Figure 3A**; **Table 3A**) and 109% in IRRI146 background (**Figure 3B**; **Figure S1D**; **Table 3A**), which was mainly explained by an increase of leaf length, whereas the width was not affected (data not presented). In the field, *qTSN* effect on the leaf area of the top leaves was already expressed from FL-3 upward for both genetic backgrounds (and significant for all these leaves only in NIL1) (**Figures 3C,D**), with 34% of increase in IR64 background (**Figure S2D**; **Table 3A** for QTL effect in NIL1) and 29% of increase in IRRI146 (**Table 3A**). In GH IRRI it was expressed and significant from FL-2 upward (**Figure 3E**). In the field and GH IRRI, the positive effect of *qTSN* on individual leaf size was mainly supported by the width (data not presented) rather than the length.

Under Low Light Conditions

Traits related to plant growth and development were more affected by the treatment in the field. This can be easily explained by the fact that treatments were established by planting density from sowing onwards in the field, while shading treatment in GH-CNRS was imposed only from PI time to heading. Accordingly, early tillering (**Figures 1A,D, 2A,D**; **Tables 2, 3B** for the rate of trait plasticity in response to light treatment) and leaf appearance rates (**Figures 1B,E, 2B,E**) were poorly affected by shading in GH-CNRS. By contrast stem length was more affected by shading in GH-CNRS, i.e., decreased, as observed in field-IRRI but only in HD treatment in IRRI146 isolines.

Peduncle and internode thickness were also reduced under low access to light in both experiments but only in the NILs, whereas they were not modified in the parents (not presented). Individual leaf size was almost not affected by the reduction of incoming light. In GH-CNRS, main tiller FL size of that parents of both backgrounds increased by about 30% under shading, whereas under the same conditions it was maintained or reduced

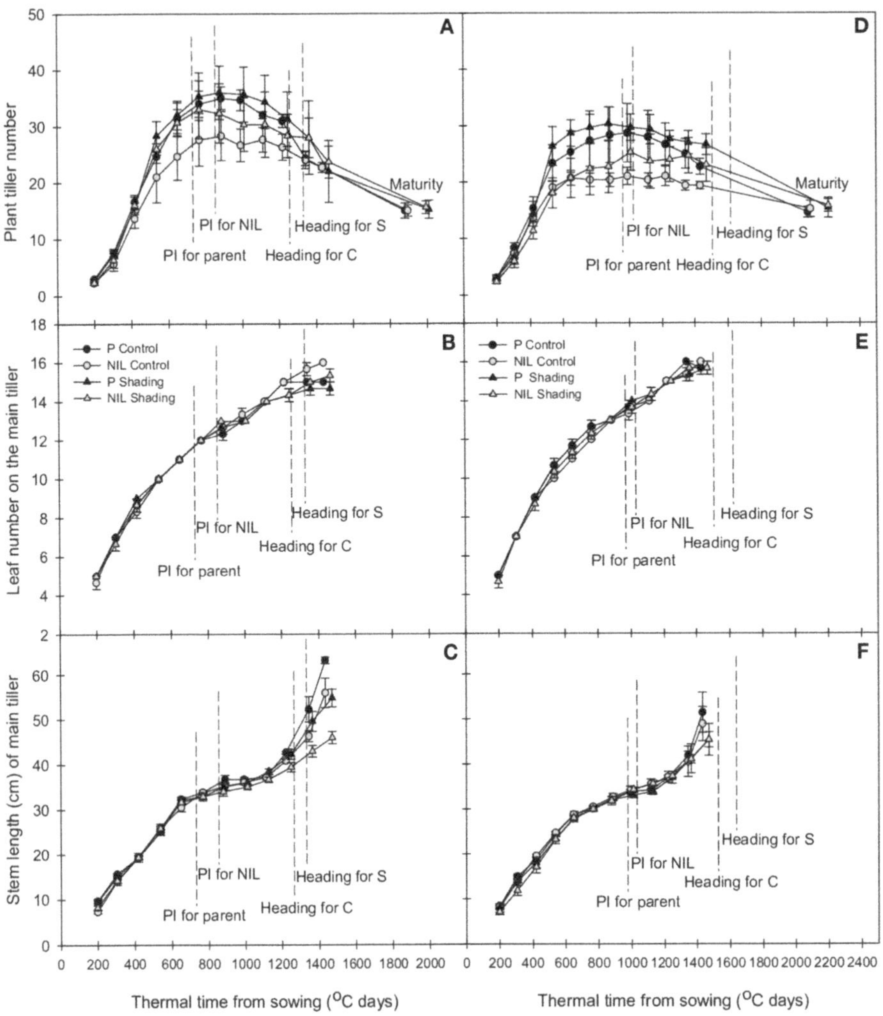

FIGURE 1 | Change with thermal time of plant tiller number (A,D), leaf number on the main tiller (B,E), and stem length of the main tiller (C,E) of parent (black) and NIL (gray) in IR64 (A–C) and IRRI146 (D–F) background, under control (C) and shading (S) in GH-CNRS trial. The values are mean ± SE. $n = 3$.

with the NILs. In the field, slight increase in FL area due to high density was observed only for the parent of IR64 background (2%) (**Figures 3A,D; Table 3B**).

As observed under full light conditions, tiller number was reduced with *qTSN* under low plant access to light (shading in GH-CNRS, HD in the field), however, the difference between NIL and parents was not as strong as observed under higher plant access to light (**Figures 1A,D, 2A,D; Table 3B**), and final tiller number was similar between NILs and parents at maturity. The *qTSN* effect on leaf appearance (positive for IR64 background in GH-CNRS; negative in the field; unchanged for IRRI146 background in both trials) and stem elongation (negative or unchanged) was similar than that observed in full light conditions (**Figures 1, 2**). The *qTSN* positive effect on individual leaf size under low plant access to light was appreciable but less pronounced than that observed under high access to light, as the low access to light increased top leaf size of the parents but not of the NILs (**Figure 3; Table 3B**). No QTL effect was observed on the traits related to peduncle and internode anatomy under low light condition (not presented).

Biomass, Leaf Area and Grain Productions

At plant level, in GH-CNRS, *qTSN4* increased plant final grain production (FGDW) which was significant in all cases (**Table 2**) except for IRRI146 background under shading (**Figure 4B; Table 3A**). Meanwhile, plant shoot biomass was not affected by *qTSN4* neither at PI and FLO (heading in GH-CNRS) (not presented) nor at physiological maturity (**Figure 4C; Tables 2, 3A**). In the field, *qTSN* poorly affected grain and straw biomass and not systematically in a positive way. An increase could be observed only in IR64 background in plant grain production but this was not significant. The *qTSN* effect was even significantly negative on plant grain production in IRRI146 under LD (**Figure 4E; Table 3A**). These contradictory results between GH-CNRS and field conditions regarding *qTSN4* effect on grain production at plant level were also observed in GH-IRRI where

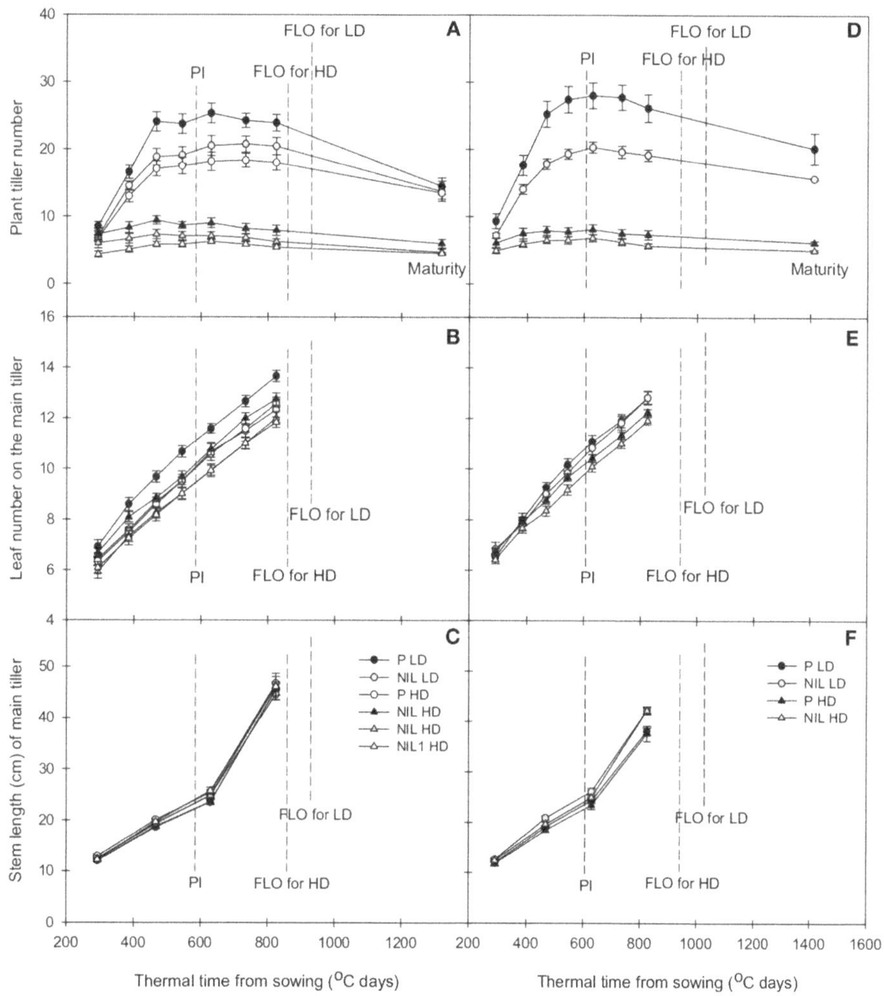

FIGURE 2 | Change with thermal time of plant tiller number (A,D), leaf number on the main tiller (B,E), and stem length of the main tiller (C,E) of parent (black) and NIL (gray) in IR64 (A–C) and IRRI146 (D–F) background, under low density (LD) and high density (HD) in field trial. The values are mean ± SE. $n = 4$.

no significant QTL effect on plant grain production was observed (results not presented). The leaf area per plant at flowering time was unaffected by *qTSN* but the distribution of leaf area was modified in a way that individual leaves were larger but fewer in the presence of *qTSN*. This was true for both genetic backgrounds in both trials (**Figures 4D,H**). However, plant leaf area was affected by light conditions in the field, where it was higher in full light (LD) compared to low light (HD) condition.

At main stem level, panicle dry weight at MAT was increased by *qTSN* in both genetic backgrounds and trials (**Figures 4B,F**; **Table 2**; $P < 0.01$). This increase was more pronounced and systematic in GH-CNRS. It was generally higher under low light conditions for IR64 background (115% under shading and 51% under HD in NIL1, but no effect observed in the NIL of IR64 in field-IRRI, **Table 3A**), whereas it was more homogenous among treatments for IRRI146 (**Table 3A**). The main stem panicle DW was systematically reduced by low access to light and no *qTSN* × treatment interactions were observed (**Tables 2, 3B**). The main stem DW at FLO was systematically increased by *qTSN* ($P < 0.01$, **Table 2**), and this was generally stronger in IR64 background compared to IRRI146 and more particularly in field-IRRI. This *qTSN* effect was not maintained until MAT as no significant difference for main stem DW between parents and NIL could be observed at that stage (**Tables 2, 3**). Low plant access to light systematically reduced main stem DW at FLO, more particularly in GH-CNRS (**Table 3**). This was maintained until maturity only in the field-IRRI as no more significant treatment effect was observed in the GH-CNRS at this time (**Tables 2, 3**). No *qTSN* × treatment interaction for main stem DW was observed, neither at MAT or FLO, except in field-IRRI at FLO (**Table 2**).

Relationship Between Tillering Dynamics and Main Stem Growth Rate

Overall, above-mentioned results pointed out two key nodes of regulation of plant phenotypes due to *qTSN* introgression,

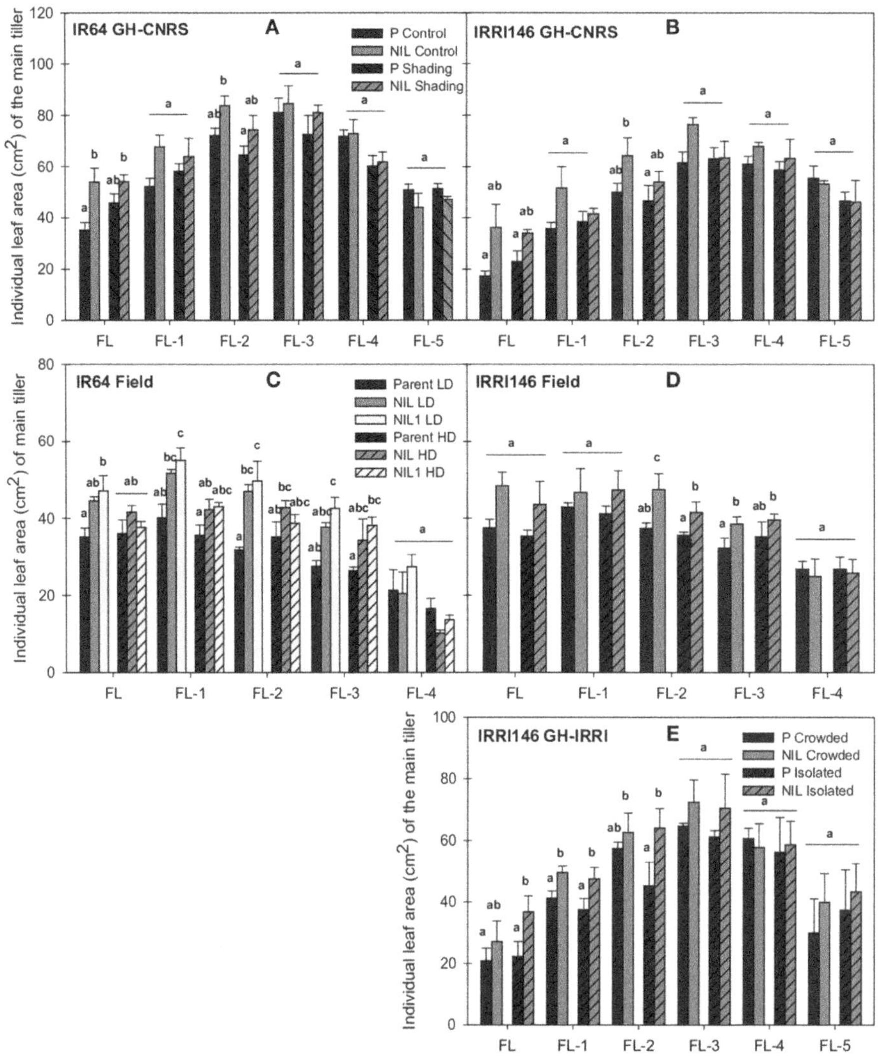

FIGURE 3 | Individual leaf area of main tiller (FL, flag leaf; FL-1–FL-5, one to five leaves below the flag leaf) at flowering (FLO) of parent (black), NIL (gray), and NIL1 (white) in IR64 background (A) and IRRI146 background (B) in GH-CNRS under control and shading, in IR64 background (C) and IRRI146 background (D) in the field under low density (LD) and high density (HD), in IRRI146 background (E) in GH-IRRI under crowded and isolated plants. The values are mean ± SE. Results of Duncan test for multiple comparisons of each genotype per treatment at 5% level are shown in the letters above the bars. $n = 4$ in GH-CNRS and filed-IRRI, $n = 3$ in GH-IRRI.

namely: (i) tillering and tiller number, generally reduced by *qTSN* and (ii) main stem biomass (either before or after FLO), generally increased by *qTSN* (**Table 3A**). In association with the opposite effect of *qTSN4* on these two traits, no additional difference was observed between parents and NILs regarding the resulting plant shoot DW at FLO and MAT. Results of plant FGDW were, however, dependent on cropping conditions (**Figures 4A,D**; **Table 3A**). In order to further explore the relationship between tillering and main stem DW, the change of main stem growth rate during panicle development (PI–FLO period) was plotted against tillering rate before PI. A negative correlation could be observed between these variables, stronger in IRRI146 ($R^2 = 0.5$) than in IR64 ($R^2 = 0.28$) backgrounds (**Figures 5A,B**). Interestingly, the average value for the NIL was systematically at a higher position on the y-axis (with reference to main stem growth rate from PI to FLO) compared to that of the parents. However, the correlation disappeared when analyzing the same relationship in each experiment separately (**Figures 5C,D**).

In order to evaluate whether this early trade-off between tillering and main stem growth rate (from PI to FLO) could impact grain production, main stem growth rate from PI to FLO was plotted against main stem panicle DW at maturity. This is presented in **Figures 6A,B**, showing a slightly positive correlation between these variables ($R^2 = 0.12$) when analyzing data from parents and NILs together, whereas there was no correlation in IRRI146 ($R^2 = 0.02$) background. This positive correlation was getting even stronger when considering trials separately

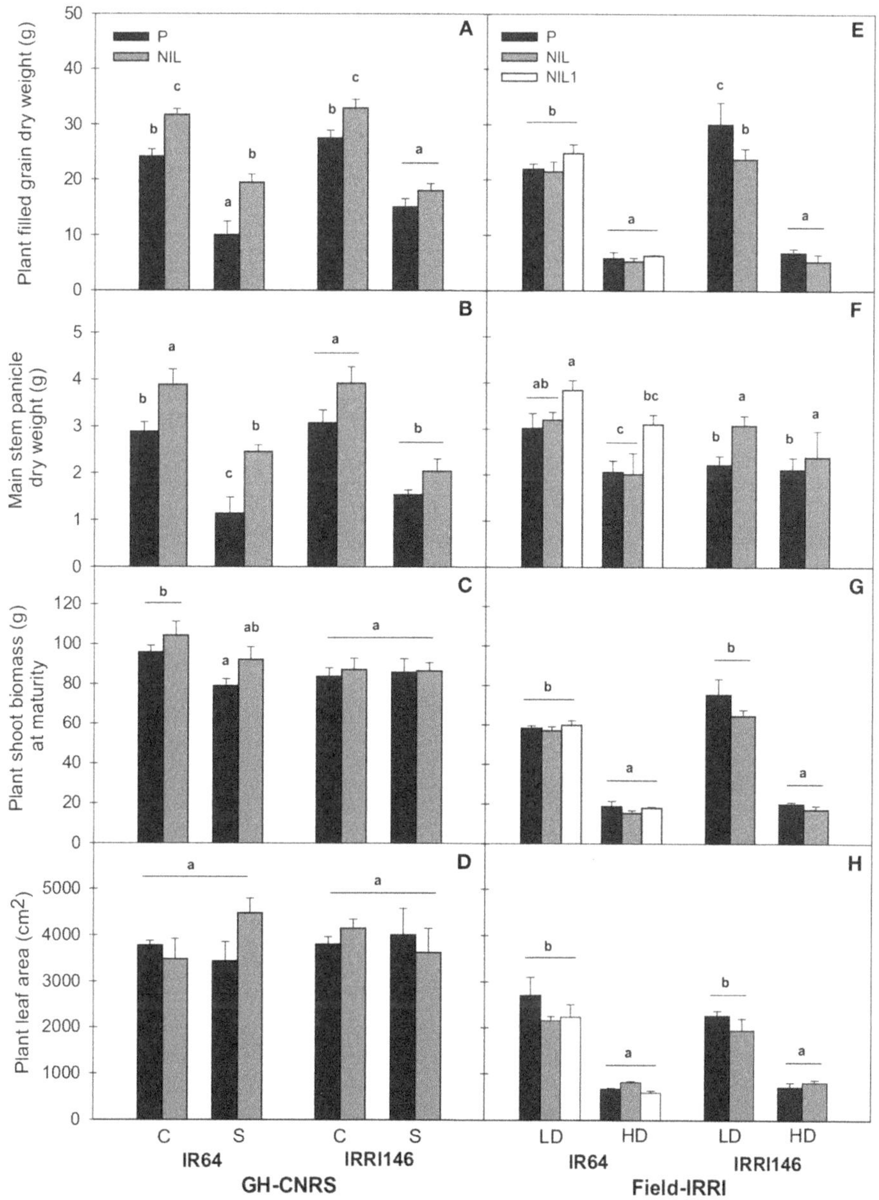

FIGURE 4 | Plant filled grain dry weight (A,E), main stem panicle dry weight at maturity (B,F), plant shoot biomass at maturity (C,G), plant green leaf area at flowering (D,H) under control (C), and shading (S) in GH-CNRS (A–D), under low density (LD), and high density (HD) in field-IRRI (E–J). The values are mean ± SE. Results of Duncan test for multiple comparisons of each genotype per treatment at 5% level are shown by the letters above error bars. $n = 5$ at maturity and $n = 4$ at flowering in GH-CNRS, $n = 4$ in field-IRRI.

(**Figures 6C,D**), in particular at GH-CNRS. In all situations, the average of NIL values showed a higher main stem growth rate from PI to FLO related to a bigger panicle DW on the main stem at maturity.

The relationship between main stem and plant shoot growth rates from PI to FLO was thereafter explored, and no significant correlation was observed (not presented). Meanwhile, plant grain production was positively correlated to the whole plant shoot growth rate from PI to FLO (**Figure 7**). This correlation was higher in IR64 ($R^2 = 0.49$) than in IRRI146 ($R^2 = 0.37$) background and in field-IRRI ($R^2 = 0.50$) than in GH-CNRS ($R^2 = 0.28$) trial. Interestingly, with respect to this correlation, the NILs performed better than parents only in GH-CNRS whatever the genetic background (**Figure 7C**), i.e., in the cropping situation where the trade-off between tillering and main stem growth rates was the lowest for the NIL and the nearest from that of parents (**Figure 5C**). It can be mentioned that plant grain production was not correlated to main stem growth rate from PI to FLO (**Figure S3**).

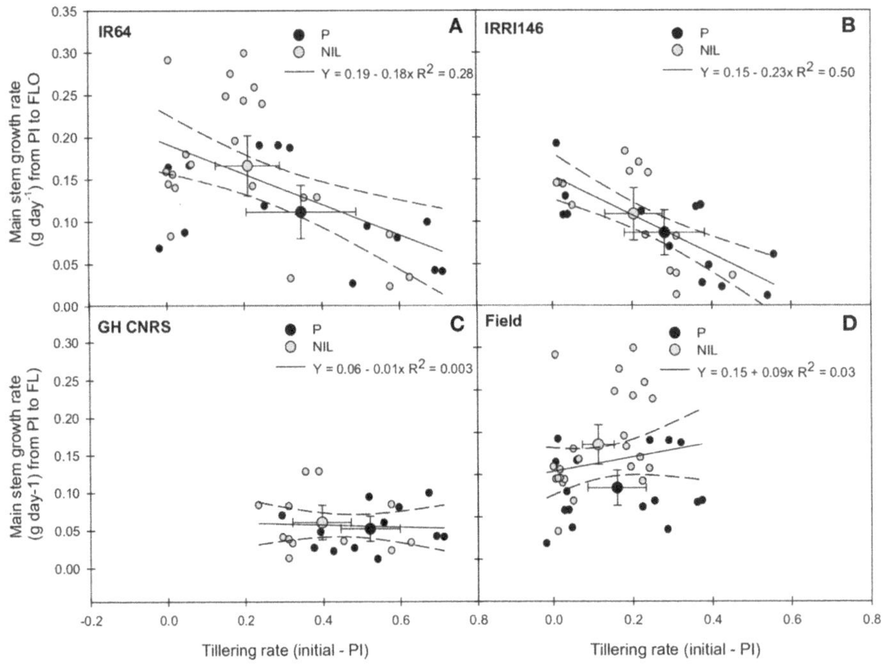

FIGURE 5 | Relationship between tillering rate from initial measurement to panicle initiation (PI) and main stem growth rate from PI to flowering (FLO) of the parent (black symbol) and the NIL (gray symbol), in IR64 background in GH-CNRS and field trials (A) in IRRI146 background in GH-CNRS and field trials (B) in GH-CNRS in IR64 and IRRI146 backgrounds (C) in field-IRRI experiment in IR64 and IRRI146 backgrounds (D). The values are mean ± SE. Regression curves are associated with confidence interval at $P = 0.05$. $n = 40$ for IR64 background and field-IRRI trial, $n = 32$ for IRRI146 background and GH-CNRS trial for regression curve.

Carbon Assimilation and Sugar Related Traits

In order to identify whether the difference in main stem growth rate was associated with a particular metabolic pattern, starch and net assimilation rate at ambient CO_2 concentration of 400 ppm were quantified during panicle development. In GH-CNRS, *qTSN4* enhanced assimilation only in shading treatment by 33% for IR64 and 24% for IRRI146 background (**Figure 8**). Shading significantly reduced assimilation in parent lines, 29 and 16% for IR64 and IRRI146 background, respectively, whereas in the NILs, assimilation was maintained under shading (**Figure 8A**). In the field, no significant *qTSN* and treatment effect was observed on assimilation, even if it was slightly decreased with *qTSN* in both backgrounds (**Figure 8B**). In GH-IRRI, *qTSN4* increased assimilation under low light (crowded plants) conditions but it was the opposite under full light (isolated plants) conditions (**Figure S4A**).

Internode starch content increased in the presence of *qTSN* for both backgrounds across the trials (**Figures 8C,D; Figure S4B**). The *qTSN4* effect was stronger under full light conditions (control in GH-CNRS, LD in the field, isolated in GH-IRRI), and significant in IRRI146 background in GH-CNRS (18%) (**Figure 8C**) and IR64 background in the field (11%) (**Figure 8D**). Similar trend was observed in leaf blade, with no significant effect of *qTSN* across the trials (not presented). However, in most cases, leaf starch was strongly reduced under low plant access to light.

DISCUSSION

The isolines (NIL) used in this study carried *qTSN4* or *qTSN12*, known to enhance leaf and panicle sizes but to reduce panicle number in some environmental situations (Fujita et al., 2013; Okami et al., 2015). The present study aimed at better characterizing this trade-off by comparing the NILs to their recurrent parents IRRI146 and IR64 regarding morphogenesis and C source-sink balance along the whole plant cycle and their behavior under low access to light.

The QTSN Affects Rice Morphogenesis and Physiology at Earlier Stage than Expected

A reduction of the rate of tiller emergence before PI (as early as 400°C days) was observed in this study in the presence of *qTSN* for both genetic backgrounds under both treatments. This was, however, not addressed in previous studies (Fujita et al., 2012, 2013) where the breeders mainly focused their attention on latter traits measured between flowering and maturity. Nevertheless, Okami et al. (2015) confirmed that tiller number was reduced with *qTSN4* under drought stress at vegetative stage as tillering rate per unit of above-ground biomass of the NIL was lower than that of the parent, which is in line with the present study. But in contrast to the present study, no difference between P and NIL was observed regarding tillering dynamics under well water conditions (Okami et al., 2015) and the ratio of main stem

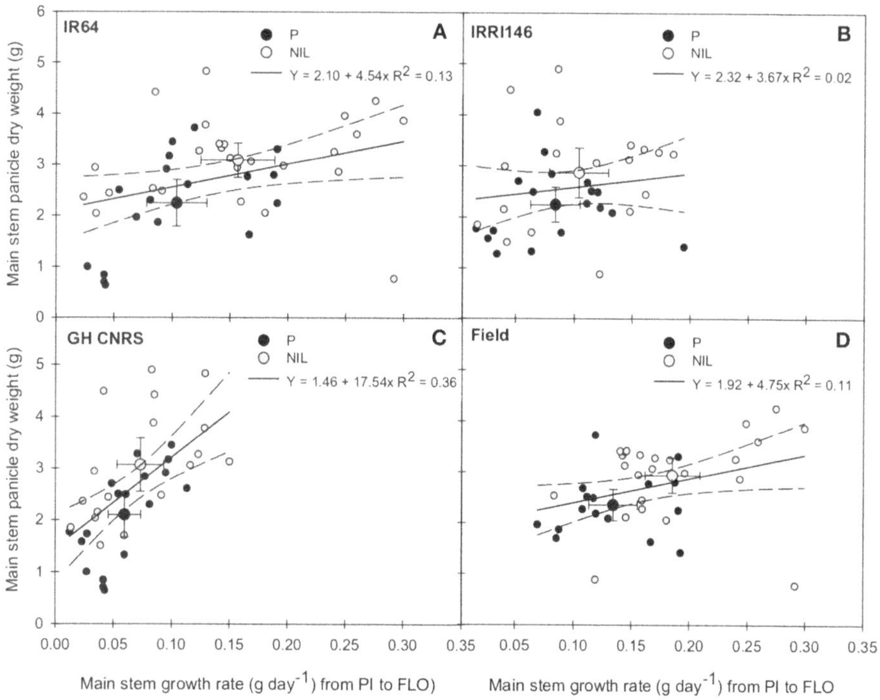

FIGURE 6 | Relationship between main stem growth rate from panicle initiation (PI) to flowering (FLO) and main stem panicle dry weight at maturity of the parent (black symbol) and the NIL (gray symbol), in IR64 background in GH-CNRS and field trials (A) in IRRI146 background in GH-CNRS and field trials (B) in GH-CNRS in IR64 and IRRI146 backgrounds (C) in field-IRRI experiment in IR64 and IRRI146 backgrounds (D). The values are mean ± SE. Regression curves are associated with confidence interval at $P = 0.05$. $n = 42$ for IR64 background and field-IRRI trial, $n = 34$ for IRRI146 background and GH-CNRS trial for regression curve.

leaf area to tiller number (Okami et al., 2012). Interestingly, in the present study, main stem growth rate between PI and FLO was inversely proportional to tillering rate before PI (**Figure 5**). Considering that reduction in tillering rate is expected to provide more assimilate available to the growing stems, the difference in main stem growth rate appears as a consequence of the change in early tillering rate. Several hypotheses subtending this correlation can be raised. On one hand, it can be hypothesized that *qTSN* implies a higher apical dominance due to hormonal signals, e.g., in relation to strigolactone (Jamil et al., 2012) or other hormones (ABA, IAA, GA3; Liu et al., 2011). This may be associated with a higher sensitivity to the red/far red ratio within the canopy maybe brought by *qTSN*. Indeed cessation of tillering in crops has been widely reported to be correlated with the increase, with crop age, in red/far-red ratio sensed by the plant within the canopy, even before any C limitation occurred within the plant (Ballaré and Casal, 2000; Ugarte et al., 2010). On the other hand, sink strength of growing stems should be higher at early stage in the presence of QTL, due to the initiation and pre-dimensioning of larger organs at meristem level, which may decrease tiller bud outgrowth. In this latter case, tillering would be reduced by a competition (or at least a signaling of competition) for resources active meristem and organs and tiller bud outgrowth, as pointed out by Rebolledo et al. (2012) across japonica rice genetic diversity. Whatever the hypothesis, the reduction in early tillering rate coincides with the appearance of leaf 8 in GH (rather leaf 7 in the field) so with the initiation of leaf 10 or 11, based on the developmental pattern established by Nemoto et al. (1995). Interestingly in GH, the leaves with larger area in the presence of QTL were the 3–4 top leaves of the main stem, so those initiated right after leaf 11.

The size of organs was enhanced by *qTSN4* on the main stem, from 3 to 4 leaves below the flag leaf up to the panicle. In addition to individual leaf area from the upper phytomers, internode length, and internode and peduncle thickness, were also enhanced with *qTSN*, while peduncle length and plant height were reduced. This finding was also reported in a recent study using the same genetic materials but comparing genotype behavior under drought and well-watered conditions (Okami et al., 2015). This is in line with Fujita et al. (2013) using the same genetic materials and Wu et al. (2011) using other genetic materials revealing that rice plants with larger culm diameter exhibited longer and wider flag leaf, more grains per panicle, as well as lower tiller number. Furthermore, Liu et al. (2008) confirmed that a thicker peduncle plays an important role in the determination of panicle size and grain yield potential.

This behavior in the presence of *qTSN* may be thus related to a stronger expression of apical dominance as the changes in organ size could not be detected at plant level: plant leaf area and shoot biomass at flowering were not different between parents and NILs, mainly due to the compensation between organ size and number. Nevertheless, the increase in C assimilation (in GH

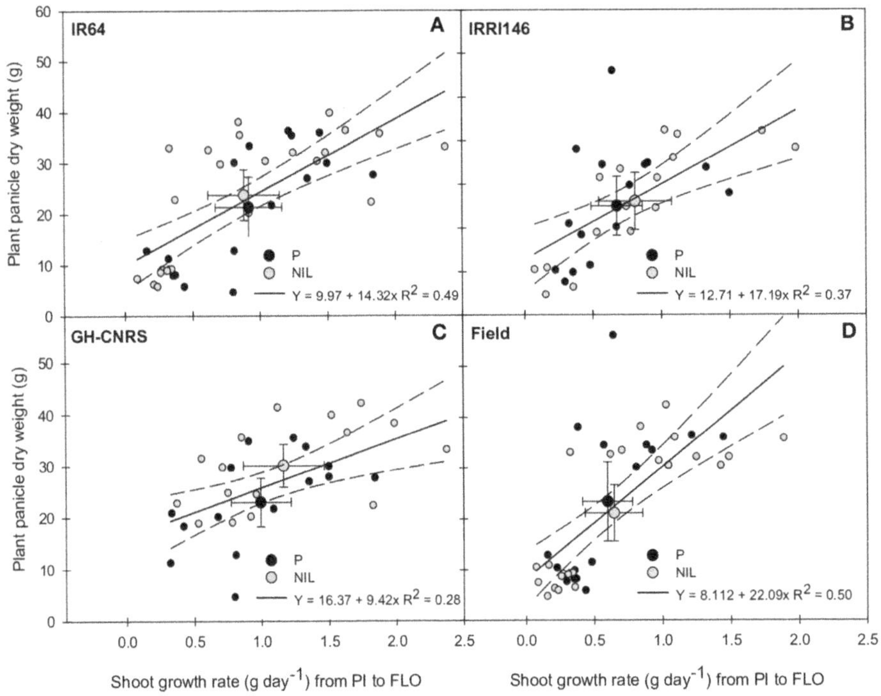

FIGURE 7 | Relationship between plant shoot growth rate from panicle initiation (PI) to flowering (FLO) and plant panicle dry weight at maturity of the parent (black symbol) and the NIL (gray symbol), in IR64 background in GH-CNRS and field trials (A) in IRRI146 background in GH-CNRS and field trials (B) in GH-CNRS in IR64 and IRRI146 backgrounds (C) in field-IRRI experiment in IR64 and IRRI146 backgrounds (D). The values are mean ± SE. Regression curves are associated with confidence interval at $P = 0.05$. $n = 42$ for IR64 background and field-IRRI trial, $n = 34$ for IRRI146 background and GH-CNRS trial for regression curve.

under shading) and in sugar (in particular starch) storage in stems (only in GH) suggested that *qTSN* enhances C availability within the main tiller at least during panicle development. Whether this is related to leaf anatomy and the elaboration of plant leaf area based on larger, thicker but fewer leaves needs to be confirmed. The fact that *qTSN4* co-localizes with NAL1, a gene involved in leaf anatomy, veining pattern and carboxylation, provides further insight to this *QTL* (Qi et al., 2008).

The QTSN Early Trade-Off Between Tiller and Main Stem Growth Partially Explains its Environment Dependent Effect on Plant Grain Production

The reduction of early tillering and its benefits for main stem growth rate in the presence of *qTSN* was dependent on the genetic background and the environment: it was more pronounced in GH-CNRS and in IR64. In addition, the positive effect of *qTSN* on plant grain production was mostly revealed in GH-CNRS while it was weak or inexistent in the field. This can be explained by the fact that in GH-CNRS fertile tiller (i.e., panicle) number was less reduced by *qTSN4* compared to that observed in the field, while its effect on panicle size was strong. The low effect in the field cannot be totally interpreted, but it is in line with a previous study (Okami et al., 2014) that revealed the absence of difference in grain yield between parents and NILs of IR64 background during three summer seasons under flooded and aerobic conditions in Japan. The same authors, however, reported some differences under aerobic conditions in one season with low nitrogen supply (90 kg N ha^{-1}). These results support the present study suggesting that a positive effect of the *qTSN* may be expressed depending on the cropping environment, where stressing environments (low light, low N, low soil exploration as in pots in greenhouse) should favor the benefits of *qTSN* on panicle size and, if ever, on grain production.

Earlier Tiller Cessation and its Implication for Yield Potential

Increasing yield potential by inducing an earlier cessation of tiller production was already proposed in previous studies comparing performances of hybrids and inbreds (Bueno and Lafarge, 2009; Lafarge et al., 2010). An earlier cessation of tillering was then related to an earlier biomass accumulation within reproductive stems and to a higher biomass remobilization from internodes to panicles, as also promoted by a higher sensitivity to the red/far red ratio within the canopy (Ballaré and Casal, 2000). Interestingly, the present study also highlighted the correlation between an earlier tillering cessation and higher main stem biomass growth, in association with a larger sink size. In the present study, however, these correlations were weak at plant level in the field particularly, highlighting the complexity

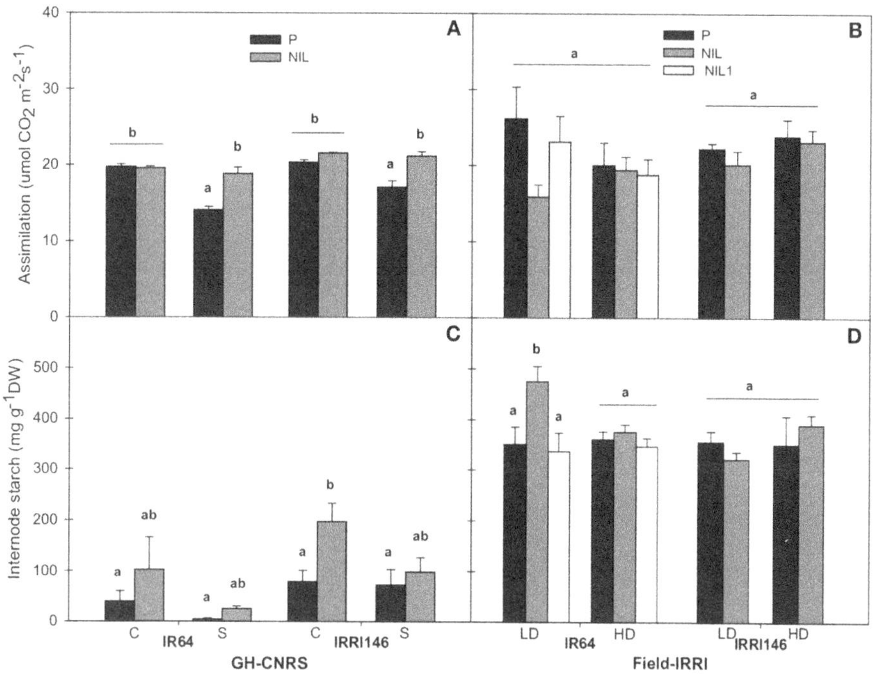

FIGURE 8 | Carbon assimilation (A,B) and internode starch concentration during panicle development (C,D) of parent (black), NIL (gray) and NIL1 (white) under control (C) and shading (S) in GH-CNRS (A,C), under low density (LD) and high density (HD) in field-IRRI (B,D). The values are mean ± SE. Results of Duncan test for multiple comparisons of each genotype per treatment at 5% level are shown in the letters above the bars. $n = 3$ in GH-CNRS, $n = 4$ in field-IRRI.

of the GxE interactions and trade-offs underlying the *qTSN* effect on plant grain production. It will be interesting to pay more attention on other fertile tillers to better understand *qTSN* impact on the whole plant growth and grain production. Nevertheless, this study reinforces the interest of developing genotypes optimizing tillering dynamics as long as yield potential is concerned.

CONCLUSION

The *qTSN* was confirmed in this study as a QTL potentially increasing panicle size due to an increase in stem growth rate, and in the size of the top leaves and internodes, at least at the main stem level. However, the trade-off between panicle size and panicle number was identified as the key node modulating the environment-dependent *qTSN* positive effect on plant grain production. This study revealed indeed that this trade-off was already visible at early stage through an earlier cessation of tiller production due to *qTSN* introgression, which coincided with the initiation at meristem level of phytomers with potentially larger leaves and internodes. Although it cannot be concluded if this early effect impacts directly tillering or organ dimensioning at meristem level, it seems worth going deeper in the understanding of the physiological regulation of this allele on plant functioning including contrasted cropping condition like limited radiation (i.e., during wet season in the tropics or N availability) in further studies.

AUTHOR CONTRIBUTIONS

TL, SY, MD, and DL participated in the design of the study; DA, AD participated in performing the research; AF and AC participated in biochemistry analysis; DA, TL, and DL participated in data analysis and wrote the manuscript. All authors read and approved the final manuscript.

ACKNOWLEDGMENTS

This work was a part of GRiSP (Global Rice Science Partnership) consortium and Yield Potential Project, in the scheme of double degree program between Bogor Agricultural University, Indonesia and Montpellier SupAgro, France. The authors would like to gratitude to Faculty of Agriculture, University of Lambung Mangkurat, Indonesia for supporting the PhD program; and to the Directorate General of Higher Education of Indonesia, Agris Mundus (Agreenium) and CIRAD for providing a PhD scholarship. Lastly we thank Nicole Sonderegger, Armelle Soutiras, and John Julius Manuben for sugar content analyses; and Denis Fabre and Bermenito Punzalan for their expertise on ecophysiological measurements.

REFERENCES

Ashikari, M., Sakakibara, H., Lin, S. Y., Yamamoto, T., Takashi, T., Nishimura, A., et al. (2005). Cytokinin oxidase regulates rice grain production. *Science* 309, 741–745. doi: 10.1126/science.1113373

Ballaré, C. L., and Casal, J. J. (2000). Light signals perceived by crop and weed plants. *Field Crops Res.* 67, 149–160. doi: 10.1016/S0378-4290(00)00090-3

Boehringer, S. A. (1984). *Methods of Enzymatic Food Analysis using Single Reagents.* Mannheim: Boehringer Mannheim GmbH.

Borràs-Gelonch, G., Rebetzke, G. J., Richards, R. A., and Romagosa, I. (2012). Genetic control of duration of pre-anthesis phases in wheat (*Triticum aestivum* L.) and relationships to leaf appearance, tillering, and dry matter accumulation. *J. Exp. Bot.* 63, 69–89. doi: 10.1093/jxb/err230

Brennan, J. P., and Malabayabas, A. (2011). "Impacts of IRRI germplasm on the Philippines since 1985," in *International Rice Research Institute's Contribution to Rice Varietal Yield Improvement in South-East Asia* (Canberra, ACT: Australian Center for International Agricultural Research), 25–43. Impact Assesment Series Report. 74.

Bueno, C. S., and Lafarge, T. (2009). Higher crop performance of rice hybrids than of elite inbreds in the tropics: 1. hybrids accumulate more biomass during each phenological phase. *Field Crops Res.* 112, 229–237. doi: 10.1016/j.fcr.2009.03.006

Chen, S., Gao, R., Wang, H., Wen, M., Xiao, J., Bian, N., et al. (2015). Characterization of a novel reduced height gene (Rht23) regulating panicle morphology and plant architecture in bread wheat. *Euphytica* 203, 583–594. doi: 10.1007/s10681-014-1275-1

Chen, S., Zeng, F. R., Pao, Z. Z., and Zhang, G. P. (2008). Characterization of high-yield performance as affected by genotype and environment in rice. *J. Zhejiang Univ. Sci. B* 9, 363–370. doi: 10.1631/jzus.B0710603

Dingkuhn, M., Laza, M. R. C., Kumar, U., Mendez, K. S., Collard, B., Jagadish, K., et al. (2015). Improving yield potential of tropical rice: achieved levels and perspectives through improved ideotypes. *Field Crops Res.* 182, 43–59. doi: 10.1016/j.fcr.2015.05.025

Dong, X., Wang, X., Zhang, L., Yang, Z., Xin, X., Wu, S., et al. (2013). Identification and characterization of *OsEBS*, a gene involved in enhanced plant biomass and spikelet number in rice. *Plant Biotechnol. J.* 11, 1044–1057. doi: 10.1111/pbi.12097

Dong, Y. J., Kamiunten, H., Ogawa, T., Tsuzuki, E., Terao, H., Lin, D. Z., et al. (2004). Mapping of QTLs for leaf developmental behavior in rice (*Oryza sativa* L.). *Euphytica* 138, 169–175. doi: 10.1023/B:EUPH.0000046799.21410.13

Dreccer, M. F., Chapman, S. C., Ogbonnaya, F. C., Borgognone, M. G., and Trethowan, R. M. (2008). Crop and environmental attributes underpinning genotype by environment interaction in synthetic-derived bread wheat evaluated in Mexico and Australia. *Aust. J. Agric. Res.* 59, 447–460. doi: 10.1071/AR07220

Fujita, D., Ebron, L. A., Araki, E., Kato, H., Khush, G. S., Sheehy, J. E., et al. (2010). Fine mapping of a gene for low-tiller number, Ltn, in japonica rice (*Oryza sativa* L.) variety Aikawa 1. *Theor. Appl. Genet.* 120, 1233–1240. doi: 10.1007/s00122-009-1251-7

Fujita, D., Santos, R. E., Ebron, L. A., Telebanco-Yanoria, M. J., Kato, H., Kobayashi, S., et al. (2009). Development of introgression lines of an Indica-type rice variety, IR64, for unique agronomic traits and detection of the responsible chromosomal regions. *Field Crops Res.* 114, 244–254. doi: 10.1016/j.fcr.2009.08.004

Fujita, D., Tagle, A. G., Ebron, L. A., Fukuta, Y., and Kobayashi, N. (2012). Characterization of near-isogenic lines carrying QTL for high spikelet number with the genetic background of an indica rice variety IR64 (*Oryza sativa* L.). *Breed. Sci.* 62, 18–26. doi: 10.1270/jsbbs.62.18

Fujita, D., Trijatmiko, K. R., Tagle, A. G., Sapasap, M. V., Koide, Y., Sasaki, K., et al. (2013). *NAL1* allele from a rice landrace greatly increases yield in modern *indica* cultivars. *Proc. Natl. Acad. Sci. U.S.A.* 110, 20431–20436. doi: 10.1073/pnas.1310790110

Fujita, K., and Yoshida, S. (1984). Partitioning of photosynthates between panicle and vegetative organs of rice under different planting densities. *Soil Sci. Plant Nutr.* 30, 519–525. doi: 10.1080/00380768.1984.10434719

Gaju, O., Reynolds, M. P., Sparkes, D. L., Mayes, S., Ribas-Vargas, G., Crossa, J., et al. (2014). Relationships between physiological traits, grain number and yield potential in a wheat DH population of large spike phenotype. *Field Crops Res.* 164, 126–135. doi: 10.1016/j.fcr.2014.05.015

Hashida, Y., Aoki, N., Kawanishi, H., Okamura, M., Ebitani, T., Hirose, T., et al. (2013). A near isogenic line of rice carrying chromosome segments containing OsSPS1 of Kasalath in the genetic background of Koshihikari produces an increased spikelet number per panicle. *Field Crops Res.* 149, 56–62. doi: 10.1016/j.fcr.2013.04.020

Huang, X., Qian, Q., Liu, Z., Sun, H., He, S., Luo, D., et al. (2009). Natural variation at the DEP1 locus enhances grain yield in rice. *Nat. Genet.* 41, 494–497. doi: 10.1038/ng.352

Ikeda, K., Ito, M., Nagasawa, M., Kyozuka, J., and Nagato, Y. (2007). Rice *ABERRANT PANICLE ORGANIZATION* 1, encoding an F-box protein, regulates meristem fate. *Plant J.* 51, 1030–1040. doi: 10.1111/j.1365-313X.2007.03200.x

Itoh, Y., and Shimizu, H. (2012). Phyllochron dynamics during the course of late shoot development might be affected by reproductive development in rice (*Oryza sativa* L.). *Dev. Genes Evol.* 222, 341–350. doi: 10.1007/s00427-012-0419-3

Jamil, M., Charnikhova, T., Houshyani, B., van Ast, A., and Bouwmeester, H. (2012). Genetic variation in strigolactone production and tillering in rice and its effect on *Striga hermonthica* infection. *Planta* 235, 473–484. doi: 10.1007/s00425-011-1520-y

Khush, G. S. (1987). Rice breeding: past, present and future. *J. Genet.* 66, 195–216. doi: 10.1007/BF02927713

Komatsu, K., Maekawa, M., Ujiie, S., Satake, Y., Furutani, I., Okamoto, H., et al. (2003). LAX and SPA: major regulators of shoot branching in rice. *Proc. Natl. Acad. Sci. U.S.A.* 100, 11765–11770. doi: 10.1073/pnas.1932414100

Lafarge, T. A., Broad, I. J., and Hammer, G. L. (2002). Tillering in grain sorghum over a wide range of population densities: identification of a common hierarchy for tiller emergence, leaf area development and fertility. *Ann. Bot.* 90, 87–98. doi: 10.1093/aob/mcf152

Lafarge, T., and Bueno, C. S. (2009). Higher crop performance of rice hybrids than of elite inbreds in the tropics: 2. does sink regulation, rather than sink size, play a major role? *Field Crops Res.* 114, 434–440. doi: 10.1016/j.fcr.2009.03.007

Lafarge, T., Seassau, C., Martin, M., Bueno, C., Clément-Vidal, A., Shreck, E., et al. (2010). Regulation and recovery of sink strength in rice plants grown under changes in light intensity. *Funct. Plant Biol.* 37, 413–428. doi: 10.1071/FP09137

Li, X., Qian, Q., Fu, Z., Wang, Y., Xiong, G., Zeng, D., et al. (2003). Control of tillering in rice. *Nature* 422, 618–621. doi: 10.1038/nature01518

Liu, G. L., Mei, H. W., Yu, X. Q., Zou, G. H., Liu, H. Y., Hu, S. P., et al. (2008). QTL analysis of panicle neck diameter, a trait highly correlated with panicle size, under well-watered and drought conditions in rice (*Oryza sativa* L.). *Plant Sci.* 174, 71–77. doi: 10.1016/j.plantsci.2007.09.011

Liu, W., Zhang, D., Tang, M., Li, D., Zhu, Y., Zhu, L., et al. (2013). THIS1 is a putative lipase that regulates tillering, plant height, and spikelet fertility in rice. *J. Exp. Bot.* 64, 4389–4402. doi: 10.1093/jxb/ert256

Liu, Y., Wang, Q., Ding, Y., Li, G., Xu, J., and Wang, S. (2011). Effects of external ABA, GA3 and NAA on the tiller bud outgrowth of rice is related to changes in endogenous hormones. *Plant Growth Regul.* 65, 247–254. doi: 10.1007/s10725-011-9594-x

Miura, K., Ikeda, M., Matsubara, A., Song, X. J., Ito, M., Asano, K., et al. (2010). *OsSPL14* promotes panicle branching and higher grain productivity in rice. *Nat. Genet.* 42, 545–592. doi: 10.1038/ng.592

Nemoto, K., Morita, S., and Baba, T. (1995). Shoot and root development in rice related to the phyllochron. *Crop Sci.* 35, 24–29. doi: 10.2135/cropsci1995.0011183X003500010005x

Okami, M., Kato, Y., Kobayashi, N., and Yamagishi, J. (2014). Agronomic performance of an IR64 introgression line with large leaves derived from New Plant Type rice in aerobic culture. *Eur. J. Agron.* 58, 11–17. doi: 10.1016/j.eja.2014.03.001

Okami, M., Kato, Y., Kobayashi, N., and Yamagishi, J. (2015). Morphological traits associated with vegetative growth of rice (*Oryza sativa* L.) during the recovery phase after early-season drought. *Eur. J. Agron.* 64, 58–66. doi: 10.1016/j.eja.2014.12.006

Okami, M., Kato, Y., and Yamagishi, J. (2012). Allometric relationship between the size and number of shoots as a determinant of adaptations in rice to water-saving aerobic culture. *Field Crops Res.* 131, 17–25. doi: 10.1016/j.fcr.2012.02.014

Okawa, S., Makino, A., and Mae, T. (2003). Effect of irradiance on the partitioning of assimilated carbon during the early phase of grain filling in rice. *Ann. Bot.* 92, 357–364. doi: 10.1093/aob/mcg147

Qi, J., Qian, Q., Bu, Q., Li, S., Chen, Q., Sun, J., et al. (2008). Mutation of the rice *narrow leaf1* gene, which encodes a novel protein, affects vein patterning and polar auxin transport. *Plant Physiol.* 147, 1947–1959. doi: 10.1104/pp.108.118778

Rebolledo, M. C., Dingkuhn, M., Clément-Vidal, A., Rouan, L., and Luquet, D. (2012). Phenomics of rice early vigour and drought response: are sugar related and morphogenetic traits relevant? *Rice* 5:22. doi: 10.1186/1939-8433-5-22

Streck, N. A., Lago, I., Bosco, L. C., de Paula, G. M., Oliveira, F. B., Gabriel, L. F., et al. (2009). Relationship between panicle differentiation and main stem leaf number in rice genotypes and red rice biotypes. *Sci. Agric.* 66, 195–203. doi: 10.1590/S0103-90162009000200008

Terao, T., Nagata, K., Morino, K., and Hirose, T. (2010). A gene controlling the number of primary rachis branches also controls the vascular bundle formation and hence is responsible to increase the harvest index and grain yield in rice. *Theor. Appl. Genet.* 120, 875–893. doi: 10.1007/s00122-009-1218-8

Tivet, F., Pinheiro, B. D. S., Raissac, M. D., and Dingkuhn, M. (2001). Leaf blade dimensions of rice (*Oryza sativa* L. and *Oryza glaberrima Steud.).* Relationships between tillers and the main stem. *Ann. Bot.* 88, 507–511. doi: 10.1006/anbo.2001.1447

Ugarte, C. C., Trupkin, S. A., Ghiglione, H., Slafer, G., and Casal, J. J. (2010). Low red/far-red ratios delay spike and stem growth in wheat. *J. Exp. Bot.* 61, 3151–3162. doi: 10.1093/jxb/erq140

Wang, W., Li, G., Zhao, J., Chu, H., Lin, W., Zhang, D., et al. (2014). DWARF TILLER1, a WUSCHEL-related homeobox transcription factor, is required for tiller growth in rice. *PLoS Genet.* 10:e1004154. doi: 10.1371/journal.pgen.1004154

Weng, X., Wang, L., Wang, J., Hu, Y., Du, H., Xu, C., et al. (2014). Grain number, plant height, and heading date7 is a central regulator of growth, development, and stress response. *Plant Physiol.* 164, 735–747. doi: 10.1104/pp.113.231308

Wu, L. L., Liu, Z. L., Wang, J. M., Zhou, C. Y., and Chen, K. M. (2011). Morphological, anatomical, and physiological characteristics involved in development of the large culm trait in rice. *Aust. J. Crop Sci.* 5, 1356–1363.

Zhang, B., and Yamagishi, J. (2010). Response of spikelet number per panicle in rice cultivars to three transplanting densities. *Plant Prod. Sci.* 13, 279–288. doi: 10.1626/pps.13.279

Conflict of Interest Statement: The authors declare that the research was conducted in the absence of any commercial or financial relationships that could be construed as a potential conflict of interest.

19

Transcriptome Analysis of *Brassica rapa* Near-Isogenic Lines Carrying Clubroot-Resistant and –Susceptible Alleles in Response to *Plasmodiophora brassicae* during Early Infection

Jingjing Chen[1†]*, Wenxing Pang*[1†]*, Bing Chen*[2]*, Chunyu Zhang*[3*] *and Zhongyun Piao*[1*]

[1] *Department of Horticulture, Shenyang Agricultural University, Shenyang, China,* [2] *Department of Characteristic Horticulture, Taizhou Institute of Agricultural Sciences, Jiangsu Academy of Agricultural Sciences, Taizhou, China,* [3] *National Key Laboratory of Crop Genetic Improvement and College of Plant Science and Technology, Huazhong Agricultural University, Wuhan, China*

Edited by:
Jaime Prohens,
Universitat Politècnica de València, Spain

Reviewed by:
Jutta Ludwig-Müller,
Technische Universität Dresden, Germany
Gary Peng,
Agriculture and Agri-Food Canada, Canada

****Correspondence:***
Chunyu Zhang
zhchy@mail.hzau.edu.cn;
Zhongyun Piao
zypiao@syau.edu.cn

† These authors have contributed equally to this work.

Although *Plasmodiophora brassicae* is one of the most common pathogens worldwide, the causal agent of clubroot disease in *Brassica* crops, resistance mechanisms to it are still only poorly understood. To study the early defense response induced by *P. brassicae* infection, a global transcriptome profiling of the roots of two near-isogenic lines (NILs) of clubroot-resistant (CR BJN3-2) and clubroot-susceptible (BJN3-2) Chinese cabbage (*Brassica rapa*) was performed by RNA-seq. Among the 42,730 unique genes mapped to the reference genome of *B. rapa*, 1875, and 2103 genes were found to be up- and down-regulated between CR BJN3-2 and BJN3-2, respectively, at 0, 12, 72, and 96 h after inoculation (hai). Functional annotation showed that most of the differently expressed genes are involved in metabolism, transport, signal transduction, and defense. Of the genes assigned to plant-pathogen interactions, 151 showed different expression patterns between two NILs, including genes associated with pathogen-associated molecular patterns (PAMPs) and effectors recognition, calcium ion influx, hormone signaling, pathogenesis-related (PR) genes, transcription factors, and cell wall modification. In particular, the expression level of effector receptors (resistance proteins), PR genes involved in salicylic acid (SA) signaling pathway, were higher in clubroot-resistant NIL, while half of the PAMP receptors were suppressed in CR BJN3-2. This suggests that there was a more robust effector-triggered immunity (ETI) response in CR BJN3-2 and that SA signaling was important to clubroot resistance. The dataset generated by our transcriptome profiling may prove invaluable for further exploration of the different responses to *P. brassicae* between clubroot-resistant and clubroot-susceptible genotypes, and it will strongly contribute to a better understanding of the molecular mechanisms of resistance genes of *B. rapa* against *P. brassicae* infection.

Keywords: clubroot, *Plasmodiophora brassicae*, disease resistance, RNA-seq, *Brassica rapa*

INTRODUCTION

Much research has focused on plant-pathogen interactions because of their importance to both agriculture and scientific research (Brown and Hovmoller, 2002). *Plasmodiophora brassicae* Wor. is a soil-borne, obligate, and biotrophic pathogen that attacks *Brassica* crops, leading to clubroot, and subsequent reductions in crop yield (Dixon, 2009). Many strategies have been proposed for controlling clubroot, among which the use of resistant cultivars is still the most cost-effective and environmentally friendly (Diederichsen et al., 2009; Donald and Porter, 2009).

Early reports suggested that both qualitative and quantitative traits were involved in clubroot resistance in *Brassica rapa*. At present, ten loci—*Crr1, Crr2, Crr3, Crr4, CRa, CRb, CRc, CRk, PbBa3.1*, and *PbBa3.3*—have been reported as being associated with clubroot resistance (CR) in *B. rapa* (Matsumoto et al., 1998; Suwabe et al., 2003, 2006; Hirai et al., 2004; Piao et al., 2004; Sakamoto et al., 2008; Chen et al., 2013). Two of these, *Crr1a* and *CRa*, have been cloned and found to be classical resistance genes that contain Toll-interleukin receptor (TIR)-nucleotide-binding (NB)-leucine-rich repeat (LRR; Ueno et al., 2012; Hatakeyama et al., 2013), but the response mechanisms to *P. brassicae* infection associated with these two CR genes have yet to be fully elucidated.

Plants have evolved two innate immune systems to combat with the attack of various pathogens (Jones and Dangl, 2006). The first mode of plant immune system is referred to as pathogen-associated molecular pattern (PAMP)-triggered immunity (PTI), which is triggered by the detection of PAMPs by pattern recognition receptor (PRR) proteins located on the external face of the host cell. However, pathogens can suppress PTI through secreting effectors into host cells. These pathogen effectors are recognized by specific resistance (*R*) genes in plants and triggers activation of the second mode of defense response, known as effector-triggered immunity (ETI). A number of defense responses are activated at the transcriptome level, including mitogen-activated protein kinase (MAPK) cascades, transcription factors, and hormone signaling (Moore et al., 2011), during PTI and ETI processes. Based on the transcriptome of *P. brassicae*, Schwelm et al. (2015) proposed that chitin might act as a PAMP of *P. brassicae* and the recognition of it can trigger defense responses in plants. However, these two processes have not been well characterized in *B. rapa* when challenged with *P. brassicae*.

Brassica crops are infected by *P. brassicae* in two distinct stages, consisting of primary infection of the root hairs followed by secondary infection of the root cortex (Kageyama and Asano, 2009). Feng et al. (2013) reported that primary and secondary infection of canola (*B. napus*) occurred 12 and 72 h, respectively, following inoculation with *P. brassicae*. The success of primary infection by *P. brassicae* was also observed in resistant genotypes of *Brassica* crops (Deora et al., 2012), indicating that in resistant *Brassica* strains *P. brassicae* is blocked at later stages of infection. Thus, studying the differentially expressed genes (DEGs) at two stages of infection helps in understanding host–*P. brassicae* interactions.

Analysis of global gene expression is one means of exploring the molecular basis of interactions between *Brassica* crops and *P. brassicae*, particularly with respect to mechanisms of resistance and the basal defense response. To date, several "-omics" approaches have been employed to examine the interactions between hosts and *P. brassicae*. Devos et al. (2006) and Cao et al. (2008), focusing on the changes of protein composition in *Arabidopsis* and *B. napus*, respectively, reported the details of the primary infection following inoculation. Their results demonstrated that most of the differentially regulated proteins were involved in plant defense, hormone metabolics, and detoxification. Siemens et al. (2006) identified more than 1000 DEGs in infected vs. control roots based on the ATH1 Affymetrix 22K microarray, which included genes associated with sugar phosphate metabolics, growth, cell cycle, and defense. A microarray analysis of *Arabidopsis thaliana* demonstrated that the number of DEGs involved in pathogen recognition and signal transduction was highest during the early stages of infection (Agarwal et al., 2011). A study using the complete *Arabidopsis* transcriptome microarray (CATMA) showed that, when compared with immune response in susceptible response, metabolic changes in the partial resistance response were reduced or delayed, and abnormal cell enlargement and proliferation were actively inhibited at 7 days post-inoculation on *A. thaliana* (Bur-0; Jubault et al., 2013). More recently, Schuller et al. (2014) confirmed the role of auxin and cytokinin metabolism and signaling in clubroot development based on microarray data and laser microdissection of *Arabidopsis* roots. They also found that brassinosteroid (BR) synthesis and signal perception were involved in clubroot development. Chu et al. (2014) reported that the signaling and metabolic activity of jasmonate acid (JA) and ethylene (ET) were up-regulated significantly in resistant populations compared to the susceptible lines at 15 days post-inoculation, while no increase in the expression of genes involved in salicylic acid (SA) metabolic and signaling pathways were detected. All of these functional studies focused on either the susceptible response or the late stage resistance response. Comparing responses to *P. brassicae* infection in susceptible and resistant plant lines is critical for understanding the defense mechanisms involved in resistance to clubroot.

In our previous studies, we identified and finely mapped a dominant *CRb* gene that confers resistance to *P. brassicae* pathotypes 2, 4, and 8 (Piao et al., 2004; Zhang et al., 2014). In addition, a pair of near-isogenic lines (NILs)—a clubroot-resistant (CR BJN3-2) line and a clubroot-susceptible (BJN3-2) line—were developed for *B. rapa* (Zhang et al., 2012a). The fact of similar genetic background and differences at the locus of a target gene between NILs would facilitate to analyze the response to pathogen in resistance and susceptible line as well as to compare the difference between the two lines on the transcription level.

RNA-Seq, a powerful approach for detecting DEGs and novel-expressed genes over a broad dynamic range (Blencowe et al., 2009; Wang et al., 2009b), was employed in this study: (i) to identify DEGs between CR BJN3-2 and BJN3-2; and (ii) to gain an insight into host–*P. brassicae* interactions for clubroot resistance during the early stages of infection by *P. brassicae*.

MATERIALS AND METHODS

Plant Material

Two Chinese cabbage NILs carrying the clubroot-resistant allele of *CRbCRb* (CR BJN3-2) and the clubroot-susceptible allele of *crbcrb* (BJN3-2) were inoculated with suspension of *P. brassicae*. CR BJN3-2 was developed by integration of the *CRb* gene from the CR Shinki DH line of Chinese cabbage into the Chinese cabbage inbred line BJN3-2 based on the marker-assisted selection (Zhang et al., 2012a). SSR genotyping of CR BJN3-2 revealed that 100% of the recurrent parent genome was recovered.

P. Brassicae Inoculation

The single-spore isolate (SSI) *P. brassicae*, which was used in the previous study (Zhang et al., 2014), was maintained and propagated in the susceptible Chinese cabbage lines. Resting spores were diluted to a density of 10^7 spores per mL in sterile distilled water after isolation from homogenized clubbed roots. Thirty-day-old plants of CR BJN3-2 and BJN3-2 were inoculated with *P. brassicae* by injecting the soil around each plant with 1 mL of SSI suspension, in accordance with Piao et al. (2004). The inoculated plants were maintained in the culture room under a 16-h photoperiod at 25°C and the soil was kept moist during the treatment period.

Microscopic Investigation

To determine the timing of primary and secondary infection of the two NILs, infection processes of *P. brassicae* in the roots were observed using an inverted microscope (Olympus IX70, Japan) and imaged with digital single-lens reflex (DSLR) camera (Nikon D5300, Japan) every 2 h after inoculation until the secondary infection was established. The method of *P. brassicae* inoculation was the same as above. The roots were washed thoroughly with distilled water to remove the spores adsorbed on the surface, and then were transferred to a new 90 mm diameter glass petri dish with 30 ml distilled water for microscope observation. At each time point, five individual plants were examined, and three fields were observed for each sample.

Tissue Sampling

Based on the observed infection stages, the primary and secondary infection were found to be happened at 12 and 72 h after inoculation (hai). Therefore, the roots of the CR BJN3-2 and BJN3-2 plants were sampled at 0, 12, 72, and 96 hai for analysis of the DEGs at the early stages of infection. Three independent biological replicates of both two NILs were performed for *P. brassicae* inoculation. Roots of 30 (three biological replications, ten plants for each replicate) plants were sampled and pooled for RNA extraction at each time point. The roots were washed in distilled water to remove the pathogen inoculum and immediately frozen in liquid nitrogen and stored at −80°C until use. To verify the success of infection, a pair of NILs was maintained in the culture room for 30 days after inoculation.

To cover the shortage caused by the lack of biological replication for RNA-seq, three independent biological replications were performed for *P. brassicae* inoculation. Inoculation process and tissue sampling were exactly the same with the preparation of samples as above for RNA-seq. The pooled RNA was used to confirm the RNA-seq data results by quantitative RT-PCR (qRT-PCR).

RNA isolation

RNA was extracted using TRIzol reagent (Invitrogen, Carlsbad, CA, USA) according to the manufacturer's instructions. RNA purity was determined using a Nanodrop spectrophotometer (Thermo Fisher Scientific Inc., Wilmington, DE, USA), 1% formaldehyde gel electrophoresis, and a 2100 Bioanalyzer (Agilent Technologies, Santa Clara, CA, USA). To eliminate the variability caused by human operation during the infection and sampling process, RNA from three biological replicates at each time point (0, 12, 72, and 96 hai) were pooled.

cDNA Library Construction and Sequencing

cDNA library preparation and sequencing were conducted by the Biomarker Technology Company in Beijing, China. mRNA was isolated using beads with Oligo (dT), and then broken into short fragments after the addition of fragmentation buffer. These short fragments of mRNA were used as templates to synthesize first-strand cDNA with random primers. Second-strand cDNA was synthesized in a reaction containing DNA polymerase I, RNase H, dNTPs, and buffer. The resulting cDNA were then subjected to end-repair polymerase and mixed with Solexa adapters. Suitable fragments were recovered from an agarose gel. Next, PCR amplification was performed to enrich the purified cDNA template. Finally, all the eight libraries (samples from 0, 12, 72, and 96 hai in two NILs) were sequenced using an Illumina HiSeq™ 2000. The raw RNA-seq data has been deposited at GEO (http://www.ncbi.nlm.nih.gov/geo/) under accession number GSE74044 (http://www.ncbi.nlm.nih.gov/geo/query/acc.cgi?acc=GSE74044).

Digital Gene Expression Analysis

Gene expression levels of all eight libraries were estimated by calculating read density as Reads Per Kilo base per Million mapped reads (RPKM; Mortazavi et al., 2008). EBseq software (Leng et al., 2013) was used to identify DEGs using pair-wise comparison. This program can estimate the variance of RNA-seq data without biological replicates by pooling similar genes together. *P*-values obtained via EBseq were further corrected using the Benjamini-Hochberg procedure (Benjamini and Hochberg, 1995), and the corrected *P*-values were used to determine the false discovery rate (FDR). Sequences were deemed to be significantly differentially expressed if the FDR was <0.01 and there was at least a two-fold change (>1 or < -1 in log2 ratio value) in RPKM between two libraries. Differential patterns of gene expression at the various time points are represented by Venn diagrams.

Sequence Data Analysis and Annotation

We mapped sequencing reads to the reference database for the *B. rapa* genome (Version 1.2, http://brassicadb.org/brad/)

using TopHat (Trapnell et al., 2009). Unigene sequences were compared with the nonredundant protein (nr) database (Deng et al., 2006), the Cluster of Orthologous Groups (COG) database (Tatusov et al., 2000), the Kyoto Encyclopedia of Genes and Genomes (KEGG) pathway database (Kanehisa et al., 2004), and Gene Ontology (GO; Ashburner et al., 2000). The Blast2GO program (Conesa et al., 2005) was used to obtain GO annotation of the unigenes. WEGO and Top GO software were then used to perform GO functional classification and enrichment analysis of all unigenes in order to view the distribution of gene functions.

Quantitative RT-PCR Analysis

To validate the RNA-seq results, we performed qRT-PCR on 40 selected DEGs. Twenty-one of the selected DEGs were involved in resistant to *P. brassicae* (including 9 receptor like proteins, 5 *R* genes, 2 calmodulin-like genes, 2 WRKY transcription factors, one beta-1,3 glucanase, and one cytochrome P450) and the rest of 19 DEGs were from the model of SA, JA, and ET signal pathway (Supplementary Table S1). Notably, 33 of the DEGs were from 0 hai.

The qRT-PCR reactions were performed using SYBR Green SuperReal PreMix Plus (TIANGEN) on a Bio-Rad CFX 96 real-time PCR system (Bio-Rad, Hercules, CA, USA). The primers were designed using Primer Premier 5.0 software and are listed in Supplementary Table S1. Analysis of gene expression was performed for all samples at 0, 12, 72, and 96 h after inoculation of CR BJN3-2 and BJN3-2 with *P. brassicae*. Each sample was divided into three independent biological and technical replicates. The comparative C_T method of quantitation was used with *B. rapa 18S rRNA* and *Actin* as internal controls. The conditions for amplification were as follows: 15 min denaturation at 95°C, followed by 40 cycles of 95°C for 10 s, 60°C for 20 s, and 72°C for 25 s. Following amplification, melting curves were performed by increasing temperatures from 55 to 95°C at intervals of 0.5°C every 10 s to confirm the specificity of the PCR amplification.

RESULTS

Comparative Analysis of *P. brassicae* Infection of CR BJN3-2 and BJN3-2 Genotypes

The differences in the infection process of *P. brassicae* between CR BJN3-2 and BJN3-2 were monitored after inoculation. There were no spores in the roots at 0 hai in both two lines, but spores were observed around the surface of the roots. Primary zoospores inside the root hairs were observed at 12 hai in both two lines. The presence of plasmodium in the cortical tissue was observed in BJN3-2 at 72 hai, indicating the beginning of secondary infection. Subsequently, numberous plasmodia were observed in the cortical tissue. In contrast, the secondary infection was not happened in CR BJN3-2 even after 72 and 96 hai. The same results were also obtained by Feng et al. (2013). Therefore, four time points (0, 12, 72, and 96 hai) were selected to investigate the differential transcript changes between two NILs at an early infection stage.

Severity of disease in both two NILs was assessed 30 days after inoculation with *P. brassicae* (Supplementary Figure S1). We found severe clubs on the roots of the susceptible line BJN3-2, but no clubroot symptoms were detected in the clubroot-resistant CR BJN3-2.

RNA-Seq Analysis

Changes in transcript levels between CR BJN3-2 and BJN3-2 at 0, 12, 72, and 96 hai with *P. brassicae* were analyzed by RNA-seq. A total of 180,185,005 reads were generated by 100-bp paired-end sequencing from the eight cDNA libraries, constituting 35.4 Gb of cDNA sequence. Of these, 88,634,677 reads were obtained from CR BJN3-2 and 91,550,328 from BJN3-2. GC% of sequence data from the eight libraries were all ~48%, and Q30% (i.e., those reads with an average quality score >30) were all >80%, an indication that the accuracy and quality of the sequencing data were sufficient for further analysis. Approximately 78% of the sequenced reads (140 million mapped reads) were in alignment with the *B. rapa* genome reference sequence (Version 1.2, http://brassicadb.org). An overview of the sequencing process is shown in Supplementary Table S2. The distribution of the expression level of the unigenes had similar patterns among the eight samples, suggesting there was no bias in the construction of the cDNA libraries (Supplementary Figure S2).

The total mapped reads were aligned to each region in the reference genome, including the exon, intergenic, and intron regions, with the percentage of reads mapped to the exon being the highest in all eight libraries (Supplementary Figure S3).

Transcriptome Analysis in Response to *P. brassicae*

A total of 3812 DEGs were identified between CR BJN3-2 and BJN3-2 (Supplementary Table S3). As shown in the Venn diagram in **Figure 1**, 1875 genes were up-regulated and 2103 were down-regulated in CR BJN 3-2, while only 83 genes were up-regulated and 321 were down-regulated at all four time points. Compared with BJN 3-2, 987 genes were up-regulated and 1069 genes were down regulated at 0 hai. Following, *P. brassicae* inoculation, 587 genes were up-regulated, and 1010 genes were down-regulated at 12 hai; 670 gens were up-regulated and 899 were down-regulated at 72 hai; 574 genes were up-regulated and 878 genes were down-regulated at 96 hai. Number of the genes that were up/down-regulated was more than any other time point in CR BJN3-2 at 0 hai (**Figure 1**, Supplementary Tables S4–S5).

Functional Annotation of DEGs

All of the 3812 DEGs were annotated by comparing their sequences against five public databases, which permitted the assignment of several functional annotations (Supplementary Table S6): 3573 (93.73%) DEGs were found to have significant matches in the nr database, 2815 (73.85%) in the Swiss-Prot database, 1034 (27.12%) in the COG, 788 (20.67%) in the KEGG, and 3355 (88.01%) in the GO.

A total of 3355 DEGs were assigned to three GO classes: biological process, cellular component, and molecular function. For the four time points, the cellular components most commonly represented were "cell" and "cell part," whereas

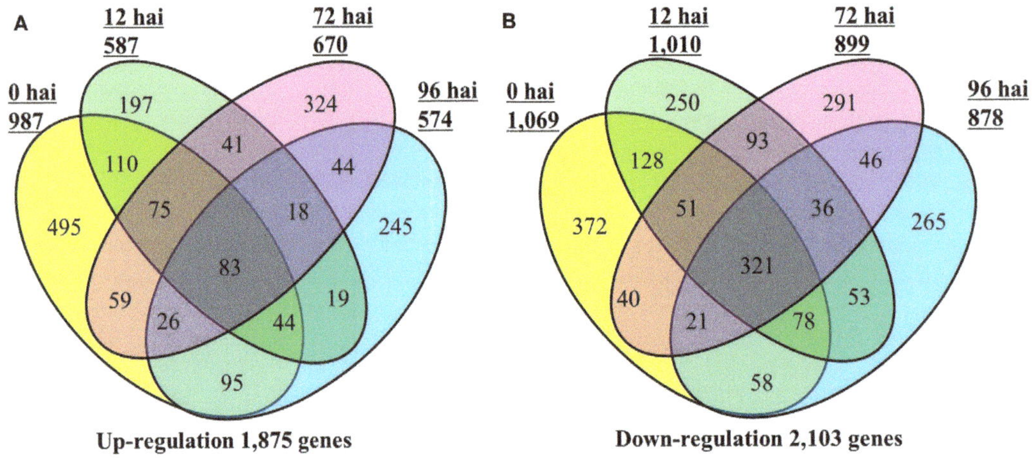

FIGURE 1 | Number of differentially expressed genes in the CR BJN3-2 that were (A) up-regulated and (B) down-regulated (FDR < 0.01 and fold change > 2.0 or < −2.0) compared with BJN3-2. Numerals inside the parentheses indicate the number of genes expressed at each hai. The total number of DEGs is noted at the bottom of each Venn diagram. hai, hours after inoculation.

"catalytic activity" and "binding" were among the most commonly represented molecular function categories. The top five subcategories in the biological process class were "metabolic process," "cellular process," "response to stimulus," "biological regulation," and "developmental process" (**Figure 2**). GO enrichment analysis of the DEGs in the biological process class show the top GO terms at each time point represented as a directed acyclic graph (Supplementary Figures S4–S7). We extracted the top ten GO terms at each time point (**Figure 3**). For instance, the GO terms "salicylic acid biosynthetic process," "type I hypersensitivity," and "focal adhesion assembly" were present at all four time points. In total, there were 6, 3, 2, and 2 specific GO terms at 0, 12, 72, and 96 hai, respectively, (**Figure 3**).

We found 1034 DEGs that had a COG classification. Among the 25 COG categories, the largest group was "General function prediction only" followed by "Tanscription," "Signal transduction mechanisms," "Amino acid transport and metabolism," and "Replication, recombination and repair." However, we did not find DEGs in the categories of W (Extracellular structure) and Y (Nuclear structure; **Figure 4**).

To identify the biological pathways that are active in *B. rapa*, we mapped the differentially expressed unigene sequences to the reference canonical pathways in KEGG. A total of 788 DEGs were annotated in the KEGG database and assigned to 111 KEGG pathways. The "Plant hormone signal transduction" was the most common term and contained 473 DEGs (12.41%), followed by "Ribosome" (448, 11.75%), "Plant-pathogen interaction" (283, 7.42%), "Oxidative phosphorylation" (265, 6.95%), and "Protein processing in endoplasmic reticulum" (258, 6.77%; **Table 1**).

TABLE 1 | Top 20 enriched KEGG pathways.

Pathway	The number of DEGs	Pathway ID
Plant hormone signal transduction	473 (12.41%)	ko04075
Ribosome	448 (11.75%)	ko03010
Plant-pathogen interaction	283 (7.42%)	ko04626
Oxidative phosphorylation	265 (6.95%)	ko00190
Protein processing in endoplasmic reticulum	258 (6.77%)	ko04141
Spliceosome	255 (6.69%)	ko03040
Purine metabolism	235 (6.16%)	ko00230
RNA transport	231 (6.06%)	ko03013
Starch and sucrose metabolism	202 (5.30%)	ko00500
Glycolysis/Gluconeogenesis	193 (5.06%)	ko00010
Amino sugar and nucleotide sugar metabolism	189 (4.96%)	ko00520
Phenylpropanoid biosynthesis	185 (4.85%)	ko00940
Ubiquitin mediated proteolysis	185 (4.85%)	ko04120
Pyrimidine metabolism	184 (4.83%)	ko00240
Phenylalanine metabolism	160 (4.20%)	ko00360
mRNA surveillance pathway	157 (4.12%)	ko03015
Ribosome biogenesis in eukaryotes	150 (3.93%)	ko03008
Carbon fixation in photosynthetic organisms	146 (3.83%)	ko00710
Pyruvate metabolism	138 (3.62%)	ko00620
Arginine and proline metabolism	137 (3.59%)	ko00330

DEGs Involved in Resistance to *P. brassicae*

To gain a deeper insight into the defense mechanisms *B. rapa* employs to prevent infection by *P. brassicae*, a list of 151 differentially regulated disease resistance genes was obtained by combining data from the literature and a keyword search in the *B. rapa* genome annotation (Supplementary Table S7). These included 23 PRRs, 15 NBS-LRR containing *R* genes, 2 respiratory burst oxidase homologs (RBOH), 6 Ca^{2+} influx coding genes, 2 MAPK, 18 pathogenesis-related (PR) proteins, 20 hormone metabolism (SA [three genes], JA [eight genes], and ET [nine genes]) related genes, 34 cell wall modification related genes, 10 chitinases, and 23 WRKY factors. To get a global view of DEGs involved in resistant to *P. brassicae*, the fold changes of these genes between the two NILs were represented via a heat map

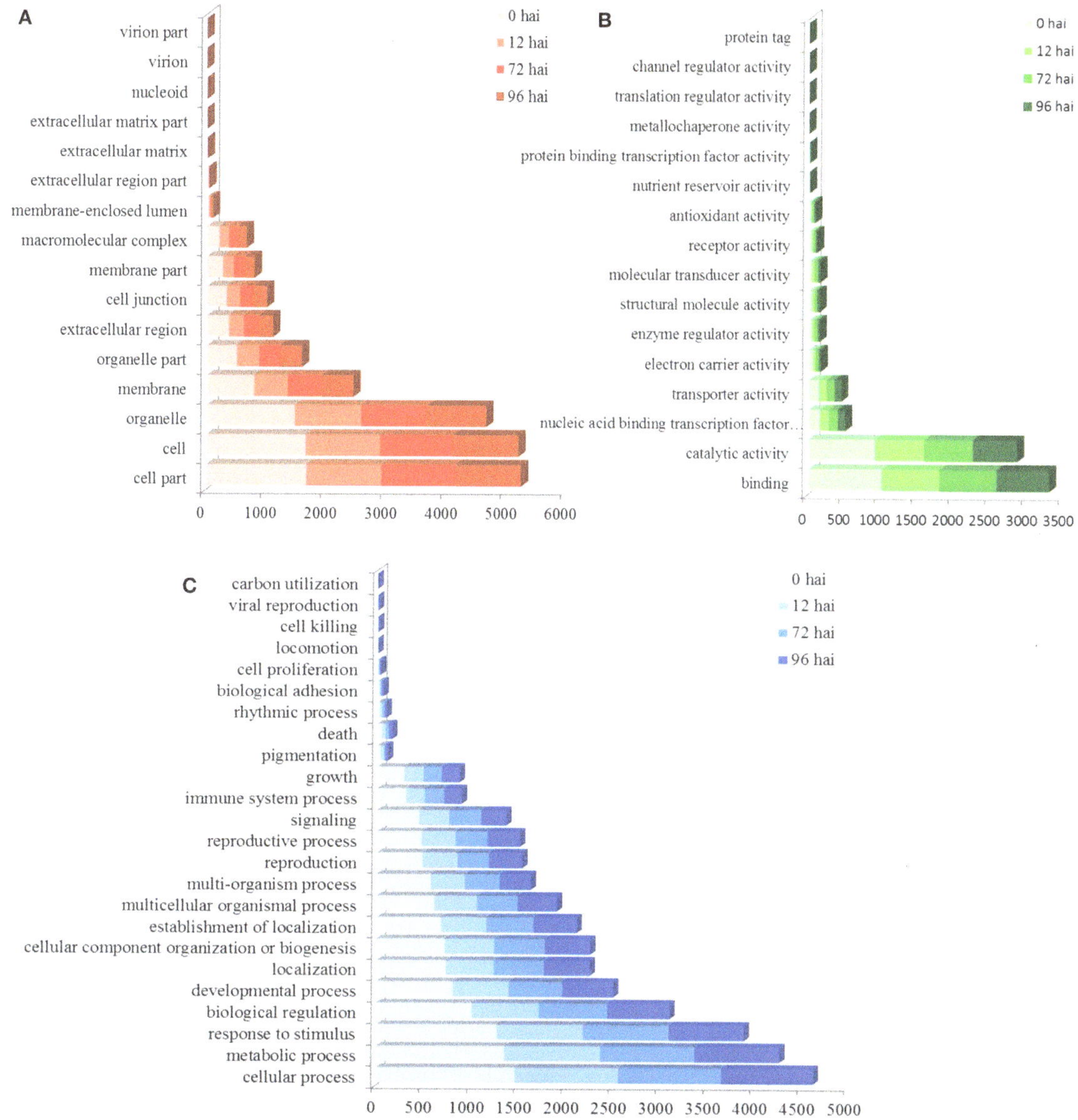

FIGURE 2 | GO assignment of differentially expressed genes (DEGs) in CR BJN3-2 and BJN3-2. The unigenes were mapped to three main categories: cellular component **(A)**, molecular function **(B)**, and biological process **(C)**. The y-axis indicates the number of annotated DEGs. hai, hours after inoculation.

(**Figure 5**). It is clearly that the *R* genes, MAPK, and PR proteins were up regulated in CR BJN3-2, while PRRs, WRKY factors and cell wall modification related genes showed an irregular regulation.

Expression Patterns of the Genes in the SA, JA, and ET Signaling Pathways

In the current study, there were three genes related to the SA pathway, including two up-regulated *NIM1-interaction 2* (*NIMIN2*) genes, that were components of this defense signaling pathway (**Figure 5**, Supplementary Table S7). In addition, a *B. rapa* gene homolog to *SUPPRESSOR OF NPR1-1* was significantly down-regulated at all four time points (**Figure 5**, Supplementary Table S7), while eight genes involved in JA biosynthesis and signaling (including 12-oxophytodienoatereductase 1[*OPR1*], lipoxygenases 3 *[LOX3]*, *LOX4*, jasmonate ZIM-domain [*JAZ*], and jasmonate-associated 1[*JAS1*]) and nine genes associated with ET biosynthesis and signaling (including 1-amino-cyclopropane-1-carboxylate synthase [*ACS*], ethylene-insensitive 3 [*EIN3*], EIN3-binding box protein [*EBF*], and ethylene response factor 2 [*ERF2*]) were down-regulated in CR BJN 3-2 at different time points (**Figure 5**, Supplementary Table S7). These results suggested an induced SA pathway and repressed JA/ET pathway during the host–*P. brassicae* interaction.

Quantitative RT-PCR Validation

To confirm the results of the RNA-seq, 40 DEGs were selected for qRT-PCR assays, including 21 genes involved in resistant to *P. brassicae* and 19 genes from the model of SA, JA, and ET signal pathway (Supplementary Table S1). For all 40 DEGs, including

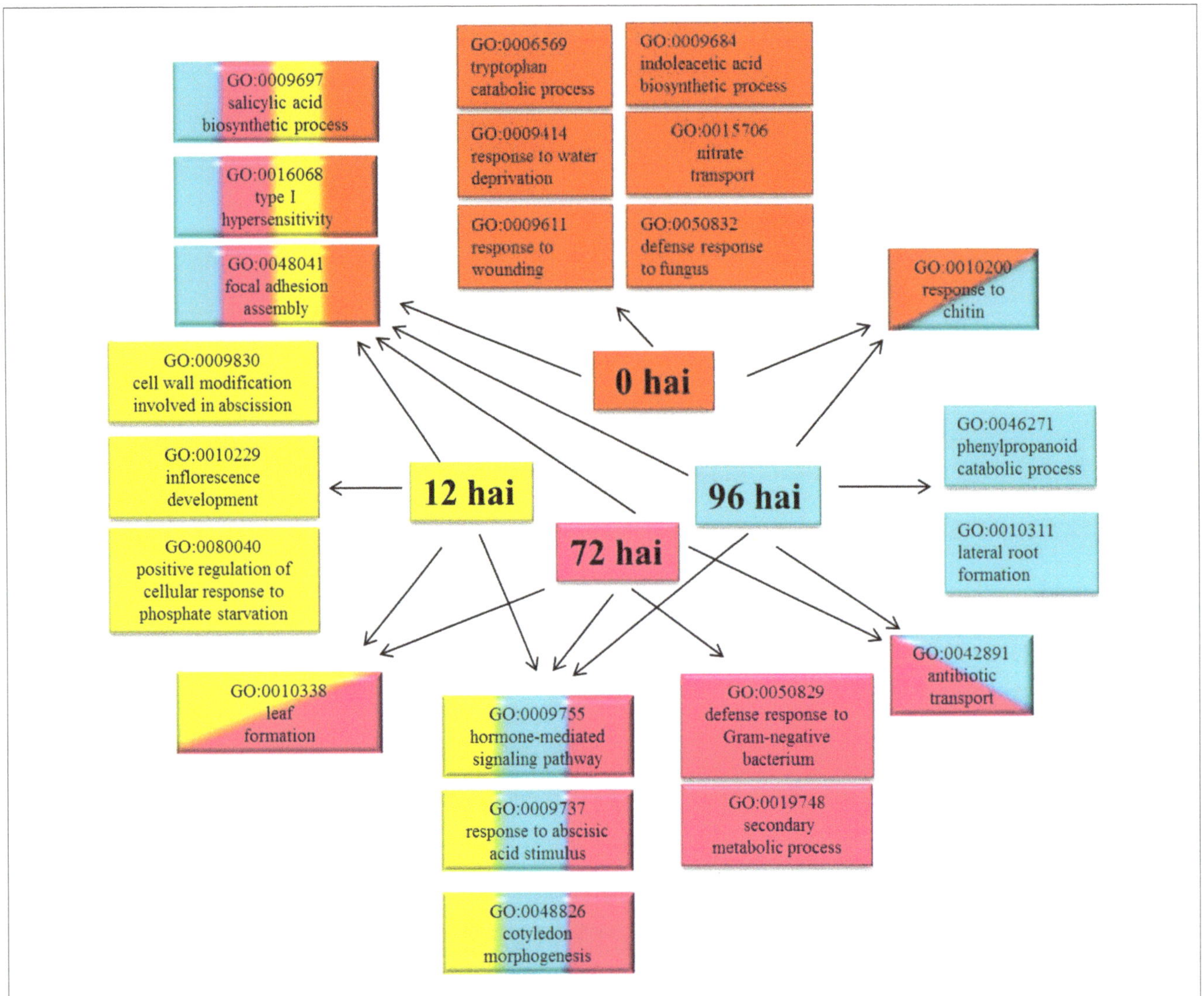

FIGURE 3 | Top ten Gene Ontology (GO) terms (biological process) that were significantly enriched at four time points. GO terms enriched only at 0 hai are shown in red, GO terms enriched only at 12 hai are shown in yellow, GO terms enriched only at 72 hai are shown in purple, GO terms enriched only at 96 hai are shown in blue, and GO terms enriched at more than one time point are indicated in combined colors. hai, hour after inoculation.

33 identified at 0 hai, qRT-PCR detected the same expression tendency as the RNA-seq analyses (**Figure 6**). It is a further indication of the high degree of reliability of the RNA-seq used in this study.

DISCUSSION

Identifying the genes that are regulated in response to *P. brassicae* infection represents a major challenge to understanding the basis of qualitative resistance. The use of a NIL pair carrying either the resistant or susceptible allele for the *CRb* locus in this study provided the opportunity to identify genes that exhibited differential transcript accumulation in the resistant or susceptible genotype. Compared to traditional genetic populations, NILs offer several advantages for transcriptional analyses due to the minimization of genetic background interference and enhancement of the sensitivity and accuracy of transcriptional analyses (Keurentjes et al., 2007). Although the RNA-seq data was analyzed without biological replicates, we could examine the *B. rapa* transcriptome at the early stages of *P. brassicae* infection, compares the response of susceptible and resistance lines of *B. rapa* to *P. brassicae* infection, and developed an integrated model for *B. rapa* during the early stages of *P. brassicae* infection.

A total of 3812 DEGs were detected between CR BJN3-2 and BJN3-2. Among them, more genes were detected to be differentially expressed at 0 hai. Detection of more DEGs at 0 hai were also found in potato and soybean when they were challenged by phytophthora and bacterial leaf pustule, respectively, (Kim et al., 2011; Gyetvai et al., 2012). However, it was not clear why more DEGs were detected at 0 hai. In the current study, DEGs related to the two main defense mechanisms (PTI and ETI) were investigated. One hundred and fifty one

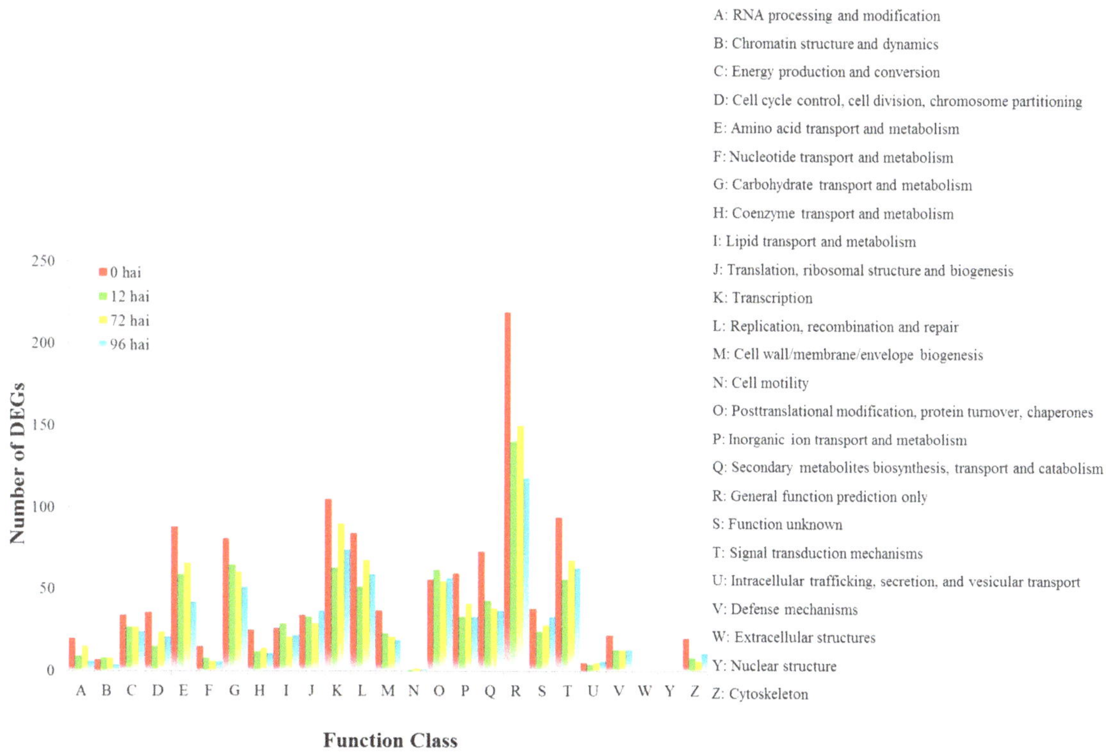

FIGURE 4 | COG functional classification of differentially expressed genes in CR BJN3-2 and BJN3-2.

DEGs involved in PTI and ETI were found by combining data from the literature and a keyword search in the *B. rapa* genome annotation (Supplementary Table S7) and discussed below.

Up-Regulation of PR Proteins in CR BJN3-2

One of the features of the defense response of plants is the production of PR proteins that are induced specifically in pathological or related situations (Van Loon and Van Strien, 1999). After inoculation with *P. brassicae*, 18 PR proteins (including 7 PR proteins, 5 thaumatin family proteins, and 6 lipid transfer proteins) were differentially regulated between CR BJN3-2 and BJN3-2 (**Figure 5**, Supplementary Table S7). All 18 PR proteins were up-regulated in CR BJN3-2 following inoculation with *P. brassicae*. This is consistent with a previous microarray analysis of *A. thaliana*, in which it was observed that several PR proteins, including thaumatin and defensins, were induced in the partially resistant line of the species (Jubault et al., 2013). Our results suggest that these PR proteins are most likely also involved in *B. rapa* resistance to *P. brassicae*.

Pathogen Perception by PRRs

The genes related to PRRs were induced in both two NILs. Among the 23 differentially expressed PRRs, 14 were induced in BJN3-2, and 9 were induced in CR BJN3-2 (**Figure 5** and Supplementary Table S7). The initial step in the defense response of plants to the presence of a pathogen is the conserved PAMPs detected by PRRs, resulting in PTI (Jones and Dangl, 2006). PRRs thus play a fundamental role in PTI (Shiu and Bleecker, 2001). Of note, key PRR genes triggered by PAMPs, such as brassinosteroid insensitive 1-associated kinase 1 (*BAK1*), flagellin sensing 2 (*FLS2*), chitin elicitor receptor kinase (CERK), and chitin elicitor-binding protein (CEBiP), which transduces signals that trigger PTI (Dodds and Rathjen, 2010), showed no difference between CR BJN3-2 and BJN3-2, a finding that was contrary to our expectation. Using NILs of BLP-susceptible and BLP-resistant soybeans, Kim et al. (2011) examined the response of PTI-related genes after inoculation with *Xanthomonas axonopodis* pv. *glycines* and found that PRRs such as *FLS2* and *EFR* were up-regulated at 0 hai in the resistant soybean line. Li et al. (2012) proposed that *BAK1*, *CEBiP*, and *CERK1*, critical genetic components of PTI, were up-regulated following inoculation with *Foc TR4* in a resistant line of banana plant. That the PTI response in *B. rapa* as a result of *P. brassicae* infection differs from the response to other pathogens leads us to speculate that perhaps PTI does not play a role in the interaction between CR BJN 3-2 and *P. brassicae*.

Pathogen Detection by NB-LRR Disease Resistance Protein

The NB-LRR containing *R* gene is a crucial component of ETI, as it detects and binds to pathogen effectors and triggers the subsequent defense response. In our study, of the 15 differentially expressed R proteins between the two NILs, only two genes encoding NBS-LRR proteins were down-regulated in CR BJN3-2 compared with BJN3-2 (**Figure 5**, Supplementary Table S7). Two transcripts homologous to *Arabidopsis* "*RESISTANT TO P. SYRINGAE 4*" (*RPS4*) and "*RESISTANT TO P. SYRINGAE 5*"

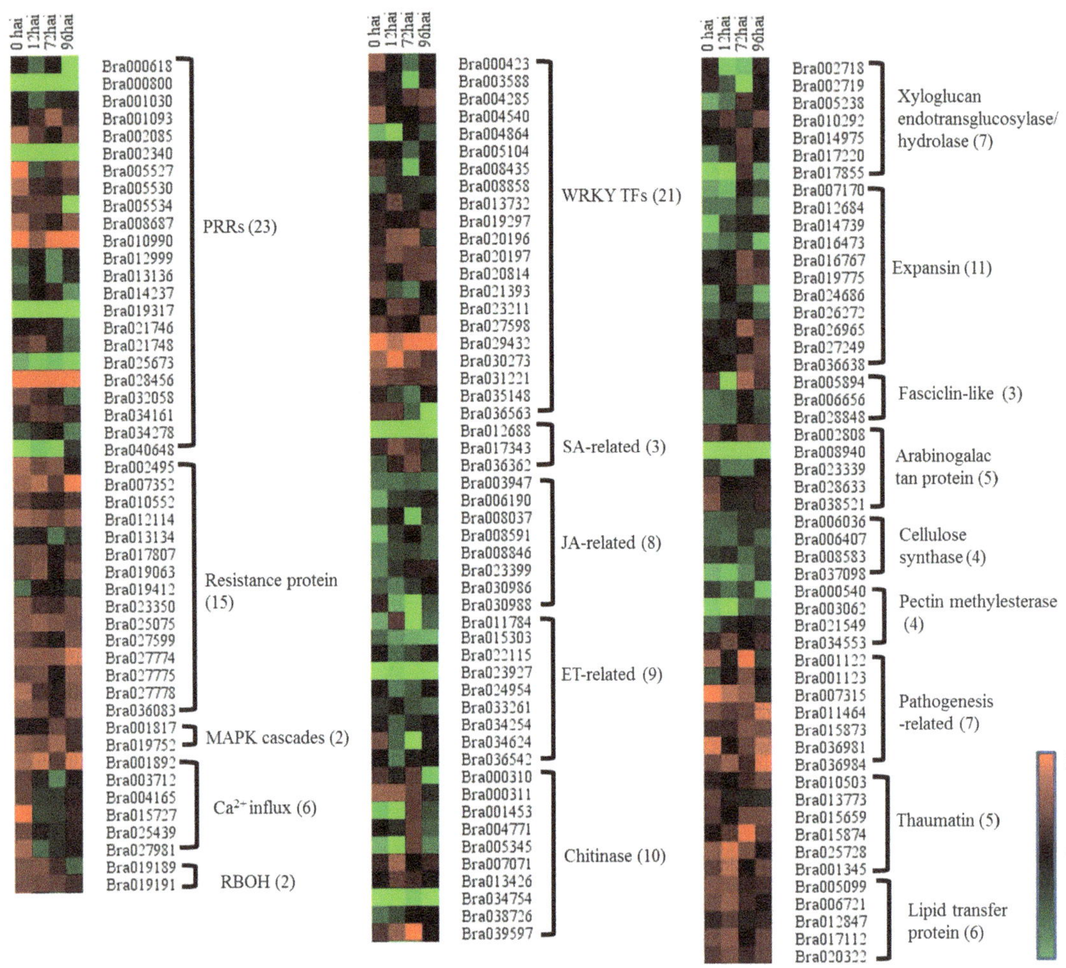

FIGURE 5 | Heat maps of differentially expressed genes for putative candidate genes assigned to clubroot-resistant genes in *B. rapa*. R and S, resistant and susceptible genotype, respectively; hai, hours after inoculation; green indicates down-regulated DEGs, and red indicates up-regulated DEGs in CR BJN3-2 when compared to BJN3-2.

(*RPS5*), respectively, were up-regulated at different time points in CR BJN3-2 (Supplementary Table S5). Previous work has shown that *RPS4* confers specific resistance to *Pseudomonas syringae* pv. *tomato* carrying the avirulence gene *AvrRPS4* (Gassmann et al., 1999). It was also reported that *RPS5* confers resistance to *P. syringae* strains that contain the avirulence gene *avrPphB* (Ade et al., 2007). Hence the *RPS4* and *RPS5* may be involved in the resistance interaction between *P. brassicae* and *B. rapa*. A *RPW8*-like gene encoding powdery mildew resistance protein (Wang et al., 2009a) was up-regulated in the *B. rapa* resistant NIL, as well. Taken together, our analysis suggests that the functions of *R* genes may be conserved across species and that the ETI response was more robust in CR BJN3-2 than in BJN3-2.

Previously, the *CRb* locus was narrowed to an interval of approximately 83.5 kb based on the "Chiifu-401-42" sequences located on chromosome A03 (Zhang et al., 2014). The target interval contained 15 putative genes and several classes of genes that could potentially be involved in resistance activities, including one TIR-NBS-LRR gene, one NBS–LRR gene, and several putative regulatory genes. There was, however, no significant difference in the expression of these genes between the two NILs (Supplementary Table S8). This may be due to the instability of the translation product and the differences in sequence between the CR regions of two NILs. Further confirmation via cloning of the *CRb* gene is needed.

Activation of MAPK Cascades and WRKY Transcription Factors

MAPK activation is one of the earliest signaling events following pathogen detection (Ichimura et al., 2002). Seven genes (*MPK3, MPK4, MPK6, MKK1, MKK2, MKK4,* and *MKK5*) were highly expressed at all time points in both NILs and no differences in the expression of these genes were found between CR BJN3-2 and BJN3-2 with the exception of *MKK5*, which was up-regulated at 72 hai in CR BJN3-2 (**Figure 5**, Supplementary Table S7). A *B. rapa* homolog of *SUPPRESSOR OF MKK1 MKK2 2* (*SUMM2*), a *R* protein encoding the coiled-coil (CC)-NB-LRR

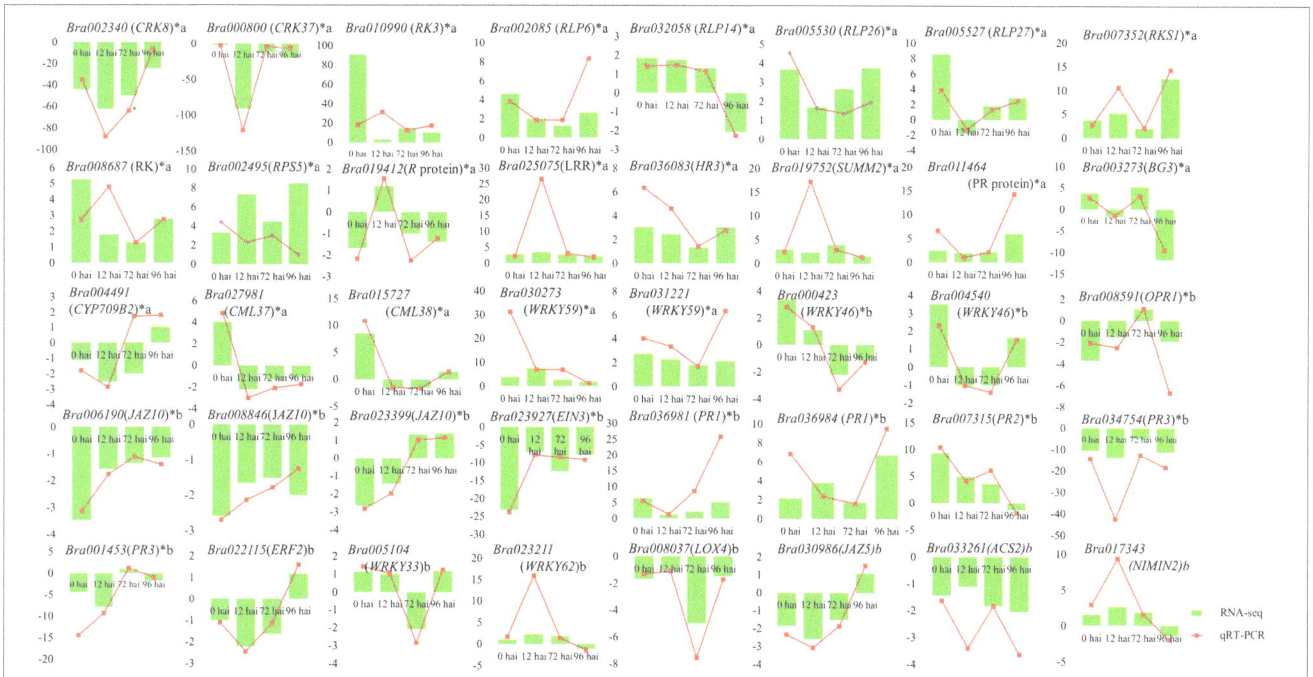

FIGURE 6 | Validation of RNA-seq data by qRT-PCR. Forty DEGs involved in clubroot resistant and SA, JA, ET signal pathway (including 33 genes from 0 hai) were selected for validation and showed the same tendency with RNA-seq. y-axis showed the fold change between the two NILs, positive value indicated up regulated in CR BJN3-2 and negative indicated down regulated. *indicated DEGs from 0 hai; a indicated DEGs involved in resistant to *P. brassicae*; b indicated DEGs from the model of SA, JA/ET signal pathway.

domain, was up-regulated at 0, 12, and 72 hai in CR BJN3-2. Zhang et al. (2012b) reported that *SUMM2* becomes active when the *MEKK1-MKK1/MKK2-MPK4* cascade is disrupted by pathogens, supporting the hypothesis that *R* proteins evolved to protect plants when microbial effectors suppress PTI. These results suggest that although PTI were suppressed in both NILs, ETI response was more robust in the resistant NIL. *WRKY*-type transcription factors, which act in a complex defense response network as both positive and negative regulators in plant immunity (Pandey and Somssich, 2009), were then activated by the MAPK cascade. In this study, some DEGs encoding *WRKY* exhibited different expression patterns between two NILs, with *WRKY 9, −16, −19, −29, −38, −46, −53, −59,* and *−62* being up-regulated in CR BJN3-2 at different time points whereas *WRKY 12, −23, −33, −40, −50,* and *−75* were down-regulated (**Figure 5**, Supplementary Table S7). It's unclear whether the *WRKY*s acted positively or negatively in response to the presence of *P. brassicae*, and thus the role of *WRKY* in clubroot resistance need further study.

Ca^{2+} Influx and Respiratory Burst Oxidase Homolog

The *B. rapa* calmodulin-like 37 (*CML 37*)-, *CML11-*, *CML24-*, and *CML38*-like genes were all up- regulated at 0 hai in CR BJN3-2 compared to BJN3-2 (**Figure 5**, Supplementary Table S7), indicating that Ca^{2+} participates in signal transduction during early-stage plant defense response. Similarly, an up-regulation of Ca^{2+}-ATPase-like genes was also observed at 0 hai in soybean bacterial leaf pustule-resistance-NIL after inoculation with *X. axonopodis* pv. *glycines* (Kim et al., 2011). Calcium is an essential second messenger in the signal transduction pathways that regulate plant response to abiotic and biotic stress (Zipfel, 2009). Ca^{2+} can activate RBOH protein which was involved in the production of reactive oxygen species *in vitro* (Sagi and Fluhr, 2001). Two Chinese cabbage genes (*Bra019189* and *Bra019191*), the putative functions of which were RBOH, exhibited up-regulation in CR BJN3-2 (**Figure 5**, Supplementary Table S7). This result was in agreement with previous observations in wheat, cotton, and cucumber after infection by the fusarium wilt fungal pathogen (Dowd et al., 2004), and indicated that there was a higher ROS level in the resistant NILs, which inhibited the colonization of the pathogen on the roots.

An Induced SA Signal Pathway and Repressed JA/ET Signal Pathway in Resistant NIL

P. brassicae belongs to an obligate biotrophic protist. It is believed that plant resistance to biotrophic pathogens is controlled largely by SA-mediated signaling pathways, while resistance to necrotrophic pathogens is mediated by the JA and ET signaling pathways (Glazebrook, 2005). To study whether SA-mediated pathway was also involved in the resistant mechanism to *P. brassicae* in *B. rapa*, we examined the expression pattern of SA-related DEGs. It is found that NIMIN2, *PR1*, and *PR2* were up-regulated in the current study. NIMIN2 along with nonexpresser of PR genes 1 (*NPR1*) is a master regulator of the SA-mediated induction of defense genes (Spoel et al., 2003; Koornneef and Pieterse, 2008). Our results indicated an induced SA signal pathway in CR BJN3-2 after inoculation with *P. brassicae*. Consistently, Lovelock et al. (2013) demonstrated that SA is capable of suppressing clubroot in broccoli (*Brassicae oleracea* var. *italica*). And several studies have reported that *P. brassicae* secretes a methyl transferase (*PbBSMT*) that reduces the accumulation of SA in infected roots and that the expression level of *PbBSMT* in spores and plasmodia was low, whereas transferase expression was high in clubroots of host plants (Ludwig-Müller et al., 2015; Schwelm et al., 2015). All these studies suggested that a low SA level in the host can facilitate colonization by *P. brassicae*.

Eight genes related to JA biosynthesis and signaling and nine genes associated with ET biosynthesis and signaling were down-regulated in CR BJN3-2 (**Figure 5**, Supplementary Table S7). For example, *PR3* (basic chitinase) controlled by JA/ET-dependent pathways (Thomma et al., 1998) were down-regulated. The *WRKY62* was up-regulated in CR BJN3-2, which was demonstrated to function downstream of *NPR1* and regulate SA-mediated inhibition of JA signaling (Mao et al., 2007). The *JAZ* proteins can repress the function of *EIN3/EIL1*, probably by suppressing the DNA binding of *EIN3* (Zhu et al., 2011). Based on our results, a schematic illustration of the interactions between SA and JA (**Figure 7**) suggested that SA signal pathway was induced in the resistant NIL. Jubault et al. (2013) also reported the occurrence of an induced SA signaling pathway and a suppressed JA/ET signaling pathway in *A. thaliana* 7 dai with *P. brassicae*. Combined with previous results, it is clear that the SA signaling pathway and not the JA/ET signaling pathway plays a critical role in host resistance to *P. brassicae* infection.

Cell Wall Modification

We also identified DEGs related to clubroot-induced abnormal cell enlargement and uncontrolled cell division of host roots.

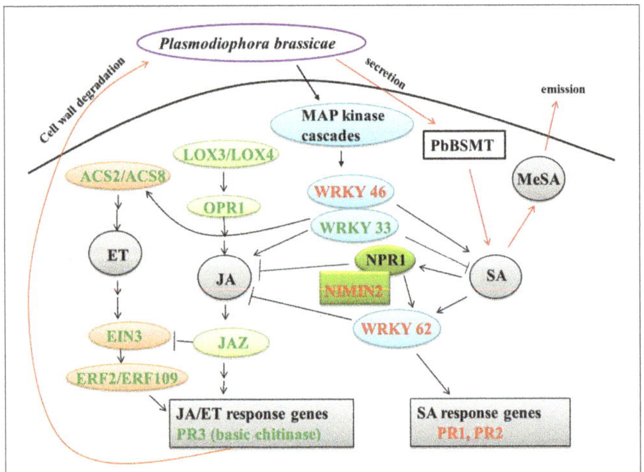

FIGURE 7 | Crosstalk between salicylic acid and jasmonate/ethylene. Green and red coloring indicate down- and up-regulated genes, respectively, in CR BJN3-2. Black arrows, indicate results derived from the current study; red arrows, indicate results derived from other studies. ACS, 1-amino-cyclopropane-1-carboxylate synthase; EIN, ethylene-insensitive; ERF, ethylene response factor; JAZ, jasmonate-domain protein; LOX, lipoxygenase; MeSA, methyl salicylate; NIMIN, NIM1-interacting; NPR, nonexpresser of PR genes; OPR, 12-oxophytodienoate reductase; PbSMT, *P. brassicae* methyltransferase; PR, pathogenesis-related.

Siemens et al. (2006) reported that genes involved in cell division and expansion, such as cell cycle genes and expansins, were up-regulated in *A. thaliana* after inoculation with *P. brassicae*. Jubault et al. (2013) reported that cell enlargement and proliferation were actively inhibited in the partial quantitative resistance response compared to the susceptible one. Consistent with these studies, our analysis showed that most of cell division and expansion related genes were down regulated in resistant NIL. Of the 34 genes involved in cell division and expansion identified in the current study, only 11 genes were up-regulated in the clubroot-resistant NIL at 0, 12, 72, and 96 hai, including xyloglucan endotransglucosylase/hydrolase 16 and 24, expansin 7, 12 and 18, arabinogalactan protein (AGP) 1, 2 and 14, pectin methylesterase (PME) 44, with the remaining 23 genes being down-regulated in the resistant NIL (**Figure 5**, Supplementary Table S7). Cell division and expansion were suppressed in the clubroot-resistant NIL, indicating that the abnormal cell division that results from *P. brassicae* infection was suppressed in CR BJN 3-2.

Chitinase in *B. rapa* Response to *P. brassicae* Infection

We identified 10 genes related to chitinase that were differentially expressed between the two NILs. Among them, 4 chitinase-related genes (*Bra007071, Bra013426, Bra000311*, and *Bra039597*) were up-regulated. Chitin is main component of the cell wall in *P. brassicae* (Moxham and Buczacki, 1983) and is known to be a kind of PAMP secreted by most pathogens (Latgé and Beauvais, 2014). It is not clear whether chitin released by *P. brassicae* is PAMP, but seven chitin synthase of *P. brassicae* were found to be highly expressed during infection process (Schwelm et al., 2015). The induced chitinase might provide the evidence for the degradation of chitin secreted by *P. brassicae* although it should be confirmed further.

Five genes were down-regulated in the CR BJN3-2 genotype compared to the BJN3-2 genotype, including three PR3-type basic chitinase (**Figure 5**, Supplementary Table S7). PR3 was controlled by JA/ET dependent pathway, and the induced PR3 exhibited the resistance to necrotrophic pathogens together with PR4 (chitinase type I and II) (Glazebrook, 2005). Down regulation of PR3 further indicated that the JA and ET pathways were not involved in the resistance to *P. brassicae* in *B. rapa*.

The chitinase gene *Bra004771* was up regulated at 72 hai, but down regulated at 96 hai in CR BJN3-2. There are also several reports showing that different types of chitinase operate against different pathogens in different plant species (Tabei et al., 1998; Takatsu et al., 1999; Datta et al., 2001; Rohini and Rao, 2001; Kishimoto et al., 2002; Takahashi et al., 2005). As such, the role that different chitinase subfamilies play in the interaction between *P. brassica* and *B. rapa* need to be investigated further.

CONCLUSION

In this study, we first investigated the transcriptome response in the roots of clubroot-resistant and clubroot-susceptible NILs of *B. rapa* during the early stages of *P. brassicae* infection. A total of 3812 differentially expressed genes were identified. The up-regulation of resistance protein-coding genes, SA-related genes and PR genes demonstrated that these components play a role in the interaction between *B. rapa* and *P. brassicae*. At the same time, our results also showed that the genes involved in the uncontrolled cell division and expansion that are induced by clubroot were suppressed to a greater extent in resistant NIL than in susceptible NIL. The present work revealed that many genes are differentially expressed between clubroot-resistant and clubroot-susceptible NILs when responding to *P. brassicae* infection, greatly expanded the vision of the genetic basis and improved our understanding of the molecular mechanisms underlying *B. rapa* resistance to *P. brassicae*.

AUTHOR CONTRIBUTIONS

JC analyzed the data, performed the experiments, and drafted the manuscript. WP participated in the data analysis and helped to draft the manuscript. BC helped analyzed the data. CZ contributed analysis tools and helped to draft the manuscript. ZP conceived the study, participated in its coordination, and helped to draft the manuscript. All authors have read and approved the final manuscript.

ACKNOWLEDGMENTS

This study was supported by grants from the National Natural Scientific Foundation of China, Project No. 31171967; the National Natural Scientific Foundation of China, Project No. 31471882; the "948" program of the Ministry of Agriculture, Project No. 2015-Z57 and the Postgraduate Innovation Cultivation Project of Shenyang Agricultural University.

REFERENCES

Ade, J., DeYoung, B. J., Golstein, C., and Innes, R. W. (2007). Indirect activation of a plant nucleotide binding site-leucine-rich repeat protein by a bacterial protease. *Proc. Natl. Acad. Sci. U.S.A.* 104, 2531–2536. doi: 10.1073/pnas.0608779104

Agarwal, A., Kaul, V., Faggian, R., Rookes, J. E., and Cahill, L. M. M. (2011). Analysis of global host gene expression during the primary phase of the *Arabidopsis thaliana-Plasmodiophora brassicae* interaction. *Funct. Plant Biol.* 36, 462–478. doi: 10.1071/FP11026

Ashburner, M., Ball, C. A., Blake, J. A., Botstein, D., Butler, H., Cherry, J. M., et al. (2000). Gene ontology: tool for the unification of biology. *Nat. Genet.* 25, 25–29. doi: 10.1038/75556

Benjamini, Y., and Hochberg, Y. (1995). Controlling the false discovery rate: a practical and powerful approach to multiple testing. *J. R. Stat. Soc. B* 57, 289–300.

Blencowe, B. J., Ahmad, S., and Lee, L. J. (2009). Current generation high-throughput sequencing: deepening insights into mammalian transcriptomes. *Genes Dev.* 23, 1379–1386. doi: 10.1101/gad.1788009

Brown, J. K., and Hovmoller, M. S. (2002). Aerial dispersal of pathogens on the global and continental scales and its impact on plant disease. *Science* 297, 537–541. doi: 10.1126/science.1072678

Cao, T., Srivastava, S., Rahman, M. H., Kav, N. N. V., Hotte, N., Deyholos, M. K., et al. (2008). Proteome-level changes in the roots of *Brassica napus* as a result of *Plasmodiophora brassicae* infection. *Plant Sci.*174, 97–115. doi: 10.1016/j.plantsci.2007.10.002

Chen, J., Jing, J., Zhan, Z., Zhang, T., Zhang, C., and Piao, Z. (2013). Identification of Novel QTLs for isolate-specific partial resistance to *Plasmodiophora brassicae* in *Brassica rapa*. *PLoS ONE* 8:e85307. doi: 10.1371/journal.pone.0085307

Chu, M., Song, T., Falk, K. C., Zhang, X., Liu, X., Chang, A., et al. (2014). Fine mapping of *Rcr1* and analyses of its effect on transcriptome patterns during infection by *Plasmodiophora brassicae*. *BMC Genomics* 15:1166. doi: 10.1186/1471-2164-15-1166

Conesa, A., Gotz, S., Garcia-Gomez, J. M., Terol, J., Talon, M., and Robles, M. (2005). Blast2GO: a universal tool for annotation, visualization and analysis in functional genomics research. *Bioinformatics* 21, 3674–3676. doi: 10.1093/bioinformatics/bti610

Datta, K., Tu, J., Oliva, N., Ona, I., Velazhahan, R., Mew, T. W., et al. (2001). Enhanced resistance to sheath blight by constitutive expression of infection-related rice chitinase in transgenic elite indica rice cultivars. *Plant Sci.* 160, 405–414. doi: 10.1016/S0168-9452(00)00413-1

Deng, Y., Li, J. Q., Wu, S. F., Zhu, Y. P., Chen, Y. W., and He, F. C. (2006). Integrated nr database in protein annotation system and its localization. *Comput. Eng.* 32, 71–72.

Deora, A., Gossen, B. D., and McDonald, M. R. (2012). Infection and development of *Plasmodiophora brassicae* in resistant and susceptible canola cultivars. *Can. J. Plant Pathol.* 34, 239–247. doi: 10.1080/07060661.2012.681071

Devos, S., Laukens, K., Deckers, P., Straeten, D. V. D., Beeckman, T., Inze, D., et al. (2006). A hormone and proteome approach to picturing the initial metabolic events during *Plasmodiophora brassicae* infection on *Arabidopsis*. *Mol. Plant Microbe Interact.* 19, 1431–1443. doi: 10.1094/MPMI-19-1431

Diederichsen, E., Frauen, M., Linders, E. G. A., Hatakeyama, K., and Hirai, M. (2009). Status and perspectives of clubroot resistance breeding in Crucifer crops. *J. Plant Growth Regul.* 28, 265–281. doi: 10.1007/s00344-009-9100-0

Dixon, G. (2009). The occurrence and economic impact of *Plasmodiophora brassicae* and clubroot disease. *J. Plant Growth Regul.* 28, 194–202. doi: 10.1007/s00344-009-9090-y

Dodds, P. N., and Rathjen, J. P. (2010). Plant immunity: towards an integrated view of plant-pathogen interactions. *Nat. Rev. Genet.* 11, 539–548. doi: 10.1038/nrg2812

Donald, C., and Porter, I. (2009). Integrated control of clubroot. *J. Plant Growth Regul.* 28, 289–303. doi: 10.1007/s00344-009-9094-7

Dowd, C., Wilson, L. W., and McFadden, H. (2004). Gene expression profile changes in cotton root and hypocotyl tissues in response to infection with *Fusarium oxysporum f.* sp *vasinfectum*. *Mol. Plant Microbe Interact.* 17, 654–667. doi: 10.1094/MPMI.2004.17.6.654

Feng, J., Hwang, S., and Strelkov, S. E. (2013). Studies into primary and secondary infection processes by *Plasmodiophora brassicae* on canola. *Plant Pathol.* 62, 177–183. doi: 10.1111/j.1365-3059.2012.02612.x

Gassmann, W., Hinsch, M. E., and Staskawicz, B. J. (1999). The *Arabidopsis RPS4* bacterial-resistance gene is a member of the TIR-NBS-LRR family of disease-resistance genes. *Plant J. Cell Mol. Biol.* 20, 265–277. doi: 10.1046/j.1365-313X.1999.t01-1-00600.x

Glazebrook, J. (2005). Contrasting mechanisms of defense against biotrophic and necrotrophic pathogens. *Annu. Rev. Phytopathol.* 43, 205–227. doi: 10.1146/annurev.phyto.43.040204.135923

Gyetvai, G., Sønderkær, M., GÖbel, U., Basekow, R., Ballvora, A., Imhoff, M., et al. (2012). The transcriptome of compatible and incompatible interactions of potato (*Solanum tuberosum*) with phytophthora infestans revealed by DeepSAGE analysis. *PLoS ONE* 7:e31526. doi: 10.1371/journal.pone.0031526

Hatakeyama, K., Suwabe, K., Tomita, R. N., Kato, T., Nunome, T., Fukuoka, H., et al. (2013). Identification and characterization of *Crr1a*, a gene for resistance to clubroot disease (*Plasmodiophora brassicae* Woronin) in *Brassica rapa* L. *PLoS ONE* 8:e54745. doi: 10.1371/journal.pone.0054745

Hirai, M., Harada, T., Kubo, N., Tsukada, M., Suwabe, K., and Matsumoto, S. (2004). A novel locus for clubroot resistance in *Brassica rapa* and its linkage markers. *Theor. Appl. Genet.* 108, 639–643. doi: 10.1007/s00122-003-1475-x

Ichimura, K., Shinozaki, K., Tena, G., Sheen, J., Henry, Y., Champion, A., et al. (2002). Mitogen-activated protein kinase cascades in plants: a new nomenclature. *Trends Plant Sci.* 7, 301–308. doi: 10.1016/S1360-1385(02)02302-6

Jones, J., and Dangl, J. L. (2006). The plant immune system. *Nature* 444, 323–329. doi: 10.1038/nature05286

Jubault, M., Lariagon, C., Taconnat, L., Renou, J., Gravot, A., Delourme, R., et al. (2013). Partial resistance to clubroot in *Arabidopsis* is based on changes in the host primary metabolism and targeted cell division and expansion capacity. *Funct. Integr. Genomics* 13, 191–205. doi: 10.1007/s10142-013-0312-9

Kageyama, K., and Asano, T. (2009). Life cycle of *Plasmodiophora brassicae*. *J. Plant Growth Regul.* 28, 203–211. doi: 10.1007/s00344-009-9101-z

Kanehisa, M., Goto, S., Kawashima, S., Okuno, Y., and Hattori, M. (2004). The KEGG resource for deciphering the genome. *Nucleic Acids Res.* 32, D277–D280. doi: 10.1093/nar/gkh063

Keurentjes, J. J., Bentsink, L., Alonso-Blanco, C., Hanhart, C. J., Blankestijn-De Vries, H., Effgen, S., et al. (2007). Development of a near-isogenic line population of *Arabidopsis thaliana* and comparison of mapping power with a recombinant inbred line population. *Genetics* 175, 891–905. doi: 10.1534/genetics.106.066423

Kim, K. H., Yang, J. K., Dong, H. K., Min, Y. Y., Moon, J. K., Kim, M. Y., et al. (2011). RNA-seq analysis of a soybean near-isogenic line carrying bacterial leaf pustule-resistant and -susceptible alleles. *DNA Res.* 18, 483–497. doi: 10.1093/dnares/dsr033

Kishimoto, K., Nishizawa, Y., Tabei, Y., Hibi, T., Nakajima, M., and Akutsu, K. (2002). Detailed analysis of rice chitinase gene expression in transgenic cucumber plants showing different levels of disease resistance to gray mold (*Botrytis cinerea*). *Plant Sci.* 162, 655–662. doi: 10.1016/S0168-9452(01)00602-1

Koornneef, A., and Pieterse, C. M. J. (2008). Cross talk in defense signaling. *Plant Physiol.* 146, 839–844. doi: 10.1104/pp.107.112029

Latgé, J., and Beauvais, A. (2014). Functional duality of the cell wall. *Curr. Opin. Microbiol.* 20, 111–117. doi: 10.1016/j.mib.2014.05.009

Leng, N., Dawson, J. A., Thomson, J. A., Ruotti, V., Rissman, A. I., Smits, B. M. G., et al. (2013). EBSeq: an empirical Bayes hierarchical model for inference in RNA-seq experiments. *Bioinformatics* 29, 1035–1043. doi: 10.1093/bioinformatics/btt087

Li, C. Y., Deng, G. M., Yang, J., Viljoen, A., Jin, Y., Kuang, R. B., et al. (2012). Transcriptome profiling of resistant and susceptible Cavendish banana roots following inoculation with *Fusarium oxysporum* f. sp. *cubense* tropical race 4. *BMC Genomics* 13:374. doi: 10.1186/1471-2164-13-374

Lovelock, D. A., Donald, C. E., Conlan, X. A., and Cahill, D. M. (2013). Salicylic acid suppression of clubroot in broccoli (*Brassicae oleracea* var. *italica*) caused by the obligate biotroph *Plasmodiophora brassicae*. *Australas. Plant Pathol.* 42, 141–153. doi: 10.1007/s13313-012-0167-x

Ludwig-Müller, J., Jülke, S., Geiß, K., Richter, F., Mithöfer, A., Šola, I., et al. (2015). A novel methyltransferase from the intracellular pathogen *Plasmodiophora brassicae* participates in methylation of salicylic acid. *Mol. Plant Pathol.* 16, 349–364. doi: 10.1111/mpp.12185

Mao, P., Duan, M., Wei, C., and Li, Y. (2007). WRKY62 transcription factor acts downstream of cytosolic NPR1 and negatively regulates jasmonate responsive gene expression. *Plant Cell Physiol.* 48, 833–842. doi: 10.1093/pcp/pcm058

Matsumoto, E., Yasui, C., Ohi, M., and Tsukada, M. (1998). Linkage analysis of RFLP markers for clubroot resistance and pigmentation in Chinese cabbage. *Euphytica* 104, 79–86. doi: 10.1023/A:1018370418201

Moore, J. W., Loake, G. J., and Spoel, S. H. (2011). Transcription dynamics in plant immunity. *Plant Cell* 23, 2809–2820. doi: 10.1105/tpc.111.087346

Mortazavi, A., Williams, B. A., McCue, K., Schaeffer, L., and Wold, B. (2008). Mapping and quantifying mammalian transcriptomes by RNA-Seq. *Nat. Methods* 5, 621–628. doi: 10.1038/nmeth.1226

Moxham, S. E., and Buczacki, S. T. (1983). Chemical composition of the resting spore wall of *Plasmodiophora brassicae*. *Trans. Br. Mycol. Soc.* 80, 297–304. doi: 10.1016/S0007-1536(83)80013-8

Pandey, S. P., and Somssich, I. E. (2009). The role of WRKY transcription factors in plant immunity. *Plant Physiol.* 150, 1648–1655. doi: 10.1104/pp.109.138990

Piao, Z. Y., Deng, Y. Q., Choi, S. R., Park, Y. J., and Lim, Y. P. (2004). SCAR and CAPS mapping of *CRb*, a gene conferring resistance to *Plasmodiophora brassicae* in Chinese cabbage (*Brassica rapa* ssp. *pekinensis*). *Theor. Appl. Genet.* 108, 1458–1465. doi: 10.1007/s00122-003-1577-5

Rohini, V. K., and Rao, K. S. (2001). Transformation of peanut (*Arachis hypogaea* L.) with tobacco chitinase gene: variable response of transformants to leaf spot disease. *Plant Sci.* 160, 889–898. doi: 10.1016/S0168-9452(00)00462-3

Sagi, M., and Fluhr, R. (2001). Superoxide production by plant homologues of the gp91(phox) NADPH oxidase. Modulation of activity by calcium and by tobacco mosaic virus infection. *Plant Physiol.* 126, 1281–1290. doi: 10.1104/pp.126.3.1281

Sakamoto, K., Saito, A., Hayashida, N., Taguchi, G., and Matsumoto, E. (2008). Mapping of isolate-specific QTL for clubroot resistance in Chinese cabbage (*Brassica rapa* L. ssp. *pekinensis*). *Theor. Appl. Genet.* 117, 759–767. doi: 10.1007/s00122-008-0817-0

Schuller, A., Kehr, J., and Ludwig-Müller, J. (2014). Laser microdissection coupled to transcriptional profiling of *Arabidopsis* roots inoculated by *Plasmodiophora brassicae* indicates a role for brassinosteroids in clubroot formation. *Plant Cell Physiol.* 55, 392–411. doi: 10.1093/pcp/pct174

Schwelm, A., Fogelqvist, J., Andrea Knaust, A., Jülke, S., Lilja, T., Bonilla-Rosso, G., et al. (2015). The *Plasmodiophora brassicae* genome reveals insights in its life cycle and ancestry of chitin synthases. *Sci. Rep.* 5:11153. doi: 10.1038/srep11153

Shiu, S. H., and Bleecker, A. B. (2001). Receptor-like kinasea from *Arabidopsis* form a monophyletic gene family related to animal receptor kinases. *Proc. Natl. Acad. Sci. U.S.A.* 98, 10763–10768. doi: 10.1073/pnas.181141598

Siemens, J., Keller, I., Sarx, J., Kunz, S., Schuller, A., Nagel, W., et al. (2006). Transcriptome analysis of *Arabidopsis* clubroots indicate a key role for Cytokinins in disease development. *Mol. Plant Microbe Interact.* 19, 480–494. doi: 10.1094/MPMI-19-0480

Spoel, S. H., Koornneef, A., Claessens, S. M. C., Korzelius, J. P., Van Pelt, J. A., Mueller, M. J., et al. (2003). NPR1 modulates cross-talk between salicylate- and jasmonate-dependent defense pathways through a novel function in the cytosol. *Plant Cell* 15, 760–770. doi: 10.1105/tpc.009159

Suwabe, K., Tsukazaki, H., Iketani, H., Hatakeyama, K., Fujimura, M., Nunome, T., et al. (2003). Identification of two loci for resistance to clubroot (*Plasmodiophora brassicae Woronin*) in *Brassica rapa* L. *Theor. Appl. Genet.* 107, 997–1002. doi: 10.1007/s00122-003-1309-x

Suwabe, K., Tsukazaki, H., Iketani, H., Hatakeyama, K., Kondo, M., Fujimura, M., et al. (2006). Simple sequence repeat-based comparative genomics between *Brassica rapa* and *Arabidopsis thaliana*: the genetic origin of clubroot resistance. *Genetics* 173, 309–319. doi: 10.1534/genetics.104.038968

Tabei, Y., Kitade, S., Nishizawa, Y., Kikuchi, N., Kayano, T., Hibi, T., et al. (1998). Transgenic cucumber plants harboring a rice chitinase gene exhibit enhanced resistance to gray mold (*Botrytis cinerea*). *Plant Cell Rep.* 17, 159–164. doi: 10.1007/s002990050371

Takahashi, W., Fujimori, M., Miura, Y., Komatsu, T., Nishizawa, Y., Hibi, T., et al. (2005). Increased resistance to crown rust disease in transgenic Italian ryegrass (*Lolium multiflorum* Lam.) expressing the rice chitinase gene. *Plant Cell Rep.* 23, 811–818. doi: 10.1007/s00299-004-0900-1

Takatsu, Y., Nishizawa, Y., Hibi, T., and Akutsu, K. (1999). Transgenic chrysanthemum (*Dendranthema grandiflorum* (Ramat.) Kitamura) expressing a rice chitinase gene shows enhanced resistance to gray mold (*Botrytis cinerea*). *Sci. Hortic.* 82, 113–123. doi: 10.1016/S0304-4238(99)00034-5

Tatusov, R. L., Galperin, M. Y., Natale, D. A., and Koonin, E. V. (2000). The COG database: a tool for genome-scale analysis of protein functions and evolution. *Nucleic Acids Res.* 28, 33–36. doi: 10.1093/nar/28.1.33

Thomma, B. P., Eggermont, K., Penninckx, I. A., Mauch-Mani, B., Vogelsang, R., Cammue, B., et al. (1998). Separate jasmonate-dependent and salicylate-dependent defense response pathways in *Arabidopsis* are essential for resistance to distinct microbial pathogens. *Proc. Natl. Acad. Sci. U.S.A.* 95, 15107–15111. doi: 10.1073/pnas.95.25.15107

Trapnell, C., Pachter, L., and Salzberg, S. L. (2009). TopHat: discovering splice junctions with RNA-Seq. *Bioinformatics* 25, 1105–1111. doi: 10.1093/bioinformatics/btp120

Ueno, H., Matsumoto, E., Aruga, D., Kitagawa, S., Matsumura, H., and Hayashida, N. (2012). Molecular characterization of the *CRa* gene conferring clubroot resistance in *Brassica rapa*. *Plant Mol. Biol.* 80, 621–629. doi: 10.1007/s11103-012-9971-5

Van Loon, L. C., and Van Strien, E. A. (1999). The families of pathogenesis-related proteins, their activities, and comparative analysis of PR-1 type proteins. *Physiol. Mol. Plant Pathol.* 55, 85–97. doi: 10.1006/pmpp.1999.0213

Wang, W., Wen, Y., Berkey, R., and Xiao, S. (2009a). Specific targeting of the *Arabidopsis* resistance protein *RPW8.2* to the interfacial membrane encasing the fungal Haustorium renders broad-spectrum resistance to powdery mildew. *Plant Cell* 21, 2898–2913. doi: 10.1105/tpc.109.067587

Wang, Z., Gerstein, M., and Snyder, M. (2009b). RNA-Seq: a revolutionary tool for transcriptomics. *Nat. Rev. Genet.* 10, 57–63. doi: 10.1038/nrg2484

Zhang, T., Wu, D., Zhao, Z., Wang, Z., and Piao, Z. Y. (2012a). Development of near isogenic lines for clubroot resistance in Chinese Cabbage and their assessment. *Mol. Plant Breed.* 10, 722–730.

Zhang, T., Zhao, Z., Zhang, C. Y., Pang, W. X., Choi, S. R., Lim, Y. P., et al. (2014). Fine genetic and physical mapping of the *CRb* gene conferring resistance to clubroot disease in *Brassica rapa*. *Mol. Breed.* 34, 1173–1183. doi: 10.1007/s11032-014-0108-1

Zhang, Z. B., Wu, Y. L., Gao, M. H., Zhang, J., Kong, Q., Liu, Y. N., et al. (2012b). Disruption of PAMP-induced MAP kinase cascade by a *Pseudomonas syringae* effector activates plant immunity mediated by the NB-LRR protein SUMM2. *Cell Host Microbe* 15, 253–263. doi: 10.1016/j.chom.2012.01.015

Zhu, Z., An, F., Feng, Y., Li, P., Xue, L. A. M., Jiang, Z., et al. (2011). Derepression of ethylene-stabilized transcription factors (EIN3/EIL1) mediates jasmonate and ethylene signaling synergy in *Arabidopsis*. *Proc. Natl. Acad. Sci. U.S.A.* 108, 12539–12544. doi: 10.1073/pnas.1103959108

Zipfel, C. (2009). Early molecular events in PAMP triggered immunity. *Curr. Opin. Plant Biol.* 12, 414–420. doi: 10.1016/j.pbi.2009.06.003

Conflict of Interest Statement: The authors declare that the research was conducted in the absence of any commercial or financial relationships that could be construed as a potential conflict of interest.

20

Varied Growth Response of Cogongrass Ecotypes to Elevated CO_2

G. Brett Runion[1]*, Stephen A. Prior[1], Ludovic J. A. Capo-chichi[2], H. Allen Torbert[1] and Edzard van Santen[3]

[1] National Soil Dynamics Laboratory, Agricultural Research Service, United States Department of Agriculture, Auburn, AL, USA, [2] Alberta Innovates Technology Futures, Vegreville, AB, Canada, [3] Department of Crop, Soil and Environmental Sciences, Auburn University, Auburn, AL, USA

Cogongrass [*Imperata cylindrica* (L.) P. Beauv] is an invasive C_4 perennial grass which is listed as one of the top ten worst weeds in the world and is a major problem in the Southeast US. Five cogongrass ecotypes [Florida (FL), Hybrid (HY), Louisiana (LA), Mobile (MB), and North Alabama (NA)] collected across the Southeast and a red-tip (RT) ornamental variety were container grown for 6 months in open top chambers under ambient and elevated (ambient plus 200 ppm) atmospheric CO_2. Elevated CO_2 increased average dry weight (13%) which is typical for grasses. Elevated CO_2 increased height growth and both nitrogen and water use efficiencies, but lowered tissue nitrogen concentration; again, these are typical plant responses to elevated CO_2. The HY ecotype tended to exhibit the greatest growth (followed by LA, NA, and FL ecotypes) whiles the RT and MB ecotypes were smallest. Interactions of CO_2 with ecotype generally showed that the HY, LA, FL, and/or NA ecotypes showed a positive response to CO_2 while the MB and RT ecotypes did not. Cogongrass is a problematic invasive weed in the southeastern U.S. and some ecotypes may become more so as atmospheric CO_2 continues to rise.

Edited by:
Richard S. Winder,
Natural Resources Canada, Canada

Reviewed by:
Gerald Moser,
Justus Liebig University Giessen, Germany
Arkadiusz Kosmala,
Institute of Plant Genetics of the Polish Academy of Sciences, Poland

***Correspondence:**
G. Brett Runion
brett.runion@ars.usda.gov

Keywords: carbon dioxide, global change, *Imperata cylindrica*, invasive weed, nitrogen use efficiency, water use efficiency

INTRODUCTION

Invasive plants are estimated to cost U.S. agricultural and forest producers 34 billion dollars annually from decreased productivity and increased cost of weed control and are considered to be a major threat to the Earth's biodiversity (Pimentel, 2002). Elevated CO_2 stimulates plant photosynthesis, resource use efficiency, and biomass production (Amthor, 1995) which may affect the physiology and competitiveness of invasive plants. However, the effects of elevated CO_2 on invasive plants remains an understudied aspect of global change research. Bright (1998) summarizes, "Fast-growing, highly invasive plants may also be able to profit directly from the atmosphere's increased carbon content...any slower-growing natives would tend to lose out to the invaders." It has been suggested that the increase in the atmospheric concentration of CO_2 since

Abbreviations: FL, Florida cogongrass ecotype; HY, Louisiana/Mobile hybrid cogongrass ecotype; LA, Louisiana cogongrass ecotype; MB, Mobile cogongrass ecotype; NA, North Alabama cogongrass ecotype; NUE, nitrogen use efficiency; OTC, Open Top Chamber; RT, red-tip ornamental cogongrass variety; WUE, water use efficiency.

the beginning of the 20th century may be a primary factor affecting the establishment and spread of some invasive species (Ziska, 2003).

Invasive plants can disrupt terrestrial ecosystems, particularly in the southeastern U.S. with its numerous ports of entry and mild climate. One example that has become a serious problem is cogongrass [*Imperata cylindrica* (L.) P. Beauv], a perennial grass native to Southeast Asia which was introduced into the southeastern U.S. in the early 1900s (Tabor, 1949) for forage, erosion control and as packing material (Bryson and Carter, 1993). It is a widespread invader to warmer regions (> 500 million ha worldwide), is tolerant of shade, poor soils, and drought and naturalizes aggressively in dense monocultures which displace native plants (Bryson and Carter, 1993). Cogongrass is one of the top ten worst weeds in the world (Holm et al., 1991) and is a Federal Noxious Weed (Miller et al., 2010). Cogongrass is a major problem in the Southeast on disturbed lands such as forest plantations and roadsides and may become problematic on agricultural lands (Patterson et al., 1980). It is present in five or more varieties including a commercially available red-tip (RT; 'Red Baron') ornamental sold by nurseries in some states (Capo-chichi et al., 2008). Sale of this RT variety is prohibited in some southern states and removal of prior plantings has been recommended since it has viable pollen that might spread to invasive cogongrass plants and has been known to revert back to the green aggressive type (Miller et al., 2010).

It has been suggested that populations introduced from several origins over an extended period of time should have higher genetic diversity than populations that were introduced only a few times or from a single source (Pappert et al., 2000). For example, cogongrass was first introduced to Alabama from Japan (Tabor, 1952), but has likely also arrived from other locations to different ports of entry. Genetic characterization of differing populations of cogongrass may help explain spread dynamics and means of establishment (Capo-chichi et al., 2008). These investigators determined that genetic variation within and between cogongrass sites in the southern U.S. was quite large given how recently it was introduced. Spread dynamics were found to be greatly influenced by anthropogenic activities (e.g., soil disturbance, canopy removal) compared to natural factors which may accelerate opportunities for bringing together cross-compatible species previously isolated by ecology and/or geography. The objective of this study was to evaluate the response of five cogongrass ecotypes collected across the southeastern U.S. plus the RT variety to ambient and elevated atmospheric CO_2. This is the first study to examine the effects of elevated CO_2 on cogongrass and the first for any weed species to look at potential differences among ecotypes.

MATERIALS AND METHODS

The study was conducted at the soil bin facilities at the USDA-ARS National Soil Dynamics Laboratory, Auburn, Alabama. The bin used for the experimental setup is 6 m wide and 76 m long and has been modified for container studies; modifications consisted of installing a geomembrane liner (20 mL) and gravel drain system to ensure a good working surface and drainage for container studies. Open top field chambers (OTC; Rogers et al., 1983a), encompassing 7.3 m × 7.3 m of ground surface area, were used to continuously deliver target CO_2 concentrations of ambient or ambient plus 200 μmol mol^{-1} (elevated) using a delivery and monitoring system described by Mitchell et al. (1995).

The six cogongrass ecotypes used in this study were from a collection maintained at Auburn University and included Louisiana, North Alabama, Florida, Mobile, a LA-MB HY, and RT. The MB ecotype was from the suspected point of introduction near Grand Bay in Mobile County, AL as described by Tabor (1952). The RT represents a commercially available variety that can be found in ornamental nurseries. Rhizomes were collected, stored in plastic bags, and transported to glasshouses at the Plant Sciences Research Center of the Alabama Agricultural Experiment Station on the campus of Auburn University for molecular analysis. An out-crossing involving the non-native and native species led to different genotypes such as the LA-MB HY. Amplified Fragment Length Polymorphism (AFLP) confirmed that the cogongrass ecotype LA-MB HY was derived from non-native (wild type cogongrass ecotype) and native (non-invasive cogongrass ecotype) species (Capo-chichi et al., 2008). The atpB-rbcL non-coding spacer of chloroplast DNA revealed that only a few nucleotide substitutions contributed to the variation among the wild cogongrass MB ecotype and the RT (data not shown).

Plants were grown in a peat-based general purpose growing medium (PRO-MIX Bx, Premier Horticulture Inc., Quakertown, PA 18951, USA) in 1.65 L tree-pots (Short One Tree-pot, 10 cm × 23 cm, Stuewe and Sons Inc., Corvallis, OR 97333, USA) in a glasshouse for establishment (~3 wk). Plants were then transplanted into 10.65 L tree pots (TPOT4 Round Tree-pot, 22 cm × 39 cm, Stuewe and Sons Inc., Corvallis, OR 97333, USA) containing the same standard growth medium described above. Forty-eight containers of each ecotype were selected for use in the study. These plants were ranked, according to size and placed into four groups of 12 containers each, representing the largest 12 first in declining order down to the smallest 12; one container from each group was randomly assigned to each of the 12 OTCs (i.e., four containers of each plant ecotypes in each chamber). The study was conducted as a randomized complete block design with the six blocks occurring along the length of the soil bin. Plants were fertilized monthly with Miracle-Gro (15:30:15, N:P:K; Scotts Products Inc., Marysville, OH, USA) according to manufacture recommendations by mixing 600 g Miracle-Gro in 130 L deionized water; each plant received 500 mL of this solution. Containers were subjected to ambient rainfall and watered once or twice a week (1 L per container) to prevent drought-induced plant mortality.

Prior to harvest, WUE was calculated from LI-6400 Portable Photosynthesis System (LI-COR, Inc., Lincoln, NE, USA) measurements. After 6 months, height was measured and the aboveground portions were harvested by severing the plant at the ground-line. Roots were separated from the growing medium using the sieve method (Bohm, 1979). Above- and belowground plant components were then dried separately in a forced-air oven at 55°C to a constant weight, and dry weights recorded. Subsamples of above- and belowground biomass (1 mm sieve) were analyzed separately for N by dry combustion using a LECO TruSpec analyzer (LECO Corp., St. Joseph, MI, USA). Prior to analyses, data from the four containers of each ecotype within each chamber were averaged making the chamber the experimental unit ($N = 6$). Statistical analyses were conducted using the Mixed Models procedure (Proc Mixed) from SAS (Littell et al., 1996). In all cases, differences were considered significant at $P \leq 0.05$ and trends were recognized at $0.05 > P \leq 0.10$.

RESULTS

Cogongrass height was significantly increased ($P < 0.001$) under elevated atmospheric CO_2 (111.3 cm) compared to ambient conditions (105.4 cm) when averaged across all ecotypes (**Figure 1**). When averaged across CO_2 concentrations, significant differences in height ($P < 0.001$) were noted among the ecotypes (i.e., LA = NA > FL > HY > MB > RT). Further, a significant interaction ($P = 0.001$) showed that height was increased by elevated CO_2 for LA, NA, FL, and HY only (**Figure 1**).

Similarly, aboveground dry weight was significantly increased ($P = 0.001$) 13% under elevated (98.5 g) compared with ambient (87.4 g) CO_2 (**Figure 2**). A significant main effect of ecotype ($P < 0.001$) was also observed (i.e., HY > LA > NA = FL > MB > RT) as was a trend for an interaction ($P = 0.095$), where dry weight was increased by elevated CO_2 for HY, LA, and FL only (**Figure 2**).

Although elevated CO_2 resulted in a slight increase (8.9%) in belowground dry weight (elevated = 155.6 g vs. ambient = 142.9 g; **Figure 3**), this effect was not statistically significant ($P = 0.118$). However, the main effect of ecotype was significant ($P < 0.001$; HY > NA > FL = LA > MB > RT. No significant CO_2 by ecotype interaction ($P = 0.487$) was noted for this measure.

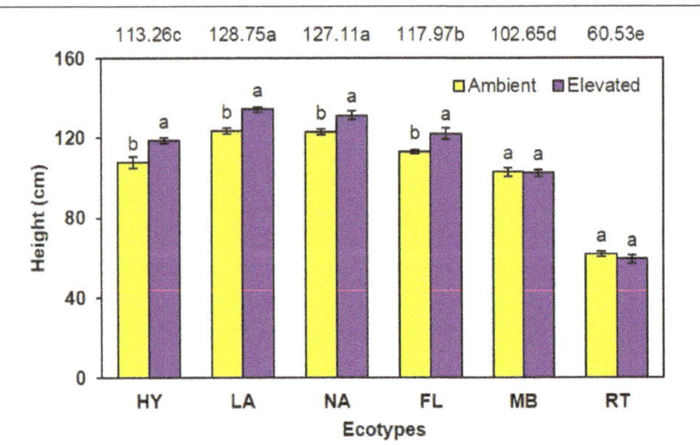

FIGURE 1 | Height of cogongrass ecotypes under ambient and elevated atmospheric CO_2 (HY, Hybrid; LA, Louisiana; NA, North Alabama; FL, Florida; MB, Mobile; RT, Red-tip). $N = 6$. Means with standard errors are shown; standard errors reflect the variability in the data and are not a means separation technique. Main effects of CO_2 ($P < 0.001$), ecotype ($P < 0.001$), and their interaction ($P = 0.001$) were significant. Bars with different letters show a significant effect of CO_2 for each ecotype; main effect ecotype means shown at the top of the graph (means followed by the same letter are not significantly different).

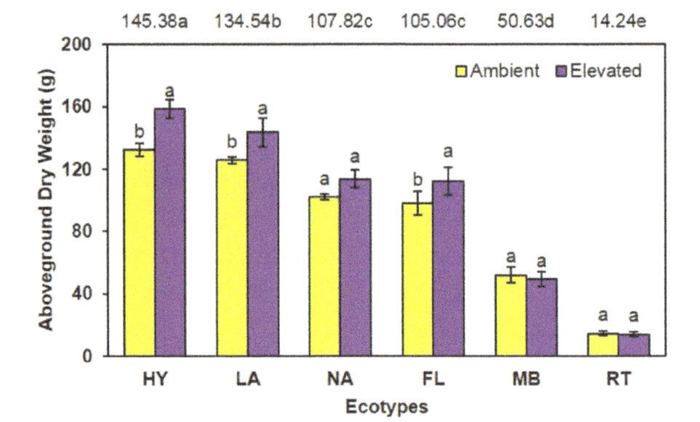

FIGURE 2 | Aboveground biomass of cogongrass ecotypes under ambient and elevated atmospheric CO_2 (HY, Hybrid; LA, Louisiana; NA, North Alabama; FL, Florida; MB, Mobile; RT, Red-tip). $N = 6$. Means with standard errors are shown; standard errors reflect the variability in the data and are not a means separation technique. Main effects of CO_2 ($P = 0.001$) and ecotype ($P < 0.001$) were significant and their interaction showed a trend ($P = 0.095$). Bars with different letters show a significant effect of CO_2 for each ecotype; main effect ecotype means shown at the top of the graph (means followed by the same letter are not significantly different).

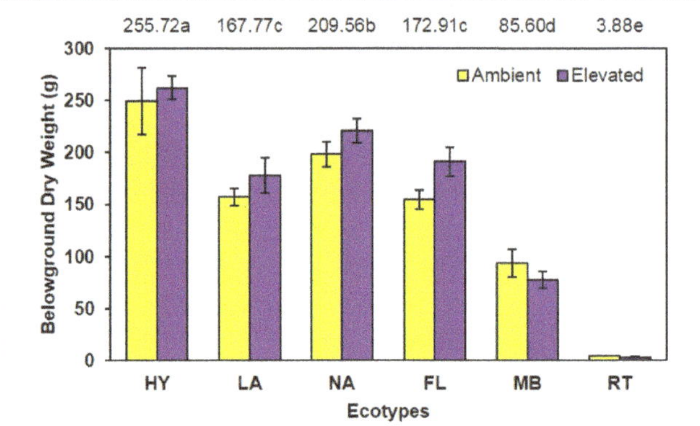

FIGURE 3 | Belowground biomass of cogongrass ecotypes under ambient and elevated atmospheric CO_2 (HY, Hybrid; LA, Louisiana; NA, North Alabama; FL, Florida; MB, Mobile; RT, Red-tip). $N = 6$. Means with standard errors are shown; standard errors reflect the variability in the data and are not a means separation technique. Main effect of ecotype ($P < 0.001$) was significant while the main effect of CO_2 showed a trend ($P = 0.118$) and their interaction ($P = 0.487$) was not significant. The omission of letters above bars indicates no effect of CO_2 for any ecotype; main effect ecotype means shown at the top of the graph (means followed by the same letter are not significantly different).

Unlike growth measurements, the main effect of CO_2 showed significantly lower ($P < 0.001$) tissue nitrogen concentration [N] under elevated (6.06 mg N/g) than ambient (6.74 mg N/g) CO_2 (**Figure 4**). A significant main effect of ecotype ($P < 0.001$) indicated that RT > FL = HY = NA > MB = LA. A significant interaction of CO_2 with ecotype ($P = 0.008$) showed that [N] was decreased by elevated CO_2 for the FL, HY, and NA ecotypes

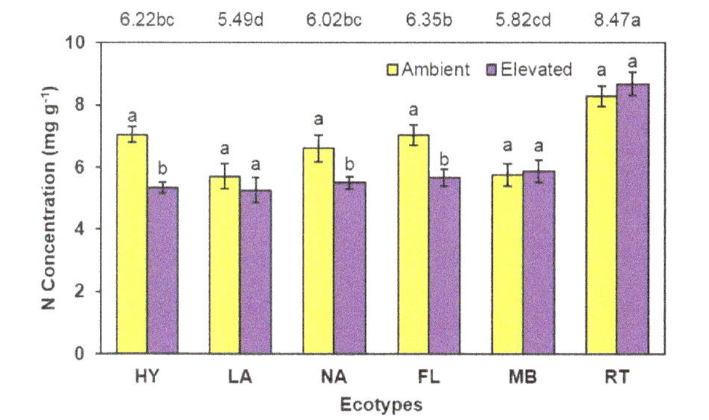

FIGURE 4 | Total plant nitrogen concentration of cogongrass ecotypes under ambient and elevated CO_2 (HY, Hybrid; LA, Louisiana; NA, North Alabama; FL, Florida; MB, Mobile; RT, Red-tip). $N = 6$. Means with standard errors are shown; standard errors reflect the variability in the data and are not a means separation technique. Main effects of CO_2 ($P < 0.001$), ecotype ($P < 0.001$), and their interaction ($P = 0.008$) were significant. Bars with different letters show a significant effect of CO_2 for each ecotype; main effect of ecotype means shown at the top of the graph (means followed by the same letter are not significantly different).

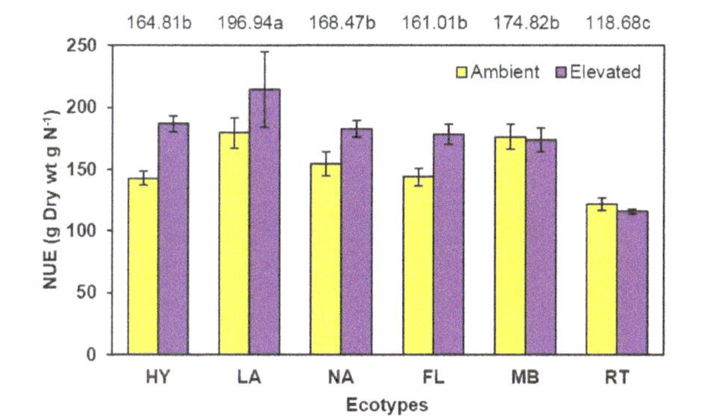

FIGURE 5 | Nitrogen use efficiency of cogongrass ecotypes under ambient and elevated CO_2 (HY, Hybrid; LA, Louisiana; NA, North Alabama; FL, Florida; MB, Mobile; RT, Red-tip). $N = 6$. Means with standard errors are shown; standard errors reflect the variability in the data and are not a means separation technique. Main effects of CO_2 ($P = 0.001$) and ecotype ($P < 0.001$) significant; their interaction ($P = 0.118$) showed a trend. The omission of letters above bars indicates no effect of CO_2 for any ecotype; main effect ecotype means shown at the top of the graph (means followed by the same letter are not significantly different).

only (**Figure 4**). Calculations of NUE as g plant biomass produced per g plant N were significant ($P = 0.001$) for the main effect of CO_2 (ambient = 153.1 vs. elevated = 175.2; **Figure 5**). Ecotype, averaged across both CO_2 treatments, significantly ($P < 0.001$) affected NUE (i.e., LA > MB = NA = HY = FL > RT). The CO_2 by ecotype interaction was not significant for NUE ($P = 0.118$); however, NUE showed a similar response as other variables in that FL, HY, LA, and NA were numerically higher under elevated CO_2 while MB and RT were actually slightly lower (**Figure 5**).

Water use efficiency, calculated from LICOR gas exchange measurements as mmol CO_2 per mol H_2O, was significantly

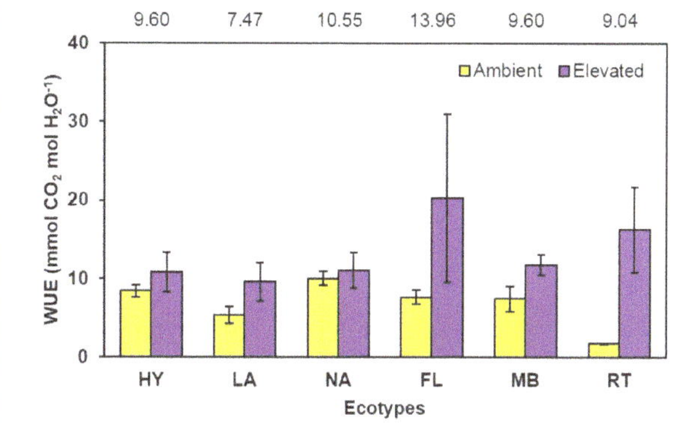

FIGURE 6 | Water use efficiency of cogongrass ecotypes under ambient and elevated CO_2 (HY, Hybrid; LA, Louisiana; NA, North Alabama; FL, Florida; MB, Mobile; RT, Red-tip). $N = 6$. Means with standard errors are shown; standard errors reflect the variability in the data and are not a means separation technique. Main effect of CO_2 ($P = 0.001$) was significant while ecotype ($P = 0.513$) and their interaction ($P = 0.226$) were not. The omission of letters above bars indicates no effect of CO_2 for any ecotype; main effect ecotype means shown at the top of the graph.

increased ($P = 0.001$) 96% by growth in elevated CO_2 (ambient = 6.8 vs. elevated = 13.3; **Figure 6**). The main effect of ecotype ($P = 0.513$) and its interaction with CO_2 ($P = 0.226$) did not affect WUE (**Figure 6**).

DISCUSSION

Cogongrass growth parameters were increased when exposed to elevated CO_2, which is typical of most plants (Rogers et al., 1994; Amthor, 1995). Although height was only slightly higher (5.6%; **Figure 1**), aboveground dry weight increase (12.7%; **Figure 2**) was in a range (10–15%) typical of C_4 plant response to CO_2 enrichment (Kimball, 1983; Prior et al., 2003). The fact that responsive ecotypes showed growth responses to elevated CO_2 typical for C_4 plants suggests that their invasive potential will not be altered as atmospheric CO_2 continues to rise, but does indicate that some ecotypes of this serious invasive weed may become more problematic.

Cogongrass ecotype also affected both height and aboveground dry weight; in general, the MB and RT were smaller than the other ecotypes. The significant interactions for these variables further showed that MB and RT were not responsive to CO_2 concentration, while the other ecotypes tended to be larger under high CO_2. It is interesting to note that HY tended to exhibit the greatest response to elevated CO_2 among ecotypes even though it was a cross from the MB ecotype which was not responsive. It is not uncommon for HYs to exhibit growth responses that either differ from or exceed their progenitors (Capo-chichi et al., 2008). This is, after all, why plant breeding programs exist for virtually all important crop species.

Despite the importance of root systems in attaining essential soil resources (i.e., water and nutrients), their response to CO_2 has received less attention than aboveground parts; however, it has been reported that roots often exhibit a larger response to elevated CO_2 than other plant organs (Rogers et al., 1994). In contrast, our study showed no increase in root dry weight under elevated CO_2 and no significant interaction with cogongrass ecotype. It is interesting to note that, despite the lack of significance, the belowground response pattern (**Figure 3**) was similar to that seen for aboveground growth (**Figure 2**). Further, ecotype effect also followed the same general pattern as aboveground dry weight in that MB and RT were smaller than the other ecotypes.

Given that weed species are more likely to have greater genetic diversity and physiological plasticity (compared to crops), they are more likely to be able to adapt to a changing environment (Ziska and Runion, 2007). This may have significant implications for developing effective weed control stategies given that elevated CO_2 may increase herbicide tolerance in some weeds due to a herbicide dilution effect caused by increased growth, as well as other potential CO_2-induced changes in plant morphology, biochemistry, and physiology (Ziska et al., 1999; Ziska and Teasdale, 2000; Archambault et al., 2001). However, increased herbicide tolerance under elevated CO_2 is not always observed (Marble et al., 2015). How cogongrass herbicide efficacy will be impacted by elevated CO_2 is not known and deserves futher study.

Total plant nitrogen concentration was reduced under elevated CO_2 (**Figure 4**). As with plant growth, this is a common response to high CO_2 (Rogers et al., 1994; Norby et al., 2001; Prior et al., 2008; Runion et al., 2009). This is a result of increased plant growth under elevated CO_2 causing a dilution effect on nutrient concentrations (Rogers et al., 1994, 1999). The RT ecotype had the highest [N] and LA was lowest (**Figure 4**). RT had the smallest growth which likely resulted in the high [N]; it is unclear why MB did not exhibit this pattern given it also had less growth. A significant CO_2 by ecotype interaction indicated that [N] was lowered by elevated CO_2 in FL, HY, and NA only. In general, these ecotypes had a larger dilution effect due to greater growth (**Figures 1** and **2**); however, why LA did not follow this pattern is not known.

Another common response to elevated CO_2 is increased NUE (Rogers et al., 1994) as observed in this study (**Figure 5**). Nutrient use efficiency (unit of biomass produced per unit of nutrient) generally increases under elevated CO_2 as plants are able to produce more biomass with available nutrients. Ecotype also affected NUE with LA being highest and RT lowest. Although the CO_2 by ecotype interaction was not significant, the response pattern was similar to other variables in that NUE was numerically higher for FL, HY, LA, and NA but not MB and RT under elevated CO_2.

As with NUE, it is also common for plants grown under elevated CO_2 to exhibit increases in water use effiency (Rogers and Dahlman, 1993). In general, C_3 plants exhibit increased photosynthesis and decreased stomatal conductance under elevated CO_2 (Amthor and Loomis, 1996), leading to increased WUE. However, the CO_2–concentrating mechanism used by C_4 species limits their photosynthetic response to CO_2 enrichment (Amthor and Loomis, 1996), but they do tend to show decreased stomatal conductance which often increases WUE (Rogers et al., 1983b). This was observed in the current study, in that photosynthesis was not significantly affected by CO_2 concentration (ambient = 2.13, elevated = 2.61 μmol CO_2 m^{-2} s^{-1}; $P = 0.14$), while stomatal conductance tended to be decreased (ambient = 0.018, elevated = 0.014 mol H_2O m^{-2} s^{-1}; $P = 0.06$) under elevated CO_2 (full data not shown). These responses led to a large increase in WUE (96%) under elevated CO_2 (**Figure 6**). Ecotype and its interaction with CO_2 did not affect WUE.

CONCLUSION

Cogongrass is one of the top ten worst weeds in the world and is listed as a Federal Noxious Weed. Since its introduction to the Southeastern U.S. it has become a major problem in forest plantations, roadsides, and agricultural systems due to its aggressive ability to develop dense monocultures which can compete with and displace desirable species. This is the first study to examine the effects of elevated CO_2 on cogongrass and the first for any weed species to look at potential differences among ecotypes. In our study, elevated CO_2 (averaged across ecotypes) increased height, biomass, and both nitrogen and water use efficiencies, but lowered tissue nitrogen concentration; again, these are typical C_4 plant responses to elevated CO_2. In general, the HY ecotype tended to exhibit the greatest growth (followed by LA, NA, and FL) while RT and MB ecotypes were smallest. Interactions of CO_2 with ecotype showed that MB and RT ecotypes did not respond to CO_2. This lack of response to CO_2 for RT (sold commerically as 'Red Baron') is significant since concerns over its potential to spread into the native landscape should not be exacerbated by the rising atmospheric CO_2 concentration. Nevertheless, it is still prohibited for sale in some states and its removal has been recommended. However, HY, LA, FL, and/or NA ecotypes responded positively to elevated CO_2, suggesting some ecotypes of this serious invasive weed may become more problematic in a future CO_2-enriched environment. These findings may influence development of future cogongrass control strategies, a subject area requiring further investigation.

AUTHOR CONTRIBUTIONS

GB conceived and conducted the research, collected and analyzed the data and co-wrote the manuscript; SP assisted with conducting the research, collecting the data, and co-wrote the manuscript; LC provided the cogongrass ecotypes, co-wrote and reviewed the manuscript; HT assisted with conducting the research and reviewed the manuscript; ES assisted with providing the cogongrass ecotypes and reviewed the manuscript.

ACKNOWLEDGMENT

The authors wish to thank Barry Dorman and Jerry Carrington for technical assistance.

REFERENCES

Amthor, J. S. (1995). Terrestrial higher-plant response to increasing atmospheric [CO_2] in relation to the global carbon cycle. *Global Change Biol.* 1, 243–274. doi: 10.1111/j.1365-2486.1995.tb00025.x

Amthor, J. S., and Loomis, R. S. (1996). "Integrating knowledge of crop responses to elevated CO_2 and temperature with mechanistic simulation models: model components and research needs," in *Carbon Dioxide and Terrestrial Ecosystems*, eds G. W. Koch and H. A. Mooney (San Diego, CA: Academic Press), 317–346.

Archambault, D. J., Li, X., Robinson, D., O'Donovan, J. T., and Klein, K. K. (2001). The effects of elevated CO_2 and temperature on herbicide efficacy and weed/crop competition. *Rept. Prairie Adapt. Res. Collab.* 1–29.

Bohm, W. (1979). *Methods of Studying Root Systems*. New York, NY: Spring-Verlag.

Bright, C. (1998). *Life Out of Bounds: Bioinvasion in a Borderless World*. New York, NY: W.W. Norton & Company.

Bryson, C. T., and Carter, R. (1993). Cogongrass, *Imperata cylindrica*, in the United States. *Weed Technol.* 7, 1005–1009.

Capo-chichi, L. J. A., Faircloth, W. H., Williamson, A. G., Patterson, M. G., Miller, J. H., and van Santen, E. (2008). Invasion dynamics and genotypic diversity of Cogongrass (*Imperata cylindrica*) at the point of introduction in the Southeastern United States. *Invasive Plant Sci. Manage.* 1, 133–141. doi: 10.1614/IPSM-07-007.1

Holm, L. G., Plucknett, D. L., Pancho, J. V., and Herberger, J. P. (1991). *The World(s Worst Weeds: Distribution and Biology*. Malabar, FL: Krieger Publishing.

Kimball, B. A. (1983). Carbon dioxide and agricultural yield: an assemblage and analysis of 430 prior observations. *Agron. J.* 75, 779–788. doi: 10.2134/agronj1983.00021962007500050014x

Littell, R. C., Milliken, G. A., Stroup, W. W., and Wolfinger, R. D. (1996). *SAS System for Mixed Models*. Cary, NC: SAS Institute, Inc.

Marble, S. C., Prior, S. A., Runion, G. B., and Torbert, H. A. (2015). Control of yellow and purple nutsedge in elevated CO_2 environments with glyphosate and halosulfuron. *Front. Plant Sci.* 6:1. doi: 10.3389/fpls.2015.00001

Miller, J. H., Manning, S. T., and Enloe, S. F. (2010). "A Management Guide for Invasive Plants in Southern Forests," in *General Technology Report SRS*-131, (Asheville, NC: U.S. Department of Agriculture Forest Service).

Mitchell, R. J., Runion, G. B., Prior, S. A., Rogers, H. H., Amthor, J. S., and Henning, S. P. (1995). Effects of nitrogen on *Pinus palustris* foliar respiratory responses to elevated atmospheric CO_2 concentration. *J. Exp. Bot.* 46, 1561–1567.

Norby, R. J., Cotrufo, M. F., Ineson, P., O'Neill, E. G., and Canadell, J. G. (2001). Elevated CO_2, litter quality, and decomposition: a synthesis. *Oecologia* 127, 153–165. doi: 10.1007/s004420000615

Pappert, R. A., Hamrick, J. L., and Donovan, L. A. (2000). Genetic variation in *Pueraria lobata* (Fabaceae), an introduced, clonal, invasive plant of the southeastern United States. *Am. J. Bot.* 87, 1240–1245. doi: 10.2307/2656716

Patterson, D. T., Flint, E. P., and Dickens, R. (1980). Effects of temperature, photoperiod, and population source on the growth of cogongrass (*Imperata cylindrica*). *Weed Sci.* 28, 505–509.

Pimentel, D. (2002). *Biological Invasions: Economic and Environmental Costs of Alien Plant, Animal, and Microbe Species*. Boca Raton, FL: CRC Press.

Prior, S. A., Runion, G. B., Rogers, H. H., and Torbert, H. A. (2008). Effects of atmospheric CO_2 enrichment on crop nutrient dynamics under no-till conditions. *J. Plant Nutr.* 31, 758–773. doi: 10.1080/01904160801928364

Prior, S. A., Torbert, H. A., Runion, G. B., and Rogers, H. H. (2003). Implications of elevated CO_2-induced changes in agroecosystem productivity. *J. Crop Prod.* 8, 217–244. doi: 10.1300/J144v08n01_09

Rogers, H. H., and Dahlman, R. C. (1993). Crop responses to CO_2 enrichment. *Vegetatio* 104/105, 117–131. doi: 10.1007/BF00048148

Rogers, H. H., Heck, W. W., and Heagle, A. S. (1983a). A field technique for the study of plant responses to elevated carbon dioxide concentrations. *Air Pollut. Control Assn. J.* 33, 42–44. doi: 10.1080/00022470.1983.10465546

Rogers, H. H., Thomas, J. F., and Bingham, G. E. (1983b). Response of agronomic and forest species to elevated atmospheric carbon dioxide. *Science* 220, 428–429. doi: 10.1126/science.220.4595.428

Rogers, H. H., Runion, G. B., and Krupa, S. V. (1994). Plant responses to atmospheric CO_2 enrichment with emphasis on roots and the rhizosphere. *Environ. Pollut.* 83, 155–189. doi: 10.1016/0269-7491(94)90034-5

Rogers, H. H., Runion, G. B., Prior, S. A., and Torbert, H. A. (1999). "Response of plants to elevated atmospheric CO_2: root growth, mineral nutrition, and soil carbon," in *Carbon Dioxide and Environmental Stress*, eds Y. Luo and H. A. Mooney (San Diego, CA: Academic Press), 215–244.

Runion, G. B., Torbert, H. A., Prior, S. A., and Rogers, H. H. (2009). "Effects of elevated atmospheric carbon dioxide on soil carbon in terrestrial ecosystems of the southeastern U.S," in *Soil Carbon Sequestration and the Greenhouse Effect*, 2nd Edn, eds R. Lal and R. F. Follett (Madison, WI: Soil Science Society of America), 233–262.

Tabor, P. (1949). Cogongrass, *Imperata cylindrica* (L.) Beauv., in the southeastern United States. *Agron. J.* 41, 270. doi: 10.2134/agronj1949.00021962004100060011x

Tabor, P. (1952). Cogongrass in Mobile County. *Alabama. Agron. J.* 44:50. doi: 10.2134/agronj1952.00021962004400010012x

Ziska, L. H. (2003). Evaluation of the growth response of six invasive species to past, present and future atmospheric carbon dioxide. *J. Exp. Bot.* 54, 395–404. doi: 10.1093/jxb/erg027

Ziska, L. H., and Runion, G. B. (2007). "Future weed, pest, and disease problems for plants," in *Agroecosystems in a Changing Climate*, eds P. C. D. Newton, R. A. Carran, G. R. Edwards, and P. A. Niklaus (Boca Raton, FL: CRC Press), 261–287.

Ziska, L. H., and Teasdale, J. R. (2000). Sustained growth and increased tolerance to glyphosate observed in a C_3 perennial weed, quackgrass (*Elytrigia repens*), grown at elevated carbon dioxide. *Aust. J. Plant Physiol.* 27, 159–166.

Ziska, L. H., Teasdale, J. R., and Bunce, J. A. (1999). Future atmospheric carbon dioxide may increase tolerance to glyphosate. *Weed Sci.* 47, 608–615.

Conflict of Interest Statement: The authors declare that the research was conducted in the absence of any commercial or financial relationships that could be construed as a potential conflict of interest.

Emission and Accumulation of Monoterpene and the Key Terpene Synthase (TPS) Associated with Monoterpene Biosynthesis in *Osmanthus fragrans* Lour

Xiangling Zeng[1], Cai Liu[1], Riru Zheng[1], Xuan Cai[1], Jing Luo[1], Jingjing Zou[1,2] and Caiyun Wang[1]*

[1] Key Laboratory for Biology of Horticultural Plants, Ministry of Education, College of Horticulture and Forestry Sciences, Huazhong Agricultural University, Wuhan, China, [2] School of Nuclear Technology and Chemistry and Biology, Hubei University of Science and Technology, Xianning, China

Edited by:
Susana M. P. Carvalho,
University of Porto, Portugal

Reviewed by:
Daniel A. Jacobo-Velázquez,
Tecnológico de Monterrey, Mexico
Hao Peng,
Washington State University, USA

***Correspondence:**
Caiyun Wang
wangcy@mail.hzau.edu.cn

Osmanthus fragrans is an ornamental and economically important plant known for its magnificent aroma, and the most important aroma-active compounds in flowers are monoterpenes, mainly β-ocimene, linalool and linalool derivatives. To understand the molecular mechanism of monoterpene production, we analyzed the emission and accumulation patterns of these compounds and the transcript levels of the genes involved in their biosynthesis in two *O. fragrans* cultivars during flowering stages. The results showed that both emission and accumulation of monoterpenes varied with flower development and glycosylation had an important impact on floral linalool emission during this process. Gene expression demonstrated that the transcript levels of terpene synthase (*TPS*) genes probably played a key role in monoterpene production, compared to the genes in the MEP pathway. Phylogenetic analysis showed that *Of*TPS1 and *Of*TPS2 belonged to a TPS-g subfamily, and *Of*TPS3 and *Of*TPS4 clustered into a TPS-b subfamily. Their transient and stable expression in tobacco leaves suggested that *Of*TPS1 and *Of*TPS2 exclusively produced β-linalool, and *trans*-β-ocimene was the sole product from *Of*TPS3, while *Of*TPS4, a predictive sesquiterpene synthase, produced α-farnesene. These results indicate that *OfTPS1*, *OfTPS2,* and *OfTPS3* could account for the major floral monoterpenes, linalool and *trans*-β-ocimene, produced in *O. fragrans* flowers.

Keywords: *Osmanthus fragrans*, MEP pathway, terpene synthase, monoterpenes, glycosylation

INTRODUCTION

Sweet *osmanthus (Osmanthus fragrans* Lour.), belonging to the Oleaceae family, is a well-known ornamental and economically important, aromatic woody plant, with the flower having a long history in China. Horticultural cultivars have spread throughout Thailand, India, and the Caucasus region (Baldermann et al., 2010). Because of its extremely powerful and unique aroma, flowers and the essential oils of *O. fragrans* are in high demand for the production of expensive perfumes, flavorings and cosmetics (Wang et al., 2009; Cai et al., 2014).

The fresh flowers are very rich in floral volatiles, including terpenoids, aromatic compounds, C6 compounds and esters. The qualitative and quantitative variability of these compounds in flowers usually depends on the cultivar and developmental stage (Li et al., 2008; Cao et al., 2009; Sun et al., 2012; Xin et al., 2013). The terpenoids, including β-ionone, β-ocimene, β-linalool, and linalool derivatives, have been detected as dominant components of fresh flower volatiles and essential oils (Wang et al., 2009; Sun et al., 2012; Xin et al., 2013), and are important in the aroma formation of *O. fragrans* (Cai et al., 2014). Of the terpenes, β-ionone is ubiquitous in *O. fragrans* flowers and its biosynthesis has been reported at the molecular level (Baldermann et al., 2010; Han et al., 2014). However, the molecular mechanism for the formation of monoterpenes such as β-ocimene and linalool in *O. fragrans* is not clear.

In plants, monoterpenes are mainly synthesized through the plastidial methylerythritol 4-phosphate (MEP) pathway, providing terpene precursors isopentenyl diphosphate (IPP) and its allylic isomer, dimethylallyl diphosphate (DMAPP; Dudareva et al., 2013). Quantitative variation in monoterpene production can be controlled by substrate flux through the MEP pathway (Munoz-Bertomeu et al., 2006; Battilana et al., 2011; Külheim et al., 2011). The first step in the MEP pathway is the condensation of pyruvate and D-glyceraldehyde 3-phosphate (G3P) to 1-deoxy-D-xylulose 5-phosphate (DXP; Dudareva et al., 2013). The first enzyme, DXP synthase (DXS), has been considered the rate-limiting enzyme for the MEP pathway flux, because of the close correlation between the gene expression and the content of plastid isoprenoids such as monoterpenes and carotenoids (Estévez et al., 2001; Xie et al., 2008; Battilana et al., 2009). The second and sixth enzyme in the pathway, DXP reductoisomerase (DXR) and 4-hydroxy-3-methylbut-2-en-1-yl diphosphate synthase (HDS), are also potential regulatory control points (Mahmoud and Croteau, 2001; Carretero-Paulet et al., 2006; Külheim et al., 2011). However, the rate-limiting role of each of these enzymes in controlling the pathway flux appears to vary among plants (Cordoba et al., 2009).

Under the catalytic action of geranyl pyrophosphate synthase (GPPS), IPP, and DMAPP are condensed head-to-tail to produce geranyl diphosphate (GPP), the monoterpene substrate (Dudareva et al., 2013). Catalysis of this linear precursor, GPP, to a broad variety of monoterpenes is by the terpene synthase (TPS) family (Degenhardt et al., 2009). TPS enzymes from different plant species have distinct phylogenetic relationships and have been classified into seven subfamilies, designated *TPS*-a to *TPS*-g (Chen et al., 2011). Despite intriguing differences between the subfamilies, there are three conserved motifs: an arginine-rich N-terminal RR(x8)W motif required for cyclization of the GPP substrate; an aspartate-rich DDxxD motif that interacts with divalent metal (usually Mg^{2+} or Mn^{2+}) ions involved in positioning the substrate for catalysis, and (N, D)Dxx(S,T)xxxE, required for second metal ion binding (Degenhardt et al., 2009). The RR(x8)W motif is involved in producing cyclic monoterpenes, and is absent in TPSs that produce acyclic products (Chen et al., 2011). Many TPSs have the ability to produce multiple terpenes from a single prenyl diphosphate substrate *in vitro* (Martin et al., 2010; Green et al., 2012; Nieuwenhuizen et al., 2013) or *in vivo* (Davidovich-Rikanati et al., 2008; Green et al., 2012). For example, the occasional complex terpene blend has been found in *Arabidopsis thaliana* and *Medicago truncatula*, often produced only by a limited number of multiproduct TPS enzymes (Tholl et al., 2005; Garms et al., 2010). To date, TPSs have been identified and characterized in many plants, including *Antirrhinum majus* (Dudareva et al., 2003; Nagegowda et al., 2008), *Actinidia* species (Nieuwenhuizen et al., 2009; Green et al., 2012) and *Vitis vinifera* (Martin et al., 2010). Despite monoterpenes making a significant contribution to the floral aroma and being rich in *O. fragrans* flowers, little is known about the *TPS* genes responsible for production of the major monoterpenes.

The biosynthetic monoterpenes are able to undergo complex processes of storage and conversion, which lead to the inconsistency between monoterpene release and gene transcript level (Chen et al., 2010; Green et al., 2012). Glycosides are a potential source of aroma and flavor compounds, and play important roles in controlling the release of the floral volatiles in flowers and fruit (Schwab et al., 2015). The glycoside volatiles are odorless and could release free aroma volatiles under the hydrolysis of β-glucosidase (Yauk et al., 2014; Schwab et al., 2015). They are highly valued in the flavor industry for enhancing the flavor and quality of grape wine and tea, and modifying the overall aroma during maturation, storage and processing in fruit (Birtić et al., 2009; Garcia et al., 2013; Yauk et al., 2014; Ohgami et al., 2015). In flowers, the organoleptic aroma contributed by free volatiles and the content of essential oil composed of non-glycoside volatiles accumulating in fresh flowers have received more attention compared to the few studies on the volatile glycoside (Picone et al., 2004; Green et al., 2012). It has been reported that glycosidically bound volatiles are more abundant than the free forms and are potentially a major source of aroma in flowers (Aurore et al., 2011; Wen et al., 2014; Yauk et al., 2014). Yang et al. (2005) found that β-D-glucosidase hydrolysis in fresh *O. fragrans* flowers enhanced the mass fractions of monoterpene volatiles. However, further research is needed on the glycoside monoterpenes in *O. fragrans*. Moreover, release of the bound volatile aglycones is dependent on flower development (Reuveni et al., 1999). Therefore, the connection between different forms of volatiles is fundamental for understanding the molecular mechanism of monoterpene biosynthesis.

Here, we give a detailed analysis of the emission and accumulation of floral monoterpenes in two *O. fragrans* cultivars during flowering. A total of 18 genes, including 13 MEP pathway genes from *DXS* to *IDI*, one *GPPS* and four *TPS* genes, were obtained by transcriptome sequencing and their expression levels were analyzed by real-time qPCR. Intriguing differences were found in the transcript levels of four *TPS*s. Furthermore, functional characterization of TPSs and their involvement in major monoterpene production in *O. fragrans* flowers are described. This work provides a better understanding of the molecular mechanism of monoterpene biosynthesis, and will also help in the biotechnological enhancement and modification of aroma in *O. fragrans*.

MATERIALS AND METHODS

Plant Materials

The 'Liuye Jingui' (abbreviated as 'Liuye') and 'Gecheng Dangui' (abbreviated as 'Gecheng') cultivars of *O. fragrans* were grown in the campus nursery of Huazhong Agriculture University in Wuhan, China. Flower opening in *O. fragrans* was divided into four stages (Zou et al., 2014): tight bud stage (S1); initial flowering stage (S2); full flowering stage (S3); and late full flowering stage (S4). Flowers were harvested at about 10 a.m. in October 2013. Each sample at each flowering stages was separated into two parts. One was directly used for headspace volatile analysis, and the other was immediately frozen in liquid nitrogen for volatile solvent and glycoside extraction, and RNA extraction.

SPME Collection and Solvent Extraction

The released floral volatiles were collected by solid-phase microextraction (SPME; Cai et al., 2014). In triplicate, 2 g fresh flowers were placed into a 20 ml capped SPME vial and incubated at 25 ± 2°C for 30 min. SPME fiber (50/30 μm DVB/CAR/PDMS on a 2 cm Stable Flex fiber, Supelco Inc. Bellefonte, PA, USA) was then exposed to the headspace of the capped vial for 30 min. The fiber was injected manually and desorbed in the injection port of the gas chromatograph (GC) with helium as the carrier gas. The fiber was desorbed for 5 min at 250°C in splitless mode. Before each set of samples was assayed, the fiber was conditioned for 1 h at 250°C in the injection port of the GC-MS and a fiber blank recorded.

The accumulated free floral volatiles were collected by solvent extraction (Green et al., 2012). In triplicate, 2 g frozen flowers, harvested at the equivalent time points to the SPME sampling, were ground to a fine powder in liquid nitrogen, transferred to a 50 ml centrifuge tube and extracted twice with 10 ml pentane/Et2O (1:1 v/v) for 30 min with gentle shaking. The two extractions were combined and stored overnight at −20°C. The following day, the upper solvent layer was carefully separated from the lower frozen water layer and reduced to 2 ml under a gentle stream of nitrogen. The concentrated extract, with 47.3 ng/μl cyclohexanone added as internal standard, was passed through a column of anhydrous $MgSO_4$ to remove any remaining water, and then injected into the GC.

Glycoside Extraction

Glycoside analysis was carried out according to Green et al. (2012) with minor modifications. In triplicate 2 g frozen flowers, harvested at the equivalent time points to the headspace sampling, were ground to a fine power in liquid nitrogen, transferred to a 50 ml centrifuge tube and resuspended in 30 ml ddH_2O. The sample was centrifuged at 8,000 *g* for 15 min at 4°C and the supernatant run on a 15 mm × 25 mm i.d. Amberlite XAD-2 column (Supelco, Bellefonte, PA, USA) according to the manufacturer's instructions, at the rate of 3 ml/min. Soluble sugars and acids were removed with 40 ml water and free volatiles by the addition of 40 ml pentane/Et2O (1:1 v/v). Bound glycosides were eluted with 20 ml methanol and evaporated to dryness in a rotary evaporator. The resulting glycoside pellet was resuspended in 2 ml deglycosylation buffer (200 mM Na_2HPO_4, 220 mM citric acid, pH 5.0) and re-extracted three times with 1 ml pentane/Et2O (1:1 v/v) to remove any remaining free compounds.

Enzymatic hydrolysis was carried out using β-glucosidase (6 u/mg; Sigma-Aldrich, Co, LLC, USA), dissolved in deglycosylation buffer, at a concentration of 10 mg/ml. The hydrolysis sample was overlaid with 1 ml Et2O and incubated at 40°C for 36 h. Following incubation, the sample was extracted a further three times with 1 ml Et2O. Prior to GC-MS analysis, the pooled extracts, with the addition of 47.3 ng/μl cyclohexanone as internal standard, were passed through a column of anhydrous $MgSO_4$ and reduced to 0.5 ml under a gentle stream of nitrogen.

GC-MS Analysis

The samples of the SPME collected, solvent extracted (1 μl) and glycoside volatiles (1 μl) were separated on a 30 m × 0.25 mm × 0.25 μm HP-5 capillary column (Thermo Scientific, Bellefonte, PA, USA). The system was a TRACE GC Ultra GC coupled to a DSQ II mass spectrometer (Thermo Fisher Scientific, Waltham, MA, USA). The GC-MS was performed according to Cai et al. (2014). The GC oven ramp for SPME collected volatiles was at 40°C for 3 min, 3°C/min to 73°C and held for 3 min, 5°C/min to 220°C and held for 1 min. The GC oven ramp for the solvent extracted and glycoside volatiles was at 40°C for 3 min, 3°C/min to 73°C and held for 3 min, 5°C/min to 240°C and held for 10 min. The flow rate of the helium (99.999%) carrier gas was 1.2 ml/min. The transfer line temperature was 280°C. For the mass detector, the ion source temperature was set at 230°C, with electronic impact (EI) mode at 70 eV over the mass range m/z 40–450 amu. A C8–C40 alkane standard solution (Sigma-Aldrich, Co., LLC., USA) was analyzed regularly to provide references for calculation of retention time (Kovats) indices (RIs) and to monitor system performance. Identification of the compounds was based on a comparison of their mass spectra and retention indices (RIs) with the authentic standards and published data, as well as standard mass spectra in the NIST05. Relative quantification of the target compounds for emission was by measuring peak areas, and for accumulation using the internal standard method.

RNA Extraction and Real-Time PCR

Total RNA was isolated from 0.20 g frozen flowers using TRIzol reagent (CoWin Biotech Co., Ltd., Beijing, China), following the manufacturer's instructions, and then treated with RQ1 DNase I (Promega, Madison, WI, USA) to remove genomic DNA. To synthesize first-strand cDNA, 3.50 μg total RNA was used with the RevertAid™ First Strand cDNA Synthesis Kit (Fermentas, Thermo Fisher Scientific Inc., USA) according to the manufacturer's instructions. The synthetic first-strand cDNAs were diluted 10-fold for gene expression analysis.

Gene expression was detected by qRT-PCR in both 'Gecheng' and 'Liuye' flowers at four flowering stages. The qRT-PCR was performed on an Applied Biosystems 7500 Fast Real-Time PCR platform with the SYBR® Premix Ex Taq™ II mix (Takara Biotechnology Co., Ltd., Dalian, China), according to the manufacturer's instructions, and the results were analyzed using the Applied Biosystems 7500 software (Applied Biosystems Life Technologies). Three biological replicates were tested, and reactions carried out in triplicate. Relative transcript levels were calculated by the $2^{-\Delta\Delta Ct}$ method using β-*actin* as the endogenous control gene for data normalization. The primers for qRT-PCR analysis are listed in Supplemental Table S2.

Isolation of *OfTPS* Genes

Based on the *TPS* unigene sequences from the transcriptome sequencing of *O. fragrans* flowers, the full-lengths of four *TPS* genes were obtained using the SMARTER^{TM} RACE method. The 5′ and 3′-RACE-Ready cDNAs were separately synthesized using the BD SMARTER^{TM} RACE cDNA Amplification Kit (Clontech, Mountain View, CA, USA). The amplified *OfTPSs* sequences were cloned into pEASY-T1 (TransGene Biotech CO., LTD, Beijing, China) and at least three independent clones were sequenced to check for PCR errors. The *OfTPSs* open reading frame (ORF) was predicted using the NCBI ORF Finder[1]. All primers used are listed in Supplemental Table S2.

Multiple Sequence Alignment and Phylogenetic Analysis

The DNAMAN 6.0 software (Lynnon Biosoft, USA) was used for multiple sequence alignment, and the phylogenetic tree constructed using the default parameters of the MEGA 6.1 software. The MEGA employed Clustal W2 software to generate multiple alignments and construction of the phylogenetic tree was based on the neighbor-joining computational method with 1000 bootstrap replicates. The bioinformatics tools ChloroP[2] and TargetP[3] were used to predict the intracellular localization of *Of*TPS proteins.

Transient and Stable Expression of *OfTPS* Genes in Tobacco

Full-length *OfTPS* ORFs were obtained from the pESAY-T1 vectors containing the target genes, using FastDigest enzymes (Fermentas, Thermo Fisher Scientific Inc., USA). *OfTPS1*, *OfTPS2*, and *OfTPS3* were digested with KpnI-XbaI, and *OfTPS4* with SmaI-XbaI. The restriction enzyme-generated inserts were cloned into the same restriction sites of the pCAMBIA 2300 binary vector to create pCAMBIA 2300::*OfTPSs* using T4 DNA ligase (Fermentas, Thermo Fisher Scientific Inc., USA). pCAMBIA 2300 contained the CaMV 35S promoter and nos-terminator. The pCAMBIA 2300::p19 was created by digesting the pGH-p19 vector using SmaI-XbaI to clone into the pCAMBIA 2300 vector, as described above. These plasmids were transformed in *Agrobacterium tumefaciens* strain EHA105 by electroporation.

Four to 6-week-old greenhouse-grown *Nicotiana benthamiana* seedlings were infiltrated with the *A. tumefaciens* strain EHA105, harboring the pCAMBIA 2300::*OfTPSs* and pCAMBIA 2300::p19, as described previously (Hellens et al., 2005). When freshly grown *Agrobacterium* cultures reached an $OD_{600\ nm}$ of 0.6–0.8, they were centrifuged and resuspended in infiltration media (10 mM MES, 10 mM $MgCl_2$, 200 μM acetosyringone). The suspensions were adjusted to an $OD_{600\ nm}$ of between 1.0 and 2.0, and incubated without shaking at 28°C for 2 h. The target gene and viral suppressor p19 *Agrobacterium* cultures were mixed in a 1:1 ratio before injection into *N. benthamiana* leaves using a syringe. After 5 days, 2 g of the treated leaves were harvested and placed in a 20 ml SPME vial for volatile analysis (as above).

Leaf disks of *N. tabacum* were transformed by co-culture with *A. tumefaciens* strain EHA105 harboring the pCAMBIA 2300::*OfTPSs* binary vector. Three to five independent transformed lines were obtained. The transformed plants, obtained after selection with kanamycin, were confirmed by semi quantitative RT-PCR with 2x Es Taq MasterMix (CoWin Biotech Co., Ltd, Beijing, China) and GAPDH as the reference gene. 2 g leaves of the transformed plants were used for volatile analysis.

RESULTS

Analysis of Monoterpene Emission and Accumulation in *O. fragrans* Flowers

Solid-phase microextraction-GC-MS analysis identified a total of 33 volatile compounds in two cultivars of 'Gecheng' and 'Liuye' at the initial flowering stage (S2), assigned to monoterpene, norisoprenoid, aromatic and fatty acid-related compounds (Supplemental Table S1). Seventeen monoterpenes were found in the two cultivars, with β-ocimene, linalool and derivatives the dominant components in both cultivars. However, the content of monoterpenes differed in the two cultivars (**Figure 1**). The relative content of total monoterpenes was higher in 'Gecheng' (70%) than 'Liuye' (5%). In particular, *trans*-β-ocimene and linalool accounted for 44 and 23% of the total volatiles in 'Gecheng', but for only 1 and 2% in 'Liuye' (Supplemental Table S1). The emission of *cis*-β-ocimene, *trans*-β-ocimene, linalool and linalool derivatives in both 'Gecheng' and 'Liuye' flowers showed a similar pattern, increasing from S1 to S3 and decreasing at S4 (**Figures 2A–D** and **3A**). The peak of β-ocimene and linalool emission occurred at S2 in 'Gecheng' and at S3 in 'Liuye,' while the peak of the linalool derivatives in both cultivars was at S3. The β-ocimene and linalool emissions at S2 were much higher in 'Gecheng' than 'Liuye.' The emissions of all linalool derivatives were higher in 'Gecheng' throughout the flowering stages.

The accumulation of free monoterpenes in flowers was analyzed by solvent extraction, combining with GC-MS, collecting samples at the same stage as for SPME (**Figures 2E,F** and **3B**). Compared with the emitted monoterpenes, β-ocimene was barely detectable, but an abundance of linalool and derivatives accumulated in *O. fragrans* flowers. *Cis*-8-hydroxylinalool and 8-hydroxylinalool, the linalolol derivatives, were not detected as emitted floral volatiles but were found in solvent extracts. The accumulation pattern of linalool was consistent with the emission pattern in both cultivars, increasing from S1 to S3 and decreasing slightly at S4. The accumulation of linalool oxides continuously increased up to S3 and slightly

[1]http://www.ncbi.nlm.nih.gov/gorf/orfig.cgi
[2]http://www.cbs.dtu.dk/services/ChloroP/
[3]http://www.cbs.dtu.dk/services/TargetP/

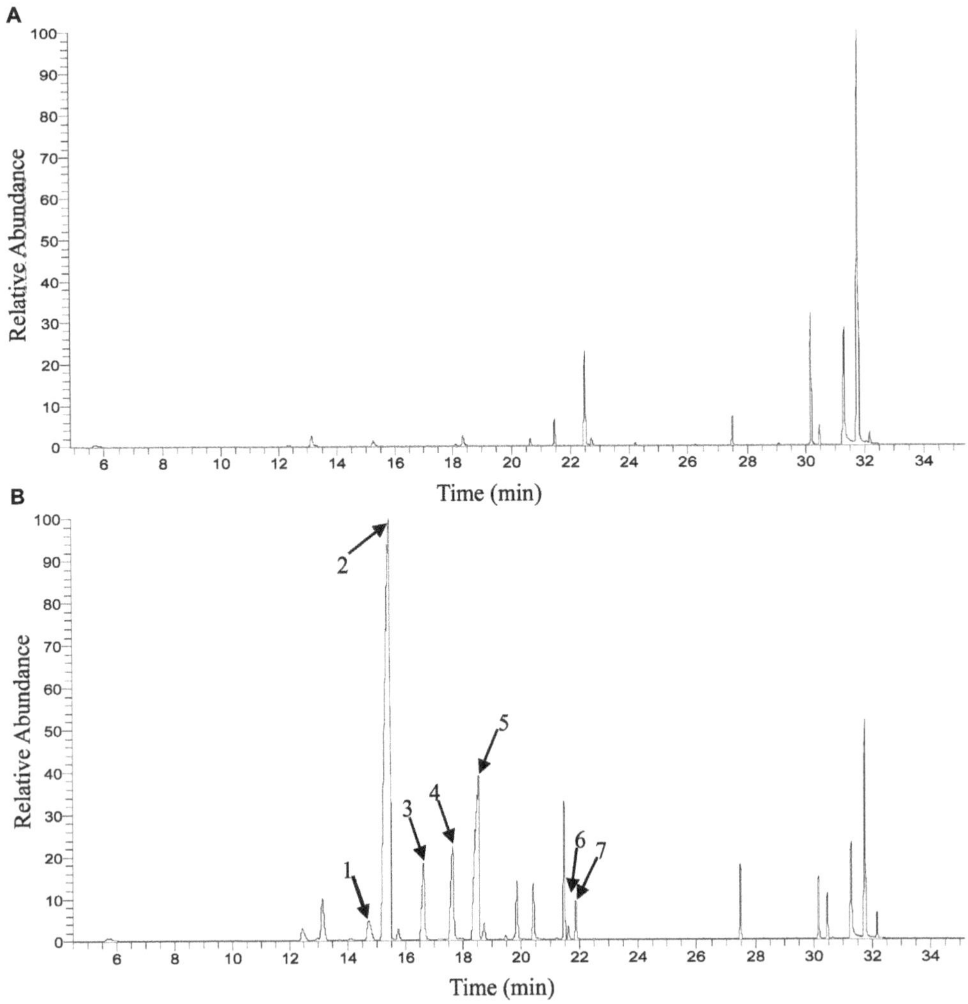

FIGURE 1 | Gas chromatograph (GC) trace of floral volatile emission in *Osmanthus fragrans* at the initial flowering stage (S2). (A) Liuye, **(B)** Gecheng. (1) *cis*-β-ocimene; (2) *trans*-β-ocimene; (3) *cis*-Linalool oxide (furanoid); (4) *trans*-linalool oxide (furanoid); (5) β-linalool; (6) *cis*-linalool oxide (pyranoid); (7) *trans*-linalool oxide (pyranoid).

decreased at S4, except for *trans*-linalool oxide(fur) in 'Gecheng,' which was still increasing at S4. The accumulation of *cis*-8-hydroxylinalool and 8-hydroxylinalool in 'Gecheng' gradually increased during the flowering stages, reaching the maximum at S4, while they basically remained steady in 'Liuye,' dropping slightly at S3. More linalool was accumulated in 'Gecheng' at S1 and S2, but in 'Liuye' at S3 and S4. The result was just the contrary to *cis*-8-hydroxylinalool and 8-hydroxylinalool. There was much more accumulation of linalool oxides in 'Gecheng' compared to 'Liuye' during flowering.

Floral glycosylated volatiles were extracted from *O. fragrans* flowers and analyzed by GC-MS after enzymatic hydrolysis (**Figures 2G,H** and **3C**; **Table 1**). Due to lack of hydoxylation in β-ocimene, only linalool and its derivatives were detected in glycosylated forms in *O. fragrans* flowers. In both cultivars, the glycosylated linalool and derivatives continuously increased during flowering, except for linalool in 'Gecheng,' which remained stable throughout the flowering stages. After S2, more glycosylated linalool accumulated in 'Liuye' than in 'Gecheng.' The glycosylated linalool derivatives as *trans*-linalool oxide (furan and pyran), *cis*-8-hydroxylinalool and 8-hydroxylinalool accumulated much more in 'Gecheng' than 'Liuye' throughout the flowering process. The percentage of glycosylation of linalool continued increasing in 'Liuye' from S1 to S4, but maintained a lower fraction in 'Gecheng.' Although the percentage of glycoside linalool oxides changed irregularly, total linalool derivatives in glycoside forms continued increasing in both cultivars from S1 to S4, due to the dominant components, *cis*-8-hydroxylinalool and 8-hydroxylinalool, were constantly increasing.

Expression Analysis of Genes Involved in Monoterpene Biosynthesis in *O. fragrans* Flowers

The MEP pathway produces IPP and DMAPP for production of monoterpenes (Munoz-Bertomeu et al., 2006) and TPS is the final

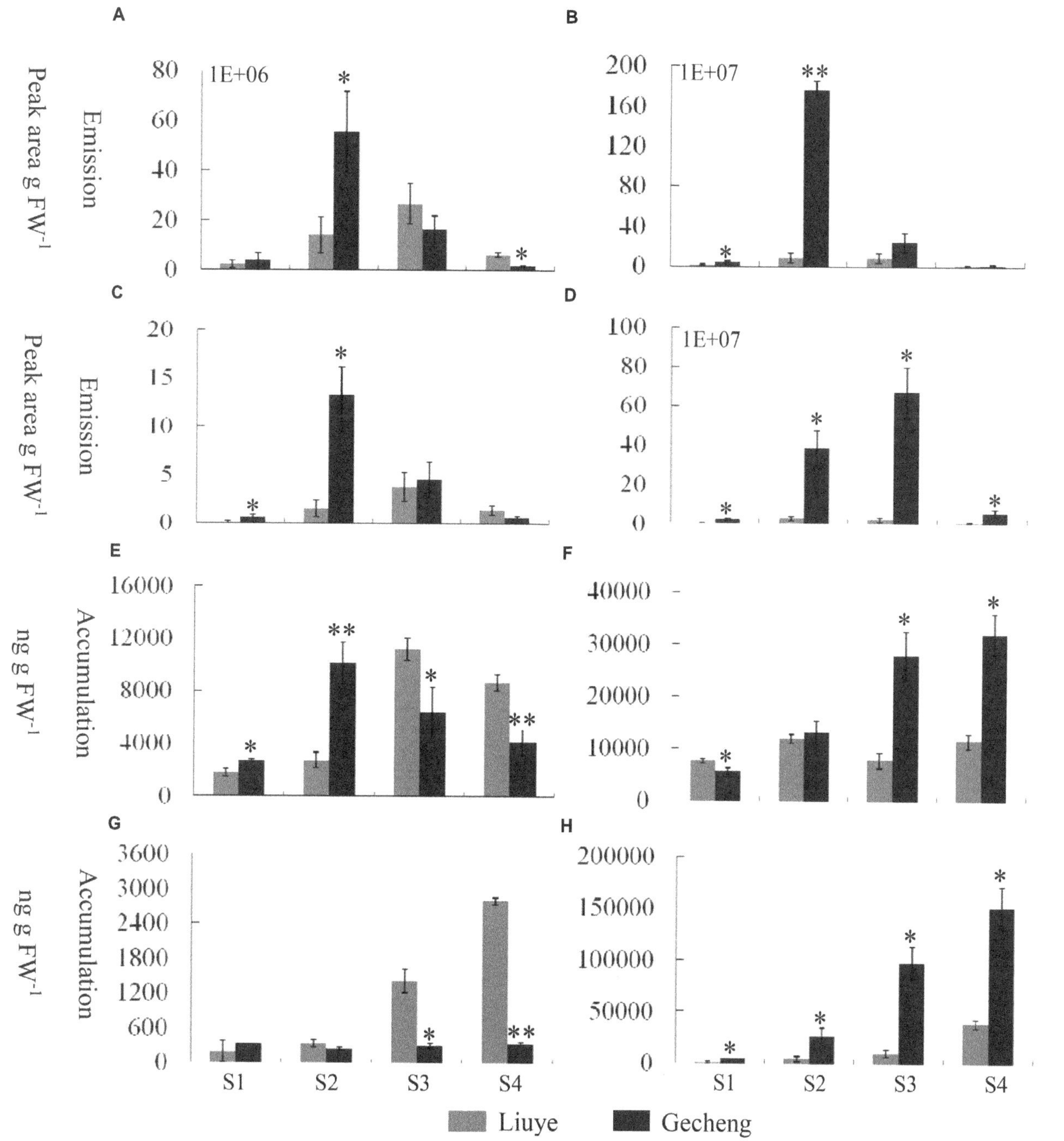

FIGURE 2 | Emission and accumulation of monoterpenes and their derivatives in *O. fragrans* flowers at four stages. (A–D) Emission abundance of monoterpenes and their derivative volatiles: **(A)** *cis*-β-ocimene; **(B)** *trans*-β-ocimene; **(C)** linalool, and **(D)** total linalool derivatives. **(E,F)** Free accumulation of linalool and its derivatives: **(E)** linalool and **(F)** total linalool derivatives. **(G,H)** Glycoside accumulation of linalool and its derivatives: **(G)** linalool and **(H)** total linalool derivatives. The four flowering stages are: (S1) tight bud stage; (S2) initial flowering stage; (S3) full flowering stage, and (S4) late full flowering stage. Data are presented as mean ± SE ($n = 3$). The asterisks indicate significant differences between the values of 'Liuye' and 'Gecheng' at a given flower stage calculated by the Student's *t*-test (*$P < 0.05$, **$P < 0.01$).

enzyme converting the precursor GPP to kinds of monoterpenes (Degenhardt et al., 2009). The transcript levels of thirteen genes involved in the eight enzyme reaction stages of the MEP pathway, one *GPPS*, and four *TPS*s were analyzed by real-time qPCR (**Figure 4**).

In the MEP pathway, the expression levels of *DXR*, *CMK1*, and *MCT1* remained stable in the two cultivars throughout the flowering stages, and only subtle differences (<2 fold) were detected in *DXS1*, *MECPS*, *HDS*, *IDS1*, *IDS2*, and *IDS3* expression between the two cultivars. There was higher expression of *CMK2*,

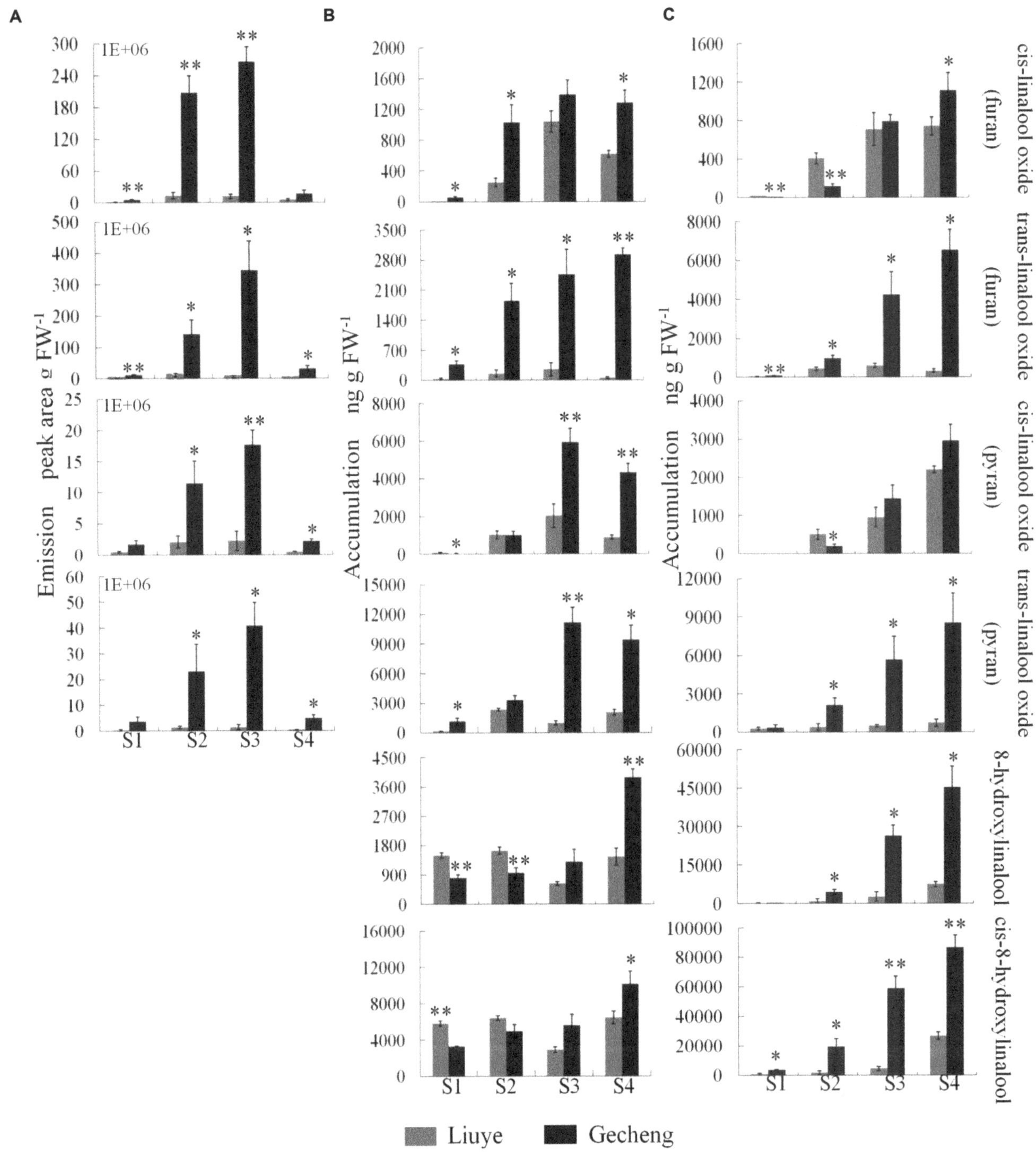

FIGURE 3 | Emission and accumulation of linalool derivatives in *O. fragrans* flowers at four stages. (A) Emission of linalool derivatives. **(B)** Accumulation of free linalool derivatives. **(C)** Accumulation of glycosylated linalool derivatives. Flower stages are: (S1) tight bud stage; (S2) initial flowering stage; (S3) full flowering stage, and (S4) late full flowering stage. Data are presented as mean ± SE ($n = 3$). The asterisks indicate significant differences between the values of 'Liuye' and 'Gecheng' at a given flower stage calculated by the Student's *t*-test (*$P < 0.05$, **$P < 0.01$).

MCT2, and *IDI* in 'Gecheng' than 'Liuye' from S1 to S4, but the difference was no more than threefold between the two cultivars, except for *MCT2* at S3. *DXS2* expression was no more than threefold higher in 'Liuye' than 'Gecheng' from S2 to S4. These results showed that, during flowering, there were minor differences between the two cultivars in the expression of most genes in the MEP pathway. Expression of *GPPS* was the same in both cultivars at S1, but three–fivefold higher in 'Liuye' from

TABLE 1 | Glycosylation of linalool and its derivatives in 'Liuye' and 'Gecheng' cultivars of *Osmanthus fragrans* flowers at four stages.

	Glycoside (% of total accumulation)							
	Liuye				Gecheng			
Compounds	**S1**	**S2**	**S3**	**S4**	**S1**	**S2**	**S3**	**S4**
β-linalool	10.03 ± 2.19	10.97 ± 0.15	11.22 ± 1.43	24.44 ± 1.22	10.96 ± 0.39	**2.36 ± 0.24**	**4.49 ± 1.05**	**7.60 ± 1.54**
Total linalool derivatives	14.89 ± 3.16	27.19 ± 3.42	55.71 ± 11.74	76.88 ± 4.93	**43.69 ± 8.50**	**67.58 ± 12.70**	77.76 ± 7.56	82.52 ± 6.10
Cis-linalool oxide (furan)	66.44 ± 5.36	61.97 ± 4.38	40.56 ± 8.36	54.65 ± 5.56	**12.79 ± 2.70**	**10.29 ± 0.37**	36.33 ± 1.07	46.56 ± 5.80
Trans-linalool oxide (furan)	51.10 ± 12.30	73.80 ± 1.38	70.51 ± 9.60	86.62 ± 10.76	**18.39 ± 2.41**	**34.81 ± 0.37**	63.21 ± 12.35	69.03 ± 8.18
Cis-linalool oxide (pyran)	15.19 ± 2.63	33.06 ± 6.81	32.04 ± 5.03	71.45 ± 1.72	26.43 ± 7.34	**16.82 ± 2.84**	**19.50 ± 4.16**	**40.49 ± 4.73**
Trans-linalool oxide (pyran)	67.29 ± 15.68	13.70 ± 2.16	33.40 ± 2.70	25.89 ± 6.30	**24.10 ± 7.80**	**38.85 ± 8.91**	33.63 ± 9.82	47.56 ± 10.48
8-hydroxylinalool	13.14 ± 3.40	37.58 ± 11.50	81.03 ± 12.90	83.73 ± 5.82	**29.65 ± 6.13**	**82.49 ± 10.55**	95.33 ± 4.34	92.08 ± 6.48
Cis-8-hydroxylinalool	11.60 ± 1.58	21.06 ± 5.45	60.09 ± 14.33	80.53 ± 4.32	**52.29 ± 12.31**	**79.63 ± 13.24**	**91.32 ± 5.09**	89.54 ± 3.64

Data are presented as mean ± SE (n = 3). The values in bold indicate significant differences to the 'Liuye' values at a given flower stage (Student's t-test $P < 0.05$).

S2 to S4. Regarding the *TPS* genes, there was no apparent difference between the two cultivars during flowering for *TPS2* and *TPS4* expression, which gradually increased from S1 to S3 followed by a dramatic increase at S4. However, it is worth noting that the expression of *TPS1* and *TPS3* was more than 100-fold higher in 'Gecheng' compared to 'Liuye' throughout flowering stage. *TPS1* expression consistently increased from S1 to S4, while *TPS3* had the highest level of transcript accumulation at S3 in the two cultivars. The differences in the levels of *TPS* gene expression between 'Gecheng' and 'Liuye' indicated their important contribution to monoterpene formation in *O. fragrans* flowers.

Sequence Characterization of *TPS* Genes from *O. fragrans* Flowers

The four full-length *TPS* genes were cloned using the RACE-PCR method. Their ORF sequences were 1746, 1668, 1776, and 1647 bp, respectively, designated *OfTPS1*, *OfTPS2*, *OfTPS3*, and *OfTPS4* (GeneBank no. KT591180 to KT591183). The sequences of these four genes were identical in the cultivars 'Gecheng' and 'Liuye.' The four encoded proteins had highly conserved elements of TPSs, the DDxxD motif and the (N,D)D(L,I,V)x(S,T)xxxE motif, implying that they had the same capacity to bind the diphosphate group substrate (**Figure 5**). The N-terminal $RR(x)_8W$ motif, involved in producing cyclic monoterpenes and absent in TPSs that produce acyclic products, only appeared in OfTPS4 (**Figure 5**). Using the 'ChloroP' and 'TargetP' programs[4], *Of*TPS1 and *Of*TPS3 were found to contain an extended N-terminus, recognized as a signal peptide, with a length of 47 and 38 amino acids, respectively. Phylogenetic analysis of the predicted amino acid sequences compared with TPS protein sequences in other species indicated that *Of*TPS1and *Of*TPS2 belong to the TPS-g subfamily, which lacks the $RR(x)_8W$ motif and mainly produce acyclic terpenes (**Figure 6**). *Of*TPS1 and *Of*TPS2 divided into two distinct clades. *Of*TPS1 grouped together with geraniol synthase and linalool synthase from *Olea europaea* and *Ocimum basilicam* in one clade, and *Of*TPS2 in the clade with linalool synthase, nerolidol/linalool synthase, myrcene synthase and ocimene synthase of *A. majus*. *Of*TPS3 and *Of*TPS4 fell into the TPS-b subfamily, covering angiosperm monoterpene synthases (**Figure 6**). *Of*TPS1, *Of*TPS3, and *Of*TPS4 showed highest amino acid sequence identity with geraniol synthase (90%), *Oe*TPS3 (90%), and *Oe*TPS2 (89%) from *O. europaea*, respectively. These results illustrate that they were the remaining uncharacterized TPSs and close to the similarly functional TPSs of *O. europaea*, also belonging to the Oleaceae family.

[4]http://www.cbs.dtu.dk/services/

Functional Identification of *OfTPS* Genes *in Planta*

To investigate the volatiles produced by *OfTPSs in planta*, transient plant expression (Hellens et al., 2005) was carried out by infiltration of *N. benthamiana* leaves using *A. tumefaciens* carrying pCAMBIA 2300::*OfTPSs*. Samples were taken 5–6 days after inoculation and emitted volatiles were analyzed by SPME headspace sampling. Over-expression of *OfTPS1* and *OfTPS2* produced large amounts of β-linalool, with no other products detected. Only *trans*-β-ocimene was detected in tobacco leaves over-expressing *OfTPS3*, and α-farnesene was the only product in those over-expressing *OfTPS4*. No related products were observed in control leaves infiltrated with empty binary vectors (**Figures 7A,C**).

Moreover, stably transformed *N. tabacum* plants were obtained through transformation with *A. tumefaciens* strain EHA105 harboring the pCAMBIA 2300::*OfTPSs*. Positive transgenic plants were further identified by semi-quantitative RT-PCR (**Figure 7D**). The volatile products of the *OfTPS* transformed plants were detected by SPME-GC-MS, and the results were the same as those from transient expression in plants (**Figure 7B**). These results suggest that both *OfTPS1* and *OfTPS2* are involved in synthesis of the same terpene product, β-linalool, and *OfTPS3* and *OfTPS4* are related to *trans*-β-ocimene and α-farnesene synthesis *in planta*, respectively.

DISCUSSION

Monoterpenes are common components of floral scent and play a role in attracting pollinators in plants (Dudareva and Pichersky,

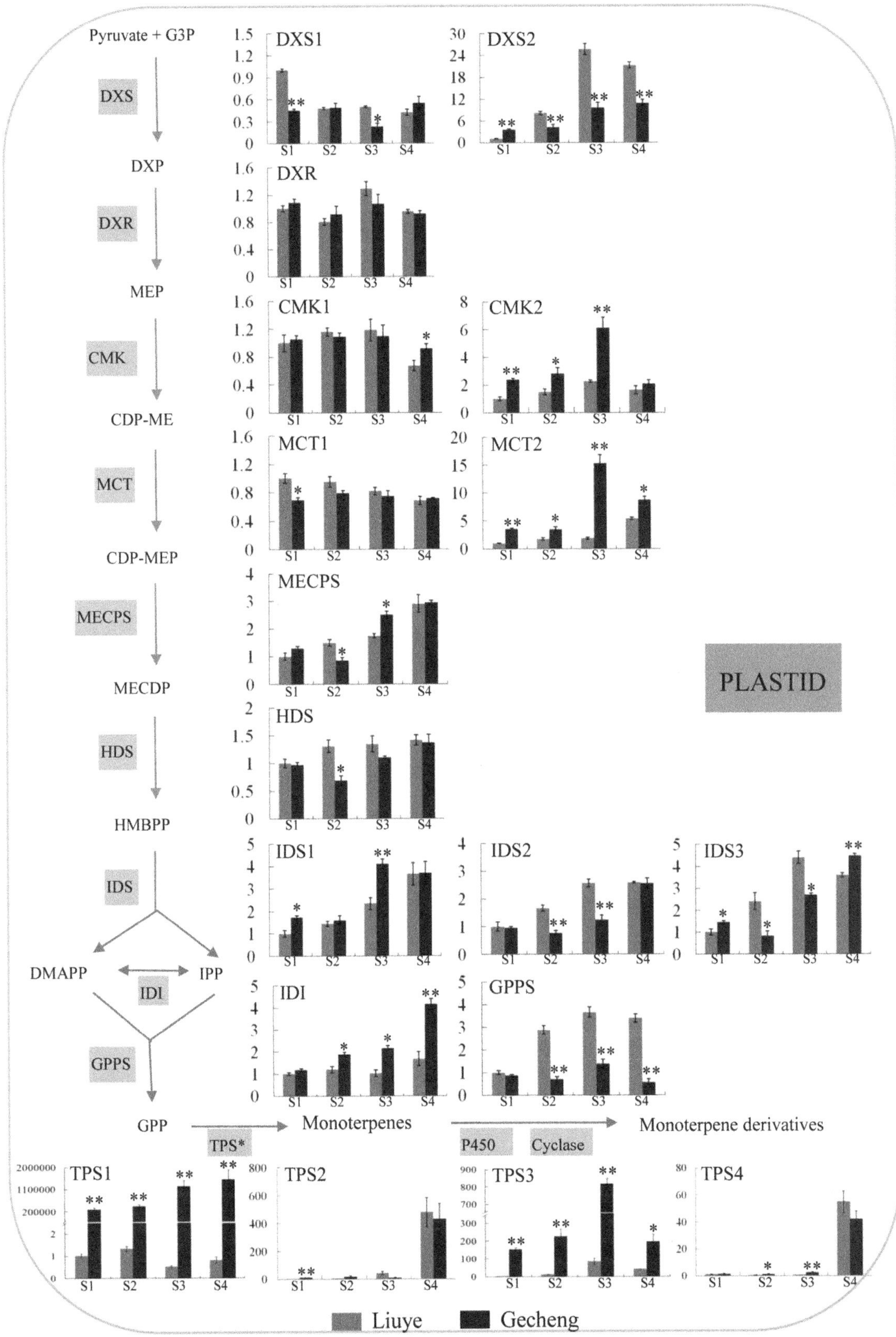

FIGURE 4 | Continued

FIGURE 4 | Continued

Expression of genes involved in monoterpene biosynthesis in *O. fragrans* flowers at four stages. Flower stages are: (S1) tight bud stage; (S2) initial flowering stage; (S3) full flowering stage, and (S4) late full flowering stage. The relative abundance was determined using a $2^{-\Delta\Delta Ct}$ method with β-actin as the reference gene. Results represent the mean ± SE of three technical repetitions and three biological replicates. The asterisks indicate significant differences between the values of 'Liuye' and 'Gecheng' at a given flower stage calculated by the Student's *t*-test (*$P < 0.05$, **$P < 0.01$). Abbreviations: G3P, glyceraldehyde 3-phosphate; DXP, 1-deoxy-D-xylulose 5-phosphate; MEP, methylerythritol phosphate; CDP-ME, 4-diphosphocytidyl-2-C-methyl-D-erythritol; CDP-MEP, CDP-ME 2-phosphate; MECPD, 2-C-methyl-D-erythritol 2,4-cyclodiphosphate; HMBPP, (E)-4-hydroxy-3-methylbut-2-en-1-yl diphosphate; IPP, isopentenyl pyrophosphate; DMAPP, dimethylallyl pyrophosphate; GPP, geranyl pyrophosphate; MVP, mevalonate 5-phosphate; FPP, farnesyl pyrophosphate; DXS, 1-deoxy-D-xylulose 5-phosphate synthase; DXR, 1-deoxy-D-xylulose 5-phosphate reductoisomerase; MCT, 2-C-methyl-D-erythritol 4-phosphate cytidylyltransferase; CMK, 4-(cytidine 5′-diphospho)-2-C-methyl-D-erythritol kinase; MECPS, 2-C-methyl-D-erythritol 2,4-cyclodiphosphate synthase; HDS, 4-hydroxy-3-methylbut-2-en-1-yl diphosphate synthase; IDS, isopentenyl diphosphate synthase; IDI, isopentenyl pyrophosphate isomerase; GPPS, geranyl pyrophosphate synthase; TPS, terpene synthase.

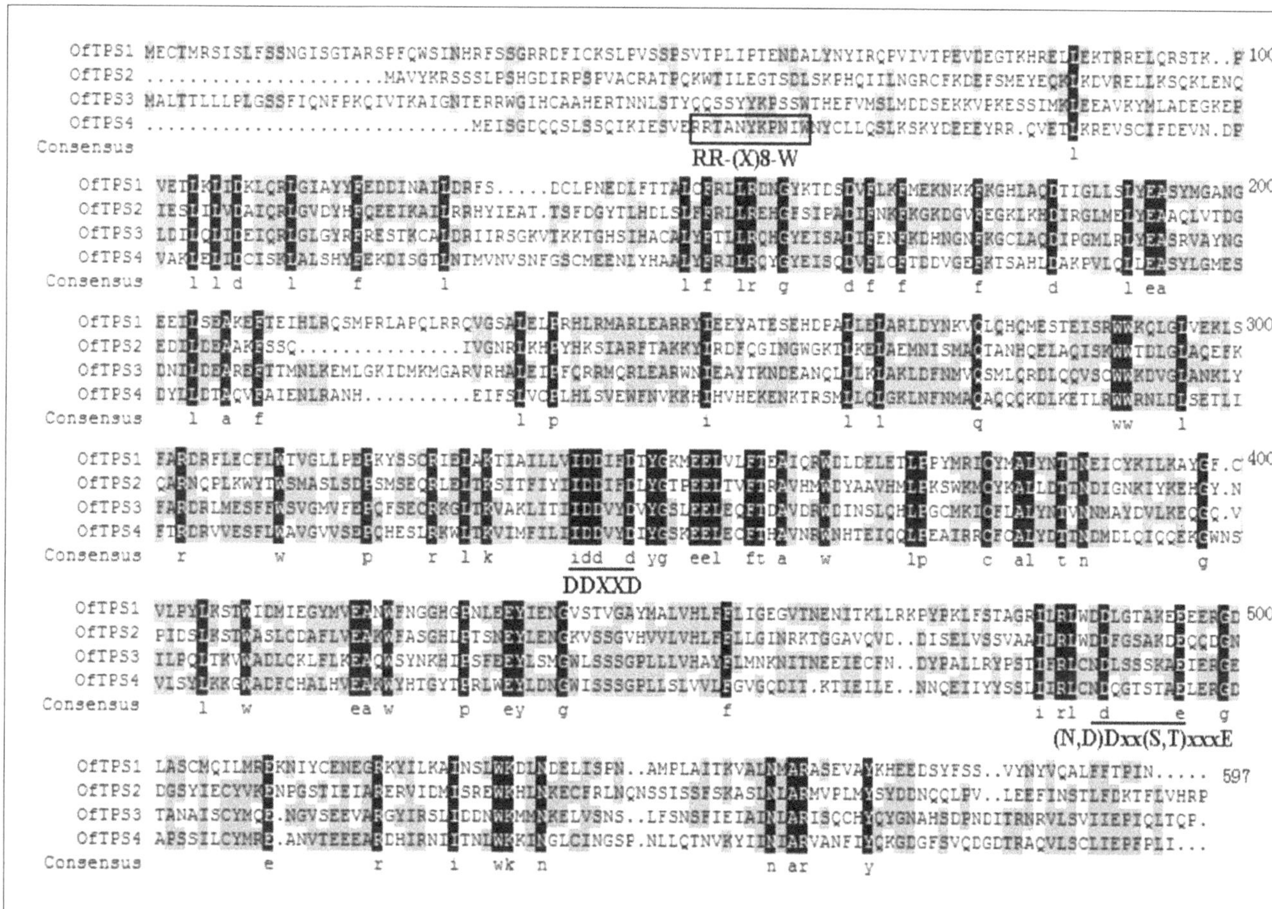

FIGURE 5 | Alignment of the amino acid sequences of *Of*TPS1, *Of*TPS2, *Of*TPS3, and *Of*TPS4. Amino acids identical in all four proteins are shaded black and those identical in three proteins are shaded gray. The three highly conserved motifs are labeled RR-$(x)_8$-W, DDxxD and (N,D)Dxx(S,T)xxxE, respectively.

2006). Previous studies focusing on floral volatiles composition from *O. fragrans* have identified about 20 monoterpenes and find that monoterpenes such as β-ocimene, linalool and linalool derivatives are the major components of floral volatiles in most cultivars (Cao et al., 2009; Sun et al., 2012; Xin et al., 2013). In an earlier study, we also identified these compounds as important aroma-active compounds in *O. fragrans* flowers, with the two cultivars of 'Gecheng' and 'Liuye' having distinct monoterpene emission profiles (Cai et al., 2014). The current study gives basically identical results: monoterpenes, especially *trans*-β-ocimene and linalool, make up a large amount of the floral volatiles in 'Gecheng' but only a trace in 'Liuye' (**Figure 1** and Supplemental Table S1). The results confirmed that 'Liuye' only produces small amounts of monoterpenes, unlike most *O. fragrans* cultivars, so this cultivar can be used as contrasting materials for our further study on the molecular mechanism of monoterpene biosynthesis.

Monoterpenes Emission and Glycosylation in *O. fragrans* Flowers

The emission of floral volatiles in *A. majus* and the content of essential oil in *Lavandula* species can vary with flower development (Dudareva et al., 2000, 2003; Lane et al., 2010). We analyzed the emission and accumulation patterns of the

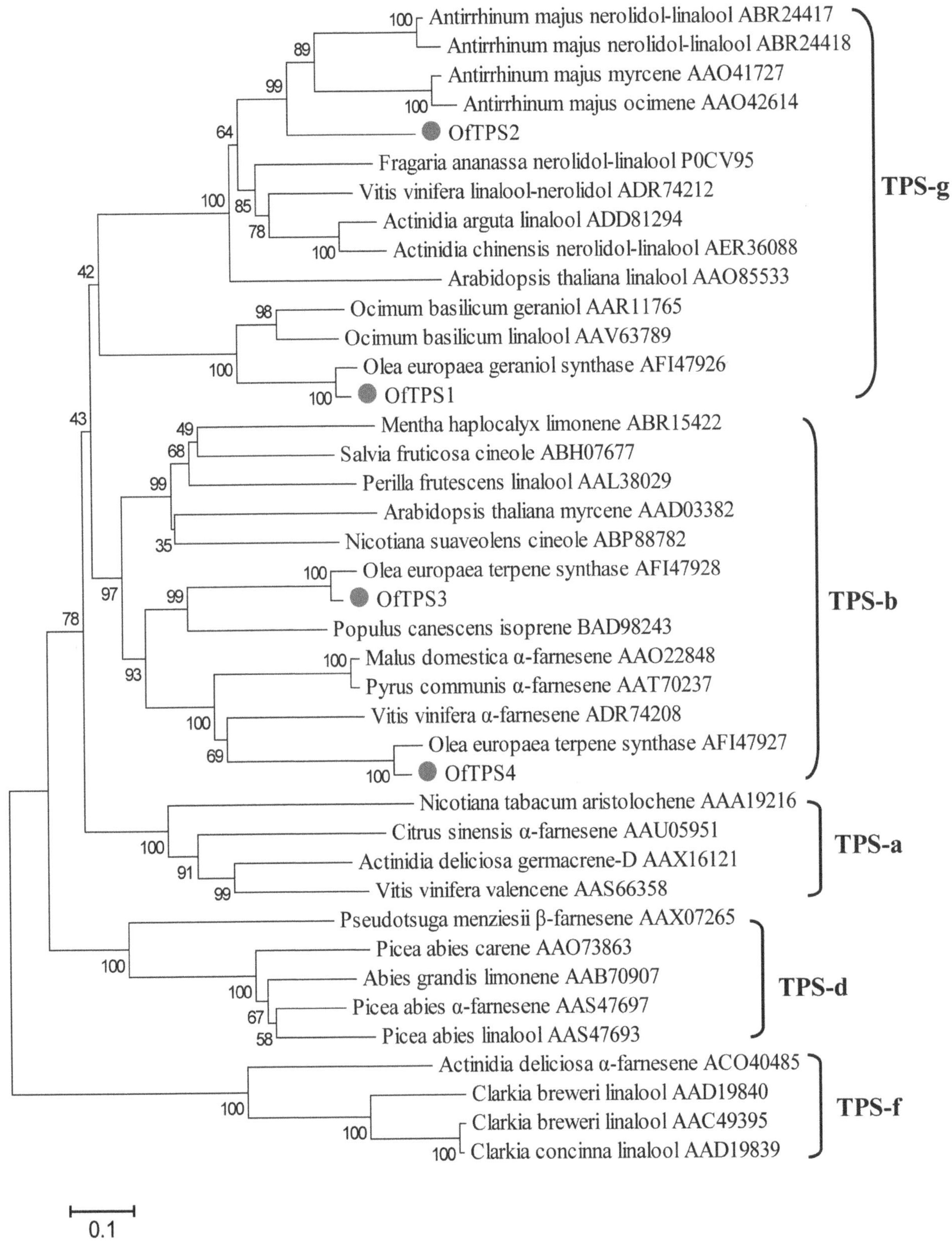

FIGURE 6 | Phylogenetic analysis of selected plant terpene synthases (TPSs) based on estimation of pair-wise distances at the amino acid level. Clusters corresponding to five TPS subfamilies (TPS-a,TPS-b, TPS-d, TPS-g, and TPS-f) are apparent. The Clustal W 2.0 algorithm was used for the alignment. Trees were inferred with the neighbor-joining (NJ) method and $n = 1000$ replicates for bootstrapping.

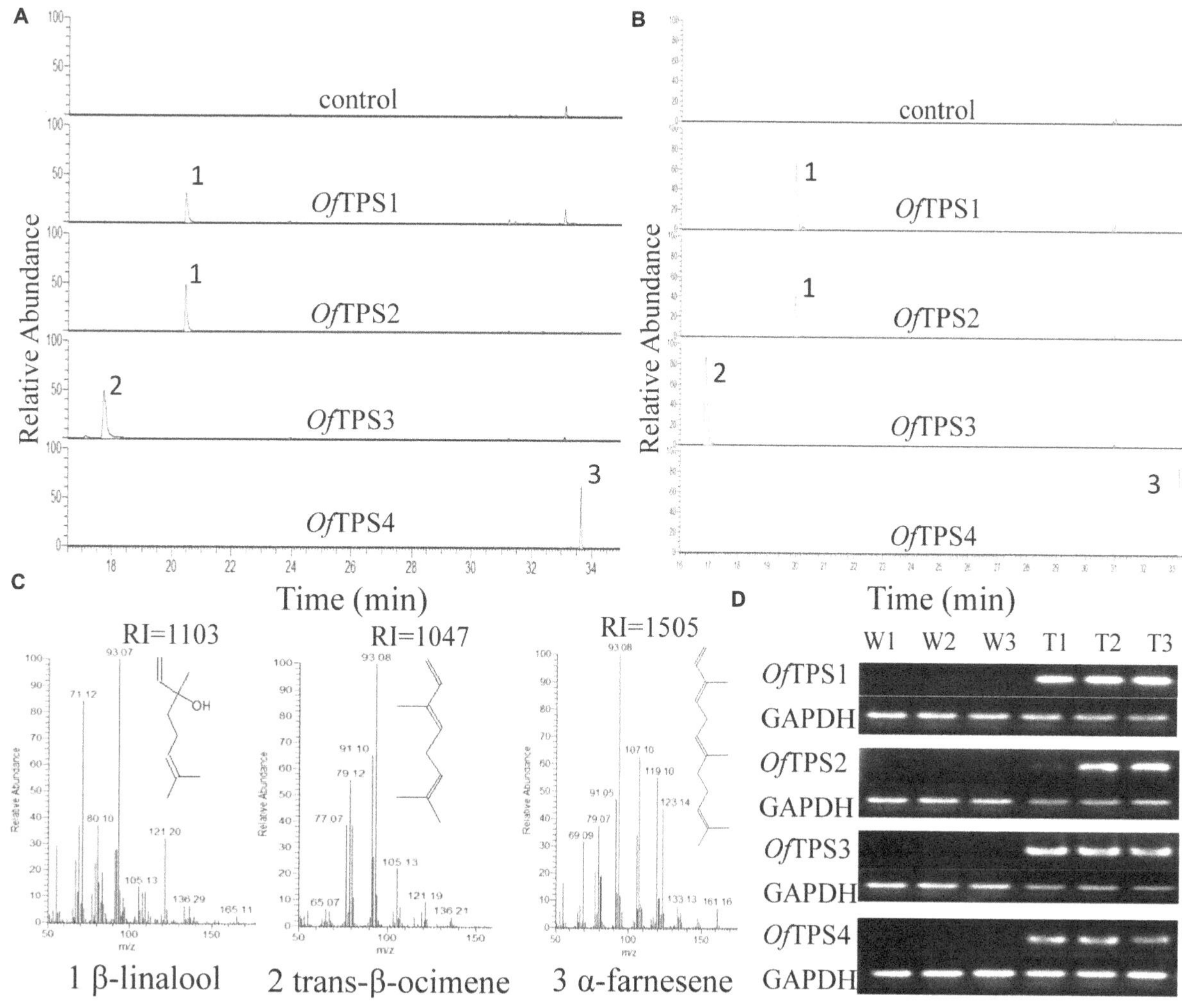

FIGURE 7 | Functional characterization of terpene synthases (TPSs) from *O. fragrans in planta.* (A) GC trace of leaves of transiently expressed *N. benthamiana* plants. (B) GC trace of leaves of transgenic *N. tabacum* plants. **(C)** Mass spectra of the products: (1) β-linalool, (2) *trans*-β-ocimene, (3) α-farnesene. Products were confirmed by comparison with authentic standards and/or GC-MS retention indices. **(D)** Semi-quantitative RT-PCR assay of *OfTPS* genes in transgenic *N. tabacum* plants.

dominant monoterpenes in two *O. fragrans* cultivars throughout the flowering stages. The highest levels of β-ocimene and linalool were found at the initial flowering stage in 'Gecheng' and at the full flowering stage in 'Liuye,' with emission of linalool derivatives reaching a peak at the initial or full flowering stage. These emission patterns were consistent with the other cultivars (Li et al., 2008) and the concentrated emission of monoterpenes coincided with the optimal time for organoleptic aroma and pollination in *O. fragrans* (Baldermann et al., 2010; Zhang, 2013). Moreover, the accumulation pattern of free linalool in the two cultivars was consistent with its emission, reaching the highest level in 'Gecheng' at the initial flowering stage and in 'Liuye' at the full flowering stage. Meanwhile, a parallel trend was observed between accumulation of free linalool oxides and their emission from the tight bud stage to the full flowering stage. These results suggested that both emission and accumulation of monoterpene are controlled by flower development, and the accumulation of free monoterpenes in flowers is directly proportional to their emission in *O. fragrans* flowers.

It has been reported that glycosylation might be involved in regulating the release of floral volatiles. For example, the decrease of glycosylated 2-phenylethanol in flowers correlating with rhythmic emission has demonstrated that glycosylation has an impact on floral 2-phenylethanol rhythmical release in *Rosa damascena* (Picone et al., 2004). But the glycosylation profile cannot account for the different rhythmic patterns of (E)-nerolidol accumulation and emission in *Actinidia chinensis* (Green et al., 2012). In this study, the accumulation and emission of linalool and its oxides in *O. fragrans* flowers shared parallel patterns from the tight bud stage to the full flowering stage. However, significant amounts of glycosylated linalool and its oxides were found during this period, and their accumulation continuously increased, especially at late full flowering stage when only a small amount of linalool and its oxides were released. These results showed that glycosylation was likely to have an impact on the linalool release in *O. fragrans* flowers at different flowering stages. Beside direct glycosylation, linalool tends to enzymatically convert to more stable linalool derivatives such as 8-hydroxylinalool and then is stored in plant tissues in glycosylated forms (Lewinsohn et al., 2001; Chen et al., 2010; Green et al., 2012). In *O. fragrans* flowers, a large amount of linalool derivatives, especially *cis*-8-hydroxylinalool and 8-hydroxylinalool, were found and depleted the pool of linalool. These results showed the complexity of linalool metabolism, and further research on the accumulation and emission of linalool and its derivative should consider the action of glucosidase, cytochrome, P450 or cyclase for further catalytic conversion.

MEP Pathway Genes Analysis of *O. fragrans* Flowers

Monoterpene biosynthesis depends on the MEP pathway and the GPPS enzyme providing the GPP substrate (Chen et al., 2011). To discover the key gene controlling monoterpenes formation, the expression of genes involving in these enzymatic reactions were monitored in 'Gecheng' and 'Liuye' during flowering stages. It has been reported that, except for *DXS* and *IDI*, the other enzymes of the MEP pathway are encoded by single copy genes in *Arabidopsis* (Cordoba et al., 2009). Two distinctive forms *of CMK* and *IDS* have been found in gymnosperms such as *Ginkgo biloba* and *Pinus taeda* (Kim et al., 2008a,b). Two different gene expression patterns of *CMK, MCT,* and *IDS* in two cultivars throughout the flowering stages indicated that at least two forms exist in *O. fragrans*. This suggested that regulation of metabolic flux by a multigene is also present at the mid-point of the MEP pathway in addition to the first step of this pathway in *O. fragrans.*

As the over-expression of DXS in transgenic *A. thaliana* and tomato result in an increase in plastid isoprenoids such as diterpenes and carotenoids, the DXS enzyme has been considered as a rate-limiting enzyme for MEP pathway flux (Enfissi et al., 2005; Carretero-Paulet et al., 2006). In glandular trichomes of *Ocimum basilicum*, the transcript abundance of *DXS* correlates with oil yield which consists mostly of monoterpenes (Xie et al., 2008). In *V. vinifera*, *DXS* co-localizes with a major QTL for the accumulation of linalool, nerol and geraniol (Battilana et al., 2009). During the berry development, *DXS* expression shows a significant correlation with the accumulation profile of these monoterpenes (Battilana et al., 2011). However, in this study, *DXS* gene expression did not agree with the emission and accumulation profiles of monoterpenes from *O. fragrans* flowers. On the contrary, there was a higher expression of *DXS2* in 'Liuye' producing less monoterpenes from flowers. In the second step of the MEP pathway, DXR has also been suggested as a rate-limiting enzyme in *A. thaliana, Mentha piperita, O. basilicum* (Wildung and Croteau, 2005; Carretero-Paulet et al., 2006; Xie et al., 2008). However, the expression of *DXR* was also inconsistent with the monoterpenes production in *O. fragrans*. *HDS* and *IDS* is associated with foliar concentrations of the monoterpene 1,8-cineole in *Eucalyptus globulus* (Külheim et al., 2011). Here, there was no connection between gene expression of *HDS* and *IDS* and the production of monoterpenes. Therefore, we concluded that these MEP pathway genes controlling the pathway flux in other species was not the rate-limiting factor for monoterpene biosynthesis in *O. fragrans*. Unexpectedly, the level of expression of *CMK*2 and *MCT*2 was higher in 'Gecheng' than in 'Liuye.' There were no reports about the connection between CMK, MCT, and monoterpene production, however, their gene expression seemed to be consistent with monoterpene emission in *O. fragrans*, the roles of *CMK* and *MCT* need further analysis.

OfTPS Genes Expression in *O. fragrans* Flowers

Transcriptional control of *TPS* genes has been shown to regulate the production of certain monoterpenes. For example, both spatial and temporal *AlstroTPS* expression correlates well with myrcene emission in scented *Alstroemeria* genotypes (Aros et al., 2012). Also, a concerted increase between expression of *LaLINS* and accumulation of linalool has been reported in the essential oil of *Lavandula angustifolia* during flower development (Lane et al., 2010). Similar results have been found in plants such as *A. majus* (Dudareva et al., 2003) and *Origanum vulgare* (Crocoll et al., 2010). In this study, *TPS2* and *TPS4* expression showed a drastic increase from full flowering stage to late full flowering stage but no difference between 'Gecheng' and 'Liuye' during flowering stages, which was not consistent with the emission and accumulation of monoterpene in the two cultivars. Interestingly, expression of *TPS1* and *TPS3* was more than 100 fold higher in 'Gecheng' as compared to 'Liuye,' and their expression was in agreement with the production of monoterpene during flowering. These results indicated a direct contribution of *TPS1* and *TPS3* to monoterpenes production in flowers of *O. fragrans.* Carotenoid and monoterpene biosynthesis share the same pathway, which indicates a possible competition for the IPP and DMAPP precursors (Botella-Pavía et al., 2004; Dudareva et al., 2013). However, although the production of monoterpenes in 'Liuye,' containing trace amounts of carotenoids, was low and 'Gecheng,' with large mounts of carotenoids still produced a lot of monoterpenes, this was only related to *TPS* genes expression. The results further demonstrated that the transcript levels of *TPS* genes on monoterpene production had a greater effect compared with the competition of carotenoid synthesis. This led to the conclusion that transcriptional regulation of *TPS* genes, especially *TPS1* and *TPS3*, was likely to be the most important mechanism for controlling production of the monoterpenes linalool or β-ocimene in *O. fragrans* flowers.

Sequence and Functional Characterization of *O. fragrans TPS* Genes

The putative functions of the four isolated TPSs from *O. fragrans* were initially predicted according to the conserved motifs and amino acid sequence similarity with known TPSs from other species. The four *Of*TPSs shared the conserved motifs DDxxD and (N,D)Dxx(S,T)xxxE) with other TPSs. *Of*TPS1 and *Of*TPS3 had the typical putative plastid targeting signal peptides but lacked the $RR(x)_8W$ motif in the N-terminal, which indicated they probably encoded acyclic monoterpene synthases. Phylogenetic analysis placed *Of*TPS1 in the TPS-g subfamily defined by monoterpene synthases, that produce acyclic scent compounds such as myrcene and ocimene in *A. majus* (Dudareva et al., 2003), and linalool in *A. thaliana* (Chen et al., 2003) and *V. vinifera* (Martin et al., 2010). Although *Of*TPS2 also belonged to the TPS-g subfamily, it was not clear whether this gene encodes a mono-TPS or sesqui-TPS, due to its unpredictable signal peptide and clustering with the bifunctional linalool/nerolidol synthase of *A. majus* (Nagegowda et al., 2008), *V. vinifera* (Martin et al., 2010), and *A. chinensis* (Green et al., 2012). *Of*TPS3 and *Of*TPS4 clustered into the TPS-b subfamily in which all characterized TPSs are either monoterpene synthases, including all *A. thaliana* monoterpene synthases except linalool synthase (Chen et al., 2003) and *V. vinifera* ocimene synthases (Martin et al., 2010), or isoprene synthases (Chen et al., 2011). But *Of*TPS4 clustered with α-farnesene synthase from *Malus domestica* (Nieuwenhuizen et al., 2013) and *V. vinifera* (Martin et al., 2010). Of the four *Of*TPS proteins, only *Of*TPS4 had the $RR(x)_8W$ motif that is present in all cyclic monoterpene synthases and most sesquiterpene and diterpene synthases (Hyatt et al., 2007), which may play a role in triggering isomerization-cyclization or act to stabilize the protein (Whittington et al., 2002a,b). Moreover, the unapparent signal peptide and high homology with α-farnesene synthase in phylogenetic analysis suggested that *Of*TPS4 was most likely a sesquiterpene synthase.

Transient and stable expression in tobacco leaves further confirm the function of *OfTPS* genes *in planta*, which could partly explain the production of major monoterpene volatiles in *O. fragrans* flowers. Over-expression of *OfTPS1* and *OfTPS3* genes in tobacco leaves resulted in the formation of only β-linalool and *trans*-β-ocimene, respectively. *Of*TPS2, that may be a bifunctional linalool/nerolidol synthase, produced β-linalool exclusively in over-expressing tobacco leaves. As previously speculated, sesquiterpene α-farnesene was the sole product emitted from tobacco leaves over-expressing *OfTPS4*. Most of the TPS enzymes from *V. vinifera* are multiple-product enzymes that can act on one substrate to produce more than one product (Martin et al., 2010). However, there was only a single product of the four *OfTPSs* in the over-expressing tobacco leaves. The TPS enzyme, that produces the same single product, has also been found in *V. vinifera* and *M. domestica* (Pechous and Whitaker, 2004; Martin et al., 2010; Nieuwenhuizen et al., 2013). *Of*TPS1, *Of*TPS3, and *Of*TPS4 showed high homology with *Oe*GES, *Oe*TPS3, and *Oe*TPS2 from *O. europaea*, respectively (**Figure 6**). The homologous TPS proteins from *O. fragrans* and *O. europaea* had entirely different enzyme activities. Heterologous expression in a bacterial system has demonstrated that OeGES is a geraniol synthase only producing geraniol, and OeTPS2 and OeTPS3 have no enzymatic activity (Vezzaro et al., 2012). Fischer et al. (2013), expressing *O. basilicum* geraniol synthase in *V. vinifera*, *A. thaliana*, *N. benthamiana* and a microbial system, has shown that the genetic background of the heterologous expression host could have an influence on minor products but not on the main product. Nevertheless the major monoterpene compounds (β-linalool and *trans*-β-ocimene) found in *O. fragrans* flowers have been detected in tobacco leaves over-expressing *OfTPS1*, *OfTPS2*, and *OfTPS3*. Moreover, gene expression of *OfTPS1* and *OfTPS3* are closely correlated to the accumulation or emission of β-linalool and *trans*-β-ocimene in *O. fragrans* flowers. Therefore, we conclude that *OfTPS1* and *OfTPS3* are separately the key genes associating with β-linalool and *trans*-β-ocimene production in *O. fragrans* flowers.

This study provided the molecular basis for the production of major monoterpenes in *O. fragrans*. The monoterpenes emission and accumulation profiles open the way to further understand the role of monoterpenes in the physiology and pollination biology of *O. fragrans* flowers. Understanding the expression of the MEP pathway genes in relation to monoterpene metabolism could generate vital information regarding the regulation of monoterpene biosynthesis in higher plants. The findings on the key TPS genes responsible for formation of the major monoterpenes would be a great advantage for breeding and manipulation of scent related-enzymes to enhance the ornamental and economic value in *O. fragrans* as well as in other plants in the future.

AUTHOR CONTRIBUTIONS

XZ, CW designed the research; XZ, CL, XC, and JZ performed the research; XZ, RZ, and CL analyzed the data; XZ wrote the article; CW, JL, and RZ critically read the article.

ACKNOWLEDGMENTS

This research was supported by the Doctoral Fund of the Ministry of Education of China (Project No 20130146110022), Fundamental Research Funds for the Central Universities (Project No 2013PY088) and Hubei Collaborative Innovation Center for the Characteristic Resources Exploitation of Dabie Mountains (Project No 2015TD02).

REFERENCES

Aros, D., Gonzalez, V., Allemann, R. K., Müller, C. T., Rosati, C., and Rogers, H. J. (2012). Volatile emissions of scented *Alstroemeria* genotypes are dominated by terpenes, and a myrcene synthase gene is highly expressed in scented *Alstroemeria* flowers. *J. Exp. Bot.* 63, 2739–2752. doi: 10.1093/jxb/err456

Aurore, G., Ginies, C., Ganou-parfait, B., Renard, C. M. G. C., and Fahrasmane, L. (2011). Comparative study of free and glycoconjugated volatile compounds of three banana cultivars from French West Indies: Cavendish, Frayssinette and Plantain. *Food Chem.* 129, 28–34. doi: 10.1016/j.foodchem.2011.01.104

Baldermann, S., Kato, M., Kurosawa, M., Kurobayashi, Y., Fujita, A., Fleischmann, P., et al. (2010). Functional characterization of a carotenoid cleavage dioxygenase 1 and its relation to the carotenoid accumulation and volatile emission during the floral development of *Osmanthus fragrans* Lour. *J. Exp. Bot.* 61, 2967–2977. doi: 10.1093/jxb/erq123

Battilana, J., Costantini, L., Emanuelli, F., Sevini, F., Segala, C., Moser, S., et al. (2009). The 1-deoxy-dxylulose 5-phosphate synthase gene co-localizes with a major QTL affecting monoterpene content in grapevine. *Theor. Appl. Genet.* 118, 653–669. doi: 10.1007/s00122-008-0927-8

Battilana, J., Emanuelli, F., Gambino, G., Gribaudo, I., Gasperi, G., Boss, P. K., et al. (2011). Functional effect of grapevine 1-deoxy-D-xylulose 5-phosphate synthase substitution K284N on *Muscat flavour* formation. *J. Exp. Bot.* 62, 5497–5508. doi: 10.1093/jxb/err231

Birtić, S., Ginies, C., Causse, M., Renard, C. M. G. C., and Page, D. (2009). Changes in volatiles and glycosides during fruit maturation of two contrasted tomato (*Solanum lycopersicum*) lines. *J. Agric. Food Chem.* 57, 591–598. doi: 10.1021/jf8023062

Botella-Pavía, P., Besumbes, O., Phillips, M. A., Carretero-Paulet, L., Boronat, A., and Rodríguez-Concepción, M. (2004). Regulation of carotenoid biosynthesis in plants:evidence for a key role of hydroxymethylbutenyl diphosphate reductase in controlling the supply of plastidial isoprenoid precursors. *Plant J.* 40, 188–199. doi: 10.1111/j.1365-313X.2004.02198.x

Cai, X., Mai, R. Z., Zou, J. J., Zhang, H. Y., Zeng, X. L., Zheng, R. R., et al. (2014). Analysis of aroma-active compounds in three sweet osmanthus (*Osmanthus fragrans*) cultivars by gas-chromatography-olfactometry and GC-mass spectrometry. *J. Zhejiang Univ. Sci. B* 15, 638–648. doi: 10.1631/jzus.B1400058

Cao, H., Li, Z., and Shen, D. (2009). GC/MS fingerprint analysis of *Osmanthus fragrans* Lour. in different varieties. *Acta Hort. Sin.* 36, 391–398.

Carretero-Paulet, L., Cairó, A., Botella-Pavía, P., Besumbes, O., Campos, N., Boronat, A., et al. (2006). Enhanced flux through the methylerythritol 4-phosphate pathway in *Arabidopsis* plants overexpressing deoxyxylulose 5-phosphate reductoisomerase. *Plant Mol. Biol.* 62, 683–695. doi: 10.1007/s11103-006-9051-9

Chen, F., Tholl, D., Bohlmann, J., and Pichersky, E. (2011). The family of terpene synthases in plants: a mid-size family of genes for specialized metabolism that is highly diversified throughout the kingdom. *Plant J.* 66, 212–229. doi: 10.1111/j.1365-313X.2011.04520.x

Chen, F., Tholl, D., D'Auria, J. C., Farooq, A., Pichersky, E., and Gershenzon, J. (2003). Biosynthesis and emission of terpenoid volatiles from *Arabidopsis* flowers. *Plant Cell* 15, 481–494. doi: 10.1105/tpc.007989

Chen, X., Yauk, Y., Nieuwenhuizen, N. J., Matich, A. J., Wang, M. Y., Perez, R. L., et al. (2010). Characterisation of an (S)-linalool synthase from kiwifruit (*Actinidia arguta*) that catalyses the first committed step in the production of floral lilac compounds. *Funct. Plant. Biol.* 37, 232–243. doi: 10.1071/FP09179

Cordoba, E., Salmi, M., and Leon, P. (2009). Unravelling the regulatory mechanisms that modulate the MEP pathway in higher plants. *J. Exp. Bot.* 60, 2933–2943. doi: 10.1093/jxb/erp190

Crocoll, C., Asbach, J., Novak, J., Gershenzon, J., and Degenhardt, J. (2010). Terpene synthases of oregano (*Origanum vulgare* L.) and their roles in the pathway and regulation of terpene biosynthesis. *Plant Mol. Biol.* 73, 587–603. doi: 10.1007/s11103-010-9636-1

Davidovich-Rikanati, R., Lewinsohn, E., Bar, E., Iijima, Y., Pichersky, E., and Sitrit, Y. (2008). Overexpression of the lemon basil α-zingiberene synthase gene increases both mono- and sesquiterpene contents in tomato fruit. *Plant J.* 56, 228–238. doi: 10.1111/j.1365-313X.2008.03599.x

Degenhardt, J., Köllner, T. G., and Gershenzon, J. (2009). Monoterpene and sesquiterpene synthases and the origin of terpene skeletal diversity in plants. *Phytochemistry* 70, 1621–1637. doi: 10.1016/j.phytochem.2009.07.030

Dudareva, N., Klempien, A., Muhlemann, J. K., and Kaplan, I. (2013). Biosynthesis, function and metabolic engineering of plant volatile organic compounds. *New Phytol.* 198, 16–32. doi: 10.1111/nph.12145

Dudareva, N., Martin, D., Kish, C. M., Kolosova, N., Gorenstein, N., Faldt, J., et al. (2003). (E)-β-ocimene and myrcene synthase genes of floral scent biosynthesis in snapdragon: function and expression of three terpene synthase genes of a new terpene synthase subfamily. *Plant Cell* 15, 1227–1241. doi: 10.1105/tpc.011015

Dudareva, N., Murfitt, L. M., Mann, C. J., Gorenstein, N., Kolosova, N., Kish, C. M., et al. (2000). Developmental regulation of methyl benzoate biosynthesis and emission in snapdragon flowers. *Plant Cell* 12, 949–961. doi: 10.2307/3871221

Dudareva, N., and Pichersky, E. (2006). "Floral scent metabolic pathways: their regulation and evolution," in *Biology of Floral Scent*, eds N. Dudareva and E. Pichersky (Boca Raton, FL: CRC/Taylor and Francis), 55–78.

Enfissi, E. M. A., Fraser, P. D., Lois, L. M., Boronat, A., Schuch, W., and Bramley, P. M. (2005). Metabolic engineering of the mevalonate and non-mevalonate isopentenyl diphosphate-forming pathways for the production of health-promoting isoprenoids in tomato. *Plant Biotechnol. J.* 3, 17–28. doi: 10.1111/j.1467-7652.2004.00091.x

Estévez, J. M., Cantero, A., Reindl, A., Reichler, S., and León, P. (2001). 1-Deoxy-D-xylulose-5-phosphate synthase, a limiting enzyme for plastidic isoprenoid biosynthesis in plants. *J. Biol. Chem.* 276, 22901–22909. doi: 10.1074/jbc.M100854200

Fischer, M. J. C., Meyer, S., Claudel, P., Perrin, M., Ginglinger, J. F., Gertz, C., et al. (2013). Specificity of *Ocimum basilicum* geraniol synthase modified by its expression in different heterologous systems. *J. Biotechnol.* 163, 24–29. doi: 10.1016/j.jbiotec.2012.10.012

Garcia, C. V., Stevenson, R. J., Atkinson, R. G., Winz, R. A., and Quek, S. Y. (2013). Changes in the bound aroma profiles of 'Hayward' and 'Hort16A' kiwifruit (*Actinidia* spp.) during ripening and GC-olfactometry analysis. *Food Chem.* 137, 45–54. doi: 10.1016/j.foodchem.2012.10.002

Garms, S., Kollner, T. G., and Boland, W. (2010). A multiproduct terpene synthase from *Medicago truncatula* generates cadalane sesquiterpenes via two different mechanisms. *J. Organ. Chem.* 75, 5590–5600. doi: 10.1021/jo100917c

Green, S. A., Chen, X., Nieuwenhuizen, N. J., Matich, A. J., Wang, M. Y., Bunn, B. J., et al. (2012). Identification, functional characterization, and regulation of the enzyme responsible for floral (E)-nerolidol biosynthesis in kiwifruit (*Actinidia chinensis*). *J. Exp. Bot.* 63, 1951–1967. doi: 10.1093/jxb/err393

Han, Y., Wang, X., Chen, W., Dong, M., Yuan, W., Liu, X., et al. (2014). Differential expression of carotenoid-related genes determines diversified carotenoid coloration in flower petal of *Osmanthus fragrans*. *Tree Genet. Genomes* 10, 329–338. doi: 10.1007/s00425-010-1222-x

Hellens, R., Allan, A., Friel, E., Bolitho, K., Grafton, K., Templeton, M., et al. (2005). Transient expression vectors for functional genomics, quantification of promoter activity and RNA silencing in plants. *Plant Methods* 1:13. doi: 10.1186/1746-4811-1-13

Hyatt, D. C., Youn, B., Zhao, Y., Santhamma, B., Coates, R. M., Croteau, R. B., et al. (2007). Structure of limonene synthase, a simple model for terpenoid cyclase catalysis. *Proc. Natl. Acad. Sci. U.S.A.* 104, 5360–5365. doi: 10.1073/pnas.0700915104

Kim, S. M., Kim, Y. B., Kuzuyama, T., and Kim, S. U. (2008a). Two copies of 4-(cytidine 5-diphospho)-2-C-methyl-D-erythritol kinase (CMK) gene in *Ginkgo biloba*: molecular cloning and functional characterization. *Planta* 228, 941–950. doi: 10.1007/s00425-008-0794-1

Kim, S. M., Kuzuyama, T., Chang, Y. J., Kobayashi, A., Sando, T., and Kim, S. U. (2008b). 1-Hydroxy-2-methyl-2-(E)-butenyl 4-diphosphate reductase (IDS) is encoded by multicopy genes in gymnosperms *Ginkgo biloba* and *Pinus taeda*. *Planta* 227, 287–298. doi: 10.1007/s00425-007-0616-x

Külheim, C., Yeoh, S. H., Wallis, I. R., Laffan, S., Moran, G. F., and Folev, W. J. (2011). The molecular basis of quantitative variation in foliar secondary metabolites in *Eucalyptus globulus*. *New Phytol.* 191, 1041–1053. doi: 10.1111/j.1469-8137.2011.03769.x

Lane, A., Boecklemann, A., Woronuk, G., Sarker, L., and Mahmoud, S. (2010). A genomics resource for investigating regulation of essential oil production in *Lavandula angustifolia*. *Planta* 231, 835–845. doi: 10.1007/s00425-009-1090-4

Lewinsohn, E., Schalechet, F., Wilkinson, J., Matsui, K., Tadmor, Y., Nam, K. H., et al. (2001). Enhanced levels of the aroma and flavor compound S-linalool by metabolic engineering of the terpenoid pathway in tomato fruits. *Plant Physiol.* 127, 1256–1265. doi: 10.1104/pp.010293

Li, Z., Cao, H., Zhu, G., Gao, J., and Shen, D. (2008). Study on chemical constituents of fragrance releases from fresh flowers of three different *Osmanthus fragrans* Lour. During different florescences. *Chem. Ind. For. Prod.* 28, 75–80.

Mahmoud, S. S., and Croteau, R. B. (2001). Metabolic engineering of essential oil yield and composition in mint by altering expression of deoxyxylulose phosphate reductoisomerase and menthofuran synthase. *Proc. Natl. Acad. Sci. U.S.A.* 98, 8915–8920. doi: 10.1073/pnas.141237298

Martin, D. M., Aubourg, S., Schouwey, M., Daviet, L., Schalk, M., Toub, O., et al. (2010). Functional annotation, genome organization and phylogeny of the grapevine (*Vitis vinifera*) terpene synthase gene family based on genome assembly, FLcDNA cloning, and enzyme assays. *BMC Plant Biol.* 10:226. doi: 10.1186/1471-2229-10-226

Munoz-Bertomeu, J., Arrillaga, I., Ros, R., and Segura, J. (2006). Up-regulation of 1-deoxy-D-xylulose-5-phosphate synthase enhances production of essential oils in transgenic spike lavender. *Plant Physiol.* 142, 890–900. doi: 10.1104/pp.106.086355

Nagegowda, D. A., Gutensohn, M., Wilkerson, C. G., and Dudareva, N. (2008). Two nearly identical terpene synthases catalyze the formation of nerolidol and linalool in snapdragon flowers. *Plant J.* 55, 224–239. doi: 10.1111/j.1365-313X.2008.03496.x

Nieuwenhuizen, N. J., Green, S. A., Chen, X., Bailleul, E. J. D., Matich, A. D., Wang, M. Y., et al. (2013). Functional genomics reveals that a compact terpene synthase gene family can account for terpene volatile production in apple. *Plant Physiol.* 161, 787–804. doi: 10.1104/pp.112.208249

Nieuwenhuizen, N. J., Wang, M. Y., Matich, A. J., Green, S. A., Chen, X., Yauk, Y. K., et al. (2009). Two terpene synthases are responsible for the major sesquiterpenes emitted from the flowers of kiwifruit (*Actinidia deliciosa*). *J. Exp. Bot.* 60, 3203–3219. doi: 10.1093/jxb/erp162

Ohgami, S., Ono, E., Horikawa, M., Murata, J., Totsuka, K., Toyonaga, H., et al. (2015). Volatile glycosylation in tea plants: sequential glycosylations for the biosynthesis of aroma β-primeverosides are catalyzed by two *Camellia sinensis* glycosyltransferase. *Plant Physiol.* 168, 464–477. doi: 10.1104/pp.15.00403

Pechous, S. W., and Whitaker, B. D. (2004). Cloning and functional expression of an (E, E)-α-farnesene synthase cDNA from peel tissue of apple fruit. *Planta* 219, 84–94.

Picone, J. M., Clery, R. A., Watanabe, N., MacTavish, H. S., and Turnbull, C. G. N. (2004). Rhythmic emission of floral volatiles from *Rosa damascena* semperflorens cv. '*Quatre Saisons.*' *Planta* 219, 468–478. doi: 10.1007/s00425-004-1250-5

Reuveni, M., Sagi, Z., Evnor, D., and Hetzroni, A. (1999). β-glucosidase activity is involved in scent production in *Narcissus* flowers. *Plant Sci.* 147, 19–24. doi: 10.1016/S0168-9452(99)00097-7

Schwab, W., Fischer, T. C., Giri, A., and Wüst, M. (2015). Potential applications of glucosyltransferases in terpene glucoside production: impacts on the use of aroma and fragrance. *Appl. Microbiol. Biotechnol.* 99, 165–174. doi: 10.1007/s00253-014-6229-y

Sun, B., Li, L., Han, Y., Zhu, Q., Yuan, W., Dong, M., et al. (2012). HS-SPME-GC-MS analysis of different *Osmanthus fragrans* cultivars from Guilin Garden in Shanghai. *J. Fujian For. Coll.* 32, 39–42.

Tholl, D., Chen, F., Petri, J., Gershenzon, J., and Pichersky, E. (2005). Two sesquiterpene synthases are responsible for the complex mixture of sesquiterpenes emitted from *Arabidopsis* flowers. *Plant J.* 42, 757–771. doi: 10.1111/j.1365-313X.2005.02417.x

Vezzaro, A., Krause, S. T., Nonis, A., Ramina, A., Degenhardt, J., and Ruperti, B. (2012). Isolation and characterization of terpene synthases potentially involved in flavor development of ripening olive (*Olea europaea*) fruits. *J. Plant Physiol.* 169, 908–914. doi: 10.1016/j.jplph.2012.01.021

Wang, L., Li, M., Jin, W., Li, S., Zhang, S., and Yu, L. (2009). Variations in the components of *Osmanthus fragrans* Lour essential oil at different stages of flowering. *Food Chem.* 114, 233–236. doi: 10.1016/j.foodchem.2008.12.014

Wen, Y. Q., He, F., Zhu, B. Q., Lan, Y. B., Pan, Q. H., Li, C. Y., et al. (2014). Free and glycosidically bound aroma compounds in cherry (*Prunus avium* L). *Food Chem.* 152, 29–36. doi: 10.1016/j.foodchem.2013.11.092

Whittington, D. A., Wise, M. L., Croteau, R., and Christianson, D. W. (2002a). Insights into monoterpene cyclization reactions in biology: crystal structure of (+)-bornyl diphosphate synthase. *Biochemistry* 41:8973.

Whittington, D. A., Wise, M. L., Urbansky, M., Coates, R. M., Croteau, R. B., and Christianson, D. W. (2002b). Bornyl diphosphate synthase: structure and strategy for carbocation manipulation by a terpenoid cyclase. *Proc. Natl. Acad. Sci. U.S.A.* 99, 15375–15380. doi: 10.1073/pnas.232591099

Wildung, M. R., and Croteau, R. B. (2005). Genetic engineering of peppermint for improved essential oil composition and yield. *Transgenic Res.* 14, 365–372. doi: 10.1007/s11248-005-5475-2

Xie, Z., Kapteyn, J., and Gang, D. R. (2008). A systems biology investigation of the MEP/terpenoid and shikimate/phenylpropanoid pathways points to multiple levels of metabolic control in sweet basil glandular trichomes. *Plant J.* 54, 349–361. doi: 10.1111/j.1365-313X.2008.03429.x

Xin, H., Wu, B., Zhang, H., Wang, C., Li, J., Yang, B., et al. (2013). Characterization of volatile compounds in flowers from four groups of sweet osmanthus (*Osmanthus fragrans*) cultivars. *Can. J. Plant Sci.* 93, 923–931. doi: 10.4141/cjps2012-333

Yang, Z., Yao, W., and Qian, H. (2005). Effect of β-D-glucosidase on aromatic compounds of sweet osmanthus. *Fine Chem.* 22, 924–926.

Yauk, Y. K., Ged, K., Wang, M. Y., Matich, A. J., Tessarotto, L., Cooney, J. M., et al. (2014). Manipulation of flavour and aroma compound sequestration and release using a glycosyltransferase with specificity for terpene alcohols. *Plant J.* 80, 317–330. doi: 10.1111/tpj.12634

Zhang, H. (2013). *Pollination Biology of Osmanthus fragrans Lour.* Ph.D. dissertation, University of Nanjing Forestry, Nanjing, 16–28.

Zou, J. J., Zhou, Y., Cai, X., and Wang, C. Y. (2014). Increase in DNA fragmentation and the role of ethylene and reactive oxygen species in petal senescence of *Osmanthus fragrans*. *Postharvest Biol. Technol.* 93, 97–105. doi: 10.1016/j.postharvbio.2014.02.015

Conflict of Interest Statement: The authors declare that the research was conducted in the absence of any commercial or financial relationships that could be construed as a potential conflict of interest.

22

Roles of Gibberellins and Abscisic Acid in Regulating Germination of *Suaeda salsa* Dimorphic Seeds Under Salt Stress

Weiqiang Li*[1,2], *Shinjiro Yamaguchi*[3], *M. Ajmal Khan*[4], *Ping An*[5], *Xiaojing Liu*[1] and *Lam-Son P. Tran*[6]

[1] Center for Agricultural Resources Research, Institute of Genetics and Developmental Biology, Chinese Academy of Science, Shijiazhuang, China, [2] Signaling Pathway Research Unit, RIKEN Center for Sustainable Resource Science, Tsurumi, Japan, [3] Department of Biomolecular Sciences, Graduate School of Life Sciences, Tohoku University, Sendai, Japan, [4] Qatar Shell Professorial Chair for Sustainable Development, Centre for Sustainable Development, College of Arts and Sciences, Qatar University, Doha, Qatar, [5] Arid Land Research Center, Tottori University, Hamasaka, Japan, [6] Plant Abiotic Stress Research Group & Faculty of Applied Sciences, Ton Duc Thang University, Ho Chi Minh City, Vietnam; Signaling Pathway Research Unit, RIKEN Center for Sustainable Resource Science, Tsurumi, Japan

Edited by:
Shabir Hussain Wani, Sher-e-Kashmir University of Agricultural Sciences and Technology, India

Reviewed by:
Jie Song, Shandong Normal University, China
Narendra Singh Yadav, Ben-Gurion University of the Negev, Israel

****Correspondence:***
Lam-Son P. Tran
sontran@tdt.edu.vn;
son.tran@riken.jp

Seed heteromorphism observed in many halophytes is an adaptive phenomenon toward high salinity. However, the relationship between heteromorphic seed germination and germination-related hormones under salt stress remains elusive. To gain an insight into this relationship, the roles of gibberellins (GAs) and abscisic acid (ABA) in regulating germination of *Suaeda salsa* dimorphic brown and black seeds under salinity were elucidated by studying the kinetics of the two hormones during germination of the two seed types with or without salinity treatment. Morphological analysis suggested that brown and black are in different development stage. The content of ABA was higher in dry brown than in black seeds, which gradually decreased after imbibition in water and salt solutions. Salt stress induced ABA accumulation in both germinating seed types, with higher induction effect on black than brown seeds. Black seeds showed lower germination percentage than brown seeds under both water and salt stress, which might be attributed to their higher ABA sensitivity rather than the difference in ABA content between black and brown seeds. Bioactive GA_4 and its biosynthetic precursors showed higher levels in brown than in black seeds, whereas deactivated GAs showed higher content in black than brown seeds in dry or in germinating water or salt solutions. High salinity inhibited seed germination through decreasing the levels of GA_4 in both seeds, and the inhibited effect of salt stress on GA_4 level of black seeds was more profound than that of brown seeds. Taken together higher GA_4 content, and lower ABA sensitivity contributed to the higher germination percentage of brown seeds than black seeds in water and salinity; increased ABA content and sensitivity, and decreased GA_4 content by salinity were more profound in black than brown seeds, which contributed to lower germination of black seeds than brown seeds in salinity. The differential regulation of ABA and GA homeostases by salt stress in dimorphic seeds might provide a strategy for *S. salsa* plants to survive adverse environmental conditions.

Keywords: germination, *Suaeda salsa*, gibberellins, abscisic acid, dimorphic seeds, salt stress

INTRODUCTION

Suaeda salsa, a leaf succulent annual herb in Chenopodiaceae family, is a main halophyte species in saline soils of China, which has economical value as being a valuable source of oil, vegetable and fodder (Zhao et al., 2002; Xu et al., 2013; Song and Wang, 2015). This species was also named as *S. maritima* subsp. *salsa* (L.) Soó or *S. liaotungensis* Kit by other reports (Li et al., 2014; Wang et al., 2015). Seed germination of halophytes is a critical stage for population establishment in saline soil and the inner conditions of seeds (Khan and Ungar, 1984; Gul et al., 2013). Like other plant species, the best germination of halophytes is obtained under non-saline conditions, and their germination decreases with the increase in level of salinity (Khan and Ungar, 1984; Gul et al., 2013). High salinity inhibits seed germination by either restricting the supply of water (osmotic effect) or causing specific injury to the metabolic machinery through disturbing the ionic balance (ionic effect) (Bajji et al., 2002; Gul et al., 2013). Thus, for the successful establishment of plants in saline environments, seeds must remain viable at high salinity and germinate when salinity decreases (Gul et al., 2013).

Abscisic acid (ABA) and gibberellins (GAs) are well-known phytohormones that are involved in regulating seed germination in *Arabidopsis*. These two hormones regulate seed germination in opposite manner. While ABA inhibits seed germination, GAs promote this biological process (Finkelstein et al., 2008; Piskurewicz et al., 2008; Yamaguchi, 2008; Nambara et al., 2010). It was reported that ABA biosynthetic mutants can germinate in the presence of salt, and fluridone (an inhibitor of ABA biosynthesis) can alleviate germination of various plants under salinity (Cheng et al., 2002; Ruggiero et al., 2004), suggesting that ABA biosynthesis might be affected by salinity during germination. With regard to GAs, bioactive GA_4 is synthesized by two enzymes GA20ox and GA3ox (GA 20-oxidase and GA 3-oxidase) from GA_{12} in *Arabidopsis*. GA20ox converts GA_{12} to GA_{15}, then to GA_{24}, and finally to GA_9 that is subsequently converted to GA_4 by GA3ox. On the contrary, GA_9 and GA_4 are catalyzed to GA_{51} and GA_{34}, respectively, by GA2ox (GA 2-oxidase) (Hedden and Thomas, 2012). Previous studies in *Arabidopsis* have showed that salt stress up-regulated genes involved in GA inactivation (e.g., *GA2ox7* gene) and suppression of GA signaling (e.g., DELLA proteins encoding genes) (Achard et al., 2006; Kim et al., 2008; Magome et al., 2008; Yuan et al., 2011; Colebrook et al., 2014). However, little is known about the effects of salt stress on the biosynthesis of the active GA_4.

It has been demonstrated that seed polymorphism provides an adaptive advantage in saline habitats through the production of multiple germination periods, which increases the chances of survival of at least some seedling cohorts (Imbert, 2002; Matilla et al., 2005). The two seed morphs of *S. salsa* (brown and black seeds) can provide multiple germination cohorts in saline habitats, providing opportunities for this species to establish next generation (Khan et al., 1998; Li et al., 2008). However, the regulatory functions of these two hormones in differential germination of dimorphic seeds under salt stress remains elusive.

Thus, in this study we examined the effects of salinity on the GA and ABA metabolism during seed germination to assess the roles of these two hormones in regulating germination of dimorphic seeds of *S. salsa* under salinity. Our findings showed that dimorphic seeds showed significant differences in the content of ABA and GAs, as well as in the sensitivity to ABA. Furthermore, our results suggest that salinity differentially inhibits germination of dimorphic seeds by differentially affecting the level of endogenous ABA and GA_4 in the seeds, as well as the sensitivity of the seeds to ABA.

RESULTS

Visible Events During Dimorphic Seed Germination

Brown seeds are enclosed by brittle outer coat, thus absorb water quickly (**Figures 1A–D,I,L**), while black seeds are enclosed by outer layer of the thin coat, thus they show hard, waxy and un-wettable appearance (**Figures 1E–H,M,P**). Both seeds have a brown inner coat on the underside (**Figures 1J,N**), and black seeds contain more endosperms than brown seeds (**Figures 1J,L,N,P**). The embryo of brown seed is bigger than black seeds with brown in color, while that of black seed is white (**Figures 1K,O**).

For black seeds, black outer coat opening (**Figure 1G**) was the first visible event, leading to radicle protrusion (germination, **Figure 1H**) of inner coat when absorbing water. For brown seeds, the process was not so clear because the outer coat was very soft after imbibition (**Figures 1C,D**). Seedlings from dimorphic seeds are similar in morphological features (**Figure 1Q**). Three-day-old seedlings germinated from black seeds were slightly smaller than those from brown seeds (**Figure 1R**). Hypocotyl and cotyledons of seedlings from black seeds were slightly redder than those from brown ones (**Figures 1Q,R**). After growing for 3 weeks in the growth chamber, there was significant difference in height and appearance between the seedlings germinated from two types of seeds (**Figure 1S**). Seedlings from brown seeds were longer (3.76 ± 0.16 cm) than those from black seeds (3.18 ± 0.22 cm).

Germination of Dimorphic Seeds in Response to ABA Under Water and Salinity

Brown seeds showed higher germination percentage than black seeds in water or NaCl solution (**Figure 2A**). Germination of brown seeds was earlier and faster than that of black seeds in both water and NaCl solutions (**Figure 2B**). With the increase in concentration of NaCl, the onset of germination delayed and the germination percentage decreased for both types of seeds (**Figure 2B**). In the presence of 600 mM NaCl, germination percentage of black seeds was almost 0%, whereas approximately 30% of brown seeds were able to germinate (**Figures 2A,B**).

To examine whether the synthesis of endogenous ABA in germinating seeds was affected by salinity which in turn influenced the germination percentage, fluridone (a potent inhibitor of ABA biosynthesis) was first used in a germination

FIGURE 1 | Morphology of dimorphic seeds (black and brown) and seedlings of *Suaeda salsa*. (A–H) Visible events during the incubation of *S. salsa* brown and black seeds in water; **(A,E)** Dry brown and black seeds; **(B,C,F,G)** Brown and black seeds in water for 10 min and 10 h; **(D,H)** Radicle emergence through the seed covering layers (seed coat and endosperm) signals the completion of germination after 24 h of imbibition in water. **(I,M)** Outer coat was partly removed from brown and black seeds after 10 min of imbibition in water. Red arrows indicate outer coat removed from seeds. **(J,N)** The embryo and endosperms are visible after outer coat of brown and black seeds was fully removed. Red arrows indicate endosperms. **(K,O)** Removal of inner coat after 10 min of imbibition in water showed coiled embryo of brown and black seeds. **(L,P)** Section through dry black and brown seeds. Red arrow indicate endosperms. **(Q)** Distinct stages of *S. salsa* brown (upper) and black (lower) seeds during germination, and developed seedlings after 0, 6, 12, 24, 36, and 48 h of imbibition in water. **(R)** 3-day-old seedlings from brown and black seeds. **(S)** 3-week-old seedlings from brown and black seeds.

assay under salt stress. As shown in **Figure 2C**, addition of 10 μM fluridone rescued the germination of brown and black seeds of *S. salsa* in the presence of 600 mM NaCl, suggesting that ABA biosynthesis might be induced by salinity in germinating seeds, which would lead to inhibition of germination. To verify this hypothesis, both fluridone and exogenous ABA were used in a germination assay of *S. salsa* seeds with and without salt stress. As shown in **Figure 2D**, treatment of seeds with 200 mM NaCl in the presence of 10 μM fluridone and various concentrations of ABAs inhibited germination of black seeds more significantly than that of brown seeds. Additionally germination percentage of black seeds in the presence of 10 μM fluridone and various concentrations of ABA without NaCl was remarkably lower than that of brown seeds. These results suggested that black seeds are more sensitive to both exogenous ABA and salt stress-induced endogenous ABA than brown seeds.

ABA Contents in Dimorphic Seed During Germination Under Water and Salinity

As a means to check whether the endogenous ABA content was indeed increased by salinity in germinating seeds which subsequently decreased the germination percentage, in the next line of our study ABA contents were determined in germinating dimorphic seeds under both water and salinity conditions. ABA levels were found to be higher in brown seeds than in black seeds during the time course of imbibition to water and salt solutions (**Figure 3**). Although the ABA contents gradually decreased in both types of germinating seeds, salinity treatments enhanced ABA levels during germination. Moreover, the induction effect of salt stress on ABA accumulation was much earlier in black seeds than brown seeds. For instance, ABA levels in black seeds treated with 200 and 600 mM

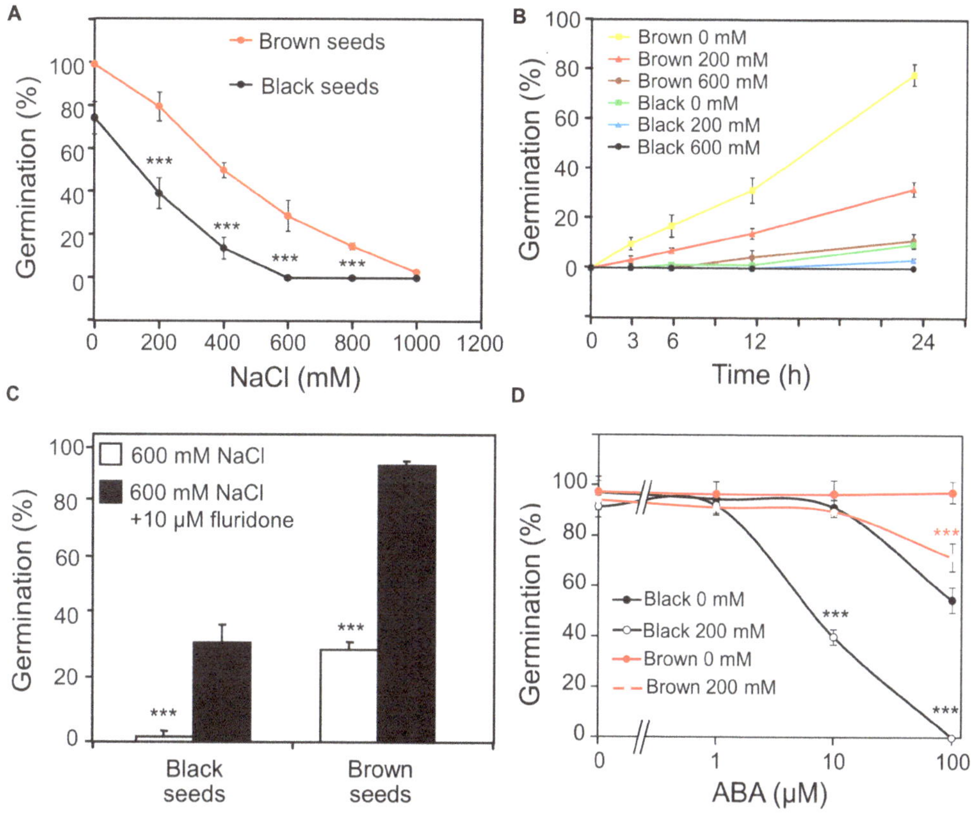

FIGURE 2 | Effect of ABA on germination of *Suaeda salsa* brown and black seeds under normal and salt stress conditions. (A) Germination percentage of brown and black seeds of *S. salsa* treated with 0, 200, 400, 600, 800, or 1000 mM NaCl solution. **(B)** Germination percentage of brown and black seeds of *S. salsa* at 3, 6, 12, and 24 h after imbibition in 0, 200, or 600 mM NaCl solution. **(C)** Germination percentage of brown and black seeds of *S. salsa* treated with 600 mM NaCl solution in the presence or absence of 10 μM fluridone. **(D)** Germination percentage of brown and black seeds of *S. salsa* treated with 0 or 200 mM NaCl solution in the presence of 10 μM fluridone and 1, 10, and 100 μM ABA. The brown and black seeds were treated with 0, 200, 400, 600, 800, or 1000 mM NaCl solution under continuous light at 22°C, and germination percentages were determined after 5 days of treatment. Percentage means and SE values (error bars) were calculated from the results of three independent experiments ($n > 100$ seeds/seed type/experiment). Asterisks indicate significant differences as determined by a Student's *t*-test (***$P < 0.001$).

NaCl were higher than that of water control after 6 h of imbibition, while ABA levels in brown seeds were induced by salt stress only after 12 h of treatment (**Figure 3**). Addition of fluridone to the germinating solutions further reduced the ABA contents in both brown and black seeds with or without salt stress, with higher ABA contents being observed in both seed types treated with salt stress + fluridone solution than with water control + fluridone solution (**Figure 3**). These results together might suggest that salinity inhibits germination of dimorphic seeds at least in part by inducing ABA accumulation, which happens earlier in black seeds than brown ones.

Germination of Dimorphic Seeds in Response to GA_1 and GA_4 Under Water and Saline Conditions

GA_1 and GA_4 are two active forms of GAs in many plants (Yamaguchi, 2008). First, to examine which form is more important for the germination of *S. salsa* seeds under water and high salinity conditions, we evaluated the promotion effects of GA_1 and GA_4 on the germination of dimorphic seeds in the presence of GA biosynthesis inhibitor paclobutrazol (50 μM). As shown in **Figures 4A,B**, 1 μM GA_4 significantly promoted brown and black seed germination, while GA_1 did not even at concentration of 50 μM. This result demonstrated that GA_4 was more active than GA_1 in promoting germination of both brown and black seeds treated with water, suggesting that both seed types are more sensitive to GA_4 than GA_1.

Thus, in the next line of our study, we investigated the promotion effect of GA_4 on the germination of the two seed types under salt stress. In the presence of NaCl and paclobutrazol, results indicated that salt stress reduced the promotion effect of GA_4 on germination percentage of both two seed types, which was found to be more severe on black seeds than brown seeds (**Figure 4C**). These results suggested that salinity decreased GA_4 response of both black and brown seeds and the negative effect of salt stress on GA_4 response of black seeds was more profound than on that of brown seeds.

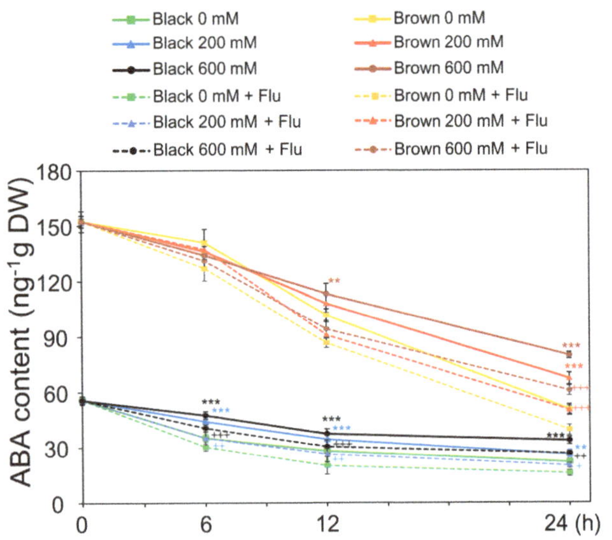

FIGURE 3 | Abscisic acid (ABA) contents in *Suaeda salsa* brown and black seeds of during germination process under normal and salt stress conditions. The brown and black seeds were treated with 0, 200, or 600 mM NaCl solution in the presence or absence of 10 μM fluridone (Flu) under continuous light at 22°C. After 0, 6, 12, and 24 h of treatment, the seed germination was examined and seed samples were harvested to determine ABA content. Means and SE values (error bars) were calculated from the results of three independent experiments. Asterisks indicate significant differences in ABA content between salt-treated and water-treated samples in the presence of Flu (**$P < 0.01$ and ***$P < 0.001$) or in the absence of Flu ($^{+}P < 0.05$; $^{++}P < 0.01$, and $^{+++}P < 0.001$) as determined by a Student's *t*-test. DW, dry weight.

GA Contents in Dimorphic Seeds During Germination Under Water and Salinity

To check whether the GA content were affected by salinity which subsequently altered the germination percentage, GA contents were determined in germinating dimorphic seeds under both water and salinity conditions. In dry seeds, the contents of the bioactive GA_4 and its biosynthetic precursors (GA_{12}, GA_{15}, GA_{24}, and GA_9) were higher in brown seeds than black seeds (**Figure 5A**; 0 h). On the contrary, the levels of deactivated GAs (such as GA_{51} and GA_{34}) were higher in black seeds than brown seeds (**Figure 5A**; 0 h).

After imbibition in water, the levels of GA_4 biosynthetic precursors (GA_{12}, GA_{15}, GA_{24}, and GA_9) in brown seeds quickly increased and reached the peak at 6 h, followed by a decrease. Addition of 600 mM NaCl into germination solution significantly inhibited GA_9 and GA_{12} syntheses in brown seeds after 3 and 6 h of treatment, respectively (**Figure 5B**). However, salt stress increased the levels of GA_{15} and GA_{24} after 3 h of treatment, as well as induced the accumulation of GA_9 after 12 h of exposure to both 200 and 600 mM NaCl (**Figures 5A,B**). On the other hand, the levels of GA_4 maintained the increasing tendency in brown seeds treated with water during the time course of treatment, whereas excessive salt of either 200 or 600 mM reduced the GA_4 accumulation (**Figures 5A,B**). Similar to what was observed in brown seeds, the levels of GA_4 biosynthetic precursors in black seeds increased in water after imbibition (**Figure 5A**). However, this increase showed slower tendency in black seeds than brown seeds. For example, content of GA_9 in black seeds reached the peak at 12 h, while that in brown seeds was maximum at 6 h after being treated in water (**Figure 5A**). Addition of NaCl into germination solution significantly inhibited GA_{12} and GA_{15} and GA_{24} syntheses in black seeds after 12 h, and 24 h of treatment, respectively (**Figures 5A,B**). However, salt stress induced the accumulation of GA_{12} after 6 h and GA_9 after 12 and 24 h of exposure to NaCl (**Figures 5A,B**). On the other hand, the levels of GA_4 maintained the increasing tendency in black seeds treated with water during the time course of treatment, whereas salt stress reduced the GA_4 accumulation. More specifically, salt stress more significantly reduced GA_4 levels in black than brown seeds. For instance, 600 mM NaCl reduced GA_4 content of brown seeds to 74% comparing with water control, while to 28% in black seeds at 12 h after imbibition (**Figure 5B**). In addition, germinating black seeds always contained lower levels of biosynthetic precursors and GA_4 than brown seeds during germination within the period of 24 h after imbibition (**Figures 5A,B**). These data collectively suggested that (i) brown seeds were more active than black seed in biosynthesis of GA_4 before and after imbibition in water, (ii) salt stress inhibited the biosynthesis of GA_4, consequently resulting in accumulation of its precursors in both black and brown seeds, and (iii) salt stress more significantly reduced the levels of GA_4 in black than brown seeds comparing with water control.

Although black seeds contained lower levels of biosynthetic precursors and GA_4 than brown seeds, the levels of deactivated GAs (GA_{51} and GA_{34}) were higher in black seeds than brown seeds after imbibition in water. High salinity significantly increased the levels of GA_{51} and GA_{34} than water in germinating black and brown seeds after 12 h of imbibition. These results suggested that black seeds were more active than brown seeds in deactivating GAs, and NaCl stress promoted deactivation of GAs in both black and brown seeds.

DISCUSSION

More than 200 species have been identified to have seed polymorphism/dimorphism with differences in seed coat properties, seed size, germination percentage and/or dormancy (Imbert, 2002; Baskin et al., 2014). Our results showed than the *S. salsa* dimorphic seeds differ not only in seed coat properties (**Figures 1I,M**) and germination ability, but also in embryo color (**Figures 1K,O**), endosperm thickness (**Figures 1J,L,N,P**) and their seedling growth (**Figures 1Q–S**). Furthermore, our study indicated the existence of the endosperms in both black and brown seeds, which was not reported by recent reports (Song et al., 2008; Song and Wang, 2015). The hard, waxy and un-wettable outer coat (**Figures 1F,M**) may prevent entry of water into embryo (Li et al., 2005; Wang et al., 2012), resulting in delay in embryo growth and thus germination (**Figures 2A,B**). Black seeds contain more endosperms than brown seeds, which might also inhibit embryo to absorb water, and thus preventing or delaying germination. In addition to the differences in seed coat morphology, embryos and seedlings form black seeds

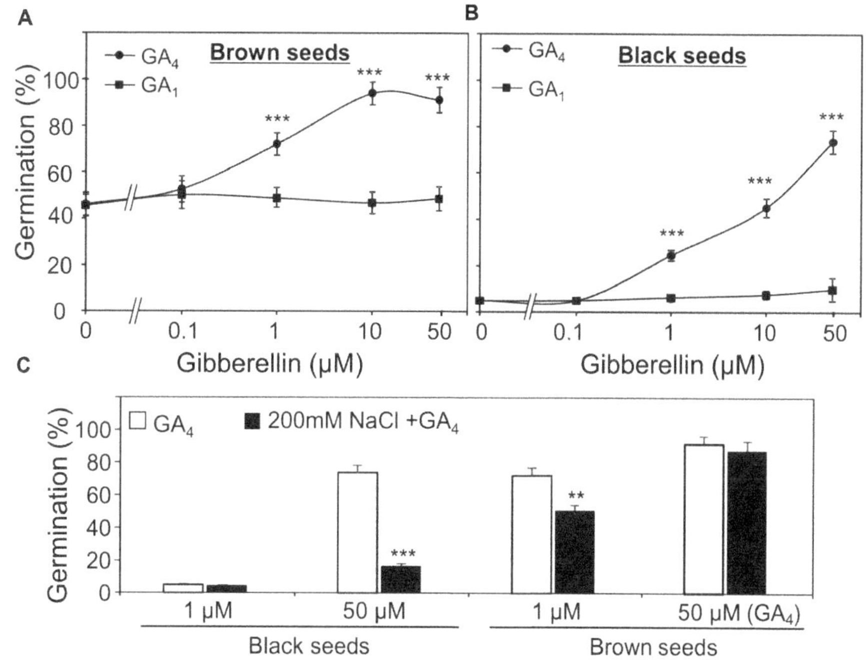

FIGURE 4 | Effect of gibberelline (GA) on germination of *Suaeda salsa* brown and black seeds under normal and salt stress conditions. Germination percentage of brown **(A)** and black **(B)** seeds in the present of 50 μm paclobutrazol and indicated concentrations of GA_4 or GA_1 under non-stressed conditions. The brown and black seeds were treated with 0, 0.1, 1, 10, or 50 μM GA_1 or GA_4 solution containing 50 μM paclobutrazol under continuous light at 22°C. Germination percentages were determined after 5 days of treatment. **(C)** Germination percentage of brown and black seeds in the present of 50 μm paclobutrazol and indicated concentrations of GA_4 under salt stress conditions. The brown and black seeds were treated with 1 or 50 μM GA_4 solution containing 50 μM paclobutrazol with or without 200 mM NaCl under continuous light at 22°C. Germination percentages were determined after 5 days of treatment. Percentage means and SE values (error bars) were calculated from the results of three independent experiments (n > 100 seeds/seed type/experiment). Asterisks indicate significant differences as determined by a Student's t-test (**$P < 0.01$ and ***$P < 0.001$).

displayed different in size and color phenotype than those from brown seeds (**Figures 1K,O,Q,R**), suggesting that the embryos and seedlings form black seeds might contain different levels of pigments as compared with those from brown seeds, which would be interesting to further investigate in future. Taken together, the observation that brown seeds had larger seed size, especially embryo size and color, and better developed embryo (especially cotyledon, **Figures 1K,O**) than black seeds implied that the black and brown seeds were in different development stage, when they were formed in their mother plants, which might result in difference in their germination ability, seedling growth, and perhaps stress tolerability as well.

In the present study, we showed that the inhibition of germination of both black and brown seeds by salt stress (**Figures 2A,B**) coincided with the enhanced ABA biosynthesis (**Figure 3**), which was evidenced by a germination assay under salt stress in the presence of fluridone with and without ABA (**Figures 2C,D**). The data indicated that fluridone alleviated the negative effect of salinity on *S. salsa* seed germination (**Figure 2C**), suggesting that salt stress inhibits germination by inducing ABA synthesis, as reported in a number of plant species, such as *Arabidopsis* and rice (Cheng et al., 2002; Xiong and Zhu, 2003; Gianinetti and Vernieri, 2007). In addition, brown seeds showed higher germination percentage and faster germination than black seeds, which might be attributed to their insensitivity to ABA in comparison with black seeds. As showed in **Figure 2D**, even 100 μM ABA could not inhibit brown seed germination in water, whereas this concentration of ABA effectively inhibited germination of black seeds by approximately 50%. Salinity increased ABA response in both black and brown seeds, with higher ABA sensitivity observed in black seeds relative to brown seeds in the present of salinity (**Figure 2D**), which might be related to differential function of protein(s) homologous to ABI5 (ABA insensitive 5) of *Arabidopsis* (Yuan et al., 2011), resulting in lower germination percentage of black seeds than brown seeds under salinity (**Figure 2D**). Our results collectively demonstrate that (i) ABA sensitivities of black and brown seeds are different, and (ii) salinity increases the differential ABA responses of these two seed types, perhaps by differentially inducing endogenous ABA level, thereby differentially affecting germination of dimorphic seeds under salinity.

Although brown seeds germinated faster and showed higher germination percentage than black seeds in water and salinity (**Figures 2A,B**), surprisingly, brown seeds were found to contain higher level of ABA than black seeds (**Figure 3**). This finding was also supported by a previous study, in which the authors used the enzyme-linked immunosorbent assay to determine the ABA content (Wang et al., 2015). One possible explanation for the faster and higher germination percentage of brown seeds relative to black seeds is that brown seeds were insensitive to ABA

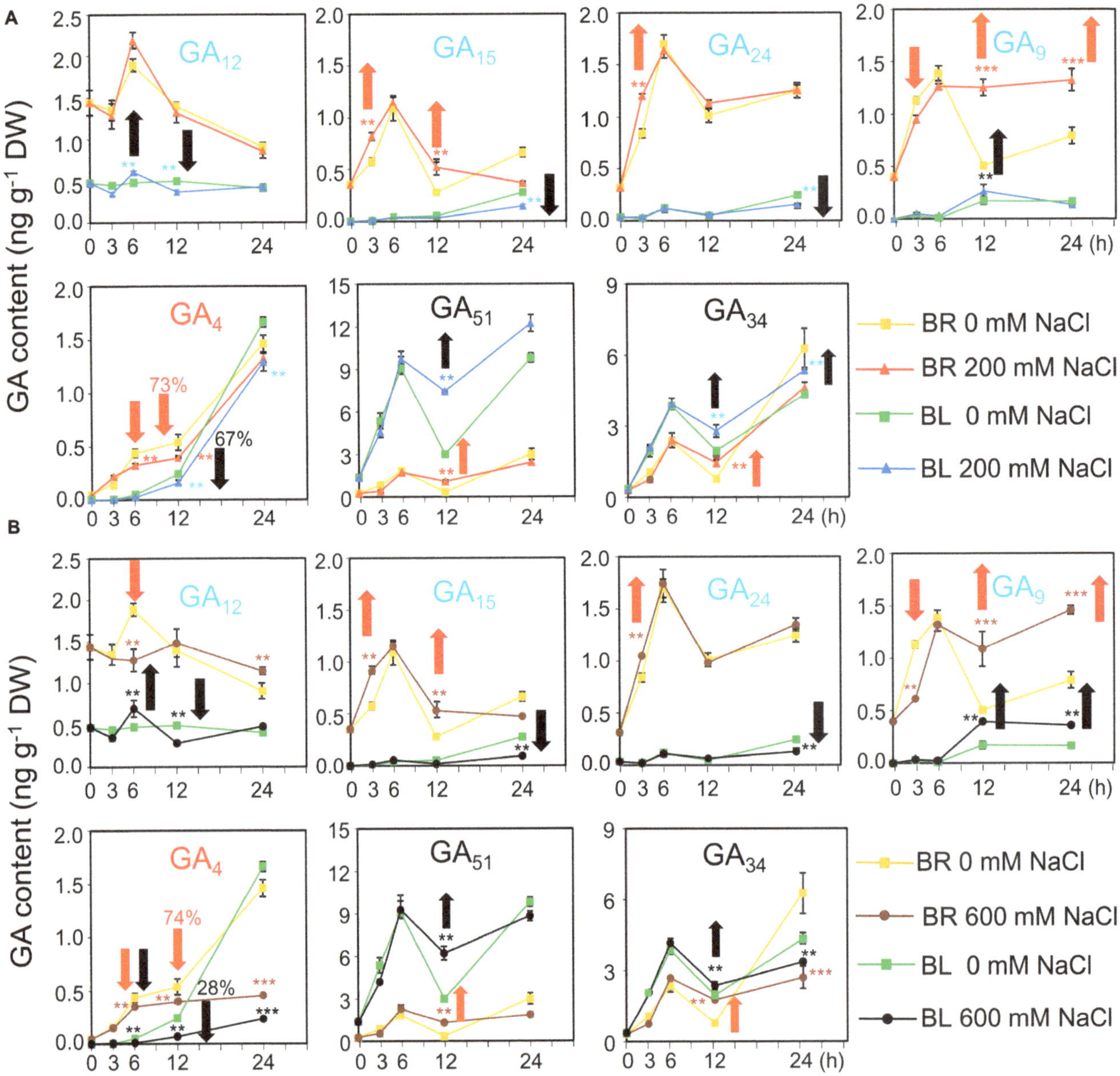

FIGURE 5 | Gibberelline (GA) contents of brown and black seeds of *Suaeda salsa* during germination process under normal and salt stress conditions. The brown (BR) and black (BL) seeds were treated with **(A)** 0 and 200 or **(B)** 0 and 600 mM NaCl solution under continuous light at 22°C. After 0, 3, 6, 12, and 24 h of treatment, seed samples were harvested to determine the GA contents. GA_{12}, GA_{15}, GA_{24}, and GA_9 (blue color) are biosynthetic precursors of bioactive GAs. GA_4 (red color) is a bioactive GA form, whereas GA_{51} and GA_{34} (black color) are deactivated forms. Red and black arrows indicate responses of brown and black seeds to salinity, respectively. Upward and downward arrows indicate increase and decrease of GA contents by salt stress, respectively. Numbers above the arrows indicate the GA reduction percentage by salt stress. Means and SE values (error bars) were calculated from the results of three independent experiments. Asterisks indicate significant differences as determined by a Student's *t*-test (**$P < 0.01$ and ***$P < 0.001$). DW, dry weight.

(**Figure 2D**). Furthermore, the higher ABA content observed in brown seed type might contribute to make the brown seeds larger than black seeds (Li et al., 2008; Wang et al., 2015), because ABA promotes synthesis of seed storage proteins and lipids (Finkelstein et al., 2002; Kim et al., 2014). This finding also strengthens that brown and black seeds are in different developmental stages on *S. salsa* plants (**Figure 1**). Additionally, measurement of endogenous ABA content indicated that salt stress more significantly increased ABA content in black than brown seeds at early germination stage (**Figure 3**), which might lead to delay or inhibition of germination of black seeds in comparison with brown seeds (**Figures 2A,B**). Our finding suggests that the induction rate of endogenous ABA content in *S. salsa* dimorphic seeds by salt stress has more effects than the absolute ABA content on dimorphic seed germination during salt stress.

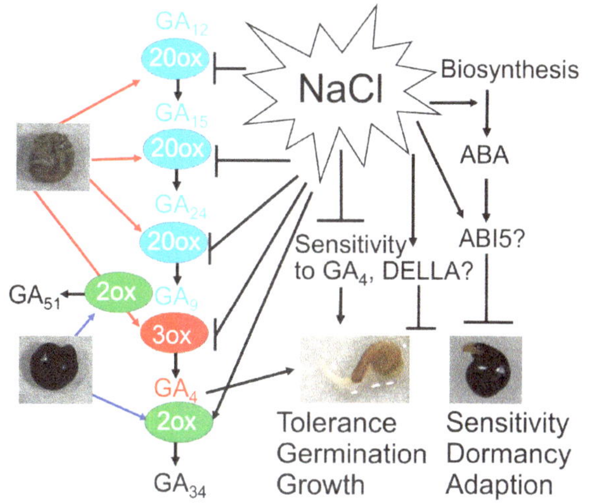

FIGURE 6 | Possible mechanisms regulating *Suaeda salsa* dimorphic seed germination under salt stress. GA_4 (red color) is synthesized from precursors GA_{12}, GA_{15}, GA_{24} and GA_9 (blue color) by GA20ox (20ox) and GA3ox (3ox) enzymes. Bioactive GA_4 is deactivated to inactive form GA_{51} and GA_{34} (black color) by GA2ox (2ox) enzyme. Brown seeds are more active in GA biosynthesis and contain higher content of GA precursors and bioactive GA_4 than black seeds. Black seeds are more active in deactivation of GA_4 and contain higher content of inactive GAs than brown seeds. Salt stress inhibits GA biosynthesis and promotes GA deactivation in both brown and black seeds. In addition, salt stress promotes ABA biosynthesis and increase ABA response. Salinity decreases sensitivity to GA_4 probably through affecting the function of DELLA proteins. Brown seeds have better salt tolerance, and faster germination and growth rates than black seeds. On the other hand, dormancy of black seeds might be an adaptation strategy that enables the seeds to avoid adverse environmental conditions. Bars indicate negative regulation, while arrows indicate positive regulation.

In contrast to ABA, GAs have been known to play positive regulatory role in seed germination (Finkelstein et al., 2008; Nambara et al., 2010; Shu et al., 2015). In the present study, by using the GA biosynthesis inhibitor paclobutrazol in the germination assays, GA_4 was shown to be the more biological active form than GA_1 in the *S. salsa* seed germination (**Figure 4**), which was supported by previous studies in which germination assays were performed without paclobutrazol (Li et al., 2005, 2008). We observed that salinity decreased GA_4 sensitivity of both black and brown seeds (**Figure 4C**) and the negative effect was more profound in black than brown seeds (**Figure 4C**), which might contribute to differential germination of dimorphic seeds under salinity. These results imply that GA response-related proteins might differentially regulate GA responses in black and brown seeds during germination under salt stress. As showed in **Figure 6**, the downstream factors of GA signaling pathway, such as *S. salsa* homologs of *Arabidopsis* DELLA proteins, might be candidate regulators as the *Arabidopsis* quadruple *della* mutant was shown to be sensitive to salt stress during germination (Achard et al., 2006; Colebrook et al., 2014). It would be interesting to carry out further genetic studies in *S. salsa* in future to gain more understanding of the roles of ABA and GA signaling pathways in regulation of the germination of black and brown seeds. Sequencing of *S. salsa* genome and development of its mutant resource should be taken to advance genetic studies in this important halophyte plant.

Analysis of the metabolism of GA_4 in both black and brown seeds during germination strengthens the positive role of GA_4 in promoting dimorphic seed germination in water and salt solutions. Dry and germinating brown seeds contained higher levels of GA_4 than black seeds (**Figure 5**), which was consistent with the finding that brown seeds displayed quicker and higher germination percentage than black seeds under water, excessive salt and excessive ABA conditions (**Figures 2A–D**). In addition, higher levels of GA_4 precursors in germinating brown than black seeds under water and salt stress treatments (**Figures 5A,B**) implied that GA biosynthesis activity is higher, perhaps due to the enhanced activities GA biosynthesis enzymes (e.g., GA20ox and GA3ox), in germinating brown than black seeds, which also supports the differential development stages of the two seed types (**Figure 1**). On the contrary, deactivated GAs forms from GA_9 and GA_4 (e.g., GA_{51} and GA_{34}) were higher in germinating black seeds than brown seeds in water and salt solutions (**Figure 6**), suggesting that germinating black seeds possess higher GA deactivated activities than brown seeds. These data collectively demonstrate that the germinating brown seeds have higher GA biosynthesis activity but lower inactivation activity than black seeds, which may contribute to higher germination percentage and faster germination rate of brown seeds in comparison with black seeds in water and salt solutions (**Figure 5**).

As GA_4 strongly binds to the GA receptor and is directly involved in GA signaling, maintenance of an appropriate level of endogenous GA_4 is the most critical for seed germination (Sun, 2008). Thus, the balance in biosynthesis and deactivation of GA_4, which is strongly affected by salinity (Hedden and Thomas, 2012), is very important in controlling the GA_4 homeostasis in the early stage of seed germination (Yamaguchi, 2008). Salinity more highly induces GA_{34} in black than brown seeds (**Figures 5A,B**), implying that salt stress promotes deactivation of GA_4 through upregulation of GA-deactivating GA2ox activity in *S. salsa* in a similar process as in *Arabidopsis*, especially in black seeds (Magome et al., 2008; Colebrook et al., 2014) (**Figure 6**). Our results showed that the effects of salt stress on the levels of some GA biosynthesis precursors in both brown and black seeds were significantly different at various time points (**Figures 5A,B**), suggesting that salt stress differentially affects the biosynthesis of active GA_4 in brown and black seeds. Salt stress decreased the content of GA_9 and increased that of their precursors GA_{15} and GA_{24} in brown seeds after 3 h of imbibition (**Figures 5A,B**), indicating that salinity inhibited the conversion of GA_{15} to GA_{24} and GA_{24} to GA_9 with the involvement of GA20ox (Yamaguchi, 2008), suggesting that salt stress might affect activity of GA20ox in brown seeds after 3 h of imbibition (**Figure 6**). However, there is no difference in GA content in black seeds after 3 h of imbibition in salt stress, implying that GA20ox activity in black might not be affected by salt stress at this time point. GA3ox is involved in the final step of synthesis of GA_4 from GA_9, and in most case, the final step is the main site for active regulation by environmental factors (Yamauchi et al., 2004). The decrease and increase in GA_4 and GA_9 contents, respectively, in both black and brown seeds at 12 h after imbibition in salt solution suggested that salinity inhibited the conversion of GA_9 to GA_4, in which GA3ox is involved (**Figure 6**). Therefore, we supposed that salinity inhibited GA3ox activity in both black and brown seeds after 12 h of imbibition in salt solution. Salinity induced lower level of GA_4 in black than brown seeds, which might contribute to reduced germination of black seed relative to brown seeds under salinity. Collectively, our results suggest that salinity inhibits GA_4 biosynthesis perhaps through reducing GA20ox and GA3ox activities, as well as promotes inactivation of GA_4 probably through increasing GA2ox activity in *S. salsa* dimorphic germination seeds, resulting in a decrease in endogenous level of GA_4 (**Figure 6**). And, this effect of salt stress is more severe on black than brown seeds.

CONCLUSION

Brown and black seeds are in different development stages, when harvested from *S. salsa* plants, and thus they show different GA and ABA homeostases, resulting in their differential germination. In addition, the differential regulation of ABA and GA kinetics by salt stress observed between brown and black seeds results in lower sensitivity of brown seeds to salinity during germination in comparison with black seeds, thereby providing different adapting strategies of dimorphic seeds to saline environment.

MATERIALS AND METHODS

Plant Materials

Seeds of *S. salsa* were collected during the fall of 2011 in the coastal saline soils of Huanghua City, Hebei Province, China. The seeds were then dried for a few days, and stored in paper bags in laboratory at approximately 20°C and 30–40% relative humidity. Black and brown seeds were manually separated from the inflorescence.

Microscopy and Image Preparation

Black and brown seeds of *S. salsa* were incubated under standard conditions, and appropriate developmental stages were selected for taking photos.

Germination Assays

Germination was assayed at 22°C under continuous light (fluorescent lamp with approximately 100 mmol $m^{-2}s^{-1}$ light intensity) in triplicates. Seeds were washed with 0.02% Triton X solution, rinsed with distilled water, then washed with respective working solutions. Subsequently, 50–100 seeds were placed on a double layer of wet filter papers within a plastic petri dish and closed. Seeds were scored as germinated when primary root protrusion was visible. To determine the effect of salinity on germination, NaCl solutions of 0, 200, and 600 mM were used for brown and black seeds. Seed germination was counted at 3, 6, 12, 24, and 120 h.

To evaluate the effect of ABA inhibitor fluridone on germination of dimorphic seeds under salinity, 10 μM fluridone was used in germination assays of brown and black seeds in 600 mM NaCl solution. To determine the response (sensitivity) of dimorphic seeds to ABA under salt stress, ABA was used in 0, 1, 10, and 100 μM concentrations in germination assays of brown and black seeds in water or 200 mM NaCl solutions in the presence of 10 μM fluridone. To determine the response (sensitivity) to GA_1 and GA_4 of dimorphic seeds in salinity, GA_1 and GA_4 solutions of 0, 0.1, 1, 10, and 50 μM were used in germination assays of brown and black seeds in water and 200 mM NaCl solutions in the presence of 50 μM GA biosynthesis inhibitor paclobutrazol. After 5 days of cultivation, the germination seeds were counted. To determine the average percentage of germination, more than 100 seeds were used for each treatment. The experiments were independently repeated three times.

Analyses of Endogenous Hormone Contents

For hormone analyses, seeds were germinated under the same environmental conditions as for the germination assays. For ABA analysis 50 mg of brown and black dry seeds were incubated in 0, 200, and 600 mM NaCl solutions in the present or absence of 10 μM fluridone. Samples were then collected at 0, 6, 12, and 24 h in three replicates, immediately frozen in liquid nitrogen and kept at −80°C until analysis. For GA analysis, 500 mg of brown and black dry seeds were incubated in 0, 200, and 600 mM NaCl solutions. Samples were then collected at 0, 3, 6, 12, and 24 h in three replicates, immediately frozen in liquid nitrogen and kept at −80°C until analysis. Purification and determination of ABA and GA contents were carried out by LC-MS/MS analysis as previously described (Saika et al., 2007; Li et al., 2011).

AUTHOR CONTRIBUTIONS

WL conceived research. WL performed the experiments and analyzed the data. SY, PA, and XL provided research materials. WL, MK, and L-ST wrote the manuscript.

ACKNOWLEDGMENTS

This work was supported in part by National High Technology R&D program of China (Project Code: 2013BAC02B01).

REFERENCES

Achard, P., Cheng, H., De Grauwe, L., Decat, J., Schoutteten, H., Moritz, T., et al. (2006). Integration of plant responses to environmentally activated phytohormonal signals. *Science* 311, 91–94. doi: 10.1126/science.1118642

Bajji, M., Kinet, J. M., and Lutts, S. (2002). Osmotic and ionic effects of NaCl on germination, early seedling growth, and ion content of *Atriplex halimus* (Chenopodiaceae). *Can. J. Bot.* 80, 297–304. doi: 10.1139/b02-008

Baskin, J. M., Lu, J., Baskin, C. C., Tan, D., and Wang, L. (2014). Diaspore dispersal ability and degree of dormancy in heteromorphic species of cold deserts of northwest China: a review. *Perspect. Plant Ecol.* 16, 93–99. doi: 10.1016/j.ppees.2014.02.004

Cheng, W. H., Endo, A., Zhou, L., Penney, J., Chen, H. C., Arroyo, A., et al. (2002). A unique short-chain dehydrogenase/reductase in *Arabidopsis* glucose signaling and abscisic acid biosynthesis and functions. *Plant Cell* 14, 2723–2743. doi: 10.1105/tpc.006494

Colebrook, E. H., Thomas, S. G., Phillips, A. L., and Hedden, P. (2014). The role of gibberellin signalling in plant responses to abiotic stress. *J. Exp. Biol.* 217, 67–75. doi: 10.1242/jeb.089938

Finkelstein, R. R., Gampala, S. S., and Rock, C. D. (2002). Abscisic acid signaling in seeds and seedlings. *Plant Cell* 14, S15–S45. doi: 10.1105/tpc.010441

Finkelstein, R., Reeves, W., Ariizumi, T., and Steber, C. (2008). Molecular aspects of seed dormancy. *Annu. Rev. Plant Biol.* 59, 387–415. doi: 10.1146/annurev.arplant.59.032607.092740

Gianinetti, A., and Vernieri, P. (2007). On the role of abscisic acid in seed dormancy of red rice. *J. Exp. Bot.* 58, 3449–3462. doi: 10.1093/jxb/erm198

Gul, B., Ansari, R., Flowers, T. J., and Khan, M. A. (2013). Germination strategies of halophyte seeds under salinity. *Environ. Exp. Bot.* 92, 4–18. doi: 10.1093/aobpla/plv112

Hedden, P., and Thomas, S. G. (2012). Gibberellin biosynthesis and its regulation. *Biochem. J.* 444, 11–25. doi: 10.1042/BJ20120245

Imbert, E. (2002). Ecological consequences and ontogeny of seed heteromorphism. *Perspect. Plant Ecol.* 5, 13–36. doi: 10.1078/1433-8319-00021

Khan, M. A., and Ungar, I. A. (1984). The effect of salinity and temperature on the germination of polymorphic seeds and growth of *Atriplex triangularis* willd. *Am. J. Bot.* 71, 481–489. doi: 10.2307/2443323

Khan, M. A., Ungar, I. A., and Gul, B. (1998). Action of compatible osmotica and growth regulators in alleviating the effect of salinity on the germination of dimorphic seeds of *Arthrocnemum indicum* L. *Int. J. Plant Sci.* 159, 313–317. doi: 10.1086/297552

Kim, S. G., Lee, A. K., Yoon, H. K., and Park, C. M. (2008). A membrane-bound NAC transcription factor NTL8 regulates gibberellic acid-mediated salt signaling in *Arabidopsis* seed germination. *Plant J.* 55, 77–88. doi: 10.1111/j.1365-313X.2008.03493.x

Kim, Y., Ahn, I., Khan, A. L., Kamran, M., Waqas, M., Lee, J., et al. (2014). Regulation of endogenous gibberellins and abscisic acid levels during different seed collection periods in *Panax ginseng*. *Hortic. Environ. Biotechnol.* 55, 166–174. doi: 10.1007/s13580-014-0146-y

Li, W., An, P., and Liu, X. (2008). Effect of storage, stratification, temperature and gibberellins on germination of dimorphic seeds of *Suaeda salsa* under saline conditions. *Seed Sci. Technol.* 36, 122–132. doi: 10.15258/sst.2008.36.1.13

Li, W., Liu, X., Hanada, A., and Khan, M. A. (2011). Effect of cold stratification, scarification and hormones on germination of dimorphic seeds of *Atriplex centralasiatica* under saline conditions. *Seed Sci. Technol.* 39, 82–92. doi: 10.15258/sst.2011.39.1.08

Li, W., Liu, X., Khan, M. A., and Yamaguchi, S. (2005). The effect of plant growth regulators, nitric oxide, nitrate, nitrite and light on the germination of dimorphic seeds of *Suaeda salsa* under saline conditions. *J. Plant Res.* 118, 207–214. doi: 10.1007/s10265-005-0212-8

Li, X. L., Yang, X., Hu, Y. X., Yu, X. D., and Li, Q. L. (2014). A novel NAC transcription factor from *Suaeda liaotungensis* K. enhanced transgenic *Arabidopsis* drought, salt, and cold stress tolerance. *Plant Cell Rep.* 33, 767–778. doi: 10.1007/s00299-014-1602-y

Magome, H., Yamaguchi, S., Hanada, A., Kamiya, Y., and Oda, K. (2008). The DDF1 transcriptional activator upregulates expression of a gibberellin-deactivating gene, GA2ox7, under high-salinity stress in *Arabidopsis*. *Plant J.* 56, 613–626. doi: 10.1111/j.1365-313X.2008.03627.x

Matilla, A., Gallardo, M., and Puga-Hermida, M. I. (2005). Structural, physiological and molecular aspects of heterogeneity in seeds: a review. *Seed Sci. Res.* 15, 63–76. doi: 10.1079/SSR2005203

Nambara, E., Okamoto, M., Tatematsu, K., Yano, R., Seo, M., and Kamiya, Y. (2010). Abscisic acid and the control of seed dormancy and germination. *Seed Sci. Res.* 20, 55–67. doi: 10.1017/S0960258510000012

Piskurewicz, U., Jikumaru, Y., Kinoshita, N., Nambara, E., Kamiya, Y., and Lopez-Molina, L. (2008). The gibberellic acid signaling repressor RGL2 inhibits *Arabidopsis* seed germination by stimulating abscisic acid synthesis and ABI5 activity. *Plant Cell* 20, 2729–2745. doi: 10.1105/tpc.108.061515

Ruggiero, B., Koiwa, H., Manabe, Y., Quist, T. M., Inan, G., Saccardo, F., et al. (2004). Uncoupling the effects of abscisic acid on plant growth and water relations. Analysis of sto1/nced3, an abscisic acid-deficient but salt stress-tolerant mutant in *Arabidopsis*. *Plant Physiol.* 136, 3134–3147. doi: 10.1104/pp.104.046169

Saika, H., Okamoto, M., Miyoshi, K., Kushiro, T., Shinoda, S., Jikumaru, Y., et al. (2007). Ethylene promotes submergence-induced expression of OsABA8ox1, a gene that encodes ABA 8 '-hydroxylase in rice. *Plant Cell Physiol.* 48, 287–298. doi: 10.1093/pcp/pcm003

Shu, K., Liu, X. D., Xie, Q., and He, Z. H. (2015). Two faces of one seed: hormonal regulation of dormancy and germination. *Mol. Plant* 5:S1674–S2052. doi: 10.1016/j.molp.2015.08.010

Song, J., Fan, H., Zhao, Y., Jia, Y., Du, X., and Wang, B. (2008). Effect of salinity on germination, seedling emergence, seedling growth and ion accumulation of a euhalophyte *Suaeda salsa* in an intertidal zone and on saline inland. *Aquat. Bot.* 88, 331–337. doi: 10.1016/j.aquabot.2007.11.004

Song, J., and Wang, B. (2015). Using euhalophytes to understand salt tolerance and to develop saline agriculture: *Suaeda salsa* as a promising model. *Ann. Bot.* 115, 541–553. doi: 10.1093/aob/mcu194

Sun, T. P. (2008). Gibberellin metabolism, perception and signaling pathways in *Arabidopsis*. *Am. Soc. Plant Biol.* 6:e0103. doi: 10.1199/tab.0103

Wang, F., Xu, Y., Wang, S., Shi, W., Liu, R., Feng, G., et al. (2015). Salinity affects production and salt tolerance of dimorphic seeds of *Suaeda salsa*. *Plant Physiol. Biochem.* 95, 41–48. doi: 10.1016/j.plaphy.2015.07.005

Wang, H., Wang, L., Tian, C., and Huang, Z. (2012). Germination dimorphism in Suaeda acuminata: a new combination of dormancy types for heteromorphic seeds. *S. Afr. J. Bot.* 78, 270–275. doi: 10.1016/j.sajb.2011.05.012

Xiong, L., and Zhu, J. K. (2003). Regulation of abscisic acid biosynthesis. *Plant Physiol.* 133, 29–36. doi: 10.1104/pp.103.025395

Xu, B., Zhang, M., Xing, C., Mothibe, K., and Zhu, C. (2013). Composition, characterisation and analysis of seed oil of *Suaeda salsa* L. *Int. J. Food Sci. Technol.* 48, 879–885. doi: 10.1111/ijfs.12040

Yamaguchi, S. (2008). Gibberellin metabolism and its regulation. *Annu. Rev. Plant Biol.* 59, 225–251. doi: 10.1146/annurev.arplant.59.032607.092804

Yamauchi, Y., Ogawa, M., Kuwahara, A., Hanada, A., Kamiya, Y., and Yamaguchi, S. (2004). Activation of gibberellin biosynthesis and response pathways by low temperature during imbibition of *Arabidopsis thaliana* seeds. *Plant Cell* 16, 367–378. doi: 10.1105/tpc.018143

Yuan, K., Rashotte, A. M., and Wysocka-Diller, J. W. (2011). ABA and GA signaling pathways interact and regulate seed germination and seedling development under salt stress. *Acta Physiol. Plant* 33, 261–271. doi: 10.1007/s11738-010-0542-6

Zhao, K., Fan, H., and Ungar, I. A. (2002). Survey of halophyte species in China. *Plant Sci.* 163, 491–498. doi: 10.1186/1756-0500-7-927

Conflict of Interest Statement: The authors declare that the research was conducted in the absence of any commercial or financial relationships that could be construed as a potential conflict of interest.

23

Corrigendum: Heat Stress during Seed Filling Interferes with Sulfur Restriction on Grain Composition and Seed Germination in Oilseed Rape (*Brassica napus* L.)

Sophie Brunel-Muguet[1], Philippe D'Hooghe[1], Marie-Paule Bataillé[1], Colette Larré[2], Tae-Hwan Kim[1,3], Jacques Trouverie[1], Jean-Christophe Avice[1], Philippe Etienne[1] and Carolyne Dürr[4]*

[1] UCBN, INRA, UMR INRA–UCBN 950 Ecophysiologie Végétale, Agronomie and Nutritions N.C.S., Normandie University, Caen, France, [2] INRA UR1268 BIA, Rue de la Géraudière, Nantes, France, [3] Department of Animal Science, Environment-Friendly Agriculture Research Center, Institute of Agricultural Science and Technology, College of Agriculture and Life Science, Chonnam National University, Gwangju, South Korea, [4] INRA, UMR 1345, Institute of Research on Horticulture and Seeds, Beaucouzé, France

Keywords: oilseed rape, sulfur, temperature, germination, grain quality, lipids, proteins, sugars

Edited by:
Laurent Deluc,
Oregon State University, USA

Reviewed by:
Richard William Bell,
Murdoch University, Australia

****Correspondence:***
Sophie Brunel-Muguet
sbmuguet@rennes.inra.fr

A corrigendum on

Heat stress during seed filling interferes with sulfur restriction on grain composition and seed germination in oilseed rape (*Brassica napus* L.)
by Brunel-Muguet, S., D'Hooghe, P., Bataillé, M. P., Larré, C., Kim, T. H., Trouverie, J., et al. (2015). Front. Plant Sci. 6:213. doi: 10.3389/fpls.2015.00213

Reason for Corrigendum:

After our article was published online, it was brought to our attention that the values of the ratio ω6/ω3 in **Table 2** were inverted between the four treatments i.e., Ctrl T-HS, HT-HS, Ctrl T-LS, and HT-LS. All of the other data related to the total FAs and the *F*-values for the FAs (in the Supplemental data, Table 2) are correct.

Corrections for the values of the ratio ω6:ω3 follow:

	Ctrl T-HS		HT-HS		Ctrl T-LS		HT-LS		T effect	S effect	T × S effect
	mean	se	mean	se	mean	se	mean	se			
C18:2/C18/3 (ω6/ω3)	1.82	0.32	2.47	0.12	1.70	0.03	3.11	0.32	17.2***	1.09 ns	2.24 ns

Therefore, high temperature increased the ratio for both S supplies condition.

The authors apologize for the mistake that changes the conclusion with regards to the effect of temperature on this ratio.

In the text and in **Figure 3**, we want to make the following modifications:

In the abstract:

Instead of "While LS had negative effects on seed composition by reducing the FA contents and increasing the ratio S-poor SSPs (12S globulins)/S-rich SSPs (2S albumins) ratio, HT had positive effects by increasing S and FA contents and decreasing the C18:2/C18:3 ratio and the 12S/2S protein ratio," we want to write: "While LS had negative effects on seed composition by reducing the FA contents and increasing the ratio S-poor SSPs (12S globulins)/S-rich SSPs (2S albumins) ratio, HT had positive effects by increasing S and FA contents and decreasing the 12S/2S protein **but it negatively affected the C18:2/C18:3 ratio by increasing it.**"

In the result section:

Page 5 line 500: Instead of "The C18:2/C18:3 (also denoted ω6/ω3, **Table 2**) ratio, which is used as an indicator of nutritional quality of oil (the lower the ratio, the higher the nutritional value), decreased significantly under heat stress under both S treatments, while no significant effect of S was observed (**Table 2**)," we want to write "The C18:2/C18:3 (also denoted ω6/ ω3, **Table 2**) ratio, which is used as an indicator of nutritional quality of oil (the lower the ratio, the higher the nutritional value), **increased** significantly under heat stress under both S treatments, while no significant effect of S was observed (**Table 2**)."

In the discussion section:

Page 8 lines 846–849: Instead of "Our results also showed that heat stress was rather beneficial because it allowed the C18:2/C18:3 ratio to be lowered, which is usually targeted to satisfy dietary requirements (Simopoulos, 2002)," we want to

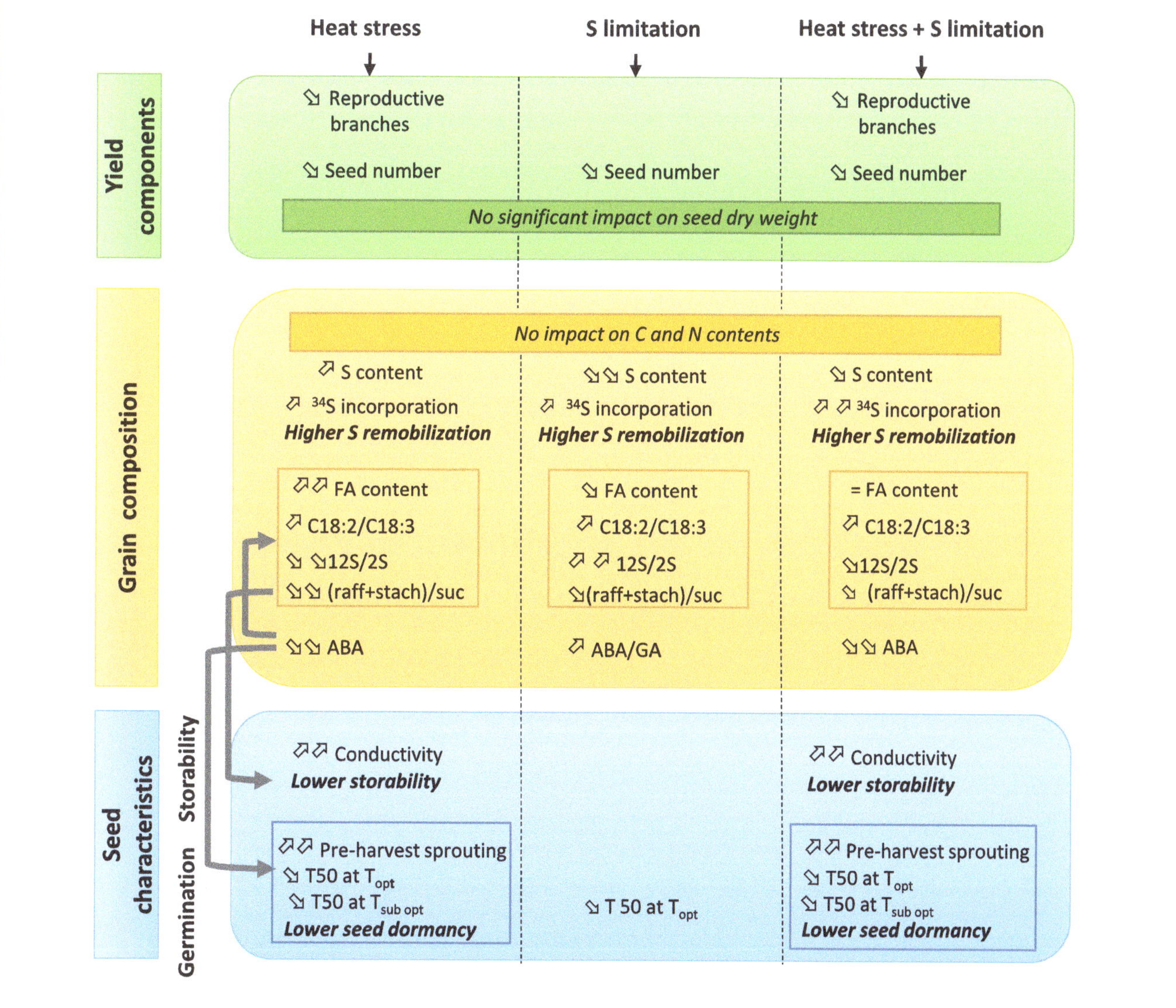

FIGURE 3 | Summary of the effects of heat stress and/or S limitation on yield components, seed composition, storability, and germination in oilseed rape. When a single effect (heat stress or S limitation) was observed, the trends (increase or decrease) are given in reference to the ANOVA in **Tables 1, 2**. When a double effect (heat stress + S limitation) is observed, the trends (increase or decrease) are given in reference to control plants (Ctrl T-HS). Grey arrows indicate known relations between variable levels. FA: fatty acids; raff: raffinose; stach: stachyose, suc: sucrose; ABA: absicic acid; GA: gibberelic acid; T50: time to reach 50% of germination; Topt: optimum temperature for germination (20°C); Tsubopt: sub-optimum temperature for germination (5°C)

write "Our results also showed that heat stress **was rather negative because it increased the C18:2/C18:3 ratio, which is usually not targeted with regards to dietary requirements** (Simopoulos, 2002)." Therefore, the following sentence is wrong "it should be noted that heat stress mitigated the negative effect of S restriction on the C18:2/C18:3 ratio, as double stress (HT-LS) had ~~lower~~ C18:2/C18:3 ratios than single stressed seed.

Page 9 line 920: Instead of "Like the C18:2/C18:3 ratio, heat stress mitigated the negative effects of S restriction (...)," we want to write "Heat stress mitigated the negative effects of S restriction...."

Page 10 line 1030: instead of "Overall, heat stress decreased yield, but improved important ratios that determine the quality of the oil and meal (lipids and proteins)," we want to write: "Overall, heat stress decreased yield, but improved an important ratio that determine the quality meal (proteins)."

In **Figure 3**

The changes deal with the arrows for the C18:2/C18:3 ratios that increased with Heat Stress (first column) and with Heat Stress+ S limitation (third column).

In addition to these points above, the authors want to correct the followings sentences:

- Page 2 line 148 (at the end of the first paragraph of the introduction): "S limitation may lead to a loss of 40% of seed yield in oilseed rape" instead of "S limitation may lead to 40% seed yield in oilseed rape."

AUTHOR CONTRIBUTIONS

SB, PD, JT, JA, and PE contributed to the experimental design, to plant growth and tissue sampling and have been involved in revising the article for important intellectual content. CD supervised the choice of relevant measurements on seeds. MB made the spectrometry analysis. CA was involved in the SSP analyses, interpretation of protein data, and revising the manuscript. TK was involved in the hormone measurements. SB performed the whole raw data analysis (including statistical analyses). SB and CD made interpretation of data and writing of the article.

ACKNOWLEDGMENTS

This work is part of the CAQ40 (CAQ40: Climate chAnge and Quality of fruit grain and seeds in the next 40 years) project, which was funded by the metaprogram ACCAF "Adaptation to Climate Change in Agriculture and Forest" by INRA.

Conflict of Interest Statement: The authors declare that the research was conducted in the absence of any commercial or financial relationships that could be construed as a potential conflict of interest.

Permissions

All chapters in this book were first published in FPLS, by Frontiers; hereby published with permission under the Creative Commons Attribution License or equivalent. Every chapter published in this book has been scrutinized by our experts. Their significance has been extensively debated. The topics covered herein carry significant findings which will fuel the growth of the discipline. They may even be implemented as practical applications or may be referred to as a beginning point for another development.

The contributors of this book come from diverse backgrounds, making this book a truly international effort. This book will bring forth new frontiers with its revolutionizing research information and detailed analysis of the nascent developments around the world.

We would like to thank all the contributing authors for lending their expertise to make the book truly unique. They have played a crucial role in the development of this book. Without their invaluable contributions this book wouldn't have been possible. They have made vital efforts to compile up to date information on the varied aspects of this subject to make this book a valuable addition to the collection of many professionals and students.

This book was conceptualized with the vision of imparting up-to-date information and advanced data in this field. To ensure the same, a matchless editorial board was set up. Every individual on the board went through rigorous rounds of assessment to prove their worth. After which they invested a large part of their time researching and compiling the most relevant data for our readers.

The editorial board has been involved in producing this book since its inception. They have spent rigorous hours researching and exploring the diverse topics which have resulted in the successful publishing of this book. They have passed on their knowledge of decades through this book. To expedite this challenging task, the publisher supported the team at every step. A small team of assistant editors was also appointed to further simplify the editing procedure and attain best results for the readers.

Apart from the editorial board, the designing team has also invested a significant amount of their time in understanding the subject and creating the most relevant covers. They scrutinized every image to scout for the most suitable representation of the subject and create an appropriate cover for the book.

The publishing team has been an ardent support to the editorial, designing and production team. Their endless efforts to recruit the best for this project, has resulted in the accomplishment of this book. They are a veteran in the field of academics and their pool of knowledge is as vast as their experience in printing. Their expertise and guidance has proved useful at every step. Their uncompromising quality standards have made this book an exceptional effort. Their encouragement from time to time has been an inspiration for everyone.

The publisher and the editorial board hope that this book will prove to be a valuable piece of knowledge for researchers, students, practitioners and scholars across the globe.

List of Contributors

Leandro H. Estornell, Manuel Talón and Francisco R. Tadeo
Centre de Genómica,Institut Valencià d'Investigacions Agràries, Montcada, Spain

Mari Wildhagen and Melinka A. Butenko
Section for Genetics and Evolutionary Biology, Department of Biosciences, University of Oslo, Oslo, Norway

Miguel A. Pérez-Amador
Instituto de Biología Moleculary Celularde Plantas, Universidad Politécnica de Valencia-Consejo Superior de Investigaciones Científicas, València, Spain

Jing Zhang
Ministry of Agriculture (MOA) Key Laboratory of Crop Ecophysiology and Farming System in the Middle Reaches of the Yangtze River, College of Plant Science and Technology, Huazhong Agricultural University, Wuhan, China
Centre for Plant Genetics and Breeding, School of Plant Biology, Faculty of Science and The UWA Institute of Agriculture, The University of Western Australia, Perth, WA, Australia

Annaliese S. Mason
Plant Breeding Department, IFZ Research Centre for Biosystems, Land Use and Nutrition, Justus Liebig University, Giessen, Germany
School of Agriculture and Food Sciences and Centre for Integrative Legume Research, The University of Queensland, Brisbane, QLD, Australia

Jian Wu and Sheng Liu
National Key Laboratory of Crop Genetic Improvement, College of Plant Science and Technology, Huazhong Agricultural University, Wuhan, China

Robert Redden
Australian Grains Genebank, Department of Economic Development Jobs Transport and Resources, Horsham, VIC, Australia

Xuechen Zhang and Guijun Yan
Centre for Plant Genetics and Breeding, School of Plant Biology, Faculty of Science and The UWA Institute of Agriculture, The University of Western Australia, Perth, WA, Australia

Tao Luo and Liyong Hu
Ministry of Agriculture (MOA) Key Laboratory of Crop Ecophysiology and Farming System in the Middle Reaches of the Yangtze River, College of Plant Science and Technology, Huazhong Agricultural University, Wuhan, China

Jacqueline Batley
Centre for Plant Genetics and Breeding, School of Plant Biology, Faculty of Science and The UWA Institute of Agriculture, The University of Western Australia, Perth, WA, Australia
School of Agriculture and Food Sciences and Centre for Integrative Legume Research, The University of Queensland, Brisbane, QLD, Australia

Fengmei Gao
Key Laboratory of Soybean Biology, Soybean Research Institute, Ministry of Education, Northeast Agricultural University, Harbin, China
National Wheat Improvement Center, Institute of Crop Science, Chinese Academy of Agricultural Sciences, Beijing, China
Keshan Sub-Academy, Heilongjiang Academy of Agricultural Sciences, Keshan, China

Weie Wen, Jindong Liu and Xianchun Xia
National Wheat Improvement Center, Institute of Crop Science, Chinese Academy of Agricultural Sciences, Beijing, China

Awais Rasheed
National Wheat Improvement Center, Institute of Crop Science, Chinese Academy of Agricultural Sciences, Beijing, China
International Maize and Wheat Improvement Center (CIMMYT) China Office, c/o Chinese Academy of Agricultural Sciences, Beijing, China

Guihong Yin
Zhoukou Academy of Agricultural Sciences, Zhoukou, China

Xiaoxia Wu
Key Laboratory of Soybean Biology, Soybean Research Institute, Ministry of Education, Northeast Agricultural University, Harbin, China

Zhonghu He
National Wheat Improvement Center, Institute of Crop Science, Chinese Academy of Agricultural Sciences, Beijing, China
International Maize and Wheat Improvement Center (CIMMYT) China Office, c/o Chinese Academy of Agricultural Sciences, Beijing, China

Mohammad G. Mostofa
Department of Biochemistry and Molecular Biology, Banga bandhu Sheikh Mujibur Rahman Agricultural University, Gazipur, Bangladesh

Daisuke Saegusa and Masayuki Fujita
Laboratory of Plant Stress Responses, Department of Applied Biological Science, Faculty of Agriculture, Kagawa University, Miki, Japan

Lam-Son Phan Tran
Plant Abiotic Stress Research Group & Faculty of Applied Sciences, Ton Duc Thang University, Ho Chi Minh City, Vietnam
Signaling Pathway Research Unit, RIKEN Center for Sustainable Resource Science, Yokohama, Japan

Zhifeng Wen, Ran Wan, Zhi Li and Xiping Wang
State Key Laboratory of Crop Stress Biology in Arid Areas, College of Horticulture, Northwest A&F University, Yangling, China
Key Laboratory of Horticultural Plant Biology and Germplasm Innovation in North west China, Ministry of Agriculture, Northwest A&F University, Yangling, China

Liping Yao
Key Laboratory of Stress Physiology and Molecular Biology for Tree Fruits of Beijing, Department of Pomology, College of Agriculture and Biotechnology, China Agricultural University, Beijing, China

Chonghuai Liu
Zhengzhou Fruit Research Institute, Chinese Academy of Agricultural Sciences, Zhengzhou, China

Xiaodong Wang and Jiefu Zhang
Key Laboratory of Cotton and Rapeseed, Ministry of Agriculture, Institute of Industrial Crops, Jiangsu Academy of Agricultural Sciences, Nanjing, China
Jiangsu Collaborative Innovation Center for Modern Crop Production, Nanjing, China

Kunjiang Yu, Hongge Li, Qi Peng, Wei Zhang, Song Chen and Maolong Hu
Key Laboratory of Cotton and Rapeseed, Ministry of Agriculture, Institute of Industrial Crops, Jiangsu Academy of Agricultural Sciences, Nanjing, China

Feng Chen
Key Laboratory of Cotton and Rapeseed, Ministry of Agriculture, Institute of Industrial Crops, Jiangsu Academy of Agricultural Sciences, Nanjing, China
Provincial Key Laboratory of Agrobiology, Jiangsu Academy of Agricultural Sciences, Nanjing, China

Francesco F. Montesano, Francesco Serio , Carlo Mininni and Angelo Parente
Institute of Sciences of Food Production, National Research Council of Italy, Bari, Italy

Angelo Signore and Pietro Santamaria
Department of Agricultural and Environmental Science, University of Bari Aldo Moro, Bari, Italy

Dandan Li, Xiaoqing Hu, Quangang Liu, Zhanchao Wang, Haizhen Zhang, Han Wang, Ming Wei, Hanzeng Wang, Haimei Liu and Chenghao Li
State Key Laboratory of Tree Genetics and Breeding, Northeast Forestry University, Harbin, China

Xuemei Xu
Library of Northeast Forestry University, Harbin, China

Suxia Yuan, Yanbin Su, Yumei Liu, Zhansheng Li, Zhiyuan Fang, Limei Yang, Mu Zhuang, Yangyong Zhang, Honghao Lv and Peitian Sun
Institute of Vegetables and Flowers, Chinese Academy of Agricultural Sciences, Beijing, China

Joonyup Kim, Ronghui Yang and Mark L. Tucker
Soybean Genomics and Improvement Lab, Agricultural Research Service, United States Department of Agriculture, Beltsville, MD, USA

Srivignesh Sundaresan
Department of Postharvest Science of Fresh Produce, Agricultural Research Organization, The Volcani Center, Bet-Dagan, Israel
The Robert H. Smith Faculty of Agriculture, Food and Environment, The Hebrew University of Jerusalem, Rehovot, Israel

Sonia Philosoph-Hadas and Shimon Meir
Department of Postharvest Science of Fresh Produce, Agricultural Research Organization, The Volcani Center, Bet-Dagan, Israel

Heshan Du, Yuqing Wang, Jingjing Yang and Wencai Yang
Beijing Key Laboratory of Growth and Developmental Regulation for Protected Vegetable Crops, Department of Vegetable Science, China Agricultural University, Beijing, China

Esayas Mendesil, Birgitta Rämert, Salla Marttila and Peter Anderson
Department of Plant Protection Biology, Swedish University of Agricultural Sciences, Alnarp, Sweden

Ylva Hillbur
Department of Plant Protection Biology, Swedish University of Agricultural Sciences, Alnarp, Sweden
International Institute of Tropical Agriculture, Ibadan, Nigeria

Kai Cao
State Key Laboratory of Crop Stress Biology for Arid Areas, Horticulture College, Northwest A&F University, Yangling, China
Laboratory of Plant Molecular Biology, Rockefeller University, New York, NY, USA

Lirong Cui, Xiaoting Zhou, Lin Ye and Zhirong Zou
State Key Laboratory of Crop Stress Biology for Arid Areas, Horticulture College, Northwest A&F University, Yangling, China

Shulin Deng
Laboratory of Plant Molecular Biology, Rockefeller University, New York, NY, USA

Jing-Hao Jin, Huai-Xia Zhang, Jun-Yi Tan, Ming-Jia Yan, Da-Wei Li, Abid Khan and Zhen-Hui Gong
College of Horticulture, Northwest A&F University, Yangling, China

Florence Auguy
Institutde Recherchepourle Développement, UMRDIADE, Equipe Rhizogenèse, Montpellier, France

Mouna Fahr and Mohamed El Mzibri
Centre National de l'Energie, des Scienceset des Techniques Nucléaires, Laboratoirede Biotechnologie des Plantes, UBRM-DSV, Rabat, Morocco

Patricia Moulin
Institut de Recherche pour le Développement, Laboratoire de Microbiologie et Biologie Moléculaire, Faculté des Sciences, Université Mohammed V-Rabat, Rabat, Morocco

Abdelaziz Smouni
Laboratoire de Physiologie et Biotechnologie Végétale, Faculté des Sciences, Université Mohammed V-Rabat, Rabat, Morocco

Abdelkarim Filali-Maltouf
Laboratoire de Microbiologie et Biologie Moléculaire, Faculté des Sciences, Université Mohammed V-Rabat, Rabat, Morocco

Gilles Béna
Institut de Recherche pour le Développement, UMR IPME, Equipe ABIP, Montpellier, France

Patrick Doumas
Institut National de la Recherche Agronomique, UMR Biochimieet Physiologie Moléculaire des Plantes, Montpellier, France

Jing Zhang, Jiantao Zhao, Fei Yan, and Zhirong Zou
State Key Laboratory of Crop Stress Biology for Arid Areas, College of Horticulture, Northwest A&F University, Yangling, China
Key Laboratory of Protected Horticultural Engineering in Northwest, Ministry of Agriculture, Yangling, China

Yao Xu
College of Forestry, Northwest A&F University, Yangling, China

Jing Liang
Shaanxi Jinpeng Seed Industry Co. Ltd., Yangling, China

Peipei Chang, Mingjun Li, Yan Liang
State Key Laboratory of Crop Stress Biology for Arid Areas, College of Horticulture, Northwest A&F University, Yangling, China

Medhin H. Kifle and Mark D. Laing
Discipline of Plant Pathology, School of Agricultural, Earth and Environmental Sciences University of Kwa Zulu-Natal, Pietermaritzburg, South Africa

Dewi E. Adriani
CIRAD, UMRAGAP, F-34398 Montpellier, France
2 Faculty of Agriculture, University of Lambung Mangkurat, Banjarbaru, Indonesia

Tanguy Lafarge, Audrey Dardou, Anne Clément-Vidal and Delphine Luquet
CIRAD, UMRAGAP, F-34398 Montpellier, France
Faculty of Agriculture, University of Lambung Mangkurat, Banjarbaru, Indonesia

Aubrey Fabro
Crop and Environment Science Division, International Rice Research Institute, Los Baños, Philippines

Sudirman Yahya
Department of Agronomy and Horticulture, Bogor Agricultural University, Bogor, Indonesia

Michael Dingkuhn
CIRAD, UMRAGAP, F-34398 Montpellier, France
Crop and Environment Science Division, International Rice Research Institute, Los Baños, Philippines

Jingjing Chen, Wenxing Pang and Zhongyun Piao
Department of Horticulture, Shenyang Agricultural University, Shenyang, China

Bing Chen
Department of Characteristic Horticulture, Taizhou Institute of Agricultural Sciences, Jiangsu Academy of Agricultural Sciences, Taizhou, China

Chunyu Zhang
National Key Laboratory of Crop Genetic Improvement and College of Plant Science and Technology, Huazhong Agricultural University, Wuhan, China

G. Brett Runion, Stephen A. Prior and H. AllenTorbert
National Soil Dynamics Laboratory, Agricultural Research Service, United States Department of Agriculture, Auburn, AL, USA

Ludovic J. A. Capo-chichi
Alberta Innovates Technology Futures, Vegreville, AB, Canada

Edzardvan Santen
Department of Crop, Soil and Environmental Sciences, Auburn University, Auburn, AL,USA

Xiangling Zeng, Cai Liu, Riru Zheng, Xuan Cai, Jing Luo and CaiyunWang
Key Laboratory for Biology of Horticultural Plants, Ministry of Education, College of Horticulture and Forestry Sciences, Huazhong Agricultural University, Wuhan, China

Jingjing Zou
School of Nuclear Technology and Chemistry and Biology, Hubei
University of Science and Technology, Xianning, China

Weiqiang Li
Center for Agricultural Resources Research, Institute of Genetics and Developmental Biology, Chinese Academy of Science, Shijiazhuang, China
Signaling Pathway Research Unit, RIKEN Center for Sustainable Resource Science, Tsurumi, Japan

Shinjiro Yamaguchi
Department of Biomolecular Sciences, Graduate School of Life Sciences, Tohoku University, Sendai, Japan

M. Ajmal Khan
Qatar Shell Professorial Chair for Sustainable Development, Centre for Sustainable Development, College of Arts and Sciences, Qatar University, Doha, Qatar

Ping An
Arid Land Research Center, Tottori University, Hamasaka, Japan

Xiaojing Liu
Center for Agricultural Resources Research, Institute of Genetics and Developmental Biology, Chinese Academy of Science, Shijiazhuang, China

Lam-Son P. Tran
Plant Abiotic Stress Research Group & Faculty of Applied Sciences, Ton Duc Thang University, Ho Chi Minh City, Vietnam; Signaling Pathway Research Unit, RIKEN Center for Sustainable Resource Science, Tsurumi, Japan

Sophie Brunel-Muguet, Philippe D' Hooghe, Marie-Paule Bataillé, , Jacques Trouverie, Jean-Christophe Avice and Philippe Etienne
UCBN,INRA,UMRINRA-UCBN950 Ecophysiologie Végétale, Agronomie and Nutritions N.C.S., Normandie University, Caen, France

Colette Larré
INRAUR1268 BIA, Ruedela Géraudière, Nantes, France

Tae-Hwan Kim
UCBN, INRA, UMRINRA-UCBN950 Ecophysiologie Végétale, Agronomie and Nutritions N.C.S., Normandie University, Caen, France

Department of Animal Science, Environment-Friendly Agriculture Research Center, Institute of Agricultural Science and Technology, College of Agriculture and Life Science, Chonnam National University, Gwangju, South Korea

Carolyne Dürr
INRA, UMR1345, Institute of Research on Horticulture and Seeds, Beaucouzé, France